31st International Symposium on Superconductivity (ISS2018)

Journal of Physics: Conference Series Volume 1293

Tsukuba, Japan
12 - 14 December 2018

Editors:

Atsutaka Maeda
Hiraku Ogino
Yoshiyuki Yoshida
Yasuhiro Iijima
Mutsuo Hidaka
Tsunehiro Hato
Naoyuki Amemiya
Tsuyoshi Wakuda
Hiroshi Miyazaki

ISBN: 978-1-5108-9963-6
ISSN: 1742-6588

Printed from e-media with permission by:

Curran Associates, Inc.
57 Morehouse Lane
Red Hook, NY 12571

Some format issues inherent in the e-media version may also appear in this print version.

Institute of Physics
Dirac House, Temple Back
Bristol BS1 6BE UK

Phone: 44 1 17 929 7481
Fax: 44 1 17 920 0979

techtracking@iop.org

Additional copies of this publication are available from:

Curran Associates, Inc.
57 Morehouse Lane
Red Hook, NY 12571 USA
Phone: 845-758-0400
Fax: 845-758-2633
Email: curran@proceedings.com
Web: www.proceedings.com

TABLE OF CONTENTS

1. PHYSICS AND CHEMISTRY

1.1 NOVEL MATERIALS

1.2 CUPRATE SUPERCONDUCTORS

1.3 IRON-BASED SUPERCONDUCTORS

1.4 VORTEX PHYSICS

1.5 THEORY

2. WIRES AND BULK

2.1 COATED CONDUCTORS

4.2 SUPERCONDUCTING CABLES AND ELECTRIC POWER DEVICES

4.3 ELECTRIC AIRCRAFTS AND ROTATING MACHINES

4.4 FUSIONS

4.5 MAGNETIC SEPARATION

4.6 MAGNETIC BEARING

IOP Conf. Series: Journal of Physics: Conf. Series **1293** (2019) 011001 doi:10.1088/1742-6596/1293/1/011001

31st International Symposium on Superconductivity

TSUKUBA INTERNATIONAL CONGRESS CENTER
IBARAKI, JAPAN
December 12–14, 2018

31st International Symposium on Superconductivity (ISS2018)

Preface

It is our pleasure to report that the Proceedings of the 31st International Symposium on Superconductivity (ISS2018) held at Tsukuba International Congress Center, Ibaraki, Japan, December 12–14, 2018, are now published in *Journal of Physics: Conference Series* (JPCS). ISS2018 was organized by the National Institute of Advanced Industrial Science and Technology (AIST), successively from the previous year. Total of 449 scientists, engineers, students, corporate executives and other participants (103 from abroad) gathered in ISS2018, and fruitful discussions were made to promote superconductor sciences and technologies. Total of 323 papers were presented, which included 6 plenary lectures, 79 invited talks, 38 contributed oral presentations and 200 poster presentations. The papers published in JPCS were categorized into relevant topics in the following four fields; (a) Physics and Chemistry, (b) Wires and Bulk, (c) Electronic Devices, (d) Large Scale System Applications.

To ensure the high publication standard mandated by JPCS, every paper was peer reviewed by a reviewer with expertise, and in some cases by two reviewers before it was accepted for publication. As editors of the Proceedings, we would like to express our sincere appreciation to all the reviewers involved in the evaluation of the papers for their invaluable contribution.

The Editors

Physics and Chemistry: Atsutaka Maeda (The University of Tokyo), Hiraku Ogino (AIST)
Wires and Bulk: Yoshiyuki Yoshida (AIST), Yasuhiro Iijima (Fujikura Ltd.)
Electronic Devices: Mutsuo Hidaka (AIST), Tsunehiro Hato (SUSTERA)
Large Scale System Applications: Naoyuki Amemiya (Kyoto University), Tsuyoshi Wakuda (Hitachi Ltd.), Hiroshi Miyazaki (Toshiba Co.)

IOP Conf. Series: Journal of Physics: Conf. Series **1293** (2019) 011001 doi:10.1088/1742-6596/1293/1/011001

Organization of ISS2018

General Chair:
Dr. Toshihiko Kanayama
Special Emeritus Advisor, National Institute of Advanced Industrial Science and Technology (AIST)

Local Organizing Committee:
Secretary-General: Michiya Okada, TIA Central Office, AIST

Members:
Haruhiko Obara (Deputy Director-General, Planning Headquarters, AIST)
Fumio Takemura, Teruhisa Horita, Mitsuho Furuse (Department of Energy and Environment, AIST)
Tetsuji Yasuda, Masataka Ohkubo, Mutsuo Hidaka, Yoshiyuki Yoshida, Shigeyuki Ishida, (Advisor) Akira Iyo (Department of Electronics and Manufacturing, AIST)
Susumu Ogawa, Hirofumi Yamasaki, Mitsuyasu Yokosuka, Haruko Okazaki (TIA Central Office, AIST)

Supported by:
Ministry of Economy, Trade and Industry (METI)
Ministry of Education, Culture, Sports, Science and Technology (MEXT)
New Energy and Industrial Technology Development Organization (NEDO)
Japan Science and Technology Agency (JST)
The City of Tsukuba
Tsukuba Tourism and Convention Association
Cryogenics and Superconductivity Society of Japan (CSSJ)
The Institute of Electrical Engineers of Japan (IEEJ)
The Japan Society of Applied Physics (JSAP)
Applied Superconductivity Constellations of Tsukuba (ASCOT)
TIA—a platform for open innovation

Committees of ISS2018

Program committee

Physics and Chemistry

Co-Chair: Atsutaka Maeda (The Univ. of Tokyo), Sub-Chair: Hiraku Ogino (AIST)

Members: Ryotaro Arita (RIKEN), Takekazu Ishida (Osaka Prefecture Univ.), Minoru Nohara (Okayama Univ.), Tsutomu Nojima (Tohoku Univ.), Takao Sasagawa (Tokyo Inst. of Technology), Takasada Shibauchi (The Univ. of Tokyo)

Advisors: Hiroshi Eisaki (AIST), Akira Iyo (AIST)

Wires and Bulk

Co-Chair: Yoshiyuki Yoshida (AIST), Sub-Chair: Yasuhiro Iijima (Fujikura Ltd.)

Members: Satoshi Awaji (Tohoku Univ.), Hiroshi Ikuta (Nagoya Univ.), Atsushi Ishihara (Railway Technical Research Inst.), Takeshi Kato (Sumitomo Electric Industries, Ltd.), Takanobu Kiss (Kyushu Univ.), Akiyoshi Matsumoto (National Inst. for Materials Science), Kaname Matsumoto (Kyushu Inst. of Technology), Hideki Tanaka (Hitachi Ltd.)

Electronic Devices

Co-Chair: Mutsuo Hidaka (AIST), Sub-Chair: Tsunehiro Hato (SUSTERA)

Members: Yoshimi Hatsukade (Kindai Univ.), Masashi Ohno (The Univ. of Tokyo), Shigeo Sato (Tohoku Univ.), Naoto Sekiya (Yamanashi Univ.), Masamitsu Tanaka (Nagoya Univ.), Hirotake Yamamori (AIST)

Advisor: Masataka Ohkubo (AIST)

Large Scale System Application

Co-Chair: Naoyuki Amemiya (Kyoto Univ.), Sub-Chair: Tsuyoshi Wakuda (Hitachi Ltd.), Sub-Chair: Hiroshi Miyazaki (Toshiba Co.)

Members: Mamoru Hamada (JASTEC), Kazuhiro Kajikawa (Kyushu Univ.), Shinji Matsumoto (National Inst. for Materials Science), Shin-ichi Mukoyama (Furukawa Electric Co., Ltd.), Naoko Nakamura (MAYEKAWA MFG. Co., Ltd.), Taketsune Nakamura (Kyoto Univ.), Tetsuo Oka (Shibaura Inst. of Technology), Shun Tonooka (Mitsubishi Electric Co.), Tomonori Watanabe (Chubu Electric Power Co., Inc.)

Advisor: Mitsuho Furuse (AIST)

IOP Conf. Series: Journal of Physics: Conf. Series **1293** (2019) 011001 doi:10.1088/1742-6596/1293/1/011001

International Advisory Committee

Physics and Chemistry

Jun Akimitsu (Okayama Univ.), Hideo Aoki (The Univ. of Tokyo), Xianhui Chen (Univ. of Science and Technology of China), Atsushi Fujimori (The Univ. of Tokyo), Zenji Hiroi (The Univ. of Tokyo), Maarit Karppinen (Aalto Univ.), Wai-Kwong Kwok (Argonne National Laboratory), Victor Moshchalkov (KU Leuven), Satoshi Okuma (Tokyo Inst. of Technology), Marina Putti (Univ. of Genoa), Hermann Suderow (Autonomous Univ. of Madrid), Setsuko Tajima (Osaka Univ.), Jeffrey L. Tallon (Victoria Univ. of Wellington), Takami Tohyama (Tokyo Univ. of Science), Dirk van der Marel (Univ. of Geneva), Hai-Hu Wen (Nanjing Univ.), Eli Zeldov (Weizmann Inst. of Science)

Wires and Bulk

Amalia Ballarino (CERN), David Cardwell (Univ. of Cambridge), Leonardo Civale (Los Alamos National Laboratory), Michael Eisterer (Vienna Univ. of Technology), Timothy J. Haugan (Air Force Research Laboratory), Bernhard Holzapfel (Karlsruhe Inst. of Technology), Teruo Izumi (AIST), Hiroaki Kumakura (National Inst. for Materials Science), David Larbalestier (Florida State Univ.), Qiang Li (Brookhaven National Laboratory), Xavier Obradors (Institut Ciencia de Materials de Barcelona), Venkat Selvamanickam (Univ. of Houston), Jun-ichi Shimoyama (Aoyama Gakuin Univ.), Michael Sumption (Ohio State Univ.), Yutaka Yoshida (Nagoya Univ.)

Electronic Devices

Pascal Febvre (Univ. of Savoie Mont Blanc), Masataka Ohkubo (AIST), Shigetoshi Ohshima (Yamagata Univ.), Horst Rogalla (Univ. of Colorado at Boulder), Keiichi Tanabe (SUSTERA), Zhen Wang (SC2 & SIMIT, Chinese Academy of Sciences)

Large Scale System Application

Bob Buckley (Victoria Univ. of Wellington), Toru Fukushima (SuperPower Inc.), Mathias Noe (Karlsruhe Inst. of Technology), Hiroyuki Ohsaki (The Univ. of Tokyo), Minwon Park (Changwon National Univ.), Christopher M. Rey (Energy-to-Power Solutions), Pascal Tixador (Grenoble-INP), Liye Xiao (Chinese Academy of Sciences)

Peer review statement

All papers published in this volume of *Journal of Physics: Conference Series* have been peer reviewed through processes administered by the proceedings Editors. Reviews were conducted by expert referees to the professional and scientific standards expected of a proceedings journal published by IOP Publishing.

Effect of Indium doping on the superconductivity of layered oxychalcogenide $La_2O_2Bi_3Ag_{1-x}In_xS_6$

Rajveer Jha[1]*, Yosuke Goto[1], Tatsuma D Matsuda[1], Yuji Aoki[1] and Yoshikazu Mizuguchi[1]

[1]Department of Physics, Tokyo Metropolitan University, 1-1 Minami-Osawa, Hachioji, Tokyo 192-0397, Japan

E-mail for corresponding author: *rajveerjha@gmail.com

Abstract. We report on the substitution effect of Indium (In) at the Ag site of layered oxychalcogenide $La_2O_2Bi_3Ag_{1-x}In_xS_6$. The T_c decreases with increasing In concentration. A hump in the normal state resistivity at an anomaly temperature (T^*) near 180 K was observed for all the samples. The anomaly in the resistivity at T^* is indicating the possible occurrence of a charge-density-wave (CDW) transition. The T^* does not markedly change by In doping. The x dependence of Seebeck coefficient suggests that carrier concentration does not change by In doping. The EDX analysis indicates small amount of Bi deficiency, which suggests that the Bi site is slightly substituted by In. The CDW transition is robust against the In substitution at Ag site, while T_c is decreasing due to the Bi site substitution by In. On the basis of those analyses, we propose that the suppression of superconductivity in the In-doped $La_2O_2Bi_3Ag_{1-x}In_xS_6$ system is caused by negative in-plane chemical pressure effect and partial substitution of In for the in-plane Bi site.

1. Introduction

The discovery of superconductivity in layered Bismuth disulphide (BiS_2–based) compounds $Bi_4O_4S_3$ with a transition temperature (T_c) of 4.5 K [1], $REO_{1-x}F_xBiS_2$ (RE = La, Ce, Nd, Yb, Pr) and $Sr_{1-x}La_xFBiS_2$ with T_c of 2-5 K [1–11] have created tremendous interests of condensed matter physics scientific community [1,2]. The highest T_c up to 11 K was observed in a high-pressure phase of $LaO_{1-x}F_xBiS_2$ [12-16]. Recently, we reported superconductivity in the oxychalcogenide $La_2O_2Bi_3AgS_6$ with T_c of 0.5 K [17]. Furthermore, we reported that T_c can be increased up to 2.4 K by Tin (Sn) substitution at the silver (Ag) site of $La_2O_2Bi_3AgS_6$ [18]. Basically, $La_2O_2Bi_3AgS_6$ has a crystal structure similar to that of $LaOPbBiS_3$, which is composed of La_2O_2 blocking layer and $Bi_2Pb_2S_6$ (or Bi_3AgS_6 for $La_2O_2Bi_3AgS_6$) conducting layer [19-21]. In those oxichalcogenide family ($La_2O_2M_4S_6$: M = metal), the NaCl-type layer (M_2S_2 layer) is inserted into the van der Waals gap of two BiS_2 layers [19]. The other factor essential for the superconductivity in the BiS_2 family is the chemical pressure effect, which suppresses in-plan disorder [22-23]. Therefore, we aimed to study the further substitution effects for the $La_2O_2Bi_3AgS_6$ superconductor. Here, we report the effect of In substitution at the Ag site of $La_2O_2Bi_3AgS_6$ compound. T_c decreases with increasing In concentration. The hump at T^* = 180 K in the normal state electrical resistivity is not changing clearly with In doping, which is clearly different to the case of Sn doping at the Ag site.

2. Experimental Section

The In-doped polycrystalline samples of $La_2O_2Bi_3Ag_{1-x}In_xS_6$ were synthesized by a solid-state reaction method. Powders of Bi_2O_3 (99.9%), La_2S_3 (99.9%), In (99.99%), and AgO (99.9%) and grains of Bi (99.999%) and S (99.99%) with a nominal composition of $La_2O_2Bi_3Ag_{1-x}In_xS_6$ were ground thoroughly using a pestle and a mortar for mixing, pelletized, sealed in an evacuated quartz tube, and heated at 725 °C for 15 h. For the homogeneity, we reground the obtained samples and heated at 725 °C for 15 h. Finally, we obtained phase-pure samples, and the phase purity was checked by laboratory X-ray diffraction (XRD) with Cu-K_α radiation. The crystal structure parameters were refined using the Rietveld method with RIETAN-FP [24]. A schematic image of the crystal structure was drawn using VESTA [see Figure. 1(c)] [25]. The energy-dispersive X-ray spectroscopy (EDX) on scanning electron microscope TM3030 (Hitachi) was used for analysing actual compositions. The Seebeck coefficient was measured by a four-probe technique on ZEM-3 (Advance RIKO). The electrical resistivity down to T = 0.4 K was measured by four probe method using ^{3}He probe platform on the Physical Property Measurement System (PPMS: Quantum Design).

3. Results and discussion

The room temperature XRD patterns of the In-doped $La_2O_2Bi_3Ag_{1-x}In_xS_6$ (x = 0-0.4) compounds are shown in figure 1(a). These compounds are crystallized in the tetragonal structure with the space group of *P4/nmm*. We present the XRD data up to x = 0.4, which is determined as the solubility limit of In for the Ag site in the $La_2O_2Bi_3Ag_{1-x}In_xS_6$ system. We checked fitting (reliability) parameter R_{wp} through the Rietveld refinement of XRD data of $La_2O_2Bi_3Ag_{1-x}In_xS_6$ compounds. We found that the R_{wp} is lower with keeping the M2 site substituted by In. While, the R_{wp} factor was comparably high when the refinement performed by considering both sites (M1 and M2) substitutc by In. The Rietveld analysis suggests that the In ions were substituted at the M2 site. The (103) peak position shifts towards the low angle side with increasing In doping level as shown in figure 1(b). Figure 1(c) shows the schematic unit cell of $La_2O_2Bi_3Ag_{0.9}In_{0.1}S_6$, in which chalcogenide layers of (Bi,Ag,In)S and $LaO(Bi,Ag)S_2$ are alternately stacking. The structural parameters have been determined by the Rietveld fitting of powder XRD pattern. The obtained lattice parameters are shown in figures 1(d, e). The estimated lattice parameters are a = 4.0561(1) Å and c = 19.345(1) Å for x = 0, which become a = 4.052(1) Å and c = 19. 32(1) Å for x = 0.1. The lattice parameter c decreases with small amount of In doping for x = 0-0.1, and then continuously increases with increasing x for higher x. The lattice parameter a and c are increasing for x>0.1 for the In doped compounds, while the lattice parameter a is decreasing for x> 0.1 for the Sn doped compounds. The lattice parameter c firstly decreases for x<0.2 then slightly increases for x>0.2, the value of c=19.458 (1) Å for the 50% Sn doped compound, we observed the clear positive in plane chemical pressure for the Sn doped $La_2O_2Bi_3Ag_{1-x}Sn_xS_6$ compounds [18]. Comparably, the lattice parameter c become larger for the 40% In doped sample, The lattice parameter c = 19. 647(1) Å for the 40% In doped sample. As well, the continuous increase in lattice parameter a suggests in-plane expansion of the unit cell, which should result the negative chemical pressure effect. The ratio of Bi, Ag and In in the $La_2O_2Bi_3Ag_{1-x}In_xS_6$ (x = 0-0.4) samples were studied by using EDX. To estimate the Bi amount, we tentatively analysed the composition according to the formula of $La_2O_2Bi_{3+y}(Ag_{1-x}In_x)_{1-y}S_6$. The analysed values of x_{EDX} and y_{EDX} are plotted in figures 1(f, g). We found that Bi is slightly deficient in the doped samples with x = 0.1, 0.2, 0.3 and 0.4, but the estimated y_{EDX} was close to zero.

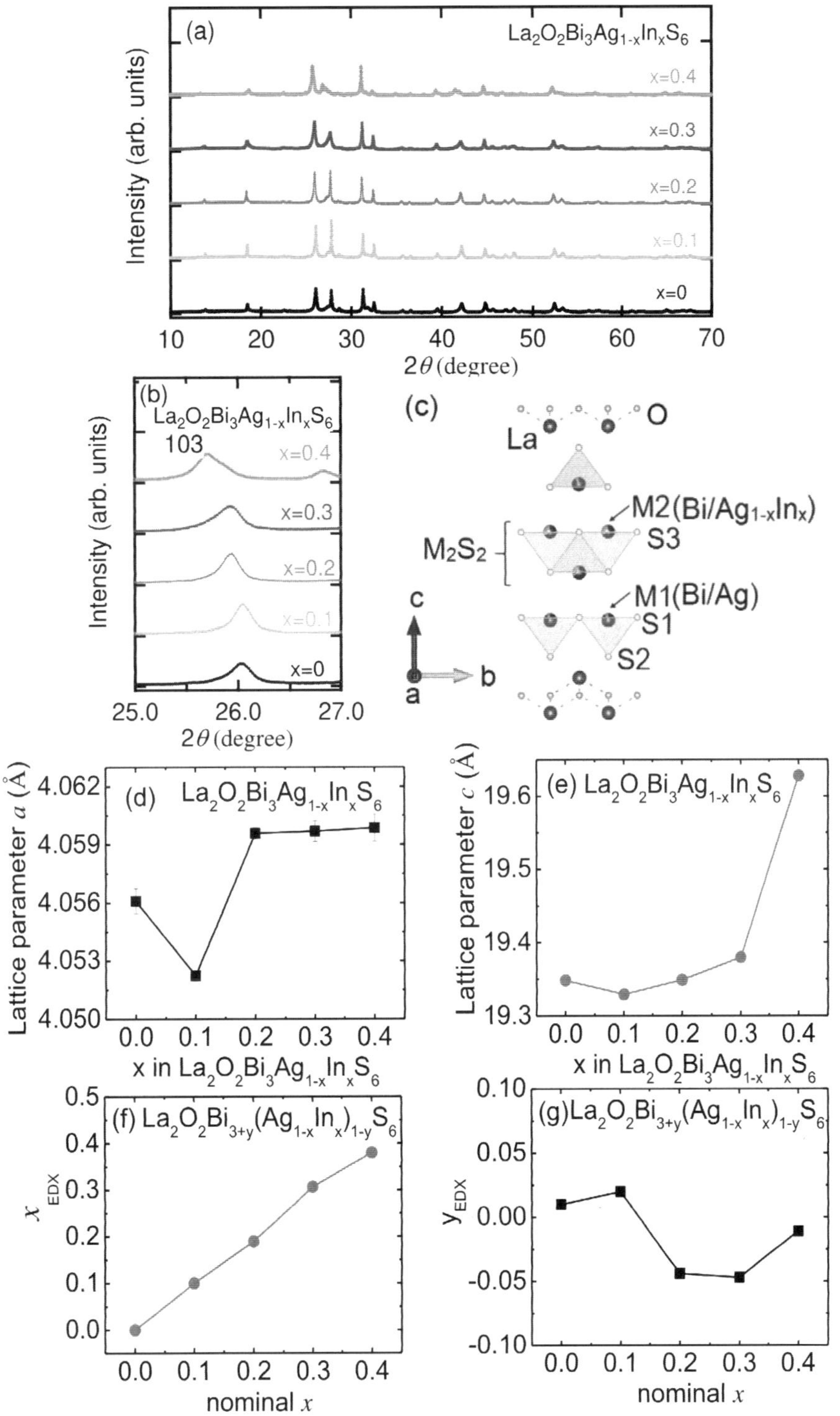

Figure. 1: (a) XRD patterns of $La_2O_2Bi_3Ag_{1-x}In_xS_6$ (x = 0-0.4) compounds scanned at room temperature. (b) Zoomed view of XRD patterns near the (103) (Miller indices) peak of tetragonal phase of the $La_2O_2Bi_3AgS_6$ structure. (c) The schematic unit cell of $La_2O_2Bi_3Ag_{0.9}In_{0.1}S_6$. (d, e) Lattice parameters a and c estimated by the Rietveld refinement. (f, g) The nominal composition dependence of x_{EDX} and y_{EDX} in $La_2O_2Bi_{3+y}(Ag_{1-x}In_x)_{1-y}S_6$ obtained by EDX.

Therefore, we assumed the composition of present samples as $La_2O_2Bi_3Ag_{1-x}In_xS_6$; hence, we name the samples using nominal composition of $La_2O_2Bi_3Ag_{1-x}In_xS_6$ in this paper. On the other hand, the In concentration is linearly increasing, which suggests that In was substituted at the Ag site as expected. Because x_{EDX} is almost the same with the nominal x and y_{EDX}<0.1, we use $La_2O_2Bi_3Ag_{1-x}In_xS_6$.

The temperature dependences of electrical resistivity [$\rho(T)$] from 300 to 0.1 K for the In-doped $La_2O_2Bi_3Ag_{1-x}In_xS_6$ (x = 0, 0.1, 0.2, 0.3, and 0.4) are shown in figure 2(a). We observed metallic behaviour with a hump at 180 K in the normal state electrical resistivity for x = 0, 0.1, 0.2, and 0.3. For x = 0.4, the normal state $\rho(T)$ shows a semiconducting-type behaviour. The anomaly temperature T^* clearly appeared for $x \leq 0.3$ and does not change with increasing x, while T^* for the x=0.4 has been confirmed from second order differentiation of normal state resistivity data. The $\rho(T)$ curves near the superconducting transition are shown in the figure 2(b). The T_c slightly shifts towards the lower temperature side with increasing In concentration in $La_2O_2Bi_3Ag_{1-x}In_xS_6$. The lowest T_c =0.4 K was observed for x = 0.4.

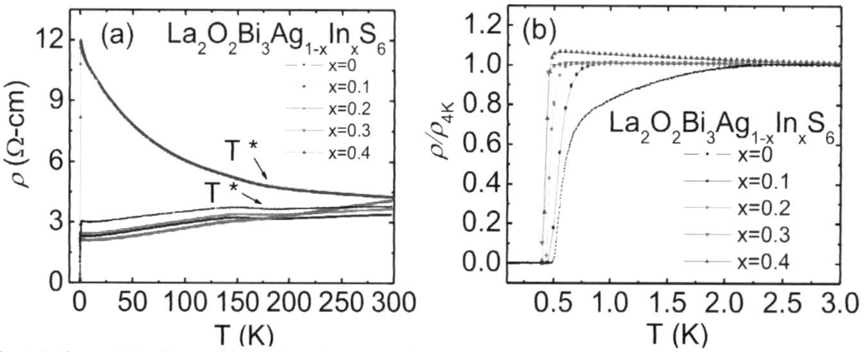

Figure. 2: (a) the $\rho(T)$ from 300 K to down to 0.1 K for the $La_2O_2Bi_3Ag_{1-x}In_xS_6$ (x=0-0.4) compounds. A hump structure appearing in the $\rho(T)$ curve is indicated by the arrow at $T = T^*$. (b) The normalized $\rho(T)/\rho(4\text{ K})$ curve in the temperature range 3.0-0.1K.

Figure 3 shows the room temperature Seebeck coefficient for $La_2O_2Bi_3Ag_{1-x}In_xS_6$ (x = 0, 0.1, 0.2, 0.3, and 0.4) samples. For all the samples, Seebeck coefficient is negative, which suggests that the dominant carrier is electron. In addition, we observed no clear change in the Seebeck coefficient by the In doping. The Seebeck coefficient measurements suggest that the In doping is not effective for carrier doping.

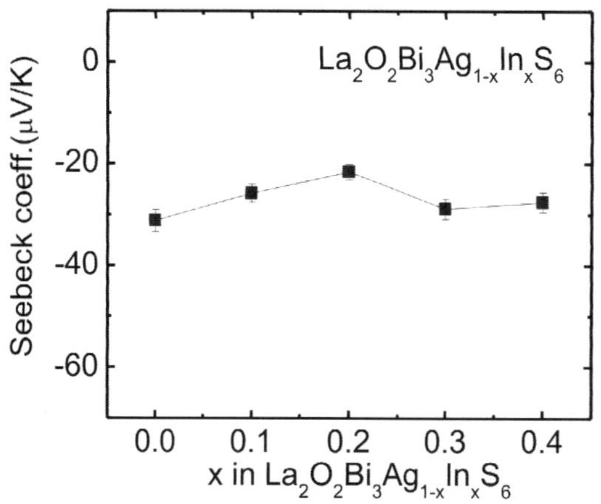

Figure. 3: The In concentration (x) dependence of Seebeck coefficient for $La_2O_2Bi_3Ag_{1-x}In_xS_6$ (x = 0-0.4) at room temperature.

Figure 4 exhibits the T vs x plot (phase diagram) of $La_2O_2Bi_3Ag_{1-x}In_xS_6$, which shows the change in superconducting transition T_c^{zero} with the In doping. The hump in the normal state resistivity (T^*) is not changing with increasing In doping. T_c slightly decreases with increasing x in $La_2O_2Bi_3Ag_{1-x}In_xS_6$. From the Seebeck coefficient, In substitution cannot increase carrier density. Therefore, the decrease in T_c for the In-doped samples might be caused by the negative in-plane chemical pressure effect caused by the in-plane expansion of the lattice by In doping. Furthermore, there is the possibility of slight substitution of In at the superconducting BiS plane (as indicated by $y_{EDX} < 0.1$). As reported in Ref. 26, $LaOInS_2$ was synthesized, but the InS_2 layer was highly disordered; split model for the In site was needed to refine the XRD data [26]. If small amount of In is substituted for the superconducting BiS plane, introduction of disorder is expected.

Although there is no evidence of charge-density-wave (CDW) ordering at T^* in the $La_2O_2Bi_3AgS_6$ system, one can assume that the CDW transition occurred at T^* from the analogy to the case of $EuFBiS_2$ [27].

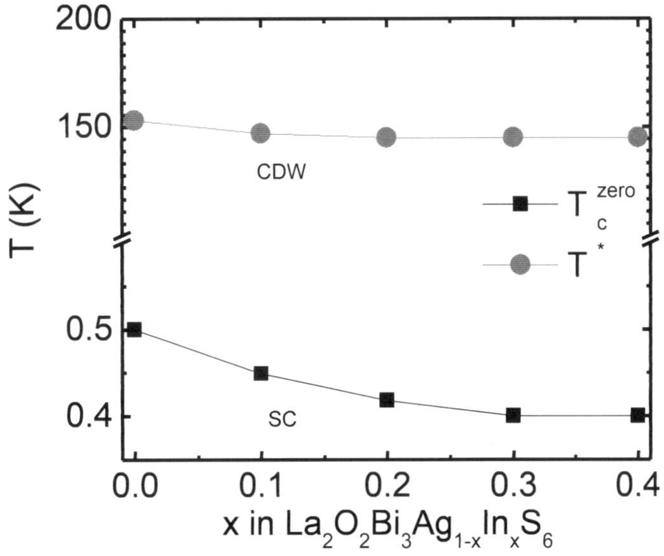

Figure. 4: The x dependence of T_c for $La_2O_2Bi_3Ag_{1-x}In_xS_6$ (x=0- 0.5) compounds and T^* scaled on the x-T phase diagram for $La_2O_2Bi_3Ag_{1-x}In_xS_6$. CDW and SC symbolises as charge density wave and superconductivity respectively.

Our latest results follow the contemporary scenario, we have observed decrease of the T^* with increasing Sn doping level and enhancement of T_c for the Sn-doped $La_2O_2Bi_3Ag_{1-x}Sn_xS_6$ system [18]. In contrast, T^* is not clearly decreasing or increasing, while T_c is decreasing marginally with increasing In doping. The interplay of CDW and superconducting orders has been reported through the high pressure study on CDW and superconductivity for the 2H-$NbSe_2$ system [28]. They observed that with increasing pressure an increase in T_c and a decrease in T_{CDW}, which shows that CDW phase and superconducting phase are the two competing phases [28]. Therefore, the contrasting evolution of superconductivity in the Sn-doped and In-doped systems should be related to the robustness of CDW ordering to the substitutions. On the basis of the discussion above, we propose that the suppression of superconductivity in In-doped $La_2O_2Bi_3Ag_{1-x}In_xS_6$ is caused by negative in-plane chemical pressure effect, introduction of disorder at the superconducting BiS plane by substitution of small amount of In. The EDX data show small amount of Bi deficiency, which suggests that the Bi site is slightly substituted by In. The decrease in T_c might be associated with the Bi site substitution by small amount of In. The Seebeck coefficient data suggest that the carrier concentration does not change sufficiently via In doping, which indicates that the CDW transition is robust against the In substitution. To precisely determine the doping site and local structure parameters, synchrotron X-ray diffraction experiment is needed.

4. Conclusion

We have investigated the substitution effect of Indium (In) at Ag site in $La_2O_2Bi_3AgS_6$. The T_c decreased with increasing In doping, and T_c^{zero} was 0.4 K for $x = 0.4$. The hump in the normal state resistivity at anomaly temperature ($T^* \sim 180$ K) remained unchanged with In doping up to $x = 0.3$. The $x = 0.4$ sample showed semiconducting-like behaviour with having CDW anomaly. The suppression of T_c in the In-doped $La_2O_2Bi_3Ag_{1-x}In_xS_6$ system is caused by negative in-plane chemical pressure effect, introduction of disorder at the superconducting BiS plane by substitution of small amount of In. The robustness of the CDW ordering to the In substitution also suggests that the carrier concentration does not change sufficiently.

Acknowledgments

This work was partly supported by Grants-in-Aid for Scientific Research (KAKENHI) (Grant Nos. 15H05886, 15H05884, 16H04493, 17K19058, 16K05454, and 15H03693).

References

[1] Mizuguchi Y, Fujihisa H, Gotoh Y, Suzuki K, Usui H, Kuroki K, Demura S, Takano Y, Izawa H, and Miura O 2012 *Phys. Rev. B* **86**, 220510
[2] Mizuguchi Y, Demura S, Deguchi K, Takano Y, Fujihisa H, Gotoh Y, Izawa H, and Miura O 2012 *J. Phys. Soc. Jpn.* **81**, 114725
[3] Singh S K, Kumar A, Gahtori B, Kirtan S, Sharma G, Patnaik S, and Awana V P S 2012 *J. Am. Chem. Soc.* **134**, 16504
[4] Demura S, Mizuguchi Y, Deguchi K, Okazaki H, Hara H, Watanabe T, Denholme S J, Fujioka M, Ozaki T, Fujihisa H, Gotoh Y, Miura O, Yamaguchi T, Takeya H, Takano Y 2013 *J. Phys. Soc. Jpn.* **82**, 033708
[5] Jha R, Kumar A, Singh S K, Awana V P S 2013 *J. Supercond. Nov. Magn.* **26**, 499
[6] Xing J, Li S, Ding X, Yang H, Wen H H 2012 *Phys. Rev. B* **86**, 214518
[7] Yazici D, Huang K, White B D, Chang A H, Friedman A J, Maple M B 2012 *Philos. Mag.* **93**, 673
[8] Yazici D, Huang K, White B D, Jeon I, Burnett V W, Friedman A J, Lum I K, Nallaiyan M, Spagna S, Maple M B 2013 *Phys. Rev. B* **87**, 174512
[9] Krzton-Maziopa A, Guguchia Z, Pomjakushina E, Pomjakushin V, Khasanov R, Luetkens H, Biswas P, Amato A, Keller H, Conder K 2014 *J. Phys.: Condens. Matter* **26**, 215702
[10] Mizuguchi Y, Omachi A, Goto A, Kamihara Y, Matoba M, Hiroi T, Kajitani J, Miura O 2014 *J. Appl. Phys.* **116**, 163915
[11] Lin X, Ni X, Chen B, Xu X, Yang X, Dai J, Li Y, Yang X, Luo Y, Tao Q, Cao G, Xu Z 2013 *Phys. Rev. B* **87**, 020504
[12] Kotegawa H, Tomita Y, Tou H, Izawa H, Mizuguchi Y, Miura O, Demura S, Deguchi K, Takano Y 2012 *J. Phys. Soc. Jpn.* **81**, 103702
[13] Wolowiec C T, Yazici D, White B D, Huang K, Maple M B 2013 *Phys. Rev. B* **88**, 064503
[14] Wolowiec C T, White B D, Jeon I, Yazici D, Huang K, Maple M B 2013 *J. Phys.: Condens. Matter* **25**, 422201
[15] Jha R, Tiwari B, Awana V P S 2014 *J. Phys. Soc. Jpn.* **83**, 063707
[16] Jha R, Tiwari B, Awana V P S 2015 *J. Appl. Phys.* **117**, 013901
[17] Jha R, Goto Y, Higashinaka R, Matsuda T D, Aoki Y, and Mizuguchi Y 2018 *J. Phys. Soc. Jpn.* **87**, 083704

[18] Jha R, Goto Y, Matsuda T D, Aoki Y, and Mizuguchi Y 2018 (Preprint arXiv:1810.08404)
[19] Sun Y L, Ablimit A, Zhai H F, Bao J K, Tang Z T, Wang X B, Wang N L, Feng C M, and Cao G H 2014 *Inorg. Chem.* **53**, 11125
[20] Mizuguchi Y, Hijikata Y, Abe T, Moriyoshi C, Kuroiwa Y, Goto Y, Miura A, Lee S, Torii S, Kamiyama T, Lee C H, Ochi M, Kuroki K 2017 *EPL* **119** 2600
[21] Mizuguchi Y, Miura A, Kajitani J, Hiroi T, Miura O, Tadanaga K, Kumada N, Magome E, Moriyoshi C, Kuroiwa Y 2015 *Sci. Rep.* **5,** 14968
[22] Hijikata Y, Abe T, Moriyoshi C, Kuroiwa Y, Goto Y, Miura A, Tadanaga K, Wang Y, Miura O, and Mizuguchi Y 2017 *J. Phys. Soc. Jpn.* **86**, 124802
[23] Sogabe R, Goto Y, Mizuguchi Y 2018 *Appl. Phys. Express* **11**, 053102
[24] zumi F and Momma K 2007 *Solid State Phenom.* **130,** 15
[25] Momma K and Izumi F 2008 *J. Appl. Crystallogr.* **41**, 653
[26] Miura A, Oshima T, Maeda K, Mizuguchi Y, Moriyoshi C, Kuroiwa Y, Meng Y, Wen X D, Nagao M, Higuchi M, and Tadanaga K 2017 *J. Mater. Chem. A* **5**, 14270
[27] Zhai H F, Tang Z T, Jiang H, Xu K, Zhang K, Zhang P, Bao J K, Sun Y L, Jiao W H, Nowik I, Felner I, Li Y K, Xu X F, Tao Q, Feng C M, Xu Z A, Cao G H 2014 *Phys. Rev. B* **90**, 064518
[28] Jerome D, Berthier C, Molinie P and Rouxel J 1976 *J. Physique Colloq.* **437C**, 125

Superconductivity in Weyl Semimetal NbP: Bulk vs. Surface

M Baenitz[1], M Schmidt[1], V Suess[1], C Felser[1] and K Lüders[1,2]

[1] Max-Planck-Institut für Chemische Physik fester Stoffe, Nöthnitzer Str. 40, 01187 Dresden, Germany

[2] Fachbereich Physik, Freie Universität Berlin, Arnimallee 14, 14195 Berlin, Germany

baenitz@cpfs.mpg.de

Abstract. Transition metal monopnictides belong to the new class of semimetals where the bulk properties are determined by the presence of pairs of nodes with different chirality formed by linear dispersive states in the k-space. Beside the anomaly in the bulk magnetotransport superconductivity is frequently found in some Weyl semimetals. We found signatures of superconductivity in ac and dc magnetization measurements of highly pure and stoichiometric NbP powder. We determined the lower and upper critical field and the Ginzburg-Landau parameter. The relative small superconducting volume fraction is related to either effect of finite grain size and/or surface superconductivity. The last mentioned may originate from either off stoichiometric (Nb-rich) surface layers or a strained surface with different electronic properties. Furthermore the intrinsic normal state susceptibility is determined taking into account a paramagnetic contribution of a few ppm of magnetic impurities.

1. Introduction

Niobium and its compounds play a major role in superconductivity including its applications to particle accelerators and superconducting magnets. Among the element superconductors niobium has with T_c = 9.5 K the highest transition temperature and belongs to the three type-II superconductors. Many Nb compounds exhibit rather high transition temperatures, like NbN (T_c = 17 K), NbC (T_c = 11.1 K) or Nb_3Ge which holds with T_c = 23.2 K the record among the binary compounds.

The system NbN belongs to the Nb monopnictides. Within these systems NbP is one of the recently discovered Weyl semimetals, a novel state of topological quantum matter with exotic transport properties [1-4]. Some of the Weyl semimetals exhibit superconductivity (e.g. $MoTe_2$) and there is a discussion if so-called type-II Weyl semimetals are the prime candidates for superconductivity [5, 6]. However, well prepared single crystals of NbP show no sign of superconductivity down to lowest temperatures [7]. Nonetheless there is strong evidence from various methods that Nb and even Ta based monopnictides are at the verge of superconductivity [8, 9]. It should also be mentioned that the magnetic counterparts MnP and CrAs show pressure induced superconductivity [10. 11].

On the other hand Nb based pnictide compounds with N, P or As show superconductivity. Well known is the system NbN which was already discovered in 1941. It shows a transition temperature of T_c = 17 K and a penetration depth of the magnetic field of λ = 200 nm. Recently it is also discussed for accelerator applications. Less known are the superconducting properties of NbP. Recently a transition temperature of T_c = 7.5 K was published for NbP powder [12]. Some transition temperatures were published earlier for sputtered Nb-P samples in the composition range from 17 to 31 at.% P with T_c-

values from 6 K up to 7.5 K for 21 at.% P [13]. Nearly nothing is known about NbAs, the third compound in this group. Some experiments seem to indicate a superconducting transition at about 10 K [14].

In this publication results of superconducting properties of NbP powder are reported. Surprisingly superconductivity occurs after milling NbP single crystals to powder samples.

2. Experimental

Several powder samples of NbP are obtained by milling NbP single crystals the preparation and properties of which are described in ref. [3]. The largest NbP powder sample has a mass of about 500 mg (sample 1) and was used for detailed Nb- and P-NMR studies [15]. All powder samples are investigated by X-ray diffraction and lattice parameters obtained are in good agreement with the literature [16]. No sign of foreign phases could be found. It is remarkable that we found well defined de Haas-van Alphen oscillations even in powder samples (see supplemental material of ref. [17]). This we take as an evidence for the presence of rather pure single crystalline grains in the powder samples.

Zero field cooled (zfc), field cooled (fc), magnetization, and ac susceptibility measurements are performed by means of commercially available magnetometers (MPMS and PPMS, Quantum Design Inc.).

3. Results and discussion

3.1. Structure and phase diagram

Figure 1 shows a scanning electron microscope (SEM) image of NbP powder. Different structure elements can be seen: small grains of about 100 nm size collected in larger conglomerates and larger blocs of parallel bars.

Figure 2 shows the superconducting phase diagram, i.e. the temperature dependence of the lower and upper critical field $B_{c1}(T)$ and $B_{c2}(T)$ determined from zfc-fc measurements and magnetization curves. The onset transition occurs at $T_c = 7.8$ K. The phase diagram exhibits a clear type-II behavior without irreversibility line. The extrapolated values of the critical fields at $T = 0$ are $B_{c1}(0) = 34$ mT and $B_{c2}(0) = 1.64$ T, respectively.

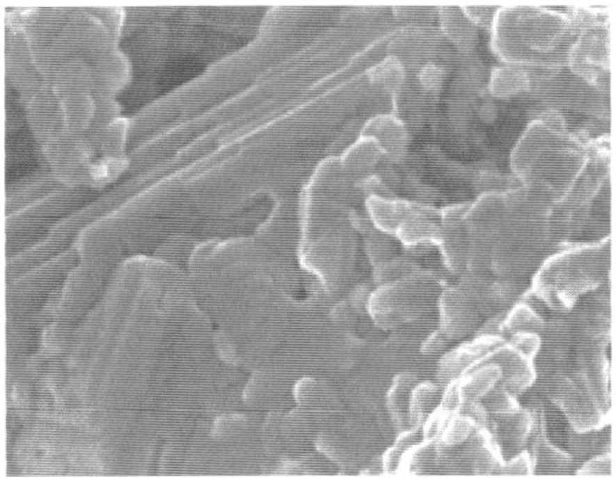

Figure 1. SEM image of NbP powder (scale: 3 x 2.25 μm)

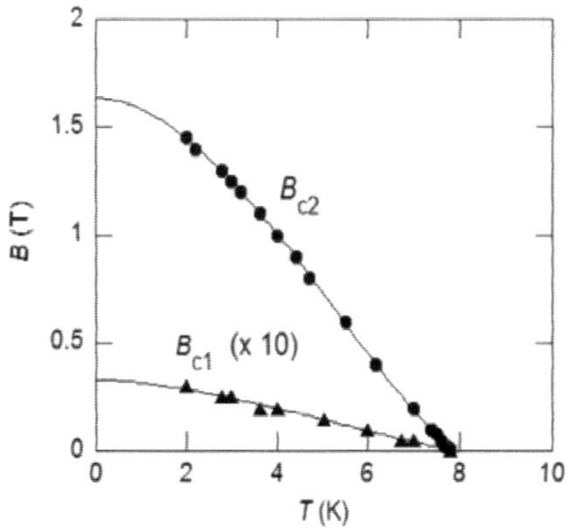

Figure 2. Field-temperature diagram of NbP powder

3.2. Superconducting behavior

The upper critical field value at $T = 0$, $B_{c2}(0) = 1.64$ T, is slightly higher than the theoretical value of the WHH theory [18, 19], $B^*_{c2}(0) = 1.5$ T, with the assumption of a clean limit and the resulting Maki parameter $\alpha = 0.15$. The reason for that might be a multi-phase grain structure of the sample or surface superconductivity. Following the expressions connecting the Ginzburg-Landau parameter κ with the ratio of the two critical fields [20], leads to a value of κ = 6.96. This κ-value clearly indicates superconducting type-II behavior and is of order of other Nb compounds. It is considerably larger than that of pure niobium (κ_{Nb} = 0.8) but well below κ = 40, the value of the A-15 compound Nb_3Sn.

The magnetic penetration depth $\lambda(0)$ is determined from $B_{c2}(0)$ via the Ginzburg-Landau coherence length $\xi_{GL}(0)$ using the relations $\lambda = \kappa \cdot \xi$ and $\xi^2 = \Phi_0/2\pi B_{c2}$ (Φ_0 = flux quantum = $h/2e$). With κ = 6.96 and $\xi_{GL}(0)$ = 14.2 nm the penetration depth results to $\lambda_{GL}(0)$ = 98.8 nm. Figure 3 shows zfc and fc curves for two magnetic fields. Plotted is the volume susceptibility which allows the determination of the superconducting volume fraction. For sample 2 which is shown in the figure it results to 6.2 %. Due to the presence of small grains (with dimensions in the order of the penetration depth) the determined fraction underestimates the real superconducting volume. An analysis of magnetization curves leads to the same value. As an example the magnetization curve for T = 2 K of sample 2 is plotted in Figure 4.

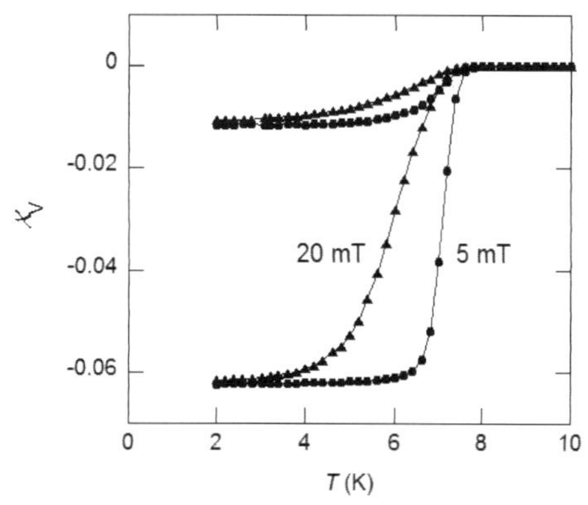

Figure 3. Zfc and fc transition curves in two magnetic fields. Plotted is the volume susceptibility calculated from the measured magnetic moment.

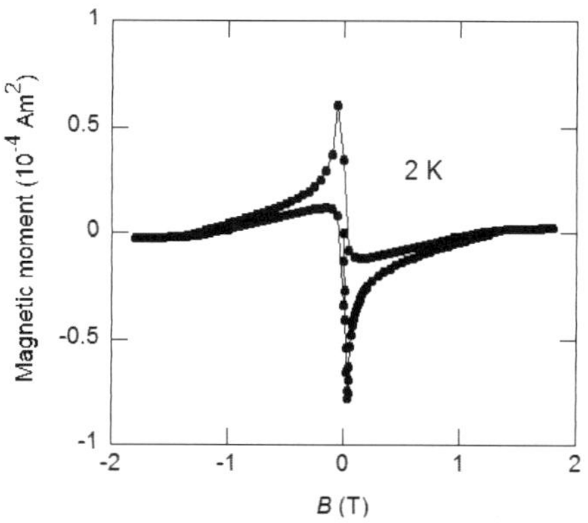

Figure 4. Magnetization curve of NbP powder obtained at a temperature of 2 K. Here the magnetic moment is plotted against the magnetic field B.

3.3. Ac susceptibility

In order to get more information about the vortex dynamics ac susceptibility measurements are performed. The temperature dependence of the real and imaginary parts of the ac susceptibility, χ' and χ'', was recorded starting at low temperatures up to temperatures well above the transition temperature and back again. Magnetic fields are applied up to B = 40 mT. Frequency and ac field amplitude were fixed to 0.1 kHz and 10 mT, respectively. The real parts are in rather good agreement with the respective zfc transition curves.

Figure 5 shows a comparison of imaginary parts of the ac susceptibility for two samples. In the case of sample 1 only one peak occurs starting at the transition temperature of T_c = 7.6 K and passing its maximum at T = 7.1 K. For sample 2, however, a second peak occurs at lower temperature with its maximum at T = 6.4 K. It originates most likely from intergranular superconducting contacts between the grains.

In Figure 6 the imaginary parts of the ac susceptibility are plotted against their real parts. The peak height of χ'' is about 14 % of the maximum values of χ'. This kind of plot can be compared with theoretical predictions [21] based either on Bean's critical state model [22] or other models based on diffusive motion of the flux lines for example [23]. In our case, however, the curves are in excellent agreement with the prediction of the critical state model with a maximum at $\chi' = -0.375$.

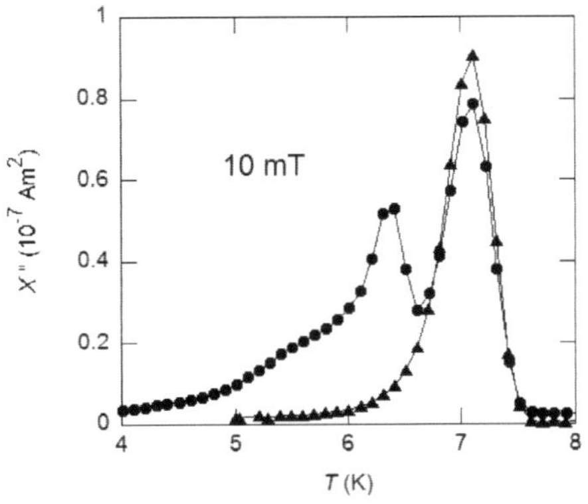

Figure 5. Comparison of the imaginary parts χ'' of the ac susceptibility for sample 1 (triangles) and 2 (circles) obtained in a magnetic field of 10 mT

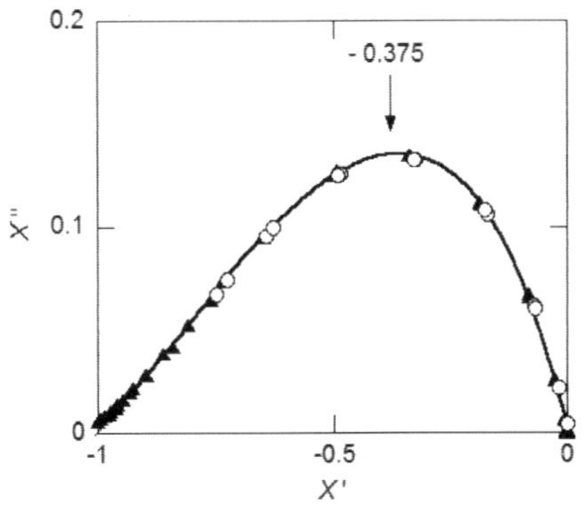

Figure 6. Imaginary parts χ'' plotted against their real parts χ' of the ac susceptibility at a magnetic field of 10 mT (sample 1 triangles and 2 open circles). The axes are normalized to the lowest value of χ'.

3.4. Magnetic behavior

Figure 7 shows as an example the fc curve of sample 1 in an applied magnetic field of $B = 0.25$ T. In order to investigate the magnetic properties of NbP powder at first the temperature independent contribution was determined. The temperature dependence of the non-superconducting part of the curve evidences the presence of i) a Curie-like paramagnetism and ii) a temperature and field independent diamagnetic contribution χ_{Dia}. A $1/T$-extrapolation of the non-superconducting part to infinite temperature leads to the diamagnetic contribution. It is represented by the horizontal line in Figure 7 whereas the Curie-like contribution is represented by the curved line. The corresponding diamagnetic volume susceptibility is $\chi_{Dia} = -4.0 \cdot 10^{-5}$ ($-6.1 \cdot 10^{-5}$ emu/mol). The diamagnetic core contribution for NbP is estimated in ref. [16] to be $-7.4\ 10^{-5}$ emu/mol. This suggests a small positive intrinsic powder susceptibility for NbP of $+1.3 \cdot 10^{-5}$ emu/mol due to a small but finite number of carriers.

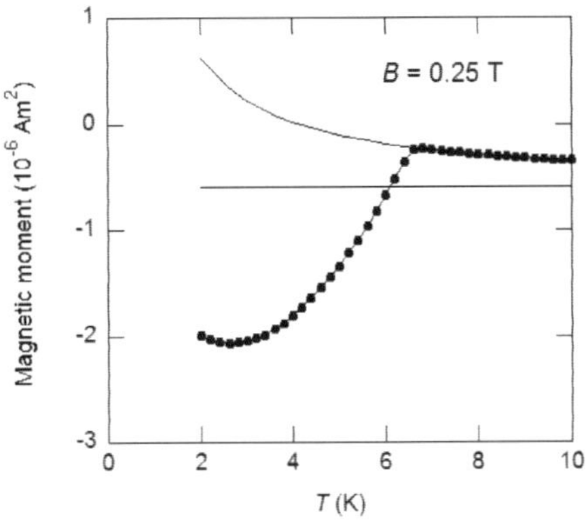

Figure 7. Diamagnetic (horizontal line) and paramagnetic (upper $1/T$-line) contributions extrapolated from the non-superconducting part of an fc curve at the magnetic field $B = 0.25$ T (sample 1)

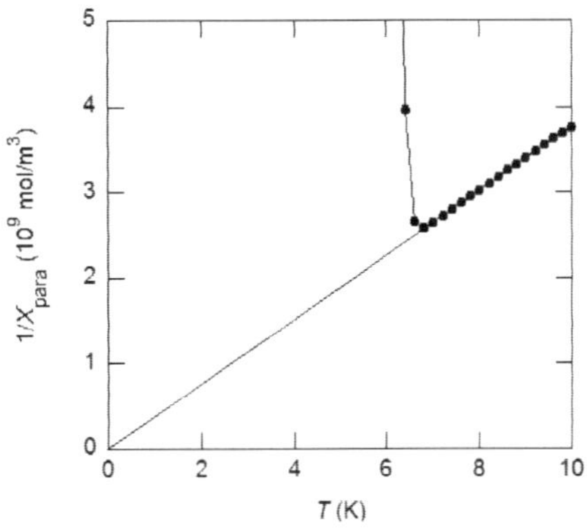

Figure 8. Plot of the same data for the determination of the paramagnetic moment

To determine the paramagnetic effective Curie moment μ the reciprocal paramagnetic volume susceptibility is plotted against the temperature (Figure 8). We obtain a value of $\mu = 0.041\ \mu_B$ (μ_B = Bohr magneton). Comparable values were found for the other samples as plotted in Table 1. This magnetism originate probably from a very tiny amount (about 60 ppm) of magnetic ions (probably Fe or Ni) which is in the specified purity range of commercially available elements used for the synthesis (see discussion in ref. [17]).

Table 1. Superconductivity and magnetic properties of NbP powder (μ_B = Bohr magneton = $9.274 \cdot 10^{-24}$ Am^2)

Sample	1	2	3
T_c (K)	7.6	7.8; 6.7	8.7; 7.2
$B_{c2}(0)$ (T)		1.64	
$B_{c1}(0)$ (T)		0.034	
superconducting volume fraction (%)	0.4	6.2	0.02
diamagnetic volume susceptibility	$-3.5 \cdot 10^{-5}$	$-4.0 \cdot 10^{-5}$	from $-2.1 \cdot 10^{-5}$ to $-6.1 \cdot 10^{-5}$
paramagnetic moment	$0.041\ \mu_B$	$0.045\ \mu_B$	$0.043\ \mu_B$

Conclusions

Powder samples of highly pure and stoichiometric NbP exhibit superconductivity with transition temperatures between 6 and 9 K in accordance with other published values for Nb-P compounds. The superconducting phase diagram shows a clear type-II behavior with a Ginzburg-Landau parameter of about $\kappa = 7$ and a penetration depth of about $\lambda_{GL}(0) = 100$ nm. Although the transition temperatures are similar to those of pure Nb the behavior in magnetic fields is quite different. The Ginzburg-Landau parameter as well as the upper critical field is larger by a factor of 6 to 7. The relatively small superconducting volume fraction of a few percent might be due to grain size effects. It might be speculated that the powder grains are strained and the system converts into a type-II Weyl semimetal for which superconductivity is more likely. A more detailed study is required to support this scenario.

Acknowledgments

The authors would like to thank H. Rave for her support in the experiments. We thank P. Werner (MPI of Microstructure Physics, Halle) for providing the SEM picture. One of us (KL) would like to thank C. Felser and A.P. Mackenzie for their kind hospitality and the possibility of doing research at the MPI-CPfS.

References

[1] Weng H, Fang C, Fang Z, Bernevig B A and Dai X 2015 *Phys. Rev.* X **5** 011029
[2] Sun Y, Wu S-C and Yan B 2015 *Phys. Rev.* B **92** 115428
[3] Shekhar C, Nayak A K, Sun Y, Schmidt M, Nicklas M, Leermakers I, Zeitler U, Skourski Y, Wosnitza J, Liu Z, Chen Y, Schnelle W, Borrmann H, Grin Y, Felser C and Yan B 2015 *Nature Physics* **11** 645
[4] Klotz J, Wu S-C, Shekhar C, Sun Y, Schmidt M, Nicklas M, Baenitz M, Uhlarz M, Wosnitza J, Felser C and Yan B 2016 *Phys. Rev.* B **93** 121105
[5] Alidoust M, Halterman K and Zyuzin A A 2017 *Phys. Rev.* B **95** 155124
[6] Rosenstein B, Shapiro B Y, Li D and Shapiro I 2018 *Phys. Rev.* B **97** 144510

[7] Stockert U, dos Reis R D, Ajeesh M O, Watzman S J, Schmidt M, Shekhar C, Heremans J P, Felser C, Baenitz M and Nicklas M 2017 *J. Phys.: Condens. Matter* **29** 325701
[8] Bachmann M D, Nair N, Flicker F, Ilan R, Meng T, Ghimire N J, Bauer E D, Ronning F, Analytis J G and Moll P J W 2017 *Sci. Adv.* **3** e160298
[9] Aggarwal L, Gayen S, Das S, Kumar R, Süß V, Felser C, Shekhar C and Sheet G 2016 *Nature Communications* **8** 13974
[10] Cheng J G, Matsubayashi K, Wu W, Sun J P, Lin F K, Luo J L and Uwatoko Y 2015 *Phys. Rev. Lett.* **114** 117001
[11] Wu W, Cheng J, Matsubayashi K, Kong P, Lin F, Jin C, Wang N, Uwatoko Y and Luo J 2014 *Nature Communications* **5** 5508
[12] Kumar P, Sudesh and Patnaik S 2016 *AIP Conf. Proc.* **1731** 140063
[13] Johnson G R and Douglass D H 1974 *Journal of Low Temperature Physics* **14** 575
[14] Baenitz M, to be published
[15] Yasuoka H et al., to be published
[16] Xu J, Greenblatt M, Emge T, Höhn P, Hughbanks T and Tian Y 1996 *Inorg. Chem.* **35** 845
[17] Yasuoka H, Kubo T, Kishimoto Y, Kasinathan D, Schmidt M, Yan B, Zhang Y, Tou H, Felser C, Mackenzie A P and Baenitz M 2017 *Phys. Rev. Lett.* **118** 236403
[18] Helfand E and Werthamer N R 1966 *Phys. Rev.* **147** 288
[19] Werthamer N R, Helfand E and Hohenberg P C 1966 *Phys. Rev.* **147** 295
[20] Brandt E H 2003 *Phys. Rev.* B **68** 054506
[21] Ling X and Budnick J I 1991 *Magnetic Susceptibility of Superconductors and other Spin Systems*, ed R A Hein *et al* (New York and London: Plenum Press) p 377
[22] Bean C P 1964 *Rev. Mod. Phys.* **34** 31
[23] Khoder A F and Couach M 1991 *Magnetic Susceptibility of Superconductors and other Spin Systems*, ed R A Hein *et al* (New York and London: Plenum Press) p 213

Effect of non-magnetic rare earth substitution for Zr on mixed anion Zr(P, Se)$_2$ superconductors II

K Iwakiri[1, 2], T Nishio[1], K Kawashima[2, 3], S Ishida[2], K Oka[2], H Fujihisa[2], Y Gotoh[2], I Hase[2], A Iyo[2], H Ogino[2], H Eisaki[2], Y Yoshida[2] and H Kitô[2]

[1]Department of Physics, Tokyo University of Science, 1-3 Kagurazaka, Shinjuku-ku, Tokyo 162-8601, Japan

[2]National Institute of Advanced Industrial Science and Technology (AIST), 1-1-1 Umezono, Tsukuba, Ibaraki 305-8568, Japan

[3]IMRA Material R&D Co., Ltd, Kariya, Aichi 448-0032, Japan

E-mail: h.kito@aist.go.jp

Abstract. In $ZrP_{1.25}Se_{0.75}$, the superconducting transition temperature T_c is increased from 5.8 K to 7.6 K when Sc atoms are partially substituted for Zr atoms having almost the same atomic radius as Sc atoms. Transport measurements show that holes are doped with substituting Sc atoms. The density of states in (Zr, Sc)$P_{1.25}Se_{0.75}$ is derived from band structure calculations and its implication in the doping behaviour and the increase of T_c is discussed.

1. Introduction

Recently, it has been reported the discovery of many new superconductors having the PbFCl-type structure such as Zr-P-Se, Zr-P-S, Hf-P-Se, and Hf-P-S intermetallic compounds and those superconducting properties [1]. Zr($P_{2-x}Se_x$) has a double block layer of Zr and ($P_{1-x}Se_x$) sandwiched between electrically conducting P-square lattice sheets along *c*-axis (Fig. 1.), which is similar to high T_c cuprates [2, 3] and iron-based superconductors [4 5]. The maximum superconducting transition temperature T_c is 6.3 K at x=0.75 [6] and changes to 7.2 K with a dome-like substitution dependence [7], as seen in high T_c cuprates and iron-based superconductors, with substituting Lu atoms for Zr atoms [8]. According to the similarities, PbFCl-type materials have been attracted much attention in terms of possibility that T_c becomes even higher [2~5].

In this paper, we report that T_c in $ZrP_{1.25}Se_{0.75}$ substituted by Sc atoms for Zr atoms reach to 7.6 K. The origin of the increase in T_c is discussed based on data obtained by transport measurements, band structure calculations, and others.

2. Experimental details

Polycrystalline samples of ($Zr_{1-y}Sc_y$)$P_{1.25}Se_{0.75}$ were synthesized by using a cubic-anvil high pressure apparatus. Powders of P (99.9999 %; Furuuchi Chemical), Se (99.999 %; Kojundo Chemical Laboratory), Zr (98%; Kojundo Chemical Laboratory), and Sc (99.9 %; Kojundo Chemical Laboratory) at a molar ratio of 1.25 : 0.75 : 1-y : y were mixed in a N_2 filled glove box and then heated

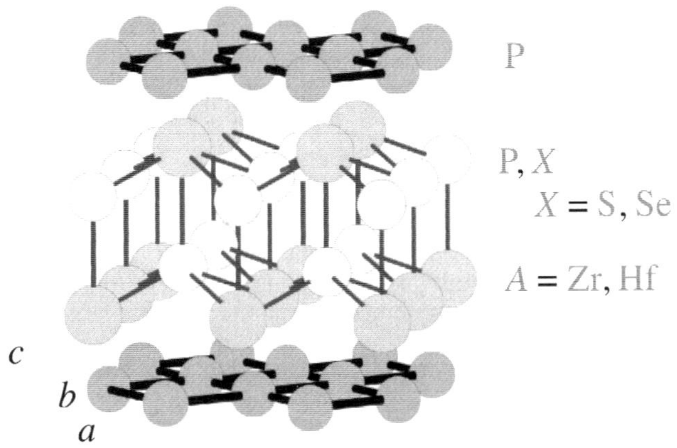

Figure 1. Schematic image of the crystal structure of $AP_{2-x}X_x$ (A = Zr, Hf; X = S, Se).

in a BN crucible at 1,425℃ under a pressure of 2 GPa for 1 hour. Finally, the mixture was quenched to room temperature. Powder X-ray diffraction (XRD) patterns were measured within the range of 5° to 80° in 2θ at room temperature. Magnetization for samples as a function of temperature was measured at 20 Oe by a SQUID magnetometer (MPMS, Quantum design). Transport measurements were carried out by a four-terminal method for rectangular samples. Scanning electron microscope (SEM) and energy dispersive X-ray spectrometry (EDX) measurements were carried out for composition analysis. Band structure calculations were made by using the WIEN2k package [9]. This calculation is based on the density functional theory (DFT) and the full-potential linearized augmented plane wave method (FLAPW), and the exchange-correlation potential is approximated by general gradient approximation (GGA) [10]. The parameter RK_{max} is chosen as 7.0. As for the self-consistent calculation we used the 1000 k-points mesh in the 1st Brillouin zone. After convergence of potential, we performed the DOS calculation using 8000 k-points mesh. Spin-orbit interaction (SOI) has been included by a second-variational approach in this calculation.

3. Results and discussion

Figure 2 shows XRD patterns of $(Zr_{1-y}Sc_y)P_{1.25}Se_{0.75}$ ($0 \leqq y \leqq 0.70$, where y shows a nominal value).

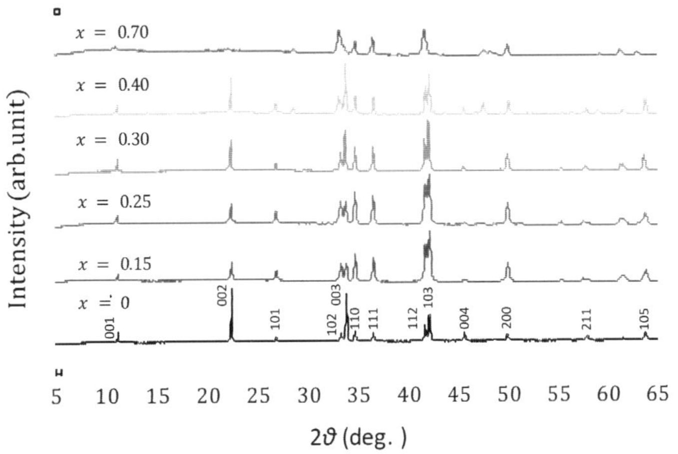

Figure 2. Powder XRD patterns of $(Zr_{1-y}Sc_y)P_{1.25}Se_{0.75}$ for y < 0.7.

Below y=0.3, $(Zr_{1-y}Sc_y)P_{1.25}Se_{0.75}$ seems not to have nontrivial impurities. Since Zr and Sc atoms are more or less the same atomic radius, Zr atoms are successfully replaced by Sc atoms. On the other hand, Y, La, and Lu substitutions for Zr atoms yield impurity phases even in the low degree of substitution [8].

Figure 3 shows lattice parameters in the a-axis and c-axis as a function of y for $(Zr_{1-y}Sc_y)P_{1.25}Se_{0.75}$.

These lattice parameters do not show a significant change with increasing y, which is consistent with the thing that Zr atoms are successfully replaced by Sc atoms.

Figure 4(a) shows magnetic susceptibility for $(Zr_{1-y}Sc_y)P_{1.25}Se_{0.75}$ as a function of temperature. Apparently, T_c increases with increasing y in the range of y<0.25 and reaches to the maximum T_c to be 7.6 K at y=0.25. On the other hand, T_c decreases in the range of y>0.25. The T_c behaviour looks like a dome-like substitution dependence, as shown in Figure 4(b). Besides T_c, the superconducting volume fraction somewhat fluctuates with increasing y and at y=0.7 it is very low. It indicates that at y=0.7 samples have a large proportion of impurity phases.

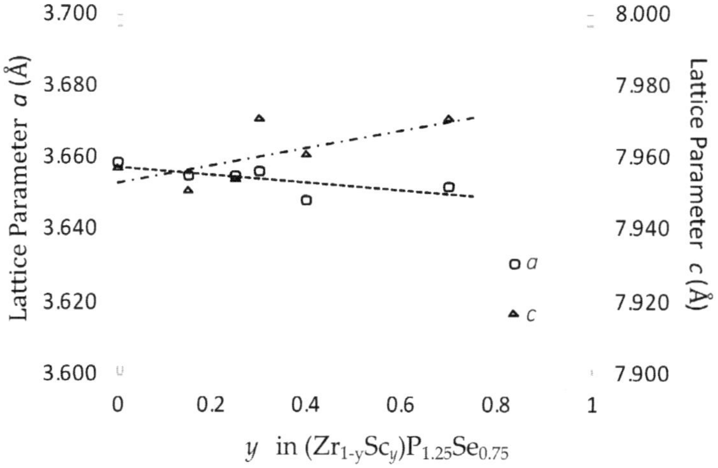

Figure 3. Lattice parameters in the a-axis and c-axis as a function of y for $(Zr_{1-y}Sc_y)P_{1.25}Se_{0.75}$.

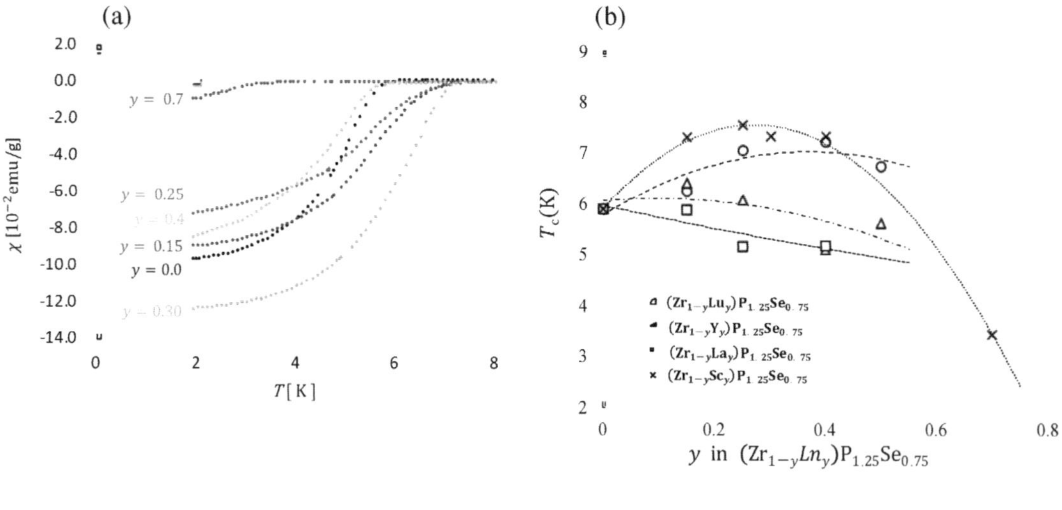

Figure 4. (a) T-dependence of magnetic susceptibility for $(Zr_{1-y}Sc_y)P_{1.25}Se_{0.75}$. (b) y-T_c phase diagram of $(Zr_{1-y}Ln_y)P_{1.25}Se_{0.75}$ (Ln =Sc, Y, La and Lu)

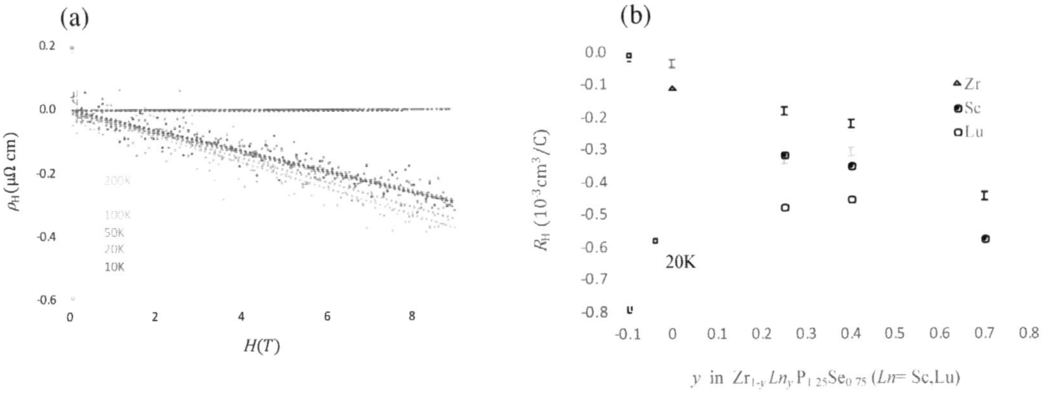

Figure 5. (a) H dependence of ρ_H for $y = 0.25$ at 200, 150, 100, 50, 20 and 10K. (b) y dependence of R_H.

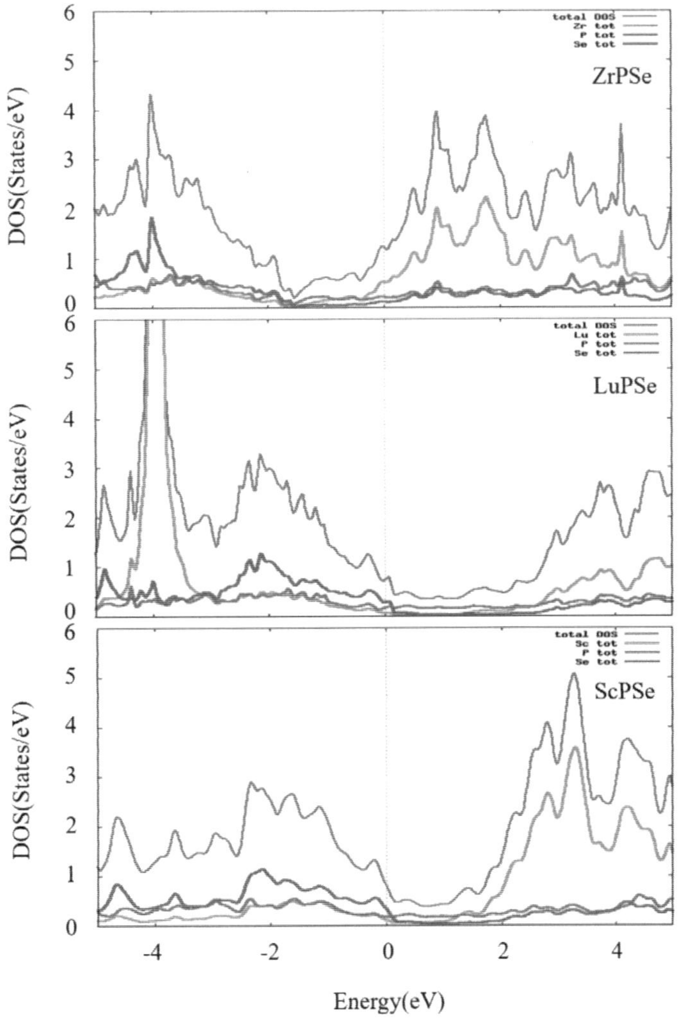

Figure 6. The density of states around the Fermi energy derived from band structure calculations for ZrPSe, LuPSe, and ScPSe.

Figure 5(a) shows the Hall resistivity ρ_H for y=0.25 obtained by Hall effect measurements at different temperatures as a function of magnetic field. The ρ_H values show a linear magnetic field dependence which becomes weaker as the temperature decreases. The same tendency has been also observed for different y values and the Lu substitution.

Figure 5(b) shows the Hall coefficient R_H taken at 20 K for the Lu and Sc substitution as a function of y. All the R_H values are negative and R_H decreases with the substitution. Thus, it is considered that electrons are dominated in charge carriers and holes are doped with the substitution.

Figure 6 shows the density of states around the Fermi energy derived from band structure calculations for ZrPSe, LuPSe, and ScPSe. Essentially, there are analogies between them. For $(Lu_{0.5}Sc_{0.5})P_{1.25}Se_{0.75}$, and $(Zr_{0.5}Sc_{0.5})P_{1.25}Se_{0.75}$, the Fermi energy shifts to lower energy in comparison with that in Zr $P_{1.25}Se_{0.75}$, indicating that holes are doped with the substitution. This is consistent with the Hall effect measurements above mentioned. It seems that a rigid band picture is valid in Zr $P_{1.25}Se_{0.75}$ with the substitution. In this picture, the density of states becomes lower with Fermi energy shifts, so that it is not likely that T_c is raised with the substitution in terms of the density of states. Although irregularity in Zr sites may concern an increase of T_c, this problem still remains unsolved.

4. Summary

In $Zr_{1-y}Sc_yP_{1.25}Se_{0.75}$, T_c increased from 6.3 K to 7.6 K as y changed from 0 to 0.25. The maximum T_c of 7.6 K is the highest value in Zr-P-Se compounds. The increase of T_c is not consistent with Hall effect measurements and band structure calculations. To solve this problem future work is needed.

Acknowledgments

This work partially supported by JSPS KAKENHI Grant-in-Aid for Scientific Research on Innovative Areas "Mixed anion" (Grant Number JP16H6439).

References

[1] Kitô H, Yanagi Y, Ishida S, Oka K, Gotoh Y, Fujihisa H, Yoshida Y, Iyo A and Eisaki H 2014 *J. Phys. Soc. Japan* **83** 074713
[2] Bednorz J G and Müller K A 1986 *Z. Phys.* B **64** 189
[3] Wu M K, Ashburn J R, Torng C J, Hor P H, Meng R L, Gao L, Huang Z J, Wang Y Q, and Chu C W 1987 *Phys. Rev. Lett.* **58** 908
[4] Kamihara Y, Watanabe T, Hirano M, and Hosono H 2008 *J. Am. Chem. Soc.* **130**, 3296-3297
[5] Rotter M, Tegel M, and Johrendt D 2008 *Phys. Rev. Lett.* **101** 107006
[6] Ishida S, Fujihisa H, Hase I, Yanagi Y, Kawashima K, Oka K, Gotoh Y, Yoshida Y, Iyo A, Eisaki H and Kitô H 2016 *Supercond. Sci. Technol.* **29** 055004
[7] Takagi H, Ido T, Ishibashi S, Uota M, Uchida S and Tokura Y 1989 *Phys. Rev.* B **40** 2254
[8] Iwakiri K, Nishio T, Kawashima K, Ishida S, Oka K, Fujihisa H, Gotoh Y, Iyo A, Ogino H, Eisaki H, Yoshida Y, and Kitô H 2018 *J. Phys. Conf. Ser.* **1054** 012002
[9] Schearz K , Blaha P, 2003 *Comput. Mater. Sci*, **28** 259
[10] Perdew J P, Burke K, Ernzerhof M, 1996 *Phys. Rev. Lett.* **77** 3865

Microscopic Study of Domain Structure in Charge Density Wave States in 2*H*-$TaS_{1.8}Se_{0.2}$

S Ohta[1], Y Fujisawa[2], S Demura[3] and H Sakata[1]
[1]Department of Physics, Tokyo University of Science[1], Shinjyuku-ku, Tokyo 162-8601, Japan
[2]Okinawa Institution of Science and Technology[2], Onna-son Kunigami-gun Okinawa 904-0495, Japan
[3]College of Science and Technology, Nihon University[3], Chiyoda-ku Tokyo 101-0062, Japan

E-mail: 1217606@ed.tus.ac.jp

Abstract. Transition metal dichalcogenides 2*H*-TaS_2 and 2*H*-$TaSe_2$ show charge density wave (CDW) state and superconductivity. It has been reported that elemental substitution of chalcogenide suppresses the CDW and enhances superconductivity. However, how the CDW order is suppressed by the substitution has not been clarified yet. To clarify this, we performed real space observation of the CDW state in the commensurate phase of 2*H*-$TaS_{1.8}Se_{0.2}$ by scanning tunneling microscopy. The existence of domains separated by walls at which the phase of the CDW shifts was observed. The change in the electronic state on the domain boundaries in 2*H*-$TaS_{1.8}Se_{0.2}$ was not as drastic as that in 1*T*-$Fe_xTa_{1-x}S_2$.

1. Introduction

Transition metal dichalcogenides (TMDCs) show various quantum phenomena such as charge density wave (CDW) or superconductivity. In some TMDCs, it has been reported that the CDW order is suppressed and superconductivity is enhanced by an elemental substitution. For example, superconductivity emerges by substitution of Ta for small amount of Fe about a few percent in 1*T*-TaS_2 [1]. In 2*H*-TaS_2, which is another polytype of 1*T*-TaS_2, when Se is substituted by S, i.e. 2*H*-$TaS_{2-x}Se_x$, CDW transition disappears between $0.35 < x < 1.48$, and superconducting transition temperature (T_C) about 4 K appears between $0.1 < x < 1.9$, whereas T_C of pristine 2*H*-TaS_2 and 2*H*-$TaSe_2$ is 0.8 K and 0.14 K, respectively [3].

When the superconductivity emerges, how the commensurate CDW order is suppressed is an interesting problem. In 1*T*-TaS_2, the appearance of the peculiar domain structure of the CDW and the drastic change of the electronic state on the CDW domain boundaries were observed when the superconductivity emerges [2].

In this study, we examined the CDW states in 2*H*-TaS_2 and 2*H*-$TaS_{1.8}Se_{0.2}$ microscopically by scanning tunnelling microscopy (STM) to clarify how the CDW order is suppressed by Se doping.

IOP Conf. Series: Journal of Physics: Conf. Series **1293** (2019) 012004 doi:10.1088/1742-6596/1293/1/012004

2. Experimental

The polycrystal samples of $2H$-$TaS_{2-x}Se_x$ (x=0 and 0.2) were prepared by heating the mixtures of Ta (Kojyundo Chemicals Co. Ltd., 99.999%), S (Kojyundo Chemicals Co. Ltd., 99.99%), and Se (Kojyundo Chemicals Co. Ltd., 99.99%), with nominal compositions in vacuumed quartz tubes at 900°C for 3 days. The single crystals were grown from the polycrystals by chemical vapor transport method using iodine (Wako Chemicals Co. Ltd., 99.5%) as transport agent [4]. The thermal process was the followings. First, the tube was heated up to 900 °C in the source zone and 800 °C in the growth zone for 8 hours, then, kept for a week, finally, cooled down to room temperature for 16 hours.

Figure 1 shows the temperature dependence of electrical resistivity for x = 0 and 0.2 samples. The CDW transition temperature (T_{CDW}) can be defined as the temperature where the slope changes abruptly, and T_c is defined as the temperature at which the resistivity becomes zero. From the results, T_{CDW} and T_c were determined as 55 K and 2.6 K for x = 0.2 sample, respectively. In x = 0 sample, T_{CDW} was 78K and T_c was below 2.2 K. The observed suppression of the CDW and the enhancement of the superconductivity in x = 0.2 sample are consistent with the results reported in a previous paper [3].

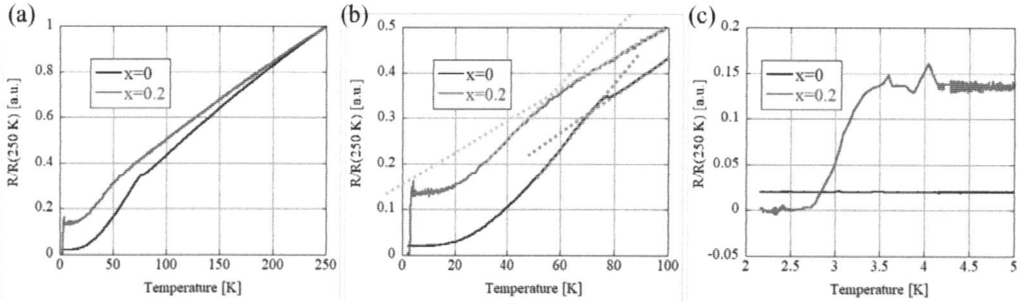

Figure 1 (a) The temperature dependence of the normalized electrical resistivity in $2H$-TaS_2 (blue line) and $2H$-$TaS_{1.8}Se_{0.2}$ (red line). (b) Resistivity below 100 K. (c) Resistivity below 5 K.

STM measurements were performed with a laboratory-built STM at 4.2 K. The electrochemically polished Au wire was used as a STM tip. The clean surface of the single crystal was prepared by cleavage at 4.2 K *in situ*. A bias voltage was applied to the sample in all measurements.

3. Result

Figure 2 (a - d) show the typical STM images and the FFT images of the samples (x = 0 and 0.2). The STM image of x = 0 sample (Fig. 2 (a)) indicates the triangular lattice of the chalcogenide atoms and 3 x 3 CDW. Corresponding spots can be seen in the FFT image shown in Fig. 2 (b). The STM image of x = 0.2 (Fig. 2 (c)) also shows the triangular lattice of the atoms and the superstructure of the CDW with rather large background roughness. The period of the superstructure is same as that of x = 0 and corresponding spots can be seen in FFT image shown in Fig. 2 (d). In addition, there are several phase shifts in x = 0.2 sample, at which the phase of the modulation wave slips off for $2\pi/3$ or $4\pi/3$, i.e. slips of one or two atoms. Figure 2 (e) shows the inverse FFT image obtained from FFT spots corresponding to atoms and one of the three CDW vectors marked in Fig. 2 (d). As indicated by white arrows, the crest of CDW shifts by one atom near the center of the figure. As can be seen from the figure, a row of such phase shifts, indicated by white broken line, composes a wall which divides the commensurate region into domains.

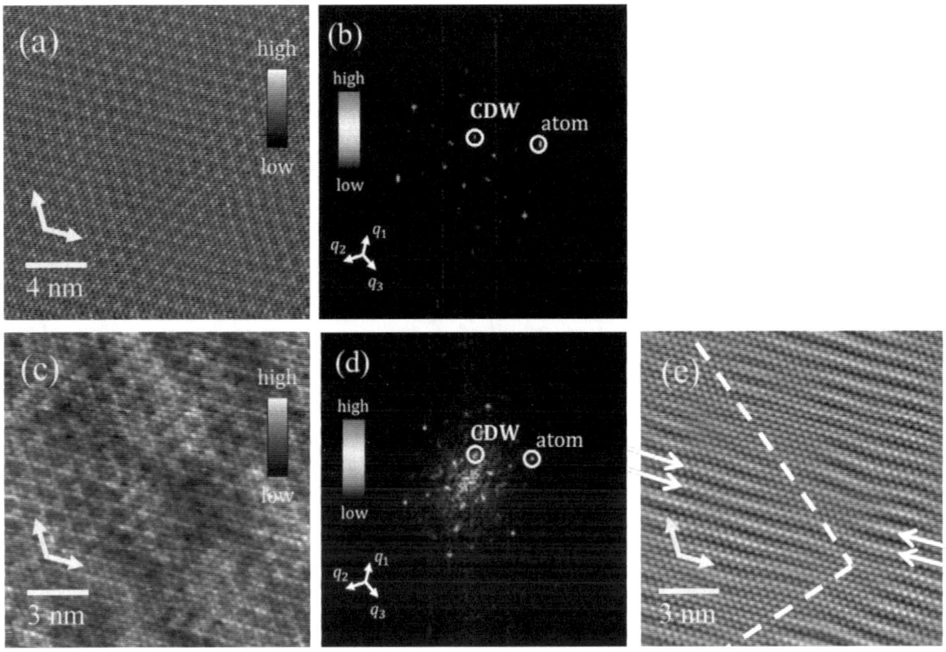

Figure 2 (a - b) STM image of $2H$-TaS_2 taken at +300 mV and 300 pA on 20 nm × 20 nm field of view and the FFT image. (c - d) STM image of the $2H$-$TaS_{1.8}Se_{0.2}$ taken at +500 mV and 200 pA on 15 nm × 15 nm field of view and the FFT image. (e) Inverse FFT image obtained from FFT spots corresponding to atoms and one of three CDW vectors marked with a white circle in Fig. 2 (d). White broken line indicates a domain wall, which consists of a row of phase shifts.

4. Discussion

From the STM measurements for $2H$-$TaS_{2-x}Se_x$ (x = 0, 0.2), we obtained the following two results. The first is the emergence of the phase shifts of the CDW with Se doping. The existence of phase shifts suppresses the coherence of the CDW. Because the contrast of the CDW in the STM images does not change with Se doping, the amplitude of the CDW does not seem to be reduced. Thus, the reduction of T_{CDW} observed in electric resistivity measurements in the sample of x = 0.2 is thought to be caused by the suppression of the coherence in the CDW.

The second is the lack of the contrast on the domain walls in STM images. In the STM image shown in Fig. 2 (c), almost no contrast at the domain wall can be seen. This indicates that the electronic state on the domain wall does not change from that in the domains. The same result was obtained in the STM images obtained with negative bias voltage. These results are quite contrasted to that observed in $1T$-$Fe_xTa_{1-x}S_2$, where the contrast of the domain wall is quite different from that in the domain in both bias voltages [2]. In $1T$-$Fe_xTa_{1-x}S_2$, the possibility that the metallic domain wall is responsible to the superconductivity has been pointed out. In $2H$-$TaS_{2-x}Se_x$, however, because the electronic state on the wall is close to that on the domains, the wall is not responsible to the superconductivity. Thus the enhancement of the superconductivity observed in $2H$ compound is responsible to the change in the electronic state in the domains by Se substitution unlike $1T$ compounds.

5. Summary

We performed STM observations in 2*H*-$TaS_{2-x}Se_x$ ($x = 0.2$), which shows reduced T_{CDW} and enhanced T_C. We found the emergence of the several phase shifts of the CDW, which reduce the coherence of the CDW. We also found the change in the electronic states on the domain boundaries in 2*H*-$TaS_{1.8}Se_{0.2}$ was not as drastic as that in 1*T*-$Fe_xTa_{1-x}S_2$.

References

[1] L J Li, W J Lu, X D Zhu, L S Ling, Z Qu, and Y P Sun 2012 *EPL* **97** 67005
[2] Fujisawa Y, Shimabukuro T, Kojima H, Kobayashi K, Ohta S, Machida T, Demura S, and Sakata H 2017 *J. Phys. Soc. Jpn.* **86** 113703
[3] L Li *et al* 2017 *npj Quantum Materials* **2** 11
[4] A J Patel, M K Bhayani, A R Jani 2009 *Chalcogenide Letters* **6** 491

IOP Conf. Series: Journal of Physics: Conf. Series **1293** (2019) 012005 doi:10.1088/1742-6596/1293/1/012005

Influence of microfabrication on superconducting properties of exfoliated thin films of layered superconductor $NbSe_2$: reactive ion etching

Hikari Tomori, Naoki Hoshi, Dai Inoue, Akinobu Kanda

Division of Physics, Faculty of Pure and Applied Sciences, University of Tsukuba, Tennodai, Tsukuba, Ibaraki 305-8571, Japan

E-mail: tomori4486@gmail.com

Abstract. We have experimentally investigated the influence of the reactive ion etching (RIE) on the superconducting properties of exfoliated thin films of layered superconductor $NbSe_2$ through transport measurement. Wires with width of 2 μm and 0.5 μm were fabricated from a single $NbSe_2$ film (thickness $\sim$ 65 nm) using RIE with CF_4 and O_2 gas mixture. While the 2-μm wires did not exhibit any distinguishable degradation in residual resistivity, residual resistance ratio, superconducting transition and current-voltage characteristics in the superconducting state, significant degradation was observed in the 0.5-μm wires.

1. Introduction

When the size of a superconductor approaches the characteristic length scales such as the coherence length and/or the magnetic penetration depth, the quantum confinement of Cooper pairs brings geometric effects. Thus, by controlling the geometry of the superconductor, one is able to manipulate mesoscopic superconducting states, which might lead to superconducting classical and quantum devices.

Mesoscopic superconducting effect manifests itself in vortex states.[1] In a mesoscopic type-II superconductor with several quantum vortices, the arrangement of the vortices depends not only on the vortex-vortex interaction but also on the interaction between a vortex and the sample edge, so that the triangular Abrikosov vortex lattice, which is characteristic of bulk superconductors, is deformed into a multivortex state (MVS) in which the vortex arrangement is strongly influenced by the sample shape, or a giant vortex state (GVS) in which the multiple vortices coalesce.[2, 3, 4, 5]

The common method of obtaining a shape-defined mesoscopic superconductor is the electron beam lithography followed by the deposition of superconducting materials in vacuum. So far, we have used electron-beam-evaporated aluminum as a material of mesoscopic superconductors to experimentally investigate the vortex states in mesoscopic superconductors. By using our original *small tunnel junction method*,[6, 7] we successfully confirmed the existence of a giant vortex and transitions between different MVSs or between an MVS and a GVS in mesoscopic circular and square thin films. These transitions were induced by an applied magnetic field [8, 9] as well as by a locally injected supercurrent.[10, 11, 12, 13] However, the controllability of the mesoscopic superconducting states was limited by the existence of defects which pin the

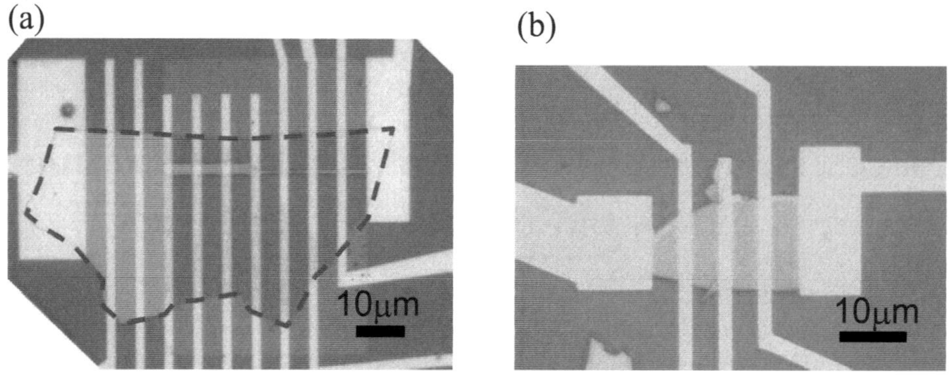

Figure 1. Optical microscope images of the sample treated with RIE (a) and the sample of a $NbSe_2$ flake without RIE treatment (b). In (a), the dashed line indicates the area of the $NbSe_2$ flake before the RIE treatment.

vortices. Actually in our aluminum samples, while the lateral dimensions are well controlled by the state-of-the-art electron beam lithography system, the control of the vertical dimension is poor because of the surface roughness, which is inevitable in evaporated aluminum films.

For this reason, we are looking for an alternative material for the study of mesoscopic superconductivity, and focusing on exfoliated thin films of layered superconductors.

Recently, the mechanical exfoliation technique, called the Scotch tape method, developed in the graphene research[14, 15], has been applied to layered superconductors, and thin superconductors with atomically flat surface can be easily obtained.[16, 17, 18] Such atomically flat superconductors are able to resolve the above defect problem. For the application of the exfoliated thin superconductors to the mesoscopic superconductivity research, one need to shape the lateral dimensions. In the graphene research, the reactive ion etching (RIE) is commonly used to shape the sample. The extremely high mobility is maintained even after the RIE process.[19] Here, we apply the RIE to the thin films of layered superconductors, and investigate the influence of the RIE on the superconducting properties.

2. Experiment

As a layered superconductor, we used $NbSe_2$. Optical microscope images of the samples are shown in Fig. 1.

For the fabrication of the RIE-treated sample (Fig. 1(a)), first an $NbSe_2$ flake with thickness of 65 nm was obtained using mechanical exfoliation of a bulk crystal on a SiO_2/Si substrate. Then RIE was performed with resist mask of ZEP-520a (Zeon Co.), gas mixture of $CF_4{:}O_2 = 24$ Pa:2.7 Pa, a power of 50 W, and for 1 min to form wires with widths of 2 μm and 0.5 μm from a single flake. Finally, Au/Cr electrodes are connected to the flake. In this sample, electron transport of two wires with width of 2 μm (samples #2 and #3) and two wires with width of 0.5 μm wires (samples #4 and #5) with the same length (4 μm) were investigated. Also, as a reference, a sample of a pristine $NbSe_2$ flake (without RIE treatment) with thickness of 65 nm (sample #1) was prepared.

Electron transport of the samples were measured with four-terminal method in vacuum in a cryostat with appropriate cryogenic lowpass filters. The differential resistance was obtained by numerically differentiating the current-voltage characteristics.

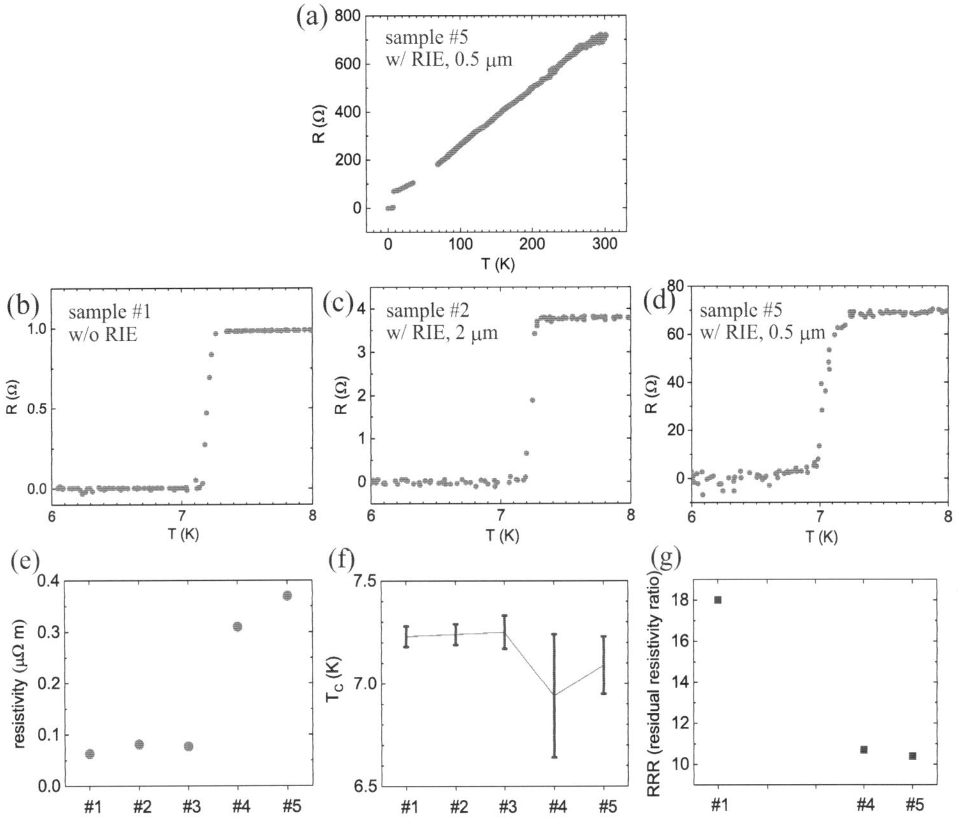

Figure 2. (a) Temperature dependence of zero-bias resistance of sample #5 (a RIE-treated wire with width of 0.5μm) between 4 K and 300 K. (b,c,d) Temperature dependence of zero-bias resistance of samples #1 (no RIE treatment), #2 (a RIE-treated wire with width of 2 μm) and #5 around the superconducting transition. Residual resistivity (e), superconducting transition temperature (T_C) and the transition width (f), and the residual resistivity ratio (g) for all samples.

3. Results and discussion

All samples exhibited the superconducting transition around 7 K. Typical temperature dependence of zero-bias resistance between 4 K and 300 K is shown in Fig. 2(a). Also, the close-ups around the superconducting transition of samples #1 (no RIE treatment), #2 (a RIE-treated wire with width of 2 μm) and #5 (a RIE-treated wire with width of 0.5μm) are shown in Figs. 2(b), (c) and (d), respectively. While the superconducting transition of wires with width of 2 μm (samples #2 and #3) was similar to that of a flake without RIE (sample #1), the transition of wires with width of 0.5 μm (samples #4 and #5) was relatively smeared.

To elucidate this point, we show the residual resistivity, superconducting transition temperature (T_C) and the transition width, and the residual resistivity ratio (RRR) for all samples in Figs. 2(e-g). Here, we define the transition width as the temperature range where the resistance are between the 5% and 95% of the normal state residual resistance, and T_C as the midpoint of the transition width. Besides, the residual resistivity ratio is the ratio of the resistance at 300 K to the residual resistance at low temperatures. In Figs. 2(e) and (f),

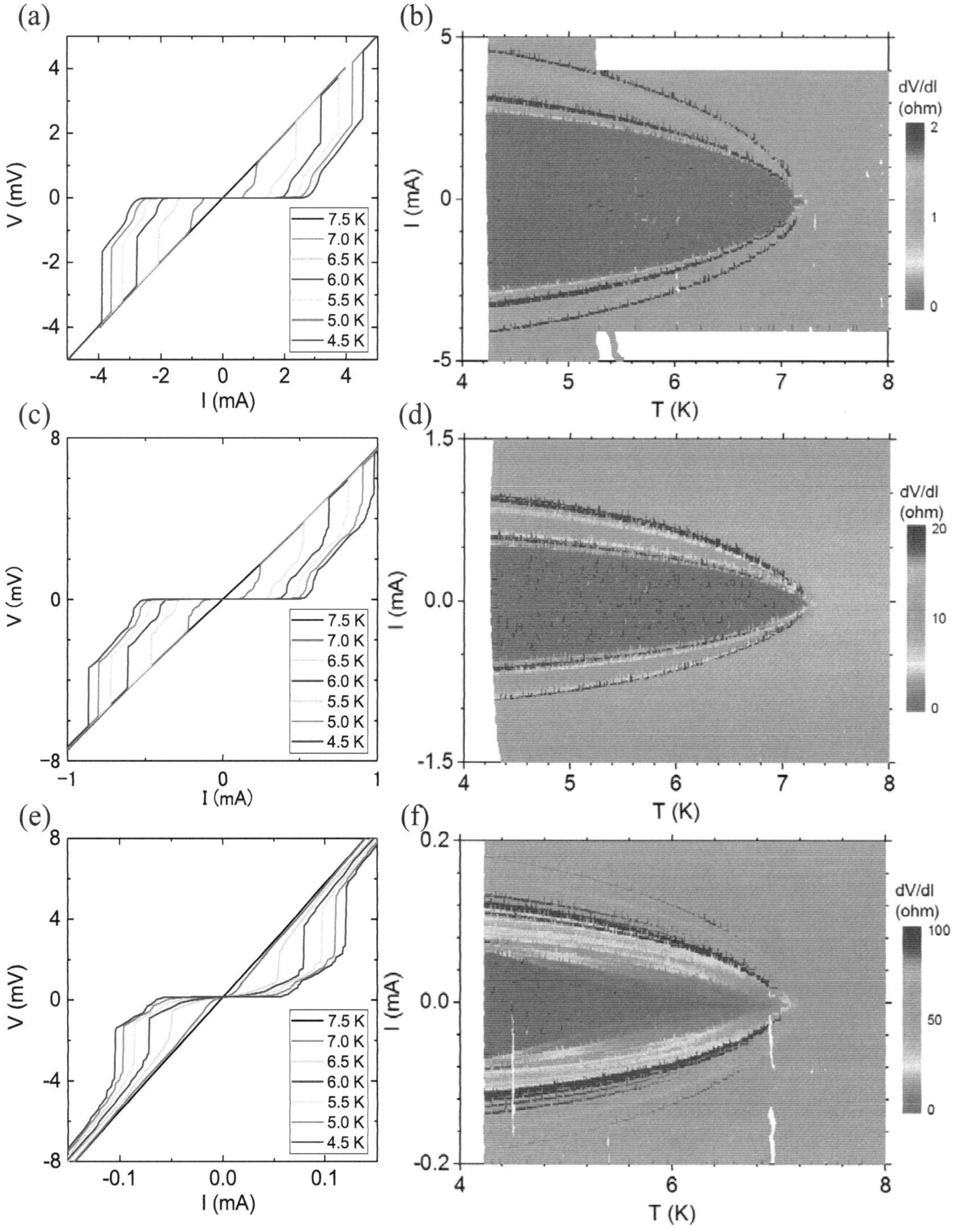

Figure 3. (a,c,e) Current-voltage characteristics of samples #1(a), #3(b) and #4(c) for several temperatures. (b, d, f) Differential resistance plotted in the current-temperature plane for samples #1(b), #3(d) and #4(f), respectively. The biasing current was swept from negative to positive values.

while the resistivity, T_C and the transition width of samples #1, #2 and #3 are almost the same, those of samples #4 and #5 exhibit clear difference from the others, indicating that the superconductivity of wires with width of 0.5 μm deteriorates, while that of wires with width of 2 μm does not. This conclusion is supported by the result of RRR. (Note that RRRs for samples #2 and #3 were not available due to missing resistance data at room temperature.)

Finally we discuss the current voltage characteristics. Figure 3 shows the current-voltage characteristics of samples #1 (Figs. 3(a) and (b)), #3 (Figs. 3(c) and (d)) and #4 (Figs. 3(e) and (f)). The biasing current was swept from negative to positive values. In Figs. 3(b,d,f), the differential resistance is plotted in the current-temperature plane. The red regions around the zero current correspond to the zero-resistance state, and thus the upper edge of a red region corresponds to the critical current. The lines around the red regions in Figs. 3(b,d,f) correspond to steep voltage change in Figs. 3(a,c,e), which appear below the critical temperature ($\approx$ 7.2 K). These are presumably due to the phase slips. Both in Figs. 3(b) and (d), the boundaries of the red regions are clear, and several lines corresponding to phase slips are seen. In this way, Figs. 3(a) and (c) (3(b) and (d)) are similar to each other except for the current scale which is due to the difference in sample width, indicating again that the superconductivity of wires with width of 2 μm does not deteriorate due to the RIE. On the other hand, in Fig. 3(f), the boundary of the red region is blurred, and more lines corresponding to phase slips are seen in comparison with Figs. 3(b) and (d), indicating that there exist many phase slip centers. These results indicate that the wires with width of 0.5 μm are damaged by the RIE treatment.

4. Conclusions

We have experimentally investigated the influence of the reactive ion etching (RIE) on the superconducting properties of exfoliated thin films of layered superconductor $NbSe_2$ through transport measurement. While the 2-μm wires did not exhibit any distinguishable degradation in residual resistivity, residual resistance ratio, superconducting transition and current-voltage characteristics in the superconducting state, significant degradation was observed in the 0.5-μm wires.

Acknowledgments
This work was supported by JSPS KAKENHI Grant Number JP15H05867 [Hybrid Quantum Systems]. The electron beam lithography was conducted at Nanotechnology Platform Facility in University of Tsukuba. The authors gratefully acknowledge fruitful discussions with M. Hayashi.

References

[1] Kanda A, Ootuka Y, Kadowaki K, Peeters F M 2010 Novel superconducting states in nanoscale superconductors *The Oxford Handbook of Nanoscience and Technology* vol 1, ed. A Narlikar and Y Y Fu (Oxford: Oxford University Press) chapter 19 pp 639-676
[2] Deo P S, Schweigert V A and Peeters F M 1997 *Phys. Rev. Lett.* **79** 4653
[3] Moshchalkov V V, Qiu X G and Bruyndoncx V 1997 *Phys. Rev. B* **55** 11793
[4] Schweigert V A and Peeters F M 1998 *Phys. Rev. B* **57** 13817
[5] Bruyndoncx V, Rodrigo J G, Puig T, Van Look L, Moshchalkov V V and Jonckheere R 1998 *Phys. Rev. B* **60** 4285
[6] Kanda A, Geisler M C, Ishibashi K, Aoyagi Y and Sugano T 2000 *Physica B* **284-288** 1870
[7] Kanda A and Ootuka Y 2003 *Physica B* **329-333** 1421
[8] Kanda A, Baelus B J , Peeters F M , Kadowaki K, Ootuka Y 2004 *Phys. Rev. Lett.* **93** 257002
[9] Baelus B J, Kanda A, Peeters F M, Ootuka Y and Kadowaki K 2005 *Phys. Rev. B* **71** 140502(R)
[10] Milosevic M V, Kanda A, Hatsumi S, Peeters F M, Ootuka Y 2009 *Phys. Rev. Lett.* **103** 217003
[11] Hatsumi S, Ootuka Y and Kanda A 2009 *Physica C* **469** 1080
[12] Hatsumi S, Kuroda Y, Ootuka Y and Kanda A 2010 *J. Phys.: Conf. Series* **232** 012003
[13] Hatsumi S, Kuroda Y, Ootuka Y and Kanda A 2010 *Physica C* **470** 1141
[14] Novoselov K S, Geim A K, Morozov S V, Jiang D, Zhang Y, Dubonos S V, Grigorieva I V and Firsov A A 2004 *Science* **306** 666

[15] Kanda A 2013 Experimental approaches to graphene electron transport for device applications *Physics and Chemistry of Graphene: Graphene to Nanographene* ed T Enoki and T Ando (Singapore: Pan Stanford) chapter 3 pp 89-205

[16] Xi X, Wang Z, Zhao W, Park J-H, Law K T, Berger H, Forro L, Shan J and Mak K F 2015 *Nat. Phys.* **12** 139

[17] Cao Y, Mishchenko A, Yu G L, Khestanova E, Rooney A P, Prestat E, Kretinin A V, Blake P, Shalom M B, Woods C, Chapman J, Balakrishnan G, Grigorieva I V, Novoselov K S, Piot B A, Potemski M, Watanabe K, Taniguchi T, Haigh S J, Geim A K and Gorbachev R V 2015 *Nano Lett.* **15** 4914

[18] Khestanova E, Birkbeck J, Zhu M, Cao Y, Yu G L, Ghazaryan D, Yin J, Berger H, Forro L, Taniguchi T, Watanabe K, Gorbachev R V, Mishchenko A, Geim A K and Grigorieva I V 2018 *Nano Lett.* **18** 2623

[19] Wang L, Meric I, Huang P Y, Gao Q, Gao Y, Tran H, Taniguchi T, Watanabe K, Campos L M, Muller D A, Guo J, Kim P, Hone J, Shepard K L, Dean C R 2013 *Science* **342** 614

Influence of focused-ion-beam microfabrication on superconducting transition in exfoliated thin films of layered superconductor $NbSe_2$

Hikari Tomori, Naoki Hoshi, Dai Inoue, Akinobu Kanda

Division of Physics, Faculty of Pure and Applied Sciences, University of Tsukuba, Tennodai, Tsukuba, Ibaraki 305-8571, Japan

E-mail: tomori4486@gmail.com

Abstract. We have experimentally studied the influence of the focused ion beam (FIB) microfabrication on the superconducting properties of exfoliated thin films of layered superconductor $NbSe_2$ through transport measurement. We observed significant decrease of the residual-resistance ratio (RRR), indicating the formation of defects by the FIB. Although clear superconducting transition was seen before the FIB microfabrication, after FIB it was blurred, and one sample exhibited insulating behavior. The possible origins of the changes in the superconducting properties are discussed.

1. Introduction

Atomic layers have attracted much attention in recent years. Rapidly developing technology of mechanical exfoliation of layered materials using sticky tapes and transfer/stacking of atomic layers, first developed in the graphene research[1, 2], have been widely used for various kinds of two-dimensional materials, which have various features such as metal, insulator, superconductor and so on.[3] Studies on thin film superconductors also benefit from these technological innovations.

Previously thin-film superconductors were mainly obtained by vacuum deposition and epitaxial growth, but in general deposited films include a lot of defects, and the substrate is limited for epitaxial layers. On the other hand, by utilizing the technology cultivated in graphene research, one is able to obtain superconducting thin films which are clean and atomically uniform. Accordingly, thin film superconductors have been studied more extensively[4, 5, 6] and also are expected as materials for next generation quantum devices. Here, note that for investigating detailed transport characteristics and making devices for applications, microfabrication is required.

There are several kinds of microfabrication techniques for layered materials. One is the reactive ion etching (RIE), which is commonly used for atomic layers such as graphene and hexagonal boron nitride. Extremely high mobility is kept in graphene even after the RIE microfabrication.[7] Another is the focused ion beam (FIB) milling, which is conventionally used for microfabrication of high-T_C superconductors.[8, 9] According to the previous study, when bulk high-T_C superconductors are microfabricated with FIB, damage is spread only over the range of several tens of nm from the surface.[10] On the other hand, the influence of FIB

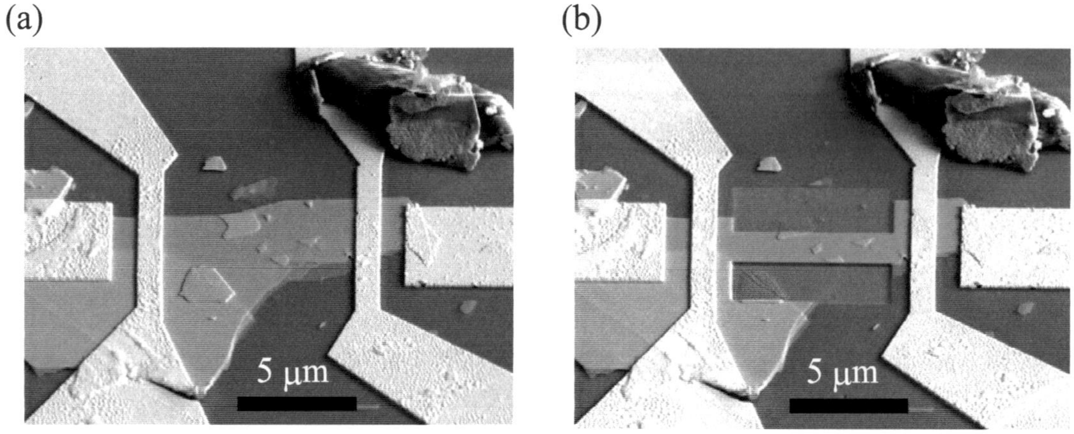

Figure 1. Optical microscope images (a) before and (b) after the FIB microfabrication of sample #3.

microfabrication on thin layered superconductors has not been studied yet. Here, we investigate the influence of FIB microfabrication on the superconducting properties of niobium diselenide ($NbSe_2$), which is a layered superconductor with a bulk T_C of ~7.2 K.

2. Experimental

We obtained $NbSe_2$ flakes on a SiO_2/Si substrate by exfoliating a bulk crystal using Scotch tape[1, 2], but due to our poor technique, flakes consisted of several stacking layers, as seen in Fig. 1(a). Electrodes (Au (100 nm)/Cr (5 nm)) were connected to a flake using conventional electron beam lithography followed by metal deposition in vacuum. For the FIB milling, we used XVision 200DB (SII Nanotechnology Inc.) with the acceleration voltage of 30 kV and a dose of $\sim 2 \times 10^9$ gallium ions/μm^2.

We made three samples, #1, #2, and #3. Nominal thicknesses are 57 nm, 57 nm, 43 nm for samples #1, #2, and #3, respectively. Optical microscope images before and after the FIB microfabrication of sample #3 are shown in Figs. 1(a) and (b), respectively. The electron transport before and after the FIB microfabrication was investigated above ~4.5 K in vacuum in a cryostat with appropriate cryogenic lowpass filters. The zero-bias resistance was obtained by numerically differentiating the linear part of current-voltage characteristics around the origin with bias voltages typically less than 10 mV. Magnetic field was applied perpendicular to the substrate.

3. Results and discussion

Figure 2(a) shows the zero-bias resistance (R) as a function of the temperature before the FIB microfabrication. As shown in Fig. 2(b), the samples exhibited the superconducting transition with $T_C = 7.0$ K and the transition width $\Delta T_C \approx 0.1$ K. Here, we define T_C as the temperature where the resistance becomes 50% of its value at 8 K (R(8 K)) and ΔT_C as the temperature width where the resistance drops from 90% to 10% of R(8 K). The residual-resistance ratio (RRR), defined as R(270 K)/R(8 K), is shown in Table 1. The FIB microfabrication induced striking change in the superconducting transition. Figure 2(c) shows the resistance variation in a wide temperature range after the FIB microfabrication. While the resistance of samples #1 and #2 slightly decreased with decreasing temperature ($dR/dT > 0$, metallic behavior), sample

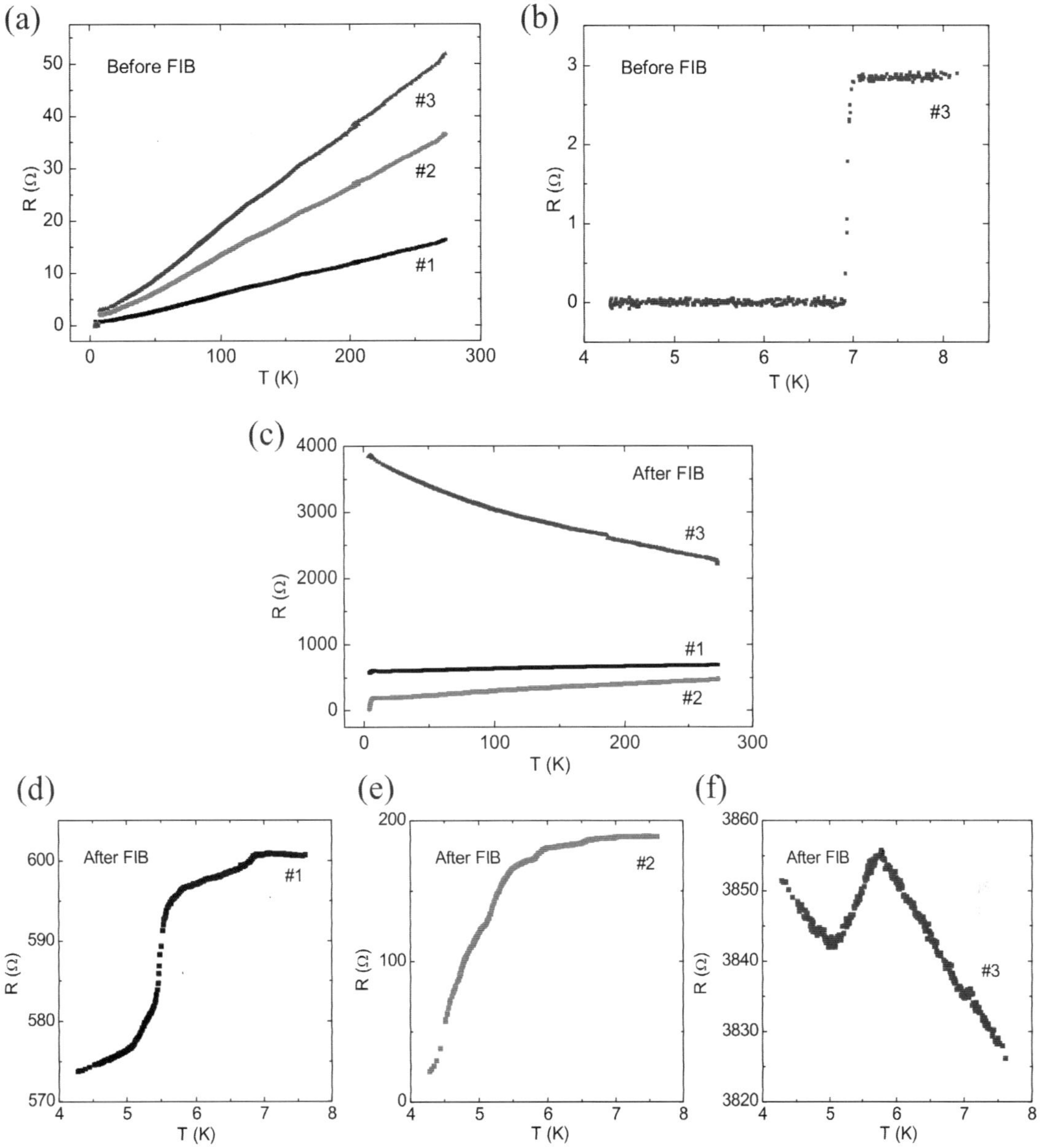

Figure 2. (a) Zero-bias resistance is plotted as a function of the temperature for samples, #1,#2, and #3 before the FIB microfabrication. (b) Superconducting transition of sample #3. (c) Zero-bias resistance is plotted as a function of the temperature for samples, #1, #2, and #3 after the FIB microfabrication. (d-f) Resistance - temperature curves between 4.3 K and 7.6 K for samples #1(d), #2(e), #3(f) after the FIB microfabrication.

#3 exhibited an insulating behavior with $dR/dT < 0$. The residual-resistance ratio (RRR), defined as $R(270\text{ K})/R(8\text{ K})$ for samples with metallic behavior, is shown in Table 1. The RRR is dramatically deteriorated by the FIB microfabrication, indicating that the gallium ion beam introduced significant defects in the $NbSe_2$ flakes. The insulating behavior in sample #3 also indicates damage due to the FIB treatment. These results present a great contrast to the above-mentioned previous work for high-T_C superconductors.[10] The possible origins of the defects will be discussed below. Figs. 2(d-f) are the close-ups of the low-temperature resistance. By the FIB microfabrication, the superconducting transition to the zero-resistance state was washed out above 4.3 K for all samples. Although the sample width after the FIB microfabrication was the same for all samples, the low-temperature transport varied sample to sample. For sample #1, resistance slightly decreased below 6.9 K and exhibited a drop at 5.5 K. However the relative amount of the resistance drop was small, with the resistance ratio $R(4.3\ \mathrm{K})/R(6.9\ \mathrm{K}) = 0.96$. For sample #2, the resistance started to drop at 7.0 K, and it took 50% of $R(7\ \mathrm{K})$ at 4.7 K. The resistance ratio $R(4.3\ \mathrm{K})/R(7.0\ \mathrm{K}) = 0.12$. For sample #3, while it displayed an insulating behavior, small resistance drop still appeared between 5.0 K and 5.8 K.

Table 1. Type of behavior and residual-resistance ratio (RRR) before and after the RIE treatment

Sample	Before RIE		After RIE	
	behavior	RRR	behavior	RRR
#1	metallic	21.8	metallic	1.15
#2	metallic	17.6	metallic	2.48
#3	metallic	18.0	insulating	

To confirm whether the resistance drops seen in all samples after the FIB microfabrication originate from superconductivity, we investigated the magnetic field dependence of the resistance. The results are shown in Fig. 3. Figures 3(a,c,e,g) are resistance of sample #3 before the FIB microfabrication, resistances of samples #1, #2, and #3 after the FIB microfabrication, respectively, plotted in the magnetic field (B) -temperature (T) plane. The resistance range of the color map is approximately between the minimum and maximum values of the resistance for each plot. Figures 3(b,d,f,h) are resistance of sample #3 before the FIB microfabrication, resistances of samples #1, #2, and #3 after the FIB microfabrication, respectively, plotted as a function of temperature for several magnetic fields. In Fig. 3(a), for sample #3 before FIB microfabrication, the boundary between the blue and red regions, indicated by dashed lines, corresponds to the critical magnetic field. The critical magnetic field increases linearly with decreasing temperature. The other samples (samples #1 and #2 before the FIB microfabrication) exhibited the same characteristics (not shown). For samples after the FIB microfabrication, in Figs. 3(c,e,g), the boundary between the blue and red regions, corresponding to a characteristic magnetic field for the resistance change, shows a trend similar to the one in Fig. 3(a), as indicated by dashed lines. That is, the characteristic magnetic field linearly increases with decreasing temperature. (Note that in Fig. 3(g), this is seen only at $T < 5.5$ K.) Also, in Fig. 3(b) before the FIB treatment, the resistance drop becomes less steep as the magnetic field increases. This trend is also seen in Figs. 3(d) and (f). Thus, we conclude that the resistance drop seen in samples after the FIB microfabrication is related to the superconductivity.

Finally, we discuss the origin of the change in the superconducting properties due to the FIB microfabrication. Here, we point out three possibilities. The first is related to the disorder. As clearly seen in RRR values in Table 1, the FIB induces many defects. Defects, or disorder, can cause broadening of the superconducting transition, and in the limit of the strong disorder, the superconductor-insulator transition, which was intensively studied theoretically and experimentally as a dissipation-driven quantum phase transition.[11, 12] The decrease of

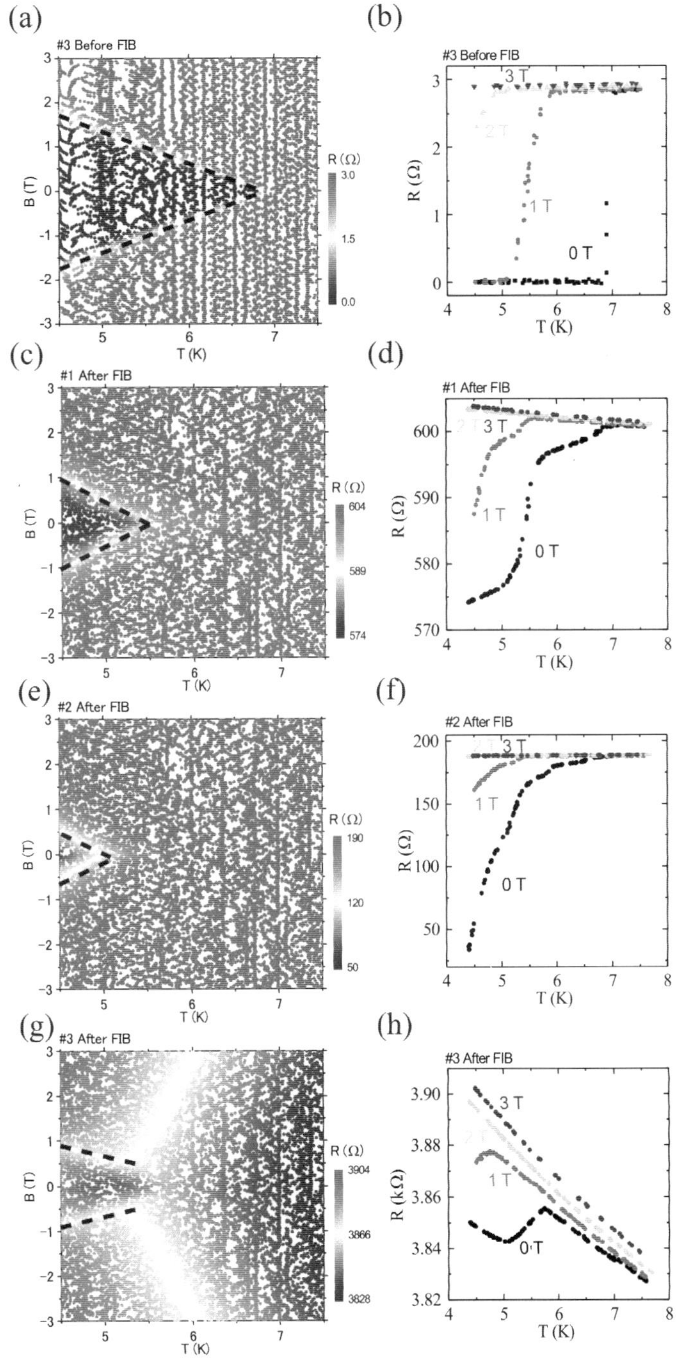

Figure 3. (a,c,e,g) Magnetic field and temperature dependence of resistance of sample #3 before the FIB microfabrication, resistances of samples #1, #2, and #3 after the FIB microfabrication are plotted, respectively. Dashed lines indicate boundaries between blue and red regions. (b,d,f,h) The resistances for several magnetic field values are plotted as a function of temperature for sample #3 before the FIB microfabrication, samples #1, #2, and #3 after the FIB microfabrication, respectively.

the superconducting transition temperature is explained as a precursor of the superconductor-insulator transition. The second possible origin is the intercalation of the gallium atoms in $NbSe_2$ crystals. It is known that the gallium-intercalated $NbSe_2$ exhibits lower T_C as compared to the pristine $NbSe_2$, depending on the amount of gallium.[13] For example, T_C of $Ga_{0.1}NbSe_2$ is 2.7 K. The FIB might prompt the non-uniform intercalation of gallium atoms, leading to spatially non-uniform superconducting transition. The third is the cleavage of $NbSe_2$ layers by the FIB. The impact of the gallium ions might widen the interlayer spacing of $NbSe_2$, forming effective monolayer and/or few layer $NbSe_2$, which are still electrically connected to each other. It is known that T_C decreases as the layer number of $NbSe_2$ decreases, and $T_C \approx 2 - 3$ K for monolayer $NbSe_2$.[4, 5, 6]

The detailed origin of the change in the superconducting properties due to the FIB microfabrication is not clear at this moment. Further experimental and theoretical researches are needed for the full understanding of the microfabrication of the layered materials.

4. Conclusions

We have experimentally studied the influence of the focused ion beam (FIB) microfabrication on the superconducting properties of exfoliated thin films of layered superconductor $NbSe_2$ through transport measurement. The sample is an FIB processed wires with width of 2 μm. We observed significant decrease of RRR and the deterioration of the superconducting properties. The origin of these effect is discussed.

Acknowledgments

This work was supported by JSPS KAKENHI Grant Number JP15H05867 [Hybrid Quantum Systems] and University of Tsukuba Basic Research Support Program Type B. The FIB microfabrication was conducted at the AIST Nano-Processing Facility, supported by "Nanotechnology Platform Program" of the Ministry of Education, Culture, Sports, Science and Technology (MEXT), Japan.

References

[1] Novoselov K S, Geim A K, Morozov S V, Jiang D, Zhang Y, Dubonos S V, Grigorieva I V and Firsov A A 2004 *Science* **306** 666
[2] Kanda A 2013 Experimental approaches to graphene electron transport for device applications *Physics and Chemistry of Graphene: Graphene to Nanographene* ed T Enoki and T Ando (Singapore: Pan Stanford) chapter 3 pp 89-205
[3] Geim A K and Grigorieva I V 2013 *Nature* **499** 419
[4] Xi X, Wang Z, Zhao W, Park J-H, Law K T, Berger H, Forro L, Shan J and Mak K F 2015 *Nat. Phys.* **12** 139
[5] Cao Y, Mishchenko A, Yu G L, Khestanova E, Rooney A P, Prestat E, Kretinin A V, Blake P, Shalom M B, Woods C, Chapman J, Balakrishnan G, Grigorieva I V, Novoselov K S, Piot B A, Potemski M, Watanabe K, Taniguchi T, Haigh S J, Geim A K and Gorbachev R V 2015 *Nano Lett.* **15** 4914
[6] Khestanova E, Birkbeck J, Zhu M, Cao Y, Yu G L, Ghazaryan D, Yin J, Berger H, Forro L, Taniguchi T, Watanabe K, Gorbachev R V, Mishchenko A, Geim A K and Grigorieva I V 2018 *Nano Lett.* **18** 2623
[7] Wang L, Meric I, Huang P Y, Gao Q, Gao Y, Tran H, Taniguchi T, Watanabe K, Campos L M, Muller D A, Guo J, Kim P, Hone J, Shepard K L, Dean C R 2013 *Science* **342** 614
[8] Inomata K, Sato S, Nakajima K, Tanaka A, Takano Y, Wang H B, Nagao M, Hatano H and Kawabata S 2005 *Phys. Rev. Lett.* **95** 107005
[9] Ozyuzer L, Koshelev A E, Kurter C, Gopalsami N, Li Q, Tachiki M, Kadowaki K, Yamamoto T, Minami H, Yamaguchi H, Tachiki T, Gray K E, Kwok W K and Welp U 2007 *Science* **318** 1291
[10] Kakizaki Y, Koyama J, Yamaguchi A, Umegai S, Ayukawa S and Kitano H 2017 *Jap. J. Appl. Phys.* **56** 043101
[11] Haviland D B, Liu Y and Goldman A M 1989 *Phys Rev Lett* **62** 2180
[12] Fisher M P A 1990 *Phys. Rev. Lett.* **65** 923
[13] Naik I and Rastogi A K 2010 *Physica B* **405** 955

IOP Conf. Series: Journal of Physics: Conf. Series **1293** (2019) 012007 doi:10.1088/1742-6596/1293/1/012007

New oxide diluted magnetic semiconductor $La_{0.75}Ca_{0.25}Cu_{0.75}Mn_{0.25}SO$ with independent spin and charge doping by first -principle study

Haoze Chen[1], Linxian Li[1], Sheng Wang[1], Yuke Li[2], Jianye Chen[1], Li Zhang[1], *

[1]China Jiliang University, Hangzhou,310018, China

[2]Hangzhou Normal University, Hangzhou,310018, China

E-mail: lzhang6688@sina.com

Abstract. Here we report the first-principle study of a bulk oxide diluted magnetic semiconductor (DMS) $La_{0.75}Ca_{0.25}Cu_{0.75}Mn_{0.25}SO$ and its parent LaCuSO. As a wide band gap p-type oxide semiconductor, LaCuSO system satisfies all the conditions forecasted theoretically to be a room temperature DMS, which is important for formation of high temperature spintronic devices. Then we compare the band gap and optical properties (such as loss function, reflectivity, absorption, the real part and the imaginary part of dieletric function among LaCuSO, $(La_{0.75}Ca_{0.25})(Cu_{0.75}Mn_{0.25})SO$, $(La_{0.75}Ba_{0.25})(Ag_{0.75}Mn_{0.25})SO$ and $(Y_{0.75}Sr_{0.25})(Cu_{0.75}Mn_{0.25})SO$. The results show that the band gap of LaCuSO system decreases most drastically with co-doped Ca and Mn (from 2.1eV to 1eV). As for the optical properties, both the position and intensity of the peaks changed most drastically, too. It illustrates that the energy band structure influences the optical properties.

I. Introduction

Diluted magnetic semiconductors have both the spin and charge properties of electrons, which are the material basis for the preparation of spin-optical devices, and are of great significance for the transmission, processing and storage of information. The earliest dilute magnetic semiconductor was II-VI diluted magnetic semiconductor. Since then, people have successively studied group III-V, II-VI and IV-VI family of diluted magnetic semiconductors and oxide dilute magnetic semiconductors. Due to the limitation of the chemical solubility of Mn, most of the DMS systems are available only as thin films. However, only the bulk crystalline specimens make it possible to perform muon spin relaxation (μSR), nuclear magnetic resonance (NMR) and even neutron scattering. Therefore, the bulk DMS materials are highly required.

Recently, a few of Mn doped pnictides such as Li(Zn,Mn)As, $(Ba,K)(Zn,Mn)_2As_2$,and (La,Sr)(Zn,Mn)SO were reported to be a bulk DMS materials[1]. Among them, LaCuSO is expected to be one of the candidate materials for optoelectronic devices in ultraviolet and/or blue regions, since it has a wide band-gap energy of 3.1 eV with an intense photoluminescence due to the interband transition at room temperature [2,3]. Therefore, in this paper, bulk DMSs based on LaCuSO (LaAgSO, YCuSO) have been investigated by first-principle calculation in order to combine electricity, light and magnetism together for synthesizing devices. The first paragraph after a heading is not indented.

II. Calculation method and model construction

The CASTEP module in Material Studio was selected for calculation. The valence electrons of O, S, La, Cu, Mn and Ca were selected as $O2s^2 2p^4$, $S3s^2 3p^4$, $La5d^1 6s^2$, $Cu3d^{10} 4s^1$, $Mn3d^5 4S^2$ and $Ca4S^2$, respectively. The convergence criterion of the optimization parameters in the calculation is single atomic energy, 1.0×10^{-5} eV / atom, Max. Force,0. 3 eV /nm, the Max. stress, 0. 05 GPa, Max. displacement, 0.0001 nm, through the convergence test, the plane energy cutoff is 380 eV, and the Monkhorst-Pack special K point is 4 × 4 × 4.

The parent LaCuSO belongs to a tetragonal layered structure of the ZrCuSiAs-type with a space groupP4/nmm, No.129, Z=2. a=b=3.99250 Å, c=8.55250 Å, α=β=γ=90◦. The atomic positions are La: 2c (1/4,1/4,0.14219); Cu: 2b (3/4,1/4,1/2); S: 2c (1/4,1/4,0.3117)and O: 2a(3/4,1/4,0).The optimized lattice constants are a=b=3.97118 Å and c=8.49978 Å. In order to calculate the electronic structure and study the optical property of $(La_{0.75}Ca_{0.25})(Cu_{0.75}Mn_{0.25})SO$, we fix the Mn atom and find out the find out the lowest energy relates with the Sr position.

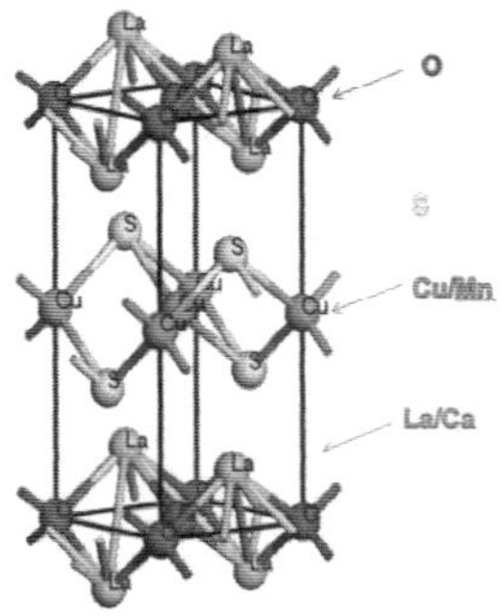

Fig.1 Crystal structure of (La,Ca)(Cu,Mn)SO which belongs to tetragonal ZrCuSiAs type.

III. Electronic structure

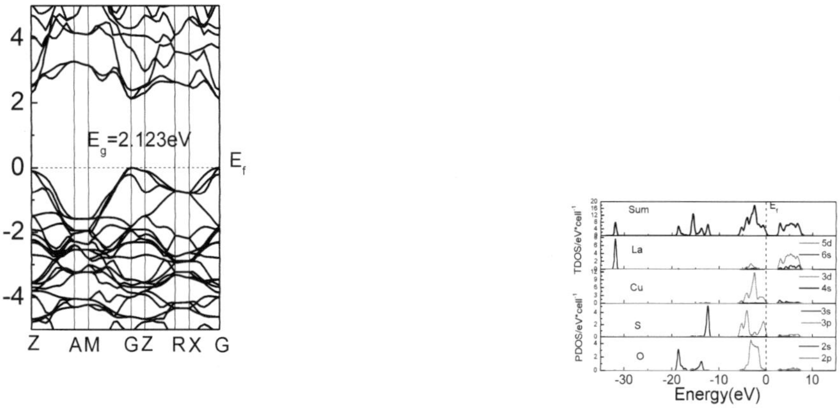

Fig.2 (a)Band structure and (b) DOS of parent LaCuSO.

We first focus on the electronic structure of its parent compound LaCuSO according to the structure of fig. 1. As Cu is a diamagnetic metal, we calculate the electronic structure without considering its spin. The DOS and band structure were depicted in fig. 2(a) and (b). In this paper, all the DOS and band figures are plotted with the Fermi level being zero. Obviously, LaCuSO is a direct semiconductor with the band gap 2.123eV at high symmetry G point in Brillouin zone.

The states range from -6.0 eV to 0 eV are mainly contributed by S3p, O2p and Cu3d orbital, which consist of valence bands. The top of valence bands contains primarily S3p state and Cu 3d, forming covalent Cu-S bonds around fermi surface. However, the conduction band is mainly composed of La5d states with a gap 2eV above E_f.

Then we move on the band structure of $(La_{0.75}Ca_{0.25})$ $(Cu_{0.75}Mn_{0.25})SO$ with spin up and down for comparison in fig.3(a). Whether the spin is up or down, the bottom of conduct band and upper of valence band locate at high-symmetry point G in Brillouin zone, which indicates $(La_{0.75}Ca_{0.25})$ $(Cu_{0.75}Mn_{0.25})SO$ is still a direct semiconductor with a band gap 0.962eV in the spin up state while only 0.88eV in the spin down state. Compared with pure LaCuSO, the band gap apparently reduces a lot after doping, particularly in the latter state.

The DOS of $(La_{0.75}Ca_{0.25})(Cu_{0.75}Mn_{0.25})SO$ are plotted in fig. 3(b). Overall, after Mn and Ca atoms are doped into the lattice, there is a polarization and hybridization between 3d states of the transition metal and the 3p of sulfur atom. The conduction bands are mainly composed of La 5d and Mn 3d band. Mn dopant plays an important role in the DOS around the fermi level. Below E_f from -4eV to 0eV, the asymmetry of PDOS of Mn irons dominates in the up state, compared with that at 0.5-3eV above E_f, it dominates in the down state.

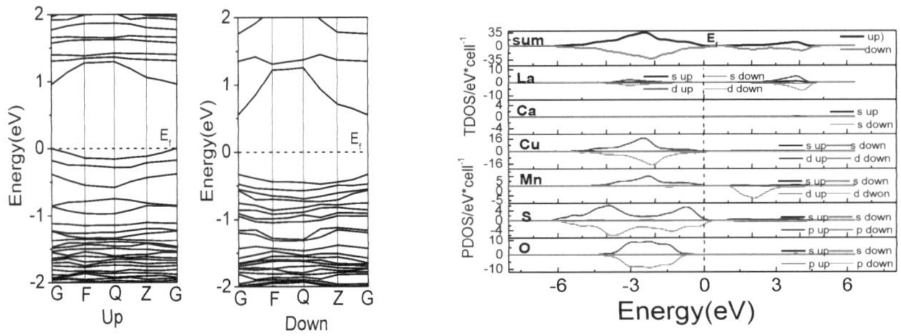

Fig.3 (a)Band structure and (b)DOS of $(La_{0.75}Ca_{0.25})$ $(Cu_{0.75}Mn_{0.25})SO$ with spin up and down.

In order to understand the influence of Ca and Cu co-dopants, we compare the band gap among the parents and co-doped materials with the same copants according to our previous results (see Table 1).

Table 1 Comparison of the band gaps among the parents and co-doped materials of LaCuSO, YCuSO and LaAgSO

Parents	Band gap(eV)	Co-doped materials	Band gap(eV)
LaCuSO	2.123	$La_{0.75}Ca_{0.25}Cu_{0.75}Mn_{0.25}SO$	Up 0.962
		$La_{0.75}Ca_{0.25}Cu_{0.75}Mn_{0.25}SO$	Down 0.88
YCuSO	1.250	$Y0_{.75}Sr_{0.25}Cu_{0.75}Mn_{0.25}SO$	Up 1.20
		$Y_{0.75}Sr_{0.25}Cu_{0.75}Mn_{0.25}SO$	Down 1
LaAgSO	1.40	$La_{0.75}Ba_{0.25}Ag_{0.75}Mn_{0.25}SO$	Up 1.25
		$La_{0.75}Ba_{0.25}Ag_{0.75}Mn_{0.25}SO$	Down 1.12

From the table, it is clear that after co-doping, the band gap of $La_{0.75}Ca_{0.25}Cu_{0.75}Mn_{0.25}SO$ decrease very quickly in both up and down states.

IV. Optical properties

First-principle calculation in investigating optical property, such as loss function, reflectivity coefficient, absorption coefficient and complex dielectric constant. These optical parameters are connected with one another and can be deduced directly from the radiative-transition-rate theory and Kramers-Kroning dispersion relation. From Fig.4(a),(b),(c),(d) and (e), it is also clear that the position of all the peaks for $(La_{0.75}Ca_{0.25})$ $(Cu_{0.75}Mn_{0.25})SO$ have the biggest "red shift" and the intensity of them decreases mostly. It indicates that co-doped Ca and Mn reduce the energy strongly, which is consistent with the change of band gap.

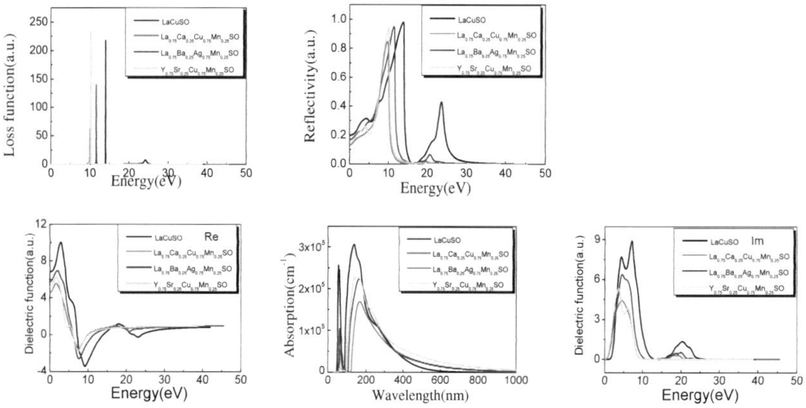

Fig.4 Comparison of (a) loss function, (b) reflectivity, (c) absorption, (d) the real part of dieletric function and (e) the imaginary part of dieletric function among parent LaCuSO, $(La_{0.75}Ca_{0.25})$ $(Cu_{0.75}Mn_{0.25})SO$, $(La_{0.75}Ba_{0.25})$ $(Ag_{0.75}Mn_{0.25})SO$ and $(Y_{0.75}Sr_{0.25})$ $(Cu_{0.75}Mn_{0.25})SO$, respectively.

Most importantly, the relative intensity of all the peaks decreases and the position of all the peaks shows "red shift" when doped. The peak intensity decreases when doping attributes to the scattering from the Ca and Mn dopant. The red shift originates from the impurity band generated by the holes to compensate the charge imbalance due to Ca doping, which agrees with the decreasing band gap when doped.

V. Conclusions

In conclusion, we found the band gap decrease after co-doping. $La_{0.75}Ca_{0.25}Cu_{0.75}Mn_{0.25}SO$ decreases most quickly result from the severe scattering from the Ca and Mn dopant .On the other hand, the position of all the peaks for $La_{0.75}Ca_{0.25}Cu_{0.75}Mn_{0.25}SO$ have the biggest "red shift" and the intensity of them decreases mostly. Thus, the difference in optical properties result from the difference in band structure (band gap).

Acknowledgments

This work was financially supported by the National Science Foundation of China (No.61376094) and China Scholarship Council (No. 201408330028).

References:

[1] K. Zhao,Z. Deng, X. C. Wang et al, 2013 Nature Communications,4, 1442.
[2] X.J.Yang, Y.K. Li, C.Y. Shen et al,2013 APL, 103,022410.
[3] K.Ueda, S.Inoue and H.Hosono et al,2001 APL, 78,2333

IOP Conf. Series: Journal of Physics: Conf. Series **1293** (2019) 012008 doi:10.1088/1742-6596/1293/1/012008

Transmission EBSD (t-EBSD) to determine grain and grain boundary properties on nanostructured superconductor samples

A. Koblischka-Veneva[1,2], M. R. Koblischka[1,2], J. Schmauch[1], M. Murakami[2]

[1] Experimental Physics, Saarland University, P.O.Box 151150, 66041 Saarbrücken, Germany
[2] Superconducting Materials Laboratory, Department of Materials Science and Engineering, Shibaura Institute of Technology, Tokyo 135-8548, Japan

E-mail: anjela@shibaura-it.ac.jp

Abstract. The knowledge of the properties of the grains and their grain boundaries is essential for the further optimization of sample performance of various high-T_c samples ($YBa_2Cu_3O_y$ (YBCO), Bi-based cuprates, MgB_2, iron-based materials). In these materials either the grains are of nanometer dimensions or the various additions of secondary phase(s) to improve the flux pinning are required to be of nanometer dimensions, so a high achievable resolution and better imaging properties are important to obtain reasonably high image quality of the recorded Kikuchi patterns to enable automated orientation mapping using the EBSD technique. Therefore, the newly developed transmission-EBSD (t-EBSD) technique is the method of choice to analyze the grain and the grain boundary properties in such nanostructured superconducting materials. Several results obtained on MgB_2, melt-textured, infiltration-growth processed YBCO and electrospun $Bi_2Sr_2CaCu_2O_8$ nanowires are presented.

1. Introduction

A proper knowledge of the microstructure of superconducting samples is essential to interpret data of the common integral measurement techniques like AC or DC magnetization or electric transport data (resistance, I/V-measurements). The electron backscatter diffraction (EBSD) technique was demonstrated in the literature by several authors to be a very useful tool on metallic and ceramic samples [1–4]. In the case of $YBa_2Cu_3O_y$ (YBCO or Y-123) superconductors, the introduction of flux pinning sites in the nanometer range demands a high-resolution microstructure analysis for melt-textured bulks as well as for tapes (coated conductors or powder-in-tube), whereas polycrystalline materials like MgB_2 or iron-based superconductors exhibit grain sizes in the nanometer range, which is also difficult to be analyzed for the standard EBSD technique. The transmission EBSD (t-EBSD) technique was very recently developed by several authors [5–8] with the goal to improve the spatial resolution of EBSD on materials with nanometer-sized grains. The samples investigated require to be transparent for the electron beam, so the sample fabrication as for transmission electron microscopy (TEM) has to be applied. As result, the spatial resolution could be improved from several tens of nanometers to ~5-10 nm, depending on the material to be studied [9, 10]. Additionally, the effects of charging in non-conducting samples are considerably reduced, which is very important for non-conducting,

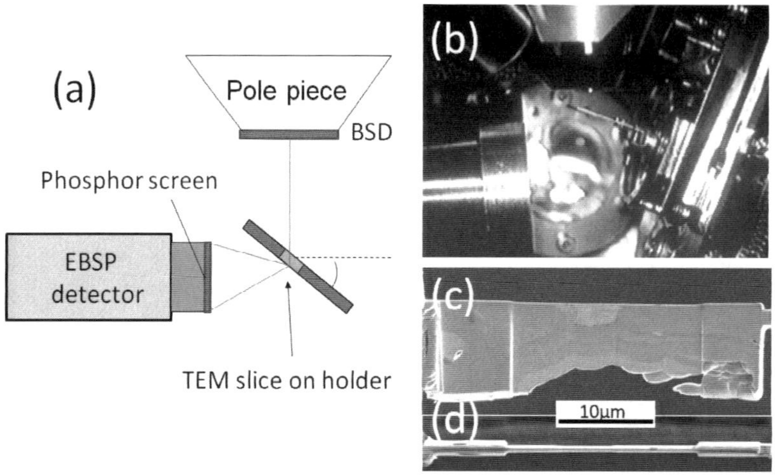

Figure 1. (a) Schematic drawing of the configuration for t-EBSD. Also indicated are the sample inclination angles required for EBSD imaging. BSD – backscatter detector, EBSP – electron backscatter pattern. (b) View of the TEM-slice sample holder for t-EBSD mounted in the SEM chamber, allowing to perform both convential EBSD and t-EBSD measurements. (c) gives a surface view of a TEM slice of melt-textured YBCO fixed on the micromanipulator, and (d) gives the cross-section of the slice.

semiconducting or even biomaterials [11]. Therefore, the t-EBSD technique is optimally suited for the analysis of crystallographic orientations of the grains or inclusions also in the various superconducting materials, especially those intended for applications [12, 13]. Up to now, the use of t-EBSD in the literature focused mainly on MgB_2, where the nanometer-sized grains and carbon additions as flux pinning sites play an important role [14,15]. In the case of sintered MgB_2 samples prepared using various reaction temperatures, t-EBSD was found to be the only way to obtain EBSD mappings on all the samples [15]. Furthermore, the crystallographic orientations of crystallites within electrospun $Bi_2Sr_2CaCu_2O_8$ (Bi-2212) nanowire fabric samples could only be investigated by employing the t-EBSD technique [16].

In this contribution, we present results of grain and grain boundary (GB) properties like the boundary density, GB misorientation and the grain size aspect ratio for spark-plasma sintered (SPS) MgB_2, melt-textured, infiltration-growth (IG)-processed YBCO and electrospun Bi-2212 nanowires. To enable a direct comparison of the results obtained, we give the same mappings and plots for each sample type, wherever possible.

2. Experimental procedures

The materials investigated comprise spark-plasma sintered (SPS) MgB_2 [17, 18], melt-textured YBCO bulk samples prepared by the infiltration growth (IG) technique [19] and Bi-2212 nanowires [20]. Details about sample preparation can be found in the references mentioned.

In a first step, all sample surfaces were subjected to mechanical polishing using SiO_2 papers, diamond pastes (3 μm, 1 μm and 0.25 μm) and Struers SiO_2 OP-S suspension (colloidal silica, particle size $\approx$ 40 nm) to a total roughness of several nanometers. Details of the polishing procedure are described in previous publications [4, 21]. The polished sample surfaces were then analyzed in the scanning electron microscope (SEM) to select the best-suited sites for the subsequent TEM slice fabrication.

In a second step, the TEM slices for the t-EBSD measurements were fabricated using a dual-beam focused-ion beam (FIB) workstation (FEI Strata DB 235) with a routine allowing

IOP Conf. Series: Journal of Physics: Conf. Series **1293** (2019) 012008 doi:10.1088/1742-6596/1293/1/012008

for reduced surface damage. After lifting-off the TEM slice from the sample with the micromanipulator, the surface is ion-polished in a separate step by 2 KeV Ga-ions to a thickness of about 80 nm. This step serves to further reduce the preparation damage of the surface area and for a further thinning of the sample to be transparent to the electron beam. Here, it should be noted that, in general, polycrystalline samples are difficult to be handled in this process. The breaking out using the micromanipulator may lead to irregular-shaped bottom sides of the slice; in the worst case, the slice can break into several pieces. To further improve the imaging quality, the TEM slices were treated by additional low-angle argon ion-polishing (5 KeV, 5 min) to increase the image quality (IQ) of the resulting Kikuchi patterns as described in [22] for the investigation of ferrite samples.

Figure 1 (a) gives a schematic drawing of the configuration for operating EBSD in the transmission mode (t-EBSD). The t-EBSD works with a transmitted electron beam in contrast to the conventional mode of EBSD. Therefore, an electron-transparent sample is required for t-EBSD. Such samples can be straightforwardly prepared using FIB milling. The TEM-slices were mounted in the SEM on a specially fabricated sample holder allowing for the correct 70° inclination of the sample required for EBSD. The stage with the sample holder is inclined to an angle of −20°, which enables, together with the sample mounting, the same detector position to be used for the EBSD detector as in the standard configuration. This is shown in Fig. 1 (b). Here, the electron beam is passing through the sample (transmission mode) and an electron cone emanating from the sample is formed on its backside [23]. When this cone is intercepted by the phosphor screen, the Kikuchi patterns can be observed. Figure 1 (c) presents the final TEM slice and its cross section (d), ready for investigation.

The EBSD analysis was performed in a JEOL 7000F SEM microscope equipped with an orientation imaging analysis unit (EDAX Inc., OIM Analysis™) [24]. The Kikuchi patterns were generated at an acceleration voltage of 15 kV (standard configuration), and at 30 kV in t-EBSD mode. The patterns were recorded by means of a DigiView camera system. For t-EBSD, the working distance is set to 5 mm. The stepsize was chosen to be 5 nm. More details of the t-EBSD measurements are described in Refs. [11, 15].

3. Results and discussion

Figure 2 present EBSD measurements and analysis on the SPS-MgB_2 sample. The original bulk sample (diameter 30 mm, thickness 15 mm) was cut twice in perpendicular direction in the center, producing a slice of 2 mm thickness. This slice was cut in several $2 \times 2 \times 2$ mm^3 cubes for various measurements. The sample surface of the TEM slice was selected to be a horizontal plane with respect to the original bulk. The uniaxial pressure (50 MPa) was applied parallel to the bulk height [17, 18], i.e., perpendicular to the selected sample surface.

Figure 2 (a) presents the EBSD image quality (IQ) mapping. We notice that there are elongated, large MgB_2 grains with high IQ values, but also may tiny grains with lower IQs, filling the spaces between the bigger ones. In some large grains, there are embedded small grains to be seen. In Fig. 2 (b) an orientation mapping in the ND-direction (i.e., normal to the sample surface) is given. The color code for the orientation is shown in the stereographic triangle below the map. We see here that the grains in SPS-sample are not fully randomly oriented, but do show a preferred texture as indicated by the dominant magenta color. Figure 2 (c) gives the corresponding pole figure. Figure 2 (d) is an IQ-map with the EBSD-detected GBs highlighted in color. The minimum GB misorientation detectable by EBSD is 2°. The colors are distributed as follows: 2° – 15° using blue lines, 20° – 55° using green lines, 60° – 75° using yellow lines, 80° – 125° using orange lines and 130° and above using red lines. The largest amount of the GBs is found in the range 20° – 55° (green), with some small angle GBs (blue) found within the large MgB_2 grains. This is further illustrated in the plot of Fig. 2 (g), giving the statistical distribution of the GB misorientation. The total length of all GBs detected in the analyzed area

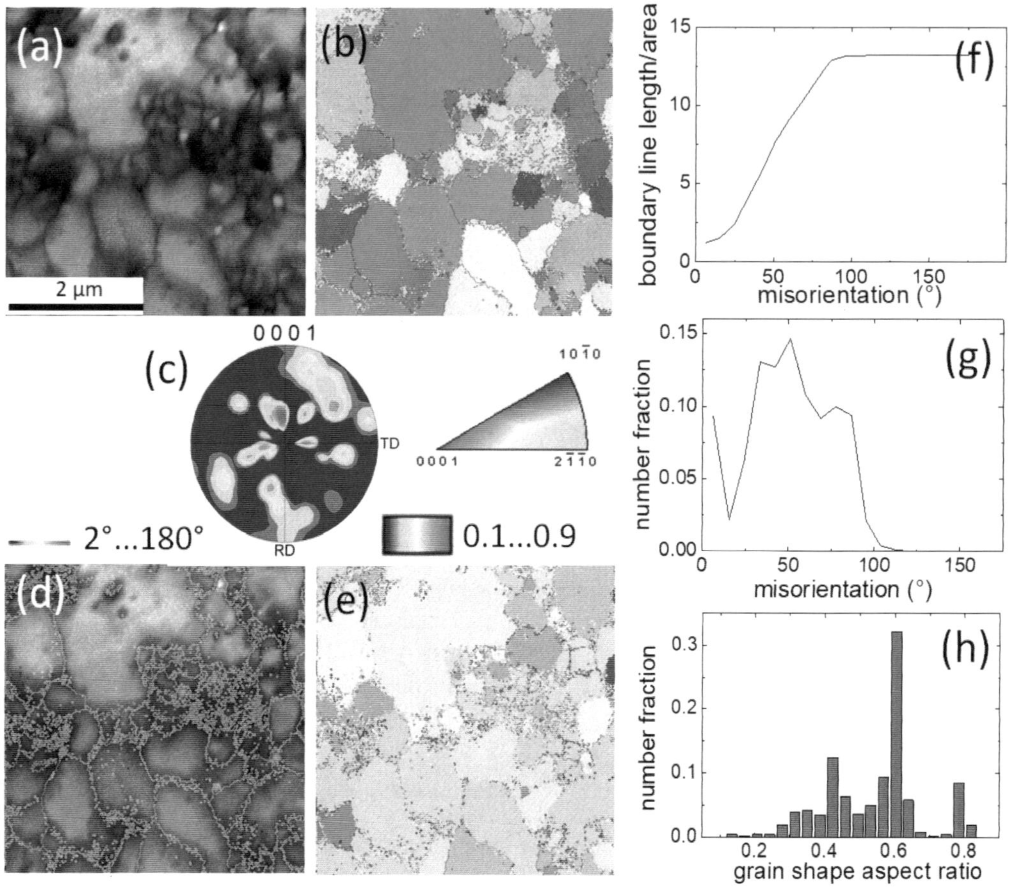

Figure 2. EBSD measurement and analysis on a SPS-MgB_2 sample. (a) IQ map, (b) orientation mapping giving the orientations normal to the sample surface (ND), (c) corresponding pole figure, (d) IQ map with the grain boundaries marked in color (— 2° – 15°, — 20° – 55°, — 60° – 75°, — 80° – 125° and — 130° and above), (e) map of the grain shape aspect ratio, (f) plot of the boundary density, (g) plot of the GB misorientation and (h) plot of the grain shape aspect ratio.

is 440.36 μm (25424 boundaries with misorientation > 2°). In Fig. 2 (e), a map of the grain shape aspect ratio, γ_{ar}, is presented. This ratio ranges from 0.1 (blue) to 0.9 (red), defined as the length of the minor axis divided by the length of the major axis. Therefore, round grains have $\gamma_{ar} \approx 1$ [24]. The plot of Fig. 2 (f) gives the GB density (boundary line length/area) as function of the misorientation; extracted from Fig. 2 (d). The high number of GBs in the investigated area is due to the MgB_2 grain size in the 100 nm range, and thus, a strong GB pinning can be provided. The plot of Fig. 2 (g) presents the statistics of the GB misorientation angles, showing a maximum at 50°. This is a consequence of the texture induced by the SPS process [15]. Finally, Fig. 2 (h) gives the distribution of the grain shape aspect ratio for all grains in the map, corresponding to the map of Fig. 2 (e). The maximum of ~0.6 indicates that the MgB_2 grains have dominantly an elliptic shape, whereas the small grains have a circular shape. This analysis yields important information about the sample studied. The GBs with misorientation angles of approx. 50° can be indeed effective flux pinning sites. A dominating pinning at GBs is also what is observed in magnetic measurements of this sample type. However, the presence

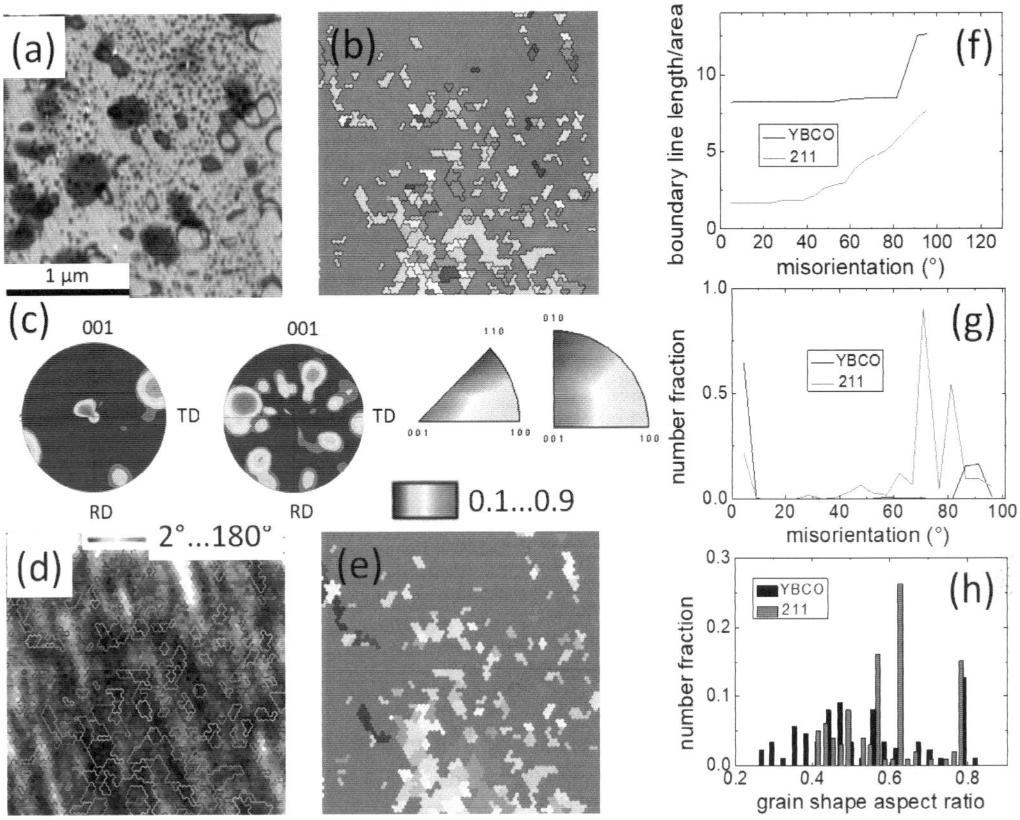

Figure 3. EBSD measurement and analysis on an IG-processed YBCO sample. (a) IQ map, (b) orientation mapping giving the orientations normal to the sample surface (ND), (c) corresponding pole figures, (left) YBCO and (right) 211, (d) IQ map with the grain boundaries marked in color (—— 2° – 15°, —— 20° – 55°, —— 60° – 75°, —— 80° – 125° and —— 130° and above), (e) map of the grain shape aspect ratio, (f) plot of the boundary density, (g) plot of the GB misorientation and (h) plot of the grain shape aspect ratio.

of the numerous small grains along the GBs provides a larger resistance to the current flow as could be expected from a sample without additional flux pinning sites. These results, therefore, corroborate the results of the resistance measurements discussed in [25].

Figure 3 shows a t-EBSD analysis on an IG-processed, YBCO bulk sample [19]. The unique properties of the IG-processed samples (i.e., extremely high irreversibility fields) are due the presence of very small Y_2BaCuO_5 (211) particles, being useful flux pinning sites as their size is comparable to 2 times the coherence length, ξ, especially at elevated temperatures. As result, these particles provide good flux pinning [26–28]. The sample selected for the investigation stems from the *c*-axis growth sector of the bulk (diameter 20 mm, height 8 mm), ~5 mm apart and 3 mm below the seed crystal. The sample surface is a typical *(a,b)*-plane of a melt-textured YBCO sample. To investigate the properties of these tiny 211 particles and their interaction with the YBCO matrix, the location for milling out the TEM slice was carefully chosen between the large 211 particles. In EBSD configuration (tilted to 70°), the SEM image clearly reveals these particles (Fig. 3 (a)) as dark spots. Figure 3 (b) gives an orientation map in ND direction for both phases, YBCO and 211. The color code for this map is given in the stereographic triangles below the map. The unit cells of YBCO and 211 are both orthorhombic, but the *c*-axis of 211 is much longer as that of YBCO, so the distinction between the two cells is simple, but some problems prevail as discussed in detail in Ref. [29]. Due to the increased IQ values of the t-EBSD images, it is possible to resolve the tiny 211 particles embedded within the YBCO matrix, which have sizes between 18 and 45 nm. From this mapping, we can learn that the larger 211 grains cause an alteration of the surrounding YBCO matrix like the big 211 grains measured in Refs. [30, 31]. Only the extremely small 211 grains do not affect the homogeneous growth of the YBCO matrix. This is a very important result for the fabrication of chemical flux pinning sites using secondary phases. In Fig. 3 (c), the corresponding pole figures are given for YBCO (left) and 211 (right). This figure reveals the strong texture of YBCO and the random-like orientation of the 211 particles. Figure 3 (d) presents an IQ map with the highlighted GBs. The total length of the GBs in the investigated section is 104.48 μm (4524 boundaries with misorientation $> 2°$). Fig. 3 (e) shows the grain aspect ratio mapping, revealing a large number of small, round grains embedded in a big, one-grain matrix, which is typical for melt-textured YBCO. The plot of Fig. 3 (f) gives the GB density in the investigated area. Here, we show the data for the two phases separately. The GB density for YBCO is nearly constant, with a sharp increase towards 90°, whereas the 211 phase exhibits an increasing GB density. Fig. 3 (g) shows the distribution of the GB misorientations, where YBCO shows a majority of low-angle GBs. The 211 phase has in contrast some GBs with angles between 40 and 60°, and a maximum at 70°. Fig. 3 (h) presents a plot of the grain boundary aspect ratio of both phases. YBCO exhibits several large, elongated grains, while the 211 phase has a peak at $\sim$0.6. For both phases, the small grains have a grain size aspect ratio of 0.8. These phase-resolved analysis of the GB and grain properties corroborates the earlier results obtained on melt-textured bulk samples [30, 31].

Finally, Fig. 4 presents t-EBSD results obtained on an individual piece of a nanowire from an electrospun, Bi-2212 nanowire network sample [20]. The dimensions of the nanowires enable t-EBSD measurements without additional sample preparation as discussed in Ref. [16]. Figure 4 (a) presents a SEM image of Bi-2212 nanowires mounted on a TEM grid. The yellow rectangle marks the analyzed area. In Fig. 4 (b), the result of the grain orientation mapping in normal orientation is presented. The color code is given in the stereographic triangle below the map. We observe that the Bi-2212 grains have a dominant orientation in [100]-direction (dominating green color in the map), which is due to the previous arrangement of the precursor material within the original polymer nanowire. In Fig. 4 (c), the corresponding pole figures in ND (left) and RD (right) directions are shown, which manifest the grain orientation along the former polymer nanowire. Figure 4 (d) presents an IQ map with the highlighted GBs. A large number of small-angle GBs is observed, but also a large number of GBs with 30 – 90° misorientation. The total length of the GBs in the investigated section is 71.85 μm (2489 boundaries with misorientation $> 2°$). On such a sample with strongly anisotropic, mostly elongated grains which is typical for the Bi-2212 or Bi-2223 compounds [32, 33], an analysis of the grain shape aspect ratio is very informative. Figure 4 (e) gives the grain shape aspect ratio mapping. The map reveals that a medium ratio of 0.6 is dominating in the selected area. This information describes the grain growth depending, e.g., on the reaction temperature and the holding time. Therefore, such EBSD analysis enables an optimization of the grain shape aspects of such an anisotropic material. Figure 4 (f) presents the GB density in the measured area, which is here monotonously increasing with increasing misorientation. Fig. 4 (g) shows the distribution of the GB misorientations, revealing a high amount of small-angle GBs and a large number of GBs in the range 40 – 90°. The plot of Fig. 4 (h) gives the distribution of the grain shape aspect ratio. The Bi-2212 grains have a compact shape, which is due to the reduced reaction temperature (800 °C) necessary to maintain the nanowire shape. The more elongated grains being typical for the Bi-2212 compound require obviously reaction temperatures around 830 °C as common for the tape fabrication [34].

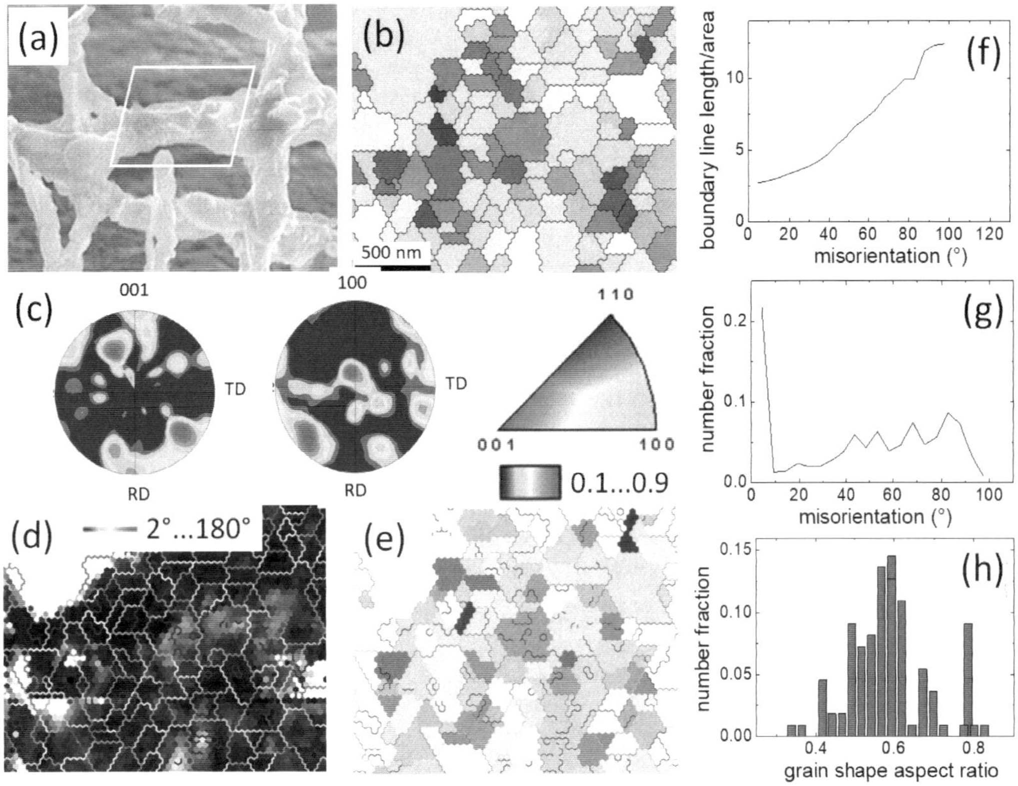

Figure 4. EBSD measurement and analysis on an IG-processed YBCO sample. (a) SEM image of a piece of a Bi-2212 nanowire network fabric sample, (b) orientation mapping giving the orientations normal to the sample surface (ND), (c) corresponding pole figures in ND and RD directions, (d) IQ map with the grain boundaries marked in color (— 2° – 15°, — 20° – 55°, 60° – 75°, — 80° – 125° and — 130° and above), (e) map of the grain shape aspect ratio, (f) plot of the boundary density, (g) plot of the GB misorientation and (h) plot of the grain shape aspect ratio.

4. Conclusions

The t-EBSD technique is demonstrated to be a very useful instrument to analyze the microstructures of various superconducting materials, especially when grains in the nanometer range or nanometer-sized particles as flux pinning sites embedded in the superconducting matrix are to be analyzed. The use of FIB-prepared TEM-slices allows proper areas for analysis to be selected, which is very important regarding inhomogeneously distributed flux pinning sites. Furthermore, t-EBSD enables automated multi-phase EBSD scans to be performed due to the increased image quality of the Kikuchi patterns obtained on ceramic samples or on samples with nanometer-sized grains. In general, the EBSD data collected on various superconducting samples provide essential input to develop specifically tailored growth techniques for superconducting materials.

Acknowledgments

We thank J. Noudem (CRISMAT, Caen, France) for providing us with the spark-plasma sintered MgB_2 sample, K. Nakazato, M. Muralidhar (SIT, Tokyo, Japan) for the IG-processed YBCO bulk, and X. L. Zeng (UdS, Saarbrücken, Germany) for the Bi-2212 nanowire fabrics and K.

Berger, Q. Nouailhetas and B. Douine (GREEN, Nancy, France) for valuable discussions. This work is supported by DFG-ANR grant Ko 2323/10, which is gratefully acknowledged.

References

[1] Humphreys F J 2004 *Scr Mater* **51** 771
[2] Dingley D 2004 *J Microsc* **213** 214
[3] Chen D, Kuo J-C and Wu W-T 2011 *Ultramicroscopy* **111** 1488
[4] Koblischka M R and Koblischka-Veneva A 2013 *Phase Transit* **86** 651
[5] Trimby P W 2012 *Ultramicroscopy* **120** 16
[6] Keller R and Geiss R 2012 *J Microsc* **245** 245
[7] Sneddon G C, Trimby P W and Cairney J M 2016 *Mat Sci Eng R* **110** 1
[8] Suzuki S 2013 JOM **65** 1254
[9] Britton B *et al* 2013 *Microsc Anal* **27** 8
[10] van Bremen R *et al* 2016 *Ultramicroscopy* **160** 256
[11] Koblischka-Veneva A *et al* 2018 *Nano Research* **11** 3911
[12] Silver T M, Dou S X, Jin J X 2001 *Europhysics News* Mai-June p. 82
[13] A. Luiz (ed.) 2011 Applications of high-T_c superconductivity *InTech Open*, DOI: 10.5772/2522
[14] Wong D C K *et al* 2015 *Scripta Mater* **101** 36
[15] Koblischka-Veneva A *et al* 2016 *Supercond Sci Technol* **29** 044007
[16] Koblischka-Veneva A *et al* 2018 *J Phys: Conf Ser* **1054** 012005
[17] Noudem J G *et al* 2014 *J Appl Phys* **116** 163916
[18] Noudem J G *et al* 2016 *Materials Today: Proceedings* **3** 545
[19] Nakazato K *et al* 2014 *Physica C* **504** 4
[20] Zeng X L *et al* 2017 *Supercond Sci Technol* **30** 035014
[21] Koblischka-Veneva A and Koblischka M R 2004 in: MO imaging (eds Johansen T H and Shantsev D V) Kluwer: Dordrecht p 242
[22] Koblischka-Veneva A *et al* 2010 *J Phys: Conf Ser* **200** 082014
[23] Reimer L 1985 Springer Science: Berlin Heidelberg
[24] Orientation Imaging Microscopy (OIM™) Software Version V8.1, User Manual, EDAX Inc., Draper, UT.
[25] Wiederhold A *et al* 2016 *J Phys: Conf Ser* **695** 012004
[26] Koblischka-Veneva A *et al* 2016 *J Phys: Conf Ser* **695** 012002
[27] Pavan Kumar Naik S, Muralidhar M and Murakami M 2019 *IEEE Trans Appl Supercond* **29** 6800305
[28] Namburi D K *et al* 2016 *Supercond Sci Technol* **29** 095010
[29] Koblischka-Veneva A *et al* 2019 *IEEE Trans Appl Supercond* **29** 6800504
[30] Koblischka-Veneva A *et al* 2005 *Supercond Sci Technol* **18** S158
[31] Ogasawara K *et al* 2003 *Physica C* **386** 225
[32] Koblischka-Veneva A *et al* 2008 *Physica C* **468** 174
[33] Koblischka-Veneva A, Koblischka M R 2008 *J Phys: Conf Ser* **94** 012011
[34] Skov-Hansen P *et al* 2001 *IEEE Trans Appl Supercond* **11** 3740

IOP Conf. Series: Journal of Physics: Conf. Series **1293** (2019) 012009 doi:10.1088/1742-6596/1293/1/012009

Porous high-T_c superconducting cuprates: Advantages and applications

M. R. Koblischka[1,2], A. Koblischka-Veneva[1,2], S. Pavan Kumar Naik[2], D. Gokhfeld[3], M. Murakami[2]

[1] Experimental Physics, Saarland University, P.O. Box 151150, 66041 Saarbrücken, Germany
[2] Superconducting Materials Laboratory, Department of Materials Science and Engineering, Shibaura Institute of Technology, Tokyo 135-8548, Japan
[3] Kirensky Institute of Physics, Federal Research Center KSC SB RAS, Krasnoyarsk, 660036, Russia

E-mail: m.koblischka@gmail.com; miko@shibaura-it.ac.jp

Abstract. Porous high-T_c superconducting cuprates are promising materials representing an alternative preparation route to enable the fabrication of large-scale, light-weight superconducting samples. There are several advantages of such samples including the much easier (and faster) oxygenation process, a simpler scalability to produce large samples, and of course, the reduced weight. Two different types of such samples were prepared in the literature: (i) Superconducting foams, prepared using polyurethane foams converted to green phase foams followed by an infiltration growth (IG) process. (ii) Superconducting nanowire networks prepared by spinning from sol-gel precursors. Such fabric-like nanowire networks are extremely light-weight, but show very interesting properties. We discuss the properties of such samples concerning both the physical parameters and the respective microstructures and give an overview about possible applications.

1. Introduction

Most research activities in the field of high-T_c superconductivity (HTSc) are focused to increase the critical current densities for various classical applications like energy transport and generation of high magnetic fields [1,2]. This forces to improve the crystallographic texture and to provide strong flux pinning. The corresponding materials are three types of HTSc samples: Tapes or wires, thin films and bulk samples. However, all of these sample types have certain drawbacks which hinder the development of some types of applications. One issue is, e.g., the scalability and shaping to obtain large superconducting samples of various shapes [3], the mechanical stability when generating large magnetic fields [4] and the problems arising when oxygenating bulk samples. In the latter case, the required time to oxygenate large, bulk samples requires weeks, and internal cracking may appear due to the phase transformation in the Y-Ba-Cu-O system ($YBa_2Cu_3O_{6...6.5} \rightarrow YBa_2Cu_3O_{6.6...7}$) [5]. As an answer to these problems, one can find in the literature several attempts to grow different types of high-T_c superconducting samples. In many current fields of research, porous metallic or ceramic samples play an increasingly important role as it becomes possible to produce light-weight samples with increased mechanical strength [6–9]. For superconducting porous samples, there are several adantages like (i) the easy and fast oxygenation process due to the open-porous structure, (ii) the scalability of the sample size, (iii) the ideal cooling process and (iv) the reduced weight and the reduced amount of (expensive) superconducting (precursor) material. Therefore, some years ago, several different types of porous material were studied [10–15], but the further development of these materials was overrun by the progress of preparing coated conductors [16, 17]. However, even up to date, the aforementioned problems are not yet overcome, and certain applications are only possible with, or still require, the use of porous superconducting materials.

2. Experimental procedures

Figure 1. YBCO foam (a,c) and Bi-2212 nanowire network fabric (b,d). The images (a) and (b) give the preforms of both types of porous HTSc samples, and (c) and (d) the fully-reacted, final products. Note the seed crystal and the YBCO cloth on the foam surface in (c), and the size of the nanowire fabric samples in (b) and (d).

In the present contribution, we focus on two distinctly different porous HTSc materials: Bulk-like $YBa_2Cu_3O_y$ (YBCO) foams and $Bi_2Sr_2CaCu_2O_{8+\delta}$ (Bi-2212) nanowire network fabrics prepared by means of electrospinning. The foam samples were prepared at RWTH Aachen, which is described in detail in Refs. [10, 11]. The superconducting transition temperature, T_c, of the foam samples is 91.1 K like in the bulk samples. Here it is important to point out that crystallographic orientation was introduced to the foam by using the infiltration growth process on a Y_2BaCuO_5 (211) foam and a seed crystal on top of the structure. This treatment introduced indeed a texture as evidenced by x-ray and neutron diffraction [18].

The Bi-2212 nanowire network fabrics were fabricated by electrospinning as described in Refs. [19–21]. The T_c of the fabric samples is just below 77 K, and previous investigations have shown that the superconducting properties are well developed below 30 K like in the case of bulk Bi-2212 samples [22]. Alternatively, another fabrication technique, the solution blow spinning, was recently developed to produce similar nanowire network fabrics. In this way, YBCO nanowires were produced as well [23].

Magnetic hysteresis loops were recorded out using SQUID magnetometry, for details see, e.g., Ref. [24]. Trapped field (TF) measurement of the foam were carried out using a homemade setup with a scanning Hall probe operating at 77 K. The sample was field cooled in a field of a Nd-Fe-B permanent magnet (0.5 T surface field) in liquid nitrogen. TF profiles were recorded after 15 min waiting time at 1.5-2 mm height above the foam sample surface, which is more rough than that of a bulk sample.

3. Results and discussion

3.1. Nanowire network fabrics

The Bi-2212 nanowire network fabrics consist of long (up to 100 μm) nanowires (diameter ~250 nm) which form a nonwoven fabric material with numerous interconnects between the polycristalline nanowires. These interconnects enable a current flow through the entire sample perimeter. As a result, the nanowire fabrics were found to perform at low temperatures even at large applied magnetic fields up to 10 T as discussed in Refs. [19–21].

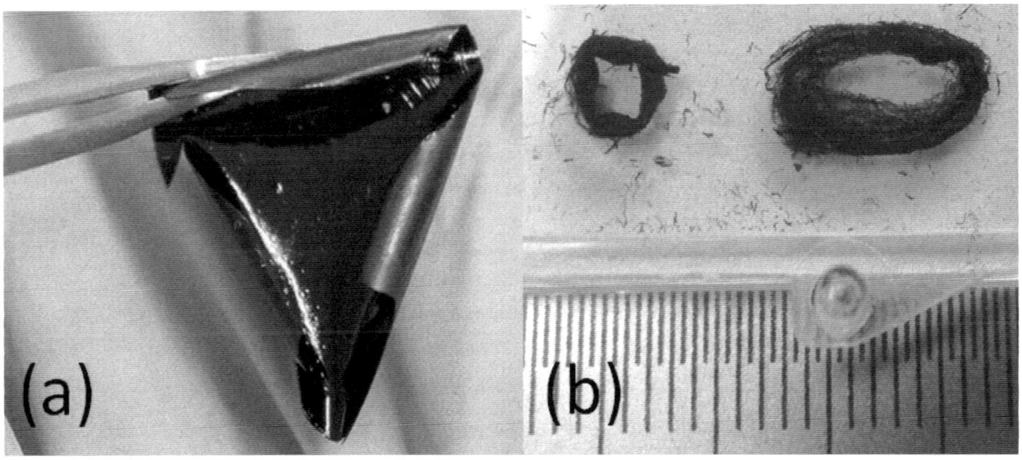

Figure 2. Applications of Bi-2212 nanowire network fabric samples: (a) fully flexible, superconducting foil and (b) completly reacted, donut-shaped nanowire network fabric samples.

Figure 1 (b) already shows that the upscaling of electrospun fabrics is very straightforward. The fully-reacted Bi-2212 nanowire network fabrics are extremely light-weight (only 0.05 g/cm^3 density, porosity 0.9928) [24] and as result, very brittle, which requires additional treatment to make the samples useful for applications. One possible route would be to embed the fully reacted nanowire fabric into a suitable polymer [25], so that a flexible superconducting foil (Fig. 2 (a)) results. Such a flexible foil could easily cover also complicated shapes when being used as a magnetic shield. Another possible application was already discussed in Ref. [26], where the fabric-sample could be used as a large-scale 'superconducting carpet', which would be placed in a thin plastic casing and cooled by streaming liquid nitrogen. In this way, a large platform to levitate magnets above it could be created.

More shaping possibilities are provided in the preform stage which consists of polymer nanowires containing the ceramic precursor material. Here, an easy shaping process is possible: One can cut a small section off and wrap it around a wire to create a donut shape. Figure 2 (b) presents such fully-reacted nanowire donut samples which could be employed as light-weight magnetic shielding.

To bring the nanowire fabric samples towards real applications, it would be a great benefit to have samples with a higher T_c and a less-2D-like behavior as in the case of Bi-2212. A superconducting carpet operating at 77 K with liquid nitrogen as a coolant could be very interesting for automation purposes [26]. The shielding applications could also work at low temperatures/high magnetic field environments, as the resistance measurements on nanowire fabric samples were possible in fields up to 10 T, which one would not expect for such polycrystalline HTSc samples.

3.2. Superconducting foams

The superconducting foams can be applied as trapped field magnets, as the upscaling enables to create foam samples of larger dimensions as bulk samples, avoiding the extremely long oxygenation times required. Furthermore, the foams have much less weight compared to the bulks. The present trapped field values are not spectacular, but here it is important to mention that the current shape of the foams was never optimized for large trapped fields. First experiments [27] had shown trapped fields of ~400 G, when field-cooling (FC) the a smaller foam sample in a field of 600 mT. In these experiments, it was noted that a broad peak and several sharp peaks resulted.

Figures 3 and 4 show the results of a trapped field experiment on the YBCO foam sample. The sample has dimensions $5 \times 2 \times 2$ cm^3, and was field-cooled in the field of a square Nd-Fe-B magnet (size $30 \times 30 \times 25$ mm^3) which had a surface field of 0.5 T (remanent magnetization 1.3 T). The magnet was located on top of the top surface (with the former seed crystal, labelled (1) – see the inset) of the foam. Figure 3 gives the measured trapped field distributions as 3D surface plots as well as contour plot for the foam sample surface (1) and the side (2), and Fig. 4 shows the measurement results for sides (3) and (4). On all sample sides, we find a trapped field peak, and characteristically, also the presence of several sharp, small peaks. The foam sides (2) and (4), which are 90° off the original field direction, exhibit a broad peak with only a small $B_{\rm trap}$. Remarkably, there are many of the small sharp peaks on side (4). The measurement on side (3) reveals a negative peak field of −100 G. The measurements clearly demonstrate the presence of two types of currents in the foam sample – one current density, which is running through the entire sample perimeter, resulting in the large, cone-shaped maxima with $B_{\rm trap,max}$, and one locally distributed current density being responsible for the small sharp peaks. Both currents are percolative, truely 3D currents flowing through the foam struts, and are not confined to planes in the sample. Repeating the scan of B_z reproduces most of these peaks, but the height may vary. We ascribe these peaks to local current loops in the sample, which were formed around local defects (e.g., large pores, pore clusters or non-superconducting struts), and are compressed due to the rearrangement of the internal field, B_i, when removing the external field. Similar observations were made already in Ref. [27], where only the surface (1) close to the seed crystal was measured. Here we have to note that the Hall probes measure only the z-component of B, which implies that current paths along the z-axis without a z-component of the generated magnetic field will not be detected. Only the (a,b)-projection of the real circulating 3D currents will be measured, and, the various contributions to it are weighted relative to their distance to the scanning Hall probe.

The measured trapped field values are still smaller than those of YBCO bulks, but regarding the difference in current density and the reduced weight of the foam samples, the results are not so bad. Also, most IG-processed bulks were found to exhibit still smaller trapped fields

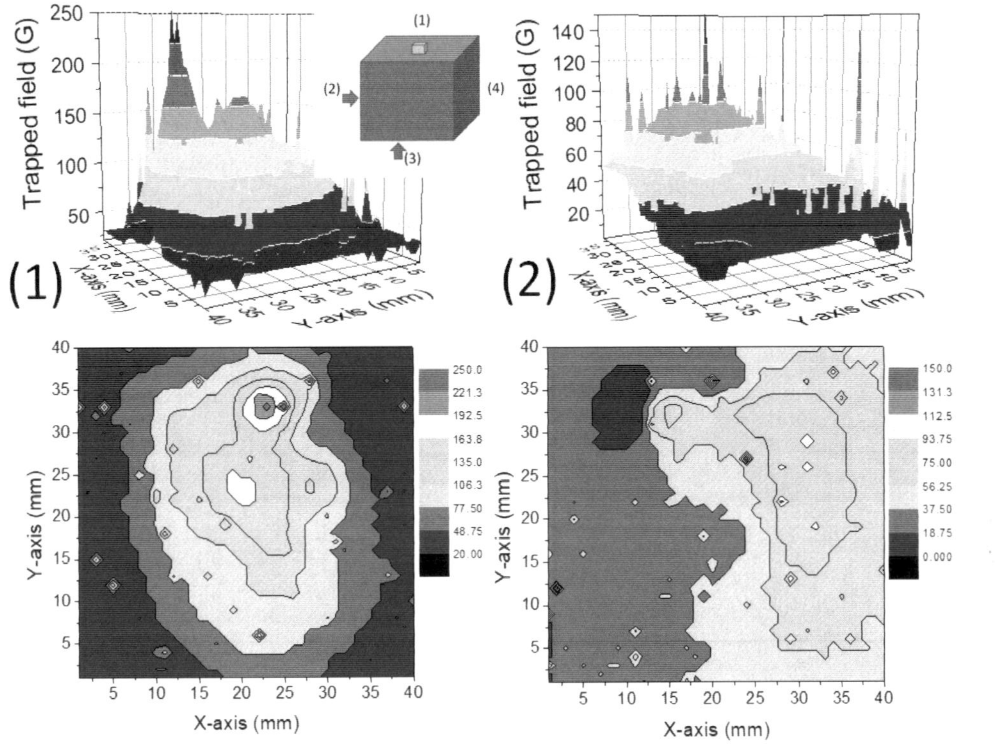

Figure 3. Trapped field measurements on a foam sample. Distribution of B_z measured by scanning Hall probe on sides (1) and (2) of the foam sample. The inset gives a schematic drawing of the foam sample and the nomenclature. The permanent magnet in the FC cooling process was placed on side (1).

as the MTMG-processed bulks. Furthermore, the present foam sample is 14 years old and has underwent many experiments as shown in Fig. 5 below. Here, it is important to point out that the present sample shape was never optimized for trapped field measurements, nor the texture of the sample, which has a dominating orientation, but even magnetization measurements on single struts still reveal signs of granularity [28]. A simple estimation of the overall transport current density in the foam sample can be done using the size of the trapped field peak ($B_{\rm trap}$) with the relation [29]

$$J_c R \propto B_{\rm trap}/\mu_0 \quad . \tag{1}$$

With the maximal trapped field of 250 G and $R = 15$ mm, we obtain $J_c = 130$ A/cm^2.

Due to the multi-peak trapped field distribution as seen in Figs. 3 and 4, this estimation gives the lower boundary for the real critical current density flowing in the sample. This observation points out that the current flow in such porous samples is more complex as in bulk samples, so we need a closer look to it.

Concerning the current flow in porous high-T_c materials, we have to be aware that these are polycrystalline materials, so we have to consider two different contributions; the intergranular ($J_{c,\rm inter}$) and intragranular ($J_{c,\rm intra}$) currents. The intragranular currents are controlled by the flux pinning and are as high as those of single crystals or bulk samples, whereas the intergranular currents are affected by the grain boundaries between the superconducting grains and by the

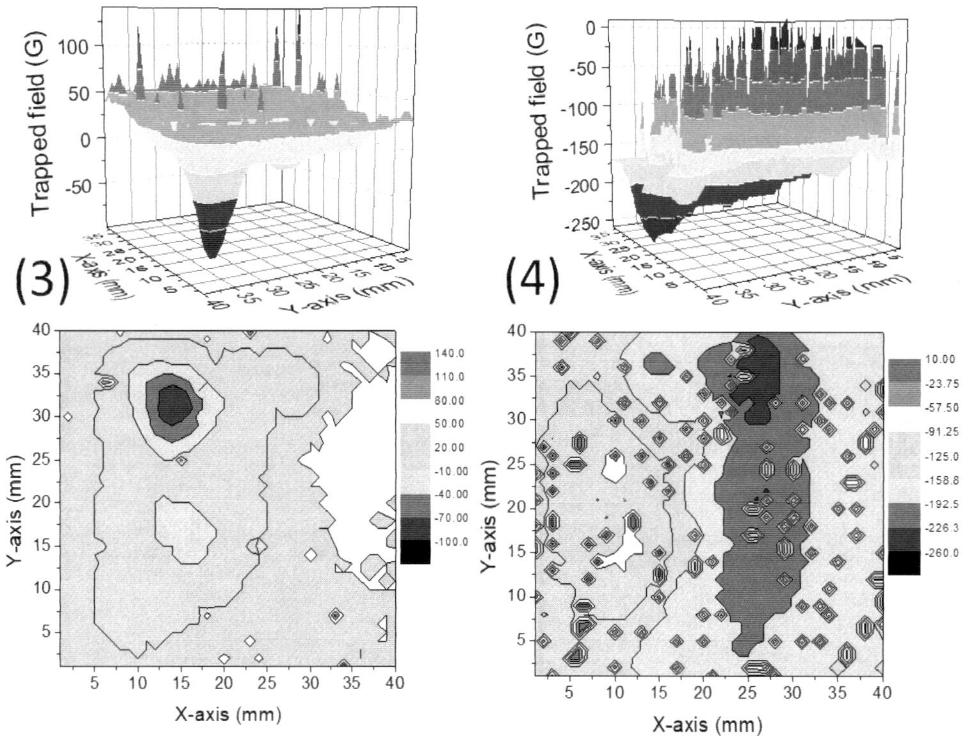

Figure 4. Distribution of B_z measured by scanning Hall probe sides (3) and (4).

relative orientations of the anisotropic granules to each other. This describes the case of dense, bulk polycrystalline samples. In porous samples, there is an additional contribution due to the percolative current flow caused by the porosity and the spacial orientation of the foam struts to the external field. In case of the nanowire networks, the contacts formed by the interconnects between the nanowires provide another limiting factor. Therefore, the overall currents flowing through the entire sample are limited by all these factors, which is the main drawback for the porous superconducting materials.

Figure 5 presents $J_{c,intra}$ for the Bi-2212 nanowire fabrics. The data for $J_{c,intra}$ (open symbols) stem from measured magnetization loops of extended pieces of the nanowire fabric, and were evaluated using the Bean model and corrected for porosity by a factor 1/0.0072. Details of this calculation using the extended critical state model (ECSM) on the measured magnetization data can be found in Ref. [24]. The Bi-2212 nanowires show relatvely high current densities only at temperatures below 30 K, which is in stark contrast to the YBCO foam, which is also shown in Fig. 5 using full symbols for the intragranular current density measured on a single foam strut at $T = 60$ K and 77 K [28]. Therefore, liquid nitrogen can be used as coolant for the YBCO foam samples. The lower limit of the overall current density determined from the trapped field data is given also in Fig. 5 for comparison. Calculating this overall current density for the nanowire fabric sample yields about 0.05 A/cm^2.

The YBCO foam shows much higher values of H_{irr} than the Bi-2212 nanowire fabric sample, being comparable to those of IG-processed bulk samples. The H_{irr} reaches 7 T already at 77 K, which renders the measurement difficult [28]. The irreversibility field is known to be strongly

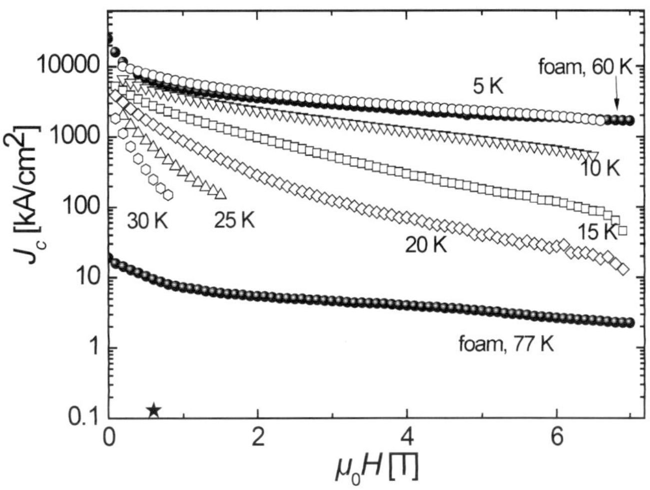

Figure 5. Critical current densities of the YBCO foam and the Bi-2212 nanowire network fabrics. The open symbols give the intragranular current density $J_{c,\mathrm{intra}}$ of the Bi-2212 nanowires; the full symbols give the intragranular current density $J_{c,\mathrm{intra}}$ measured on an individual YBCO foam strut piece at 60 K and 77 K for comparison. The star indicates the overall current density of the foam sample determined from the trapped field plots at 77 K.

dependent on the pinning strength and on the scale of the current circulation [30, 31]. As also texture is achieved in the YBCO foam samples [18], this is very promising for future improvement of the overall critical current densites of such foam samples.

The superconducting YBCO foams could see many different applications which make use of the efficient cooling process, e.g., for fault current limiters, which was already discussed in the first publications on foams [10, 11]. Cooling of superconducting foam structures from room temperature to 77 K proceeds about ten times faster compared to respective bulk materials of the same mass and composition [6]. This can directly be demonstrated when cooling a foam sample in liquid nitrogen, e.g., to test the levitation properties on a magnetic rail consisting of 3 Nd-Fe-B magnets in a row (S-N-S configuration, [32]) as shown in Fig. 6. When operating on such a magnetic rail, the foam sample is cooled so well that it takes ~5 min to warm up again. Furthermore, the lift height of the foam sample is much higher than that of an YBCO bulk sample due to the reduced weight. Another interesting possibility would be to turn around the situation of the levitation experiment: Using long pieces of foam would create an easily coolable superconducting rail, where magnetic objects can levitate upon. Such a superconducting rail system could be cooled very effectively by liquid nitrogen being pumped through.

The examples presented are not yet all possibilities for superconducting foam samples. Superconducting foams with small open porosity can easily be continuously reinforced, e.g. with resins, to improve their mechanical properties and, thus, to overcome the forces encountered in levitation and quasi-permanent magnet applications [33]. Also an electroplating process to further improve the mechanical strength as done for metallic Al foams [9] could be carried out here, as well as impregnation using resins. This will contribute to avoid the cracking problems of bulk superconductor samples prevailing when trapping large magnetic fields at lower temperatures. The large amount of possible applications of foam samples suggest that further development of this sample type is an important task for the future. Combining the

Figure 6. Superconducting foam levitating above a magnetic rail. (a) shows the foam sample at full speed running on the rail, and (b) gives the foam sample when being close to warm up.

latest developments concerning the IG processing of HTSc materials, the optimization of the sample shape for the given task and improved understanding of the open porous structure can lead to unique types of superconducting samples to fulfill different demands of applications. This will be the goal for future experimental work concerning a second generation of superconducting foam samples.

4. Conclusions

To conclude, porous high-T_c superconducting cuprates are very promising materials for applications wherever the sample weight or the cooling efficiency counts. This is given, e.g., in space experiments, for fault current limiters, and trapped field applications in rotating machinery. The achieved trapped field values are still small, but promising and the cooling efficiency can be demonstrated in simple levitation experiments. The future work will require also a deeper understanding of the current flow in such samples, combining already existing modelling approaches of mechanical properties of metallic foams with modelling of the superconducting current flow.

Acknowledgments

We thank X. L. Zeng, T. Karwoth for the work on electrospinning, G. Schmitz (ACCESS, RWTH Aachen, Germany) for the foam sample and K. Berger, B. Douine and Q. Nouailhetas (GREEN, Nancy, France) for valuable discussions concerning the foam materials. This work is supported by DFG-ANR grant Ko 2323/10 ("Superfoam"), SIT start-up grant and Volkswagen foundation, which is gratefully acknowledged.

References

[1] Larbalestier D C, Gurevich A, Feldmann D M and Polyanskii A 2001 *Nature* **414** 368
[2] Gurevich A 2011 *Nature Mater* **10** 255
[3] Noudem J G 2011 *J Supercond* **24** 105
[4] Johansen T H 2000 *Supercond Sci Technol* **13** R121
[5] Diko P 2004 *Supercond Sci Technol* **17** R45
[6] Nettleship I 1996 *Key Eng. Mater.* **122** 305
[7] Colombo P 2002 *Key Eng Mater* **206-213** 1913
[8] Hill Ch and Eastoe J 2017 *Adv Colloid Interface Sci* **247** 496

[9] Jung A, Diebels S, Koblischka-Veneva A, et al. (2013) *Adv Eng Mater* **16** 15
[10] Reddy E S and Schmitz G J 2002 *Am Ceram Soc Bull* **81** 35
[11] Reddy E S, Herweg M and Schmitz G J 2003 *Supercond Sci Technol* **16** 608
[12] Petrov M I, Tetyueva T N, Kveglis L I, Efremov A A, Zeer G M, Shaikhutdinov, K A, Balaev D A, Popkov S I and Ovchinnikov S G 2003 *Tech Phys Lett* **29** 986
[13] Fiertek P and Sadowski W 2006 *Material Science Poland* **24** 1103
[14] Fiertek P, Andrzejewski B and Sadowski W 2010 *Rev Adv Mater Sci* **23** 52
[15] Terent'ev K Yu, Gokhfeld D M, Popkov S I, Shaikhutdinov K A and Petrov M I (2011) *Phys Solid State* **53** 2409
[16] Eisterer M, Moon S H and Freyhardt H C 2016 *Supercond Sci Technol* **29** 060301
[17] Durrell J H, Ainslie M D, Zhou D, Vanderbemden P, Bradshaw T, Speller S, Filipenko M and Cardwell D A 2018 *Supercond Sci Technol* **31** 103501
[18] Noudem J G, Guilmeau E, Chateigner D, Lambert S, Reddy E S, Ouladdiaf B and Schmitz G J 2004 *Physica C* **408410** 655
[19] Zeng X L, Koblischka M R and Hartmann U 2015 *Mat Res Express* **2** 095022.
[20] Koblischka M R, Zeng X L, Karwoth T, Hauet T and Hartmann U 2016 *IEEE Trans Appl Supercond* **26** 1800605
[21] Koblischka M R, Zeng X L, Karwoth T, Hauet T and Hartmann U 2016 *AIP Adv* **6** 035115
[22] Koblischka M R and Sosnowski J 2005 *Eur J Phys B* **44** 277
[23] Rotta M, Zadorosny L, Carvalho C L, Malmonge J A, Malmonge L F and Zadorosny R 2016 *Ceram Int* **42** 16230
[24] Zeng X L, Karwoth T, Koblischka M R, Hartmann U, Gokhfeld D, Chang C and Hauet T 2017 *Phys Rev Mater* **1** 044802
[25] Haupt S G, Riley G R, and McDevitt J T (1993) *Adv Mater* **5** 755
[26] Koblischka M R and Koblischka-Veneva A 2018 *AIMS Material Sci* **5** 1199
[27] Bartolomé E, Granados X, Puig T, Obradors X, Reddy E S and Schmitz G J 2004 *Phys Rev* B **70** 144514
[28] Koblischka M R, Koblischka-Veneva A, Chang C, Hauet T, Reddy E S and Schmitz G J 2018 *IEEE Trans. Appl. Supercond.* DOI: 10.1109/TASC.2018.2880334.
[29] Bean C P 1962 *Phys Rev Lett* **8** 250
[30] Ihara N and Matsushita T 1996 *Physica* C **257** 223
[31] Matsushita T, Otabe E S, Wada H, Takahama Y and Yamauchi H 2003 *Physica* C **397** 38
[32] Santosh M and Koblischka M R 2014 *Eur J Phys Educ* **5**(4) 1
[33] Tomita M and Murakami M 2003 *Nature* **421** 517

Uniform hole doping in $HgBa_2Ca_2Cu_3O_{8+\delta}$ studied by ^{63}Cu NMR

Yutaka Itoh[1], **Akihiro Ogawa**[2] **and Seiji Adachi**[3]

[1]Department of Physics, Graduate School of Science, Kyoto Sangyo University, Kamigamo-Motoyama, Kika-ku, Kyoto 603-8555, Japan
[2]Chugoku Electric Power Company Inc. Energia Research Institute, 3-9-1 Kagamiyama, Higashi Hiroshima, Hiroshima 739-0046, Japan
[3]Superconducting Sensing Technology Research Association, 2-11-19 Minowa, Kohoku, Yokohama, Kanagawa 223-0051, Japan

E-mail: `yitoh@cc.kyoto-su.ac.jp`

Abstract. We studied local magnetic properties of triple-CuO_2-layer superconductors $HgBa_2Ca_2Cu_3O_{8+\delta}$ with rich ^{63}Cu isotope (underdoped T_c = 124 K and optimally doped T_c = 134 K) by ^{63}Cu NMR spin-echo techniques. The temperature dependences of the ^{63}Cu nuclear spin-lattice relaxation rate $^{63}(1/T_1)$ and the spin Knight shift $^{63}K_{\rm spin}$ of the inner plane Cu(1) site were nearly the same as those of the outer plane Cu(2) site in the normal states. The pseudogap temperature T^* defined as the maximum temperature of $^{63}(1/T_1T)$ was 190 K for the underdoped T_c = 124 sample and 143 K for the optimally doped T_c = 134 K sample. The NMR results indicate the uniform hole distribution in the three layers, which is consistent with a coalescence of three-derived Fermi surfaces predicted by a band theory. The room temperature in-plane $^{63}K^{ab}_{\rm spin}$ depends on whether the number of CuO_2 planes per unit cell is even or odd for the Hg-based multilayer cuprate superconductors.

1. Introduction

The optimally hole doped $HgBa_2Ca_2Cu_3O_{8+\delta}$ (Hg1223) has the highest superconducting transition temperature T_c = 134 K among the ever reported cuprate superconductors [1, 2]. Hg1223 has three CuO_2 planes per unit cell. Crystallographically inequivalent Cu sites are the 4-oxygen coordinated inner plane (IP) Cu(1) site and the 5-oxygen coordinated outer plane (OP) Cu(2) site. The left panel in Fig. 1 is an illustration of the crystal structure of Hg1223. One of the issues on the multilayer cuprate superconductors is how the doped holes distribute in the multiple planes per unit cell. NMR spin-echo techniques enable us to observe selectively the inequivalent Cu sites.

Transport measurements for single crystals Hg1223 with T_c = 134 K tell us the effects of antiferromagnetic spin fluctuations and a pseudogap [3]. The pseudogap temperature of T^* is defined as the peak temperature of the ^{63}Cu nuclear spin-lattice relaxation rate divided by temperature $1/T_1T$. The spin pseudogap opens below T^*. The Cu nuclear spin-lattice relaxation studies revealed the pseudogap for underdoped single crystals with T_c = 115 K (T^* = 230 K) [4], an underdoped Re-substituted powder sample with T_c = 126 K (T^* = 172 K) [5], and an optimally doped powder sample with T_c = 133 K (T^* = 160 K) [6, 7].

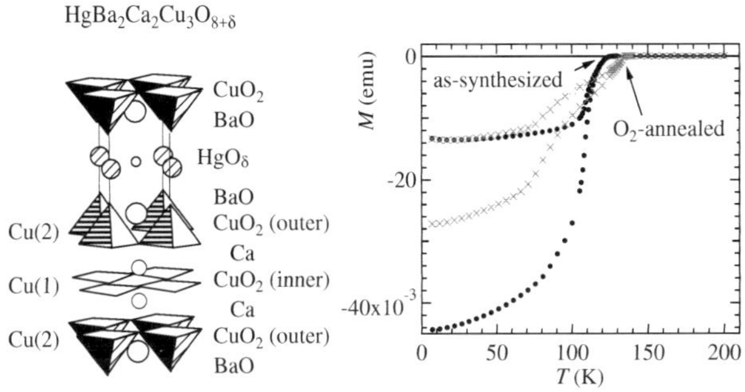

Figure 1. Illustration of crystal structure of Hg1223 (left panel). Magnetizations of the oriented powders for a field of 100 Oe perpendicular to the aligned c axis (right panel). Closed circles and crosses are for as-synthesized and O_2-annealed samples, respectively.

In this paper, we report the site-selective ^{63}Cu NMR measurements for two serial samples of the triple-CuO_2-layer high-T_c superconductors Hg1223, in which the isotope ^{63}Cu was enriched to improve a NMR signal-to-noise ratio. We found uniformly hole-doped CuO_2 planes and an in-plane spin Knight shift $^{63}K^{ab}_{\rm spin}$ associated with the number parity of CuO_2 planes per unit cell.

2. Experiments

Polycrystalline powder samples of Hg1223 were prepared by a solid state reaction of mixed powders of high-purity HgO, BaO, CaO, and ^{63}Cu-rich CuO (the isotope ^{63}Cu of concentration $> 90\%$) [8]. The mixed powder was pelletized and encapsulated in a quartz tube under high vacuum. The pellets sealed in the evacuated tube were fired at 775 °C for 4 days. As-synthesized samples are underdoped superconductors. A part of the as-synthesized samples was annealed in O_2 gas at 370 °C for 12 hours and then cooled for 10 hours. Two serial samples with as-synthesized and O_2-annealed were confirmed to be in single phase by measurements of powder X-ray diffraction patterns. The powder samples mixed in epoxy (Stycast 1266) were oriented and cured in a magnetic field of ~7.5 T at room temperature. NMR experiments were performed for the magnetically c-axis aligned powder samples of Hg1223.

The right panel in Fig. 1 shows zero-field cooled and field cooled magnetizations of the aligned powder samples of Hg1223, which were measured by a superconducting quantum interference device (SQUID) magnetometer (QUANTUM Design, MPMS). The onset temperature of superconductivity was T_c = 124 K for the as-synthesized underdoped sample and T_c = 134 K for the O_2-annealed optimally doped sample [9]. The magnetization for the O_2-annealed sample shows a weak temperature dependence around 90 K. Although the weak temperature dependence might be due to an inhomogeneous oxygen distribution, an overdoped Hg1223 with T_c = 90 K is unprecedented.

A phase-coherent-type pulsed spectrometer was utilized for the ^{63}Cu NMR (a nuclear spin $I = 3/2$, the nuclear gyromagnetic ratio $^{63}\gamma_n/2\pi = 11.285$ MHz/T) experiments at an external magnetic field of $B_0 = 7.4847$ T and a reference Larmor frequency of $\nu_{\rm L} = {}^{63}\gamma_n B_0 = 84.465$ MHz. The NMR frequency spectra were obtained from the integration of the ^{63}Cu nuclear spin-echoes. ^{63}Cu nuclear spin-lattice relaxation curves were obtained by an inversion recovery technique. The nuclear spin-lattice relaxation time $^{63}T_1$ was estimated by using the theoretical curve for the transition line of $I_z = 1/2 \leftrightarrow$ -1/2 of a spin $I = 3/2$. The subscript indices cc or ab of Knight shifts ^{63}K's and nuclear spin-lattice relaxation rates $(1/^{63}T_1)$'s indicate the direction of a static magnetic field applied along the c axis or in the ab plane.

3. Experimental results

3.1. ^{63}Cu NMR spectra

Figure 2 shows the central transition lines ($I_z = 1/2 \leftrightarrow$ -1/2) of ^{63}Cu NMR spectra with $B_0 \parallel c$ (left panels) and $B_0 \perp c$ (right panels) for the as-synthesized Hg1223 (upper panels) and the O_2-annealed Hg1223 (lower panels). Site assignment to Cu(1) and Cu(2) conforms to the previous work for a triple-layer superconductor [10]. Figure 2 shows higher orientation of the present powder samples than that in [7].

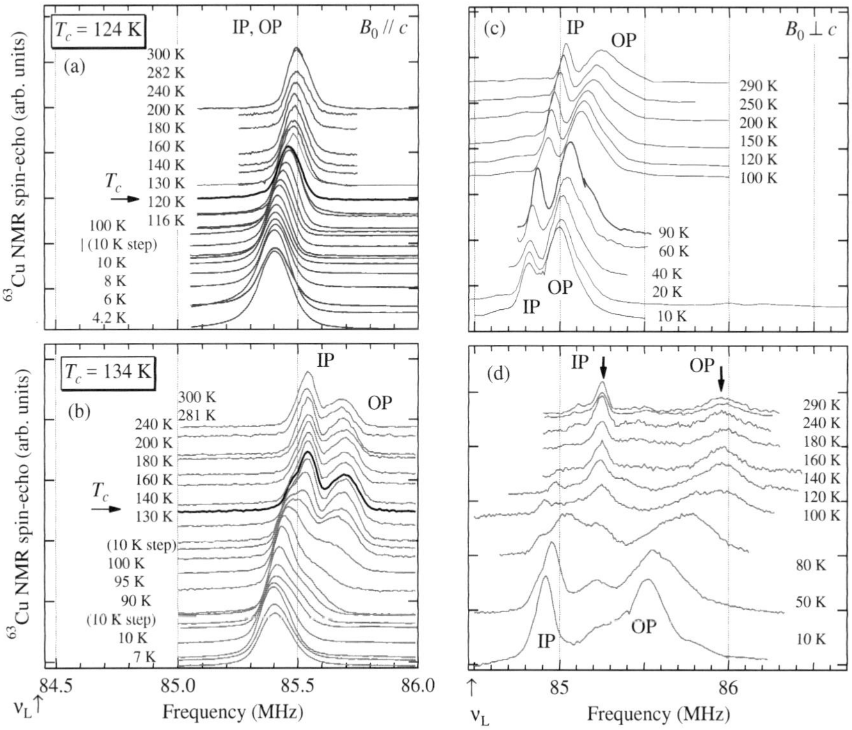

Figure 2. Central transition lines ($I_z = 1/2 \leftrightarrow$ -1/2) of ^{63}Cu NMR frequency spectra with $B_0 \parallel c$ [left panels (a) and (b)] and $B_0 \perp c$ [right panels (c) and (d)] for the as-synthesized Hg1223 [upper panels (a) and (c)] and the O_2-annealed Hg1223 [lower panels (b) and (d)]. IP stands for the inner plane Cu(1) site and OP stands for the outer plane Cu(2) site.

We focus on the central transition line ($I_z = 1/2 \leftrightarrow$ -1/2) in the second order perturbation of a nuclear quadrupole interaction with uniaxial electric field gradients [11]. The c axis is the maximum principal axis of the electric field gradient tensor. Then, the resonance frequency ν^{cc} for $B_0 \parallel c$ is expressed by the Knight shift K^{cc} as

$$\nu^{cc} = (1 + K^{cc})\nu_{\mathrm{L}}. \quad (1)$$

The resonance frequency ν^{ab} for $B_0 \perp c$ is expressed by the Knight shift K^{ab} and the second order quadrupole shift as

$$\nu^{ab} = (1 + K^{ab})\nu_{\mathrm{L}} + \frac{3}{16}\frac{\nu_Q^2}{(1 + K^{ab})\nu_{\mathrm{L}}}, \quad (2)$$

where $^{63}\nu_Q$ is the nuclear quadrupole resonance frequency along the maximum principal axis. The NQR frequencies $^{63}\nu_Q$'s were estimated to be 9.0 MHz at the inner Cu(1) and 12.0 MHz at the outer Cu(2) in the as-synthesized Hg1223, and 10.4 MHz at the inner Cu(1) and 19.5 MHz at the outer Cu(2) in the O_2-annealed Hg1223.

The ^{63}Cu Knight shift $^{63}K^{\alpha\beta}$ ($\alpha\beta = cc$ and ab) is the sum of the orbital shift $K^{\alpha\beta}_{\rm orb}$ and the spin shift $K^{\alpha\beta}_{\rm spin}$ as $K^{\alpha\beta} = K^{\alpha\beta}_{\rm orb} + K^{\alpha\beta}_{\rm spin}$. The spin shift is expressed by on-site hyperfine coupling constants $A_{\alpha\beta}$, a supertransfered coupling constant B, and a uniform spin susceptibility χ_s as $K^{\alpha\beta}_{\rm spin} = (A_{\alpha\beta} + 4B)\chi_s$ [12, 13, 14, 15].

Single transition lines in Fig. 2(a) were observed both in the normal state and the superconducting state of the as-synthesized Hg1223 for $B_0 \parallel c$. Hence, not only the spin shift $K^{cc}_{\rm spin}$ but also the orbital shift $K^{cc}_{\rm orb}$ at Cu(1) are identical to those at Cu(2). From the as-synthesized Hg1223 [Fig. 2(a)] to the O_2-annealed Hg1223 [Fig. 2(b)] for $B_0 \parallel c$, the heat treatment with oxygen annealing leads to separate the single lines into two lines and shift them toward the higher frequency side in the normal state. Between 90 and 134 K in Fig. 2(b), triple Gaussian functions reproduce the NMR spectra, so that three Knight shifts for $B_0 \parallel c$ were estimated.

Two lines in Fig. 2(c) are separated mainly due to the difference in the quadrupole shifts at Cu(1) and Cu(2) for the as-synthesized Hg1223 with $B_0 \perp c$. Two broad lines in Fig. 2(d) are separated due to the double differences in the quadrupole shifts and the spin Knight shifts at Cu(1) and Cu(2) for the O_2-annealed Hg1223 with $B_0 \perp c$. More than two lines of Cu(2) spectra with $B_0 \perp c$ in Fig. 2(d) indicate multiple $\nu_Q(2)$'s for the O_2-annealed H1223, which are consistent with multiple zero-field NQR spectra [16].

3.2. ^{63}Cu Knight shifts

Figure 3 shows ^{63}Cu Knight shifts $^{63}K^{\alpha\beta}(1, 2)$ ($\alpha\beta = cc$ and ab) in Hg1223 from underdoped to optimally doped, which were estimated from Fig. 2 by using (1) and (2). For the as-synthesized Hg1223, $K^{\alpha\beta}(1)$ is nearly the same as $K^{\alpha\beta}(2)$. All the Knight shifts $K^{\alpha\beta}(1, 2)$ decrease with cooling in the normal state, which indicates the opening of a large pseudogap. We estimated $\Delta K^{cc}_{\rm spin}/\Delta K^{ab}_{\rm spin} = 0.42$. The inner plane spin susceptibility $\chi_s(1)$ is nearly the same as the outer plane $\chi_s(2)$. The distribution of the doped holes is uniform in the three layers. The uniform hole distribution in the three layers is consistent with a coalescence of the three-derived Fermi surfaces predicted by a band theory [17]. The uniform hole distribution in the inequivalent CuO_2 planes has also been found in the previously reported Hg1223 [4, 6] and the underdoped triple-layer $Tl_2Ba_2Ca_2Cu_3O_{10-\delta}$ [18].

For the O_2-annealed Hg1223, all the Knight shifts $K^{\alpha\beta}(1, 2)$ are nearly independent of temperature in the normal state and decrease below 90 K but not 134 K. The differences between $K^{\alpha\beta}(1)$ and $K^{\alpha\beta}(2)$ in the normal state are nearly independent of temperature above 134 K. The temperature independent spin susceptibility is typical for the optimally and overdoped superconductors. The lowest frequency component in the triple Gaussian functions for $B_0 \parallel c$ starts to decrease below 134 K.

We consider three possible reasons why no drop at 134 K in $K^{cc,ab}(1,2)$ but the drop at 90 K is found for the O_2-annealed Hg1223, (i) the inhomogeneous oxygen distribution with overdoped T_{c2} = 90 K, (ii) coexistence of vortex solid and liquid states [19, 20], and (iii) bifurcation of superconducting order parameters in the inner and outer CuO_2 planes [21]. A slightly overdoped Hg1223 is known to be synthesized [8]. An overdoped T_{c2} = 90 K could be the first in Hg1223. Irreversibility lines of Hg1223 are lower than those of Tl-layer cuprate superconductors [22]. Then, a vortex liquid state may be broad in a magnetic filed vs temperature diagram of Hg1223. Different superconducting order parameters at Cu(1) and Cu(2) could result from a magnetic bilayer coupling.

IOP Conf. Series: Journal of Physics: Conf. Series **1293** (2019) 012010 doi:10.1088/1742-6596/1293/1/012010

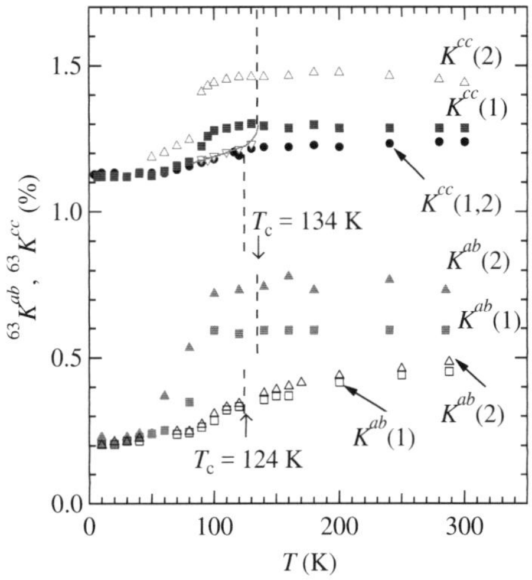

Figure 3. ^{63}Cu Knight shifts $^{63}K^{cc,ab}(1, 2)$ for Hg1223 from underdoped (black symbols) to optimally doped (red and blue symbols). Squares and upward triangles indicate the inner plane Cu(1) site and the outer plane Cu(2) site, respectively. Downward open triangles at 90−134 K are the lowest frequency component in triple Gaussian fitting functions for the O_2-annealed Hg1223 with $B_0 \parallel c$.

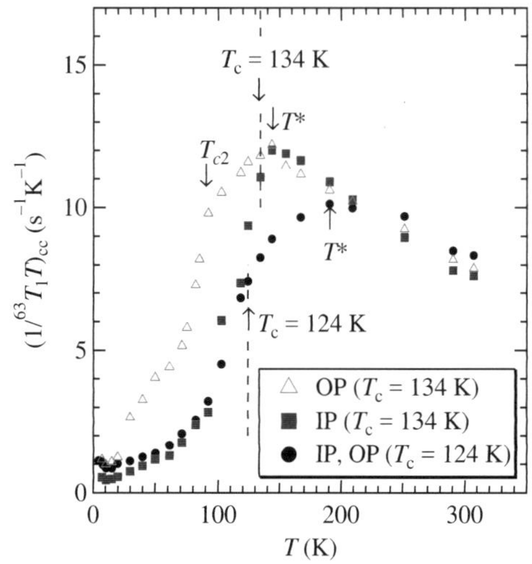

Figure 4. ^{63}Cu nuclear spin-lattice relaxation rates divided by temperature $(1/^{63}T_1T)_{cc}$ with $B_0 \parallel c$ for Hg1223 from underdoped to optimally doped. Black symbols are for the as-synthesized Hg1223. Red and blue symbols are for the O_2-annealed Hg1223. Squares and upward triangles indicate the inner plane Cu(1) site and the outer plane Cu(2) site, respectively.

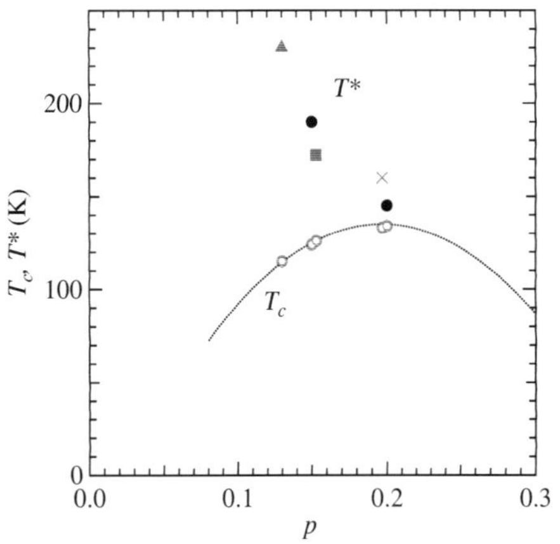

Figure 5. Phase diagram as a function of the hole concentration p for Hg1223. The hole concentration p is estimated from [8]. A triangle [4], a cross [6], a square [5], and closed circles (the present works) indicate pseudogap temperatures T^*'s. Open circles are T_c's. A dotted curve is guide for the eye.

3.3. ^{63}Cu nuclear spin-lattice relaxation rates divided by temperature

Figure 4 shows ^{63}Cu nuclear spin-lattice relaxation rates divided by temperature $(1/^{63}T_1T)_{cc}$ with $B_0 \parallel c$ for Hg1223 from underdoped to optimally doped. For the as-synthesized Hg1223, $(1/^{63}T_1T)_{cc}$ increases with decreasing temperature and starts to decreases at the pseudogap temperature $T^* = 190$ K. For the O_2-annealed Hg1223, both $(1/^{63}T_1T)_{cc}$'s at Cu(1) and Cu(2) increase with decreasing temperature in nearly the same Curie-Weiss temperature dependence

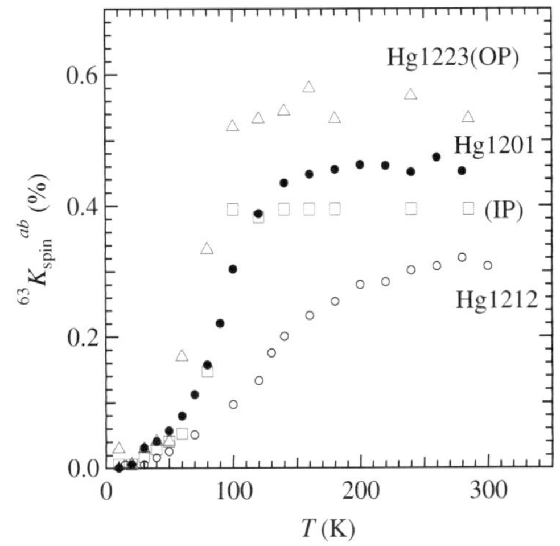

Figure 6. T dependences of ^{63}Cu spin Knight shifts $^{63}K_{spin}^{ab}$'s for a single-layer $HaBa_2CuO_{4+\delta}$ (Hg1201) with T_c = 98 K [23], a double-layer $HaBa_2CaCu_2O_{6+\delta}$ (Hg1212) with T_c = 127 K [24], and the present triple-layer Hg1223 with T_c = 134 K at the optimally doping level.

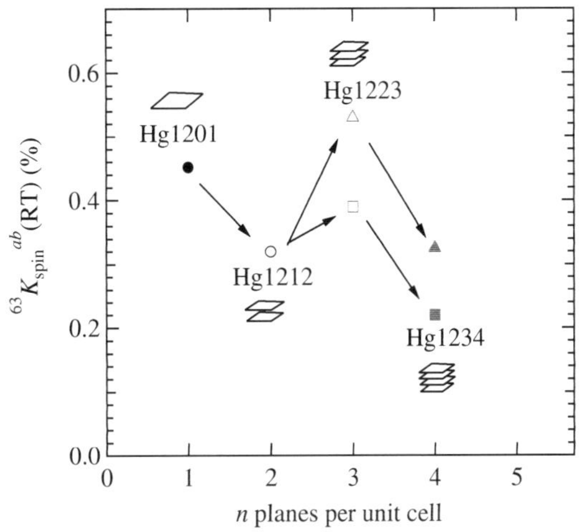

Figure 7. ^{63}Cu spin Knight shifts $^{63}K_{spin}^{ab}$'s (280−300 K) at the optimally doping level plotted against the number n of CuO_2 planes per unit cell. Data for Hg1234 [25] are included. Parallelograms illustrate the number parity of CuO_2 planes per unit cell. Arrows are guide for the eye.

and start to decrease separately at $T^* = 143$ K. Below $T^* = 143$ K and in the superconducting state below 134 K, the temperature dependence in $(1/^{63}T_1T)_{cc}$ of Cu(2) is different from that of Cu(1). $(1/^{63}T_1T)_{cc}$ of Cu(2) shows a secondary rapid decrease below $T_{c2} = 90$ K, which may be associated with the drop of the Knight shift at 90 K. Nearly the same T dependences above 143 K and the large difference below 134 K at Cu(1) and Cu(2) have been found in $(1/^{63}T_1T)_{ab}$ for $B_0 \perp c$. $(1/T_1)_{ab}/(1/T_1)_{cc} \sim 1.7$ around 200 K was estimated for both samples.

Figure 5 shows a phase diagram of Hg1223 with respect to the hole concentration p [8]. The pseudogap temperatures T^*'s and T_c's are plotted against the hole concentration p. T^* monotonically decreases as p increases both for Cu(1) and Cu(2) in Hg1223. The uniform pseudogap temperatures T^*'s characterize Hg1223.

4. Discussions

4.1. Even and odd number of CuO_2 planes per unit cell

Figure 6 shows temperature evolution of ^{63}Cu spin Knight shifts $^{63}K_{spin}^{ab}$'s for a single-layer $HaBa_2CuO_{4+\delta}$ (Hg1201) with $T_c = 98$ K [23], a double-layer $HaBa_2CaCu_2O_{6+\delta}$ (Hg1212) with $T_c = 127$ K [24], and the present triple-layer Hg1223 with $T_c = 134$ K at the optimally doping level. $^{63}K_{spin}^{ab}$ of Hg1212 is smaller than that of Hg1201. $^{63}K_{spin}^{ab}$ of Hg1223 is larger than that of Hg1212. Figure 6 indicates that the value of the ^{63}Cu spin Knight shift $^{63}K_{spin}^{ab}$ at 280−300 K depends on whether the number of CuO_2 planes per unit cell is even or odd.

Figure 7 shows the room temperature ^{63}Cu spin Knight shifts $^{63}K_{spin}^{ab}$(RT) at the optimally doping level against the number n of CuO_2 planes per unit cell. The values $^{63}K_{spin}^{ab}$'s at 280−300 K are adopted as the room temperature shifts. Data for optimally doped Hg1234 ($HgBa_2Ca_3Cu_4O_{10+\delta}$) are from [25]. The relation of $^{63}K_{spin}^{ab}$(RT) vs p for the single layers is different from that for the double layers [26], which breaks a single universal relation of

$^{63}K^{ab}_{\rm spin}$(RT) vs p in [27]. The value of $^{63}K^{ab}_{\rm spin}$(RT) for the even number of CuO_2 planes per unit cell is smaller than that for the odd number of CuO_2 planes per unit cell. The number parity of the CuO_2 planes plays a significant role in $^{63}K^{ab}_{\rm spin}$(RT).

The value of $^{63}K^{ab}_{\rm spin}$(RT) at the optimally doped level tends to decrease as the number n of the CuO_2 planes increases. All the inner and outer $^{63}K^{ab}_{\rm spin}$(RT)'s in an optimally doped five-layer superconductor Hg1245 [28] are larger than those in the four-layer Hg1234 [25], but the single $^{63}K^{ab}_{\rm spin}$(RT) for the two inner Cu sites in the five-layer Hg1245 is smaller than that in the double-layer Hg1212 [24].

An intrabilayer coupling is no chief ingredient for the pseudogap, because the pseduogap is in the single layer Hg1201 [23, 29]. However, the intrabilayer superexchange coupling constant J_b between the double CuO_2 planes in antiferromagnetic insulators $YBa_2Cu_3O_{6+\delta}$ is estimated to be $J_b \sim 0.1J_{\|}$ with an in-plane superexchange coupling constant $J_{\|}$ from observation of an optical magnon [30, 31, 32]. The intrabilayer coupling may induce spin pairing between adjacent CuO_2 planes and then may enhance the spin pseudogap, so that the uniform spin susceptibility of the even number of CuO_2 layers may be smaller than that of the odd number of CuO_2 layers [33, 34, 35, 36].

The O_2-annealed Hg1223 shows $^{63}K^{\alpha\beta}_{\rm spin}(2) > {}^{63}K^{\alpha\beta}_{\rm spin}(1)$ in the normal state. We consider two possible reasons for their difference with the uniform hole distribution. One is a constant second spin susceptibility, which was inferred from anisotropic Knight shifts of single crystals Hg1201 [37, 38]. Such a second spin susceptibility of Cu(2) may be different from that of Cu(1). The other is the intrabilayer coupling effect with a doping dependent J_b. The number of the nearest neighbor CuO_2 plane is 2 at Cu(1) and 1 at Cu(2). Then, we speculate the mean field expressions of $\chi_s(1) = \chi_s/(1+2J_b\chi_s)$ and $\chi_s(2) = \chi_s/(1+J_b\chi_s)$ which lead to $\chi_s(2) > \chi_s(1)$.

4.2. ^{63}Cu hyperfine coupling constant ratios

The ratios of the ^{63}Cu hyperfine coupling constants A_{cc}, A_{ab}, and B are expressed by the experimental anisotropies $r_u = \Delta K^{cc}_{\rm spin}/\Delta K^{ab}_{\rm spin}$ and $r_A = \sqrt{2r_{AF}-1}$ [$r_{AF} = (1/T_1)_{ab}/(1/T_1)_{cc}$] [39] as

$$\frac{A_{cc}}{4B} \approx -\frac{r_A + r_u - 2r_A r_u}{r_A - r_u}, \quad \frac{A_{ab}}{4B} \approx \frac{r_A + r_u - 2}{r_A - r_u}, \tag{3}$$

and

$$\frac{A_{ab}}{A_{cc}} \approx -\frac{r_A + r_u - 2}{r_A + r_u - 2r_A r_u}. \tag{4}$$

Substituting $r_u = 0.42$ and $r_{AF} = 1.7$ into (3) and (4), we estimated the ratios of the ^{63}Cu hyperfine coupling constants for the as-synthesized Hg1223 as $A_{cc}/4B = -0.59$, $A_{ab}/4B = -0.03$ and $A_{ab}/A_{cc} = +0.05$. Assuming $A_{ab} + 4B = 182$ kOe/μ_B [39], we obtained the values of $A_{cc} = -110$, $A_{ab} = -5.6$, and $B = 47$ kOe/μ_B for the underdoped Hg1223. Cancellation of a relatively large core polarization and a spin dipole field from the $3d_{x^2-y^2}$ orbital can result in the small negative A_{ab} [40].

5. Conclusions

The site-selective ^{63}Cu NMR results indicate uniform hole distribution in the three CuO_2 planes of Hg1223 near the optimally and less doping levels. For the Hg-based multilayer cuprate superconductors at the optimally doping level, the room temperature spin shift $^{63}K^{ab}_{\rm spin}$ in the even number of CuO_2 planes per unit cell is smaller than that in the odd number of CuO_2 planes per unit cell.

References

[1] Schilling A, Cautoni M, Guo J D and H.R. Ott H R 1993 *Nature* **363** 56
[2] Antipov E V, Loureiro S M, Chaillout C, Capponi J J, Bordet P, Tholence J L, Putilin S N and Marezio M 1993 *Physica* C **215** 1
[3] Carrington A, Colson D, Dumont Y, Ayache C, Bertinotti A, Marucco J F 1994 *Physica* C **234** 1
[4] Julien M H, Carretta P, Horvatić M, Berthier C, Berthier Y, Ségransan P, Carrington A and Colson D 1996 *Phys. Rev. Lett.* **76** 4238
[5] Goto A, Shimizu T, Sastry P V P S S and Schwartz J 1999 *Phys. Rev.* B **59** 14169
[6] Magishi K, Kitaoka Y, Zheng G q, Asayama K, Tokiwa K, Iyo A and Ihara H 1995 *J. Phys. Soc. Jpn.* **64** 4561
[7] Magishi K, Kitaoka Y, Zheng G q, Asayama K, Tokiwa K, Iyo A and Ihara H 1996 *Phys. Rev.* **53** R8906
[8] Fukuoka A, Tokiwa-Yamamoto A, Itoh M, Usami R, Adachi S and Tanabe K 1997 *Phys. Rev.* **55** 6612
[9] Colson D, Bertinotti A, Hammann J, Marucco J F and Pinatel A 1994 *Physica* C **233** 231
[10] Statt B W and Song L M 1993 *Phys. Rev.* **48** 3536
[11] Takigawa M, Hammel P C, Heffner R H, Fisk Z, Smith J L and Schwarz R B 1989 *Phys. Rev.* B **39** 300
[12] Mila F and Rice T M 1989 *Physica* C **157** 561
[13] Millis A J, Monien H and Pines D 1990 *Phys. Rev.* B **42** 167
[14] Monien H, Pines D and Slichter C P 1990 *Phys. Rev.* B **41** 11120
[15] Imai T 1990 *J. Phys. Soc. Jpn.* **59** 2508
[16] Breitzke H, Eremin I, Manske D, Antipov E V, Lüders K 2004 *Physica* C **406** 27
[17] Singh D J and Pickett W E 1994 *Phys. Rev. Lett.* **73** 476
[18] Piskunov Y V, Mikhalev K N, Zhdanov Yu I, Gerashenko A P, Verkhovskii S V, Okulova K A, Medvedev E Yu, Yakubovskii A Yu, Shustov L D, Bellot P V and Trokiner A 1998 *Physica* C **300** 225
[19] Reyes A P, Tang X P, Bachman H N, Halperin W P, Martindale J A and Hammel P C 1997 *Phys. Rev.* B **55** R14737
[20] Itoh Y, Michioka C, Yoshimura K, Hayashi A and Ueda Y 2005 *J. Phys. Soc. Jpn.* **74** 2404
[21] Suhl H, Matthias B T and Walker L R 1959 *Phys. Rev. Lett.* **3** 552
[22] Maignan A, Putilin S N, Hardy V, Simon Ch and Raveau B 1996 *Physica* C **266** 173
[23] Itoh Y, Machi T, Adachi S, Fukuoka A, Tanabe K and Yasuoka H 1998 *J. Phys. Soc. Jpn.* **67** 312
[24] Itoh Y, Tokiwa-Yamamoto A, Machi T and Tanabe K 1998 *J. Phys. Soc. Jpn.* **67** 2212
[25] Tokunaga Y, Ishida K, Magishi K, Oshugi S, Zheng G q, Kitaoka Y, Asayama K, Iyo A, Tokiwa K and Ihara H 1999 *Physica* B **259-261** 571
[26] Itoh Y 2015 *Physics Procedia* **65** 41
[27] Mukuda H, Shimizu S, Iyo A and Kitaoka Y 2012 *J. Phys. Soc. Jpn.* **81** 011008
[28] Mukuda H, Yamaguchi Y, Shimizu S, Kitaoka Y, Shirage P and Iyo A 2008 *J. Phys. Soc. Jpn.* **77** 124706
[29] Itoh Y, Machi T, Fukuoka A, Tanabe K and Yasuoka H 1996 *J. Phys. Soc. Jpn.* **65** 3751
[30] Rossat-Mignot J, Regnault L P, Jurgens M J, Vettier C, Burlet P, Henry J Y and Lapertot G 1990 *Physica* B **163** 4
[31] Reznik D, Bourges P, Fong H F, Regnault L P, Bossy J, Vettier C, Milius D L, Aksay I A and Keimer B 1996 *Phys. Rev.* B **53** R14741
[32] Hayden S M, Aeppli G, Perring T G, Mook H A and Doğan F 1996 *Phys. Rev.* B **54** R6905
[33] Millis A J, Ioffe L B, Monien H 1995 *J. Phys. Chem. Solids* **56** 1641
[34] Norman B, Kohno H and Fukuyama H 1995 *J. Phys. Soc. Jpn.* **64** 3903
[35] Kishine L 1996 *Preprint* arXiv:cond-mat/9604128
[36] Kishine J 1997 *Physica* C **282-287** 1771
[37] Rybicki D, Kohlrautz J, Haase J, Greven M, Zhao X, Chan M K, Dorow C J and Veit M J 2015 *Phys. Rev.* B **92** 081115
[38] Haase J, Jurkutat M, Kohlrautz J 2017 *Condense. Matter* **2** 16
[39] Itoh Y, Machi T and Yamamoto A 2018 *J. Phys. Conf. Series* **1054** 012006
[40] Bleaney B, Bowers K D and Pryce M H L 1955 *Proc. Roy. Soc. London*, Ser. A **228** 166

Optimization of $YbBa_2Cu_3O_y$ Thick Film Formation on MgO Substrates using Yb_2BaO_5 and liquid phase

A. Hattori, M. Muralidhar*, A. Sai Srikanth and M. Murakami

Superconducting Material Laboratory, Department of Material Science and Engineering, Shibaura Institute of Technology, 3-7-5 Toyosu, Koto-ku, Tokyo 135-8546, Japan

*miryala1@shibaura-it.ac.jp

Abstract. The production of best thick $YbBa_2Cu_3O_y$ (Yb123) films on MgO substrates are crucial for the utilization of this technology in silver sheeted wire as similar to Bi-2223. In this work, in order to optimize $YbBa_2Cu_3O_y$ films on MgO substrates, a mixture of alpha-terpineol and 2-ethyl acetate was utilized as a solvent to which 100 nm sized stoichiometric powder mixture of Yb_2BaO_5, $BaCuO_2$ and CuO was added and grinded for several hours to form a highly dense paste. Thick film precursors were prepared by spreading the paste over MgO substrates with a screen printing technique and various sintered and quench studies are done. Growth temperature is systematically varied between 800 and 950 °C for 30 minutes in air and quenched in liquid nitrogen. Similarly, sintering durations are varied systematically from 15 minutes to 8 hours. XRD reveals conformed that large Yb123 grains are formed when the thick films are sintered at 825 °C for 1 hour on MgO substrate. SEM micrographs are also in congruence with the XRD results depicting the giant Yb123 crystals. The present results clearly demonstrate that optimization of the growth temperature and time is crucial to obtain the large thick $YbBa_2Cu_3O_y$ films on MgO substrates.

1. Introduction

In the current day trend, most of the research carried on High T_c superconductor technology is focused on developing high performance, good quality cheaper superconducting tapes and power cables operating at 77 K [1]. Since the first generation Bi-based silver coated cables and tapes, the $REBa_2Cu_3O_y$ (RE – rare earth element) type coated conductors represent the second generation superconductors [2,3]. Few challenges in thin films along sides production cost is the alignment of grains and grain connectivity issues. But these can be addressed with high temperature biaxial texturing which gives alignment and improves grain connectivity. But, practical applications also include thick films, where the impact of biaxial texturing decreases with increase in thickness. Our aim is to process $YbBa_2Cu_3O_y$ coated conductors cheaply, for commercial applications while countering the difficulties. In past few years, we have fabricated good quality RE-123 (RE – Gd, Er) thick films on MgO and Ag substrates using RE-210, $BaCuO_2$ and CuO mixture [4,5]. The Ba-Cu-O precursor in the mixture is a low melting compound (900 °C), which improves the density as well as quality of the thick film and quench studies pointed out that the corresponding 123 phase can be synthesized below silver's melting point [4]. Also, since the melting point is below Ag, it is easy to use for applications alongside with Ag. Because of this low melting nature, it is worth studying the properties of Yb-123, which is a low melting RE-123 material as this promotes the application based attributes.

In our present work, we processed Yb-123 thick film on MgO substrate and characterized it after quenching. We have optimized the temperature, dwelling time and other parameter for good growth of Yb-123 thick film. XRD, DTA and SEM analysis have been done to explain the trend and other properties of thick films synthesized.

2. Experimental details

Homemade Yb_2BaO_5 powders were made by using the solid state reaction technique and then mixed with high purity commercial powders of $BaCuO_2$ and CuO in stoichiometric propositions to obtain of $YbBa_2Cu_3O_y$. Later this mixture is ball milled for 2 hours to promote the number of nanoscale particles (around 100 nm) [4]. These powders were air dried for 24 hours, then vacuum dried at 200 °C for 6 hours, and finally cooled to room temperature. A mixture of alpha-terpinol and 2-ethyl acetate was used as a solvent to which the ball milled powder mixture was added and grinded for several hours to form a highly dense paste. Thick films (about 0.5 mm thick) were then fabricated on commercial MgO single crystal substrates using screen printing technique, where the dense paste is spread on the substrate manually using printing head through a 165-mesh screen.

Two types of quench studies were done to optimize the temperature as well as dwelling time. In the first experiment, to optimize the temperature profile, the films were heated to and maintained at particular growth temperatures, which are varied from 800 to 950 °C in a box furnace. After attaining the required temperatures, the samples were held at that temperature for 30 minutes and then rapidly quenched in liquid nitrogen (77 K). Subsequently, the samples were cleaned in an ultrasonic cleaner and dried in a drying box, to be ready for characterization. In the second experiment, similar quenching studies are done to optimize the dwelling time required for maximum Yb123 growth. The structure of the samples was determined by means of a high-resolution automated X-ray powder diffractometer RINT2200 with a step size of 0.01° from 5° to 90°, using Cu-Kα radiation generated at 40 kV and 30 mA. The microstructure of these films was studied with a scanning electron microscope (SEM).

3. Results and discussion

3.1) Quench studies for temperature optimization to obtain good quality Yb-123 thick films on MgO substrate:

After depositing the mixture paste and preparing the thick film samples using screen-printing, they are subjected to the specific heat cycle as mentioned in experimental section. The growth temperatures are systematically varied such as 800, 825, 850, 875, 900, 925 and 950 °C (see figure 1-(a)).

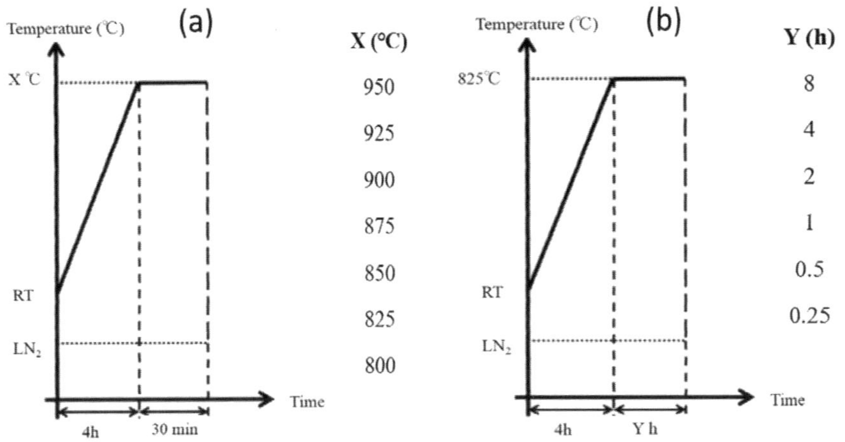

Figure 1. Sintering pattern; (a)- Temperature optimization, (b)- Dwell time optimization.

The thick films are characterized using XRD for identifying the phases formed. From the XRD results, it can be seen that 825 °C is the best temperature condition for optimum Yb123 growth. At high temperatures, the dominant phase is Yb211 as well as liquid phase as we can observe the intensity of Yb123 dropping as well as disappearing (at 950 °C), while the peak intensity of Yb211 and liquid phase starts increasing (see figure 2), maximum at 950 °C.

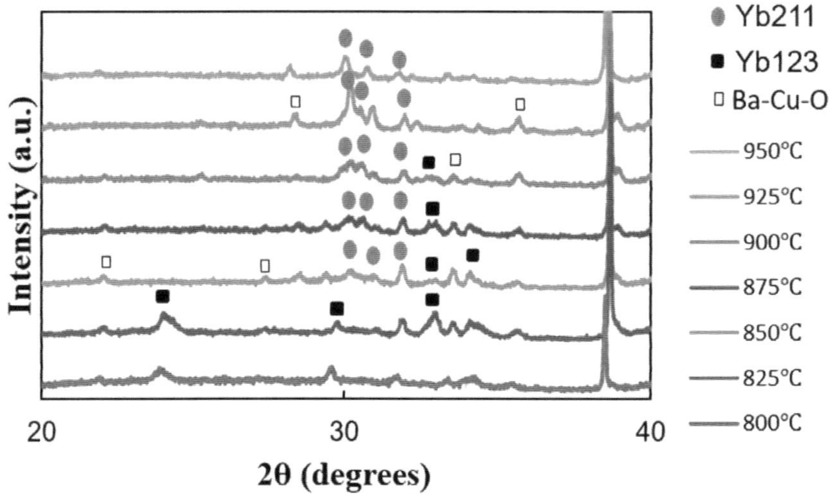

Figure 2. XRD of Yb-123 thick film samples sintered at various temperatures.

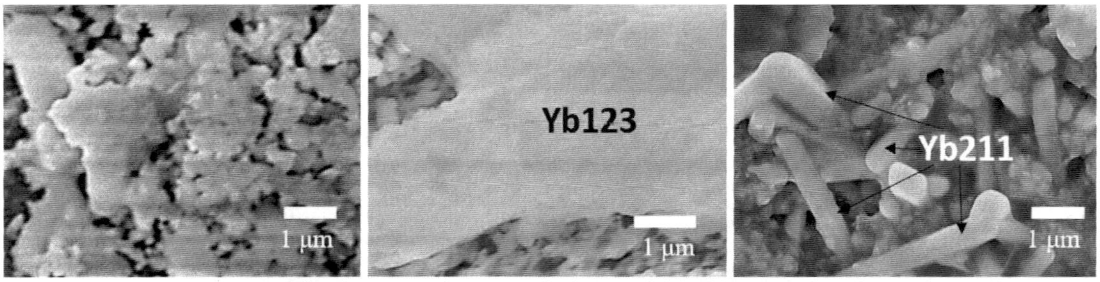

Figure 3. Low magnification SEM images of Yb-123 thick film samples sintered at various temperatures; left-800 °C, middle-825 °C and right-950 °C.

Microstructural studies from FESEM showed that large crystals of Yb123 are formed at 825 °C which is in accordance with our XRD results. When the temperature is 800 °C small grains of Yb123 are formed (see figure 3-left) and large crystals are formed at the temperature 825 °C (see figure 3-middle). While at high temperatures such as 950 °C Yb211 particles can be seen miner amount of Yb123 (see figure 3-right) which can also be observed from XRD. Hence, when the growth temperature is around 825 °C the growth of crystalline Yb123 is optimum in Yb-123 thick films grown on MgO subtracts.

3.2) Quench studies for dwell time optimization to obtain good quality Yb-123 thick films on MgO substrate:

It is also equally important to optimize the dwelling time or soaking time for optimum for large crystal growth. In order to optimize the dwelling time, we fabricated Yb-123 thick films using 825 °C as growth temperature and vary the sintering duration or dwell time. In this study, we systematically varied the dwell time such as 0.25, 0.5, 1, 2, 4 and 8 hours (see figure 1-(b)). XRD analysis reveals that, highest Yb123 is formed when the dwelling time is 1 hour as the peak intensity corresponding to the Yb123 is highest of all the other samples (see figure 4). It can be seen that the Yb123 peak intensity reduces with increase in sintering duration after 1 hour, while the Yb211 peak intensity increases with sintering duration.

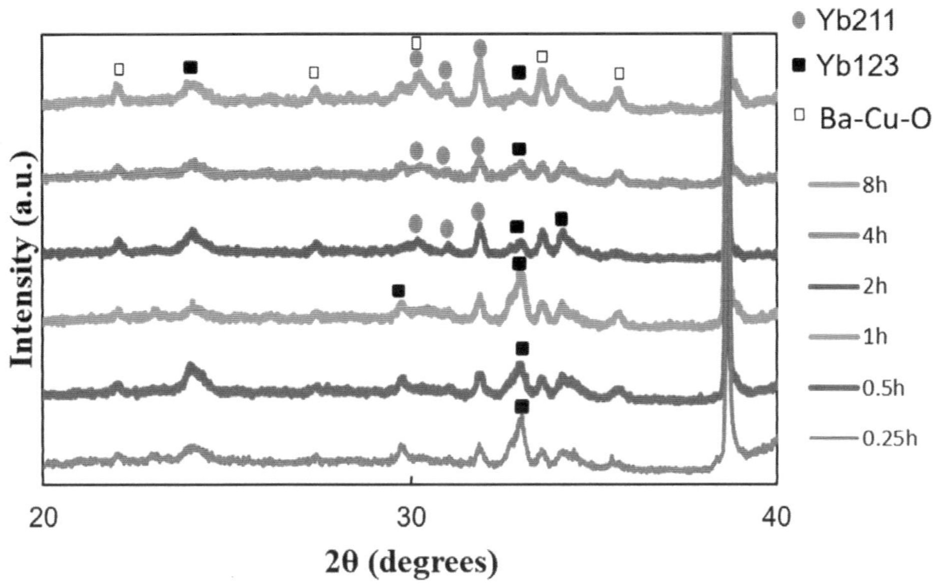

Figure 4. XRD of thick film sample sintered at various durations.

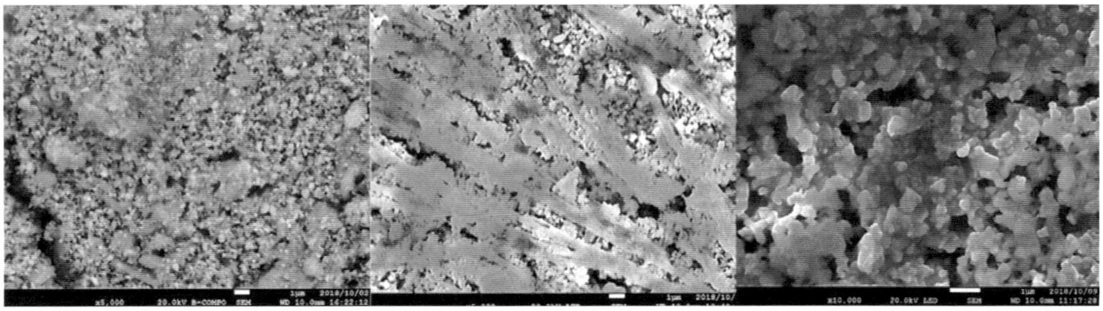

Figure 5. Low magnification SEM images of thick film samples sintered at various sintering durations; left-15 minutes, middle- 1 hour and right-8 hours.

Microstructural studies revealed that large and long crystal of Yb123 are observed in thick film, which was subjected to 1-hour sintering duration (see figure 5). Hence we can get obtain huge Yb123 crystals in thin films when the optimum sintering conditions such as 825 °C and 1 hour dwelling time are followed.

4. Conclusions

In this study, we have successfully optimized the sintering conditions required to fabricate the Yb123 large crystal thick films on MgO substrate. From XRD and SEM microstructural studies, we found that when the films are sintered at 825 °C for 1-hour optimum content of large Yb123 crystals are formed. The present results indicate that stoichiometric powder mixture of Yb_2BaO_5, $BaCuO_2$ and CuO can help to prepare the large–area thicker Yb-123 films in a short processing time.

Acknowledgments
The paper was partly supported by Grant-in-Aid FD research budget code: 112261, SIT (Shibaura Institute of Technology).

References

[1] Dusastre V, Larbalestier D, Gurevich A, Feldmann D M and Polyanskii A 2001 High-T c superconducting materials for electric power applications *Nature* **414** 368–77
[2] Haugan T, Barnes P N, Wheeler R, Meisenkothen F and Sumption M 2004 Addition of nanoparticle dispersions to enhance flux pinning of the YBa2Cu3O7-x superconductor *Nature* **430** 867–70
[3] Chen Z, Kametani F, Chen Y, Xie Y, Selvamanickam V and Larbalestier D C 2009 A high critical current density MOCVD coated conductor with strong vortex pinning centers suitable for very high field use *Supercond. Sci. Technol.* **22**
[4] Muralidhar M, Sakai N, Jirsa M and Tanaka S 2009 Solid phase (RE 2BaO 4)-Liquid phase (BaCuO 2) reaction: The way to highly oriented ErBa 2Cu 3O y superconducting thick films on commercial silver substrates *Cryst. Growth Des.* **9** 2404–8
[5] Muralidhar M, Sakai N, Murakami M, Hirabayashi I and Tanaka S 2006 RE-Ba2Cu3Oy formation mechanism via (RE)2BaO4 *Phys. C Supercond. its Appl.* **445–448** 282–5

IOP Conf. Series: Journal of Physics: Conf. Series **1293** (2019) 012012 doi:10.1088/1742-6596/1293/1/012012

Effects of vicinal substrates on the orientation of $Bi_2Sr_2CaCu_2O_{8+x}$ thin films when the metal-organic decomposition method is used

Y Yamada[1], T Okamoto[2]

[1] Department of Innovative Electrical and Electronic Engineering, National Institute of Technology, Oyama College, 771 Nakakuki, Oyama, Tochigi 323-0806, Japan
[2] Electrical, Electronics and Information Engineering, Nagaoka University of Technology, 1603-1 Kamitomioka, Nagaoka, Niigata 940-2188, Japan

yyamada@oyama-ct.ac.jp

Abstract. We have prepared (010) (or (100)) oriented $Bi_2Sr_2CaCu_2O_{8+x}$ (Bi2212) thin films by metal-organic decomposition (MOD) method which is one kind of solution method. The flat and vicinal (100) substrate (the miscut angle $\varphi = 5°$ toward the direction [001]) of $NdGaO_3$ (NGO) was used. When using vicinal substrates, the Bi2212 (020) (or (200)) peak appeared clearly in the XRD patterns. From the SEM image, it was found that elongated plate-like crystal grains of Bi2212 were grown. From the viewpoint of lattice matching, this elongated crystal grain is considered to be (010) (or (100)) oriented Bi2212 crystal grains.

1. Introduction

$Bi_2Sr_2CaCu_2O_{8+x}$ (Bi2212) has a layered perovskite structure and forms an intrinsic Josephson junction (IJJ). When a voltage is applied in the *c*-axis direction, an alternating current with a frequency proportional to the voltage is generated. A vibration mode excited by matching the frequency of the alternating current with a cavity resonance frequency can be used for a terahertz oscillator, which is considered to be a promising device that fills a frequency domain called the "terahertz gap". It has already been proved that a μW class continuous coherent terahertz wave can be oscillated [1-4].

The major preparation methods thus far have been reported on *c*-axis oriented Bi2212 that requires complicated procedures and dry etching, such as a precisely controlled etching process in depth direction in order to form the *c*-axis current paths. If a non-*c*-axis oriented thin film of which the *c*-axis parallel or incline to the substrate surface can be prepared, planar type IJJ devices can be fabricate simply by forming current paths parallel to the substrate [5,6]. In preparing such non-*c*-axis oriented thin films, it is important that selection of substrates focusing on lattice matching. If such substrates are selected and solution or powder material is coated on that substrate by a printing method, it is considered that an IJJ devices can be fabricated only by performing a heat treatment.

We are attempting to prepare (010) (or (100)) oriented Bi2212 thin films [7-9] for planar type intrinsic Josephson oscillator devices that are not required etching processes. These new type devices can be prepared by a combination of orientation control technique of Bi2212 thin film by metal-organic decomposition (MOD) method and application of raw material solution by printing method. In order to prepare planar type devices, it is necessary to form a current path parallel to the substrate, so that the *c*-axis of Bi2212 is required to be parallel or incline to the substrate. In addition, if the *c*-axis is parallel to the substrate, no *c*-twinning structure is formed (*c*-twinning structure is not preferable for preparing IJJ devices).

Bi2212 is an orthorhombic crystal having a lattice constant of a = 5.414 Å, b = 5.418 Å, c = 30.6 ~ 30.9 Å [10,11]. $NdGaO_3$ (NGO) is also orthorhombic with lattice constants a = 5.427 Å, b = 5.497 Å, c = 7.707 Å [12]. The length of the *a*-axis of Bi2212 is approximately equal to the length of the *a*-axis of NGO, and that of *c*-axis of Bi2212 is approximately equal to four times that of *c*-axis of NGO. Since these values are very close, it is expected that lattice matching occurs well and Bi2212 will be epitaxially grown. Actually, we succeeded in growing (010) (or (100)) oriented Bi2212 crystal grains [9]. However, (11*n*) oriented crystal grains are also formed, which is a problem to be solved.

In this study, we report fabrication Bi2212 thin films by the MOD method using vicinal NGO (100) substrates with the size of 10 × 10 × 0.5 mm and the miscut angle φ = 5° toward the direction [001]. Evaluation of the thin film was carried out through X-ray diffraction (XRD) pattern, observation of surface morphology by scanning electron microscope (SEM) image.

2. Experimental Procedure

Bi2212 thin films under study were prepared by the MOD method using a stoichiometric BSCCO metal organic (MO) solution (supplied by Kojundo Chemical Lab. Co., Ltd. SK-BSCCO008). Substrates used were flat and vicinal NGO (100) substrates with the size of 10 × 10 × 0.5 mm and the miscut angle φ = 5° toward the direction [001]. In the following, the flat substrates were expressed as φ = 0°. Preparation procedures were as follows [8]:

(1) An amount of 6μℓ BSCCO MO solution was dropped onto substrates using a digital micropipette.
(2) The dropped solution was spin-coated by a 2-step process with 500 rpm for 5 sec and 3000 rpm for 1 min.
(3) The spin-coated films were dried at 120 °C for 40 min.
(4) The dried films were annealed at temperature T_a for time t_a in O_2 or Air atmosphere.
(5) Furnace cooling were carried out.

The samples were taken into the box furnace at room temperature. Then, the samples were taken out after being cooled to room temperature. Figure 1(a) shows the temperature profile of a box furnace corresponding to the above steps (3) ~ (5). In this paper, annealing time t_a is 60 min. When setting the O_2 atmosphere, 100% O_2 gas was kept flowing to the box furnace from the start of annealing. The preparation condition of the samples is shown in Table 1. The crystal structures of the samples were investigated by θ-2θ XRD patterns with CuKα radiation (λ = 1.54 Å). In addition, the XRD pattern of the flat and vicinal NGO (100) substrates were also measured for reference.

The coordinate system was defined as shown in the figure 1(b). The *x*- and *z*-axis were parallel to the NGO [010] direction and the normal direction of the NGO substrate, respectively. X-rays were irradiated along with the *x*-axis, which was within the *xz* plane. The surface morphologies were observed by SEM.

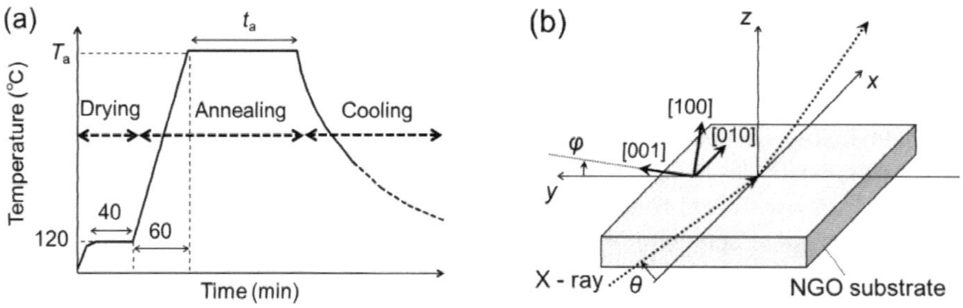

Figure 1. (a) Temperature profile of a box furnace. (b) Schematic drawing of NGO substrate and X-ray irradiate direction. φ is the miscut angle toward the direction [001].

Table 1. Summary of sample labels and their annealing conditions.

Sample label	Miscut angle φ (°)	Annealing temperature T_a (°C)	Atmosphere
N0-780-60	0	780	O_2
N5-780-60	5	780	O_2
N0-810-60	0	810	O_2
N5-810-60	5	810	O_2
N0-810-60-a	0	810	Air
N5-810-60-a	5	810	Air

3. Result and Discussion

Figure 2 shows a comparison of the observed surface morphologies by SEM with different conditions ((a) N0-780-60, (b) N5-780-60, (c) N0-810-60, (d) N5-810-60, (e) N0-810-60-a, (f) N5-810-60-a). From the viewpoint of lattice matching, elongated plate-like crystal grains (as surrounded by broken line in the figure (a)) are considered to be (010) (or (100)) oriented crystal grains. From these figures, it can be seen that the (010) (or (100)) oriented crystal grains are most abundant when annealed at 780 °C using the vicinal substrate.

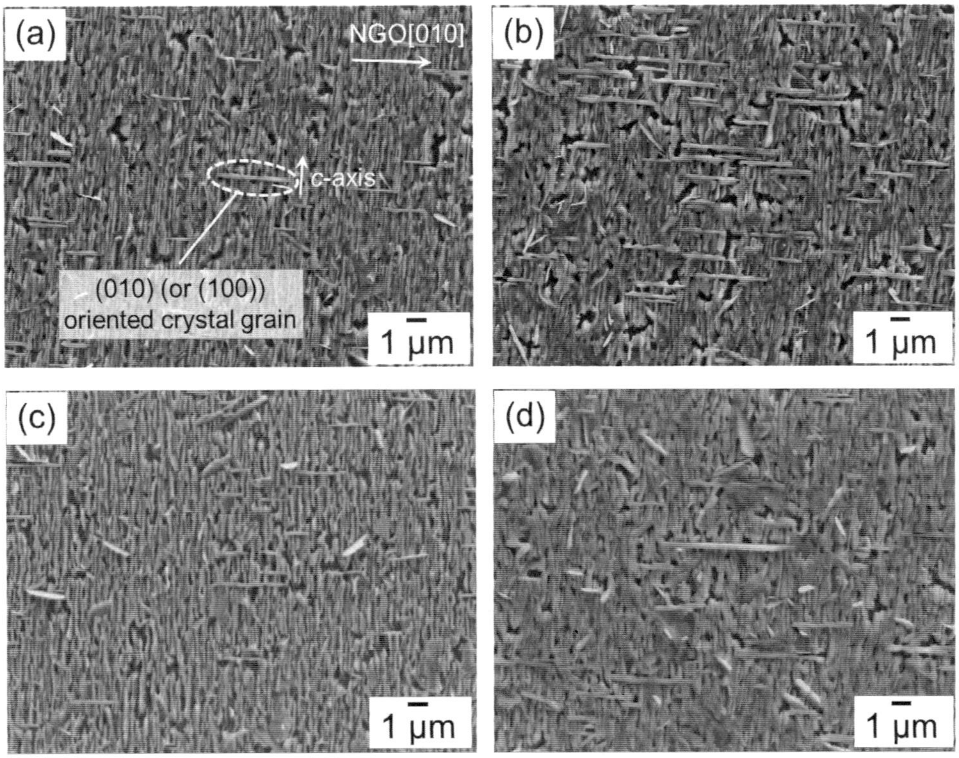

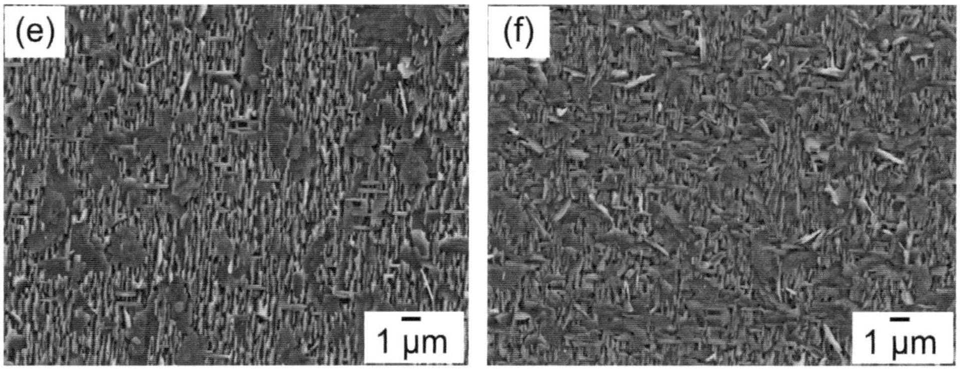

Figure 2. The observed surface morphologies by SEM of samples with different conditions ((a) N0-780-60, (b) N5-780-60, (c) N0-810-60, (d) N5-810-60, (e) N0-810-60-a, (f) N5-810-60-a). In figure (a), elongated crystal grains in the lateral direction in the part surrounded by the white broken line is considered to be the Bi2212 (010) (or (100)) oriented grain. The white solid arrow indicates the *c*-axis direction of the grain.

Figure 3 shows comparison of XRD patterns of samples with different conditions. (a) ~ (f) of figure 3 correspond to (a) ~ (f) of figure 2, respectively. Only peaks of Bi2212 phase and NGO substrates appeared in these XRD patterns. In comparison of XRD patterns of samples with different substrates types, the (11*n*) peak of Bi2212 tends to become smaller in the case of using vicinal NGO (100) substrates. In the XRD patterns of flat NGO (100) substrates, NGO (200) peaks appeared on $2\theta \approx 33°$ whereas no peak appeared on the XRD pattern of the vicinal NGO (100) substrates.

In the figure (a), (c) and (e), the peaks appeared on $2\theta \approx 33°$. They are considered to overlap NGO (200) and Bi2212 (020) (or (200)) peaks. In contrast, the peaks appeared on $2\theta \approx 33°$ in the figure (b), (d) and (f), are considered to be Bi2212 (020) (or (200)) peaks. Because in the XRD pattern of the vicinal NGO (100) substrates no peaks appeared. From these figures, it can be seen that the (010) (or (100)) oriented crystal grains are most abundant when annealed at 810 °C in O_2 atmosphere using the vicinal substrate.

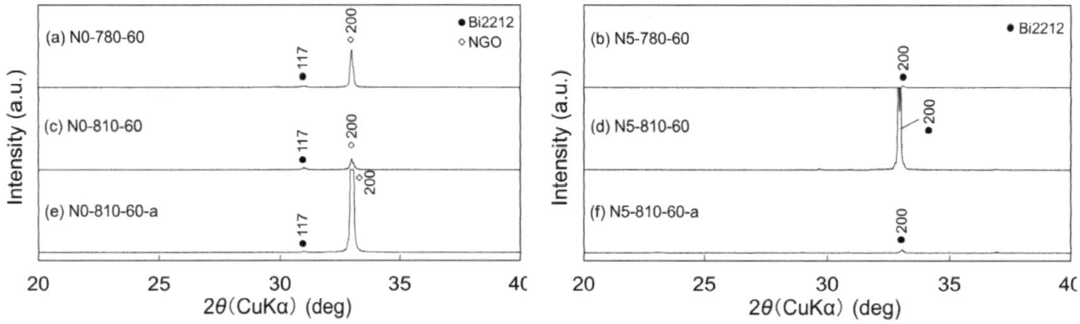

Figure 3. θ-2θ XRD patterns of samples with different conditions ((a) N0-780-60, (b) N5-780-60, (c) N0-810-60, (d) N5-810-60, (e) N0-810-60-a, (f) N5-810-60-a).

Figure 4 show schematic views (imaginary views) of lattice matching for Bi2212 and NGO substrate. The lengths of the *a*-axis and the *b*-axis of Bi2212 were set to the same average length (= 5.41 Å) in the figure. (b) is a view from NGO [0$\bar{1}$0] direction of a thin film prepared on a vicinal substrate. Since the crystal grains of Bi2212 grow to extend in the *ab* plane direction, it is considered that the (010) (or (100)) oriented grains become elongated in the *a*-axis (or *b*-axis) direction along step edges of vicinal substrate.

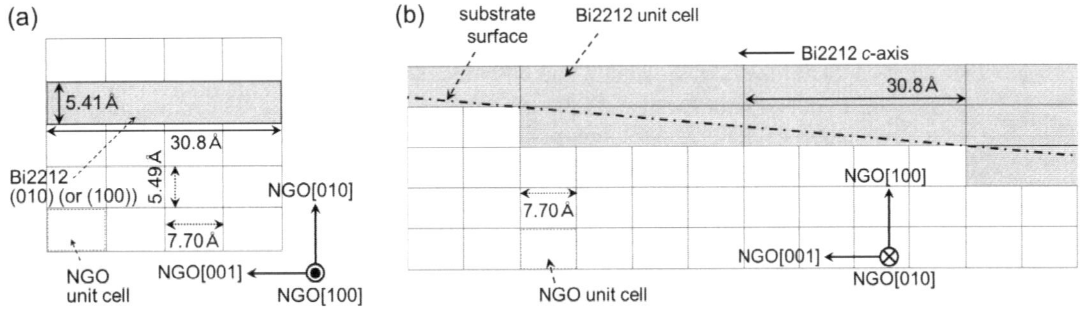

Figure 4. Schematic views (imaginary views) of lattice matching for Bi2212 and NGO substrate ((a) from NGO [100] direction, (b) from NGO [0$\bar{1}$0] direction).

4. Conclusion

In order to realize fabrication of (010) (or (100)) oriented thin film that can be used for Bi2212 IJJ device by printing method, we attempted to fabricate (010) (or (100)) oriented thin film by MOD method. For that purpose, focusing on lattice matching, NGO (100) substrates was used. The crystallinity and orientation were evaluated by the θ-2θ pattern of XRD. As a result, the followings were clarified.

(1) In comparison of XRD patterns of samples with different substrates types, when using vicinal substrates, the (11n) peak of Bi2212 tends to become smaller, and the (020) (or (200)) peak of that appeared clearly.

(2) From the viewpoint of lattice matching, elongated plate-like crystal grains in SEM images are considered to be (010) (or (100)) oriented crystal grains.

In this report, we have succeeded in growing Bi2212 (010) (or (100)) crystal grains occupying larger proportion by using vicinal NGO (100) substrates. However, the conditions under which the (010) (or (100)) oriented crystal grains became the most abundant were different from those suggested from the SEM images and the XRD patterns. Furthermore, the SEM images show that (11n) oriented crystal grains still occupy a large proportion. They are problems to be clarified in the future.

Acknowledgments
This work was supported by JSPS KAKENHI Grant Number JP17K06377.

References

[1] Ozyuzer L, Koshelev A E, Kurter C, Gopalsami N, Li Q, Tachiki M, Kadowaki K, Yamamoto T, Minami H, Yamaguchi H, Tachiki T, Gray K E, Kwok W -K, and Welp U 2007 *Science* **318** 1291
[2] Benseman T M, Gray K E, Koshelev A E, Kwok W -K, Welp U, Minami H, Kadowaki K and Yamamoto T 2013 *Appl. Phys. Lett.* **103** 022602
[3] Kashiwagi T, Yamamoto T, Kitamura T, Asanuma K, Watanabe C, Nakade K, Yasui T, Saiwai Y, Shibano Y, Kubo H, Sakamoto K, Katsuragawa T, Tsujimoto M, Delfanazari K, Yoshizaki R, Minami H, Klemm R A and Kadowaki K 2015 *Appl. Phys. Lett.* **106** 092601
[4] Kashiwagi T, Sakamoto K, Kubo H, Shibano Y, Enomoto T, Kitamura T, Asanuma K, Yasui T, Watanabe C, Nakade K, Saiwai Y, Katsuragawa T, Tsujimoto M, Yoshizaki R, Yamamoto T, Minami H, Klemm R A and Kadowaki K 2015 *Appl. Phys. Lett.* **107** 082601
[5] Tsai J S, Fujita J and Kupriyanov M Yu 1995 *Phys. Rev.* B **51** 16267
[6] Kawahara T, Ishibashi T, Kaneko H and Sato K 1999 *IEEE Trans. Appl. Supercond.* **9** 4519
[7] Yamada Y, Mori N, Atsumi T, Kato T and Ishibashi T 2015 *Physics Procedia* **65** 165
[8] Yamada Y, Kato T, Ishibashi T, Okamoto T and Mori N 2018 *AIP Advances* **8**, 015101

[9] Yamada Y, Okamoto T 2018 *Journal of Physics: Conference Series* **1054**, 012009
[10] Sunshine S A, Siegrist T, Schneemeyer L F, Murphy D W, Cava R J, Batlogg B, van Dover R B, Fleming R M, Glarum S H, Nakahara S, Farrow R, Krajewski J J, Zahurak S M, Waszczak J V, Marshall J H, Marsh P, Rupp L. W Jr. and Peck W F 1988 *Phys. Rev.* B **38** 893
[11] Horiuchi S, Maeda H, Tanaka Y and Matsui Y 1988 *Jpn. J. Appl. Phys.* **27**, L1172
[12] Vasylechko L, Akselrud L, Morgenroth W, Bismayer U, Matkovskii A and Savytskii D 2000 *J. Alloys Compd.* **297** 46

Effects of Swift-Particle Irradiations on Critical Current Density in $CaKFe_4As_4$

A Takahashi[1], S Pyon[1], S Okayasu[2], S Ishida[3], A Iyo[3], H Eisaki[3], M Imai[4], H Abe[4], T Terashima[4] and T Tamegai[1]

[1]Department of Applied Physics, The University of Tokyo, 7-3-1 Hongo, Bunkyo-ku, Tokyo 113-8656, Japan
[2]Advanced Science Research Center, Japan Atomic Energy Agency, Tokai, Ibaraki 319-1195, Japan
[3]National Institute of Advanced Industrial Science and Technology, 1-1-1 Umezono, Tsukuba, Ibaraki 305-8568, Japan
[4]National Institute for Materials Science, 1-2-1 Sengen, Tsukuba, Ibaraki 305-0047, Japan

E-mail: ayumu-takahashi141@g.ecc.u-tokyo.ac.jp

Abstract. Introduction of columnar defects to superconductors through swift-particle irradiation enhances their critical current density (J_c). Iron-based superconductors (IBSs) have been investigated as promising materials for practical applications because of their large J_c at high magnetic fields and temperatures. Recently, another promising IBS $CaKFe_4As_4$ (1144-type IBS) was found, and attracts much interest due to its high J_c in the pristine sample. We compare effects of 800 MeV Xe, 3 MeV proton, and 320 MeV Au irradiations on the critical temperature (T_c) and J_c of $CaKFe_4As_4$ single crystals, and compare them with irradiation effects in $Ba_{0.6}K_{0.4}Fe_2As_2$.

1. Introduction

Introduction of columnar defects to superconductors through swift-particle irradiation enhances their critical current density (J_c) [1-4]. It has been demonstrated that the maximum value of J_c and the corresponding dose depend on ion species and its energy [5]. The difference in the optimum dose among ion species may originate from the different diameters and lengths of created defects.

Iron-based superconductors (IBSs) have been investigated as promising materials for practical applications because of their large J_c at high magnetic fields and temperatures. In previous studies, remarkable effects have been demonstrated in IBSs by irradiating heavy ions and protons into Co or K doped $BaFe_2As_2$ (Ba-122) single crystals [3, 5]. Recently, another promising IBS $CaKFe_4As_4$ (1144-type IBS) was found [6]. Its crystal structure is similar to 122-type IBSs. $CaKFe_4As_4$ has a tetragonal structure (P4/mmm), where Ca and K layers stack alternatively along the *c*-axis [6, 7]. $CaKFe_4As_4$ shows similar superconducting properties, such as critical temperature (T_c) or upper critical field (H_{c2}), to those of optimally K-doped Ba-122-type IBS $Ba_{0.6}K_{0.4}Fe_2As_2$ [8-10]. However, alternate stacking of Ca and K along the *c*-axis may lead to different physical properties.

Here, we compare effects of 800 MeV Xe, 3 MeV proton, and 320 MeV Au irradiations on the T_c and J_c of $CaKFe_4As_4$ single crystals, and compare them with irradiation effects in $Ba_{0.6}K_{0.4}Fe_2As_2$.

IOP Conf. Series: Journal of Physics: Conf. Series **1293** (2019) 012013 doi:10.1088/1742-6596/1293/1/012013

2. Experimental Methods

$CaKFe_4As_4$ single crystals were synthesized by FeAs self-flux method. Ca granules (99.5%), K ingots (99.5%), and FeAs powder were used as starting materials. FeAs was prepared by sealing stoichiometric amounts of As grains (7N) and Fe powder (99.9%) in an evacuated quartz tube and reacting them at 700 °C for 40 h after heating at 500 °C for 10 h. A mixture with a ratio of Ca : K : FeAs =1: 1.2: 10 was placed in an alumina crucible in an argon-filled glove box. The alumina crucible was sealed in a niobium tube using arc melting method. The niobium tube was sealed in an evacuated quartz tube. The whole assembly was heated for 5 h at 1180 °C after a preliminary heating at 650 °C for 5 h, and cooled to 1050 °C at a rate of 26 °C /h, followed by cooling to 930 °C at a rate of 1.5 °C/h for the crystal growth.

800 MeV Xe and 3 MeV proton irradiations were performed at NIRS-HIMAC, and 320 MeV Au irradiation was performed using the tandem accelerator in JAEA. The irradiation dose is counted by the dose-equivalent magnetic field called "matching field", where each defect is occupied by single vortex

$$B_\Phi = n\Phi_0. \quad (1)$$

Here, n is the areal density of defects and Φ_0 is a flux quantum.

Magnetization of the crystal was measured by a superconducting quantum interference device (SQUID) magnetometer (MPMS-5XL, Quantum Design). The single crystal was placed in a quartz sample holder and fixed with Apiezon N grease. T_c was obtained from zero-field cooling (ZFC) and field-cooling (FC) magnetizaition measurements for field perpendicular to the *ab*-plane. J_c was calculated from the results of the magnetization measurements using extended Bean model.

3. Results and Discussion

3.1. 800 MeV Xe irradiation

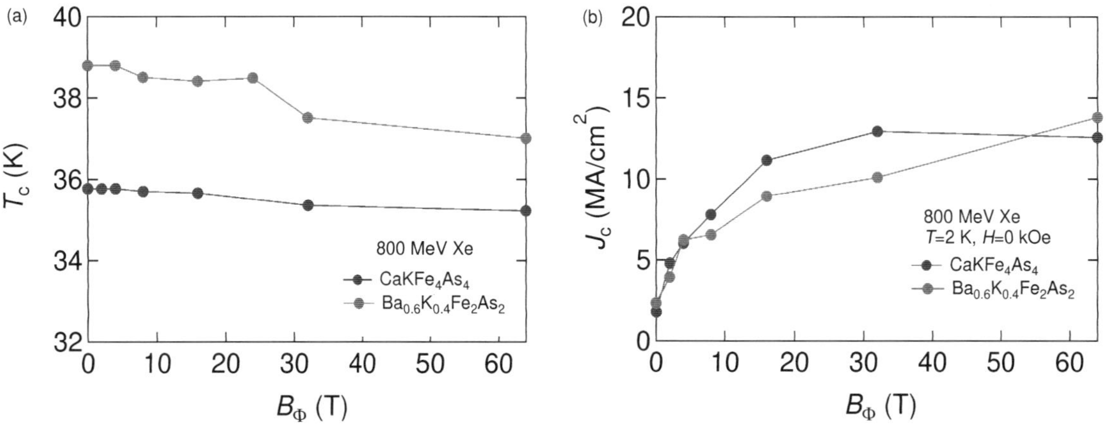

Figure 1. B_Φ dependences of (a) T_c and (b) J_c at 2 K under zero field of 800 MeV Xe-irradiated $CaKFe_4As_4$ and $Ba_{0.6}K_{0.4}Fe_2As_2$.

Figure 1 (a) shows B_Φ dependences of T_c of 800 MeV Xe-irradiated $CaKFe_4As_4$ and $Ba_{0.6}K_{0.4}Fe_2As_2$. It has been demonstrated that the suppression of T_c of 800 MeV Xe-irradiated $Ba_{0.6}K_{0.4}Fe_2As_2$ is 0.028 K/T, and is smaller than that of samples irradiated with other species [5]. It can be said that this tendency holds in the case of $CaKFe_4As_4$ since the suppression of T_c of 800 MeV Xe-irradiated $CaKFe_4As_4$ is 0.009 K/T, which is even smaller than that of $Ba_{0.6}K_{0.4}Fe_2As_2$. Irradiation-resilient feature in $CaKFe_4As_4$ could be related to the presence of novel planar defects in this material [11]. Figure 1 (b) shows B_Φ dependences of J_c of 800 MeV Xe-irradiated $CaKFe_4As_4$ and $Ba_{0.6}K_{0.4}Fe_2As_2$ at 2 K under zero field. In the case of 800 MeV Xe irradiation, B_Φ dependences of J_c in both samples are similar, where J_c is enhanced up to ~15 MA/cm^2.

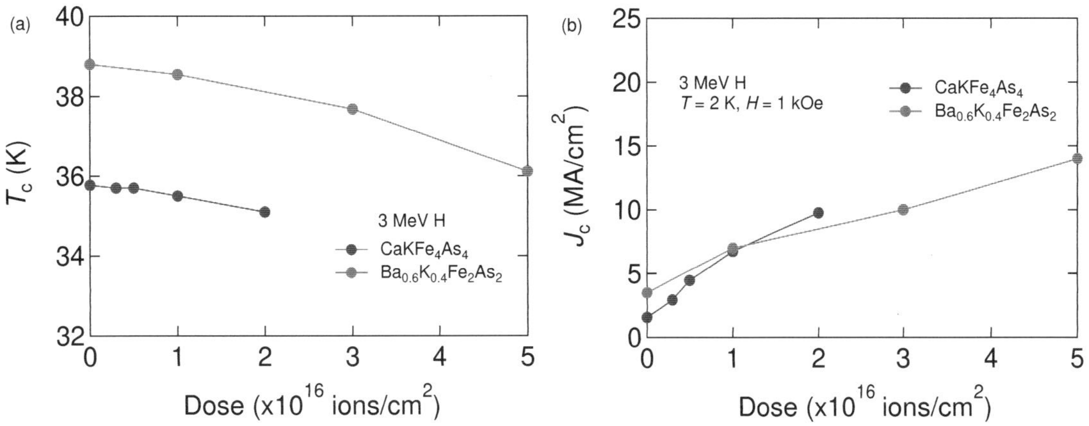

Figure 2. Dose dependences of (a) T_c and (b) J_c at 2 K at H = 1 kOe of 3 MeV proton-irradiated $CaKFe_4As_4$ and $Ba_{0.6}K_{0.4}Fe_2As_2$.

3.2. 3 MeV proton irradiation

Figure 2 (a) shows B_Φ dependences of T_c of 3 MeV proton-irradiated $CaKFe_4As_4$ and $Ba_{0.6}K_{0.4}Fe_2As_2$. At least in low-dose region, the suppression of T_c of 3 MeV proton-irradiated $CaKFe_4As_4$ is 0.34 K/ (1x10^{16} ions/cm^2), and is similar to that of $Ba_{0.6}K_{0.4}Fe_2As_2$. Figure 2 (b) shows B_Φ dependences of J_c of 3 MeV proton-irradiated $CaKFe_4As_4$ and $Ba_{0.6}K_{0.4}Fe_2As_2$ at 2 K at H = 1 kOe. In the case of 3 MeV proton irradiation, B_Φ dependences of J_c in both samples are also similar, at least in low-dose region. It is expected that higher J_c is obtained from 3 MeV proton-irradiated $CaKFe_4As_4$ with higher dose (> 2x10^{16} ions/cm^2), similar to the case of $Ba_{0.6}K_{0.4}Fe_2As_2$.

3.3. 320 MeV Au irradiation

Figure 3 (a) shows B_Φ dependences of T_c of 320 MeV Au-irradiated $CaKFe_4As_4$ and $Ba_{0.6}K_{0.4}Fe_2As_2$. Suppressions of T_c of both samples irradiated with 320 MeV Au are similar (0.11 K/T). Figure 3 (b) shows B_Φ dependences of J_c of 320 MeV Au-irradiated $CaKFe_4As_4$ and $Ba_{0.6}K_{0.4}Fe_2As_2$ at 2 K under zero field. In the case of 320 MeV Au irradiation, J_c in both samples show similar B_Φ dependences,

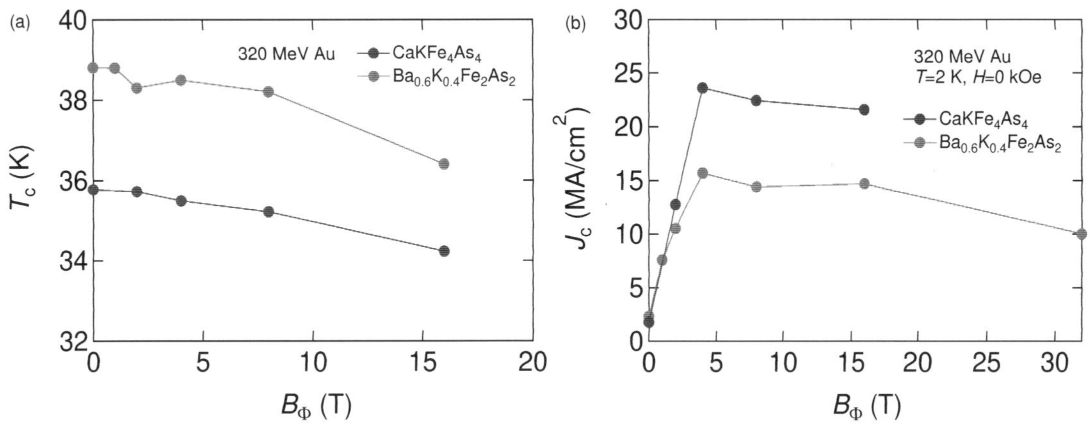

Figure 3. B_Φ dependences of (a) T_c and (b) J_c at 2 K under zero field of 320 MeV Au-irradiated $CaKFe_4As_4$ and $Ba_{0.6}K_{0.4}Fe_2As_2$. The thicknesses of $CaKFe_4As_4$ corresponding to B_Φ=2 T, 4 T, 8 T, and 16 T are 9.2 μm, 6.4 μm, 5.5 μm, and 5.4 μm.

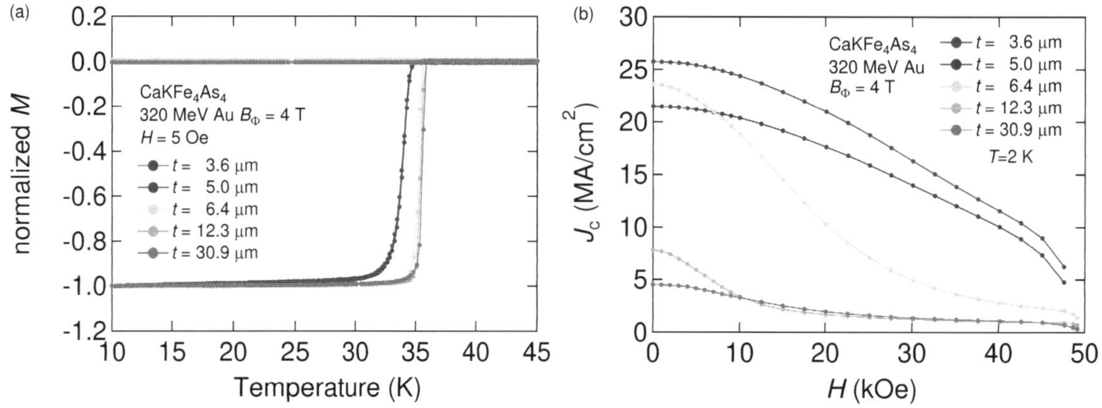

Figure 4. (a) Temperature dependence of normalized magnetization (M) at 5 Oe and (b) magnetic field dependence of J_c at 2 K of 320 MeV Au-irradiated $CaKFe_4As_4$ with the same dose (B_Φ = 4 T) but with different thicknesses.

where J_c is enhanced with increasing B_Φ up to B_Φ = 4 T and almost retains the maximum value up to B_Φ = 16 T. However, the maximum value of J_c of 320 MeV Au-irradiated $CaKFe_4As_4$ is higher than that of $Ba_{0.6}K_{0.4}Fe_2As_2$. It is expected that higher J_c close to 30 MA/cm^2 could be obtained by introducing splayed columnar defects or coexisting of columnar and point defects, which enhance J_c in $Ba_{0.6}K_{0.4}Fe_2As_2$ [12, 13].

We compare the value of T_c and the maximum value of J_c of 320 MeV Au-irradiated $CaKFe_4As_4$ with different thicknesses. Figure 4 (a) shows temperature dependence of normalized magnetization (M) at 5 Oe of 320 MeV Au-irradiated $CaKFe_4As_4$ with the same dose (B_Φ = 4 T) but with different thicknesses. The value of T_c of thin samples irradiated with 320 MeV Au tends to be lower than that of thick samples. Figure 4 (b) shows magnetic field dependence of J_c at 2 K of 320 MeV Au-irradiated $CaKFe_4As_4$ with the same dose (B_Φ = 4 T) but with different thicknesses. Although the relative error in estimating the thickness of samples becomes larger in thin samples (t < 7 μm), it can be said from Fig. 4 (b) that the maximum value of J_c of thin $CaKFe_4As_4$ samples irradiated with 320 MeV Au tends to be higher than that of thick samples (t > 12 μm). The projected range of 320 MeV Au in $CaKFe_4As_4$ is about 17 μm. Thus, in thick $CaKFe_4As_4$ samples over 10 μm irradiated with 320 MeV Au, defects may not be created through the whole crystal, and J_c in the bottom side of the crystal can be much lower than that in the top side of the crystal. Thickness dependence of T_c suppression can be also explained by the inhomogeneous creation of defects. It should be noted that similar thickness dependences of T_c suppression and J_c distribution due to inhomogeneous creation of defects by low-energy 300 MeV Xe irradiation have been reported for $Ba(Fe,Co)_2As_2$ [14].

4. Summary

We compare effects of 800 MeV Xe, 3 MeV proton, and 320 MeV Au irradiations on the T_c and J_c of 1144-type IBS $CaKFe_4As_4$ and 122-type IBS $Ba_{0.6}K_{0.4}Fe_2As_2$ single crystals. With respect to the suppression of T_c, 3 MeV proton and 320 MeV Au irradiations have similar effects on $CaKFe_4As_4$ and $Ba_{0.6}K_{0.4}Fe_2As_2$. 800 MeV Xe irradiation suppresses T_c of $CaKFe_4As_4$ a little less than that of $Ba_{0.6}K_{0.4}Fe_2As_2$. With respect to B_Φ dependences of J_c, irradiated $CaKFe_4As_4$ and $Ba_{0.6}K_{0.4}Fe_2As_2$ show similar dependences in all cases. However, in the case of 320 MeV Au irradiation, the maximum value of J_c of $CaKFe_4As_4$ is higher than that of $Ba_{0.6}K_{0.4}Fe_2As_2$. In addition, the maximum value of J_c of $CaKFe_4As_4$ irradiated with 320 MeV Au strongly depends on the thickness of samples and thin samples below 10 μm tend to show high J_c over 20 MA/cm^2.

References

[1] Civale L, Marwick A D, Worthington T K, Kirk M A, Thompson J R, Krusin-Elbaum L, Sun Y, Clem J R and Holtzberg F 1991 *Phys. Rev. Lett.* **67** 648

[2] Nakajima Y, Tsuchiya Y, Taen T, Tamegai T, Okayasu S and Sasase M 2009 *Phys. Rev.* B **80** 012510

[3] Tamegai T, Taen T, Yagyuda H, Tsuchiya Y, Mohan S, Taniguchi T, Nakajima Y, Okayasu S, Sasase M,Kitamura H, Murakami T, Kambara T and Kanai Y 2012 *Supercond. Sci. Technol.* **25** 084008

[4] Sun Y, Park A, Pyon S, Tamegai T, Kambara T and Ichinose A 2017 *Phys. Rev.* B **95** 104514

[5] Ohtake F, Taen T, Pyon S, Tamegai T, Okayasu S, Kambara T and Kitamura H 2015 *Physica* C **518** 47

[6] Iyo A, Kawashima K, Kinjo T, Nishio T, Ishida S, Fujihisa H, Gotoh Y, Kihou K, Eisaki H and Yoshida Y 2016 *J. Am. Chem. Soc.* **138** 1410

[7] Meier W R, Kong T, Kaluarachchi U S, Taufour V, Jo N H, Drachuck G, Böhmer A E, Saunders S M, Sapkota A, Kreyssig A, Tanatar M A, Prozorov R, Goldman A I, Balakirev F F, Gurevich A, Bud'ko S L and Canfield P C 2016 *Phys. Rev. B* **94** 064501

[8] Rotter M, Tegel M and Johrendt D 2008 *Phys. Rev. Lett.* **101** 107006

[9] Yuan H Q, Singleton J, Balakirev F F, Baily S A, Chen G F, Luo J L and Wang N L 2009 *Nature* **457** 565

[10] Altarawneh M M, Collar K, Mielke C H, Ni N, Bud'ko S L and Canfield P C 2008 *Phys. Rev.* B **78** 220505

[11] Pyon S, Takahashi A, Veshchunov I, Tamegai T, Ishida S, Iyo A, Eisaki H, Imai M, Abe H, Terashima T and Ichinose A (*to be published*)

[12] Park A, Pyon S, Tamegai T and Kambara T 2016 *Physica* C **530** 58

[13] Kihlstrom K J, Fang L, Jia Y, Shen B, Koshelev A E, Welp U, Crabtree G W, Kwok W K, Kayani A, Zhu S F and Wen H H 2013 *Appl. Phys. Lett.* **103** 202601

[14] Taen T, Nakajima Y, Tamegai T, Okayasu S and Sasase M 2012 *J. Phys. Conf. Ser.* **400** 022119

IOP Conf. Series: Journal of Physics: Conf. Series **1293** (2019) 012014 doi:10.1088/1742-6596/1293/1/012014

On the growth of Co-doped $BaFe_2As_2$ thin films on CaF_2

Marco Langer[1], Sven Meyer[1], Kai Ackermann[1], Lukas Grünewald[2], Sandra Kauffmann-Weiss[1], Saicharan Aswartham[3], Sabine Wurmehl[3], Jens Hänisch[1] and Bernhard Holzapfel[1]

[1] Karlsruhe Institute of Technology, Institute for Technical Physics, 76344 Eggenstein-Leopoldshafen, Germany

[2] Karlsruhe Institute of Technology, Laboratory for Electron Microscopy, 76131 Karlsruhe, Germany

[3] Leibniz Institute for Solid State and Materials Research Dresden, Institute for Solid State Research, 01171 Dresden, Germany

E-mail: marco.langer@kit.edu

Abstract. The competition between phase formation of BaF_2 and $Ba(Fe_{1-x}Co_x)_2As_2$ on CaF_2 single crystals has been analysed. $Ba(Fe_{0.92}Co_{0.08})_2As_2$ thin films have been deposited by pulsed laser deposition. X-ray diffraction, atomic force microscopy and scanning electron microscopy studies have revealed that the formation of secondary phases and misorientations as well as the growth modes of the $Ba(Fe_{0.92}Co_{0.08})_2As_2$ thin films strongly depend on the growth rate. At high growth rates, formation of BaF_2 is suppressed. The dependency of the $Ba(Fe_{0.92}Co_{0.08})_2As_2$ lattice parameters supports the idea of fluorine diffusion into the crystal structure upon suppression of BaF_2 formation similar as was proposed for $FeSe_{1-x}Te_x$ thin films on CaF_2. Furthermore, a growth mode transition from a layer growth mechanism to a three-dimensional growth mode at high supersaturation has been found, suggesting similarities between the growth mechanism of iron-based superconductors and high-T_c cuprate thin films.

1. Introduction

CaF_2 single crystalline substrates have been proven to yield *high-quality* thin films of Fe-based superconductors (FeSC), i.e. in terms of high critical temperature T_c and high critical current density J_c, grown either by pulsed laser deposition (PLD) or molecular beam epitaxy (MBE) [1]. This is especially true in the case of Co-doped $BaFe_2As_2$ (Ba122) thin films, for which $T_{c0} \sim 26$ K has been reported on CaF_2 [2], i.e. similar to values of annealed single crystals [3], whereas thin films grown on other substrates such as MgO, $SrTiO_3$ (STO) or $(LaAlO_3)_{0.3}(Sr_2TaAlO_6)_{0.7}$ (LSAT) often show a reduction of T_c by several Kelvin [4]. However, the phase purity of FeSC thin films remains a major challenge even ten years after the first reports [5]. The growth of FeSC thin films is often accompanied by formation of secondary phases such as FeAs or other Fe-rich phases and misoriented Ba122 grains on the sample surface [6] which impedes the development of many possible applications such as planar junction technology or sub-micrometer single-photon nanowire detectors. Additionally to these precipitates usually obtained on all kind of substrates, Ba122 thin films grown on CaF_2 often show the formation of BaF_2 [7, 8, 9]. This can be understood if one considers the thermal instability of CaF_2 under vacuum conditions at typical deposition temperatures, T_d, of $\sim 700\,°C$ which are necessary for the growth of epitaxial Ba122 thin films [10]. Fluorine diffusion and other interface reactions are regularly reported already at moderate $T_d \sim 300\,°C$ [11, 12, 13], and a considerably high density of free fluorine adatoms on the substrate surface in case of $T_d \sim 700\,°C$ is therefore likely. Contrary to the occasional reports of BaF_2 formation during the growth of Ba122 thin films, others, e.g. Yuan *et al.* [14] or Lee *et al.* [15], have reported the growth of films without apparent x-ray diffraction reflections of BaF_2. At present, the competition of the phase formation between BaF_2 and $Ba(Fe_{1-x}Co_x)_2As_2$ has never been discussed and it is not clear which film growth parameters most strongly influence the formation of BaF_2. Furthermore, the surface morphology and the growth modes of Ba122 films on CaF_2 has not been addressed apart from reports by Sobota *et al.* [16] and Daghero *et al.* [17], who found a granular three-dimensional growth mode surface on 500 nm scale [16] together with rectangular outgrowths ("terraces") of 100 nm to 200 nm length on 2 µm scale [17] for films with nominal Co content $x = 0.1$ and 0.8 respectively grown with KrF excimer laser at 700 °C and 7 Hz.

In this report, we discuss the influence of the growth rate, controlled in this study by the laser pulse energy E, on the growth of Ba122 thin films on CaF_2. It was found that the formation of BaF_2 predominantly occurs at low growth rates, whereas at high growth rates the formation of BaF_2 is suppressed. Furthermore, the suppression of BaF_2 formation coincides with pronounced lattice parameter changes, promoting the idea of fluorine diffusion into the Ba122 crystal structure, similar to what was proposed [11] for $FeSe_{1-x}Te_x$ thin films grown on CaF_2. Additionally, growth mode transitions from a $c_\perp$-oriented layer growth mechanism with terraces and a typical ledge height of $c/2$ at low supersaturation occurs to a $c_\parallel$-oriented three-dimensional growth mode with (110)-oriented grains at high supersaturation.

2. Experimental

The $Ba(Fe_{1-x}Co_x)_2As_2$ thin films with a nominal Cobalt content $x = 0.08$ have been grown by conventional on-axis pulsed laser deposition with a frequency-tripled Nd:YAG laser ($\lambda = 355$ nm, Quanta-Ray Indi-40-10, Spectra-Physics) at 10 Hz onto commercially available CaF_2 single crystals (CrysTec GmbH) with dimensions of 10 mm × 5 mm × 0.5 mm. A polycrystalline target was prepared by standard solid-state sintering as described in more detail in Ref. [18]. The substrates were radiantly heated by a resistive boron nitride heater, and the deposition temperature was calibrated by a thermocouple near the heater plate within the thermal shielding of the heater. The laser focus spot had an ellipsoidal shape with an area $A \approx 1\,\mu m^2$. The target was mounted on a target carousel and was scanned around three axes (x, y, rotation) to ensure homogenous ablation. The distance between substrate and target was 50 mm. The base pressures at room temperature and deposition temperature were 4×10^{-9} and 1×10^{-8} mbar respectively. The laser energy was varied between 12 mJ and 55 mJ (i.e. the laser fluence ε between 1.2 and 5.5 Jcm^{-2}) and was measured before and after each deposition behind the chamber through a laser exit window in order to confirm absence of an energy gradient during deposition.

X-ray diffraction was performed on a Rigaku SmartLab 3 kW (40 kV, 30 mA) 5-axis diffractometer with a goniometer radius of 300 mm, parallelized Cu-K_α radiation and a HyPix-3000 2D detector. Film thickness t for thin films ($t < 50$ nm) was calculated by analysing Kissing fringes obtained in specular x-ray reflectivity (XRR). The thickness of thicker films was estimated by assuming a constant growth rate under similar deposition parameters. Out-of-plane lattice constants c were calculated by extrapolation of measured lattice constants, obtained in symmetric $2\theta/\theta$ scans (line focus, 1D detector mode), plotted against the Nelson-Riley function $f_{NR} = 0.5(\cos\theta^2/\sin\theta + \cos\theta^2/\theta)$. Reciprocal space maps (RSM) were obtained with a point focus (PB-0.3 mm pinhole CBO and 0.3 mm-collimator) with the detector in 2D mode. In-plane orientation was proven by ϕ-scans and pole figures of the (103) reflection on selected samples. $R(T)$ dependencies of unstructured films were measured in van der Pauw

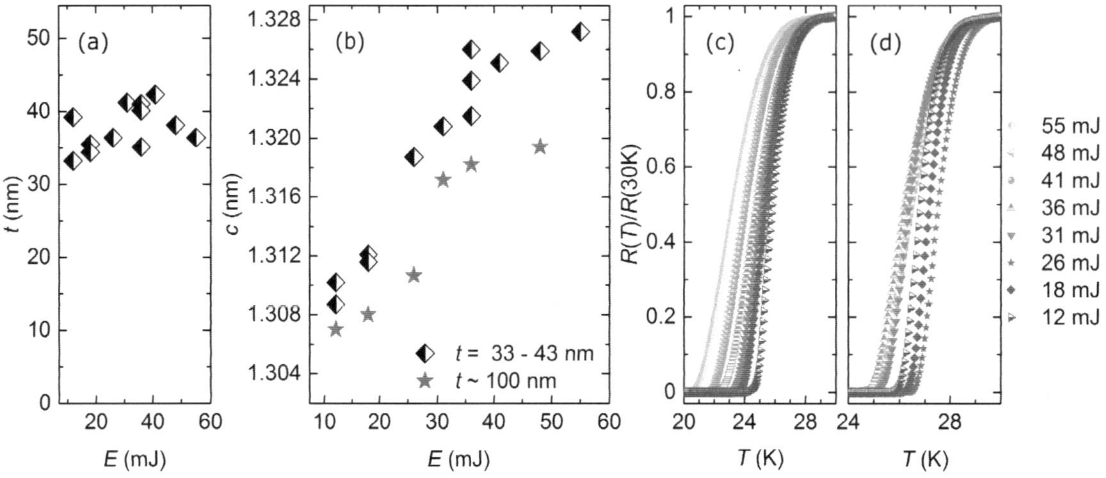

Figure 1. (a) XRR-determined film thickness vs. laser energy for the sample series with a nominal thickness $t < 50$ nm. (b) c-axis vs. laser energy for both sample series. Normalized resistance vs. temperature for thin (c) and (d) thick samples. The highest T_c of $T_{c90} = 28.5$ K and $T_{c0} = 26.1$ K was measured for a film with a nominal thickness of 100 nm grown at $E = 26$ mJ ($\varepsilon = 2.6\,\mathrm{Jcm^{-2}}$) (stars in (d)). At higher laser energies (i.e. growth rates), T_c is decreased by several Kelvin for thin and thick samples.

geometry on a Quantum Design physical properties measurement system (PPMS 14 T) with a heating rate of $1\,\mathrm{K\,min^{-1}}$. A Keithley 2460 SourceMeter in combination with a Keithley 2182A nanovoltmeter were employed for pulsed current measurements in order to reduce the voltage noise level compared to the built-in PPMS system bridge. T_{c0} was defined as the temperature at which the voltage signal exceeded the noise level. This resistively measured T_{c0} coincided with inductively measured T_c as confirmed on selected samples by a two-coil mutual inductance method, measured with a custom-designed PPMS puck mounted with the coils. Atomic force microscopy (AFM) height and adhesion maps were obtained on a Bruker Dimension Edge AFM with ScanAsyst-Air Si/Si_3N_4 tips ($f = 70$ kHz, $l = 0.4\,\mathrm{N\,m^{-1}}$) in PeakForce Tapping mode and a typical scan rate of 0.1 Hz. Scanning electron microscopy (SEM) images were taken on a Zeiss Leo 1350 equipped with a field emission gun, operated at a working distance of 5 mm and an acceleration voltage of 20 kV.

A sample series with varying laser energy at $T_d = 700\,°\mathrm{C}$ with $t < 50$ nm has been grown in order to obtain the growth rate via x-ray reflectivity. Based on this growth rate, a second series of thicker samples with a nominal thickness of $t \sim 100$ nm has been grown by adjusting the number of laser pulses.

3. Results

The thin samples grown for determination of the growth rate have a thickness ranging from 33 nm to 45 nm (figure 1(a)). The growth rate varied between $\sim 0.4\,\mathrm{Å\,s^{-1}}$ for films grown at a laser pulse energy of 12 mJ and $\sim 0.9\,\mathrm{Å\,s^{-1}}$ grown at 55 mJ. The out-of-plane lattice parameter c of both sample series shows a pronounced dependency on the laser fluence with an onset of a plateau at high laser fluence (figure 1(b)). All samples have lattice parameters bigger than the lattice parameters of single crystals with the same nominal Co content, indicating an elongation of the tetragonal unit cell along the c direction. Furthermore, the lattice parameters depend on the film thickness, where in general thicker films have lattice parameters closer to bulk values.

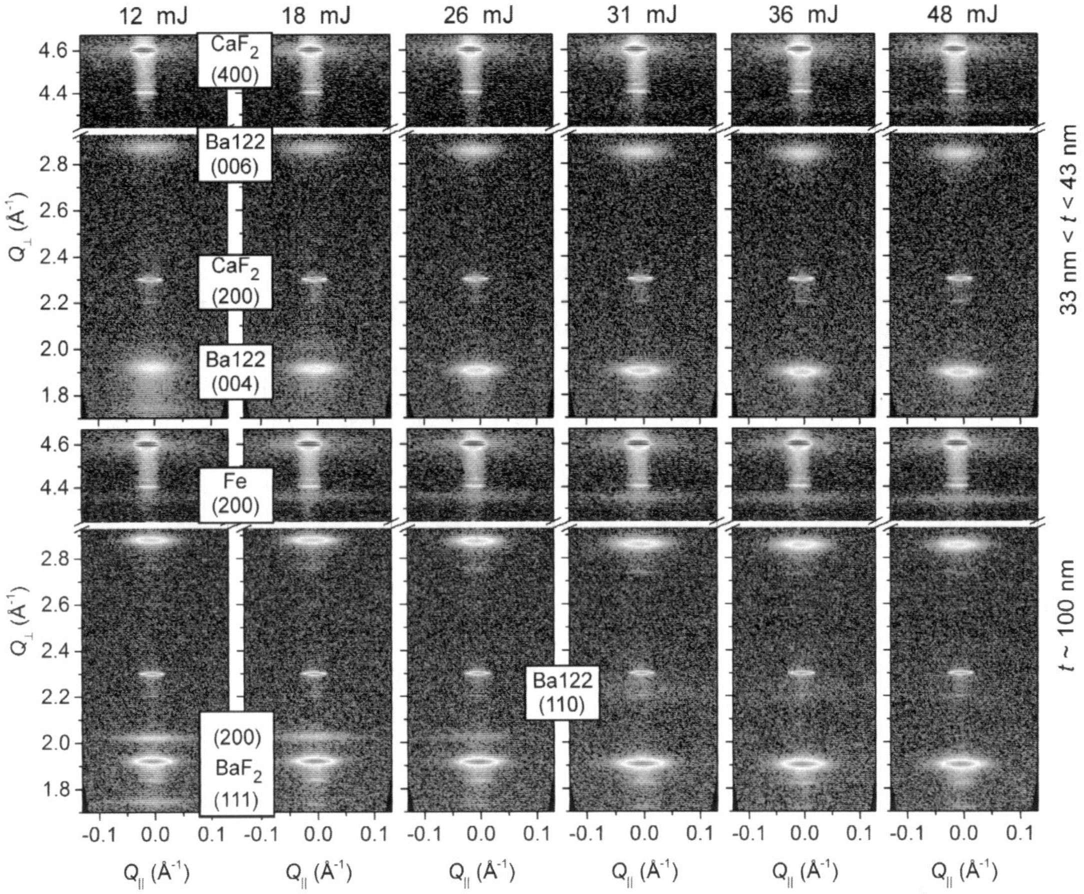

Figure 2. Reciprocal space maps of symmetric $2\theta/\theta$ scans around Ba122(004) ($Q_\perp \sim 1.9\,\text{Å}^{-1}$), Ba122(006) ($Q_\perp \sim 2.85\,\text{Å}^{-1}$) and CaF_2(400) ($Q_\perp \sim 4.6\,\text{Å}^{-1}$) of thin ($33\,\text{nm} < t < 43\,\text{nm}$, top) and thicker ($t \sim 100\,\text{nm}$, bottom) samples grown at laser energies of 12 mJ, 18 mJ, 26 mJ, 31 mJ, 36 mJ and 48 mJ (from left to right).

T_{c0} varied between 19.6 K and 23.7 K for thin and between 23.5 K and 26.1 K for thick samples with a nominal thickness of $t \sim 100\,\text{nm}$ (figure 1(c) and (d)). The highest T_c values were obtained on films grown at moderate laser energies of 12 mJ to 30 mJ, i.e. at laser fluences of $1.2\,\text{Jcm}^{-2}$ to $3.0\,\text{Jcm}^{-2}$. For both sample series, films grown at $E > 30\,\text{mJ}$ showed a reduction in T_c by several Kelvin.

X-ray diffraction using reciprocal space maps (RSM) (figure 2) revealed that all samples contain one or more secondary phases or orientations, such as BaF_2 ((111) at $Q_\perp \sim 1.74\,\text{Å}^{-1}$ and (200) at $Q_\perp \sim 2.02\,\text{Å}^{-1}$), (110)-oriented Ba122 ($Q_\perp \sim 2.19\,\text{Å}^{-1}$) and Fe-rich precipitates (Fe(200) at $Q_\perp \sim 4.34\,\text{Å}^{-1}$). BaF_2 reflections are visible for laser energies $E \leq 31\,\text{mJ}$, whereas the formation of Fe-rich precipitates is increased at high laser energy. The formation of (110)-oriented Ba122 is only apparent for samples with a sufficient thickness ($t \sim 100\,\text{nm}$), as indicated by a broad, low-intensity reflection in case of laser energies $E \geq 36\,\text{nm}$.

Films grown at low laser energy, e.g. 12 mJ, contain granular precipitates at their surface with a maximal lateral size of $\sim 170\,\text{nm}$ and a height of $\leq 40\,\text{nm}$ (figure 3(a), (d), (g) and (j)).

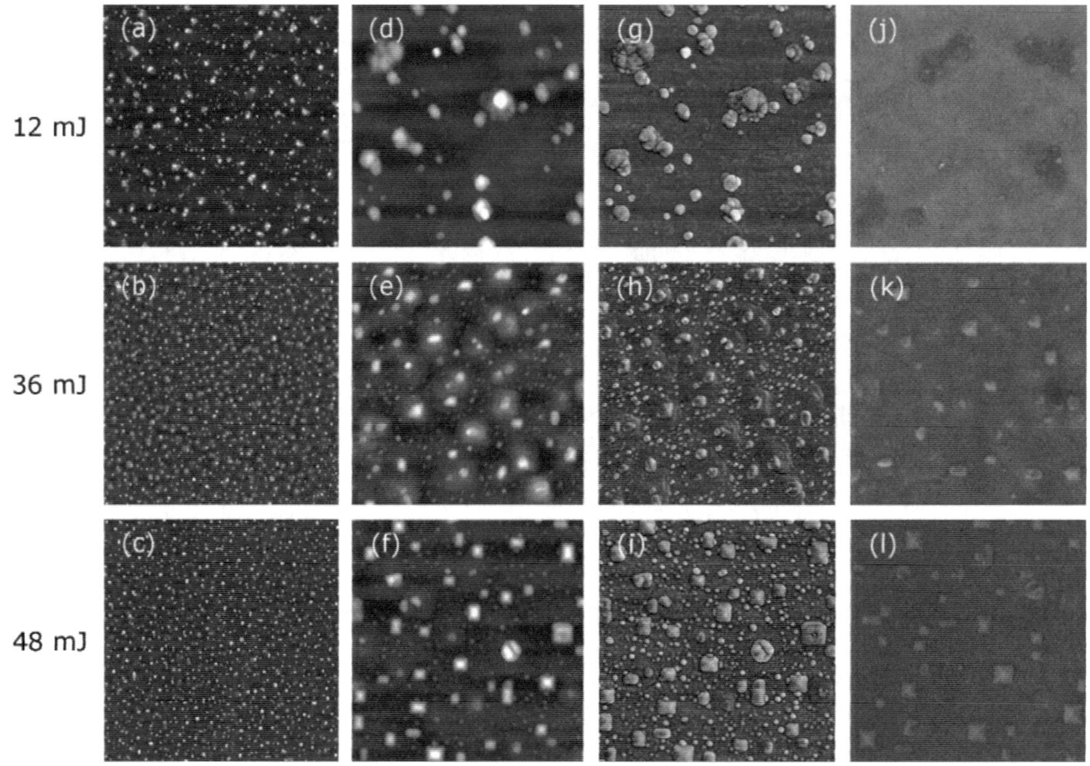

Figure 3. Representative surface morphologies of three thin samples (33 nm $< t <$ 43 nm) grown at 12 mJ (top), 36 mJ (middle) and 48 mJ (bottom). (a) - (c) 5 µm × 5 µm AFM height maps. (d) - (f) 1 µm × 1 µm AFM height maps. (g) - (i) 1 µm × 1 µm AFM adhesion maps. (j) - (l) 1 µm × 1 µm SEM micrographs.

Apart from these precipitates, the film surface is rather flat and a layer growth mechanism with a ledge height of $c/2$ is visible (figure 3(g)). A detailed determination of ledge width and shapes is often complicated due to the high density of the granular precipitates. Films grown at high laser energy, e.g. 48 mJ, show a completely different surface morphology (figure 3(c), (f), (i) and (l)). The granular precipitates are absent but a high density of square pyramids or rectangular platelets with a lateral size of $\sim$ 50 nm × 50 nm and a height of $\leq$ 20 nm, predominantly oriented along CaF_2[100], are visible. Films with a thickness of $t <$ 50 nm often show a higher density of pyramids compared to the elongated platelets as can be seen in figure 3, whereas thicker films $t \geq$ 100 nm show a similar or even higher density of the elongated flat platelets. Similar to the form of the terraces in case of low laser energy, the shape of these CaF_2[100]-oriented precipitates is often better visible in the AFM adhesion maps rather than in the AFM height maps. In contrast to the $c/2$ terraces observed on films grown at low laser energy, the layer growth mechanism is absent for films grown at high laser energy and the surface of such films contains a high density of three-dimensional islands (figure 3(i)). Films grown at an intermediate laser energy, e.g. 36 mJ, exhibit a mixture of both kinds of precipitates discussed above (figure 3(b), (e), (h) and (k)).

The growth mode transition from a layer growth mechanism with terraces to a three-dimensional growth mode without terraces occurs within a rather small energy range, as can be seen exemplary in the AFM adhesion maps for two thick samples grown at E = 31 mJ (figure

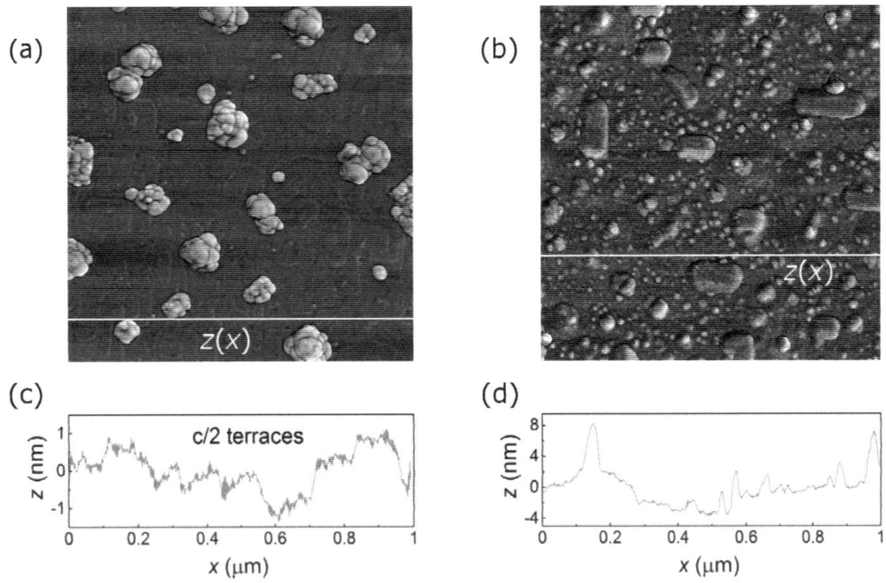

Figure 4. Surface morphology of two thick films ($t \sim 100\,\mathrm{nm}$) grown at $E = 31\,\mathrm{mJ}$ (a, c) and $E = 36\,\mathrm{mJ}$ (b, d). (a, b) $1\,\mu\mathrm{m} \times 1\,\mu\mathrm{m}$ AFM adhesion maps. (c, d) Extracted line scans of the associated AFM height maps as indicated by the white lines in (a, c). A transition between quasi-2D terraced growth and 3D island growth is clearly visible.

4(a)) and $E = 36\,\mathrm{mJ}$ (figure 4(b)). Whereas the former film exhibits terraces with a ledge height of $c/2$ and a ledge width of $\sim 50 - 100\,\mathrm{nm}$ (figure 4(c)), the terraces are absent in the latter case and the surface roughness increased significantly (figure 4(d)).

4. Discussion

The highest $T_{c0} \sim 26.1\,\mathrm{K}$ obtained on films with $t \sim 100\,\mathrm{nm}$ is comparable to the highest reported T_{c0} for Ba122 thin films by Tarantini *et al.* [2]. The reduction of T_c by several Kelvin for films with a thickness of $33\,\mathrm{nm} < t < 43\,\mathrm{nm}$ is in good agreement with the thin films grown by Yuan *et al.* [19]. The strong dependency of the lattice parameters, secondary phase formation, surface morphology and critical temperature T_c on the laser fluence, i.e. growth rate, has not been discussed so far by others, but may serve as an explanation for the contradictory published results [7, 9, 15, 19], namely the strong variation in lattice parameters and formation of BaF_2 in some cases. The c-axis lattice parameter in the range $1.307\,\mathrm{nm} \leq c \leq 1.327\,\mathrm{nm}$ of this study provides an explanation for the strong variation of reported lattice parameters, i.e. from $1.303\,\mathrm{nm}$ [19] to $1.319\,\mathrm{nm}$ [7]. In addition to the dependency on the film thickness and growth rate, we have found that the c-axis lattice parameter also strongly depends inversely on the deposition temperature, i.e. it decreases at increasing deposition temperature. These results will be published elsewhere. Films without BaF_2 precipitates show a stronger elongation of the unit cell along the c-axis compared to films which contain BaF_2 precipitates. This correlation between BaF_2 formation and changes in lattice parameters in case of BaF_2 phase formation suppression at high growth rates supports the idea of a fluorine diffusion into the Ba122 crystal structure similar as was already discussed for $FeSe_{1-x}Te_x$ thin films grown on CaF_2 [11]. The granular precipitates found on film surfaces for which x-ray analysis shows the formation of BaF_2 have similar lateral dimensions as the BaF_2 precipitates found in Ni-doped $BaFe_2As_2$

films on CaF_2 [8] and are thereby most likely BaF_2 precipitates as well. Ongoing TEM analysis will clarify the different fluorine distribution on selected films grown at different growth rates. The CaF_2[100]-oriented pyramidal precipitates are most likely (110)-oriented Ba122 grains as suggested by x-ray analysis. This interpretation is further supported by measurements of (110)-pole figures (not shown here) as well as previous TEM analysis on Ba122 films grown on MgO, for which these characteristic pyramids were also detected and identified as misoriented Ba122.

At present, it is not fully clear whether the growth mode transition from a layer growth mechanism with $c/2$ ledge heights to a three-dimensional growth with (110)-oriented precipitates is misfit-driven, supersaturation-driven or a mixture of both. Since the suggested fluorine diffusion into the Ba122 crystal structure upon increase of the growth rate changes the lattice parameters, the misfit between the substrate and the film most likely varies for the samples grown at different laser energies. Contrariwise, a supersaturation-driven growth mode transition from $c_\perp$- to $c_\|$-oriented growth at high supersaturation is well known from the growth of high-T_c cuprate thin films [20]. This transition is commonly attributed to the anisotropy of the surface Gibbs energies [21]. At low supersaturation, the critical nuclei reduce their surface energies by assuming c-axis orientation [22]. At present, films on other, fluorine-free, substrates are deposited and analysed in order to disentangle these two possible explanations. The formation of (110)-oriented Ba122 grains on the film surface depends, similar to the lattice parameters, inversely on the growth temperature, as will be discussed elsewere. At $T_\mathrm{d} \geq 720\,°\mathrm{C}$, no (110)-reflection was found in XRD, again in good agreement with the suppression of a-axis growth in case of cuprate films at high deposition temperature, i.e. at low supersaturation [23].

5. Conclusions

In summary, we have shown that the formation of BaF_2, the Ba122 lattice parameters and the surface morphology for films grown on CaF_2 strongly depend on the growth rate. Suppression of BaF_2 phase formation is achieved at high growth rates and coincides with an elongation of the unit cell along the c-axis, supporting the idea of fluorine diffusion into the Ba122 crystal structure being one of the main reasons for the commonly reported increased c-axis parameters for films grown on CaF_2. Since all films in this study were grown at the same deposition temperature, the influence of the thermal mismatch between the film and the CaF_2 substrate, usually attributed as the cause of the distortion of the Ba122 unit cell, has to be considered with care. A growth mode transition from a $c_\perp$-oriented layer growth mechanism with ledge heights of $c/2$ to a three-dimensional growth mode with $c_\|$-oriented Ba122 grains on the film surface occurs almost simultaneously at high supersaturation. Ongoing work will clarify the origin of this transition, whether it is linked to fluorine diffusion into the Ba122 crystal structure upon suppression of BaF_2 or can be explained similarly as the supersaturation-driven transitions well known for cuprate high-T_c thin films.

References

[1] Imai Y, Nabeshima F and Maeda A 2017 *Condensed Matter* **2** 25
[2] Tarantini C, Kametani F, Lee S, Jiang J, Weiss J D, Jaroszynski J, Hellstrom E E, Eom C B and Larbalestier D C 2014 *Scientific Reports* **4**
[3] Sun D, Xiao J and Lin C 2011 *Journal of Crystal Growth* **321** 55–59
[4] Haindl S, Kidszun M, Onken F, Mietke A and Thersleff T 2013 *International Journal of Modern Physics B* **27** 1330001
[5] Katase T, Hiramatsu H, Yanagi H, Kamiya T, Hirano M and Hosono H 2009 *Solid State Communications* **149** 2121–2124
[6] Katase T, Hiramatsu H, Kamiya T and Hosono H 2012 *Superconductor Science and Technology* **25** 084015
[7] Kurth F, Reich E, Hänisch J, Ichinose A, Tsukada I, Hühne R, Trommler S, Engelmann J, Schultz L, Holzapfel B and Iida K 2013 *Applied Physics Letters* **102** 142601
[8] Richter S, Kurth F, Iida K, Pervakov K, Pukenas A, Tarantini C, Jaroszynski J, Hänisch J, Grinenko V, Skrotzki W, Nielsch K and Hühne R 2017 *Applied Physics Letters* **110** 022601

[9] Iida K, Grinenko V, Kurth F, Ichinose A, Tsukada I, Ahrens E, Pukenas A, Chekhonin P, Skrotzki W, Teresiak A, Hühne R, Aswartham S, Wurmehl S, Mönch I, Erbe M, Hänisch J, Holzapfel B, Drechsler S L and Efremov D V 2016 *Scientific Reports* **6**
[10] Haindl S, Hanzawa K, Sato H, Hiramatsu H and Hosono H 2016 *Scientific Reports* **6**
[11] Ichinose A, Nabeshima F, Tsukada I, Hanawa M, Komiya S, Akiike T, Imai Y and Maeda A 2013 *Superconductor Science and Technology* **26** 075002
[12] Zarraga-Colina J, Nix R M and Weiss H 2005 *The Journal of Physical Chemistry B* **109** 10978–10985
[13] Qiu W, Ma Z, Patel D, Sang L, Cai C, Hossain M S A, Cheng Z, Wang X and Dou S X 2017 *ACS Applied Materials & Interfaces* **9** 37446–37453
[14] Yuan P, Xu Z, Wang D, Zhang M, Li J and Ma Y 2016 *Superconductor Science and Technology* **30** 025001
[15] Lee J, Jiang J, Kametani F, Oh M J, Weiss J D, Collantes Y, Seo S, Yoon S, Tarantini C, Jo Y J, Hellstrom E E and Lee S 2017 *Superconductor Science and Technology* **30** 085006
[16] Sobota R, Plecenik T, Gregor M, Truchly M, Satrapinskyy L, Vidis M, Secianska K, Kurth F, Holzapfel B, Iida K, Kus P and Plecenik A 2014 *Applied Surface Science* **312** 182–187
[17] Daghero D, Pecchio P, Laviano F, Gonnelli R, Kurth F, Grinenko V and Iida K 2014 *Applied Surface Science* **312** 23–29
[18] Kurth F, Iida K, Trommler S, Hänisch J, Nenkov K, Engelmann J, Oswald S, Werner J, Schultz L, Holzapfel B and Haindl S 2013 *Superconductor Science and Technology* **26** 025014
[19] Yuan P, Xu Z, Li C, Quan B, Li J, Gu C and Ma Y 2017 *Superconductor Science and Technology* **31** 025002
[20] Pennycook S, Chisholm M, Jesson D, Feenstra R, Zhu S, Zheng X and Lowndes D 1992 *Physica C: Superconductivity* **202** 1–11
[21] Granozio F M and di Uccio U S 1997 *Journal of Alloys and Compounds* **251** 56–64
[22] Granozio F M, Salluzzo M, di Uccio U S, Maggio-Aprile I and Fischer Ø 2000 *Physical Review B* **61** 756–765
[23] Jeschke U, Schneider R, Ulmer G and Linker G 1995 *Physica C: Superconductivity* **243** 243–251

IOP Conf. Series: Journal of Physics: Conf. Series **1293** (2019) 012015 doi:10.1088/1742-6596/1293/1/012015

Transport properties of $FeSe_{1-x}S_x$ and $FeSe_{1-y}Te_y$ epitaxial thin films under magnetic fields

N Shikama[1], T Ishikawa[1], F Nabeshima[1], A Maeda[1]

[1] Department of Basic Science, The University of Tokyo, 3-8-1 Komaba, Meguro-ku, Tokyo 153-8902, Japan

E-mail: n-shikama@maeda1.c.u-tokyo.ac.jp

Abstract. We comparatively investigated transport properties of $FeSe_{1-x}S_x$ and $FeSe_{1-y}Te_y$ epitaxial thin films under magnetic field. The contrastive behavior between Te and S substituted materials was observed in the carrier density estimated from the Hall effect as a function of x (or y), which is similar to that in T_c. This means that the effect of the disappearance of the structural transition is not universal for superconductivity in Fe chalcogenides. T_c was found to have a strong correlation with carrier concentration, which is in agreement with our recent result in FeSe film samples where the strain was changed systematically from compressive to tensile.

1. Introduction

Since the discovery of superconductivity in LaFeAs(O,F), iron based superconductors (FeSCs) have attracted much attention[1]. FeSe is a suitable material for understanding the origin of the superconductivity in FeSCs because it has the simplest crystal structure among FeSCs. The superconducting transition temperature, T_c, is about 9 K at ambient pressure[2]. Unlike other FeSCs, FeSe exhibits the tetragonal-orthorhombic structural phase transition (SPT) at about 90 K without antiferromagnetic (AFM) order. The SPT is considered to have an electronic origin and is often called as the nematic transition[3]. $FeSe_{1-y}Te_y$ has been investigated intensively from the early stage[4], and Te substitution increased T_c up to $\sim$ 14K. However there is a miscibility gap at $0.1\leq y \leq 0.4$ in bulk crystals[4], and the relation between the superconductivity and the SPT was unclear. We have successfully fabricated $FeSe_{1-y}Te_y$ thin films for all y values[5]. Interestingly. T_c rapidly increased with increasing Te content at the composition where the SPT disappears[6]. This may suggest that the SPT suppress the superconductivity. On the other hand, $FeSe_{1-x}S_x$ bulk samples do not show such a sudden increase of T_c at the SPT[7]. We recently found that $FeSe_{1-x}S_x$ films also show a gradual decrease of T_c with increasing x even at the SPT[8]. These results indicate that the relation between the superconductivity and the SPT is not universal in iron chalcogenides. To clarify the relation between the superconductivity and the SPT, we investigate the magneto-transport phenomena (Hall effect and magnetoresistance) in these samples. It is found that the carrier densities, n and p, increase as Te content increases, while they remain almost unchanged for S substituted samples. These mean that T_c has a strong correlation with carrier concentration, which is in agreement with our recent result in FeSe films where the strain was changed systematically from compressive to tensile. Therefore, it is essential to increase the carrier concentration for the T_c enhancement of iron chalcogenides.

IOP Conf. Series: Journal of Physics: Conf. Series **1293** (2019) 012015 doi:10.1088/1742-6596/1293/1/012015

2. Method

$FeSe_{1-x}S_x$ and $FeSe_{1-y}Te_y$ epitaxial films were grown on LAO substrate by the pulse laser deposition (PLD) method with a KrF laser[6, 8, 10]. The back pressure of the growth chamber is approximately $\sim 10^{-8}$ Torr. For $FeSe_{1-x}S_x$ growth, $Fe_{1.1}Se$ and FeS polycrystalline pellets were used as targets[8]. Single-phase $FeSe_{1-x}S_x$ films were obtained by ablating each target alternatively[9]. On the other hand, stoichiometric $FeSe_{1-y}Te_y$ target were used for $FeSe_{1-y}Te_y$. The a-axis and the c-axis lengths of the films were determined from the 204 or the 111 reflection peaks and the 00l reflection peaks in the X-ray diffraction (XRD) measurements. The thickness of the films was 40 $\sim$ 100 nm, which was estimated from X-ray reflectivity (XRR) measurements or by a Dektak 6 M stylus profiler. These films were deposited in a six-terminal shape using a metal mask for transport measurements. The measured area was 1.2 mm long and 1.0 mm wide. The electrical resistivity and the transport properties were measured using a Physical Property Measurement System (PPMS, Quantum Design, Inc.).

3. Result and Discussion

Figure 1 shows the temperature dependence of the resistivity of the grown films. All films show the metallic behavior and the superconducting transition. $FeSe_{1-x}S_x$ with $x \leq 0.14$ and $FeSe_{1-y}Te_y$ with $y \leq 0.20$ show a kink anomaly which corresponds to the SPT. On the other hand, $FeSe_{1-x}S_x$ with $x = 0.19$ and 0.25 and $FeSe_{1-y}Te_y$ with $y = 0.30$ and 0.40 do not show the SPT. The $x = 0.25$ S-substituted sample shows another kink anomaly in the $\rho - T$ curve at low temperatures. This kink structure is characteristic of film samples, and is considered to originate from the AFM transition[8]. Figures 2(a) and (b) show the temperature dependence of the Hall

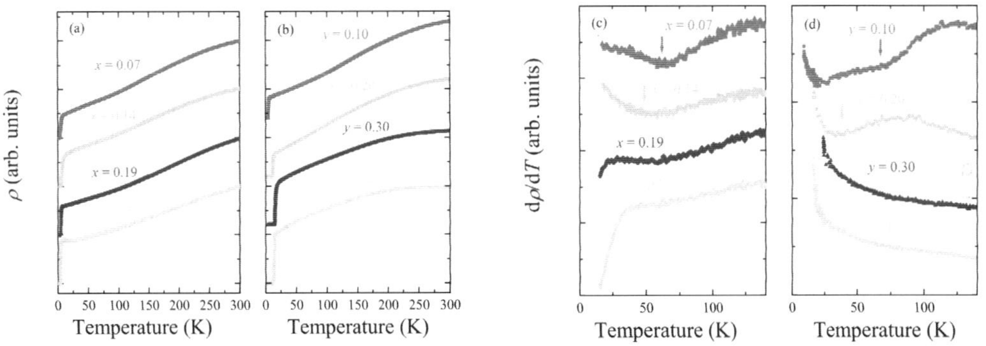

Figure 1. Temperature dependence of the resistivity, ρ, of (a) $FeSe_{1-x}S_x$ thin films and (b) $FeSe_{1-y}Te_y$ thin films, and the temperature derivative of the resistivity of (c) $FeSe_{1-x}S_x$ thin films and (d) $FeSe_{1-y}Te_y$ thin films. Arrows indicate the SPT temperature T_s.

coefficient R_H of the $FeSe_{1-x}S_x$ and $FeSe_{1-y}Te_y$ films. The obtained behavior is very contrastive between the Te-substituted system and the S-substituted system, in particular in the tetragonal phase. In both systems, R_H increases rapidly with decreasing temperature in the orthorhombic phase. In the samples without the SPT (all in the tetragonal phase), however, the R_H of the Te substituted samples stay very small and even becomes negative at low temperatures (essentially the same behavior was observed in the Te substituted films grown on CaF_2[10]), whereas the R_H of the S-substituted samples increases at low temperatures. This behavior is similar to the one in the orthorhombic phase. These suggest that the effect of the SPT on the electronic structure is rather different between the Te-substituted system and the S-substituted system.

Figure 3 shows the magnetic field-dependence of the magnetoresistance of the typical two representative films (one Te-substituted and the other S-substituted). Both films show the

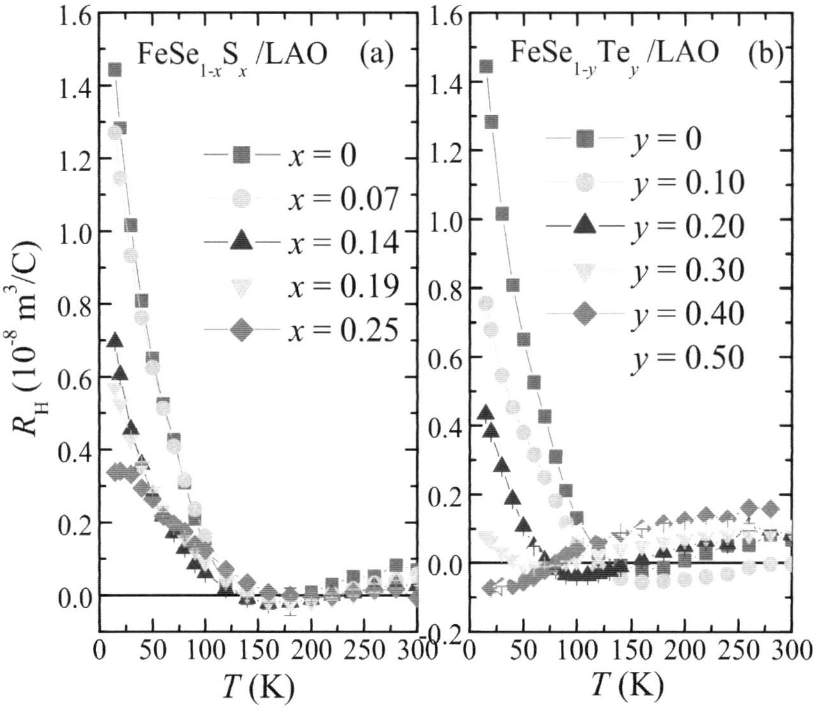

Figure 2. Temperature dependence of the Hall coefficient R_H of (a) $FeSe_{1-x}S_x$ thin films and (b) $FeSe_{1-y}Te_y$ thin films grown on LAO substrate.

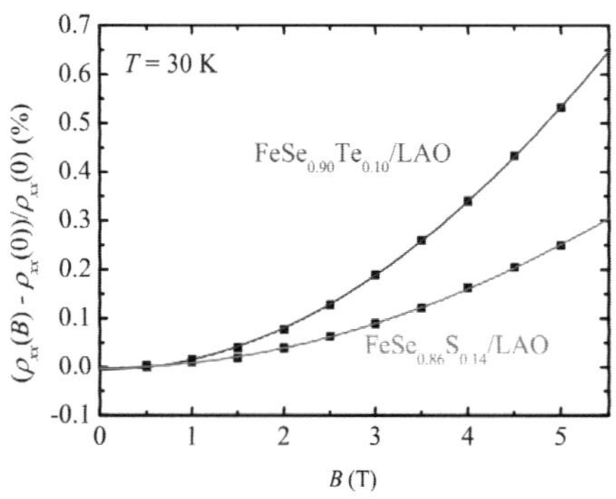

Figure 3. The magnetoresistance of the $FeSe_{0.86}S_{0.14}$ and $FeSe_{0.90}Te_{0.10}$ thin films on LAO substrate as a function of the magnetic field at T =30 K.

quadratic dependence. This suggests that we can assume a simple two-band picture of an electron and a hole. Indeed, it is well established that in iron-chalcogenides, the multiple bands contribute to the electrical transport properties such as the Hall coefficient and the magnetoresistance. As the simplest approach, the two band picture of the elctron and the hole is often worked with. Therefore, using a two band model of one hole band and one electon band, we further investigate the nature of the charge carriers.

The resistivity tensor is written as

$$\rho_{xx}(0) = \frac{1}{e(n_{\rm e}\mu_{\rm e} + n_{\rm h}\mu_{\rm h})} \tag{1}$$

$$\rho_{yx}(B) = \frac{-n_{\rm e}\mu_{\rm e}^2 + n_{\rm h}\mu_{\rm h}^2}{e(n_{\rm e}\mu_{\rm e} + n_{\rm h}\mu_{\rm h})^2}B \tag{2}$$

$$\frac{\rho_{xx}(B) - \rho_{xx}(0)}{\rho_{xx}(0)} = \frac{n_{\rm e}n_{\rm h}\mu_{\rm e}\mu_{\rm h}(\mu_{\rm e} + \mu_{\rm h})^2}{(n_{\rm e}\mu_{\rm e} + n_{\rm h}\mu_{\rm h})^2}B^2 \tag{3}$$

where $\mu_{\rm h}$, $\mu_{\rm e}$, $n_{\rm h}$, $n_{\rm e}$ are the hole mobility, the electron mobility, the hole density and the electron density, respectively. To solve these three equations on four parameters, we need to assume that $n_{\rm h}$ is proportional to $n_{\rm e}$. For simplicity, we assume $n_{\rm h} = n_{\rm e}$. The different choice for the magnitude of the proportionality constant does not affect the result largely[13]. Magnetoresistance is more remarkable at lower temperatures. However, avoiding the effect of superconductivity, we work with the data at 30 K.

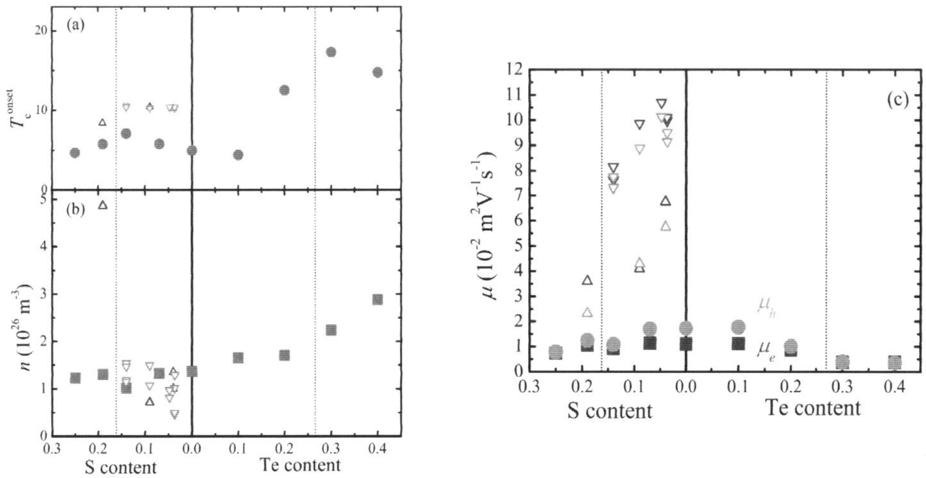

Figure 4. (a) $T_{\rm c}^{\rm onset}$ and (b) $n_{\rm e}$ and $n_{\rm h}$, and (c) $\mu_{\rm e}$ and $\mu_{\rm h}$ at 30 K as functions of S content x and Te content y. The dashed lines show the boundary between the orthorhombic and tetragonal phases. The data of the bulk samples of $\rm FeSe_{1-x}S_x$ [11, 12] are also plotted. The upper triangle is the data from[11] at 25 K, and the lower triangle is the data from[12]at 15 K.

Figures 4 (a), (b) and (c) show $T_{\rm c}^{\rm onset}$, the carrier densities, $n_{\rm e}$ and $n_{\rm h}$, and the mobilities, $\mu_{\rm e}$ and $\mu_{\rm h}$, as a function of S content and Te content, respectively. The contrastive behavior of $T_{\rm c}^{\rm onset}$ at the SPT between S-substituted samples and Te-substituted samples is observed. What is remarkable is that the similar contrastive behavior is observed in the carrier density as a function of substitution amount. In $\rm FeSe_{1-x}S_x$, $n_{\rm e}$ (and $n_{\rm h}$) does not depend largely, and is rather almost independent of x both in the orthorhombic and in the tetragonal phase. On the other hand, in $\rm FeSe_{1-y}Te_y$, $n_{\rm e}$ (and $n_{\rm h}$) increases in the tetragonal phase. These behaviors of $n_{\rm e}$ (and $n_{\rm h}$) correspond well with the characteristic dependence of $T_{\rm c}$ on $n_{\rm e}$ (and $n_{\rm h}$). Such remarkable features were not observed in the mobility. Both of $\mu_{\rm e}$ and $\mu_{\rm h}$ decrease systematically with increasing substitution amount in both cases of $\rm FeSe_{1-x}S_x$ and $\rm FeSe_{1-y}Te_y$, and no significant change was observed at the SPT. We previously observed the similar remarkable correlation of $T_{\rm c}$ with the carrier density in FeSe films with various degrees of strain introduced systematically[13].

These results suggest the close relation between $T_{\rm c}$ and the carrier density in iron chalcogenides. We already mentioned, based on our results, that the electron nematicity does not play a universally significant role on superconductivity in Fe chalcogenides. The results presented here suggests that the SPT affects on the electronic structure differently between Te substituted and S substituted samples. This is the direct origin of the difference of T_c behavior at the SPT, manifested in the close correlation between T_c and the carrier density.

4. Conclusion

We comparatively investigated transport properties of $FeSe_{1-x}S_x$ and $FeSe_{1-y}Te_y$ epitaxial thin films under magnetic field. The contrastive behavior between Te and S substituted materials was observed in the carrier density estimated from the Hall effect as a function of x (or y), which is similar to that in T_c. This means that the effect of the disappearance of the structural transition is not universal for superconductivity in Fe chalcogenides. T_c was found to have a strong correlation with carrier concentration, which is in agreement with our recent result in FeSe films where the strain was changed systematically from compressive to tensile.

References

[1] Kamihara Y, Watanabe T, Hirano M and Hosono H 2008 *J. Am. Chem. Soc.* **130** 2396
[2] Hsu F C , Luo J Y, Yeh K W, Chen T K, Huang T W, Wu P M, Lee Y C, Huang Y L, Chu Y Y, Yan D C and Wu M K 2008 *Proc. Natl. Acad. Sci. U.S.A.* **105** 14262
[3] Nakayama K, Miyata Y, Phan G N, Sato T, Tanabe Y, Urata T, Tanigaki K and Takahashi T 2014 *Phys. Rev. Lett.* **113** 237001
[4] Fang M H, Pahm H M, Qian B, Liu T J, Vehstedt E K, Liu Y, Spinu L and Mao Z Q 2008 *Phys. Rev. B* **78** 224503
[5] Imai Y, Sawada Y, Nabeshima F and Maeda A 2015 *Proc. Natl. Acad. Sci. U.S.A.* **112** 1937
[6] Imai Y, Sawada Y, Nabeshima F, Asami D, Kawai M and Maeda A 2017 *Sci. Rep.* **7** 46653
[7] Hosoi S, Matsuura K, Ishida K, Hao Wang , Mizkami Y, Watashige T, Kasahara S, Matsuda Y and Shibauchi T 2016 *Proc. Natl. Acad. Sci. U.S.A.* **113** 8139
[8] Nabeshima F, Ishikawa T, Oyanagi K, Kawai M and Maeda A 2018 *J. Phys. Soc. Jpn.* **87** 073704
[9] Fujiwara K, Shiogai J and Tsukazaki A 2017 *Jpn. J. Appl. Phys.* **56** 100308
[10] Sawada Y, Nabeshima F, Imai Y, Maeda A 2016 *Jpn. J. Appl. Phys.* **85** 073703
[11] Ovchenkov Y A, Chareev D A, Kulbachinskii V A, Kytin V G, Presnov D E, Volkova O S and Vasiliev A N 2017 *Supercond. Sci. Technol.* **30** 035017
[12] Ovchenkov Y A, Chareev D A, Kulbashinskii V A, Kytin V G, Mishikov S V, Presnov D E, Volkova O S and Vasiliev A N 2018 *arXiv* 1808.03551v2
[13] Nabeshima F, Kawai M, Ishikawa T, Shikama N and Maeda A 2018 *Jpn. J. Appl. Phys.* **57** 120314

Response of a superconductor $NbSe_2$ flake to magnetic field detected with small tunnel junctions

Naoki Hoshi, Dai Inoue, Hiroki Sonoda, Daisuke Yabe, Hikari Tomori, Akinobu Kanda

Division of Physics, Faculty of Pure and Applied Sciences, University of Tsukuba, Tennodai, Tsukuba, Ibaraki 305-8571, Japan

E-mail: kanda.akinobu.gm@u.tsukuba.ac.jp

Abstract. We have experimentally studied response of a superconductor $NbSe_2$ flake to perpendicular magnetic fields using small tunnel junctions. The $NbSe_2$ film (lateral size ~ several μm, thickness ~ 50 nm) was obtained using the mechanical exfoliation of a bulk crystal and was transfered on top of predefined Au electrodes, forming an Au/$NbSe_2$/Au double SIN tunnel junction system. At 4.2 K, between the lower and upper critical fields (H_{C1} and H_{C2}), the voltage of the Au/$NbSe_2$/Au structure under constant current of 1 μA exhibited random-telegraph-noise-like fluctuations as a function of the magnetic field which was applied perpendicular to the $NbSe_2$ flake. The origin of the voltage fluctuations is discussed.

1. Introduction

In a mesoscopic superconductor with sizes comparable to the superconducting coherence length and/or the magnetic penetration depth, the quantum confinement of Cooper pairs dramatically modifies the superconducting properties.[1] The resultant mesoscopic superconducting states are attractive not only from the viewpoint of fundamental physics but also from the viewpoint of application to classical and quantum devices.

Peculiar mesoscopic effects appear in vortex states: When a mesoscopic type-II superconductor contains several quantum vortices, the arrangement of the vortices depends not only on the vortex-vortex interaction but also on the interaction between a vortex and the sample edge, so that the triangular Abrikosov vortex lattice, which is characteristic of bulk superconductors, is deformed into a multivortex state (MVS) in which the vortex arrangement is strongly influenced by the sample shape, or a giant vortex state (GVS) in which the multiple vortices coalesce.[2, 3, 4, 5]

So far, we have experimentally investigated the vortex states in mesoscopic superconductors using our original *small tunnel junction method*,[6, 7] in which the arrangement of the vortices was estimated from the transport properties of the small tunnel junctions attached to edges of the superconductor. The superconductors were made from electron-beam evaporated aluminum. We successfully confirmed the existence of a giant vortex and transitions between different MVSs or between an MVS and a GVS in mesoscopic circular and square thin films. These transitions were induced by an applied magnetic field [8, 9] as well as by a locally injected supercurrent.[10, 11, 12, 13] Such controllability of the vortex states is important for the future realization of classical and quantum devices based on mesoscopic vortex states. During a

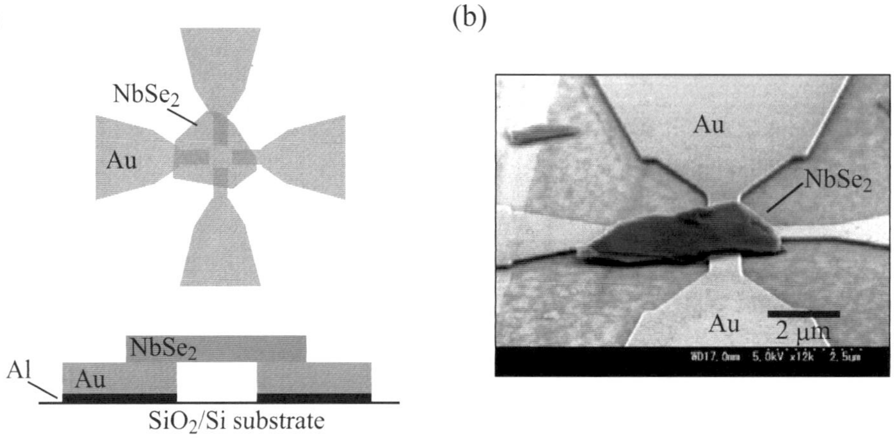

Figure 1. (a) Schematic top and side views of the sample. An $NbSe_2$ flake is placed on top of Au electrodes. (b) A typical SEM image of a sample, but this sample is different from the one reported in this paper.

series of experiments, however, we noticed a serious problem that the vortex arrangement was strongly influenced by the existence of defects. For example, in the single vortex state (vorticity $L = 1$) in a circular sample, the vortex did not situate at the center of the circle.[8] Also, the magnitude of the magnetic field/injected current which induced a transition between different vortex states varied sample to sample even when the nominal sample shape and sizes were the same. We speculate that these incidents are due to defects caused by the surface roughness of the sample. The surface roughness is inevitable in the vacuum-evaporated aluminum films. We note that we have also investigated the effect of artificially introduced defects on vortex state transitions.[14, 15, 16]

Recently, the mechanical exfoliation technique, developed in the graphene research [17, 18], has been applied to layered superconductors, and thin superconductors with atomically flat surface can be easily obtained.[20, 21] Such atomically flat superconductor might resolve the above defect problem. Here, as the first step toward the precise control of the mesoscopic vortex states in thin layered superconductors, we investigate the response of a thin layered superconductor flake to magnetic field using small tunnel junctions.

2. Experimental

As a layered superconductor, we used $NbSe_2$. The device structure is shown in Fig. 1. An $NbSe_2$ flake (lateral area ~ 21 $(\mu\text{m})^2$, thickness ~ 50 nm) was placed on top of four predefined Au electrodes (thickness: 60 nm, width: 1 μm) with an adhesive layer of Al (5 nm). No intentional tunnel barrier was inserted between Au and $NbSe_2$. The junction areas for electrodes A and B (see the inset of Fig. 2(a)) are ~ 2.1 and $2.4(\mu\text{m})^2$, respectively.

In the sample fabrication, first $NbSe_2$ flakes were obtained using the mechanical exfoliation of a bulk crystal on a SiO_2/Si substrate.[17] One of the flakes was then transfered to a substrate with Au electrodes by a microscope and a micro-manipulator using a transparent elastomer stamp (a poly dimethyl siloxane (PDMS) film) covered with poly-propylene carbonate (PPC) layer.[19] Finally the PPC layer, which was left on the $NbSe_2$ flake, was dissolved with acetone.

Electron transport between two electrodes were measured at 4.2 K in vacuum in a cryostat with appropriate cryogenic lowpass filters. The differential conductance was obtained by numerically differentiating the current-voltage characteristics. Magnetic field was applied

IOP Conf. Series: Journal of Physics: Conf. Series **1293** (2019) 012016 doi:10.1088/1742-6596/1293/1/012016

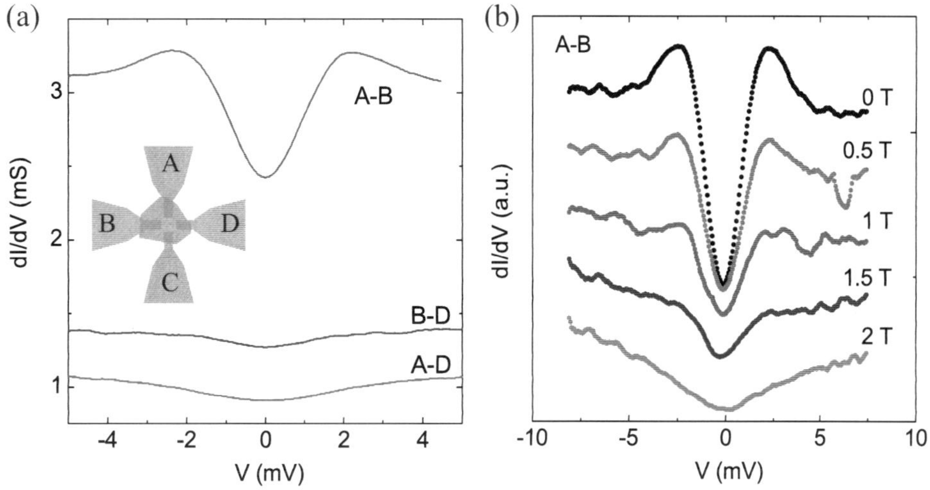

Figure 2. (a) Differential conductance (dI/dV) as a function of the bias voltage (V) for electrode pairs, A-B, A-D and B-D. The electrode C did not work in the measurement. The inset indicates the label of each electrode. (b) $dI/dV - V$ curves of electrode pair A-B for several magnetic field values.

perpendicular to the substrate.

The superconducting transition temperature of $NbSe_2$ flakes was ~7.0 K and the upper critical field at 4.2 K was ~ 2.0 T from the resistance measurement of other flakes (not shown in this paper). The literature value of the lower critical field is ~ 0.02 T.[22] The superconducting energy gap Δ was 1.1 meV calculated from the BCS theory.

3. Results and discussion

Figure 2(a) shows the differential conductance (dI/dV) as a function of the bias voltage (V) for electrode pairs, A-B, A-D and B-D. The label of each electrode is indicated in the inset of Fig. 3(a), in which the electrode C did not work in the measurement. The current-voltage characteristics were almost linear. However, there appears a distinct structure only in the $dI/dV - V$ curve of electrode pair A-B. Because the separation of the dI/dV peaks agrees with 4Δ and the structure fades away as the magnetic field approached H_{C2}, as shown in Fig. 2(b), we can conclude that tunnel barriers were formed for electrodes A and B and the structure in the $dI/dV - V$ curve reflects the superconducting energy gap of the $NbSe_2$ flake. The tunnel barrier might be due to the naturally formed oxides on the surface of the $NbSe_2$ flake. The reason for the absence of the structure for the other electrode pairs is not clear at this moment.

To elucidate the magnetic field dependence of the transport of the junctions, we fixed the current from electrode A to B to 1 μA, and measured the voltage between electrodes A and B as a function of the magnetic field. The result is shown in Fig. 3 for sweeping-up (Figs. 3(a)(c)(e)) and sweeping-down (Figs. 3(b)(d)(f)) the magnetic field. As shown in Figs. 3(a) and (b), a lot of random-telegraph-noise-like voltage fluctuations are seen below the upper critical field $H_{C2} \approx 2.0$ T both for sweeping-up and sweeping-down the magnetic field. Also, as shown in Figs. 3(c) and 3(d), the steep voltage change is less pronounced around $B = 0$ T. For increasing the magnitude of the magnetic field, $|B|$, the first steep voltage change appears at ≈ 0.05 T.

IOP Conf. Series: Journal of Physics: Conf. Series **1293** (2019) 012016 doi:10.1088/1742-6596/1293/1/012016

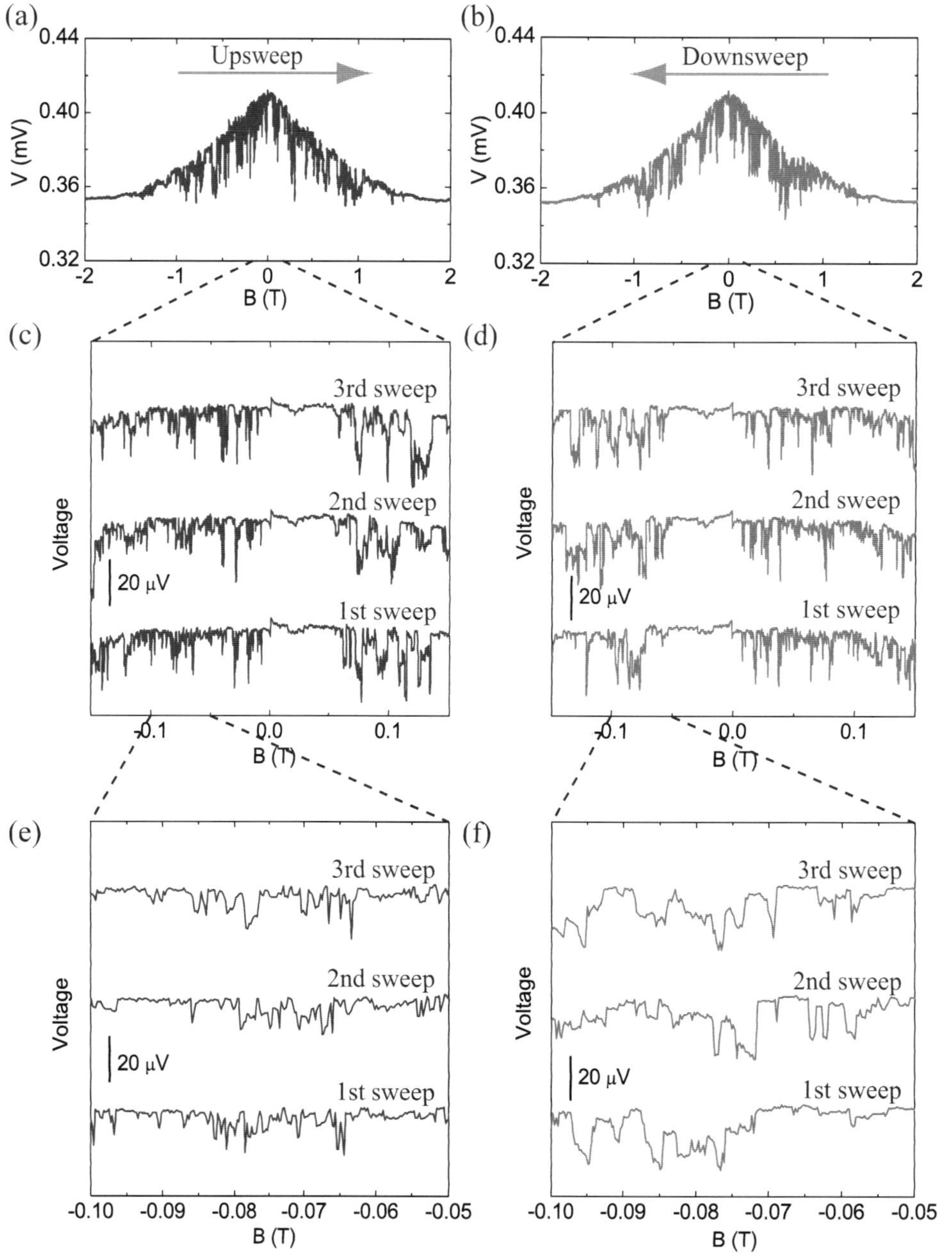

Figure 3. Magnetic field dependence of the voltage between electrodes A and B under a constant current of 1 μA in sweeping-up (a,c,e) and sweeping-down (b,d,f) the magnetic field (B). In (c)-(f), the data for three successive B-sweeps are shown but the data for each sweep are offset vertically for clarity. The noise level of the voltage measurement was ≈ 2 μV.

Taking into account that generally the penetration field of the first vortex is larger than the lower critical field (H_{C1}) due to the surface barrier[23], we speculate that the random-telegraph-noise-like voltage fluctuations occur in the vortex state (under a magnetic field between H_{C1} and H_{C2}).

Generally, the shape of the current-voltage curve of a superconductor/insulator/normal metal tunnel junction is sensitive to the superconducting energy gap at the interface, which is a function of the local supercurrent and the magnetic field.[24, 25] In the present case, under a magnetic field larger than H_{C1}, not only a shielding supercurrent is flowing near the periphery of the sample ($NbSe_2$), but circular current around each vortex exists inside the sample. Since the junction size ($\sim \mu$m) is larger than the penetration depth, $\lambda_\perp = \lambda_L^2\xi_0/d^2 \approx 140$ nm, where λ_L is the London penetration depth, ξ_0 is the coherence length and d is the thickness,[24] both the shielding supercurrent and vortices exist at the junction interface, as illustrated in Fig. 4(a). Thus, the local superconducting energy gap at the interface exhibits spatial variation. The junction voltage would reflect the averaged value of the local superconducting energy gap, which is affected both by the shielding supercurrent and by the vortex arrangement. In particular, for the electrode pair A-B, the junction size is much larger than $\lambda_\perp$, so that the gradual change in the shielding supercurrent due to the magnetic field sweep would not result in a noticeable change in the junction voltage.

Here, we consider the origin of the random-telegraph-noise-like voltage fluctuations. The magnetic field for one flux quantum per the sample area and per the junction area are 0.08 mT and $\sim$ 1 mT, respectively, both of which are much smaller than the typical separation of the steep voltage changes, as seen in Fig. 3(e) and 3(f). Thus, the voltage jumps are not caused by individual vortex entrance/expulsion in the sample or in the junction interfaces.

On the other hand, magneto-optical imaging of vortices in $NbSe_2$[26] revealed that under magnetic field sweep, the vortex motion is non-uniform and occurs in isolated jumps as well as collective motion involving many vortices. While the collective motion of vortices would lead to continuous change in the junction voltage, isolated jumps of vortices would correspond to steep voltage jumps. Here, we note that nominally tens or hundreds of vortices are situated in the junction area, depending on the magnetic field. For example, an applied magnetic field of 0.05 T corresponds to 50 vortices in the junction interface with area of ~ 2 $(\mu\text{m})^2$. Thus, we can conclude that the random-telegraph-noise-like steep voltage change is caused by vortex jumps in and near the junction interface.

Finally, we would like to comment on the difference of the present result to our previous observations of the vortex states in aluminum mesoscopic superconductors. In our previous works,[8, 9, 10, 11, 12, 13, 14, 15, 16] we were able to detect the individual vortex entrance/expulsion. This is because the junction size (~ 0.2 μm) was smaller than the magnetic penetration depth of the aluminum thin films ($\lambda_\perp \approx 420$ nm), so that there existed no vortices at the junction interface and the junction voltage mainly reflected the change in the shielding supercurrent, as illustrated in Fig. 4(b).

There would be two methods for the detection of individual vortex penetration/expulsion in mesoscopic $NbSe_2$ samples. The first is to make the junction smaller. However, this might be challenging because of the spatial resolution of the transfer process of a $NbSe_2$ flake. The other is to make the flake thinner. Since even a monolayer film can be obtained in the mechanical exfoliation, the penetration depth can be largely enhanced.

4. Conclusions

We have experimentally investigated the response of a superconductor $NbSe_2$ flake to magnetic field using small tunnel junctions. At 4.2 K, we observed random-telegraph-noise-like voltage fluctuations under a magnetic field sweep, which was attributed to vortex jumps around the junction interface. The difference of the present results from our previous results on

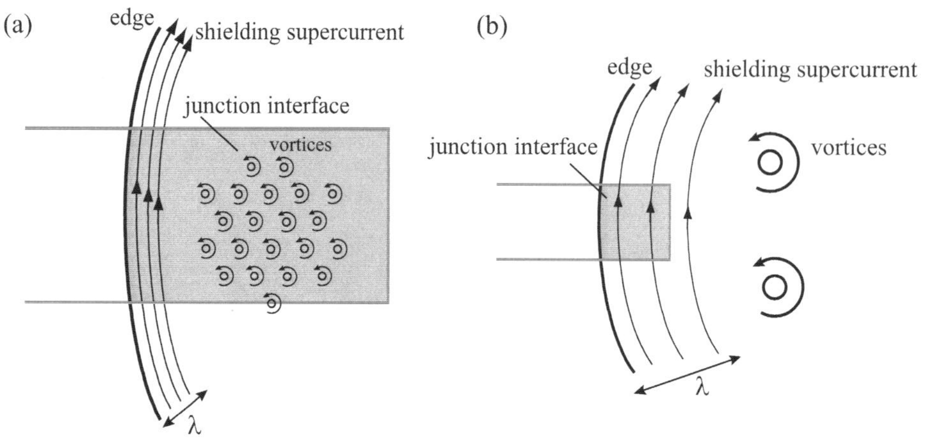

Figure 4. The junction area is compared with the magnetic penetration depth for the present $NbSe_2$ sample (a), and for the previous aluminum samples (b).

aluminum mesoscopic superconductors comes from the junction size compared with the magnetic penetration depth.

Acknowledgments

This work was supported by JSPS KAKENHI Grant Number JP15H05867 [Hybrid Quantum Systems]. The authors gratefully acknowledge fruitful discussions with M. Hayashi.

References

[1] Kanda A, Ootuka Y, Kadowaki K, Peeters F M 2010 Novel superconducting states in nanoscale superconductors *The Oxford Handbook of Nanoscience and Technology* vol 1, ed. A Narlikar and Y Y Fu (Oxford: Oxford University Press) chapter 19 pp 639-676
[2] Deo P S, Schweigert V A and Peeters F M 1997 *Phys. Rev. Lett.* **79** 4653
[3] Moshchalkov V V, Qiu X G and Bruyndoncx V 1997 *Phys. Rev. B* **55** 11793
[4] Schweigert V A and Peeters F M 1998 *Phys. Rev. B* **57** 13817
[5] Bruyndoncx V, Rodrigo J G, Puig T, Van Look L, Moshchalkov V V and Jonckheere R 1998 *Phys. Rev. B* **60** 4285
[6] Kanda A, Geisler M C, Ishibashi K, Aoyagi Y and Sugano T 2000 *Physica B* **284-288** 1870
[7] Kanda A and Ootuka Y 2003 *Physica B* **329-333** 1421
[8] Kanda A, Baelus B J , Peeters F M , Kadowaki K, Ootuka Y 2004 *Phys. Rev. Lett.* **93** 257002
[9] Baelus B J, Kanda A, Peeters F M, Ootuka Y and Kadowaki K 2005 *Phys. Rev. B* **71** 140502(R)
[10] Milosevic M V, Kanda A, Hatsumi S, Peeters F M, Ootuka Y 2009 *Phys. Rev. Lett.* **103** 217003
[11] Hatsumi S, Ootuka Y and Kanda A 2009 *Physica C* **469** 1080
[12] Hatsumi S, Kuroda Y, Ootuka Y and Kanda A 2010 *J. Phys.: Conf. Series* **232** 012003
[13] Hatsumi S, Kuroda Y, Ootuka Y and Kanda A 2010 *Physica C* **470** 1141
[14] Furugen R, Hatsumi S, Ootuka Y and Kanda A 2008 *Physica C* **468** 1301
[15] Hatsumi S, Kanda A, Furugen R, Ootuka Y and Hayashi M 2009 *J. Phys.: Conf. Series* **150** 022024
[16] Kuroda Y, Hatsumi S, Ootuka Y and Kanda A 2008 *Physica C* **470** 1145
[17] Novoselov K S, Geim A K, Morozov S V, Jiang D, Zhang Y, Dubonos S V, Grigorieva I V and Firsov A A 2004 *Science* **306** 666
[18] Kanda A 2013 Experimental approaches to graphene electron transport for device applications *Physics and Chemistry of Graphene: Graphene to Nanographene* ed T Enoki and T Ando (Singapore: Pan Stanford) chapter 3 pp 89-205
[19] Wang L, Meric I, Huang P Y, Gao Q, Gao Y, Tran H, Taniguchi T, Watanabe K, Campos L M, Muller D A, Guo J, Kim P, Hone J, Shepard K L, Dean C R 2013 *Science* **342** 614

[20] Xi X, Wang Z, Zhao W, Park J-H, Law K T, Berger H, Forro L, Shan J and Mak K F 2015 *Nat. Phys.* **12** 139
[21] Cao Y, Mishchenko A, Yu G L, Khestanova E, Rooney A P, Prestat E, Kretinin A V, Blake P, Shalom M B, Woods C, Chapman J, Balakrishnan G, Grigorieva I V, Novoselov K S, Piot B A, Potemski M, Watanabe K, Taniguchi T, Haigh S J, Geim A K and Gorbachev R V 2015 *Nano Lett.* **15** 4914
[22] Boaknin E, Tanatar M A, Paglione J, Hawthorn D, Ronning F, Hill R W, Sutherland M, Taillefer L, Sonier J, Hayden S M and Brill J W 2003 *Phys. Rev. Lett.* **90** 117003
[23] Bean C P and Livingston J D 1964 *Phys. Rev. Lett.* **12** 14
[24] Tinkham M 2004 *Introduction to Superconductivity* (New York: Dover)
[25] Bardeen J 1962 *Rev. Mod. Phys.* **34** 667
[26] Goa P E, Hauglin H, Baziljevich M, Il ′ yashenko E, Gammel P L and Johansen T H 2001 *Supercond. Sci. Technol.* **14** 729

IOP Conf. Series: Journal of Physics: Conf. Series **1293** (2019) 012017 doi:10.1088/1742-6596/1293/1/012017

Evaluation of Layer Thickness Dependence of Critical Current Density using Longitudinal Magnetic Field Effect in Superconducting Coated Conductors

T Yonenaka[1], E S Otabe[1], V Vyatkin[2], S Lee[2], T Akune[3] and T Nishizaki[3]

[1] Department of Computer Science and Electronics, Kyushu Institute of Technology, 680-4 Kawazu, Iizuka 820-8502, Japan
[2] SuperOx Japan LLC, 1880-2 Kamimizo, Chuo-ku, Sagamihara 252-0243, Japan
[3] Department of Electrical Engineering, Kyushu Sangyo University, 2-3-1 Matsukadai, Higashi-ku, Fukuoka 813-8503, Japan

yonenaka.tomohiro783@mail.kyutech.jp

Abstract. When the critical current density J_c for a superconductor is constant, the critical current I_c increases in proportion to the superconducting cross-sectional area. However, in the case of a superconducting coated conductor, when the thickness of the superconducting layer exceeds about 2 μm, there is no further increase in I_c. This is because J_c decreases due to the structural degradation of the superconducting layer. In this study, J_c for samples with superconducting layers of different thicknesses was measured using the longitudinal magnetic field effect in order to investigate the optimum layer thickness for maximizing I_c. Differences in the J_c characteristics due to the layer thickness appeared more clearly under a longitudinal magnetic field than a transverse magnetic field. This was particularly true at low temperatures, and the optimum layer thickness was found to be approximately 1 μm.

1. Introduction

The critical current I_c for a superconducting coated conductor initially increases with increasing superconducting layer thickness, but then fails to increase any further. This is because the critical current density J_c decreases due to the structural degradation of the superconducting layer [1]. The layer thickness dependence of J_c has previously been investigated under a transverse magnetic field, and J_c was found to exhibit a maximum for a superconducting layer thickness of about 2 μm [2-3].

An external magnetic field was applied to the superconductor either in the transverse direction (perpendicular to the current-flow direction) or the longitudinal direction (parallel to the current-flow) direction). It is known that when the current and the magnetic field are parallel, J_c increases with increasing magnetic field strength. This is known as the longitudinal magnetic field effect [4-7], and is associated with a reduction in the Lorentz force. As a practical application of this effect, a force-free superconducting electrical transmission cable has been proposed [8]. Therefore, a superconducting wire with a high critical current due to the longitudinal magnetic field effect is desirable.

By examining the layer thickness dependence of the longitudinal magnetic field, it is possible to determine whether the current and the magnetic field are parallel. It has been reported that for a bulk rare-earth based superconductor, J_c is larger under a longitudinal magnetic field than a transverse magnetic field at low temperatures [9]. A similar situation was found for rare-earth based superconducting thin films [10-12]. It is reported that the decrease of the critical current density in the longitudinal magnetic field means that the current and the magnetic field are not parallel, i.e., the structure of the superconducting layer is degraded [11-12].

In the present study, the dependence of J_c on the superconducting layer thickness was investigated using the longitudinal magnetic field effect. Furthermore, the crystal orientation of the superconducting coated conductor was investigated to determine the optimum thickness of the superconducting layer.

2. Experimental Methods

The sample used in the present study was a Gd-based coated conductor obtained by forming an intermediate layer on a Hastelloy substrate by ion beam assisted deposition (IBAD), and producing a superconducting layer using pulsed laser deposition (PLD) [13].

Figure 1 shows the dependence of I_c on the position in the sample, measured using the TapeStar™ system. The thickness of the superconducting layer varied depending on position, and four sample regions (labelled A to D) were selected for use in the measurements. The thickness of the layer was estimated based on the number of PLD shots. Table 1 shows the superconducting layer thickness and I_c for each sample region. Since these samples were difficult to measure in the case of high I_c, the width of the superconducting layer was reduced from 12 mm to 0.3 mm by wet etching. Table 2 shows the calculated and measured values for I_c for the samples after etching. Given that all the measured values were within the expected range, etching was considered to be successful.

The J_c-B characteristics of the samples were measured under an applied magnetic field B of up to 0.5 T at 65.0, 70.0, and 77.3 K using a four-terminal method in liquid nitrogen. A superconducting magnet was used to apply the magnetic field [14].

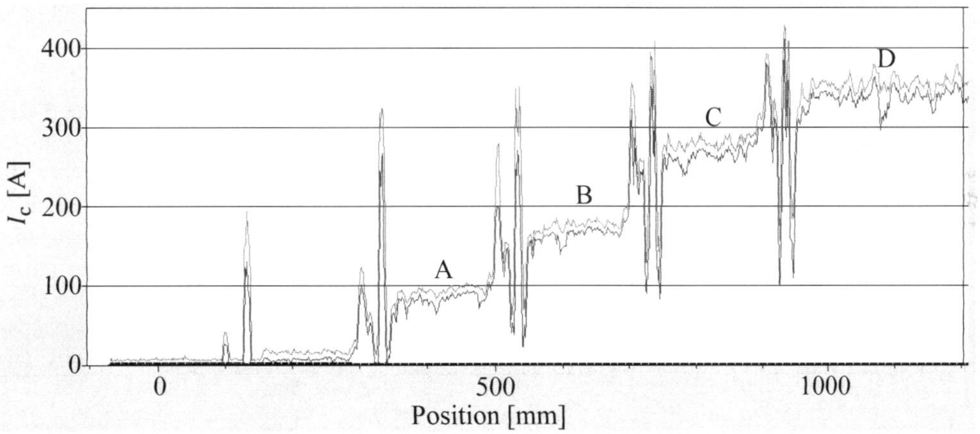

Figure 1. Dependence of minimum (blue) and maximum (red) values of I_c on sample position for superconducting coated conductor.

Table 1. Superconducting layer thickness, minimum critical current $I_{c\,min}$, and maximum critical current $I_{c\,max}$ for four 12-mm-wide samples.

Sample	Thickness [μm]	$I_{c\,min}$ [A]	$I_{c\,max}$ [A]
A	0.6	60	100
B	0.9	140	190
C	1.2	240	300
D	1.5	300	380

Table 2. Calculated and measured I_c values for each sample after etching to a width of 0.3 mm.

Sample	$I_{c\ min}$ [A]	$I_{c\ max}$ [A]	I_c [A]
A	1.50	2.50	**1.77**
B	3.50	4.75	**4.66**
C	6.00	7.50	**6.20**
D	7.50	9.50	**7.67**

3. Results and Discussion

Figure 2 shows the J_c-B characteristics at 77.3 K under longitudinal and transverse magnetic fields. J_c for sample A was smaller than that for the other samples, probably because the superconducting layer was too thin in the former sample and crystal growth was insufficient. As the magnetic field increased, the difference in J_c under longitudinal and transverse magnetic fields increased, indicating that the longitudinal magnetic field effect occurred. J_c for samples B, C, and D decreased in a similar manner under longitudinal and transverse magnetic fields, indicating no layer-thickness dependence. Therefore, it was difficult to determine the optimal thickness of the superconducting layer.

Figure 3 shows the J_c-B characteristics at 65.0 K under longitudinal and transverse magnetic fields. Under a longitudinal magnetic field, J_c for sample D decreased with increasing field strength, unlike the case for samples B and C. This is likely due to increased structural degradation of the superconducting layer as it becomes thicker, as previously reported [9-12]. In this situation, the current and magnetic field are no longer parallel, which weakens the longitudinal magnetic field effect. Therefore, the superconducting layer thickness in sample D is above optimal, whereas that in samples B and C is optimal.

Figure 4 shows the J_c characteristics for different superconducting layer thicknesses t at 0.5 T under longitudinal and transverse magnetic fields at various temperatures. There was no difference among samples B, C, and D at 77.3 K. However, J_c for sample D decreased under a longitudinal magnetic field at lower temperatures of 65.0 and 70.0 K. This means the longitudinal effect is weakened and the structure of the superconducting layer is degraded [11-12]. On the other hand, J_c for sample D did not decrease under a transverse magnetic field at low temperatures, and no clear layer thickness dependence was observed. Thus, by using the longitudinal magnetic field effect, the difference in the J_c-B characteristics due to layer thickness became more clear. From these results, it can be concluded that the optimum layer thickness under a longitudinal magnetic field was 0.9 – 1.2 μm.

For the force-free superconducting electrical transmission cable described in the introduction, the operating temperature is assumed to be 77.3 K. Therefore, even if a superconducting layer thickness of 1.5 μm is used, as in the case of sample D, there should be no problem. Thus, it will be necessary to measure thicker samples in the future. Finally, in order to determine the peak J_c value more accurately, samples should be prepared with even finer thickness intervals.

IOP Conf. Series: Journal of Physics: Conf. Series **1293** (2019) 012017 doi:10.1088/1742-6596/1293/1/012017

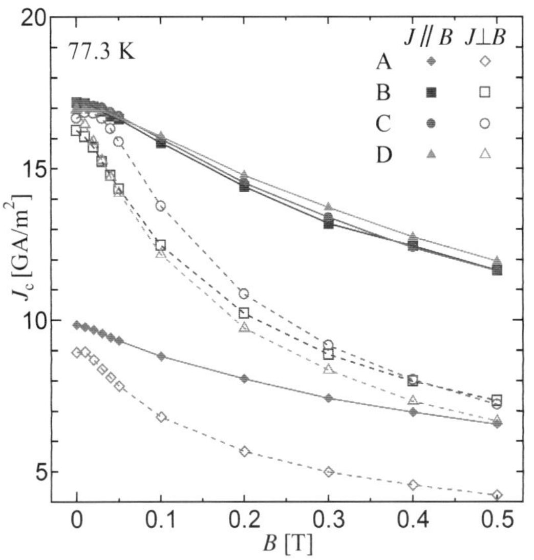

Figure 2. J_c-B characteristics at 77.3 K. The solid lines are for a longitudinal magnetic field, and the broken lines are for a transverse magnetic field. J_c for sample A is smaller than that for the other samples. J_c behaves differently under longitudinal and transverse magnetic fields due to the longitudinal magnetic field effect.

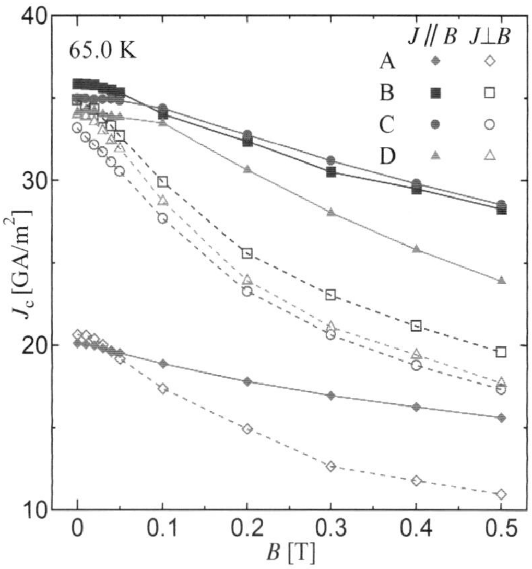

Figure 3. J_c-B characteristics at 65.0 K. As the longitudinal magnetic field increases, J_c for sample D decreases more than that for samples B and C. When the superconducting layer is thick, the structural degradation of the superconducting layer occurs. As the transverse magnetic field increases, J_c for samples B, C, and D decreases in a similar manner.

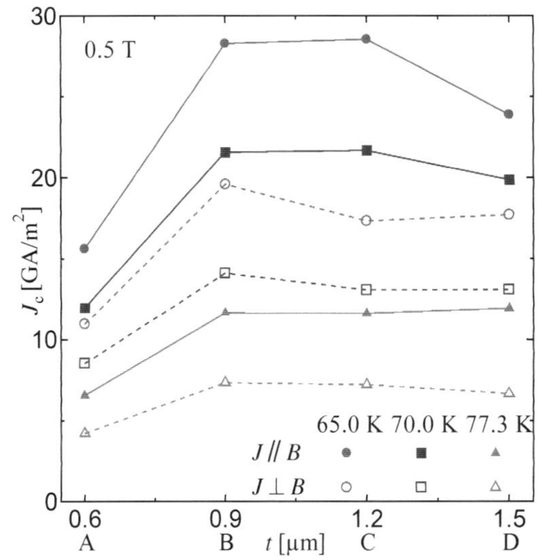

Figure 4. Dependence of J_c on superconducting layer thicknesses t at 0.5 T and various temperatures under longitudinal and transverse magnetic fields. There is no difference in J_c for $t = 0.9 - 1.5$ μm at 77.3 K. However, J_c exhibits a peak for $t = 0.9 - 1.2$ μm at 65.0 K and 70.0 K under a longitudinal magnetic field.

4. Conclusion

The effect of the superconducting layer thickness on the J_c-B characteristics was investigated using the longitudinal magnetic field effect. When the layer was too thin, crystal growth was insufficient and J_c was small. When it was too thick, J_c decreased under high magnetic fields due to the structural degradation of the superconducting layer. This did not occur under a transverse magnetic field. Using the longitudinal magnetic field effect, the optimal superconducting layer thickness for achieving the highest J_c was found to be about 1 μm at 65.0 K. However, to obtain more precise results, it is necessary to measure even thicker samples using smaller thickness intervals.

References

[1] Kato T, Yoshida R, Chikumoto N, Lee S, Tanabe K, Izumi T, Hirayama T and Shiohara Y 2012 *Physica C* **471** 1012-16
[2] Tobita H, Notoh K, Higashikawa K, Inoue M, Kiss T, Kato T, Hirayama T, Yoshizumi M, Izumi T and Shiohara Y 2012 *Supercond. Sci. Technol.* **25** 062002
[3] Fujita S, Muto S, Hirata W, Adachi Y, Yoshida T, Igarashi M, Kakimoto K, Iijima Y, Naoe K, Kiss T, Okada T and Awaji S 2018 *IEEE Trans. Appl. Supercond.* **28** No.4 6600604
[4] Bychkov Yu F, Vereshchagin V G, Zuev M T, Karasik V R, Kurganov G B and Mal'tsev V A 1969 *JETP Lett.* **9** 404
[5] Matsushita T 2011 *TEION KOGAKU* **46** 558–68
[6] Matsushita T 2011 *TEION KOGAKU* **46** 608–14
[7] Yamamoto A, Tanaka H, Shimoyama J, Ogino H, Kishio K and Matsushita T 2012 *Jpn. J. Appl. Phys.* **51** 010105
[8] Matsushita T, Kiuchi M and Otabe E S 2012 *Supercond. Sci. Technol.* **25** 125009
[9] Matsushita T, Otabe E S, Fukunaga T, Kuga K, Yamafuji K, Kimura K and Hashimoto M 1993 *IEEE Trans. Appl. Supercond.* **3** No.1 1045
[10] Safar H, Coulter J Y, Maley M P, Foltyn S, Arendt P, Wu X D and Willis J O 1995 *Phys. Rev. B* **52** R9875-78
[11] Tsuruta A, Watanabe S, Ichino Y and Yoshida Y 2014 *Jpn. J. Appl. Phys.* **53** 078003
[12] Sugihara K, Ichino Y, Tsuchiya Y, Ichinose A and Yoshida Y 2019 *Supercond. Sci. Technol.* **32** 035004.
[13] Lee S, Petrykin V, Molodyk A, Samoilenkov S, Kaul A, Vavilov A, Vysotsky V and Fetisov S 2014 *Supercond. Sci. Technol.* **27** 044022
[14] Otabe E S, Nemoto S, Kiuchi M, Matsushita T, Hayashi T, Fujino K and Ni B 2010 *J. Phys.: Conf. Ser.* **234** 032046

IOP Conf. Series: Journal of Physics: Conf. Series **1293** (2019) 012018 doi:10.1088/1742-6596/1293/1/012018

TDGL Simulation on Angular Dependence of Critical Current Density in Superconductors with Columnar Defects

Rina Yonezuka[1], Yusei Hamada[1], Kazunori Kamiji[1], Kenta Tanimura[1], Takaki Yoshihara[1], Edmund Soji Otabe[1], Yasunori Mawatari[2] and Tetsuya Matsuno[3]

[1] Department of Computer Science and Electronics, Kyushu Institute of Technology, 680-4 Kawazu, Iizuka, Fukuoka 820-8502, Japan
[2] Electronics and Photonics Research Institute, National Institute of Advanced Industrial Science and Technology, 1-1-1 Umezono, Tsukuba, Ibaraki 305-8568, Japan
[3] National Institute of Technology, Ariake College, 150 Higashihagio-Machi, Omuta, Fukuoka 836-8585, Japan

yonezuka.rina707@mail.kyutech.jp

Abstract. We numerically solved the three dimensional time-dependent Ginzburg-Landau (TDGL) equations to visualize the motion of the quantized magnetic flux lines in the superconductor under the transverse magnetic field B. We investigated the angular dependence of the critical current density J_c in the columnar pins using TDGL equations. As a result of simulations, it was confirmed that J_c decreases as increasing the angle θ between the columnar pins and the quantized magnetic flux lines. In particular at $B = 0.4$, J_c sharply decreased at θ between 20° and 30°. This result was compared with the calculation result of the overlapped volume between the quantized magnetic flux and the pins. The calculated result by TDGL equations was in good agreement with the calculated result by the overlapped volume. In addition, J_c was calculated using various pins such as the sphere, the columnar and the plane pins. It was found that plane pins are more effective than pins of other shapes. In addition, although it is experimentally difficult to fabricate, J_c of the star-shaped pin was obtained by the TDGL simulations. It was found that the J_c of the star-shaped pin shows the highest performance among all the kinds of configurations for the columnar pins in high magnetic fields.

1. Introduction

It is well known that the critical current density J_c varies depending on various conditions for pins trapping quantized magnetic flux lines in the type-II superconductors under the transverse magnetic fields [1–3]. The angular dependencies of J_c have been observed by experiments where the columnar pins were introduced by the heavy ion beams injected along the c-axis of oxide superconductors [4]. However, few cases of these dependencies were theoretically described. According to Moore's Law, computer performance has developed rapidly and has become very high, the computer numerical calculation becomes very important in addition to recent theories and experiments. It was reported that the two dimensional numerical calculation based on the time-dependent Ginzburg-Landau (TDGL) equations with a simple point pin [5, 6] in about 30 years ago. Recently, there are some reports on the pinning simulations based on the three dimensional TDGL equations [7, 8]. Further investigations using the TDGL equations with various kinds of shape and structure of pins are needed to understand the pinning effect on the quantized magnetic flux lines. Here, we focused on the pin shape and the angular dependence. Numerical simulations are expected to provide effective methods to characterize the pinning effect concerned. Instead of executing complicated and time-consuming experiments, we adopted to use the numerical simulation because it is desired to obtain some useful scheme efficiently to improve the superconducting properties for J_c.

In this study, we numerically solved the TDGL equations to visualize the movement of the quantized magnetic flux lines of the superconductor under the transverse magnetic field and to find angular dependence of J_c by the columnar pins. In addition, we investigated various kinds of pins of a sphere, a columnar, a plane and a star-shaped defects to obtain the J_c characteristics dependent on various electromagnetic and materials parameters.

2. Methods

The critical current densities J_c in various conditions of the pinning in a small superconducting cube exposed to a transport current and a transverse magnetic field were investigated. The TDGL equations for the superconducting cube was numerically solved by using the Euler method. The dimensions of the cubic material were set to be smaller than the London penetration depth [7]. In this case, the vector potential $\boldsymbol{A}$ depends only on the external magnetic field $\boldsymbol{B}$. This approximation is known as the thin wire approximation, where the GL parameter κ is expected to be sufficiently high. We show the three-dimensional dynamics of the quantized magnetic flux lines by plotting the contour surfaces of the superconducting electron density $|\Psi|^2$, where Ψ is the order parameter.

In this study, the parameters using in the original TDGL equations were normalized using the coherence length ξ and the upper critical field B_{c2} and so on for reducing the number of the constants in the TDGL equations. Hence, the calculation of the order parameter Ψ and scalar potential V are easier compared with the original TDGL equations, which are obtained by solving a set of two differential equations. The reduced TDGL equations are given as follows [8]:

$$\frac{\mathrm{d}\Psi}{\mathrm{d}t} + \mathrm{i}V\Psi + (\mathrm{i}\boldsymbol{\nabla} + \boldsymbol{A})^2\Psi - \Psi + |\Psi|^2\Psi = 0 \quad (1)$$

$$\sigma\nabla^2 V = \frac{1}{2i}(\Psi^*\nabla^2\Psi - \Psi\nabla^2\Psi^*) - \nabla\cdot(|\Psi|^2\boldsymbol{A}) \quad (2)$$

where, σ is the normal conductivity.

We considered a superconducting cube of which side length is 10ξ in the vacuum. In addition, 4 columnar pins of diameter ξ were introduced with the distance d of pins as shown in Figure 1. Here, we define the order parameter Ψ as 0 inside of the pins [5]. This approximation $\Psi = 0$ seems to be too strong setting for the pinning. However, the calculation result of J_c was not so large compared with that taken into account of the proximity effect in the normal conducting region.

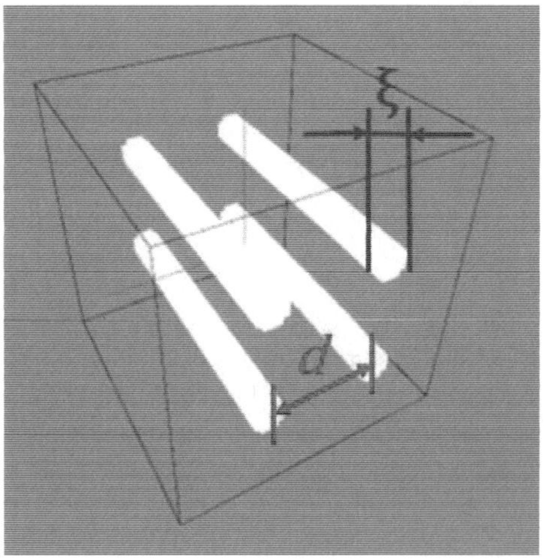

Figure 1. Geometry of the superconducting cube and the columnar pins.

The boundary condition for the electric current density $\boldsymbol{J}$ is that its normal component is zero at the surfaces of the cube. The current density and the magnetic field are applied to the y-axis and the z-axis, respectively. That is $(0, J, 0)$ and $(0, 0, B)$, respectively. Hence, the vector potential can be given by $(A_x, A_y, A_z) = (0, xB, 0)$ for the transverse magnetic field. J and B at each time are kept constant at a normalized value. Figure 2 shows the example of the quantized magnetic flux lines each of which is trapped by the 4 columnar pins, where pixels of $|\Psi|^2 < 0.1$ is colored. The hue of the color is corresponding to the phase of the order parameter. It is to be noted that almost all parameters such as B, J and Ψ are normalized. The magnetic field is normalized by upper critical field, B_{c2}. The normalized current density of 0.385 means the pair breaking current density. And the order parameter is normalized by the order parameter of equilibrium point.

Figure 3 shows the superconducting cube with different angle of the columnar pins. It is assumed that $\theta = 0°$ represents for the case when the pin and the magnetic field are in parallel, as shown in Figures 3(b), (c), and (d). Instead of rotating the magnetic field, the pin is rotated in this study to suppress the surface effects. Simulations were made with the external magnetic field $B = 0.1, 0.2, \cdots, 0.5$, the electric current density $J = 0.01, 0.02, \cdots, 0.30$, and angle $\theta = 0, 10, \cdots, 90°$.

In addition, we calculate the overlapped volume between the quantized magnetic flux and the pins. As shown in Figure 4, the quantized magnetic flux and the pin are assumed to be the quadrangular prisms. That is, the quantized magnetic flux has a cross-sectional area of $2\xi \times 2\xi$ and the pin has a cross-sectional area of $2r \times 2r$. And, we start to consider the case where the length of superconducting region along z-axis, $L_z = \infty$. Since the bold line is $r/\sin\theta$, the area of the red parallelogram is $4r\xi/\sin\theta$. In the case of $r < \xi$, the overlapped volume between the quantized magnetic flux is $8r^2\xi/\sin\theta$. Then, the case where L_z is finite is also analytically considered. Since, the detail of the calculation is complicated, equations are omitted for the case of the finite length of the superconducting region.

In order to investigate the pinning effects on J_c for different shapes of the pins, we investigated for a sphere pin as a point pin, and a plane pin. Figure 5 shows the superconducting cube with sphere pins. Figures 5(b), (c), and (d) show the number of pins for 1, 4 and 8, respectively. Figure 6 shows the superconducting cube with a plane pin. Figures 6(a) and (b) show the thickness of the plane pin for the thickness $r = 0.6$, $r = 1.0$, respectively. The pin thickness r is normalized with ξ. Figure 7 shows the superconducting cube with a star-shaped pin. Although it is experimentally difficult to fabricate the shape of the star-shaped pin, it is relatively easy to execute simulations with that kind of pins. This is one of the advantages of the TDGL numerical study. A number of simulations were made with various external magnetic fields $B = 0.05, 0.10, \cdots, 0.60$, the electric current densities $J = 0.01, 0.02, \cdots, 0.30$.

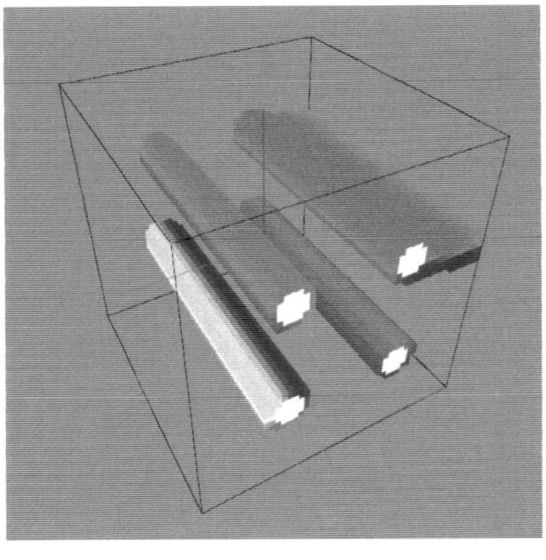

Figure 2. Simulation example: the flux lines are trapped by the columnar pins. Color represents the phase of the order parameter.

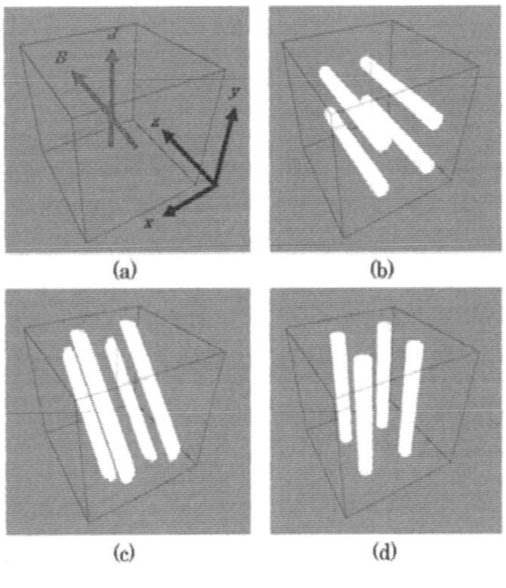

Figure 3. Superconducting cube with different angle of the columnar pins: (b) $\theta = 0°$, (c) $\theta = 45°$, (d) $\theta = 90°$.

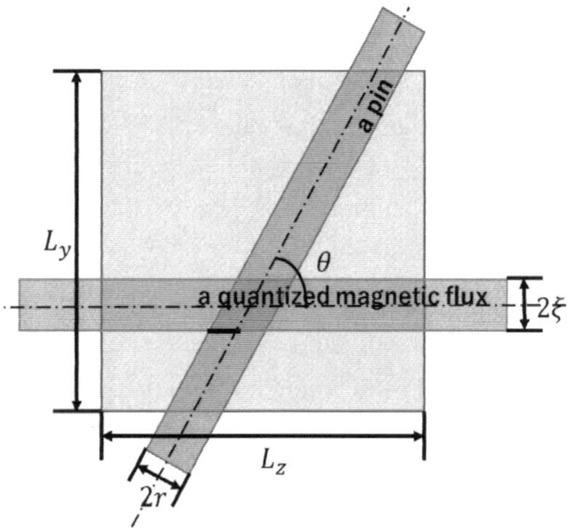

Figure 4. Overlapped volume between a quantized magnetic flux and a pin for the finite length of the superconducting region.

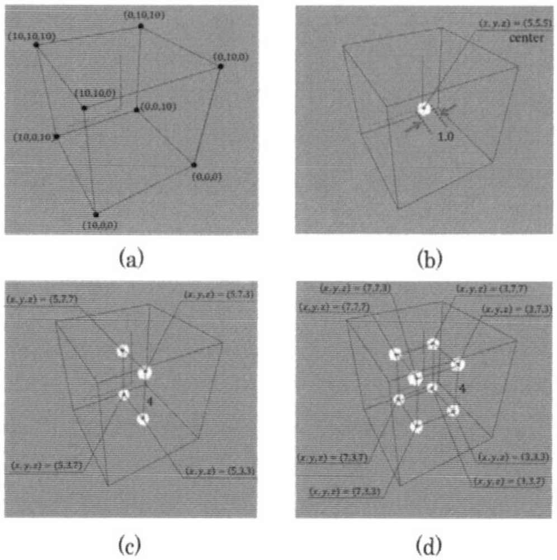

Figure 5. Superconducting cube (a) with the sphere pins: (b) 1 pin, (c) 4 pins, (d) 8 pins.

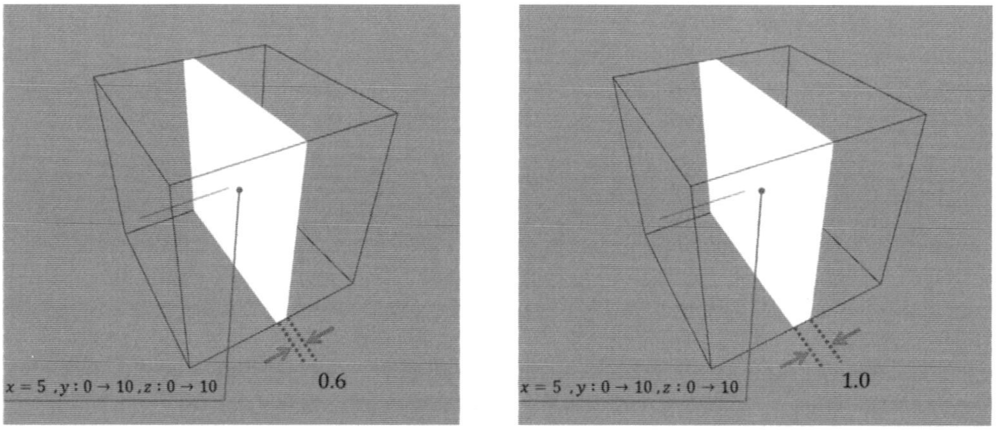

(a) (b)

Figure 6. Superconducting cube with the plane pin: (a) $r = 0.6$, (b) $r = 1.0$.

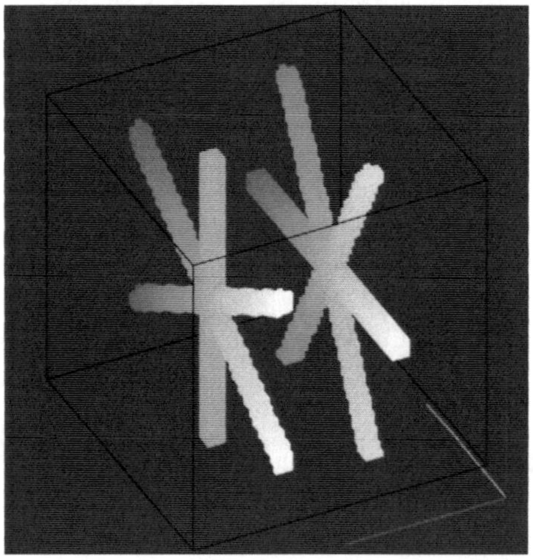

Figure 7. Superconducting cube with the star-shaped pins.

3. Results and discussion

Figure 8 shows the numerical results of the angular dependence of J_c at high magnetic fields: $B = 0.3, 0.4, 0.5$ for the columnar pins. The critical current density J_c decreases as increasing the angle. J_c at $B = 0.4$ abruptly decreases when θ is between 20° and 30°. Although the quantized magnetic flux lines could be trapped by the columnar pins for $\theta = 0 - 20°$, the quantized magnetic flux lines could not be trapped and slip through at high magnetic fields. We also examined the case of low magnetic fields: $B = 0.1, 0.2$, then it was found that the result that J_c increases as increasing the angle. These results of high and low cases seem to be a contradiction, which can be solved by the following explanation. At low magnetic fields, the space between the quantized magnetic flux lines is too large and only one flux line appeared in the small superconducting cube. Then the probability that the quantized magnetic flux lines slip through at $\theta = 0°$ was high. On the other hand, the quantized magnetic flux lines is likely to be trapped at $\theta = 90°$, since the pin and the quantized magnetic flux lines always cross each other. Therefore, we found that J_c increases as increasing the angle at low magnetic fields. This result is due to the limitation of the present calculation with the small superconducting cube.

Figure 9 shows the numerical results of the angular dependence of J_c at $B = 0.4$ and the calculation result of the overlapped volume between the quantized magnetic flux and the pins. The calculation result of the overlapped volume sharply decreases at $10° \leq \theta \leq 30°$ and then gently decreases. Thus, the simulated result based on the TDGL equations is in good agreement with the analytically calculated result by the overlapped volume. Therefore, it is concluded that the angular dependence of J_c is determined by the overlapped volume between the quantized magnetic flux and the pins.

Figure 10 shows the J_c-B property for the sphere pins. In the range of $0.35 \leq B \leq 0.6$, it is found that J_c is relatively large in the case of 8 sphere pins. This can be due to the fact that as increasing B, the number of the quantized magnetic flux lines increases by entering the superconducting region. In addition, J_c of the model with 4 sphere pins became the largest at $B = 0.3$, because the number of the quantized magnetic flux lines is 2 and the flux line lattice spacing is closed to the distance between the sphere pins, and the matching effect occurs. This effect is known as the matching effect with periodic modulations of the pins [9]. Thus, the J_c shows the peak around $B = 0.3$.

Figure 11 shows the J_c-B property for the plane pin. It is found both results $r = 0.6$ and 1.0 are the same. Therefore, the pinning works effectively both at $r = 0.6$ and at $r = 1.0$.

Those results of J_c-B property are reasonable as expected by theories and experiments. The advantage for using numerical calculation is easy to change the shape of pins compared with the theories and experiments. Figure 12 shows the J_c-B property for the star-shaped pin, the sphere pins and the plane pin. It is found that the sphere pin shows the highest J_c at low magnetic fields, and the plane pin shows the highest J_c at middle magnetic fields. At high magnetic fields, the star-shaped pin shows the highest J_c. This is because the volume of the star-shaped pin is lower than that of the plane pin, and the degradation of superconductivity of the star-shaped pin is less than that in the case of the plane pin. Therefore, it is concluded that the numerical calculation by using TDGL has high efficiency to find a new type of shape or configuration for pins to design high J_c superconducting materials.

Figure 13 shows the angular dependence of the critical current density at $B = 0.4$. It is confirmed that the angular dependence is almost flat. And the critical current density slightly negative drops at $\theta = 30, 90°$. This is due to the 6-fold rotational symmetry of the star-shaped pin. It was reported that Gd-based coated superconductor with the crossed columnar defects produced by heavy-ion irradiation shows better angular dependence of the critical current density [4]. Therefore, if the number of crossed columnar defects increases, the angular dependence of the critical current density becomes more flat according to the present work.

On the other hand, it is also true that there is a limit to the present calculation. For example, in Figs. 10 to 12, the value of the normalized upper critical field is about 0.7 – 0.8, which is not unity. The main reason is the limitation of calculation space as $10\xi \times 10\xi \times 10\xi$ due to the calculation time and cost. Hence, the surface effect is remarkable in the present calculation. In addition, it is difficult to discuss quantitatively with the present numerical calculation. Therefore, it is desired to obtain large calculation space by using such as a parallel calculation method and so on.

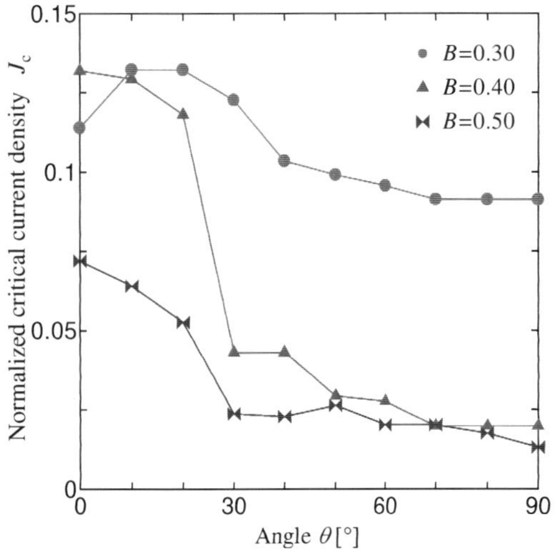

Figure 8. Numerical results of the angular dependence of J_c at $B = 0.3, 0.4, 0.5$ of columnar pins.

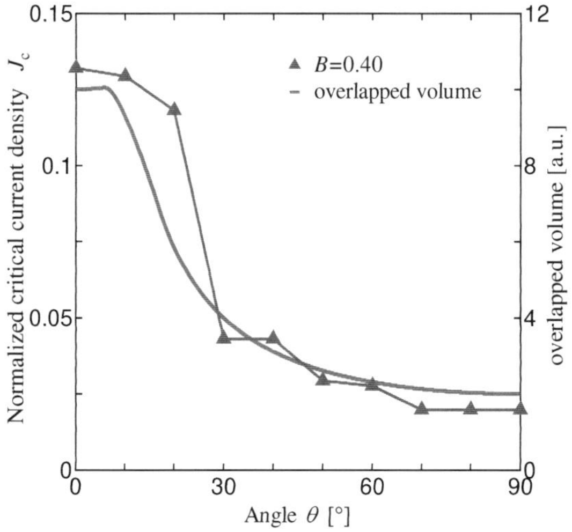

Figure 9. Numerical results of the angular dependence of J_c at $B = 0.4$ and the calculation result of overlapped volume between the quantized magnetic flux and the columnar pins.

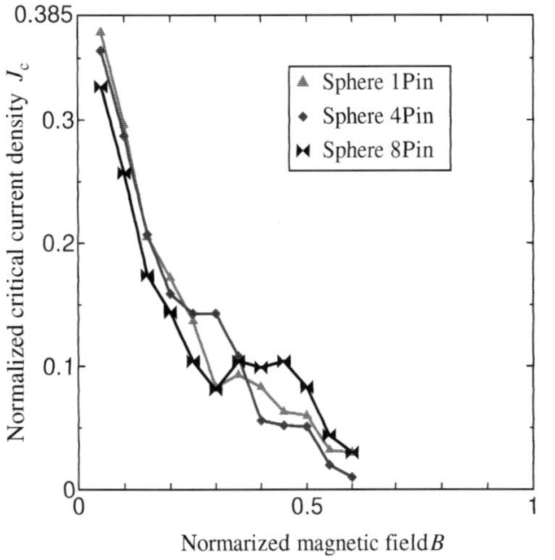

Figure 10. Magnetic field dependence of critical current density in the case of the sphere pin.

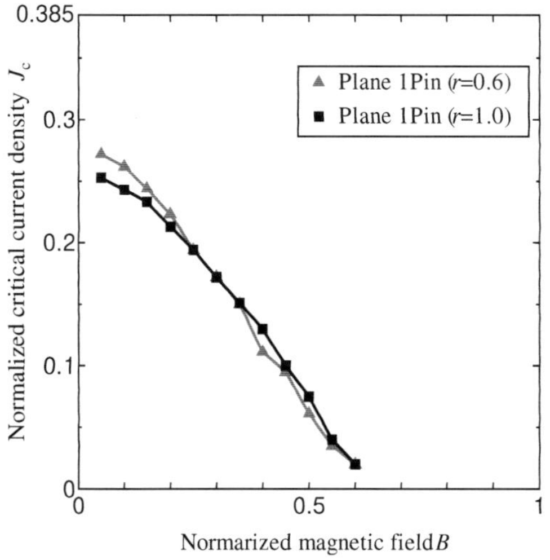

Figure 11. Magnetic field dependence of critical current density in the case of the plane pin.

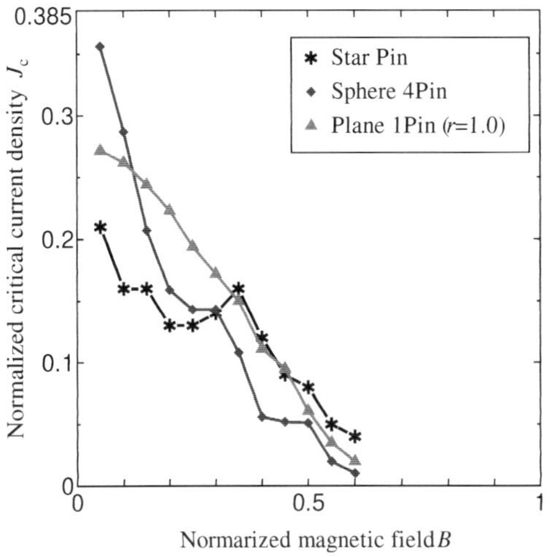

Figure 12. Magnetic field dependence of critical current density in the case of the star-shaped pin, the sphere pins and the plane pin.

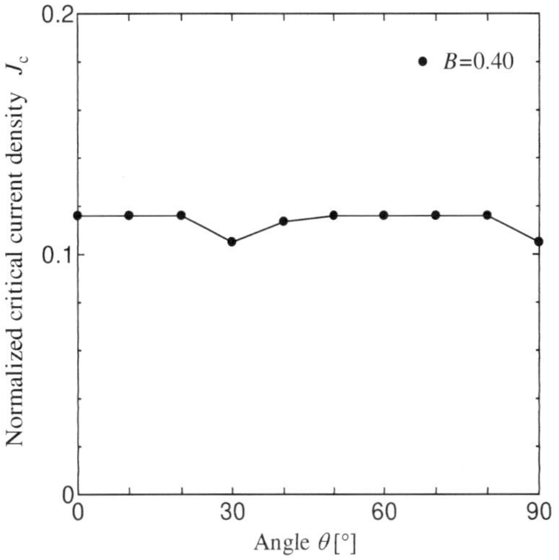

Figure 13. Angular dependence of critical current density at $B = 0.4$.

4. Conclusion

We investigated the angular dependence of J_c in the columnar pins using simulations based on the three dimensional time-dependent Ginzburg-Landau (TDGL) equations. The critical current density J_c decreases as increasing the angle. Especially J_c at $B = 0.4$ abruptly decreases in the range of $\theta = 20° - 30°$. This numerical result was compared with the calculation result of the overlapped volume between the quantized magnetic flux and the pins. And, the simulation result by TDGL agrees well with the calculated result by the overlapped volume. In this way, experimentally observed angular dependence of J_c was reproduced by the simulations in this study.

The numerical simulations based on the TDGL equations were also applied for the sphere pins and the plane pin. In the sphere pins, it is found that J_c in the model of 8 sphere pins is relatively large in the range of $0.35 \leq B \leq 0.6$. This can be due to the fact that as increasing B, the number of the quantized magnetic flux lines increases by entering the superconducting region. In addition, when $B = 0.3$, J_c of the model of 4 sphere pins became the largest. Furthermore, J_c in the model of 4 sphere pins became the largest at $B = 0.3$, because the number of the quantized magnetic flux lines is 2, and the flux line lattice spacing is closed to the distance between the sphere pins, and the matching effect occurs. Thus, the J_c shows the peak around $B = 0.3$. In the plane pin, both $r = 0.6$ and $r = 1.0$ show similar characteristics. Therefore, the pinning works effectively both at $r = 0.6$ and at $r = 1.0$. At high magnetic fields, the star-shaped shape pin shows the highest J_c. The TDGL simulation was shown to have considerable advantages to find a new type of pin shape to design engineering superconducting materials of high J_c with good properties with respect to various kinds of parameter dependence. That is, we predicted the J_c-B characteristics with the defects that have been neither fabricated nor tested by real experimental measurements. On the other hand, because of computational limitation that the computation space is small, unnatural results were obtained. Therefore, it is necessary to extend the computation space using parallel computation and so on.

References

[1] Tinkham M 1996 *Introduction to Superconductivity 2nd ed.* (New York: McGraw-Hill)
[2] Matsushita T 2014 *Flux Pinning in Superconductors* (Springer)
[3] Blatter G, Feigel'man M V, Geshkenbein V B, Larkin A I and Vinokur V M 1994 *Rev. Mod. Phys.* **66** 1125
[4] Sueyoshi T, Iwanaga Y, Fujiyoshi T, Takai Y, Mukaida M, Kudo M, Yasuda K and Ishikawa N 2017 *IEEE Trans. Appl. Super.* **27** 8001305
[5] Kato R, Enomoto Y and Maekawa S 1991 *Phys. Rev. B* **44** 6916
[6] Machida M and Kaburaki H 1993 *Phys. Rev. Lett.* **71** 3206–09
[7] Sadovskyy I A, Koshelev A E, Phillips C L, Karpeyev D A and Glatz A 2015 *J. Com. Phys.* **294** 639–654
[8] Sadovskyy I A, Koshelev A E, Glatz A, Ortalan V, Rupich M W and Leroux M 2016 *Phys. Rev. Applied* **5** 014011
[9] Haag L T, Zechner G, Lang W, Dosmailov M, Bodea M A and Pedarnig J D 2014 *Phys. C* **503** 75

IOP Conf. Series: Journal of Physics: Conf. Series **1293** (2019) 012019 doi:10.1088/1742-6596/1293/1/012019

Dynamics of a vortex system in a layered high-temperature superconductor under a pulsed current impact

I A Rudnev, A N Maksimova, A N Moroz and V A Kashurnikov

National Research Nuclear University MEPhI (Moscow Engineering Physics Institute), Kashirskoe sh. 31, 115409 Moscow, Russia

iarudnev@mephi.ru

Abstract. By using the Monte Carlo method, electric field relaxation has been numerically studied in a 2D layered HTS with defects. Time dependences of the electric field strength inside the sample for an instantaneous rise of electric current have been calculated for different temperatures. The influence of temperature on the shape of such dependences has been demonstrated. The impact of a rectangular current pulse on the vortex system has been studied. The value of the response field strength and its dependence on the amplitude and duration of the pulse have been calculated as well as the impact of pulse duration on the shape of the response.

1. Introduction

One of the application areas of both low-temperature and high-temperature superconductors are the development of high-speed current-limiting devices as well as current switches. In current limiters it is assumed to limit the sharp impulse of current during a short circuit. High-speed current switches can be used, for example, for impulse output of power from the energy storage systems. It is obvious that during a sharp pulse of current flowing through a superconductor, a highly non-equilibrium magnetic flux distributions in the superconductors (that is non-equilibrium Abrikosov vortices states) are created. Nowadays, great attention is paid to the study of processes of creation and relaxation of such non-equilibrium vortex systems. For example, in paper [1], transport characteristics of a 2*H*-$NbSe_2$ superconductor have been measured, and a history dependence associated with the so-called "peak effect" has been demonstrated. The existence of metastable vortex lattice states has also been shown. Paper [2] is dedicated to studying the vortex lattice in the Bragg glass state. An exponential time dependence of voltage in the sample under the impact of pulsed current has been demonstrated with the exponent characterizing the deviation from equilibrium. In paper [3], energy dissipation induced in the sample by the vortex motion has been studied by using short and long current pulses.

A significant number of studies have been dedicated to the influence of current rise-rate dI/dt on the current-voltage (*I-V*) curve. Some research has been done on MgB_2 samples (e.g. in [4]), as well as on high-temperature BSCCO [5] and YBCO [6] superconductors (HTS). The *I-V* curve irreversibility and staircase behaviour have been demonstrated. As has been demonstrated in theoretical and experimental papers [4–7], the *I-V* curve irreversibility becomes generally more prominent with the growth of dI/dt and may be due to the granular structure of samples, the vortex dynamics under strong pinning conditions [3], and different pinning regimes in a polycrystalline sample. In paper [5], voltage relaxation has been observed after an instantaneous stop in changing the current during *I-V* curve measurements. It has been shown that the sign of such relaxation depends on the sign of d*I*/d*t*. Magnetic field has been demonstrated to suppress the *I-V* curve hysteresis.

Since the effects considered in the mentioned papers can generally be associated with the dynamics of the vortex system of the sample in the mixed state and the vortex interactions with pinning cites, a detailed study of the vortex structure dynamics is an important task. The use of the Monte Carlo method is rather convenient in this case. It has already been successfully used to calculate magnetic and transport characteristics of HTS materials under applied transport current or external magnetic field (*H*) in papers [8–11]. For example, an S-like *I-V* curve peculiarity has been observed for a superconductor with ferromagnetic defects in magnetic field [10,11]. The mentioned peculiarity could not be reproduced when reversing the current rise, which is an example of hysteresis. The method developed by the authors of papers [8–11] can also be used to study the relaxation effects in a vortex system under the impact of instantaneous rise and fall of electric current through the sample. The purpose of this paper is to numerically study the vortex dynamics under the impact of a pulsed current depending on temperature and the parameters of the pulse. We assume for simplicity that $H=0$, i.e. the vortices are generated by the self-field of transport current.

2. Method for calculations. The critical current of HTS with defects at different temperatures

In this paper, the electric field strength – current density (*E-j*) curves and energy losses in the superconductor have been calculated by using the method developed in [8,9]. Transport losses are due to the movement of vortex lines under the impact of transport current and can be obtained by calculating the number of opposite-sign vortex pairs that have annihilated in the sample centre. By using the value of energy released in those annihilations (which equals the work of Lorenz force during vortex-antivortex movement to the centre of the superconductor) and knowing the value of transport current, one can calculate the electric field strength *E* in the superconductor. Since there is no real time in the simulations, a normalization of the calculated *E-j* curve to an experimental one is needed.

The electric field strength *E* is related to the pair annihilation energy and the magnetic field H_I created by the transport current on the edge of the sample by the following equation: $E = \alpha \frac{Q}{H_I}$, where α is the proportionality coefficient. It is assumed that a certain number of annihilated pairs corresponds to the start of the flux flow regime and to the critical electric field strength of 1 μV/cm (used in most experiments). Based on this, the coefficient α can be determined. During calculations of the $E(t)$ dependence, the released energy is memorized regularly over a certain number of Monte Carlo steps and it is assumed that the number of annihilated pairs is proportional to the calculation time. Accordingly, the proportionality coefficient α is recalculated as $\alpha \to \alpha \frac{\Delta T}{\Delta t}$, where Δt is the time interval between two neighbouring points on the $E(t)$ dependence, and ΔT is the calculation time of one point on the *E-j* curve. The time equivalent of one Monte Carlo step has been estimated as $\sim 10^{-10}$ – 10^{-8} s from calculations of relaxation of residual magnetization in HTS. For further analysis, we need to know the values of current density *j* that corresponds to the values of electric field strength *E* of 1 μV/cm, 2 μV/cm, and 3 μV/cm in the sample. We also need to know how those values depend on temperature and the pinning parameters (namely the defect potential well depth) in the HTS. To reduce the influence of calculation error, all simulations have been carried out for the same defect configuration with density $n_d = 6.7 \cdot 10^9$ cm^{-2} (the total number of defects per sample is 2020). The sample width $d = 6$ μm.

Figure 1(a) shows a series of *E-j* curves for a sample with defects of potential well depth $c = 0.01$ eV at different temperatures *T*. It can be seen that with the increasing temperature, the slope of the *E-j* curves decreases, which is in a qualitative agreement with experiment. The thermal dependences of critical current density j_c (determined by the 1 μV/cm criterion) are presented in figure 1(b) for different values of defect potential well depth. It can be seen that the critical current density decreases when the potential well depth becomes comparable to *T*. From this point on, the potential well depth was kept constant at $c = 0.01$ eV for convenience.

3. The time dependences of electric field strength at instantaneous rise of current

Let us consider the case when a current of critical density j_c (which we have determined from the *E-j* curves) is momentarily switched on (which is equivalent to a rectangular current pulse of infinite duration). Figure 2(a) shows the superconductor's response curve in the form of time dependences of the electric field inside the sample at different temperatures. These dependences have peculiarities: a sharp increase in the field strength up to values significantly exceeding 1 μV/cm in the initial section of the plot. Then the electric field slowly saturates to an average value that also exceeds 1 μV/cm, at least for the considered calculation time. The sharp increase in the initial section was also observed on the time dependences of the number and energy of vortices in the sample so it can be explained by an avalanche-like process of vortices penetrating the sample and further annihilating their counterparts in the sample centre. These peculiarities disappear at temperatures higher than 10 K (in figure 2(a), they are already not visible as the electric field saturates almost immediately) since the vortices can no longer be effectively pinned by defects at this temperature, and there is no avalanche-like process.

Figure 2(b) shows the main response parameters: E_{pulse} is the average electric field value for large time values, t_f is the front duration, i.e. the time passed from the moment of switching on the current until the electric field saturated. For each time dependence, t is shown as the number of Monte Carlo steps.

Figures 3(a) and (b) show the thermal dependences of the mentioned response parameters. Additionally, three different pulse amplitudes have been considered: the values of current density that correspond to $E = 1$ μV/cm, 2 μV/cm, and 3 μV/cm on the pre-calculated *E-j* curves (figure 1(a)). It can be seen that in all three cases, the thermal dependence of t_f (figure 3(a)) is descending in average, and the sharpest decrease can be seen at $T = 10$ K, i.e. the temperature value at which the $E(t)$ peculiarities disappear (figure 2(a)). In figure 3(b), E_{pulse} shows a weak temperature dependence, and the value of electric field jump dE decreases rather sharply with the increasing temperature.

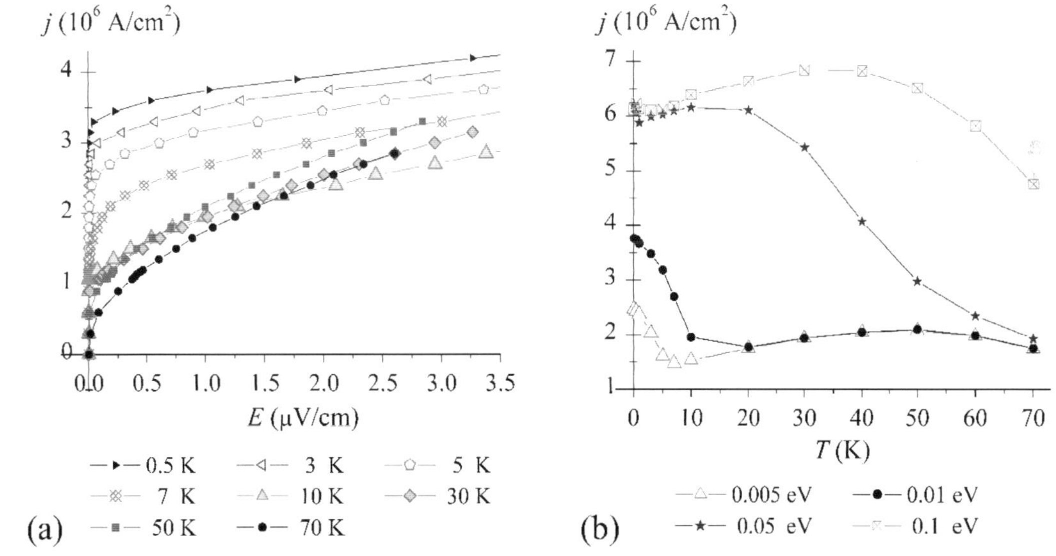

Figure 1. Transport characteristics of HTS samples with defects of density $n_d = 6.7 \cdot 10^9$ cm^{-2}: (a) *E-j* curves of a sample with a fixed defect potential well depth $c = 0.01$ eV at different temperatures from 0.2 K to 70 K; (b) The j_c versus T dependence for different defect potential well depths from 0.005 eV to 0.1 eV.

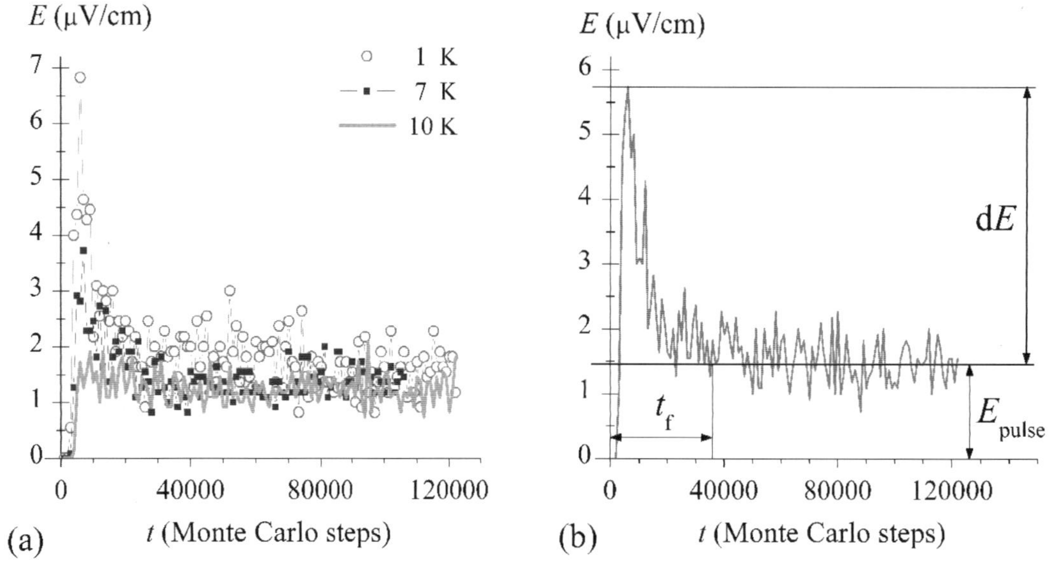

Figure 2. Time dependences of electric field strength in the sample for an instantaneous rise of current (a rectangular pulse of infinite duration): (a) for different temperatures from 0.1 K to 10 K; (b) for $T = 0.1$ K with the main response parameters shown: t_f is the front duration (the time necessary for the electric field to achieve a constant value), E_{pulse} is the average electric field value, dE is the value of field jump during rise of current (relative to E_{pulse}).

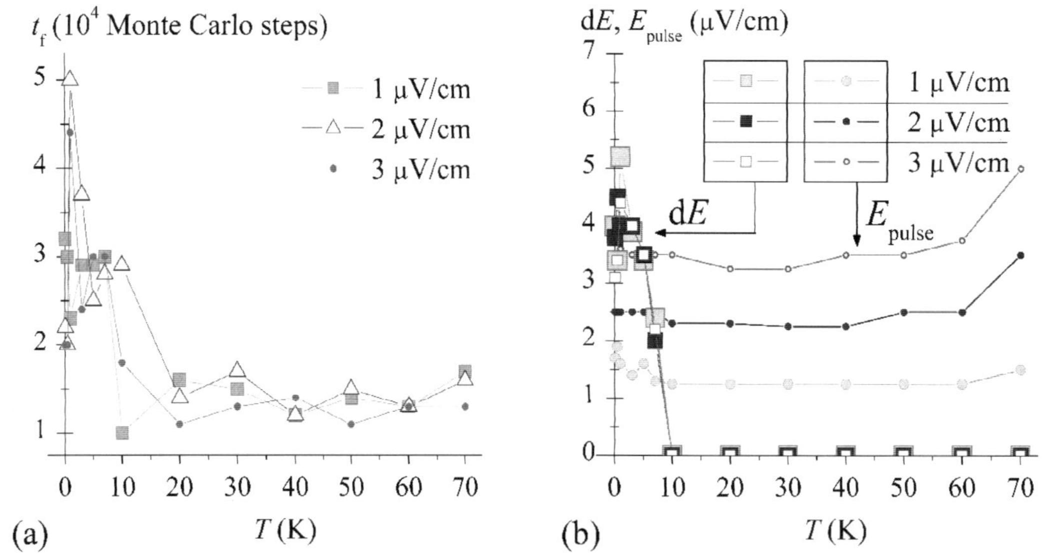

Figure 3. Thermal dependences of the response parameters for the instantaneous rise of current of densities corresponding to the electric field values of 1 μV/cm, 2 μV/cm, and 3 μV/cm obtained from E-j curves: (a) the front duration t_f; (b) the average field strength E_{pulse} and field jump dE.

4. Finite rectangular current pulses

Let us now consider the case when a rectangular current pulse of finite duration is acting on the sample. For that, we have calculated the response curves for pulses of 3 different amplitudes (same as before) and 2 different duration times: a short pulse (duration time equivalent of 7000 Monte Carlo steps) and a long pulse (time duration of 200000 steps). The temperature was kept constant this time at T = 1 K. The response curve shapes are shown in figures 4 (a), (b) and (c) for each pulse amplitude in ascending order. The thick dashed lines in all figures are for short pulses and the thin solid lines are for long pulses. The insets in each figure show the scaled parts of the responses to short pulses. If we compare the shapes of the response curves for long pulses to the ones presented in figure 2, it can be seen that the initial parts of the dependences match.

It can also be seen that in the case of short current pulse there is no saturation on the response curve, which means that the pulse duration is less than the front duration. It should be noted that increasing the amplitude of the pulse only changes the peak value of electric field strength and has no effect on the time it takes for the field to drop to zero when the current is switched off (fall time). This can be clearly seen in the insets in each figure. In the case of a long pulse, the electric field saturation can be seen throughout the pulse duration, and after the current is switched off, the field strength drops to zero. Just as for the case of short pulse, the fall time does not depend on the amplitude of the pulse. Moreover, it appears to be of the same value as for the short pulse. It should be noted that, for now, the nonzero fall time is merely a peculiarity occurring in our simulations which requires further research.

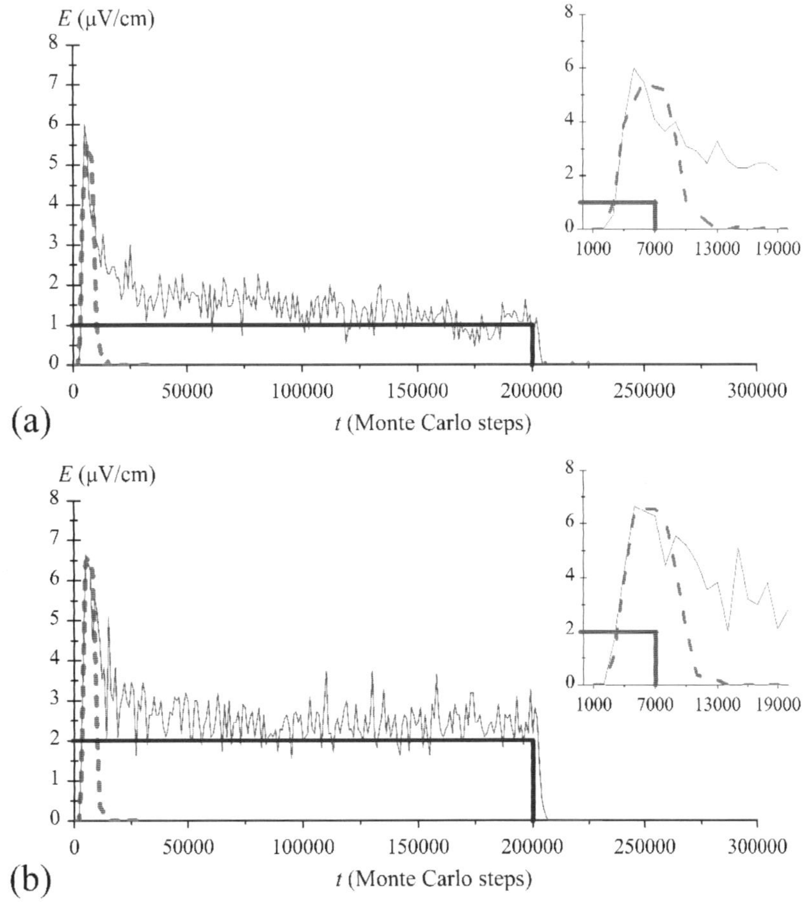

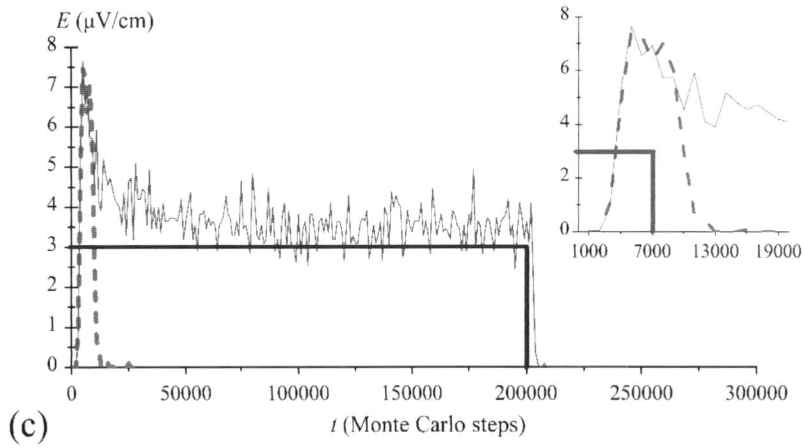

Figure 4. $E(t)$ curves (response curves) for finite rectangular current pulses of 3 different amplitudes and 2 duration times. The thick dashed lines denote the response to the short pulse (7000 Monte Carlo steps long), and the thin solid lines denote the response to the long pulse (200000 steps long). The scaled insets show the response to short pulses in more detail. Solid horizontal lines schematically show the current pulses of each duration time.

5. Conclusion

By using the Monte Carlo method, time dependences of the electric field strength in the superconductor have been calculated for an instantaneous rise of current of a given value. These dependences have demonstrated a sharp increase right after the current is switched on with a further saturation to an equilibrium value that exceeds the one that corresponds to this current of E-j curves. The sharp field jumps have been shown to disappear with the increase of temperature. Thermal dependences of the response parameters of HTS to the instantaneous rise of current have been obtained for different current values. The response curves of the superconductor to rectangular current pulses of different amplitudes and durations have been calculated. It has been shown that for pulse durations exceeding the front duration, the response curve represents the initial section of the electric field versus time dependence for an instantaneous rise of current. It has also been shown that the relaxation time does not depend on the duration of the initial pulse for the considered amplitudes and durations of pulses. The results provided by our simulations can be useful for understanding of physical processes occurring in non-equilibrium vortex systems as well as for picking out the most optimal parameters (such as temperature regimes and maximum current loads) in the design of superconducting high-speed current limiters and switchers.

Acknowledgements
The research was done under support of Russian Foundation for Basic Research (grant 17-29-10024).

References
[1] Henderson W, Andrei E Y, Higgins M J and Bhattacharya S 1996 *Phys. Rev. Lett.* **77** 2077–80
[2] Du X, Li G, Andrei E Y, Greenblatt M and Shuk P 2007 *Nat. Phys.* **3** 111–4
[3] Xiao Z L, Andrei E Y, Shuk P and Greenblatt M 2001 *Phys. Rev.* B **64** 094511
[4] Olutaş M, Kiliç A, Kiliç K and Altinkok A 2012 *J. Supercond. Nov. Magn* **25** 753–71
[5] Ušák P, Janšák L' and Polák 2000 *Physica* C **350** 139–46
[6] Altinkok A, Kiliç K, Olutaş M and Kiliç A 2012 *J. Supercond. Nov. Magn* **27** 651–65
[7] Zhang P, Ren C, Ding S Y, Ding Q, Lin F Y, Zhang Y H, Luo H and Yao X X 1999 *Supercond. Sci. Technol.* **12** 571–6

[8] Odintsov D S, Rudnev I A and Kashurnikov V A 2006 *JETP* **103** 66–76
[9] Rudnev I A, Odintsov D S and Kashurnikov V A 2008 *Phys. Lett.* A **372** 3934–6
[10] Kashurnikov V A, Maksimova A N and Rudnev I A 2014 *JETP Lett.* **100** 439–45
[11] Kashurnikov V A, Maksimova A N and Rudnev I A 2014 *Phys. Sol. State* **56** 894–911

IOP Conf. Series: Journal of Physics: Conf. Series **1293** (2019) 012020 doi:10.1088/1742-6596/1293/1/012020

Critical states in two overlapped rectangular superconducting plates

Shinsuke Ooi and Masaru Kato

Department of Mathematical Sciences,
Osaka Prefecture University, 1-1,
Gakuencho, Nakamozu, Sakai, Osaka 599-8531, Japan

E-mail: sxb01036@edu.osakafu-u.ac.jp

Abstract. We study vortex penetration into two-layer structures of superconducting plates under a perpendicular magnetic field. We solve the heat transport equation and the Maxwell equations with the current-voltage relation for superconductor, simultaneously, and obtain magnetic flux and current densities. We show how magnetic flux structure depends on the structure, especially distance of two-layer of superconductors.

1. Introduction

When an external perpendicular magnetic field applied to a superconducting plate is increased, magnetic flux penetrates into the superconducting plate from its edges. Depending on a ramping rate of increased applied field, there are two kinds of vortex penetration: a critical state and a vortex avalanche [1]. If the ramping rate of the applied magnetic field is small, the critical state appears. How this critical state appears is explained as follows. When vortices penetrate into superconducting plate, the Lorentz force from the Meissner shielding current, which flow along an edge of the superconducting plate, drive vortices into the inside of the superconducting plate. Then, if there are microscopic defects in the superconductor, vortices may be pinned by these defects. This pinning force from defects acts against the driving Lorenz force. The local balance between two forces creates a metastable equilibrium state, which is the critical state. In the critical state, magnitude of the shielding current density is adjusted to the critical current density.

On the other hand, the ramping rate is large, the vortex avalanche occurs. This phenomenon is caused by thermal fluctuation at finite temperature. The vortex, which is pinned in the pinning center, jumps to another pinning center because of the driving force and the thermal fluctuation. This movement generates heat, and temperature locally rises. Then superconductivity is weakened. Therefore, vortices further penetrate into the superconductor. Repeating this process, there appears the vortex avalanche.

The critical states and the vortex avalanches in complicated three-dimensional structures, which are composed of several stacked superconducting strips, were investigated experimentally [2,3]. In a two-layer structure, unlike single plate case, vortices mainly penetrate from overlap regions between the two superconductors, causing some dotted or linear vortex avalanches. Like the conventional vortex avalanche, there are heat generation by vortex motion and heat conduction, and therefore shapes of avalanches depend on the distance and overlap of superconducting strips, and the size of the superconducting strip.

In this study, we investigate the critical states rather than the vortex avalanche in a two-layer structure of two superconducting plates, when the ramping rate of magnetic field is not large, and heat generation and temperature rise are small. We solve the heat transport equation and the Maxwell equation with a current-voltage relation for superconductors with three-dimensional finite element method. We obtain distribution of the magnetic flux density and the current density.

2. Method

We solve following three equations. The first one is the heat transport equation,

$$C\frac{\partial T}{\partial t}=\frac{\partial X}{\partial x}+\frac{\partial Y}{\partial y}+\frac{\partial Z}{\partial z}-\alpha(T-T_0)+\boldsymbol{\mathcal{J}}\cdot\boldsymbol{E} \tag{1}$$

Here, C is a specific heat, $T(x,y,z,t)$ is temperature of the superconductor, t is a time, and $X(T)$, $Y(T)$, and $Z(T)$ are the heat flow in x, y, and z direction, respectively. The forth term on the right hand side is the heat transfer from the superconductor to the substrate surrounding the superconductor, where α is a heat transfer rate, and T_0 is temperature of the substrate. The fifth term on the right hand side is the heat generation by the vortex creep. Here $\boldsymbol{\mathcal{J}}$ is a current density, and $\boldsymbol{E}$ is an electric field. $X(T), Y(T)$, and $Z(T)$ are given as follows.

$$\begin{cases} X(T)=\kappa_{11}\frac{\partial T}{\partial x}+\kappa_{12}\frac{\partial T}{\partial y}+\kappa_{13}\frac{\partial T}{\partial z} \\ Y(T)=\kappa_{21}\frac{\partial T}{\partial x}+\kappa_{22}\frac{\partial T}{\partial y}+\kappa_{23}\frac{\partial T}{\partial z} \\ Z(T)=\kappa_{31}\frac{\partial T}{\partial x}+\kappa_{32}\frac{\partial T}{\partial y}+\kappa_{33}\frac{\partial T}{\partial z} \end{cases} \tag{2}$$

where $\kappa_{\mathrm{ij}}(\mathrm{i,j}=1,2,3)$ are heat conductivity tensors.

The second equations are the Maxwell equations,

$$\begin{cases} \mathrm{rot}(\mathrm{rot}\boldsymbol{A}-\boldsymbol{H}_a)=\frac{4\pi}{c}\boldsymbol{\mathcal{J}} \\ \boldsymbol{E}=-\frac{1}{c}\frac{\partial\boldsymbol{A}}{\partial t} \end{cases} \tag{3}$$

where $\boldsymbol{A}$ is a magnetic vector potential, $\boldsymbol{H}_a$ is an external magnetic field, c is the speed of light. The first Maxwell equation is Ampere's law, in which we ignore displacement current, and the second Maxwell equation is Faraday's law.

The third equation is the current-voltage relation for the superconductor [4],

$$\boldsymbol{\mathcal{J}}=\frac{\boldsymbol{E}}{\rho(\mathcal{J})} \tag{4}$$

Here $\rho(\mathcal{J})$ is a resistivity given by,

$$\rho(\mathcal{J})=\begin{cases} \rho_0(\mathcal{J}/\mathcal{J}_c)^{n-1} & \mathcal{J}\le\mathcal{J}_c, T\le T_c \\ \rho_0 & \mathcal{J}>\mathcal{J}_c, T\le T_c \\ \rho_n & \mathcal{J}>\mathcal{J}_c, T>T_c \end{cases} \tag{5}$$

where n is a creep exponent, ρ_0 is a resistivity constant of superconducting state, ρ_n is a resistivity of normal state, $\mathcal{J}_c$ is a critical current density, and T_c is the critical temperature. The flux creep exponent is given by $n-1=n_0(1-T/T_c)$, where n_0 is a constant, and it depends on the temperature. This current-voltage relation represents the fact that the resistance gradually rises due to vortices creeps as the current increases. The current and temperature dependences of the resistivity in the superconductor are shown in figure 1.

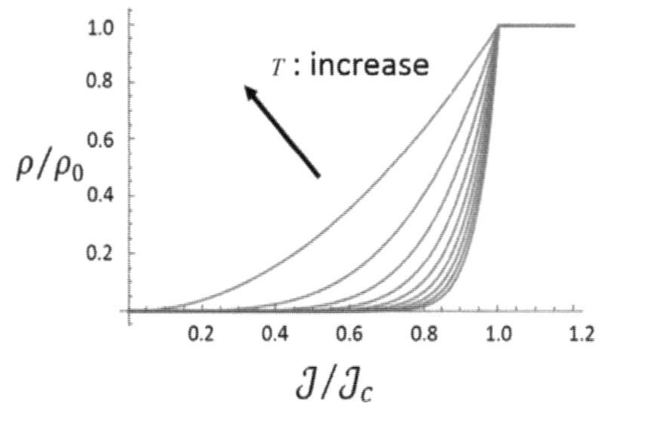

Figure 1. $\mathcal{J}$ and T dependences of resistivity

3. Results and Discussion

We consider two-layer structures of two superconducting plates, and investigate how the critical state depends on distance of two superconducting plates. The size of each superconducting plate is $30 \times 60 \times 8\ \mu m^3$. Each superconducting plate is surrounded by the substrate shown in figure 2; distance between two superconducting plates is d_i, and the overlap of two superconducting plates is 10 μm. The total system size is $70 \times 80 \times (64 + d_i)\ \mu m^3$. We change the distance from 0 to 56 μm.

We use material parameters corresponding to MgB_2, $C = 35\ kJ/K\,m^3 \times (T/T_c)^3$, $\alpha = 220\ kW/K\,m^2 \times (T/T_c)^3$, $\kappa = 0.17\ kW/K\,m \times (T/T_c)^3$, $\rho_n = \rho_0 = 7\ \mu\Omega cm$, $T_c = 39\ K$, $\mathcal{J}_c = 3600\ kA/m^2$, $n_0 = 19$ [4]. The external applied magnetic field along z axis is ramped from $H_a = 0$ kG at a constant rate, $\dot{H}_a = 3 \times 10^{-4}\ mT/s$. The initial temperature of the total system is 10 K. Although the temperature of the superconductors rises, the temperature of the substrate is kept at 10 K.

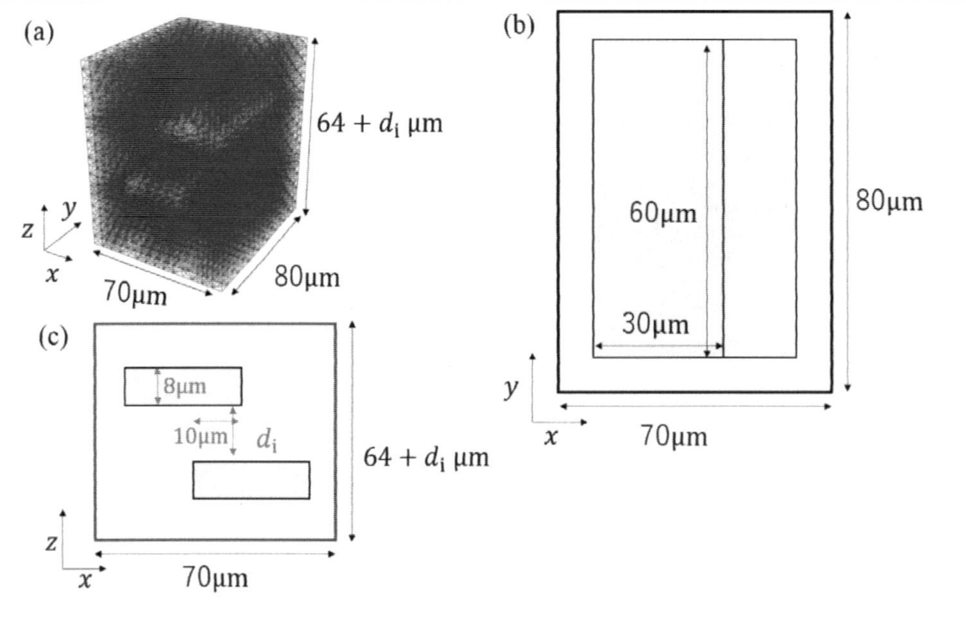

Figure 2. Three-dimensional model for a two-layer structure of superconducting plates (a), a cross-section in *x*-*y* plane (b), and a cross-section in *x*-*z* plane (c).

First, we show the critical state for $d_{\mathrm{i}} = 24\ \mu\mathrm{m}$ when the external magnetic field is $H_a = 3 \times 10^{-6}$ mT. Figures 3(a) and (b) show distribution of magnitude of magnetic flux density B along x direction at $y = 40\ \mu\mathrm{m}$, and $z = 24\ \mu\mathrm{m}, 28\ \mu\mathrm{m}$ and $32\ \mu\mathrm{m}$ in the lower superconducting plate (a) and $z = 56\ \mu\mathrm{m}, 60\ \mu\mathrm{m}$ and $64\ \mu\mathrm{m}$ in the upper superconducting plate (b). Figures 3(c) and (d) show distribution of the current density $\mathcal{J}_y/\mathcal{J}_c$ along x direction at $y = 40\ \mu\mathrm{m}$, and $z = 24\ \mu\mathrm{m}, 28\ \mu\mathrm{m}$ and $32\ \mu\mathrm{m}$ in the lower superconducting plate (c) and $z = 56\ \mu\mathrm{m}, 60\ \mu\mathrm{m}$ and $64\ \mu\mathrm{m}$ in the upper superconducting plate (d). From figures 3(a) and (b), we can see distributions of B at $z = 24$ and $64\ \mu\mathrm{m}$ have peaks at both sides of plane and distributions of B at $z = 32$ and $56\ \mu\mathrm{m}$ have a peak only one side of plate and show strong asymmetry. Total distribution of B is symmetric under spatial inversion about $z = 0$ plane. And current density distributions in figures 3(c) and (d) show similar properties as those of B. These asymmetries come from small magnetic field around the overlap regions of two superconducting plates. This is because magnetic energy becomes large if magnetic field go through the gap of two superconductors.

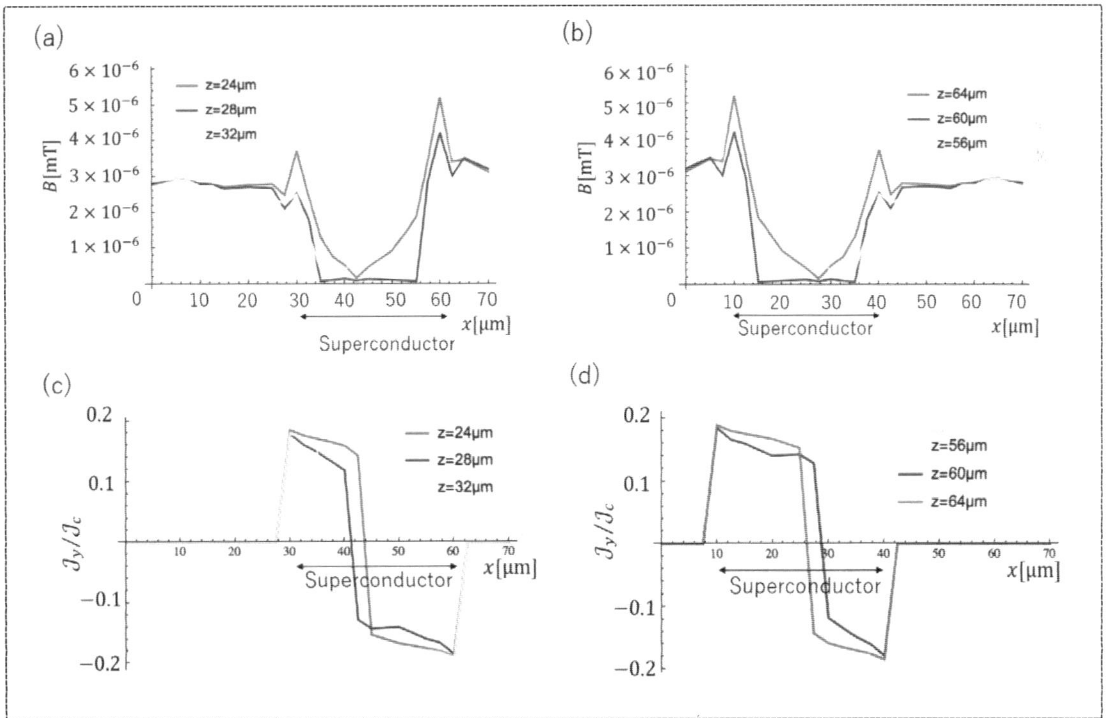

Figure 3. (a) and (b) are magnetic flux density and (c) and (d) are current density.

Next, we show results for changing d_{i} from 8 to 56. Figures 4(a) and (b) show distributions of the magnetic flux density (a) and the current density (b) at $z = 32\ \mu\mathrm{m}$ and $y = 40\ \mu\mathrm{m}$ along x direction. From figure 4(a), we can see that a peak around $x = 30\ \mu\mathrm{m}$ appears and becomes large as increasing the distance d_{i}. Correspondingly, the minimum position of the magnetic flux density moves toward inside of the superconductor, and also distribution of the current density becomes wider as shown by an arrow in figure 4(b). In figure 5, we show the dependence of the flux density at $x = 30\ \mu\mathrm{m}$, $y = 40\ \mu\mathrm{m}$ and $z = 32\ \mu\mathrm{m}$ on the distance d_{i}. We also show max flux density B_{max} at another peak at $x = 60\ \mu\mathrm{m}$, $y = 40\ \mu\mathrm{m}$ and $z = 32\ \mu\mathrm{m}$. From this figure, we can see B is saturated to B_{max} when the distance is increased. Therefore, we can expect that in the limit $d_{\mathrm{i}} \to \infty$, the magnetic flux density distribution becomes that of a single superconducting plate where two symmetric peaks at the edges of plate appear [5].

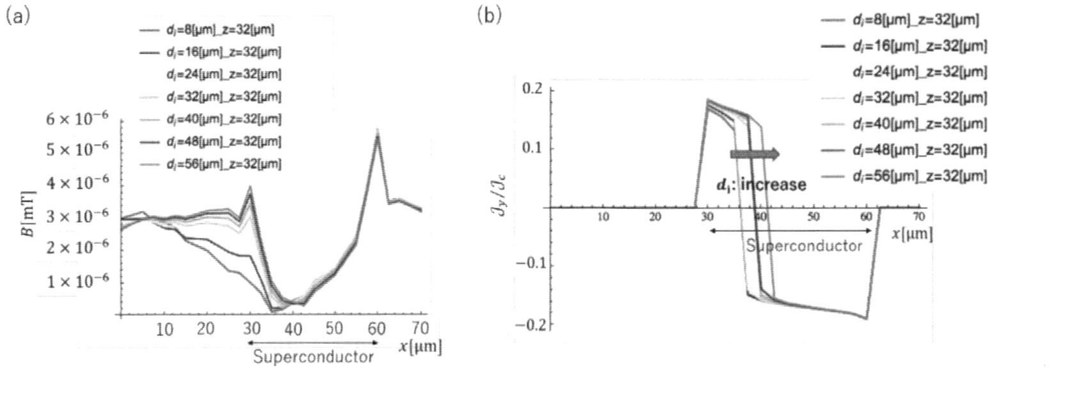

Figure 4. Distribution of magnetic flux density (a) and current density (b) for various values of distance d_i.

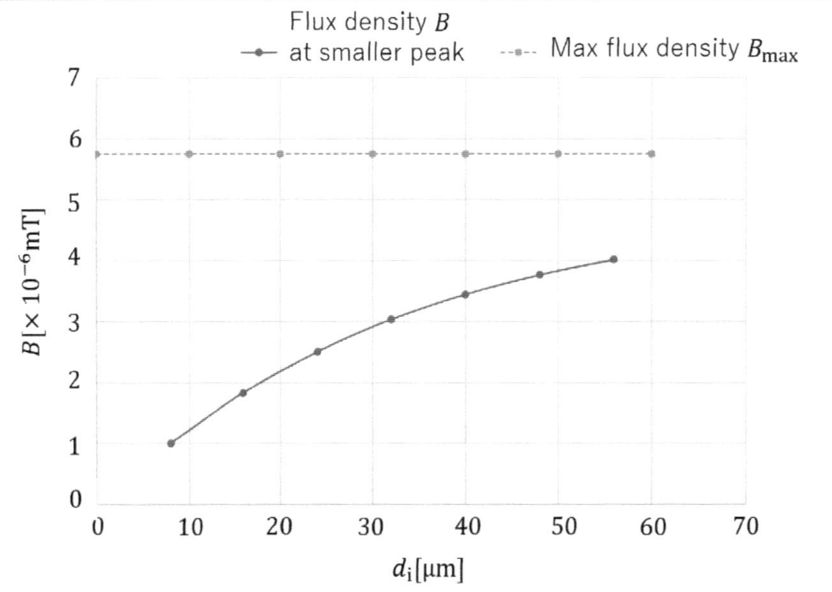

Figure 5. Dependence of peak values of the flux density on the distance between superconducting plates.

4. Summary and Future

We have studied the critical state for two-layer structure of two superconducting plates. Solving the heat transport equation and the Maxwell equations with the current-voltage relation for superconductor with three-dimensional finite element method, we have obtained distributions of the magnetic flux density and the current density. We have found asymmetric distributions of the magnetic flux density and the current density in each superconducting plate in the two-layer structure. Moreover, there is the structure dependence of the magnetic flux density. The larger the distance, the more magnetic flux penetrate from the overlapped part of two superconducting plates.

In future, we will simulate magnetic field distribution more realistically, using the boundary element method for the substrate region surrounding superconductors.

References

[1] Altshler E and Jhohansen T H 2004 *Rev. Mod. Phys.* **76** 471
[2] Tsuchiya Y, Mawatari Y, Ibuka J, Tada S, Pyon S, Nagasawa S, Hidaka M, Maezawa M and Tamegai T 2013 *Supercond. Sci. Technol.* **26** 095004
[3] Tamegai T, Mine A, Tsuchiya Y, Miyano S, Pyon S, Mawatari Y, Nagasawa S, and Hidaka M 2017 *Phys.* C. **533** 74-79
[4] Vestgården J I, Shantsev D V, Galperin Y M and Johansen T H 2011 *Phys. Rev.* B. **84** 054537
[5] Brandt E H 1992 *Phys. Rev.* B. **46** 8628

IOP Conf. Series: Journal of Physics: Conf. Series **1293** (2019) 012021 doi:10.1088/1742-6596/1293/1/012021

Simulation of vortex lattice melting in a dirty square superconducting plate

Takashi Kusafuka[1], Masaru Kato[1] and Osamu Sato[2]

[1]Osaka Prefecture University, 1-1, Gakuencho, Nakaku, Sakai, Osaka 599-8531, Japan

[2]Osaka Prefecture University College of Technology, 26-12, Sawaicho, Neyagawa, Osaka 572-8572, Japan

E-mail: syb01059@edu.osakafu-u.ac.jp

Abstract. Vortex lattice melting in a dirty mesoscopic square superconducting plate is studied using the molecular dynamics (MD) method. We include an impurity potential in the MD method and investigate vortex number and impurity number dependences of the melting transition temperature. We find strong vortex number dependence and weak pinning potential dependence of the melting temperature.

1. Introduction

A single vortex has a quantized flux ($\Phi_0 = ch/2e$) and there is screening current around it. It appears between the lower critical field (H_{c1}) and the upper critical field (H_{c2}) in a type II superconductor. Above H_{c2}, the superconductivity becomes unstable and vortices disappear. In the intermediate field regime $H_{c1} < H < H_{c2}$, vortices form a triangular lattice which is known as an Abrikosov lattice. Increasing the temperature, the vortex lattice becomes unstable. Then vortex lattice melts into a vortex liquid state. In $H-T$ phase diagram, there is a melting line of the vortex lattice. Because of large thermal and quantum fluctuation, this melting transition clearly appears in the cuprate high temperature superconductors [1,2].

Recently, Ooi et al [3] found that melting transition temperature of vortex lattices in a square superconducting plate (Bi2212) oscillates with increasing magnetic field, and it becomes maximum when the vortex number is a square number. Then, using the molecular dynamics (MD) method [4-6], Kato and Kitago [7] investigated the vortex lattice melting transition in a square superconducting plater without impurities. They found the vortex motion becomes large as temperature increases and showed the melting transition temperature can be determined from the rapid increase of the standard deviation of the position of the vortices

In this study, we investigate vortex number dependence and impurity dependence of vortex lattice melting in a dirty square superconducting plate, using the MD method.

2. Method

In the MD method, we consider following equations of motion of vortices [4-6],

$$\eta \frac{d\boldsymbol{r}_i}{dt} = \boldsymbol{f}_i^{\mathrm{imp}} + \boldsymbol{f}_{\mathrm{v}i} + \boldsymbol{f}_{\mathrm{f}i} \tag{1}$$

where $\boldsymbol{r}_i$ is a position of an i-th vortex, and η is the viscosity. $\boldsymbol{f}_i^{\mathrm{imp}}$ is a sum of pinning forces from impurities, which is given as,

$$\boldsymbol{f}_i^{\mathrm{imp}} = \sum_j^{\mathrm{pinningsite}} \frac{f_{\mathrm{p}}}{r_{\mathrm{p}}} \left|\boldsymbol{r}_i - \boldsymbol{r}_j^{\mathrm{p}}\right| \Theta\left(\frac{r_{\mathrm{p}} - \left|\boldsymbol{r}_i - \boldsymbol{r}_j^{\mathrm{p}}\right|}{\lambda}\right) \hat{\boldsymbol{r}}_{ij} \tag{2}$$

Here f_{p} and r_{p} are a strength and a range of the pinning force, and $\boldsymbol{r}_j^{\mathrm{p}}$ is the position of j-th impurity. Θ is the step function. λ is the penetration depth. $\boldsymbol{f}_{\mathrm{v}i}$ is a sum of vortex-vortex interaction forces from other vortices.

$$\boldsymbol{f}_{\mathrm{v}i} = \sum_j \boldsymbol{f}_{ij}^{\mathrm{vv}} \tag{3}$$

$$\boldsymbol{f}_{ij}^{\mathrm{vv}} = f_0 \sum_j K_1\left(\frac{r_{ij}}{\lambda}\right) \hat{\boldsymbol{r}}_{ij} \tag{4}$$

$$f_0 = \frac{\Phi_0^2}{8\pi^2\lambda^3} \tag{5}$$

where f_0 is the vortex-vortex interaction force, Φ_0 is the quantized flux, and $K_1(k)$ is the MacDonald function. This force depends on temperature T through temperature dependence of $\lambda = \lambda_0/\sqrt{1 - T/T_c}$. λ_0 is the penetration depth at $T = 0$ and $T_c$is the critical temperature, at which the penetration depth diverges. $\boldsymbol{f}_{\mathrm{f}i}$ is the thermal fluctuation force, which satisfies following equation,

$$\left\langle \boldsymbol{f}_{\mathrm{f}i}(t_1) \cdot \boldsymbol{f}_{\mathrm{f}j}(t_2) \right\rangle = 2\eta k_B T \delta_{ij} \delta(t_1 - t_2) \tag{6}$$

where k_B is the Boltzmann constant. Temperature dependence of vortex motion comes from temperature dependences of the thermal fluctuation $\boldsymbol{f}_{\mathrm{f}}$ and the vortex-vortex interaction.

3. Numerical results

We consider a mesoscopic $(3\lambda_0 \times 3\lambda_0)$ square superconducting plate with impurities under the periodic boundary condition. For investigation of the experiments in Ref.3, we should impose the fixed boundary condition. However, with the periodic boundary condition in the square shape, vortices are also confined in the square shape and they are forced to form lattices that match with the square shape, as shown below.

First, we focus on cases where vortex number is from 4 to 18 and number of impurities is 100. We set radius and strength of the impurities as $r_p = 0.03\lambda_0$ and $f_p = -0.1f_0$. The number of vortices may correspond to the external magnetic field. As a measure of vortex motion, we use an averaged variance σ^2 of a vortex position $\boldsymbol{r}_i$

$$\sigma^2 = \frac{1}{N} \sum_{i=1}^{N} \left\langle \left(\boldsymbol{r}_i - \left\langle \boldsymbol{r}_i \right\rangle\right)^2 \right\rangle \tag{7}$$

where $\langle \ldots \rangle$ is a time average. In figure 1, we show σ as a function of temperature when vortex number N is from 4 to 18. Here σ is normalized by an average of vortex distance l, where $l = \sqrt{(3\lambda_0 \times 3\lambda_0)/N}$ for N vortices. Around $T/T_c \approx 0.99$, σ/l is larger for larger N. This behavior is explained as follows. For larger N, l is small but vortex number dependence of the vortex motion in liquid phase is rather small, and therefore σ/l becomes relatively larger. But for $0.1 < T/T_c < 0.8$, σ is smaller for larger

N, except $N = 4$ and 16. This is because for smaller N, vortices moves rather freely in the solid phase and therefore σ/l becomes larger. However, for $N = 4$ and 16, σ becomes small. Especially σ is smallest for $N = 16$. This result shows vortex lattice is relatively stable for 4 and 16 vortices. In order to clarify this point, we plot σ/l as a function of vortex number N in Figure 2.

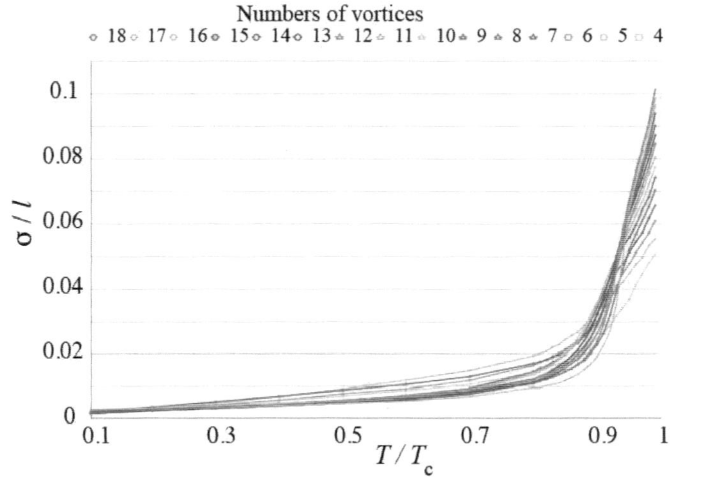

Figure 1. Standard deviation of the vortex position as function of temperature in the $3\lambda_0 \times 3\lambda_0$ square superconducting plate with 100 impurities. N is from 4 to 18.

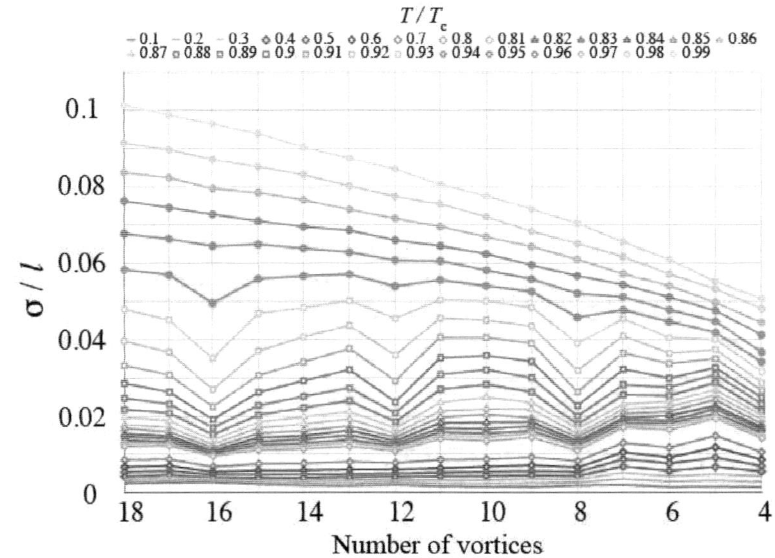

Figure 2. The standard deviation of vortex position as a function of the number of vortices.

In figure 2, when the number of vortices is 4, 8, 12, and 16, σ/l becomes small for lower temperature where the vortex lattice is formed and therefore vortex lattices for these vortex numbers becomes relatively stable. In order to investigate the stable lattice, we show trajectories of vortices when $N = 16$ at $T/T_c = 0.1$, 0.5, 0.9, and 0.98 in figure 3. For lower temperature vortices form a 4×4

IOP Conf. Series: Journal of Physics: Conf. Series **1293** (2019) 012021 doi:10.1088/1742-6596/1293/1/012021

square lattice. As we can see from figure 1, the melting transition occurs between $T/T_c = 0.9$ and $T/T_c = 0.98$.

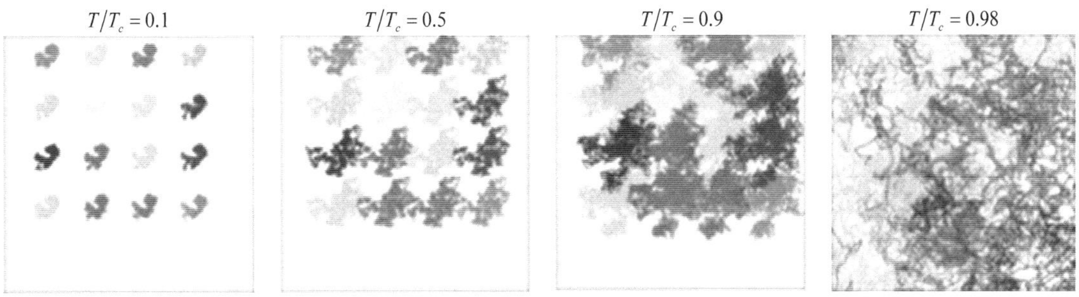

Figure 3. Trajectories of vortices at T/T_c = 0.1, 0.5, 0.9 and 0.98.

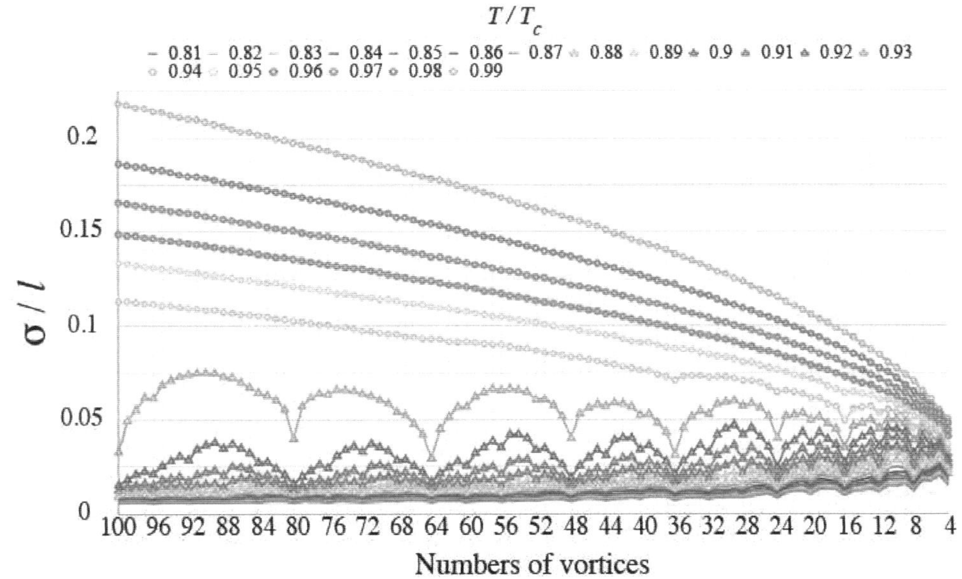

Figure 4. The standard deviations of 4 to 101 vortices in the $3\lambda_0 \times 3\lambda_0$ square superconducting plate including 100 impurities as a function of the number of vortices.

For larger number of vortices, we show σ/l as functions of vortex number for various temperatures in figure 4. At low temperature, there are large dips of σ/l when the number of the vortices is 4, 8, 16, 24, 36, 48, 64, 80 or 100. Also, there are small dips at N=12, 20, 28, 32, 40, 44, 52, 56, 60, 68, 72, 76, 84, 88, 92, and 96. All of these numbers are multiple of 4. And for large dip numbers are close to square numbers. We can say when the vortex number is a multiple of 4, the vortex lattice becomes rather stable. This result shows that vortex lattices with $N = 4m$ (m is an integer) relatively stable. This may come from our periodic boundary condition, which is suitable for square and rectangular lattices. For more realistic calculation, we must impose a fixed boundary condition.

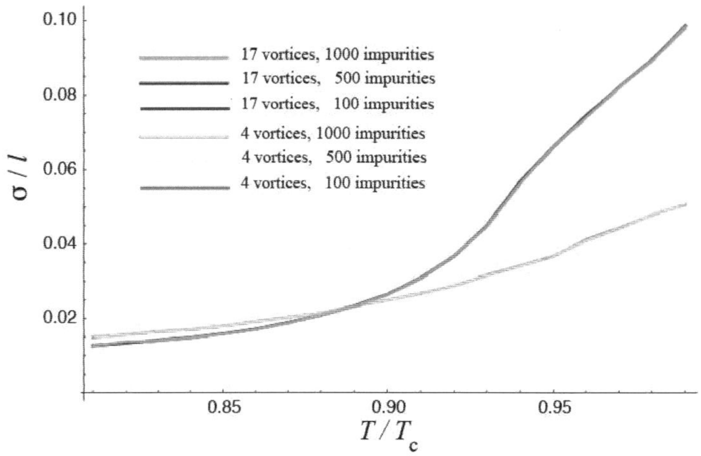

Figure 5. Standard deviation of vortex position as a function of temperature of 4 and 17 vortices in the $3\lambda_0 \times 3\lambda_0$ square superconducting plate with 100, 500, 1000 impurities.

Next, in order to investigate the effect of impurities on the vortex motion, we compare vortex motions for $N = 4$ and 17 vortices in the superconducting plate with 100, 500 and 1000 impurities. Here, $r_p = 0.03\lambda_0$ and $f_p = -0.1f_0$. In figure 5, we show σ as a function of temperature for $N = 4$ and 17 and $N_{\text{imp}} = 100$, 500 and 1000. In this figure, we cannot see any difference between $N_{\text{imp}} = 100$, 500 and 1000 cases. However, in the higher temperature region, there are small differences, which are shown in figure 6. From figure 6, we can see vortex motion becomes a little bit smaller for larger impurity number.

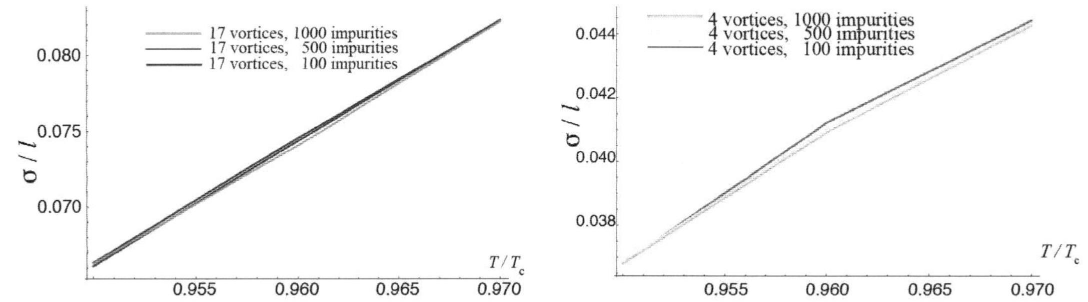

Figure 6. σ/l as a function of temperature for 17 vortices (a) and 4 vortices (b) in the $3\lambda_0 \times 3\lambda_0$ square superconducting plate with 100, 500, and 1000 impurities.

4. Summary

We have investigated vortex lattice melting in dirty superconductors, using the molecular dynamics method. We have found that when vortex number N is a multiple of 4 and close to a square number, vortex state become rather stable and melting temperature becomes higher.

In future, we will investigate vortex motion in a dirty square superconducting plate under the fixed boundary condition that is more realistic compare to present periodic boundary condition. In the fixed boundary condition, we might use a mirror image method because of the condition for the supercurrent. In this study, effects of impurities on vortex motion is small. This may come from that our parameters for impurities rather are small. Impurity potential dependence on the vortex motion is also a future problem.

References

[1] Blatter G et al 1994 *Rev. Mod. Phys.* **68** 1125
[2] Blatter G et al 2003 *The Physics of Superconductors* ed K. H. Bennemann, J. B. Ketterson chapter 10
[3] Ooi S et al. 2015 *Phys. Rev. Lett.* **114** 087001
[4] Reichhardt C, Olson C J, Groth J, Field S, and Nori F 1996 *Phys. Rev.* **53** 8898
[5] Misko V R and Peeters F M 2006 *Phys. Rev.* **74** 174507
[6] Kato M and Fujibayashi D E 2013 *Physics Procedia* **45** 133
[7] Kato M and Kitago H 2017 *J. Phys. Conf. Ser.* **871** 012028

IOP Conf. Series: Journal of Physics: Conf. Series **1293** (2019) 012022 doi:10.1088/1742-6596/1293/1/012022

Detection of the vortex-liquid phase in superconducting films by Nernst effect

K Ienaga, T Arai, T Hayashi, S Kaneko and S Okuma

Department of Physics, Tokyo Institute of Technology, 2-12-1 Ohokayama, Meguro-ku, Tokyo 152-8551 Japan

E-mail: ienaga.k.aa@m.titech.ac.jp

Abstract. To clearly detect the vortex-liquid state of a superconducting film, we constructed an experimental system for measuring the Nernst signals in a film sample. We measured Nernst signals N in an amorphous Mo_xGe_{1-x} film with a thickness of 300 nm. As the field is increased in the vortex-solid phase, N rises from zero at the boundary between the vortex-solid state and the vortex-liquid state, showing a peak in the middle of the liquid phase, and finally decreases. From a linear extrapolation of N after the peak to $N = 0$, we roughly estimated the upper bound of the liquid state. These results imply that the Nernst measurement could be an alternative to a resistivity measurement for detecting the vortex-liquid state. It would be helpful when the resistivity measurement is inconclusive in detecting vortices such as in an insulator phase in the vicinity of the superconductor-insulator transition in two dimensions.

1. Introduction

In the vortex-liquid state, freely mobile vortices give rise to non-zero resistivity in a superconductor. Using the superconducting order parameter with the phase θ and the amplitude Ψ, the liquid state is characterized by large fluctuations of θ due to the free vortex motion induced by thermal or quantum fluctuations. In the vortex-solid state, by contrast, both θ and Ψ have global coherence and the resistivity falls to zero. Both states are clearly distinguished from resistivity. On the other hand, the vortex-liquid state shows a crossover to the normal state with increasing temperature T or magnetic field B. Both states are thermodynamically equivalent and the crossover field B_{c2} is empirically estimated from resistivity. In the normal state, the resistivity decreases with decreasing temperature because of the superconducting fluctuations of both θ and Ψ, which is described as the Gaussian fluctuations [1, 2]. This behaviour is smoothly connected to the resistivity drop in the vortex-liquid state. Such a gradual change is also seen in the magnetic susceptibility.

A Nernst effect measurement is an alternative tool to detect the vortex-liquid state. When the temperature gradient $\nabla_x T$ is applied to a superconductor, vortices are driven by a thermal force and a voltage V_{xy} is generated transversely to the vortex motion, which is called a vortex Nernst effect [3-10]. The Nernst signal is defined as $N = E_{xy}/\nabla_x T$ using a transverse electric field E_{xy}. When the pinning effect is negligible, the contribution of vortices is described by

$$N = E_{xy}/\nabla_x T = B_z v_x/\nabla_x T = \rho_F S_\phi/\Phi_0, \quad (1)$$

where B_z is magnetic field, v_x average velocity of vortices, ρ_F flux-flow resistivity, S_ϕ transport

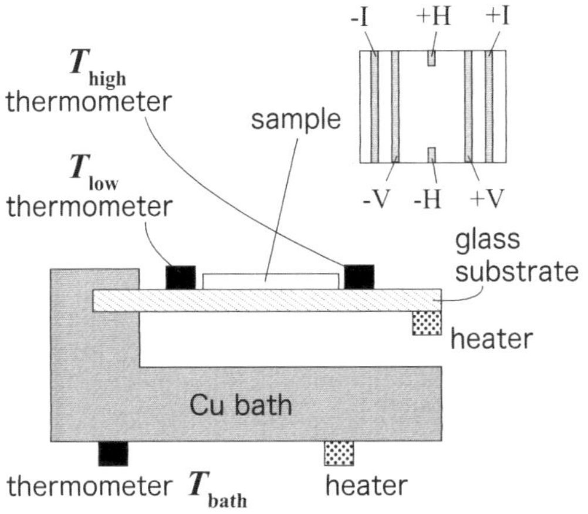

Figure 1. Schematic illustration of the measurement setup for the Nernst effect in a film sample formed on a glass substrate. One side of the substrate was attached to a Cu thermal bath and the other side was connected to a heater to supply heat current. The upper right diagram depicts the Ag electrodes on the sample.

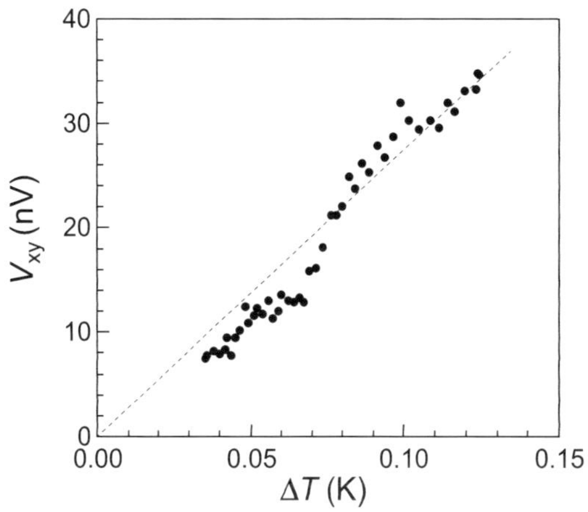

Figure 2. A typical response of V_{xy} as a function of ΔT at 5.00 K and 3.5 T in the vortex-liquid state. The linear behaviour between V_{xy} and ΔT is clearly observed.

entropy per unit length of a vortex, and Φ_0 quantized magnetic flux. For the last substitution, the balance between the thermal force and the viscous force is taken into account.

Previously, large vortex Nernst signals have been observed in the vortex-liquid state of high-T_c superconductors [5-8], where most of studies have been carried out on bulk samples. For thin films, on the other hand, very limited measurements have been performed so far, especially in the low temperature regime. At low temperature, an applied heat current must be small to prevent a thermal bath from warming up, which makes a resulting $\nabla_x T$ small and Nernst measurement difficult. Pourret *et al.* have succeeded in Nernst measurements for amorphous NbSi films with thicknesses of 12.5 and 35 nm below 1 K and clearly detected vortex Nernst signals [8-10]. They have roughly estimated B_{c2} from a linear extrapolation of N [10]. However, a boundary between the vortex-solid state and the liquid state, where N falls to zero, was not identified because of the broaden liquid state due to the two dimensionality of the sample.

In this work, we constructed an experimental system for the Nernst effect measurement in a film sample. The measurements were performed on an amorphous Mo_xGe_{1-x} film with a thickness of 300 nm, which is thicker than the amorphous thin films in the previous study [9, 10]. Our sample shows three-dimensional properties and the vortex-liquid state estimated from linear resistivity ρ is narrow [12]. We observed a clear peak of N in the liquid state and estimated B_{c2} in the same manner as the previous studies. Moreover, the solid-liquid boundary was clearly detected from a sharp rise of N as well as from the measurement of ρ.

2. Experimental

An amorphous Mo_xGe_{1-x} film with a thickness of 300 nm was sputtered onto a glass substrate mounted on a rotating stage kept at 10°C with water cooling. Since the pinning strength is weak and the pinning sites are randomly distributed in our amorphous sample, we can detect intrinsic vortex motion [13, 14]. Nernst signals from quasiparticles are also negligible because

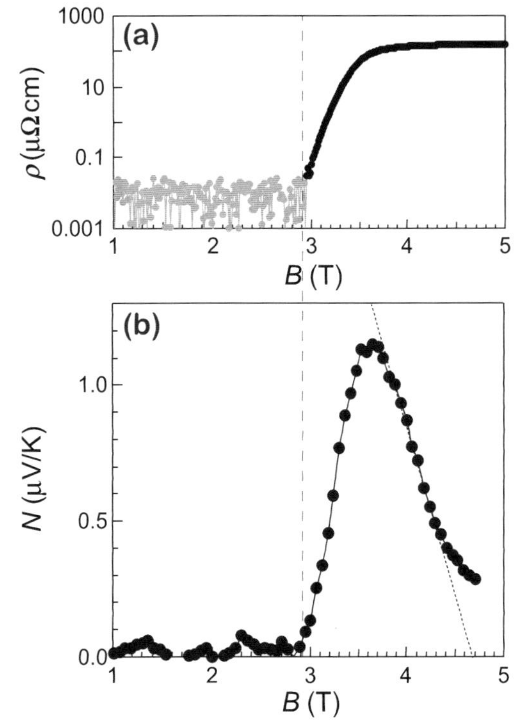

Figure 3. Field dependences of (a) the resistivity at 10 μA and (b) the Nernst signal at $\Delta T = 0.10$ K. Both were obtained at 5.00 K. The gray plots in (a) contain both positive and negative values, indicating that they are below the sensitivity of our experiment. The Nernst signal rises at the boundary between the vortex-solid and the liquid states marked by the dashed line, which is determined by the sharp rise of the resistivity from zero. The dotted line shows an extrapolation of the linear part of the Nernst signal after showing the peak.

a mean free path of the quasiparticles is the order of the interatomic distance. The transition temperature where ρ falls to zero is 6.75 K at zero field.

A measurement system for the Nernst effect in a film sample is schematically illustrated in figure 1. Using this apparatus, the Nernst effect measurement and a standard four-terminal measurement of ρ were performed. One side of the substrate was attached to a Cu thermal bath and the other side was connected to a heater to supply heat current. We used a glass substrate of low thermal conductivity with a thickness of 0.15 mm. A temperature difference, $\Delta T = T_{\mathrm{high}} - T_{\mathrm{low}}$, was measured by two RuO_2 thermometers spaced by 4.4 mm ($= L$), which were mounted on the substrate in the vicinity of the high-temperature side and the low-temperature side of the sample film. An additional thermometer and heater were also glued on the Cu bath to control the bath temperature. V_{xy} was measured between Ag electrodes (+H and -H) spaced by 1.7 mm ($= l$) using a nanovoltmeter. The magnetic field was applied perpendicular to the film plane. Nernst signals were obtained from $N = \frac{E_{xy}}{\nabla_x T} = \frac{V_{xy}/l}{\Delta T/L}$.

3. Results and discussions

First, we performed an experiment to test the linear relationship between V_{xy} and ΔT, which is necessary to obtain N from the measured V_{xy} and ΔT. Figure 2 shows a typical response of V_{xy} as a function of ΔT at 5.00 K and 3.5 T, where the large N is observed. A linear behaviour between the V_{xy} and ΔT is clearly visible. At each measurement, the background thermoelectric voltage generated at $\Delta T = 0$ is subtracted.

Next, we show ρ measured as a function of field at 5.00 K and 10 μA in figure 3(a). ρ rises from the background level at 2.9 T as marked by the dashed line, which corresponds to the boundary between the vortex-solid state and the vortex-liquid state. With further increasing field, ρ steeply increases and above 3.5 T gradually reaches the normal-state resistivity $\rho_{\mathrm{n}} = 160$ $\mu\Omega$ cm.

Figure 3(b) shows a field dependence of N at $T = 5.00$ K and $\Delta T = 0.10$ K ($T_{\rm high} = 5.05$ K and $T_{\rm low} = 4.95$ K). To subtract the thermoelectric voltage caused by misalignment of Hall electrodes, we extracted antisymmetric contributions with respect to magnetic field reversal. With increasing field, N also rises from zero at 2.9 T and then decreases after showing a peak around 3.6 T. The behaviour is consistent with what is expected from the vortex Nernst effect in the vortex-liquid state [3-10]. The rapid growth of N from the solid-liquid boundary results from the reduced pinning effects due to thermal fluctuations. The decrease of N is attributed to a decrease of S_ϕ.

According to the theory, S_ϕ is proportional to $(B_{c2} - B)$ near B_{c2} and vanishes at B_{c2} [15, 16]. To roughly estimate a value of B_{c2} from N, we extrapolated the linear part of the data after showing the peak as shown with the dotted line [7, 10]. The deviation of N from the dotted line becomes pronounced above 4.3 T, which probably originates from the Nernst signals due to the Gaussian fluctuations [8-11]. We can estimate B_{c2} to be around 4.7 T from the extrapolation to $N = 0$. This value corresponds to the field at which $\rho = 0.96\ \rho_{\rm n}$. From these results, we consider that we detected the vortex-liquid state in the thick amorphous $\text{Mo}_x\text{Ge}_{1-x}$ film using the Nernst measurements. This technique can be applied to thinner (2D) films, for example, undergoing the superconductor-insulator transition (SIT). It is of interest to explore the vortices on an insulating side of the SIT, the so-called Bose-insulator phase, which is a long-standing problem [17].

Acknowledgements

This work was supported by a Grant-in-Aid for Scientific Research (Grant Nos. 17K14337, 15K13516 and 17H02919) from the Japan Society for the Promotion of Science.

References

[1] Tinkham M 2004 *Introduction to Superconductivity* (New York: Dover) 2nd ed.
[2] Skocpol W J and Tinkham M, 1975 *Rep. Prog. Phys.* **38** 1049
[3] Solomon P R and Otter Jr. F A 1967 *Phys. Rev.* **164** 608
[4] Huebener R P and Seher A 1969 *Phys. Rev.* **181** 701
[5] Ri H -C , Gross R, Gollnik F, Beck A, Huebener R P, Wagner P and Adrian H 1994 *Phys. Rev.* B **50** 3312
[6] Xu Z A, Ong N P, Wang Y, Kakeshita T and Uchida S 2000 *Nature* **406** 486
[7] Wang Y, Li L and Ong N P 2006 *Phys. Rev.* B 73 024510
[8] Behnia K and Aubin H 2016 *Rep. Prog. Phys.* **79** 046502
[9] Pourret A, Aubin H, Lesueur J, Marrache-Kikuchi C A, Bergé L, Dumoulin L and Behnia K 2006 *Nature Phys.* **2** 683
[10] Pourret A, Aubin H, Lesueur J, Marrache-Kikuchi C A, Bergé L, Dumoulin L and Behnia K 2007 *Phys. Rev.* B **76** 214504
[11] Ussishkin I, Sondhi S L and Huse D A 2002 *Phys. Rev. Lett.* **89** 287001
[12] Okuma S, Imaizumi H, Kokubo N 2009 *Phys. Rev.* B **80** 132503
[13] Dobroka M, Kawamura Y, Ienaga K, Kaneko S and Okuma S 2017 *New J. Phys.* **19** 053023
[14] Ienaga K, Dobroka M, Shirahata Y, Kawamura Y, Kaneko S and Okuma S 2017 *J. Phys.: Conf. Series* **871** 012020
[15] Caroli C and Maki K 1967 *Phys. Rev.* **164** 591
[16] Vidal F 1973 *Phys. Rev.* B **8** 1982
[17] Goldman A M and Marković N 1998 *Physics Today* **51** 39

IOP Conf. Series: Journal of Physics: Conf. Series **1293** (2019) 012023 doi:10.1088/1742-6596/1293/1/012023

Time evolution of the vortex configuration associated with dynamic ordering detected by dc drive

S Maegochi, M Dobroka, K Ienaga, S Kaneko and S Okuma

Department of Physics, Tokyo Institute of Technology, 2-12-1, Ohokayama, Meguro-ku, Tokyo 152-8551, Japan

E-mail: sokuma@o.cc.titech.ac.jp

Abstract. When a periodic shear force with a small amplitude d_{inp} is applied to vortex assemblies having a random distribution, the vortices gradually self-organize to avoid future collisions and transform into an organized configuration. We showed recently that this random-organization or dynamic-ordering process can be detected from the time-evolution of voltage $V(t)$ that increases to a steady-state voltage. We also showed from the subsequent readout experiment of $V(t)$ using various ac amplitudes d that the transient vortex configuration during random organization is not microscopically homogeneous but consists of the disordered and organized regions. In this work, we develop an alternative readout method using a *dc* drive. It is found that the dc method gives the same results as obtained from the ac one, which further supports our view of the coexistence regions. It is expected that both methods will be applied complementarily to detect the vortex configuration over a wide range of disorder.

1. Introduction

When an ac shearing force with a small shear amplitude d_{inp} is applied to many particle assemblies having a random distribution, the particles gradually self-organize to avoid future collisions and change into an ordered configuration. This is called random organization or dynamic ordering by ac drive [1, 2, 3, 4]. In previous studies, we used a vortex system in amorphous Mo_xGe_{1-x} films to investigate these general phenomena. First, we performed measurements of the time evolution of the voltage $V(t)$, i.e., average velocity of vortices, associated with dynamic ordering by ac drive [5, 6, 7]. We call this experiment an input experiment. After freezing the vortex configuration at a desired time by turning off the ac driving current, we subjected the vortex assembly to ac drives with various amplitudes d that give rise to dynamic ordering and/or disordering. We name this experiment a readout experiment. We found unexpectedly that the transient vortex configuration created during the ac dynamic ordering is not microscopically homogeneous but consists of disordered region (DR) and ordered region (OR) [7].

When the particles with an ordered initial configuration are driven in a random pinning potential by a small dc drive, they gradually transform into a disordered configuration. This is called dynamic disordering by dc drive [5, 8, 9, 10]. In contrast to the case of the ac dynamic ordering, the transient vortex configuration created during the dc dynamic disordering seemed to be homogeneous [11]. This leads to an interesting question: what is the origin of the coexistence regions? The results mentioned above [7] suggest that the use of the ac drive and/or the dynamic ordering process would be the necessary condition for the emergence of the coexistence regions.

It is of interest to clarify the origin of the coexistence regions. For this purpose, we must conduct additional experiments, such as the input experiment exhibiting the dynamic ordering by *dc* drive and the readout experiment detecting the vortex configuration during the dc dynamic-ordering process. Note, however, that in the dynamic-ordering experiment by dc drive, the initial and final vortex configurations are, in general, more disordered than those in the dynamic-ordering experiment by ac drive [5, 6, 12]. When the configuration in the input experiment is disordered, it is difficult to use the *ac* drive in the readout experiment. This is because the ac drive, which readily orders the disordered vortex configuration, cannot *disorder* the highly disordered configuration. For the precise determination of the vortex configuration, such as the determination of the area ratio of OR, however, the readout experiment that attains both dynamic disordering and dynamic ordering is strongly desired [7]. To overcome the problem, here, we explore an alternative readout method using a dc drive.

2. Experimental

The a-Mo_xGe_{1-x} film with thickness of 330 nm was prepared by rf sputtering on a silicon substrate mounted on a water cooled rotating copper stage [5, 6, 10]. The superconducting transition temperature at zero field ($B = 0$) where the resistivity vanishes is T_c =6.3 K. The width of the sample is 0.3 mm, along which the vortices are driven. The voltage contacts spaced at 1.2 mm were used to measure the voltage V generated by vortex motion. The linear resistivity was measured using a standard four-terminal method. We also measured the time evolution of the voltage $V(t)$ just after the ac current I_{ac} of square waveform or the dc current I with a sharp rise was applied to the vortex system at $t = 0$. The voltage $V(t)$ enhanced with a preamplifier was recorded using a fast-Fourier transform spectrum analyzer with a time-resolution of up to 40 kHz. The amplitude of I_{ac} was adjusted to yield an ac voltage with a desired amplitude $V^{\infty}(\equiv |V(t \to \infty)|)$ at $t \to \infty$. The sample was directly immersed into the liquid ^{4}He. The magnetic field B was applied perpendicular to the plane of the film. All the data were acquired at 4.1 K in 3.5 T, corresponding to the peak-effect regime [13, 14, 15, 16, 17, 18, 19].

3. Results and discussion

We have known from previous studies that in our weak-pinning a-Mo_xGe_{1-x} films highly disordered vortex configuration is created in the field-cooled process, where many vortices are pinned to random pinning sites [6, 10]. Thus, in the input experiment, we prepared an initial vortex configuration by cooling the sample from 11 K ($> T_c$) to 4.1 K in a fixed magnetic field of 3.5 T. Then, we applied the square ac current I_{ac} with a frequency of 12 kHz, yielding $d_{inp} = 1$ μm and $V^{\infty} = 100$ μV, to the disordered initial vortex assembly at $t = 0$. The t-dependent voltage (divided by V^{∞}), $V(t)/V^{\infty}$, generated by periodic motion of vortices is shown in figure 1, where t is expressed as a number n of cycles and the location of $V(t)/V^{\infty} = 1$ is indicated by the horizontal dashed line. The amplitude of the voltage, $|V(t)|$, shows a monotonic increase and relaxation to a steady-state voltage $V^{\infty} = 100$ μV at many cycles ($n \geq 4000$). This indicates the dynamic ordering from the highly disordered vortex configuration to the relatively organized configuration characterized by $d_{inp} = 1$ μm and $V^{\infty} = 100$ μV.

Figure 2(a) shows the "readout" voltage $V(t)/V^{\infty}$ for three different input vortex configurations, generated responding to the ac drive with a frequency of 4 kHz yielding $d = 3$ μm and $V^{\infty} = 100$ μV. The red line represents $V(t)/V^{\infty}(\equiv V_{n=0}(t)/V^{\infty})$ for the input configuration prepared by the input ac drive with $d_{inp} = 1$ μm and $V^{\infty} = 100$ μV for $n = 0$ cycles, thus, corresponding to the most disordered initial configuration. The blue line indicates $V(t)/V^{\infty}(\equiv V_{n\to\infty}(t)/V^{\infty})$ for the input configuration prepared by the ac drive yielding d_{inp} =1 μm for many ($n >$4000) cycles, thus, corresponding to most organized initial configuration. The green line represents $V(t)/V^{\infty}(\equiv V_{n=100}(t)/V^{\infty})$ for the input configuration prepared by the ac

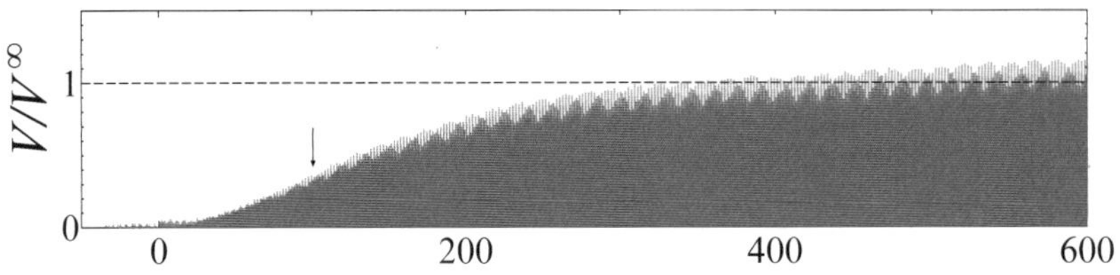

Figure 1. Input experiment of the dynamic ordering by ac drive: The time evolution of the voltage $V(t)/V^\infty$ for the highly disordered initial vortex configuration, generated in response to the ac drive with a frequency of 12 kHz yielding $d_{inp} = 1$ μm and $V^\infty = 100$ μV. An arrow indicates $n = 100$ cycles. The horizontal dashed line marks the position of $V(t)/V^\infty$ =1.

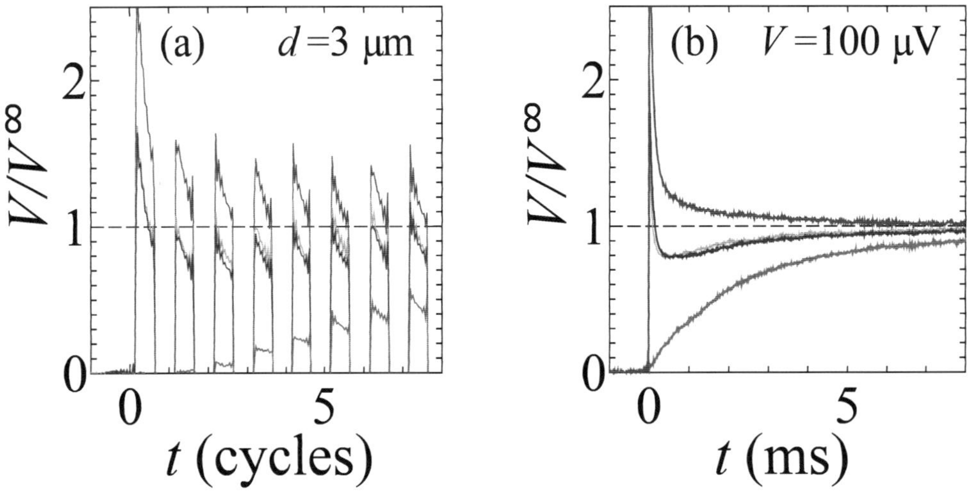

Figure 2. (a) Readout experiment by ac drive: The readout voltage $V(t)/V^\infty$ for the input configuration created by the ac drive with $d_{inp} = 1$ μm and $V^\infty = 100$ μV after $n = 0$ (red line), many cycles (n >4000) (blue line), and $n = 100$ cycles (green line), generated responding to the ac drive with a frequency of 4 kHz yielding $d = 3$ μm and $V^\infty = 100$ μV. The purple line represents the fit of the data shown with the green line to Eq. (1) with α=0.60±0.04. (b) Readout experiment by dc drive: The readout voltage $V(t)/V^\infty$ for the same input configuration as in (a); i.e., the configuration created by the ac drive with $d_{inp} = 1$ μm after $n = 0$ (red line), many cycles (n >4000) (blue line), and $n = 100$ cycles (green line), generated responding to the *dc* drive yielding $V(t \to \infty) = 100$ μV. The purple line is the fit of the data for the green line to Eq. (1) with α=0.57±0.01. The horizontal dashed lines in (a) and (b) denote the location of 1.

drive with $d_{inp} = 1$ μm for $n = 100$ cycles, thus, corresponding to the moderately disordered initial configuration. The horizontal dashed line denotes the location of $V(t)/V^\infty$ =1.

The monotonic increase and decrease in $|V(t)|/V^\infty$ followed by a saturation to $|V(t \to \infty)|/V^\infty = 1$, indicative of dynamic ordering and disordering, respectively, are clearly observed as the red and blue lines. By contrast, a nonmonotonic fast decrease in $|V(t)|/V^\infty$ followed by a slow increase clearly seen as the green line indicates the emergence of the coexistence regions of OR and DR, as observed previously [7]. It is found that $V_{n=100}(t)/V^\infty$ is well reproduced by a weighted average of $V_{n=0}(t)/V^\infty$ and $V_{n\to\infty}(t)/V^\infty$ with a suitable weight parameter α, which is expressed as:

$$V_{n=100}^{fit}(t) = \alpha V_{n\to\infty}(t) + (1-\alpha)V_{n=0}(t). \quad (1)$$

The purple line shows the best fit of the data for the green line to Eq. (1), using α=0.60±0.04. It was shown in our recent paper [7] that α corresponds to the area ratio of OR in the sample.

In figure 2(b) we show the readout voltage $V(t)/V^{\infty}$ generated responding to the *dc* drive yielding $V(t \to \infty)(\equiv V^{\infty}) = 100$ μV for the same input vortex configuration as studied above (figure 2(a)); i.e., the configuration formed by the ac drive with $d_{inp} = 1$ μm and $V^{\infty} = 100$ μV after $n = 0$ (red line), many cycles (n >4000) (blue line), and $n = 100$ cycles (green line). The horizontal dashed line again marks the position of 1. We observe qualitatively the same behavior as $|V(t)|/V^{\infty}$ in figure 2(a), namely, a monotonic increase (red line), a monotonic decrease (blue line), and a nonmonotonic fast decrease followed by a slow increase (green line) to 1. Again, we find that the green line is reproduced by a weighted average of the red and blue lines. The purple line represents the fit of the green line to Eq. (1) with α=0.57±0.01. This value is close to α=0.60±0.04 obtained above. The result strongly suggests that we have indeed detected the intrinsic properties related to the vortex configuration not only from the ac readout, but also from the dc readout.

The characteristic time at which V/V^{∞} shown by the green line in figure 2(a) takes a minimum is around t =2 cycles, thus, corresponding to about 0.5 ms. This time roughly coincides with $t =$ 0.5 ms around which V/V^{∞} shown by the green line in figure 2(b) takes a minimum. Therefore, the characteristic time scales seen between figures 2(a) and (b) are nearly identical, insensitive to whether the readout is performed by ac drive or dc drive.

Obviously, a better fit is obtained by dc drive. However, this does not immediately mean that the ac readout experiment can be thoroughly replaced by the dc one. This is because the readout experiment by dc drive is applicable only when the initial and/or final configuration in the input experiment are highly disordered. This can be understood considering the fact that when both the initial and final configurations are relatively ordered, we are unable to perform the readout experiment that shows the dynamic ordering by using any dc drive. Conversely, the readout experiment by ac drive is applicable only when the initial and/or final configuration in the input experiment are relatively ordered. This is because when both the initial and final configurations are highly disordered, it is difficult to conduct the readout experiment that shows the dynamic disordering by using any ac drive [20, 21, 22], even with large d or small V^{∞}. In the present work, we have carefully selected the particular regions of the initial and final vortex configurations in the input experiment where both the dynamic ordering and disordering can be realized in the subsequent dc as well as ac readout experiment.

Now that we have established the readout technique by dc drive in addition to the ac one developed recently [7], we expect that both techniques will be applied complementarily to detect the vortex configuration over a wide range of disorder. The study using both techniques includes the determination of the vortex configuration during the dynamic ordering by dc drive, as mentioned in the introduction.

4. Conclusions

We study the random organization or dynamic ordering by ac drive using a vortex system of an a-Mo_xGe_{1-x} film with weak random pinning. We showed recently that the random organization by ac drive can be detected from the t-dependent voltage $|V(t)|$ that increases and saturates to a steady-state value [7]. From the subsequent readout experiment of $V(t)$ using various ac amplitudes d, it was shown that the transient vortex configuration during the ac dynamic ordering is not microscopically homogeneous but consists of DR and OR. The area ratio of OR, $\alpha(n)$, increases monotonically from 0 to 1 with increasing n [7].

In this work, we have developed an alternative readout method using a *dc* drive. It is found that the dc readout method gives the same results (e.g., the same value of α) as obtained from the ac drive. This finding provides an additional support for our view of the coexistence of OR and DR, and the interpretation of α [7]. We expect that these two methods will be used complementarily to detect the vortex configuration over a wide range of disorder, as well as to clarify the origin of the coexistence regions.

Acknowledgments
This work was supported by a Grant-in-Aid for Scientific Research from the Ministry of Education, Culture, Sports, Science, and Technology of Japan.

References

[1] Pine D J, Gollub J P, Brady J F and Leshansky A M 2005 *Nature* **438** 997
[2] Gollub J and Pine D 2006 *Phys. Today* **59** 8
[3] Corté L, Chaikin P M, Gollub J P and Pine D J 2008 *Nature Phys.* **4** 420
[4] Mangan N, Reichhardt C and Olson Reichhardt C J 2008 *Phys. Rev. Lett.* **100** 187002
[5] Okuma S, Tsugawa Y and Motohashi A 2011 *Phys. Rev.* B **83** 012503
[6] Okuma S, Kawamura Y and Tsugawa Y 2012 *J. Phys. Soc. Japan* **81** 114718
[7] Dobroka M, Kawamura Y, Ienaga K, Kaneko S and Okuma S 2017 *New J. Phys.* **19** 053023
[8] Reichhardt C and Olson Reichhardt C J 2009 *Phys. Rev. Lett.* **103** 168301
[9] Okuma S and Motohashi A 2012 *New J. Phys.* **14** 123021
[10] Okuma S, Motohashi A and Kawamura Y 2013 *Phys. Lett.* A **377** 2990
[11] Dobroka M, Kawamura Y, Ienaga K, Kaneko S and Okuma S, unpublished.
[12] Peréz Daroca D, Pasquini G, Lozano G S and Bekeris V 2011 *Phys. Rev.* B **84** 012508
[13] Kes P H and Tsuei C C 1983 *Phys. Rev.* B **28** 5126
[14] Bhattacharya S and Higgins M J 1993 *Phys. Rev. Lett.* **70** 2617
[15] Banerjee S S, Patil N G, Ramakrishnan S, Grover A K, Bhattacharya S, Mishra P K, Ravikumar G, Rao T V C, Sahni V C, Higgins M J, Tomy C V, Balakrishnan G and Paul D M 1999 *Phys. Rev.* B **59** 6043
[16] Xiao Z L, Andrei E Y and Higgins M J 1999 *Phys. Rev. Lett.* **83** 1664
[17] van der Beek C J, Colson S, Indenbom M V and Konczykowski M 2000 *Phys. Rev. Lett.* **84** 4196
[18] Bermúdez M M, Eskildsen M R, Bartkowiak M, Nagy G, Bekeris V and Pasquini G 2015 *Phys. Rev. Lett.* **115** 067001
[19] Guillamón I, Córdoba R, Sesé J, De Teresa J M, Ibarra M R, Vieira S and Suderow H 2014 *Nat. Phys.* **10** 851
[20] Valenzuela S O 2002 *Phys. Rev. Lett.* **88** 247003
[21] Pautrat A, Goupil C, Simon C, Charalambous D, Forgan E M, Lazard G, Mathieu P and Brûlet A 2003 *Phys. Rev. Lett.* **90** 087002
[22] Nojima T, Takahashi K, Chotoku M, Ochiai A, Aoki H, Lee H G and Lee S I 2009 *J. Phys.: Conf. Ser.* **150** 052189

31st International Symposium on Superconductivity (ISS2018) IOP Publishing
IOP Conf. Series: Journal of Physics: Conf. Series **1293** (2019) 012024 doi:10.1088/1742-6596/1293/1/012024

Impurity induced insulator-to-metal transitions in half-filled Mott insulators

Hisatoshi Yokoyama[1], Ryo Sato[1] and Kenji Kobayashi[2]

[1]Department of Physics, Tohoku University, Aoba-ku, Sendai, 980-8578, Japan
[2]Department of Natural Science, Chiba Institute of Technology, Narashino, 275-0023, Japan

E-mail: yoko@cmpt.phys.tohoku.ac.jp

Abstract. In view of cuprate superconductors, effects of point-type impurity potential (V) on half-filled antiferromagnetic and paramagnetic states are studied for a strongly correlated square-lattice Hubbard model ($U/t = 12$) with a diagonal transfer (t'), using a variational Monte Carlo method. In the trial states, we introduce a Rice-Brinkman-type one-body projector for V, which enables us to treat the whole range of V ($-\infty < V/t < \infty$) with small statistical errors. Filling-control-type Mott (insulator-to-metal) transitions are found to occur at $V = V_U^{(\pm)}$ ($\sim \pm U$). In the metallic states ($|V| > |V_U^{(\pm)}|$), charge carriers are holes (electrons) in the case of attractive (repulsive) potential. The mechanism of the transitions is intuitively understood.

1. Introduction

Cuprate superconductors (SCs) are in the regime of Mott physics, and have inherent disorder originating in the carrier dopants. Furthermore, the impurity effect of apical oxygen in the T'-structure cuprates (typical for electron-doped systems) has been widely noticed [1] in connection with the so-called non-carrier-doped SCs [2]. Thus, it is important to clarify the impurity effects on strongly correlated systems.

In the previous reports, we studied the effects of point-type impurity potentials V on antiferromagnetic (AF) and paramagnetic (PM) states using a variational Monte Carlo (VMC) method. We mainly discussed U/t dependence of the impurity effect and also V/t dependence for large values of U/t in attractive potential ($V < 0$) [3], and metal-to-insulator (Mott) transitions occurring in partially-filled systems for $\delta_{\rm imp} = \delta$ ($\delta_{\rm imp}$: impurity density, δ: doping rate) in weakly repulsive potential ($V = V_{\rm M} = 0 - 2t$) [4]. However, the VMC optimization with the trial states used in these studies, $\Psi(\#x)$ ($x = 1 - 6$), often brought about uncontrollable large statistical fluctuations for $V > V_{\rm M}$, which prevented us from making a systematic study in the repulsive potential. From the VMC data, we presume the cause of this shortcoming of $\Psi(\#x)$ to be the fact that many-body correlation factors, which distinguish the impurity and host (non-impurity) sites, are used for controlling the one-body impurity potential prior to the original role.

To remedy such shortcomings, we consider a new trial state $\Psi(\theta)$, in which a one-body projector for V ($\mathcal{P}_\theta$) [5] is introduced and many-body correlation factors do not depend on the attribute of sites (impurity or host). This trial state has another merit, namely, simple structure with respect to an often-used electron-hole transformation. Consequently, the properties of repulsive potential, especially at half filling, are expected from those of attractive potential, and

IOP Conf. Series: Journal of Physics: Conf. Series **1293** (2019) 012024 doi:10.1088/1742-6596/1293/1/012024

vice versa. $\Psi(\theta)$ basically exhibits behavior similar to $\Psi(\#x)$ for $V < V_{\rm M}$. In this article, using $\Psi(\theta)$, we discuss insulator-to-metal transitions occurring in strong impurity potential and other properties at half filling.

2. Formalism

2.1. Model

In this study, we consider a Hubbard model ($U \geq 0$) with a point-type spin-independent impurity potential (V) on extended square lattices:

$$\mathcal{H} = \mathcal{H}_{\rm kin} + \mathcal{H}_U + \mathcal{H}_{\rm imp} = -\sum_{(i,j),\sigma} t_{ij}\left(c^\dagger_{i\sigma}c_{j\sigma} + {\rm H.c.}\right) + U\sum_j n_{j\uparrow}n_{j\downarrow} + V\sum_{\ell=1}^{N_{\rm imp}}\sum_\sigma n_{\ell\sigma}, \quad (1)$$

where $n_{j\sigma} = c^\dagger_{j\sigma}c_{j\sigma}$, (i,j) indicates the pairs on sites i and j, and $N_{\rm imp}$ is the number of impurity sites, which we will explain shortly. Here, we set the hopping integral t_{ij} as t for nearest neighbors (≥ 0), t' for diagonal neighbors, and 0 otherwise ($\mathcal{H}_{\rm kin} = \mathcal{H}_t + \mathcal{H}_{t'}$). We use t and the lattice spacing as the units of energy and length, respectively. Here, we focus on the cases at half filling ($n = N/N_{\rm s} = 1$ or $\delta = |1-n| = 0$, N: number of electrons, $N_{\rm s}$: number of sites) and of typical impurity densities $\delta_{\rm imp} = N_{\rm imp}/N_{\rm s} \sim 0.08$ in strong correlation ($U/t = 12$). We assume that the impurity potential V comes from a single species, for example, apical oxygen atoms in a T'-structure system [1], and act on a nearby single site ("impurity site") for simplicity. If we grant this simplification, the index ℓ in $\mathcal{H}_{\rm imp}$ runs over the impurity sites $\mathbf{r}_\ell$ ($\ell = 1, \cdots, N_{\rm imp}$). Furthermore, we assume that $\mathbf{r}_\ell$ are randomly distributed. We widely varies V (even for $V < 0$) to grasp overall features of impurity effects. Although the above setting of impurity potential is possibly simplistic, we believe that fundamentals of impurity effects with strong correlation can be captured.

2.2. Method

We apply a many-body variation theory to the above model. In computing expectation values, we employ a VMC method [6, 7, 8], which enables us to exactly treat many-body wave functions. As a trial wave function, we use a Jastrow type: $\Psi = \mathcal{P}\Phi$. In the previous studies [3, 4], we distinguished the impurity and host sites in the (many-body) projection factor $\mathcal{P}$ in $\Psi(\#x)$. We found, however, that this leads to the uncontrollable fluctuation in the VMC optimization for $V > V_{\rm M}$, as mentioned. Therefore, in the new trial state $\Psi(\theta)$, we do not distinguish the impurity and host sites in the many-body projection factors, $\mathcal{P} = \mathcal{P}_{\rm H}\mathcal{P}_\sigma\mathcal{P}_Q\mathcal{P}_{\rm G}$, where $\mathcal{P}_{\rm H}$, $\mathcal{P}_\sigma$, $\mathcal{P}_Q$, and $\mathcal{P}_{\rm G}$ are a nearest-neighbor charge [4] and spin ($\mathcal{P}_\sigma = \beta^\sigma$ with β being a parameter and $\sigma = 4\sum_{\langle i,j\rangle} S^z_i S^z_j$) correlators, a nearest-neighbor doublon-holon binding factor, and onsite (Gutzwiller) factor, respectively, where doublon (holon) indicates a doubly occupied (empty) site. For $\mathcal{P}_Q$ and $\mathcal{P}_{\rm G}$, we adopt the same forms as those used for the uniform systems [8]. Instead of controlling the effect of impurity potential using many-body correlators, we introduce a one-body projector [5]

$$\mathcal{P}_\theta = \theta^\kappa, \quad \text{with} \quad \kappa = \sum_{\ell,\sigma} n_{\ell\sigma}, \quad (2)$$

and θ being a parameter. Thus, the new trial state becomes $\Psi(\theta) = \mathcal{P}\mathcal{P}_\theta\Phi$.

To obtain the Slater-determinantal part Φ, we diagonalize the following mean-field Hamiltonian of spin σ,

$$\mathcal{H}^{(\sigma)}_{\rm MF} = \mathcal{H}^{{\rm BR}(\sigma)}_{\rm kin} + U\sum_j \langle n_{j\sigma}\rangle n_{j-\sigma} + \mathcal{H}^{(\sigma)}_{\rm imp}. \quad (3)$$

Table 1. Relationship of elements in Hamiltonian Eq. (1), related quantities (defined later), and variational (and fixed) parameters in trial wave function $\Psi(\theta)$ through unitary transformation Eq. (4) at half filling. In the column of physical quantities, the quantity in the left[right]-hand side indicates, say, $n_{\rm imp}(V,t')$ $[2-n_{\rm imp}(-V,-t')]$.

Hamiltonian	Physical quantities	Variational (fixed) parameters	
$\mathcal{H}_t \leftrightarrow \mathcal{H}_t$	$n_{\rm imp} \leftrightarrow 2-n_{\rm imp}$	$g \leftrightarrow g$	$\delta_{\rm AF} \leftrightarrow \delta_{\rm AF}$
$\mathcal{H}_{t'} \leftrightarrow -\mathcal{H}_{t'}$	$n_{\rm hst} \leftrightarrow 2-n_{\rm hst}$	$\zeta_{\rm d} \leftrightarrow \zeta_{\rm h}$	$t_1 \leftrightarrow -t_1$
$\mathcal{H}_U \leftrightarrow \mathcal{H}_U$	$m \leftrightarrow m$	$\zeta_{\rm h} \leftrightarrow \zeta_{\rm d}$	$t_2 \leftrightarrow -t_2$
$\mathcal{H}_V \leftrightarrow -\mathcal{H}_V + 2VN_{\rm imp}$	$d \leftrightarrow h$	$\eta \leftrightarrow \eta$	$t_3 \leftrightarrow t_3$
		$\beta \leftrightarrow \beta$	$t_4 \leftrightarrow t_4$
		$(V \leftrightarrow -V)$	$\theta \leftrightarrow \theta^{-1}$

In the kinetic part $\mathcal{H}_{\rm kin}^{{\rm BR}(\sigma)}$, a band-renormalization effect is introduced in the same manner as in Refs. [3, 4, 8]. In the second term, we take account of an AF long-range order of the nesting vector $\mathbf{Q}=(\pi,\pi)$ also in the same manner as in the previous studies. In the last term, we use the raw impurity term in the Hamiltonian to avoid the redundancy of the screening effect, that is, because $V_{\rm eff}/V$ [in $\Psi(\#x)$] and $\mathcal{P}_\theta$ play similar roles. Thus, V in $\Psi(\theta)$ is a constant given in Eq. (1). Diagonalizing $\mathcal{H}_{\rm MF}^{(\sigma)}$ of $N_{\rm s}\times N_{\rm s}$ matrices, we obtain eigenenergies ε_n and eigenvectors $\phi_{n\sigma}$ with corresponding quasiparticle operators $\alpha_{n\sigma}^\dagger$, which are linear combinations of $c_{j\sigma}^\dagger$. Using $\alpha_{n\sigma}^\dagger$, one-body part is constructed as $\Phi=\prod_{n\ (\varepsilon_n\le\varepsilon_{\rm F})}\prod_\sigma \alpha_{n\sigma}^\dagger|0\rangle$, where $\varepsilon_{\rm F}$ indicates the energy of the highest occupied level (HOMO). $\Psi(\theta)$ is reduced to the exact noninteracting ground state of Eq. (1) for $U=0$ and to the corresponding uniform states for $V=0$. The AF state $\Psi_{\rm AF}=\mathcal{P}\mathcal{P}_\theta\Phi_{\rm AF}$ is reduced to the PM state $\Psi_{\rm PM}=\mathcal{P}\mathcal{P}_\theta\Phi_{\rm PM}$ for $\delta_{\rm AF}\to 0$ [4] and $\beta\to 1$.

We compute expectation values with respect to $\Psi(\theta)$ using a VMC method similar to that in Ref. [8]. We use systems of $N_{\rm s}=L\times L$ sites with $L=10-16$ and the periodic-antiperiodic boundary conditions. In a VMC sweep, we use 20,000 samples, because the convergence of $\Psi(\theta)$ is much better than $\Psi(\#x)$. A configuration of impurities $\{\mathbf{r}_\ell\}$ is randomly chosen and fixed throughout a single sweep of VMC calculation (optimization and measurement). In most cases, we will display data of 4 kinds of $\{\mathbf{r}_\ell\}$ simultaneously.

2.3. Electron-hole transformation

The electron-hole (unitary) transformation $\mathcal{U}$ defined below is often used to treat a more-than-half-filled system, especially in cases with $\mathcal{H}_{t'}$, in a less-than-half-filled band:

$$\mathcal{U}^{-1}c_{j\sigma}^\dagger\mathcal{U}\to e^{i\mathbf{Q}\cdot\mathbf{r}_j}\tilde{c}_{j\sigma},\quad\text{or}\quad \mathcal{U}^{-1}c_{\mathbf{k}\sigma}^\dagger\mathcal{U}\to\tilde{c}_{\mathbf{Q}-\mathbf{k}\sigma}. \tag{4}$$

The expectation value of an observable O with respect to a wave function $|\Psi\rangle$ is transformed as

$$O_{\rm avr}=\langle O\rangle=\frac{\langle\Psi|O|\Psi\rangle}{\langle\Psi|\Psi\rangle}=\frac{\langle\tilde\Psi|\tilde O|\tilde\Psi\rangle}{\langle\tilde\Psi|\tilde\Psi\rangle}=\widetilde{\langle O\rangle}, \tag{5}$$

with $\tilde O=\mathcal{U}^{-1}O\mathcal{U}$ and $|\tilde\Psi\rangle=\mathcal{U}^{-1}|\Psi\rangle$. At half filling, the original forms and transforms of entities often have simple relations (Table 1), which are useful for connecting the cases of attractive and repulsive potentials and for checking the correctness of calculations. The trial wave function $\Psi(\theta)$ is transformed by simply substituting the variational parameters as in the right column of Table 1, in contrast with the former wave function $\Psi(\#6)$. As for the Hamiltonian Eq. (1),

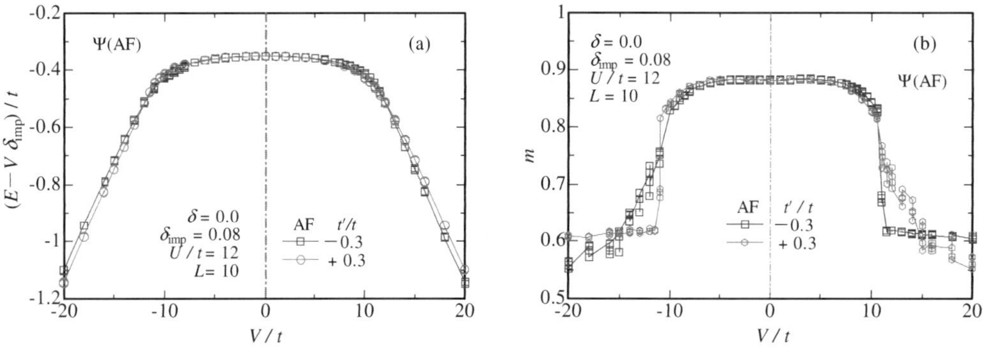

Figure 1. (a) Total energy per site shifted by $-V\delta_{\rm imp}$ (i.e. $\mathcal{E}$) and (b) staggered magnetization as functions of V/t at half filling for AF state. Both quantities are computed using the VMC method. The PM state shows similar symmetric behavior in $\mathcal{E}$ as in (a) (not shown).

$\mathcal{H}_t$ and $\mathcal{H}_U$ are invariant, whereas the signs of $\mathcal{H}_V$ and $\mathcal{H}_{t'}$ are reversed with and without a constant, respectively. Thus, the total energy per site E satisfies the relation,

$$E(V,t') - V\delta_{\rm imp}\ \left[\equiv \mathcal{E}(V,t')\right] = E(-V,-t') - (-V\delta_{\rm imp})\left[= \mathcal{E}(-V,-t')\right], \tag{6}$$

where the model parameters other than V and t' are fixed. As another example, the staggered magnetization (an AF order parameter),

$$m = \frac{2}{N_{\rm s}}\left|\sum_j e^{i\mathbf{Q}\cdot\mathbf{r}_j}\langle S_j^z\rangle\right|, \tag{7}$$

satisfies the relation

$$m(V,t') = m(-V,-t'). \tag{8}$$

Thus, $\mathcal{E}$ and m are symmetric with respect to $V = 0$, if the sign of t' is reversed along with the sign of V. Figs. 1(a) and 1(b) show the V/t dependence of $\mathcal{E}(V,t')$ and $m(V,t')$, respectively, actually calculated by the VMC method for the AF state. The cases for $t'/t = -0.3$ $(+0.3)$ are simultaneously plotted with blue (red) symbols and lines. We can confirm that the above symmetry, namely, Eqs. (6) and (8) are satisfied, indicating the VMC calculations are correct.

3. Insulator-to-metal transitions

In the uniform cases ($V = 0$), the AF and PM ($U > U_{\rm c} \sim 8t$ [9], $U_{\rm c}$: Mott transition point) states are insulating. In Fig. 1(a), E exhibits rapid variations of slope at $V = V_U^{(\pm)} \sim \pm 11t$; correspondingly, in Fig. 1(b), m rapidly drops at $V = V_U^{(\pm)}$ as $|V|$ increases. In particular for $V > 0$ and $t'/t = -0.3$ (equivalently for $V < 0$ and $t'/t = +0.3$), the behavior at $V = V_U$, namely discontinuity of m, suggests first-order transitions.

To study these anomalies, let us consider the momentum distribution function,

$$n(\mathbf{k}) = \frac{1}{2}\sum_\sigma \langle c_{\mathbf{k}\sigma}^\dagger c_{\mathbf{k}\sigma}\rangle. \tag{9}$$

Quantities defined in the wave number space, such as $n(\mathbf{k})$ and the charge-density structure factor $N(\mathbf{q})$ [Eq. (10)], are basically not well-defined in systems without translational symmetry.

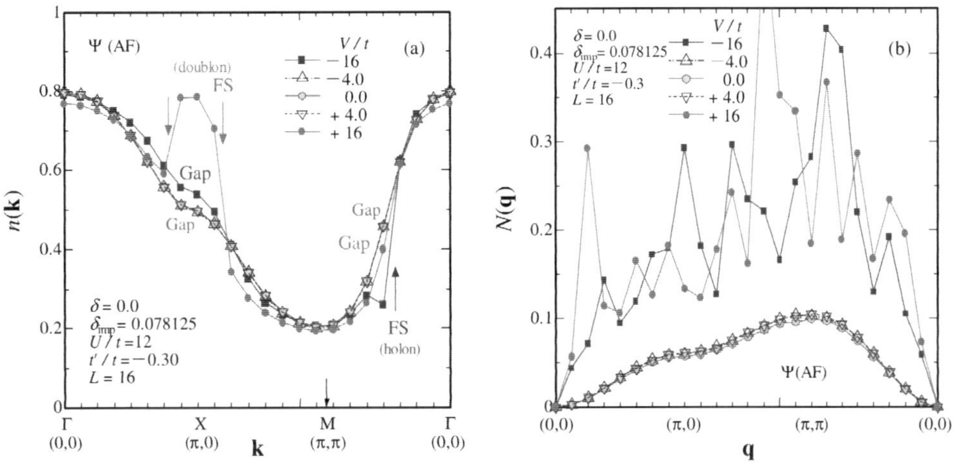

Figure 2. (a) Momentum distribution function and (b) charge-density structure factor for AF state along path of $\mathbf{k}$ (or $\mathbf{q}$), $(0,0)$-$(\pi,0)$-(π,π)-$(0,0)$ for five different values of V/t at half filling. The data for a value of V/t are computed for a fixed impurity configuration, which is different for different V/t.

Nevertheless, they provide us with useful information as far as the electronic states preserve some homogeneous nature, or information as to how the randomness affects the electronic states, as discussed below. In Fig. 2(a), we show $n(\mathbf{k})$ for five different values of V/t on the indicated path of $\mathbf{k}$. For $V_U^{(-)} < V < V_U^{(+)}$, $n(\mathbf{k})$ is smooth and does not have Fermi surface (FS) (discontinuity) at any point on this path, meaning that the state is insulating. Note that the values for $V/t = -4$, 0, and 4 almost coincide; the feature of the homogeneous case ($V = 0$) as an insulator is preserved within this regime. This is understood by the fact that the screening of V is almost complete there. Because it is not easy to directly see the screening effect in $\Psi(\theta)$, we show in Fig. 3 the screening parameter $V_{\rm eff}/V$ obtained for a similar wave function $\Psi(\#6)$ [4], in which the screening becomes complete for $V_{\rm eff}/V = 0$ and vanishes for $V_{\rm eff}/V = 1$. The screening of V is almost complete for $V_U^{(-)} < V < 0$, whereas the effect of V somewhat remains for $V < V_U^{(-)}$. This difference of screening effect is marked in the charge-density structure factor

$$N(\mathbf{q}) = \frac{1}{N_{\rm s}} \sum_{i,j} e^{i\mathbf{q}\cdot(\mathbf{r}_i - \mathbf{r}_j)} \langle n_i n_j \rangle - n^2, \qquad (10)$$

as shown in Fig. 2(b). It shows that for $|V| < |V_U^{(\pm)}|$, the electron configuration is hardly affected by V, whereas for $|V| > |V_U^{(\pm)}|$ the electron configuration largely depends on the impurity configuration.

Returning to Fig. 2(a), for $V < V_U^{(-)}$, we find that a pocket FS appears around $(\pi/2, \pi/2)$ in the nodal direction $(0,0)$–(π,π), as seen for $V/t = -16$. It means that an insulator-to-metal transition occurs at $V = V_U^{(-)}$ and a metallic state is realized for $V < V_U^{(-)}$. Because $n(\mathbf{k})$ decreases in the pocket compared to the insulating cases, the carriers are expected to be holes. On the other hand for $V > V_U^{(+)}$, $n(\mathbf{k})$ exhibits a pocket FS around the antinodal point $(\pi, 0)$, as seen for $V/t = +16$. A similar Mott transition occurs at $V = V_U^{(+)}$ and the state becomes metallic for $V > V_U^{(+)}$. In this case, the carriers are considered electrons.

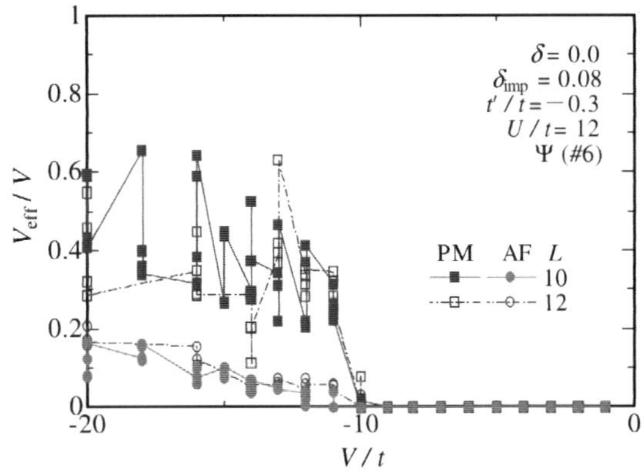

Figure 3. Optimized screening parameter obtained in the former wave function $\Psi(\#6)$ [3, 4] as a function of V/t (< 0) for $U/t = 12$. The properties of $\Psi(\#6)$ and $\Psi(\theta)$ are basically the same. The AF and PM cases are plotted together. Calculations for $V > 0$ by $\Psi(\#6)$ is not easy, but the values of $V_{\rm eff}/V$ for $V > 0$ are guessed from that for $V < 0$, because $V_{\rm eff}/V$ should be broadly symmetric with respect to $V/t = 0$.

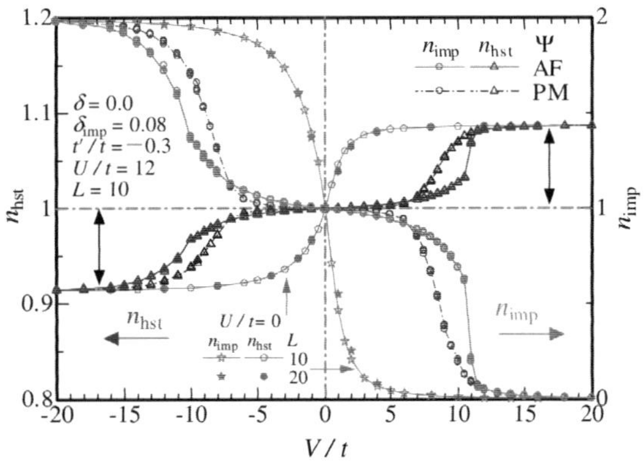

Figure 4. Average electron densities at impurity sites ($n_{\rm imp}$, right axis) and host sites ($n_{\rm hst}$, left axis) as functions of impurity potential at half filling. $n_{\rm imp}$ and $n_{\rm hst}$ have the relations shown in Table 1. The parts of charge carriers in $n_{\rm hst}$ are indicated by double-headed arrows. In this case, $N_{\rm imp}/N_{\rm hst} \sim 0.087$ for $|V|/t \to \infty$. As a reference, the numerically calculated exact values for $U = 0$ ($L = 10$ and 20) are also plotted with stars and pentagons.

To clarify the nature of this Mott transition, we consider the average electron densities on the impurity sites ($n_{\rm imp}$) and the host sites ($n_{\rm hst}$) and the average doublon density (d) calculated as

$$n_{\rm imp} = \frac{1}{N_{\rm imp}} \sum_{\ell=1}^{N_{\rm imp}} \sum_{\sigma} \langle n_{\ell\sigma} \rangle, \qquad n_{\rm hst} = \frac{N - N_{\rm imp} n_{\rm imp}}{N_{\rm s} - N_{\rm imp}}, \qquad d = \frac{1}{N_{\rm s}} \sum_j \langle n_{j\uparrow} n_{j\downarrow} \rangle. \tag{11}$$

In Fig. 4, $n_{\rm imp}$ and $n_{\rm hst}$ are shown as functions of V/t. In the insulating area ($V_U^{(-)} < V < V_U^{(+)}$), $n_{\rm hst}$ is nearly 1, namely, half filling is realized in the host sites. On the other hand for $V < V_U^{(-)}$ ($V > V_U^{(+)}$), $n_{\rm hst}$ approaches $1 - (+) N_{\rm imp}/N_{\rm hst}$ ($N_{\rm hst} = N_{\rm s} - N_{\rm imp}$: number of host sites), because $|V| \gtrsim U$ and the impurity sites are almost occupied by doublons (holons). Therefore, for $V < V_U^{(-)}$ ($V > V_U^{(+)}$), the charge carriers (indicated by double-headed arrows in Fig. 4) are holons (doublons), which is consistent with the feature of $n(\mathbf{k})$ discussed above. This feature is reflected in the average doublon density shown in Fig. 5. In the metallic regimes, d increases up

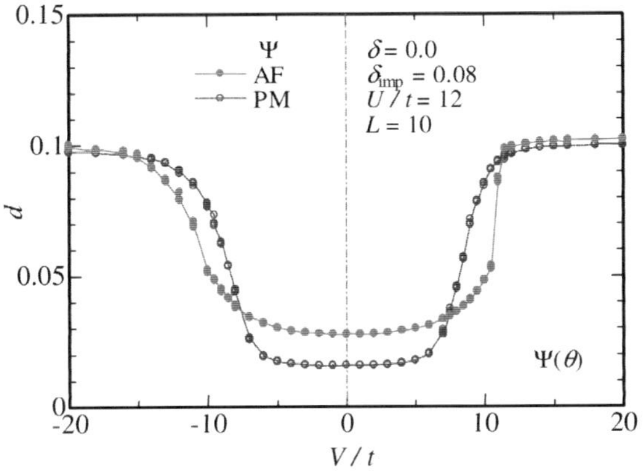

Figure 5. Average doublon (doubly occupied sites) densities in the whole system as a function of impurity potential for AF and PM cases. The contribution owing to that all impurity sites are occupied by doublons or holons is $N_{\rm imp}/N_{\rm hst} \sim 0.087$ in this case. The residual part is the contribution from local doublon-holon-pair creation.

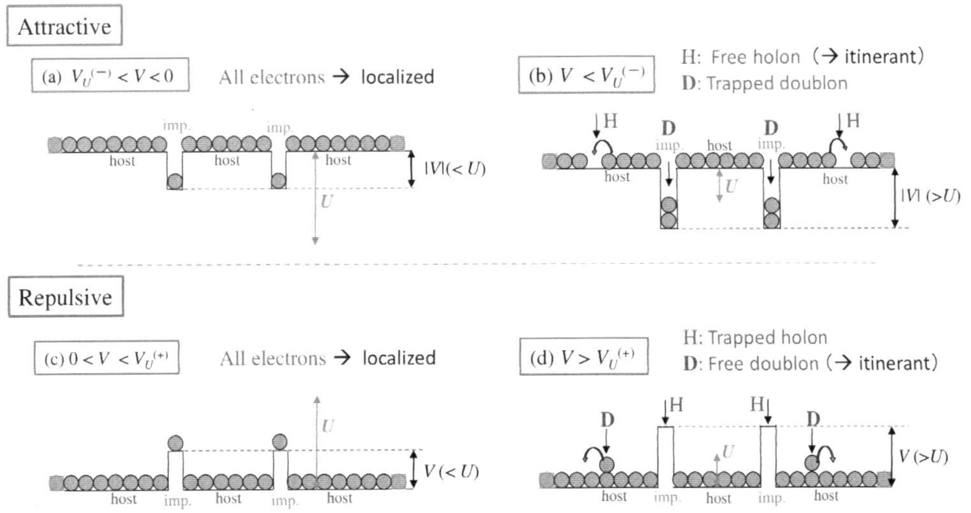

Figure 6. Explanatory drawing of impurity-induced insulator-to-metal transition at half filling in Mott regime ($U > U_{\rm c}$). (a) Weakly and intermediately attractive cases ($|V| \lesssim U$), (b) strongly attractive cases ($|V| \gtrsim U$), (c) weakly and intermediately repulsive cases ($V \lesssim U$), and (d) strongly repulsive cases ($V \gtrsim U$).

to 0.1; for attractive V, doublons are created and trapped in the impurity sites, and the pair-produced holons are in the host sites and become the carriers. For repulsive V, every impurity site is occupied by a holon and the pair-produced doublon freely moves in the host sites.

Figure 6 schematically shows this mechanism of the Mott transition. In the cases of weak and intermediate potential ($|V| \lesssim U$) shown in (a) and (c), every site is occupied by an electron except when a local D-H pair is produced. These cases are basically in the same situation as the pure case ($V = 0$), in which an ordinary Mott insulating state is realized. On the other hand in the case of strong potential ($|V| \gtrsim U$), almost all impurity sites are occupied by doublons or holons. As a result, the counter holons (+-charged carriers) or doublons (−-charged carriers) are created in the host sites and move around freely. From such a intuitive picture, the Mott transition point is roughly estimated to be $|V_U^{(\pm)}| \sim U - |E_{\rm kin}^{(\rm D/H)}|$, where $E_{\rm kin}^{(\rm D/H)}$ $(\sim -t)$ is the kinetic energy of free carriers (doublon/holon) in the host sites. This estimation seems appropriate for the AF case ($|V_U^{(\pm)}| \sim 11t$), but somewhat larger for the PM case ($|V_U^{(\pm)}| \sim 7t$).

4. Conclusion

We studied the effects of point-type impurity potentials on a strongly correlated AF and PM states at half filling, using a VMC method. We showed that a filling-control-type Mott transitions take place for strong impurity potential ($V = V_U^{(\pm)} \sim \pm U$). The mechanism is intuitively expected. In fact, we have found that such insulator-to-metal transitions are not restricted to the present case of $\delta = 0$ but manifest themselves widely for $\delta_{\rm imp} > \delta > 0$ with $V > 0$. We will deliver a more systematic report elsewhere. Similar Mott transitions have been also found in a superconducting state [10]. It is important to consider how the Mott transitions discussed here are related to the behavior of cuprates, or more generally to phenomena in other materials.

Acknowledgments

The authors thank Tsutomu Watanabe for useful discussions. This study is supported in part by Grants-in-Aid from the Ministry of Education, Culture, Sports, Science and Technology, Japan.

References

[1] Adachi T, Mori Y, Takahashi A, Kato M, Nishizaki T, Sasaki T, Kobayashi N and Koike Y 2013 *J. Phys. Soc. Japan* **82** 063713
[2] Matsumoto O, Utsuki A, Tsukada A, Yamamoto H, Manabe T and Naito M 2009 *Physica C* **469** 924
[3] Yokoyama H, Sato R and Kobayashi K 2017 *J. Phys. Conf. Ser.* **871** 012032
[4] Yokoyama H, Sato R and Kobayashi K 2018 *J. Phys. Conf. Ser.* **1054** 012014
[5] Rice T M and Brinkman W F 1972 *Phys. Rev. B* **5** 4350
[6] Yokoyama H and Shiba H 1987 *J. Phys. Soc. Japan* **56** 1490.
[7] Umrigar C J, Wilson K G and Wilkins J W 1988 *Phys. Rev. Lett.* **60** 1719
[8] Sato R and Yokoyama H 2016 *J. Phys. Soc. Japan* **85** 074701
[9] Yokoyama H, Ogata M and Tanaka Y 2006 *J. Phys. Soc. Japan* **75** 114706
[10] Sato R, Yokoyama H to be publised in *J. Phys. Conf. Ser.* (This volume)

IOP Conf. Series: Journal of Physics: Conf. Series **1293** (2019) 012025 doi:10.1088/1742-6596/1293/1/012025

Relationship between superconductivity and anisotropy in two-dimensional Hubbard model

Kenji Kobayashi[1] **and Hisatoshi Yokoyama**[2]

[1]Department of Natural Science, Chiba Institute of Technology, Shibazono, Narashino 275-0023, Japan

[2]Department of Physics, Tohoku University, Aoba-ku, Sendai, 980-8578, Japan

E-mail: kenji.kobayashi@it-chiba.ac.jp

Abstract. In connection with the symmetry-breaking phenomena found in cuprate superconductors, we check whether an anisotropy spontaneously appears or not in strongly correlated electrons that exhibit superconductivity and/or antiferromagnetism. A variational Monte Carlo method is applied to the square-lattice Hubbard model ($U/t = 12$) with diagonal transfer t'. In the trial wave function used, we introduce band renormalization factors including adjustable parameters for the anisotropy corresponding to the fourfold rotational symmetry breaking. It is confirmed that the anisotropy appears in the overdoped regime, where the Fermi-liquid features are predominant. In the underdoped area, the anisotropy does not appear for $t'/t = 0$ and $+0.3$, whereas it possibly appears in the underdoped regime for $t'/t = -0.3$.

1. Introduction

Recently, symmetry-breaking phenomena have been successively found in cuprate superconductors; electronic nematic order breaking the rotational symmetry [1] and charge density wave breaking the translational symmetry [2] were experimentally discovered along with superconductivity (SC). These symmetry breakings attract much interests because they may provide important insights into the relationship between SC and the enigmatic pseudogap state. Pomeranchuk instability, a spontaneous breaking of fourfold rotational symmetry of the Fermi surface without lattice distortion in this case, is a noteworthy candidate for the nematicity observed in cuprate superconductors [3, 4, 5].

In this article, we will check whether the x–y anisotropy spontaneously appears in the two-dimensional strongly correlated Hubbard model using a variational Monte Carlo (VMC) method, and clarify the relationship between the anisotropy and SC when the model parameters are varied. In order to accurately determine the x–y anisotropy/isotropy in the electronic state, we employ the periodic boundary conditions in both x and y directions on a slightly tilted square lattice. In the trial wave function, we introduce the following features:

(i) We adopt a mixed state of antiferromagnetic (AF) and SC orders, by which we can treat a continuous description from their coexistence to the mutual exclusivity.

(ii) The band renormalization effect owing to electron correlation is independently introduced into AF and SC orders by adjusting the parameters of hopping integrals, some of which control the anisotropy in x and y directions.

(iii) As multi-body correlation factors, a doublon-holon binding factor important for Mott physics and an on-site Gutzwiller factor, are used to capture the essence of strong correlation.

By applying the VMC method with this wave function to a square Hubbard (t-t'-U) model, we will elucidate the relationship between SC and anisotropy in electronic states in the strongly correlated regime.

2. Method

We consider the Hubbard model on a square lattice with the next-nearest-neighbor transfer t', as shown in Fig. 1(a),

$$\begin{aligned}\mathcal{H} &= \mathcal{H}_t + \mathcal{H}_{t'} + \mathcal{H}_U \\ &= -t \sum_{\langle \mathbf{r}\mathbf{r}' \rangle, \sigma} \left(c^\dagger_{\mathbf{r},\sigma} c_{\mathbf{r}',\sigma} + \text{h.c.} \right) - t' \sum_{(\mathbf{r}\mathbf{r}'), \sigma} \left(c^\dagger_{\mathbf{r},\sigma} c_{\mathbf{r}',\sigma} + \text{h.c.} \right) + U \sum_{\mathbf{r}} n_{\mathbf{r}\uparrow} n_{\mathbf{r}\downarrow},\end{aligned} \tag{1}$$

where $c^\dagger_{\mathbf{r},\sigma}$ creates an electron of spin σ at site $\mathbf{r}$, $n_{\mathbf{r}\sigma} = c^\dagger_{\mathbf{r},\sigma} c_{\mathbf{r},\sigma}$, and $\langle \mathbf{r}\mathbf{r}' \rangle$ and $(\mathbf{r}\mathbf{r}')$ denote indices of nearest-neighbor and diagonal-neighbor pairs, respectively. In order to improve the accuracy in distinguishing between isotropy and anisotropy, we use periodic boundary conditions in both x and y directions. In addition, in order to increase the number of data points satisfying closed-shell conditions, under which VMC calculations are stabilized, we use a slightly tilted square lattice with edge vectors $\mathbf{a}_1 = (m, n)$ and $\mathbf{a}_2 = (-n, m)$ [Fig. 1(a)], where the unit of distance is the lattice constant. For this lattice, reciprocal primitive vectors are given by $\mathbf{b}_1 = \frac{2\pi}{N_s}(m, n)$ and $\mathbf{b}_2 = \frac{2\pi}{N_s}(-n, m)$, and the number of sites is $N_s = m^2 + n^2$. Corresponding reciprocal lattice points in the first Brillouin zone are generated by Wigner-Seitz construction, as shown in Fig. 1(b). Note that this lattice has a slight Pomeranchuk instability because the rotational symmetry is strictly reduced from C_4 to C_2 in a finite system. Thus, VMC calculations are performed on the tilted square lattices of $m = 9$–13 and $n = 1$, and we compare results for different lattice sizes to distinguish between isotropy and anisotropy.

As a variational wave function, we use a Gutzwiller-Jastrow type, $\Psi = \mathcal{P}\Phi$. For a many-body correlation factor $\mathcal{P}$, we assume the form $\mathcal{P} = \mathcal{P}_G\, \mathcal{P}_Q$, where $\mathcal{P}_G$ is a usual on-site (Gutzwiller) projection [6], $\mathcal{P}_G = \prod_{\mathbf{r}}[1 - (1-g) n_{\mathbf{r}\uparrow} n_{\mathbf{r}\downarrow}]$, and $\mathcal{P}_Q$ is an extended projection between a doubly occupied site (doublon, D) and an empty site (holon, H), which controls the probability weight of D-H binding classified by the surrounded configuration, namely, the number of H neighboring D and *vice versa* [7, 8].

For the one-body part Φ, we adopt a mixed state of $d_{x^2-y^2}$-wave SC (dSC) and AF orders for a continuous description of their interplay. For N_e electrons, $d_{x^2-y^2}$-wave singlet state Φ with a nearest-neighbor-pairing gap Δ_d is written as [9],

$$\Phi = \left(\sum_{\mathbf{k}} \varphi_{\mathbf{k}} b^\dagger_{\mathbf{k}\uparrow} b^\dagger_{-\mathbf{k}\downarrow} \right)^{\frac{N_e}{2}} |0\rangle\,, \tag{2}$$

where

$$\varphi_{\mathbf{k}} = \frac{\Delta(\mathbf{k})}{\varepsilon^{\mathrm{SC}}(\mathbf{k}) - \zeta + \sqrt{(\varepsilon^{\mathrm{SC}}(\mathbf{k}) - \zeta)^2 + |\Delta(\mathbf{k})|^2}}\,, \tag{3}$$

$\Delta(\mathbf{k}) = \Delta_d(\cos k_x - \cos k_y)$, and ζ is a parameter corresponding to the chemical potential for $U/t \to 0$. An AF gap Δ_{AF} is introduced through a quasi-particle operator $b_{\mathbf{k}\sigma}$ of a mean-field theory;

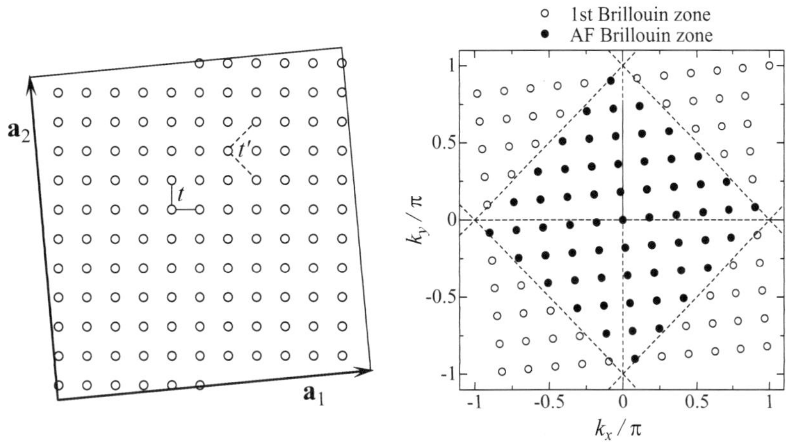

(a) Tilted square lattice (b) 1st Brillouin zone

Figure 1. (a) Tilted square lattice, spanned by edge vectors $\mathbf{a}_1 = (m, n)$ and $\mathbf{a}_2 = (-n, m)$ with $m = 11$ and $n = 1$. (b) Corresponding reciprocal lattice in the first Brillouin zone by Wigner-Seitz construction with the periodic-periodic boundary conditions. Closed circles are reciprocal lattice points in the folded AF Brillouin zone.

$$b^{\dagger}_{\mathbf{k}\sigma} = \alpha_{\mathbf{k}} c^{\dagger}_{\mathbf{k}\sigma} + \lambda_{\sigma} \beta_{\mathbf{k}} c^{\dagger}_{\mathbf{k}+\mathbf{Q}\sigma} \tag{4}$$

and

$$b^{\dagger}_{\mathbf{k}+\mathbf{Q}\sigma} = -\lambda_{\sigma} \beta_{\mathbf{k}} c^{\dagger}_{\mathbf{k}\sigma} + \alpha_{\mathbf{k}} c^{\dagger}_{\mathbf{k}+\mathbf{Q}\sigma} \, , \tag{5}$$

where $\lambda_{\sigma} = +1(-1)$ for $\sigma = \uparrow (\downarrow)$, and

$$\alpha_{\mathbf{k}} \; (\beta_{\mathbf{k}}) = \left[\frac{1}{2} \left(1 - (+) \frac{\varepsilon^{\mathrm{AF}}(\mathbf{k})}{\sqrt{|\varepsilon^{\mathrm{AF}}(\mathbf{k})|^2 + |\Delta_{\mathrm{AF}}|^2}} \right) \right]^{\frac{1}{2}} , \tag{6}$$

with $\mathbf{Q} = (\pi, \pi)$ and $\mathbf{k} \in$ AF Brillouin zone.

To improve accuracy of the wave function, a band-renormalization effect owing to electron correlation is individually introduced into the dSC and AF orders by adjusting the parameters of hopping integrals. To this end, we assume that $\varepsilon^{\mathrm{SC}}(\mathbf{k})$ in Eq. (3) and $\varepsilon^{\mathrm{AF}}(\mathbf{k})$ in Eq. (6) are separately fitted by a tight-binding band composed of transfers up to fifth-neighbor sites as shown in Fig. 2,

$$\begin{aligned} \varepsilon^{\mathrm{X}}(\mathbf{k}) \;\; = \;\; & -2(t^{\mathrm{X}}_{1x} \cos k_x + t^{\mathrm{X}}_{1y} \cos k_y) \\ & -2t^{\mathrm{X}}_{2\mathrm{p}} \cos(k_x + k_y) - 2t^{\mathrm{X}}_{2\mathrm{m}} \cos(k_x - k_y) \\ & -2t^{\mathrm{X}}_{3} \left[\cos(2k_x) + \cos(2k_y)\right] \\ & -4t^{\mathrm{X}}_{4} \left[\cos(2k_x)\cos(k_y) + \cos(k_x)\cos(2k_y)\right] \\ & -2t^{\mathrm{X}}_{5} \left[\cos(3k_x) + \cos(3k_y)\right] \, , \end{aligned} \tag{7}$$

IOP Conf. Series: Journal of Physics: Conf. Series **1293** (2019) 012025 doi:10.1088/1742-6596/1293/1/012025

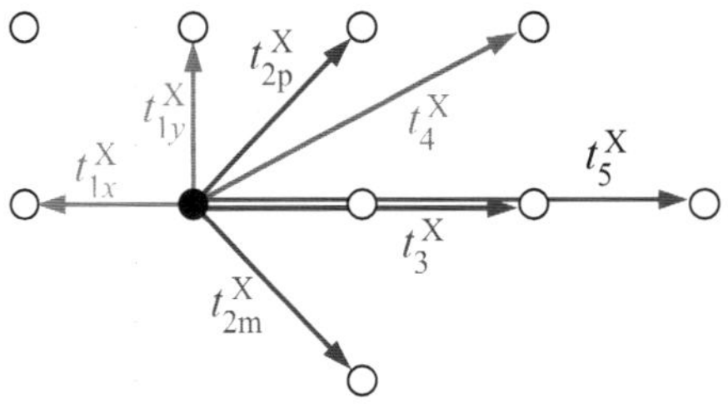

Figure 2. Band parameters $t_\ell^{\rm X}$ for fitting $\varepsilon^{\rm X}(\mathbf{k})$ with X = SC or AF, which are optimized to adjust the band-renormalization effect owing to electron correlation.

with X = SC or AF, where asymmetries in $t_1^{\rm X}$ between x and y directions and $t_2^{\rm X}$ between upward and downward diagonals are also introduced to deal with the Pomeranchuk instability. Here, $t_{1x}^{\rm X}$ (X = SC, AF) is fixed to t in Eq. (1) as the unit of energy.

As a result, variational parameters to be optimized in Φ amount to Δ_d, $\Delta_{\rm AF}$, ζ, $t_{1y}^{\rm X}$, $t_{2{\rm p}}^{\rm X}$, $t_{2{\rm m}}^{\rm X}$, $t_3^{\rm X}$-$t_5^{\rm X}$ with X = SC, AF. Besides these fifteen in Φ, variational parameters in the correlation factors are g in $\mathcal{P}_{\rm G}$ and 10 parameters of for the nearest-neighbor D-H binding in $\mathcal{P}_{\rm Q}$. According to a standard procedure of VMC using the correlated measurement for efficient optimization [10], we obtain the optimal set of the 26 parameters by minimizing the energy. Using the optimized wave functions, we calculate various quantities. Among them, we mainly discuss the following quantities, here. As an order parameter of dSC, we use the real-space dSC correlation function for the nearest-neighbor pairing defined by,

$$P_d(\mathbf{r}) = \frac{1}{N_{\rm s}} \sum_{\mathbf{r}'} \sum_{\boldsymbol{\tau},\boldsymbol{\tau}'} \lambda_{xy}(\boldsymbol{\tau}) \lambda_{xy}(\boldsymbol{\tau}') \langle \Delta^\dagger(\mathbf{r}'+\mathbf{r},\boldsymbol{\tau}) \Delta(\mathbf{r}',\boldsymbol{\tau}') \rangle , \tag{8}$$

with

$$\Delta(\mathbf{r},\boldsymbol{\tau}) = \frac{1}{\sqrt{2}} \left(c_{\mathbf{r}\uparrow} c_{\mathbf{r}+\boldsymbol{\tau}\downarrow} - c_{\mathbf{r}\downarrow} c_{\mathbf{r}+\boldsymbol{\tau}\uparrow} \right) , \tag{9}$$

$\lambda_{xy}(\boldsymbol{\tau}) = +1$ for $\boldsymbol{\tau} = (\pm 1, 0)$, and $\lambda_{xy}(\boldsymbol{\tau}) = -1$ for $\boldsymbol{\tau} = (0, \pm 1)$. Actually, we show $P_d(\mathbf{r})$ at the farthermost distance $\mathbf{r}_{\rm max}$ as a measure of dSC order. The staggered magnetization,

$$m_s = \frac{1}{N_{\rm s}} \left| \sum_{\mathbf{r}} e^{i\mathbf{Q}\cdot\mathbf{r}} \langle S_{\mathbf{r}}^z \rangle \right| , \tag{10}$$

with $\mathbf{Q} = (\pi, \pi)$ is also calculated as an order parameter of AF state.

To check whether the x–y anisotropy spontaneously appears, we will calculate a discrepancy between electron-transfers in different directions, A_ℓ for the ℓ-th neighbor hoppings, defined by

$$\begin{aligned} A_1 &= |h(1,0) + h(-1,0) - h(0,1) - h(0,-1)| , \\ A_2 &= |h(1,1) + h(-1,-1) - h(1,-1) - h(-1,1)| , \end{aligned} \tag{11}$$

where

$$h(\boldsymbol{\tau}) = \frac{1}{N_{\rm s}} \sum_{\mathbf{r},\sigma} \langle c_{\mathbf{r},\sigma}^\dagger c_{\mathbf{r}+\boldsymbol{\tau},\sigma} + {\rm h.c.} \rangle \tag{12}$$

is the expectation value of bare electron-hoppings without hamiltonian parameters, and $\boldsymbol{\tau}$ is a vector connecting neighbor sites; $\boldsymbol{\tau} = \pm(1,0), \pm(0,1)$ for the nearest-neighbors, or $\boldsymbol{\tau} = \pm(1,1), \pm(1,-1)$ for the second-nearest-neighbors. For $A_\ell \sim 0$, hoppings are isotropic whereas for finite values of A_ℓ, hoppings become anisotropic between the orthogonal directions. In addition, we discuss the anisotropy using the momentum distribution function,

$$n(\mathbf{k}) = \frac{1}{2}\sum_\sigma \langle c^\dagger_{\mathbf{k},\sigma} c_{\mathbf{k},\sigma}\rangle \ . \tag{13}$$

As pilot calculations, we have fixed the value of U/t at 12 as a typical case of strongly correlated regime, and compare the results of $t'/t = 0$ and $t'/t = \pm 0.30$ for the tilted square lattices of $m = 9$–13 and $n = 1$ ($N_\mathrm{s} = 82$–170) with periodic-periodic boundary conditions.

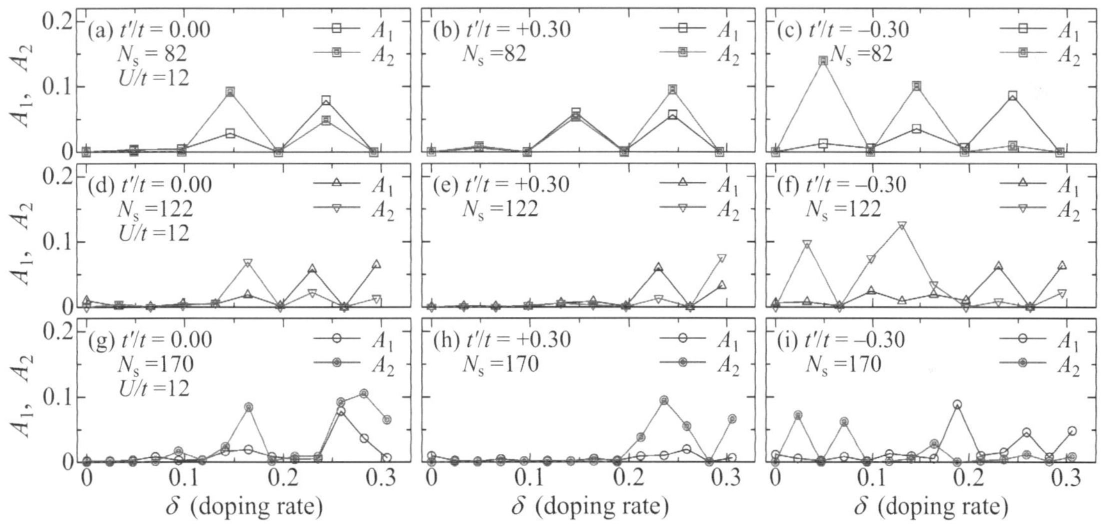

Figure 3. Anisotropy indicator A_1 (in x–y directions) and A_2 (lattice-diagonal directions) defined in Eq. (11) as functions of δ. The lattice size is $N_\mathrm{s} = 82$ ($m = 9$) in the first row [(a)–(c)], $N_\mathrm{s} = 122$ ($m = 11$) in the second row [(d)–(f)], and $N_\mathrm{s} = 170$ ($m = 13$) in the last row [(g)–(i)], with $n = 1$. From left to right, the value of t'/t is varied: $t'/t = 0$, $+0.3$, and -0.3.

3. Results and discussion

Let us begin with the x–y anisotropies of hoppings. In Fig. 3, the anisotropy indicator A_ℓ [Eq. (11)] is plotted as a function of doping rate δ for $N_\mathrm{s} = 82$, 122, and 170 and for $t'/t = 0$, ± 0.3. The δ dependence of A_1 and A_2 often exhibits zigzag behavior in every panel. We find that at most data points where A_1 and A_2 have an appreciable finite value, the one-body state becomes a degenerate (open-shell) state. Therefore, whether or not the state becomes anisotropic is a subtle problem that sensitively depends on the system size and the boundary conditions. In such a premise, we argue that x–y anisotropy hardly appears in the underdoped regime ($\delta \lesssim 0.16$) where the AF order develops, for $t'/t = 0$, and $+0.3$ (the first and second

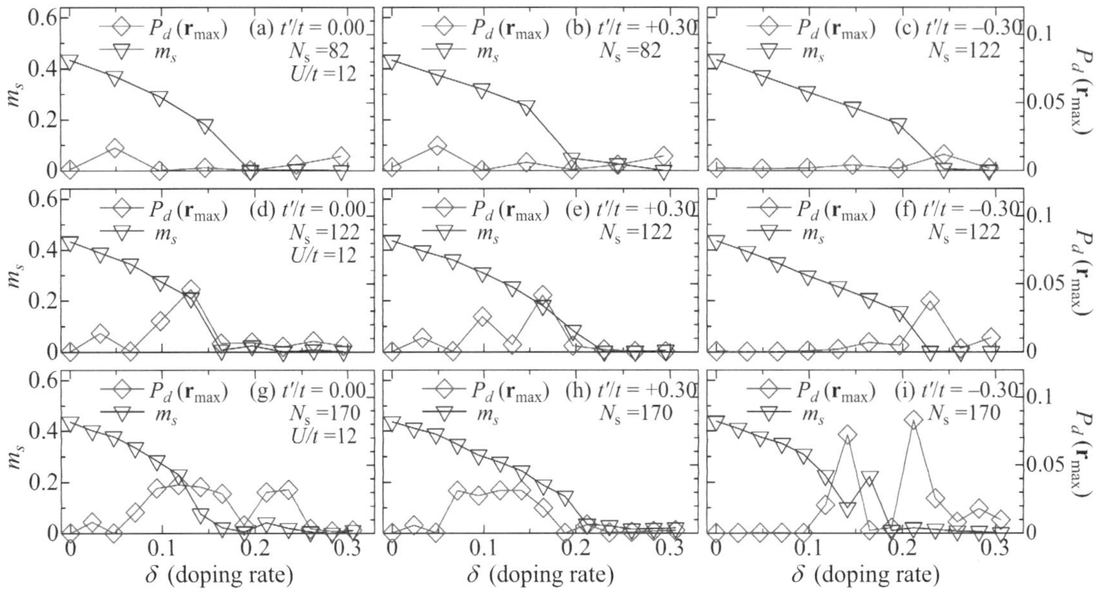

Figure 4. δ dependence of order parameters, namely, $P_d(\mathbf{r}_{\max})$ for dSC (diamonds, right axis) and m_s for AF (triangles, left axis). The system sizes and values of t'/t are the same as those in Fig. 3.

columns of panels), but the anisotropy appreciably appears in the overdoped regime, where the Fermi-liquid (metallic) behavior becomes marked. This is consistent with the previous study [5]. In contrast, for $t'/t = -0.3$ (the last column in Fig. 3), the anisotropy appears even in the underdoped regime. In this case, however, the magnitude of A_1 and A_2 tends to decrease as $N_{\rm s}$ increases. Thus, we would like to leave the conclusion as to the underdoped regime for future studies.

Plotted in Fig. 4 are the expectation values of $P_d(\mathbf{r}_{\max})$ and m_s estimated with the optimized Ψ for the systems corresponding to Fig. 3. Comparing $P_d(\mathbf{r}_{\max})$ and m_s in Fig. 4 with those previously obtained in the periodic-antiperiodic boundary conditions [11, 12], the data of $P_d(\mathbf{r}_{\max})$ largely fluctuate, suggesting that dSC is sensitive to the underlying Fermi surface. Comparing Figs. 3 and 4, we find that if the orders (AF or dSC), especially both, are robust, the anisotropy is suppressed. We will discuss the competition between each order and the anisotropy again in a coming publication.

Lastly, we show contour maps of the momentum distribution function $n(\mathbf{k})$ for $N_{\rm s} = 122$ in Fig. 5. For clarity, the axes of the wave vector $\mathbf{k}$ in the reciprocal space are reversely rotated by $-\theta$, where $\theta = \tan^{-1}(n/m)$ is a tilt angle for tilted square lattice with edge vectors $\mathbf{a}_1 = (m, n)$ and $\mathbf{a}_2 = (-n, m)$. Panels in the first column [Figs. 5(a), 5(e), and 5(i)] show the cases at half filling for $t'/t = 0$, and ± 0.3. In fact, the fourfold rotational symmetry in the occupied $\mathbf{k}$ points is subtly broken even at half filling for finite $N_{\rm s}$. This weak anisotropy will disappears as m increases. Anyway, the symmetricity seems broadly preserved at half filling. However, the Pomeranchuk instability becomes marked in some cases, for example in Fig. 5(k), corresponding to the large values of A_ℓ in Fig. 3. We will check the size dependence again in a future study.

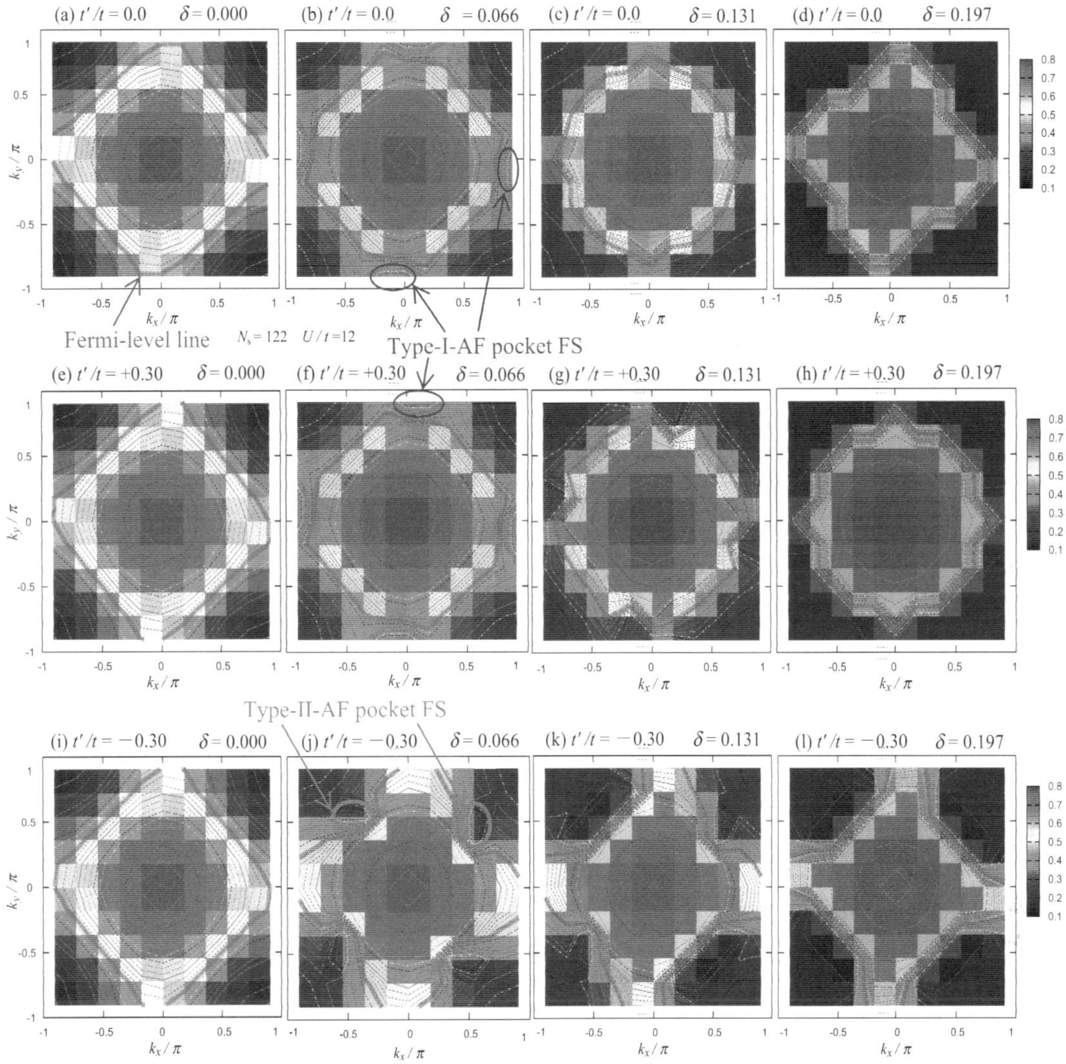

Figure 5. Contour plots of the momentum distribution function $n(\mathbf{k})$. From left to right, the doping rate is increased: $\delta = 0$ (half-filling), 0.066, 0.131, and 0.197. The values of t'/t are 0 [first row, (a)–(d)], +0.3 [second row, (e)–(h)], and −0.3 [third row, (i)–(l)]. The lattice size is $N_{\rm s} = 11^2 + 1^2 = 122$, and $U/t = 12$. The thick pink line in each panel indicates Fermi-level line drawn by connecting the points that take the half of electron density, $(1-\delta)/2$.

4. Summary

We have studied the spontaneous appearance of the fourfold rotational anisotropy in the two-dimensional strongly correlated t-t'-U model on the bases of VMC calculations. Using an improved wave function on a lattice with the fourfold rotational symmetry, we confirmed that the anisotropy appears in the overdoped area, where the state is almost Fermi liquid [5]. In the underdoped regime, the anisotropy is suppressed in the cases of type-I AF ($t'/t = 0$, $+0.3$), whereas in the case of the type-II AF ($t'/t = -0.3$) the anisotropy possibly appear. More systematic calculations are left for future studies.

References

[1] Sato Y, Kasahara S, Murayama H, Kasahara Y, Moon E -G, Nishizaki T, Loew T, Porras J, Keimer B, Shibauchi T and Matsuda Y 2017 *Nat. Phys.* **13** 1074
[2] Kawasaki S, Li Z, Kitahashi M, Lin C T, Kuhns P L, Reyes A P and Zheng G -q 2017 *Nat. Commun.* **8** 1267
[3] Yamase H, Oganesyan V and Metzner W 2005 *Phys. Rev. B* **72** 035114
[4] Yamase H and Metzner W 2006 *Phys. Rev. B* **73** 214517
[5] Edegger B, Muthukumar V N and Gros C 2006 *Phys. Rev. B* **74** 165109
[6] Gutzwiller M C 1963 *Phys. Rev. Lett.* **10** 159
[7] Kobayashi K and Yokoyama H 2007 *Physica C* **463-465** 141
[8] Kobayashi K and Yokoyama H 2008 *J. Phys. Chem. Solids* **69** 3274
[9] Giamarchi T and Lhuillier C 1991 *Phys. Rev. B* **43** 12943
[10] Umrigar C J, Wilson K G and Wilkins J W 1988 *Phys. Rev. Lett.* **60** 1719
[11] Yokoyama H, Ogata M, Tanaka Y, Kobayashi K and Tsuchiura H 2013 *J. Phys. Soc. Jpn.* **82** 014707
[12] Sato R and Yokoyama H 2016 *J. Phys. Soc. Jpn.* **85** 074701

IOP Conf. Series: Journal of Physics: Conf. Series **1293** (2019) 012026 doi:10.1088/1742-6596/1293/1/012026

The coexisting state of the staggered flux and *d*-wave superconducting order in a *t*-*J* type model

S Fukuda[1], K Yamazaki[1], H Tsuchiura[1,2] and M Ogata[3]

[1] Department of Applied Physics, Tohoku University, Sendai, Japan
[2] Center for Spintronics Research Network, Tohoku University, Sendai, Japan
[3] Department of Physics, University of Tokyo, Bunkyo, Japan

E-mail: tsuchi@solid.apph.tohoku.ac.jp

Abstract. We study the quasiparticle structures in the coexisting phase of the d-wave superconductivity (dSC) and the staggered flux state using the renormalized mean-field theory for the t-J type model with additional nearest-neighbor Coulomb repulsion V, based on the Gutzwiller approximation. We find that this coexisting state exhibits the dome-like behavior of dSC in the phase diagram and also the so-called two-gap structure found in ARPES experiments, both of which are quite consistent with the experimental results.

1. Introduction

A series of experimental studies over the recent decade has provided important information previously unexpected on the pseudogap (PG) state in cuprate superconductors: it is accompanied by some phase transitions with the broken rotational or translational symmetry, which implies that there can be nematic order or charge order. It has also been suggested that the PG phase and d-wave superconductivity (dSC) can coexist roughly in the first half of the superconducting dome. One of the experimental evidences of the coexisting pseudogap phase with dSC is a gap-structure in the ARPES data opening with broken particle-hole symmetry found in Bi-2201, which is consistent with the so-called two-gap structure in the quasiparticle excitation spectra [1, 2].

Quite recently, we have shown that the observed gap-structure can be well described if we assume that the staggered flux (SF) state exists in the PG phase [3]. We have also shown that the gap-structure cannot be described by assuming antiferromangetic order. The SF state is a normal, or non-superconducting state with broken symmetries, and has been studied by many groups since the early stage of research on cuprate superconductors [4, 5, 6, 7, 8, 9]. This state was proposed firstly as a candidate of the ground state for the CuO_2 plain, then has been studied in the context of the pseudogap phase. Theoretically, it has been known that the SF state has higher variational energy than that of dSC for all hole-doping $\delta > 0$ in the standard models, such as the t-J model and the Hubbard model [9, 10]. However, at least within a mean-field theory, the SF state can be stabilized by introducing the nearest-neighbor Coulomb repulsion V to the t-J model, and even can coexist with dSC [8] in the underdoped region of the phase diagram. Thus we can believe that the SF is still a possible candidate for the PG.

One prominent objection to the idea that the SF is the origin of the PG is that the SF gap opens always at the center of the tight-binding band due to the two-sublattice nature of the

SF state. The resulting gap structure in the quasiparticle densities of states (DOS) reported in previous studies [3, 7, 8] are highly electron-hole asymmetric even in the coexisting phase of SF and dSC, which has never been observed in STS experiments. Thus it is necessary to clarify if the coexisting phase of SF and dSC can exhibit a plausible quasiparticle spectra. In this paper, we carry out such a study using the renormalized mean-field theory for the t-J model with additional nearest-neighbor Coulomb repulsion V.

2. Model

We start with the t-J model with both the *next-nearest-neighbor* hopping term and the *nearest-neighbor* Coulomb interaction term, given as

$$\begin{aligned}\mathcal{H} &= -t\sum_{\langle i,j\rangle,\sigma}\left(c^\dagger_{i\sigma}c_{j\sigma}+\text{H.c.}\right) - t'\sum_{(i,j),\sigma}\left(c^\dagger_{i\sigma}c_{j\sigma}+\text{H.c.}\right) - \mu\sum_{i,\sigma}c^\dagger_{i\sigma}c_{i\sigma} \\ &\quad +J\sum_{\langle i,j\rangle}\boldsymbol{S}_i\cdot\boldsymbol{S}_j + V\sum_{\langle i,j\rangle}n_in_j \end{aligned} \tag{1}$$

in the standard notation, where $n_j=\sum_\sigma c^\dagger_{j\sigma}c_{j\sigma}$, and $\langle i,j\rangle$ and (i,j) mean the summation over nearest-neighbor and next-nearest-neighbor pairs. Throughout this paper, we take $J/t=0.2$, $t'/t=-0.3$, and $V/t=0.1$. This Hamiltonian operates only in the subspace where there is no doubly occupied sites, which can be implemented by a Gutzwiller projection.

To deal with the projection effects, we can use the so-called renormalized mean-field theory in which the transfer integral t and the superexchange coupling J are renormalized as $t_{\text{eff}}=g_tt$ and $J_{\text{eff}}=\frac{3}{4}g_sJ$ by using doping dependent factors

$$g_t=\frac{2\delta}{1+\delta}\ ,\quad g_s=\frac{4}{(1+\delta)^2}\ , \tag{2}$$

respectively [11]. Since we need further mean-field approximation to treat the renormalized Hamiltonian, we assume the particle-particle pairing amplitude

$$\langle c^\dagger_{\boldsymbol{k}\uparrow}c^\dagger_{-\boldsymbol{k}\downarrow}\rangle = \Delta_d(\cos k_x-\cos k_y)\ , \tag{3}$$

and also the particle-hole pairing amplitude

$$\langle c^\dagger_{\boldsymbol{k}\sigma}c_{\boldsymbol{q}\sigma}\rangle = \xi_1(\cos k_x+\cos k_y)\delta_{\boldsymbol{k},\boldsymbol{q}} + i\xi_2(\cos k_x-\cos k_y)\delta_{\boldsymbol{k}+\boldsymbol{Q},\boldsymbol{q}}\ , \tag{4}$$

with $\boldsymbol{Q}=(\pi,\pi)$. It should be noted here that ξ_2 in (4) is the order parameter of SF. Then we obtain the set of the self-consistent equations, with $s=\pm1$, as follows:

$$\xi_1 = \frac{1}{2N_s}\sum_{\boldsymbol{k},s}\tanh\left(\frac{E_{\boldsymbol{k}s}}{2T}\right)\frac{A_{\boldsymbol{k}}\gamma_{\boldsymbol{k}}}{E_{\boldsymbol{k}s}}\left[1+s\frac{4g_tt'\varphi_{\boldsymbol{k}}+\tilde{\mu}}{\varepsilon_{\boldsymbol{k}}}\right]\ , \tag{5}$$

$$\xi_2 = \frac{1}{2N_s}\sum_{\boldsymbol{k},s}\tanh\left(\frac{E_{\boldsymbol{k}s}}{2T}\right)\frac{B_{\boldsymbol{k}}\eta_{\boldsymbol{k}}}{E_{\boldsymbol{k}s}}\left[1+s\frac{4g_tt'\varphi_{\boldsymbol{k}}+\tilde{\mu}}{\varepsilon_{\boldsymbol{k}}}\right]\ , \tag{6}$$

$$\Delta_d = \frac{1}{2N_s}\sum_{\boldsymbol{k},s}\tanh\left(\frac{E_{\boldsymbol{k}s}}{2T}\right)\frac{D_{\boldsymbol{k}}\eta_{\boldsymbol{k}}}{E_{\boldsymbol{k}s}}\ , \tag{7}$$

$$\delta = -\frac{1}{N_s}\sum_{\boldsymbol{k},s}\tanh\left(\frac{E_{\boldsymbol{k}s}}{2T}\right)\frac{s\left[\varepsilon_{\boldsymbol{k}}+s\left(4g_tt'\varphi_{\boldsymbol{k}}+\tilde{\mu}\right)\right]}{E_{\boldsymbol{k}s}}\ , \tag{8}$$

where

$$A_{\boldsymbol{k}} = \left(2g_t t + J^{+}_{\text{eff}}\xi_1\right)\gamma_{\boldsymbol{k}}\ ,\quad B_{\boldsymbol{k}} = J^{+}_{\text{eff}}\xi_2\eta_{\boldsymbol{k}}\ ,\quad D_{\boldsymbol{k}} = J^{-}_{\text{eff}}\Delta_d\eta_{\boldsymbol{k}}\ , \tag{9}$$

with

$$\varepsilon_{\boldsymbol{k}} = \sqrt{A^2_{\boldsymbol{k}} + B^2_{\boldsymbol{k}}}\ ,\quad E_{\boldsymbol{k}s} = \sqrt{(s\varepsilon_{\boldsymbol{k}} + 4g_t t'\varphi_{\boldsymbol{k}} + \tilde{\mu})^2 + D^2_{\boldsymbol{k}}}\ , \tag{10}$$

and

$$\gamma_{\boldsymbol{k}} = \cos k_x + \cos k_y\ ,\quad \eta_{\boldsymbol{k}} = \cos k_x - \cos k_y\ ,\quad \varphi_{\boldsymbol{k}} = \cos k_x \cos k_y\ . \tag{11}$$

The additional renormalized parameter $J^{\pm}_{\text{eff}}$ in the above equations is given as

$$J^{\pm}_{\text{eff}} = J_{\text{eff}} \pm V\ . \tag{12}$$

We note here that Δ_d and ξ_2 are the order parameters of dSC and SF, respectively. Thus, the effective coupling constant for dSC is $J^{-}_{\text{eff}} = J_{\text{eff}} - V$, while that for SF is $J^{+}_{\text{eff}} = J_{\text{eff}} + V$. This implies that the nearest-neighbor Coulomb repulsion V enhances SF, but suppresses dSC.

3. Results and Discussion

First we show the doping dependences of the order parameters Δ_d and ξ_2 given in eqs. (3) and (4) in Fig. 1. The dSC gradually emerges only when $\delta \gtrsim 0.05$ by hole doping, while the SF state monotonically decreases and vanishes at $\delta \simeq 0.2$. Here we can clearly see that these two states coexist within the region of $0.05 \lesssim \delta \lesssim 0.2$. It is worth emphasizing here that dSC does not appear between $\delta = 0$ and 0.05, which is consistent with the typical experimental phase diagram, and at the same time, cannot be captured within the previous mean-field or variational scenario. For example, when we consider the coexisting phase of dSC and antiferromagnetism, it is known that the dSC order parameter appears even at half-filling [12]. Thus the dome-like behavior of the dSC order parameter shown in Fig. 1 can be a peculiarity of the coexisting phase of SF and dSC.

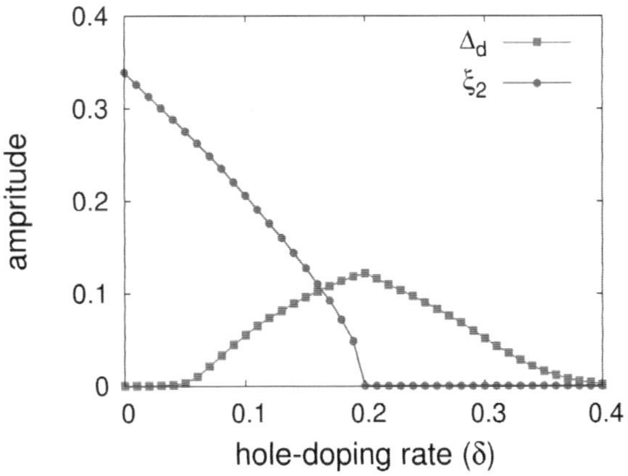

Figure 1. The order parameters Δ_d and ξ_2 versus hole doping at $T = 0$.

Next, let us look at the quasiparticle states in the coexisting state. Figure 2 shows the density of states (DOS) for the coexisting state (the red line) and the pure SF state (the blue line). The pure SF state shows a highly electron-hole asymmetric spectra, which is consistent with previous studies [3, 7]. In the coexisting state, however, the DOS exhibits two different gap-structures, one is highly asymmetric and another is symmetric around the Fermi energy ($E/t = 0$), as

IOP Conf. Series: Journal of Physics: Conf. Series **1293** (2019) 012026 doi:10.1088/1742-6596/1293/1/012026

seen in Fig. 2. Comparing this DOS with that for pure SF state, we easily notice that the larger asymmetric gap is due to SF state, and the smaller symmetric gap is a consequence of the coexisting dSC. We note here that the gap amplitudes found in the DOS roughly correspond to $J^{+}_{\rm eff}\xi_2$ and $J^{-}_{\rm eff}\Delta_d$. In this situation, we find that the dSC gap is formed using the residual quasiparticle states around the Fermi energy in the SF state.

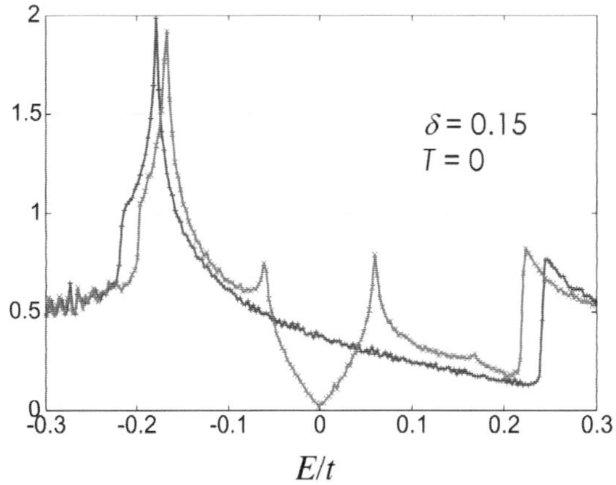

Figure 2. The density of states (DOS) for pure SF state (blue), and the coexisting phase of SF and dSC (red).

Finally, let us discuss the doubly gapped structure found in Fig. 2 in some more detail. Since the $\boldsymbol{k}$-dependences of the quasiparticle states are integrated out in the DOS, it is interesting to see here the angle resolved gap structure in the coexisting state. Fig. 3 shows the gap functions for the coexisting state, and also the pure dSC state as functions of the so-called Fermi angle. Here, $\theta = \pi/4$ corresponds to the nodal direction of the dSC gap in $\boldsymbol{k}$-space. We can clearly see the two-gap structure which is quite consistent with the ARPES experimental data [13]. Therefore, we can conclude that the coexisting state of SF and dSC can be a possible candidate of the pseudogap phase.

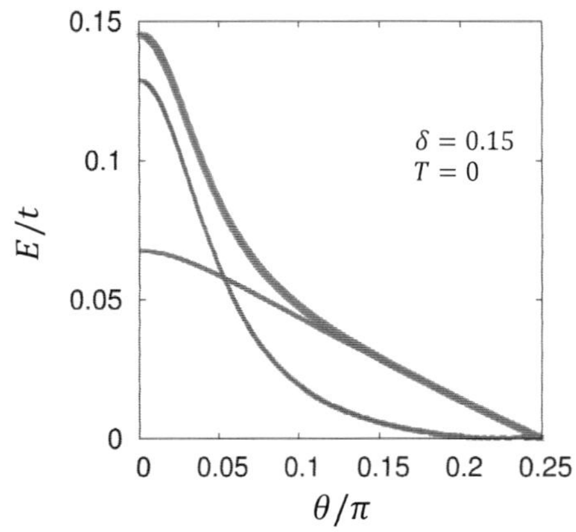

Figure 3. Gap functions for the coexisting phase of SF and dSC (red),pure SF state (blue) and pure dSC state (green) as functions of the Fermi angle.

4. Summary

We have studied the quasiparticle states in the coexisting phase of dSC and SF states within the renormalized mean-field theory based on the t-J model. The present analysis has shown that the dome-like structure of dSC in the phase diagram and the two-gap structure can be well described by the coexisting state, and thus also shown that the SF phase can be a possible candidate of symmetry-breaking pseudogap states coexisting with the dSC.

5. Acknowledgments

This work was supported by JSPS KAKENHI Grant Number JP17K05528. The numerical computations were carried out at the Cyber-science Center, Tohoku University.

References

[1] Hashimoto M, He R -H, Tanaka K, Testaud J -P, Meevasana W, Moore R G, Lu D, Yao H, Yoshida Y, Eisaki H, Devereaux T P, Hussain Z and Shen Z -X 2010 *Nature Phys.* **6** 414
[2] Hashimoto M, Vishik I M, He R -H, Devereaux T P and Shen Z -X 2014 *Nature Phys.* **10** 483
[3] Tsuchiura H, Ogata M, Yamazaki K and Asaoka R 2018 *J. Phys.: Conf. Ser.* **969** 012066
[4] Affleck I and Marston J B 1988 *Phys. Rev.* B **37** 3774
[5] Lee P A and Wen X -G 2001 *Phys. Rev.* B **63** 224517
[6] Chakravarty S, Laughlin R B, Morr D K and Nayak C 2001 *Phys. Rev.* B **63** 094503
[7] Morr D K 2002 *Phys. Rev. Lett.* **89** 106401
[8] Zhou S and Wang Z 2004 *Phys. Rev.* B **70** 020501(R)
[9] Yokoyama H, Tamura S and Ogata M 2016 *J. Phys. Soc. Jpn.* **85** 124707
[10] Zhang F C 1990 *Phys. Rev. Lett.* **64** 974
[11] Zhang F C, Gros C, Rice T M and Shiba H 1988 *Supercond. Sci. Technol.* **1** 36
[12] Ogata M and Himeda A 2003 *J. Phys. Soc. Jpn.* **72** 374
[13] Kaminski A, Kondo T, Takeuchi T and Gu G 2015 *Philos. Mag.* **95** 453

IOP Conf. Series: Journal of Physics: Conf. Series **1293** (2019) 012027 doi:10.1088/1742-6596/1293/1/012027

Antiferromagnetism, superconductivity and phase diagram in the three-band model of high-temperature cuprates

Takashi Yanagisawa

Electronics and Photonics Research Institute, National Institute of Advanced Industrial Science and Technology (AIST), Tsukuba Central 2, 1-1-1 Umezono, Tsukuba 305-8568, Japan

E-mail: t-yanagisawa@aist.go.jp

Abstract. We investigate the ground state of the two-dimensional three-band d-p model for for high-temperature superconductors on the basis of a variational Monte Carlo method. We employ an optimization variational Monte Carlo method that we have developed recently. The many-body wave function is improved beyond the Gutzwiller ansatz by adopting the wave function in the form $\psi = \exp(-S)\psi_G$ where ψ_G represents the Gutzwiller function and S is a kinetic operator. The strong magnetic correlation and also superconductivity (SC) are induced by the on-site Coulomb repulsive interaction. It is important to clarify the phase diagram that includes superconductive phase and antiferromagnetic phase. We show the phase diagram to show the antiferromagnetic region in the parameter space. High-temperature superconductivity may occur in the strongly correlated region near the antiferromagnetic boundary.

1. Introduction

The physics of high-temperature superconductivity has been studied intensively for about 30 years since the discovery of cuprate high-temperature superconductors[1]. It has been established that the Cooper pairs of cuprate superconductors have the *d*-wave symmetry. This indicates that the electron correlation plays an important role for the appearance of high-temperature superconductivity. It is also known that cuprate parent materials without carries are Mott insulators.

The CuO_2 plane is commonly contained in high-temperature cuprates. The CuO_2 plane consists of oxygen atoms and copper atoms and the electronic model for this plane is the model with d and p orbitals called the d-p model (or three-band Hubbard model)[2, 3, 4, 5, 6, 7, 8, 9, 10, 11, 12, 13, 14, 15, 16, 17, 18]. It is very hard to elucidate the phase diagram of the d-p model because of strong correlation between electrons. We often use simplified models such as the two-dimensional single-band Hubbard model[19, 20, 21, 22, 23, 24, 25, 26, 27, 28, 29, 30, 31, 32, 33, 34, 35, 36] or ladder model[37, 38, 39, 40] to make clear the phase diagram of correlated electron systems.

A variational Monte Carlo method is a useful method to investigate electronic properties of strongly correlated electron systems where we calculate the expectation values numerically. It is necessary to improve variational wave functions to obtain reliable results for correlated electrons. We proposed wave functions by multiplying an initial wave function by $\exp(-S)$-type operators[36, 41], where S is a correlation operator with variational parameters. An optimization process is performed in a systematic way by multiplying by the exponential-type operators repeatedly[41]. It has been shown that the ground-state energy is lowered considerably by using this type of wave functions[36].

2. Hamiltonian

The three-band model that explicitly includes oxygen p and copper d orbitals contains the parameters U_d, U_p, t_{dp}, t_{pp}, t'_d, ϵ_d and ϵ_p. Our study is within the hole picture and the Hamiltonian is written as

$$\begin{aligned} H_{dp} &= \epsilon_d \sum_{i\sigma} d^{\dagger}_{i\sigma} d_{i\sigma} + \epsilon_p \sum_{i\sigma} (p^{\dagger}_{i+\hat{x}/2\sigma} p_{i+\hat{x}/2\sigma} + p^{\dagger}_{i+\hat{y}/2\sigma} p_{i+\hat{y}/2\sigma}) \\ &+ t_{dp} \sum_{i\sigma} [d^{\dagger}_{i\sigma}(p_{i+\hat{x}/2\sigma} + p_{i+\hat{y}/2\sigma} - p_{i-\hat{x}/2\sigma} - p_{i-\hat{y}/2\sigma}) + \text{h.c.}] \\ &+ t_{pp} \sum_{i\sigma} [p^{\dagger}_{i+\hat{y}/2\sigma} p_{i+\hat{x}/2\sigma} - p^{\dagger}_{i+\hat{y}/2\sigma} p_{i-\hat{x}/2\sigma} \\ &- p^{\dagger}_{i-\hat{y}/2\sigma} p_{i+\hat{x}/2\sigma} + p^{\dagger}_{i-\hat{y}/2\sigma} p_{i-\hat{x}/2\sigma} + \text{h.c.}] \\ &+ t'_d \sum_{\langle\langle ij \rangle\rangle \sigma} \epsilon_{ij} (d^{\dagger}_{i\sigma} d_{j\sigma} + \text{h.c.}) + U_d \sum_i d^{\dagger}_{i\uparrow} d_{i\uparrow} d^{\dagger}_{i\downarrow} d_{i\downarrow}. \end{aligned} \tag{1}$$

$d_{i\sigma}$ and $d^{\dagger}_{i\sigma}$ represent the operators for the d hole. $p_{i\pm\hat{x}/2\sigma}$ and $p^{\dagger}_{i\pm\hat{x}/2\sigma}$ denote the operators for the p holes at the site $R_{i\pm\hat{x}/2}$, and in a similar way $p_{i\pm\hat{y}/2\sigma}$ and $p^{\dagger}_{i\pm\hat{y}/2\sigma}$ are defined. t_{dp} is the transfer integral between adjacent Cu and O orbitals and t_{pp} is that between nearest p orbitals. $\langle\langle ij \rangle\rangle$ denotes a next nearest-neighbor pair of copper sites. t'_d was introduced (as shown in Fig.1) to reproduce the Fermi surface[42] reported in several cuprate superconductors such as $Bi_2Sr_2CaCu_2O_{8+\delta}$[43] and $Tl_2ba_2CuO_{6+\delta}$[44]. ϵ_{ij} takes the values ± 1 according to the sign of the transfer integral between next nearest-neighbor d orbitals. U_d is the strength of the on-site Coulomb repulsion between d holes. In this paper we neglect U_p among p holes because U_p is small compared to U_d[45, 46, 47, 48, 49]. U_p is not important in the low-doping region where the p-hole correlation effect is small. The values of band parameters were estimated as, for example, $U_d = 10.5$, $U_p = 4.0$ and $U_{dp} = 1.2$ in eV[46] where U_{dp} is the nearest-neighbor Coulomb interaction between holes on adjacent Cu and O orbitals. We neglect U_{dp} because U_{dp} is small compared to U_d. We use the notation $\Delta_{dp} = \epsilon_p - \epsilon_d$. The number of sites is denoted as N, and the total number of atoms is $N_a = 3N$. The energy unit is given by t_{dp}.

The single-band Hubbard model can be regarded as a simplified version of the three-band d-p model. This model has been studied more intensively. The Hamiltonian is given by

$$H = \sum_{ij\sigma} t_{ij} c^{\dagger}_{i\sigma} c_{j\sigma} + U \sum_i n_{i\uparrow} n_{i\downarrow}, \tag{2}$$

where $\{t_{ij}\}$ are transfer integrals and U is the on-site Coulomb energy. The transfer integral t_{ij} for nearest-neighbor pairs $\langle ij \rangle$ is given as $t_{ij} = -t$ and that for next-nearest neighbor pair $\langle\langle ij \rangle\rangle$ is $t_{ij} = -t'$. Otherwise, t_{ij} vanishes. The number of sites is denoted as N and the number of electrons is as N_e. The energy unit is given by t. $n_{i\sigma}$ is the number operator: $n_{i\sigma} = c^{\dagger}_{i\sigma} c_{i\sigma}$. The second term in the Hamiltonian represents the on-site repulsive interaction between electrons.

IOP Conf. Series: Journal of Physics: Conf. Series **1293** (2019) 012027 doi:10.1088/1742-6596/1293/1/012027

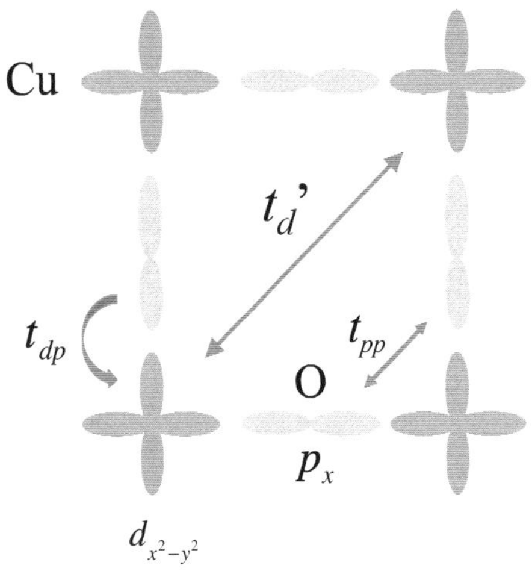

Figure 1. The transfer integral t'_d in the CuO_2 plane. t'_d is the transfer integral between next nearest-neighbor copper sites. We can also consider the transfer integral t'_{pp} between next nearest-neighbor oxygen atoms.

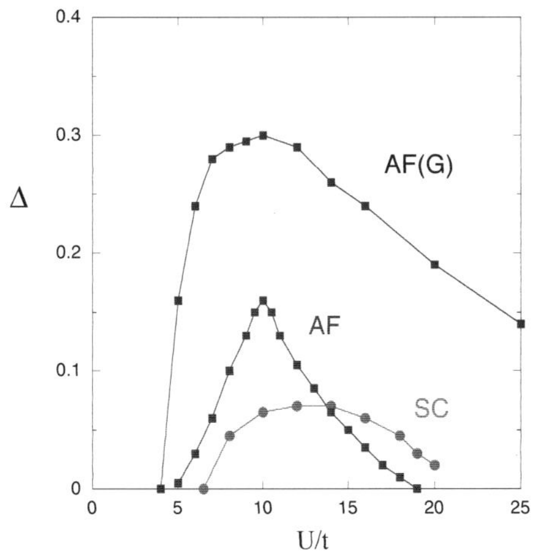

Figure 2. AF and SC order parameters as a function of U/t for the two-dimensional Hubbard model on a 10×10 lattice[36] (some data have been updated). The upper curve AF(G) indicates the result obtained by the Gutzwiller function, and those for AF and SC are obtained using ψ_λ.

3. Wave functions with strong electron correlation

The starting wave function for correlated electrons is Gutzwiller-projected wave function given by $\psi_G = P_G\psi_0$, where P_G is the Gutzwiller operator to control the double occupancy of d holes:

$$P_G = \prod_i [1 - (1-g)n_{di\uparrow}n_{di\downarrow}] = \exp\left(-\alpha \sum_i n_{di\uparrow}n_{di\downarrow}\right), \tag{3}$$

where $\alpha = \ln(1/g)$. g is a variational parameter in the range from 0 to unity. $n_{di\sigma} = d^\dagger_{i\sigma}d_{i\sigma}$ is the number operator for d holes. ψ_0 is the Fermi sea where the lowest band is occupied up to the Fermi energy μ. For the d-p model, ψ_0 contains the variational parameters $\tilde{t}_{dp}$, $\tilde{t}_{pp}$, $\tilde{t}'_d$, $\tilde{\epsilon}_p - \tilde{\epsilon}_d$:

$$\psi_0 = \psi_0(\tilde{t}_{dp}, \tilde{t}_{pp}, \tilde{t}'_d, \tilde{\epsilon}_p - \tilde{\epsilon}_d). \tag{4}$$

In the non-interacting case, parameters $\tilde{t}_{dp}$, $\tilde{t}_{pp}$, $\tilde{t}'_d$ and $\tilde{\epsilon}_p - \tilde{\epsilon}_d$ coincide with those in the Hamiltonian, respectively. We fix $\tilde{t}_{dp} = t_{dp}$ as energy unit.

We consider the following wave function that is improved from the Gutzwiller wave function[36, 41, 50, 51, 52, 53, 54]:

$$\psi_\lambda = \exp(-\lambda K)\psi_G, \tag{5}$$

where λ is a variational parameter and K is the kinetic part of the Hamiltonian. The expectation values are evaluated by using the auxiliary field method[41, 55]. The kinetic part K also contains the band parameters t_{pp}, t'_d and $\epsilon_p - \epsilon_d$ as variational parameters:

$$K = K(\hat{t}_{pp}, \hat{t}'_d, \hat{\epsilon}_p - \hat{\epsilon}_d). \tag{6}$$

We take $\hat{t}_{pp} = \tilde{t}_{pp}$, $\hat{t}'_d = \tilde{t}'_d$ and $\hat{\epsilon}_p - \hat{\epsilon}_d = \tilde{\epsilon}_p - \tilde{\epsilon}_d$, for simplicity. Thus we have g, $\tilde{t}_{pp}$, $\tilde{t}'_d$, $\hat{\epsilon}_p - \hat{\epsilon}_d = \tilde{\epsilon}_p - \tilde{\epsilon}_d$, and λ as variational parameters. The expectation values for this type of wave function are calculated by using the variational Monte Carlo method. We can evaluate the expectation value correctly within statistical errors.

The correlated superconducting (SC) state is represented by the projected-BCS wave function. This is given as

$$\psi_S = P_{N_e} P_G \psi_{BCS}, \tag{7}$$

where ψ_{BCS} indicates the BCS wave function:

$$\psi_{BCS} = \prod_{\mathbf{k}} (u_{\mathbf{k}} + v_{\mathbf{k}} \alpha^{\dagger}_{\mathbf{k}\uparrow} \alpha^{\dagger}_{-\mathbf{k}\downarrow})|0\rangle. \tag{8}$$

$\alpha^{\dagger}_{\mathbf{k}\sigma}$ indicates the creation operator of the state in the lowest band with the momentum $\mathbf{k}$ which is represented by a linear combination of d and p electron operators $d^{\dagger}_{\mathbf{k}\sigma}$ and $p^{\dagger}_{\mathbf{k}\sigma}$. The BCS parameters $u_{\mathbf{k}}$ and $v_{\mathbf{k}}$ are given by the ratio $v_{bfk}/u_{\mathbf{k}} = \Delta_{\mathbf{k}}/(\xi_{\mathbf{k}} + \sqrt{\xi^2_{\mathbf{k}} + \Delta^2_{\mathbf{k}}})$ where $\xi_{\mathbf{k}}$ is the dispersion relation of the lowest band. We assume the d-wave symmetry $\Delta_{\mathbf{k}} = \Delta(\cos k_x - \cos k_y)$ and we regard Δ as the superconducting gap. Δ is a variational parameter that is optimized to give the lowest ground energy. P_{N_e} is a projection operator which extracts only the states with a fixed total hole number.

It is convenient to perform the particle-hole transformation for down-spin holes[36, 51] in multiplying by the operator $\exp(-\lambda K)$:

$$\psi_\lambda = \exp(-\lambda K) P_G \psi_{BCS} = \exp(-\lambda K) P_G \prod_{\mathbf{k}} (u_{\mathbf{k}} \beta^{\dagger}_{\mathbf{k}} + v_{\mathbf{k}} \alpha^{\dagger}_{\mathbf{k}})|\tilde{0}\rangle, \tag{9}$$

where $\beta^{\dagger}_{\mathbf{k}} = \alpha_{-\mathbf{k}\downarrow}$ and $\alpha^{\dagger}_{\mathbf{k}} = \alpha^{\dagger}_{\mathbf{k}\uparrow}$. $|\tilde{0}\rangle$ denotes the vacuum for newly defined α and β particles satisfying $\alpha_{\mathbf{k}}|\tilde{0}\rangle = \beta_{\mathbf{k}}|\tilde{0}\rangle = 0$.

4. Antiferromagnetic phase and t'_d

In the two-dimensional Hubbard model, the antiferromagnetic (AF) correlation can be controlled by the Coulomb interaction U[36]. In the strongly correlated region where U is larger than the bandwidth, the antiferromagnetic correlation is suppressed and as a result superconducting state can be stable. We would expect a similar phase diagram for the d-p model. It is not, however, easy to have control of the antiferromagnetic state in the three-band d-p model. It appears that the AF state is considerably stable in the d-p model, compared to that in the single-band Hubbard model.

We show that the parameter t'_d plays an important role regarding a stability of the AF state. In Fig.3 the antiferromagnetic region is exhibited in the plane of t'_d and the level difference $\epsilon_p - \epsilon_d$ where we have 72 holes on 8×8 lattice with 192 atoms in total. There is a paramagnetic region for negative t'_d. When the level difference is small, the paramagnetic state can be stable for small $|t'_d|$.

There is a possibility of high-temperature superconductivity near the boundary where the AF correlation suppressed. We have three parameters U_d, $\epsilon_p - \epsilon_d$ and t'_d. For $t'_d = 0$, the AF phase exists as shown in Fig.4 where the doping rate is $\delta = 0.25$. AF order exists in a wide range of parameters. The strongly correlated d-wave superconducting state, however, exists when d and p level difference is small. The pure d-wave superconductivity may be realized in this region.

5. Superconducting wave function

We perform the electron-hole transformation for down-spin electrons to multiply the projected BCS function by the operator $\exp(-\lambda K)$. The electron (hole) number is not fixed as in the BCS

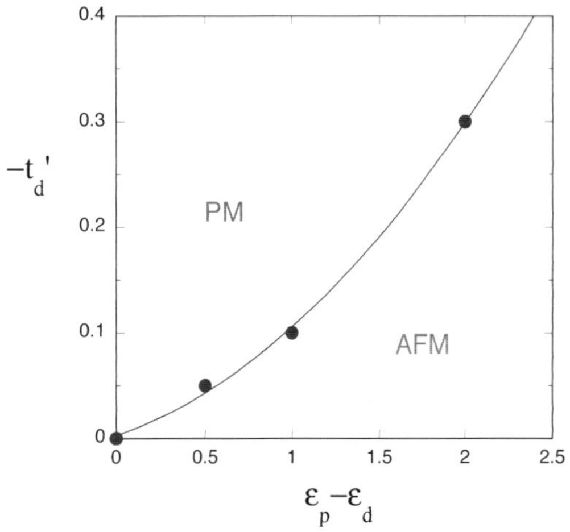

Figure 3. Antiferromagnetic region in the plane of t'_d and $\epsilon_p - \epsilon_d$ of the three-band d-p model. The energy unit is given by t_{dp}. The hole density is $\delta = 0.125$ for 72 holes on 8×8 lattice We used $t_{pp} = 0.4t_{dp}$, $U_d = 10t_{dp}$ and $U_p = 0$.

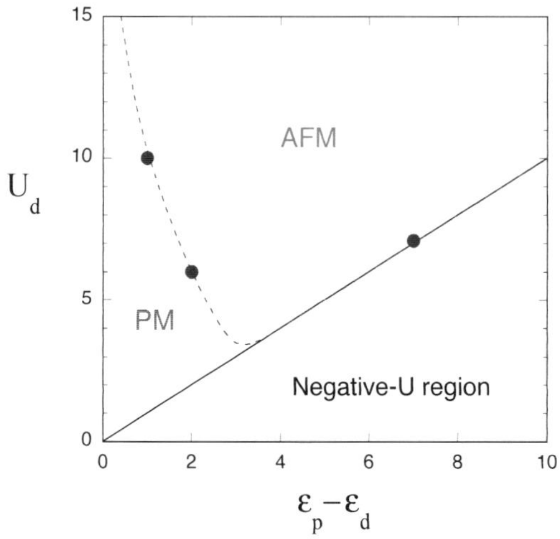

Figure 4. Antiferromagnetic region in the plane of U_d and $\epsilon_p - \epsilon_d$ in units of t_{dp} for 8×8 lattice with 76 holes. The dashed line is a guid for eyes. The band parameters are $t_{pp} = 0.4t_{dp}, t'_d = 0$ and $U_p = 0$.

wave function. The chemical potential is not a variational parameter since it is used to adjust the total electron number.

We show the result of calculations in Fig.5 where the total number N_h is evaluated as the expectation value by varying the chemical potential. In this result in slightly overdoped region, the optimized SC order parameter is finite indicating that the d-wave SC state is indeed realized. We expect that the SC state is stabilized in the region where $\epsilon_d - \epsilon_p$ is small and near the boundary of the AF region. There is a possibility of high-temperature superconductivity in this region.

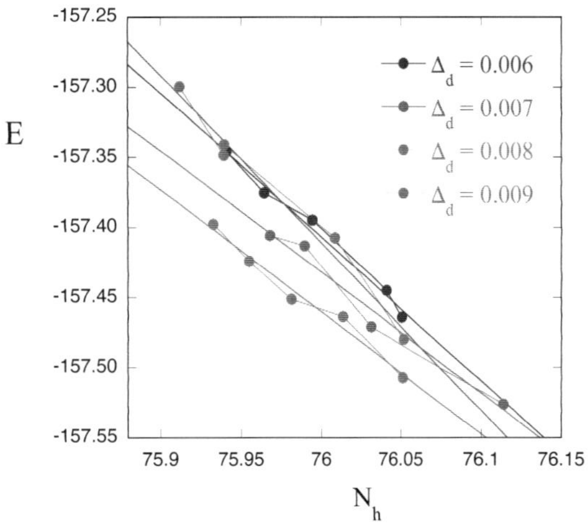

Figure 5. Total energy as a function of hole number N_h on 8×8 lattice. We set $U_d = 10$, $t_{pp} = 0$, $t'_d = 0$, $\epsilon_d = -0.5$ and $\epsilon_p = 0$. The gap parameter $\Delta = \Delta_d = \Delta_p$ is introduced for the BCS wave function. At $N_h = 76$, the energy has a minimum at finite value of Δ.

6. Discussion and Summary

It is important to determine the phase diagram in the study of correlated electron systems. In particular, we need to clarify the stability of magnetic order in electronic models of high-temperature superconductors. For this purpose, it is necessary to improve and optimize the wave function to include electron correlation. The exponential kinetic operator $e^{-\lambda K}$ used in our work has an effect to move electrons to get the kinetic-energy gain. This will increase the double occupancy, and the variational parameters are determined to lower the ground-state energy by balancing the kinetic energy against the Coulomb repulsive energy. We can lower the ground-state energy further by multiplying the wave function by exponential kinetic operator.

The importance of multiplying by $e^{-\lambda K}$ lies in that the AF phase narrows since AF correlation is suppressed. In the 2D Hubbard model, the AF correlation is considerably suppressed when U is large being greater than the bandwidth where the kinetic-energy gain is dominant over the AF energy lowering. This also holds for the d-p model. By using the Gutzwiller function, the optimized value of the AF order parameter is very large and is larger than that of the superconducting order parameter more than by one order of magnitude in the d-p model. The AF order parameter is reduced and becomes of the same order as the superconducting one on the basis of our wave function. This would give a possibility of superconductivity.

We investigated the ground state of the two-dimensional d-p (three-band Hubbard) model by employing the optimization variational Monte Carlo method. We introduced long-range transfer intergrals t'_d to conquer the antiferromagnetism in the d-p model. We have shown the phase diagram in the plane of U_d and $\epsilon_p - \epsilon_d$. We expect that there also occurs a crossover between strongly correlated and weakly correlated regions by controlling the antiferromagnetic correlation in the d-p model. There is a crossover in the two-dimensional Hubbard model[36]. A large fluctuation presumably exists in the crossover region. This indicates a possibility of high-temperature superconductivity. A crossover from weakly to strongly coupled systems is universal phenomenon that exists ubiquitously in the world as in the Kondo effect[56, 57, 59, 60, 61, 62, 63].

References

[1] Bednorz J G and Müller K A 1986 *Z. Phys.* B **64** 189.
[2] Emery V J 1987 *Phys. Rev. Lett.* **58** 2794.
[3] Hirsch J E, Loh E Y, Scalapino D J and Tang S 1989 *Phys. Rev.* B **39** 243.
[4] Scalettar R T, Scalapino D J, Sugar R L, and White S R 1991 *Phys. Rev.* B **44** 770.
[5] Oguri A, Asahata T and Maekawa S 1994 *Phys. Rev.* B **49** 6880.
[6] Koikegami S and Yamada K 2000 *J. Phys. Soc. Jpn.* **69** 768.
[7] Yanagisawa T, Koike S and Yamaji K 2001 *Phys. Rev.* B **64** 184509.
[8] Koikegami S and Yanagisawa T 2001 *J. Phys. Soc. Jpn.* **70** 3499.
[9] Yanagisawa T, Koike S and Yamaji K 2003 *Phys. Rev.* B **67** 132408.
[10] Koikegami S and Yanagisawa T 2003 *Phys. Rev.* B **67** 134517.
[11] Koikegami S and Yanagisawa T 2006 *J. Phys. Soc. Jpn.* **75** 034715.
[12] Yanagisawa T, Miyazaki M and Yamaji K 2009 *J. Phys. Soc. Jpn.* **78** 013706.
[13] Weber C, Lauchi A, Mila F and Giamarchi T 2009 *Phys. Rev. Lett.* **102** 017005.
[14] Lau B, Berciu M and Sawatzky G A 2011 *Phys. Rev. Lett.* **106** 036401.
[15] Weber C, Giamarchi T and Varma C M 2014 *Phys. Rev. Lett.* **112** 117001.
[16] Avella A, Mancini F, Paolo F and Plekhanov E 2013 *Euro. Phys. J.* B **86** 265
[17] Ebrahimnejad H, Sawatzky G A and Berciu M 2016 *J. Phys. Cond. Matter* **28** 105603.
[18] Tamura S and Yokoyama H 2016 *Phys. Procedia* **81** 5.
[19] Hubbard J 1963 *Proc. Roy. Soc. London* **276** 238.
[20] Hubbard J 1963 *Proc. Roy. Soc. London* **281** 401.
[21] Gutzwiller M C 1963 *Phys. Rev. Lett.* **10** 159.
[22] Zhang S, Carlson J and Gubernatis J E 1997 *Phys. Rev.* B **55** 7464.
[23] Zhang S, Carlson J and Gubernatis J E 1997 *Phys. Rev. Lett.* **78** 4486.
[24] Yanagisawa T and Shimoi Y 1996 *Int. J. Mod. Phys.* B **10** 3383.
[25] Nakanishi T, Yamaji K and Yanagisawa T 1997 *J. Phys. Soc. Jpn.* **66** 294.
[26] Yamaji K, Yanagisawa T, Nakanishi T and Koike S 1998 *Physica* C **304** 225.
[27] Yamaji K, Yanagisawa T and Koike S 2000 *Physica* B **284-288** 415.
[28] Bulut N 2002 *Advances in Phys.* **51** 1587.
[29] Yokoyama H, Tanaka Y, Ogata M and Tsuchiura H 2004 *J. Phys. Soc. Jpn.* **73** 1119.
[30] Yokoyama H, Ogata M and Tanaka Y 2006 *J. Phys. Soc. Jpn.* **75** 114706.

IOP Conf. Series: Journal of Physics: Conf. Series **1293** (2019) 012027 doi:10.1088/1742-6596/1293/1/012027

[31] Aimi T and Imada M 2007 *J. Phys. Soc. Jpn.* **76** 113708.
[32] Miyazaki M, Yanagisawa T and Yamaji K 2004 *J. Phys. Soc. Jpn.* **73** 1643.
[33] Yanagisawa T 2008 *New J. Phys.* **10** 023014.
[34] Yanagisawa T 2013 *New J. Phys.* **15** 033012.
[35] Yanagisawa T, Miyazaki M and Yamaji K 2014 *J. Mod. Phys.* **4** 33.
[36] Yanagisawa T 2016 *J. Phys. Soc. Jpn.* **85** 114707.
[37] Noack R M, White S R and Scalapino D J 1995 *Europhys. Lett.* **30** 163.
[38] Yamaji K, Shimoi Y and Yanagisawa T 1994 *Physica* C **235** 2221.
[39] Yanagisawa T, Shimoi Y and Yamaji K 1995 *Phys. Rev.* B **52** R3860.
[40] Nakano T, Kuroki K and Onari S 2007 *Phys. Rev.* B **76** 014515.
[41] Yanagisawa T, Koike S and Yamaji K 1998 *J. Phys. Soc. Jpn.* **67** 3867.
[42] Yanagisawa T and Miyazaki M 2014 *Europhys. Lett.* **107** 27004.
[43] McElroy K, Simmonds R W, Hoffman J E, Lee D H, Orenstein J, Eisaki H, Uchida S and Davis J C 2003 *Nature* **422** 592.
[44] Hussey N E, Abdel-Jaead M, Carrington A, Mackenzie A P and Balicas L 2003 *Nature* **425** 814.
[45] Weber C, Haule K and Kotliar G 2008 *Phys. Rev.* B **78** 134519.
[46] M. Hybertsen S, Schluter M and Christensen N E 1989 *Phys. Rev.* B **39** 9028.
[47] Eskes H, Sawatzky G A and Feiner L F 1989 *Physica* C **160** 424.
[48] McMahan A K, Annett J F and Martin R M 1990 *Phys. Rev.* B **42** 6268.
[49] Eskes H and Sawatzky G 1991 *Phys. Rev.* B **43** 119.
[50] Otsuka H 1992 *J. Phys. Soc. Jpn.* **61** 1645.
[51] Yanagisawa T, Koike S and Yamaji K 1999 *J. Phys. Soc. Jpn.* **68** 3608.
[52] Eichenberger D and Baeriswyl D 2007 *Phys. Rev.* B **76** 180504.
[53] Baeriswyl D, Eichenberger D and Menteshashvii M 2009 *New J. Phys.* **11** 075010.
[54] Baeriswyl D 2011 *J. Supercond. Novel Magn.* **24** 1157.
[55] Yanagisawa T 2007 *Phys. Rev.* B **75** 224503.
[56] Kondo J 2012 *The Physics of Dilute Magnetic Alloys* (Cambridge University Press, Cambridge).
[57] Nagaoka Y 1965 *Phys. Rev.* **138** A1112.
[58] Yanagisawa T 2012 *J. Phys. Soc. Jpn.* **81** 094713.
[59] Yanagisawa T 2015 *J. Phys. Soc. Jpn.* **84** 074705.
[60] Solyom J 1979 *Adv. Phys.* **28** 201.
[61] Yanagisawa T 1991 *J. Phys. Soc. Jpn.* **60** 29.
[62] Yanagisawa T 2016 *Europhys. Lett.* **113** 41001.
[63] Yanagisawa T 2019 *Prog. Theor. Exp. Phys.* **2019** 023A01.

Electronic Structure of Novel Non-centrosymmetric Superconductor Mg_2Rh_3P

I Hase[1], T Yanagisawa[1], A Iyo[1], H Fujihisa[1], Y Goto[1], H Eisaki[1] and K Kawashima[2]

[1] National Institute of Advanced Industrial Science and Technology (AIST), Tsukuba, 305-8568, Japan
[2] IMRA Material R&D Co. Ltd.

E-mail: i.hase@aist.go.jp

Abstract. We calculated the electronic structure of a novel non-centrosymmetric superconductor $Mg_{2-x}Rh_3P$ from first-principles. Stoichiometric Mg_2Rh_3P shows a pseudogap at the Fermi level. This is consistent with the experimental fact that the stoichiometric sample Mg_2Rh_3P does not show superconductivity and the Mg-deficient sample $Mg_{1.9}Rh_3P$ shows superconductivity with $T_c \sim 3.6$ K. The band dispersion of Mg_2Rh_3P shows a characteristic band splitting due to the anti-symmetric spin-orbit coupling (ASOC). The magnitude of ASOC of Mg_2Rh_3P is comparable to Li_2Pd_3B. We also show that its 5d-counterpart Mg_2Ir_3P may have quite a large ASOC, and is a good candidate of parity-mixed superconductor likewise Li_2Pd_3B.

1. Introduction

The non-centrosymmetric (i.e. without space inversion symmetry) superconductors have been paid much attention because they can exhibit anomalous parity-mixed superconducting state [1,2]. Among them, Li_2M_3B (M = Pd, Pt) are one of the most energetically studied compounds [3-12]. NMR experiment shows that the superconducting state of Li_2Pt_3B includes considerable spin-triplet component [5], while in Li_2Pd_3B it seems to be spin-singlet. However, for both compounds the symmetry of the order parameter is still controversial. The large upper critical magnetic field ($H_{c2} \sim 4$ T) of Li_2Pt_3B also indicates that the Cooper pair forms the spin-triplet state [6]. This large H_{c2} is also advantageous for various applications. The difference between Li_2Pd_3B and Li_2Pt_3B has been explained as follows: In the non-centrosymmetric crystals, spin-orbit interaction (SOI) has an asymmetric component, and it causes a characteristic parity mixing. Since Pt-5d orbitals has larger SOI than Pd-4d orbitals, Li_2Pt_3B has a large parity mixing while Li_2Pd_3B does not have such anomaly.

Recently we have discovered a new ternary superconductor $Mg_{2-x}Rh_3P$, which has the non-centrosymmetric crystal structure similar to Li_2M_3B [13]. The detail of the symmetry of the order parameter has not been clear yet. Interestingly, stoichiometric sample Mg_2Rh_3P does not show superconductivity above 1.8K, and about 5% deficiency of Mg leads to the superconducting phase with $T_c \sim 3.6$ K. These experimental results immediately invoke us two questions: First, is the parity mixing in $Mg_{2-x}Rh_3P$ large or not? How about the 5d-counterpart compound, Mg_2Ir_3P (which is not synthesized yet)? Second, why the stoichiometric sample Mg_2Rh_3P does not show superconductivity while Mg-deficient sample $Mg_{1.9}Rh_3P$ show superconductivity?

In order to answer these questions, we have performed an *ab-initio* band structure calculation of Mg_2Rh_3P and Mg_2Ir_3P. This paper is organized as follows: The details of calculations are described in Section 2. The calculated results and the discussions are shown in Section 3. Finally we give some conclusions in Section 4.

We found that Mg_2Rh_3P has a pseudogap at the Fermi level (E_F). About 10% Mg deficiency shifts E_F and increases DOS drastically, which can explain the above-mentioned experimental result. On the other hand, Mg_2Ir_3P has a large DOS at E_F in the stoichiometric form.

2. Methods

We have calculated the electronic structure of Mg_2Rh_3P and Mg_2Ir_3P from first-principles. This calculation is based on the density functional theory (DFT) and the full-potential linearized augmented plane wave method (FLAPW), and the exchange-correlation potential is approximated by general gradient approximation (GGA) [14]. The whole calculation is implemented by the computer code WIEN2k [15]. The unit cell includes four formula unit, i.e. $Z = 4$. We used the parameter RK_{max}=7.0, and the muffin-tin radii are set as r(Mg) = 2.37, r(Rh/Ir) = 2.36 and r(P) = 1.83 a.u. As for Mg_2Rh_3P, we used the experimentally determined lattice parameter (a = 7.0905 Å) and space group (#213, $P4_132$). This space group is enantiomorphic with the space group of Li_2Pt_3B, which is denoted as $P4_332$ (#212). These two space groups are the same except the helicity. We have also optimized the internal parameter x of Mg(x, x, x) and y of Rh(1/8, y, y+1/4) within the above restrictions. We obtained x = 0.04062 and y = 0.17794, which agrees well with the experimental value (x_{exp} = 0.0397 and y_{exp} = 0.1818). As for hypothetical compound Mg_2Ir_3P, we assumed the same lattice constant and space group as Mg_2Rh_3P, and performed an optimization of the internal parameter x and y. We obtained x = 0.04009 and y = 0.17465, which are very close to those of Mg_2Rh_3P. As for the structure optimization and the self-consistent calculation we used the 1000 k-points mesh in the 1st Brillouin zone. Hereafter we show the results for the optimized structure. After convergence of potential, we performed the DOS calculation using 8000 k-points mesh. Spin-orbit interaction (SOI) has been included by a second-variational approach in this calculation.

3. Results and Discussions

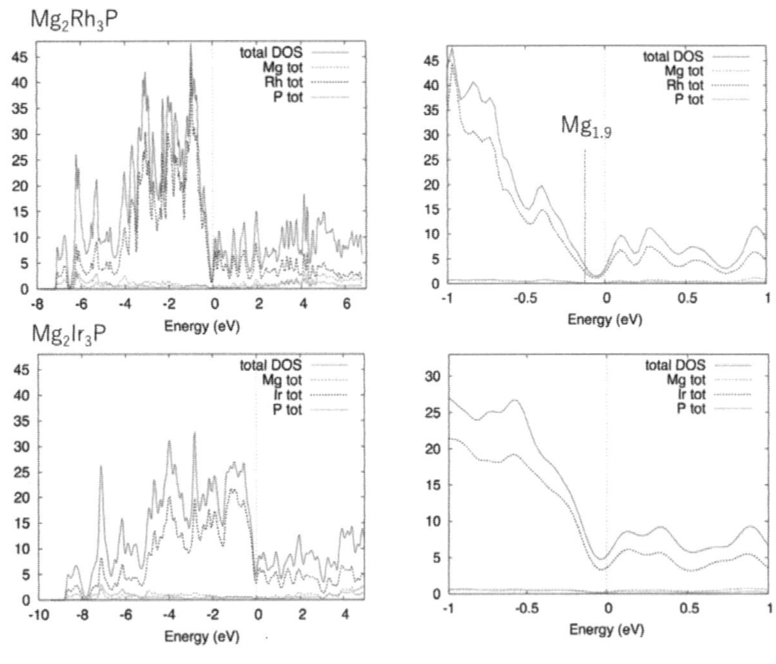

Figure 1. DOS curves of Mg_2Rh_3P and Mg_2Ir_3P. The right panels are blow-ups near E_F. The units of vertical axis are eV^{-1} per unit cell (i.e. 4 Mg_2Rh_3P).

First we show the DOS curve of Mg_2Rh_3P and Mg_2Ir_3P in figure 1. Since they are isovalent, the shape of DOS curves is very similar. The valence bands mostly consist of Rh/Ir-d orbitals, and P-p orbitals are strongly hybridized with them. The width of the valence band (below E_F) is 7.1 eV for Mg_2Rh_3P and 8.8 eV for Mg_2Ir_3P. This is mainly because Rh4d orbitals are more localized than Ir5d orbitals. This behavior is also seen in Li_2Pd_3B and Li_2Pt_3B. The width of the valence band of Li_2Pt_3B is about 20% larger than that of Li_2Pd_3B [4].

The right panels are the blow-up near E_F. We can clearly see that Mg_2Rh_3P has a pseudogap near E_F. This extremely small DOS is apparently disadvantageous for superconductivity.

Next we discuss the electronic structure of $Mg_{1.9}Rh_3P$. As discussed above, the valence bands mostly consist of Rh-d and P-p orbitals, and Mg-s and Mg-p orbitals have a negligible contribution. Therefore, we can look Mg_2Rh_3P as a Zintl phase, like $[Mg^{2+}{}_2]^{4+}[Rh_3P]^{4-}$. According to this picture, Mg-deficient phase $Mg_{1.9}Rh_3P$ can be described as $[Mg^{2+}{}_{1.9}]^{3.8+}[Rh_3P]^{3.8-}$. That is, Mg deficiency gives a $0.2e$ hole doping per formula unit to the valence band. In the rigid-band picture, this effect is described as a shift of E_F. The shifted E_F is shown by the dotted vertical line in the upper-right panel. This shift drastically increases DOS as twice or more. This behavior gives a natural explanation of the experimental fact that stoichiometric Mg_2Rh_3P does not show superconductivity, and Mg-deficient $Mg_{1.9}Rh_3P$ shows superconductivity.

On the other hand, Mg_2Ir_3P may have different properties. Similar to Mg_2Rh_3P, the DOS of Mg_2Ir_3P has a large dip at E_F. However, there is still a large amount of the residual DOS. Therefore, we can expect that Mg_2Ir_3P will be superconducting in the stoichiometric composition.

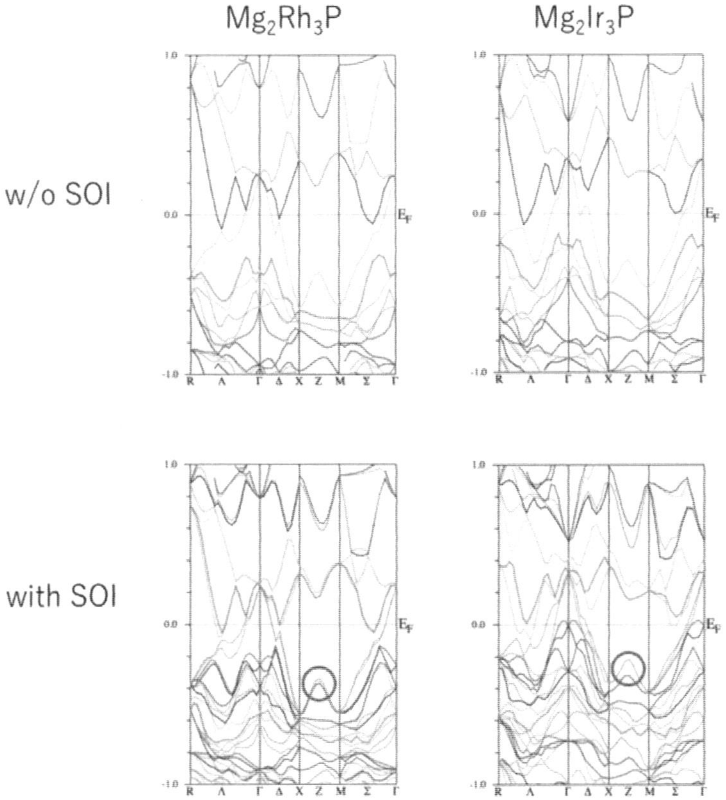

Figure 2. Energy dispersion curves of Mg_2Rh_3P and Mg_2Ir_3P near E_F. The units of vertical axis are eV. The upper (lower) panels are the calculated results without (with) including SOI. Circles show a typical band splitting due to ASOC.

Figure 2 is the band dispersion of Mg_2Rh_3P and Mg_2Ir_3P. The upper panels are the band dispersions without SOI, and the lower panels are with SOI. First we focus on the case without SOI. In this case, all bands are doubly degenerated due to the spin degree of freedom. Moreover, due to the high symmetry and non-symmorphic nature of the space group, there are some additional degeneracy along some symmetry points and axes. For example, along M–Z-R axis all the bands are doubly (if we include spin, 4-fold) degenerated. At R point all bands are 4-fold (8-fold) degenerated. These properties are useful for counting the number of the bands.

Next we discuss the effect of SOI. Since the space group lacks space inversion symmetry, the spin-orbit interaction induces so-called asymmetric spin-orbit coupling (ASOC) [16]. ASOC lifts the double degeneracy due to the spin degree of freedom at most of the k-points. Moreover, ASOC lifts part of the higher degeneracy discussed above. For example, the 4-fold degeneracy (including spin) along M-Z-R axis is partially lifted with 2+2 fold degeneracy by SOI (see the circle in the lower panels of Figure 2). This ASOC can play a crucial role to induce the parity mixing of the superconducting order parameter [16,17]. In Li_2Pt_3B, the band splitting due to ASOC near E_F is ~200 meV at X point [4]. On the other hand, the corresponding band splitting (near E_F and at X point) is only ~50 meV. However, it is not easy to compare the magnitude of SOI between Li_2Pt_3B and Mg_2Ir_3P, because the band structure and the number of valence electrons are different. For example, the splitting at Z(1/2,1/4,0) is ~100 meV both for Li_2Pt_3B and Mg_2Ir_3P, as seen in ref.4 and the circle in Figure 2. We can estimate the strength of SOI in Mg_2Ir_3P is comparable to Li_2Pt_3B, or a little bit smaller than Li_2Pt_3B on average.

The comparison between Mg_2Rh_3P and Mg_2Ir_3P is much easier. As seen in Figure 2, the effect of SOI in Mg_2Rh_3P is much smaller, at least less than half of that in Mg_2Ir_3P. Similar estimation has been obtained as for the effect of SOI in Li_2Pd_3B and Li_2Pt_3B, i.e. the splitting by SOI in Li_2Pd_3B is less than half of that in Li_2Pt_3B on average. Considering that Li_2Pt_3B seems to have quite a large component of spin-triplet pairing, Mg_2Ir_3P may also have such component if it is synthesized. And we can also expect that the observed superconductivity in Mg_2Rh_3P is a rather conventional one, likewise in Li_2Pd_3B.

4. Conclusion

We have calculated the band structure of non-centrosymmetric superconductor Mg_2Rh_3P and its 5d-counterpart compound Mg_2Ir_3P from first-principles. Mg_2Rh_3P has a pseudogap at the Fermi energy, which seems to be consistent that the stoichiometric Mg_2Rh_3P does not show superconductivity. We found that Mg_2Ir_3P has much larger spin-orbit interaction than Mg_2Rh_3P, and is a potential candidate of the parity-mixed superconductor.

Acknowledgments

This work was partially supported by KAKENHI (Grant No. JP26400379) from Japan Society for the Promotion of Science (JSPS).

References

[1] Bauer E *et al.* 2004 *Phys. Rev. Lett.* **92** 027003
[2] Bauer E and Sigrist M 2011 *Non-centrosymmetric Superconductors* (Amsterdam: Springer-Verlag)
[3] Togano K, Badica P, Namkamori Y, Orimo S, Takeya H and Hirata K 2004 *Phys. Rev. Lett.* **93** 247004
[4] Lee K -W and Pickett W, 2005 *Phys. Rev.* B **72** 174504

[5] Mochiku T, Takeya H, Wuernisha T, Mori K, Ishigaki T, Kamiyama T, Fujii H and Hirata K 2006 *Physica* C **445** 57
[6] Badica P, Kondo T, Kudo T, Nakamori Y, Orimo S and Togano K 2004 *Appl. Phys. Lett.* **85** 4433
[7] Khasanov R, Landau I L, Baines C, La Mattina F, Maisuradze A, Togano K and Keller H 2007 *Phys. Rev.* B **76** 104506
[8] Yuan H Q, Agterberg A F, Hayashi N, Badica P, Vandervelde D, Togano K, Sigrist M and Salamon M B 2006 *Phys. Rev. Lett.* **97** 017006
[9] Doria M M, Salem-Sugui M, Jr., Badica P and Togano K 2006 *Phys. Rev.* B **73** 184524
[10] Nishiyama M, Inada Y and Zheng G -Q 2007 *Phys. Rev. Lett.* **98** 047002
[11] Shimamura A, Furukawa Y, Kumagai K, Takeya H and Hirata K 2007 *Physica* C **460** 663
[12] Tsuda S *et al.* 2009 *J. Phys. Soc. Jpn.* **78** 034711
[13] Iyo A *et al.* in preparation.
[14] Perdew J P, Burke K, Ernzerhof M 1996 *Phys. Rev. Lett.* **77** 3865
[15] Blaha P, Schwarz K, Madsen G K H, Kvasnicka D and Luitz J 2001 WIEN2k, *An Augmented Plane Wave + Local Orbbitals Program for Calculating Crystal Properties* (Wienna: Vienna Univ. of Technology)
[16] Frigeri P A, Agterberg D F, Milat I and Sigrist M 2006 *Eur. Phys. J.* B **54** 435
[17] Samokhin K V and Mineev V P 2008 *Phys. Rev.* B **77** 104520

IOP Conf. Series: Journal of Physics: Conf. Series **1293** (2019) 012029 doi:10.1088/1742-6596/1293/1/012029

Impurity effects on $d_{x^2-y^2}$-wave superconducting state in strongly correlated Hubbard model

Ryo Sato and Hisatoshi Yokoyama

Depertment of Physics, Tohoku University, Sendai 980-8578, Japan

E-mail: satoryo@cmpt.phys.tohoku.ac.jp

Abstract. With cuprate superconductors in mind, we study the effects of point-type impurity potential (V) on a $d_{x^2-y^2}$-wave superconducting (d-SC) state in a strongly correlated square-lattice Hubbard model ($U/t = 12$) with a diagonal transfer (t'), using a variational Monte Carlo method. At half filling, insulator-to-conductor (filling-control-type Mott) transitions occur at $V \equiv V_U^{(\pm)} \sim \pm U$. The phase realized for $V < V_U^{(-)}$ is robust d-SC and the charge carriers are holes, whereas for $V > V_U^{(+)}$ a metallic (or weak d-SC) state appears with charge carriers of electrons. For a partially-filled case where the impurity density is equal to the doping rate, a d-SC-to-insulator transition arises at $V \equiv V_{\rm M} \sim 4t$. In the weak-$|V|$ phases, the impurity effects in the electron distribution are smeared out by the screening effect, whereas in the strong-$|V|$ phases, the electron distribution directly depend on the impurity configurations.

1. Introduction

Cuprate superconductors (SCs) [1] are inherently inhomogeneous systems in that the charge carriers are necessarily introduced with charged dopants, which randomly enter into the block layers and affect the carriers, in addition to other disorders specific to individual compounds. In most of relevant theoretical studies, such disorder potential has been treated with Anderson-Hubbard-type models [2], in which the local one-body potential V in each site is randomly distributed in the range $-D \geq V \geq D$ (D: a constant), instead of the (more simple) impurity Hubbard model [3], in which the impurity sites with a one-body potential of fixed V are randomly chosen. Since the understanding of disorder in the strongly correlated systems is still insufficient, it seems better to start with primitive cases in the impurity Hubbard model. Actually, the effects of positive and negative potential are found distinct.

In previous reports [4, 5, 6], we considered the impurity effects on the antiferromagnetic (AF) and paramagnetic (PM) or normal states in a strongly correlated Hubbard model, using a variational Monte Carlo (VMC) method. We found that in weak or intermediate attractive potential the impurity effects are almost completely screened out in a strongly correlated regime. And that a metal-to-insulator (Mott) transition occurs for weak repulsive potential. However, the trial states used in Ref. [4, 5] are insufficient for treating strong repulsive potential.

In this study, we modify the trial state to be applicable to any model parameter set [8, 6]; in addition, we formulate the VMC scheme applicable to the $d_{x^2-y^2}$-wave superconducting (SC) state. In this report, we represent some preliminary results at half filling and for a special partially-filled case, where the impurity density is equal to the doping rate.

2. Model and Method

In this series of research [4, 5, 6], we consider a Hubbard model ($U \geq 0$) with a point-type spin-independent impurity potential (V) on extended square lattices:

$$\mathcal{H} = \mathcal{H}_{\rm kin} + \mathcal{H}_U + \mathcal{H}_{\rm imp} = -\sum_{(i,j),\sigma} t_{ij} \left(c^\dagger_{i\sigma} c_{j\sigma} + {\rm H.c.} \right) + U \sum_j n_{j\uparrow} n_{j\downarrow} + V \sum_{\ell=1}^{N_{\rm imp}} \sum_\sigma n_{\ell\sigma}, \quad (1)$$

where $n_{j\sigma} = c^\dagger_{j\sigma} c_{j\sigma}$, (i,j) indicates the pairs on sites i and j, and $N_{\rm imp}$ is the number of impurity sites, which we will explain shortly. Here, we set the hopping integral t_{ij} as t for nearest neighbors (≥ 0), t' for diagonal neighbors, and 0 otherwise ($\mathcal{H}_{\rm kin} = \mathcal{H}_t + \mathcal{H}_{t'}$). We use t and the lattice spacing as the units of energy and length, respectively. We define the electron density n and the doping rate δ as $n = N/N_{\rm s}$ and $\delta = |1 - n|$ (N: number of electrons, $N_{\rm s}$: number of sites). For simplicity, we assume that the impurity potential V comes from a single species and act on a nearby single site ("impurity site"). If we grant this simplification, the index ℓ in $\mathcal{H}_{\rm imp}$ runs over the impurity sites $\mathbf{r}_\ell$ ($\ell = 1, \cdots, N_{\rm imp}$). Furthermore, we assume that $\mathbf{r}_\ell$ are randomly distributed. In this report, we concentrate on the cases of impurity densities, $\delta_{\rm imp} = N_{\rm imp}/N_{\rm s} \sim 0.08$ in strong correlation ($U/t = 12$). We widely varies V (even for $V < 0$) to grasp overall features of impurity effects. Although the above setting of impurity potential is possibly simplistic, we believe that the essence of impurity effects with strong correlation can be captured.

To the above model, we apply a variational Monte Carlo (VMC) method [7], which enables us to exactly treat many-body wave functions. As a trial wave function, we use a Jastrow type: $\Psi = \mathcal{P}\Phi$. In the previous studies [4, 5], we distinguished the impurity and host sites in the (many-body) projection factor $\mathcal{P}$. We found, however, that this leads to the uncontrollable fluctuation in the VMC optimization for $V \gtrsim 2t$. Therefore, in the new trial state here, we do not distinguish the impurity and host sites in the many-body projection factors, $\mathcal{P} = \mathcal{P}_{\rm H}\mathcal{P}_\sigma\mathcal{P}_Q\mathcal{P}_{\rm G}$, where $\mathcal{P}_{\rm H}$, $\mathcal{P}_\sigma$, $\mathcal{P}_Q$, and $\mathcal{P}_{\rm G}$ are a nearest-neighbor charge [5] and spin ($\mathcal{P}_\sigma = \beta^\sigma$ with β being a parameter and $\sigma = 4\sum_{\langle i,j\rangle} S^z_i S^z_j$) correlators, a nearest-neighbor doublon-holon binding factor, and onsite (Gutzwiller) factor, respectively. For $\mathcal{P}_Q$ and $\mathcal{P}_{\rm G}$, we adopt the same forms as those used for the uniform systems [7]. Instead of indirectly controlling the effect of V using many-body correlators, we introduce a one-body projector [8]

$$\mathcal{P}_\theta = \theta^\kappa, \quad \text{with} \quad \kappa = \sum_{\ell,\sigma} n_{\ell\sigma}, \quad (2)$$

and θ being a parameter. The new trial state becomes $\Psi(\theta) = \mathcal{P}\mathcal{P}_\theta\Phi$, which is found to be well convergent for any V.

To obtain the Slater-determinantal part Φ, we diagonalize the following mean-field Hamiltonian,

$$\mathcal{H}_{\rm MF} = \mathcal{H}^{\rm BR}_{\rm kin} + \mathcal{H}_{d\text{-SC}} + \mathcal{H}_{\rm imp}. \quad (3)$$

In the kinetic part $\mathcal{H}^{\rm BR}_{\rm kin}$, a band-renormalization effect is introduced in the same manner as in Refs. [4, 5, 6, 7]. In the impurity term, we use the raw form in the Hamiltonian [a constant $V_{\rm imp}$ ($=V$) instead of a parameter $V_{\rm eff}$] to avoid the redundancy of the screening effect, which is also introduced by $\mathcal{P}_\theta$. The second term is the mean-field pair potential term for d-SC,

$$\mathcal{H}_{d\text{-SC}} = -\sum_{\mathbf{k}} \left(\Delta_{\mathbf{k}} c^\dagger_{\mathbf{k}\uparrow} c^\dagger_{-\mathbf{k}\downarrow} + {\rm h.c.} \right), \quad (4)$$

$$\Delta_{\mathbf{k}} = \Delta_d \left(\cos k_x - \cos k_y \right). \quad (5)$$

IOP Conf. Series: Journal of Physics: Conf. Series **1293** (2019) 012029 doi:10.1088/1742-6596/1293/1/012029

To treat the site-dependent SC state, we follow a general mean-field scheme of the Bogoliubov-de Gennes (BdG) equation [9]. We rewrite $\mathcal{H}_{\rm MF}$ for the BdG equation as

$$\mathcal{H}_{\rm BdG} = \sum_{i,j} \begin{pmatrix} c_{i\uparrow}^\dagger & c_{i\downarrow} \end{pmatrix} \begin{pmatrix} H_{ij} & F_{ij} \\ F_{ji}^* & -H_{ij} \end{pmatrix} \begin{pmatrix} c_{j\uparrow} \\ c_{j\downarrow}^\dagger \end{pmatrix}. \tag{6}$$

In diagonalizing $\mathcal{H}_{\rm BdG}$, we solve the equation,

$$\sum_j \begin{pmatrix} H_{ij} & F_{ij} \\ F_{ji}^* & -H_{ij} \end{pmatrix} \begin{pmatrix} U_{\alpha j} \\ V_{\alpha j} \end{pmatrix} = E^\alpha \begin{pmatrix} U_{\alpha i} \\ V_{\alpha i} \end{pmatrix} \tag{7}$$

with $U_{\alpha i}$ and $V_{\alpha i}$ being the coefficients of two sets of eigenfunctions corresponding to E^α and

$$H_{ij} = -t\sum_{\tau} \delta_{j,i+\tau} - \sum_{m=1}^{4}\sum_{\tau_m} t_m \delta_{j,i+\tau_m} - \mu\delta_{i,j} + V_{\rm imp}\sum_{\ell=1}^{N_{\rm imp}} \delta_{i,j}\delta_{i,\ell}, \tag{8}$$

$$F_{ij} = -\frac{2}{N_{\rm s}}\Delta_d \sum_{\mathbf{k}\,(k_y>0)} (\cos k_x - \cos k_y)\cos\left[\mathbf{k}\cdot(\mathbf{r}_i - \mathbf{r}_j)\right], \tag{9}$$

where τ and τ_m (m = 1–4) run over the first to fifth nearest-neighbor sites, respectively, and ℓ the impurity sites. After diagonalization, the operators corresponding to the eigenvalue E^α are written as

$$\gamma_{\alpha\uparrow} = \sum_i U_{\alpha i} c_{i\uparrow} + \sum_i V_{\alpha i} c_{i\downarrow}^\dagger, \tag{10}$$

$$\gamma_{\alpha\downarrow} = -\sum_i V_{\alpha i} c_{i\uparrow}^\dagger + \sum_i U_{\alpha i} c_{i\downarrow}. \tag{11}$$

Using $\gamma_{\alpha\sigma}$, the ground-state wave function within the BdG equation is given as

$$|\Phi_{\rm BdG}\rangle = \prod_{\alpha\,(\alpha>N_{\rm s})} \gamma_{\alpha\uparrow}\gamma_{\alpha\uparrow}|0\rangle = ({\rm const.}) \times \exp\left[\sum_{i,j} \left(U^{-1}V\right)_{ij} c_{i\uparrow}^\dagger c_{j\downarrow}^\dagger\right]|0\rangle, \tag{12}$$

which is written for a fixed-electron-number form, which we use, as [10, 11]

$$|\Phi_{d\text{-SC}}\rangle = \left[\sum_{i,j} \left(U^{-1}V\right)_{i,j} c_{i\uparrow}^\dagger c_{j\downarrow}^\dagger\right]^{\frac{N_{\rm e}}{2}} |0\rangle. \tag{13}$$

Equation (13) is reduced to the usual BCS form for $V_{\rm imp} \to 0$. The d-SC state $\Psi_{d-{\rm SC}} = \mathcal{P}\mathcal{P}_\theta\Phi_{d-{\rm SC}}$ is reduced to the normal state $\Psi_{\rm N} = \mathcal{P}\mathcal{P}_\theta\Phi_{\rm FS}$ ($\Phi_{\rm FS}$: Fermi sea) for $\Delta_d \to 0$ and the bare chemical potential.

We compute expectation values with respect to $\Psi_{d-{\rm SC}}$ using a VMC method similar to that in Ref. [7]. The optimization is performed using the stochastic reconfiguration method [12]. In calculating physical quantities, we average 2.5×10^5 samples. We use systems of $N_{\rm s} = L \times L$ sites with $L = 10 - 20$ and the periodic-antiperiodic boundary conditions. A configuration of impurities $\{\mathbf{r}_\ell\}$ is randomly chosen and fixed throughout a single sweep of VMC calculation (optimization and measurement). In most cases, we will display data of a typical $\{\mathbf{r}_\ell\}$.

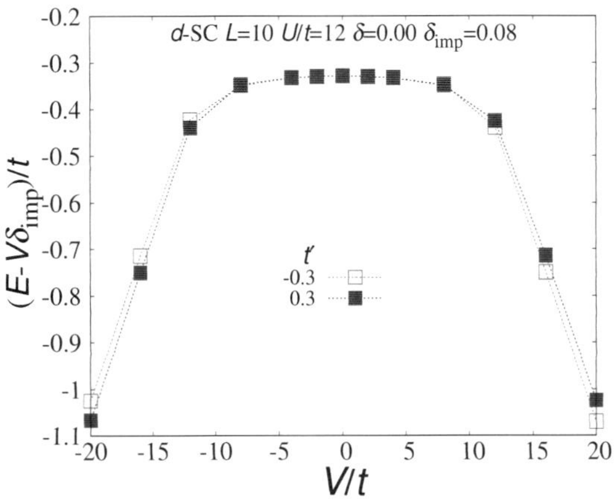

Figure 1. The total energy per site shifted by $-V\delta_{\rm imp}$, namely, $\mathcal{E}(V,t')$ at half filling is shown as a function of impurity potential V/t. The data were obtained by VMC calculations. The data for $t'/t = \pm 0.3$ are simultaneously plotted.

3. Results

Before discussing physics, we touch on a celebrated electron-hole (unitary) transformation

$$\mathcal{U}^{-1} c_{j\sigma}^\dagger \mathcal{U} \to e^{i\mathbf{Q}\cdot\mathbf{r}_j} \tilde{c}_{j\sigma}, \quad \text{or} \quad \mathcal{U}^{-1} c_{\mathbf{k}\sigma}^\dagger \mathcal{U} \to \tilde{c}_{\mathbf{Q}-\mathbf{k}\sigma}, \tag{14}$$

to confirm the correctness of VMC calculations, as well as to make good use of the symmetrical relations quantities have, especially, at half filling. Since some details are described in another report [6], here we summarize resultant relations relevant to the present subject. If we define a shifted energy as $\mathcal{E}(V,t') = E - V\delta_{\rm imp}$ (E: total energy per site) and average electron densities on the impurity and host sites as,

$$n_{\rm imp} = \frac{1}{N_{\rm imp}} \sum_{\ell=1}^{N_{\rm imp}} \sum_\sigma \langle n_{\ell\sigma} \rangle, \qquad n_{\rm hst} = \frac{N - N_{\rm imp} n_{\rm imp}}{N_{\rm s} - N_{\rm imp}}, \tag{15}$$

they have symmetrical relations,

$$\mathcal{E}(V,t') = \mathcal{E}(-V,-t'), \tag{16}$$
$$n_{\rm imp}(V,t') = 2 - n_{\rm imp}(-V,-t'), \tag{17}$$
$$n_{\rm hst}(V,t') = 2 - n_{\rm hst}(-V,-t'), \tag{18}$$

at half filling. Similarly, because the present system has inversion symmetry, the $d_{x^2-y^2}$-wave SC correlation function $P_d(\mathbf{r})$ and the optimized $d_{x^2-y^2}$-wave pairing parameter $\Delta_{\rm k}$ should satisfy the relation, $P_d(\mathbf{r})(V,t') = P_d(\mathbf{r})(-V,-t')$ and $\Delta_d(V,t') = \Delta_d(-V,-t')$, respectively, at half filling. $P_d(\mathbf{r})$ is the real-space d-SC correlation function for the nearest-neighbor-site pairing

$$P_d(\mathbf{r}) = \frac{1}{N_{\rm s}} \sum_i \sum_{\tau,\tau'=\hat{\mathbf{x}},\hat{\mathbf{y}}} (-1)^{1-\delta(\tau,\tau')} \left\langle \Delta_\tau^\dagger(\mathbf{R}_i) \Delta_{\tau'}(\mathbf{R}_i + \mathbf{r}) \right\rangle, \tag{19}$$

$\hat{\mathbf{x}}$ ($\hat{\mathbf{y}}$) denotes the lattice vector in the x (y) direction, $\delta(\tau,\tau')$ is the Kronecker delta, and $\Delta_\tau^\dagger(\mathbf{R}_i)$ is the creation operator of a nearest-neighbor singlet pair at site $\mathbf{R}_i$,

$$\Delta_\tau^\dagger(\mathbf{R}_i) = (c_{i\uparrow}^\dagger c_{i+\tau\downarrow}^\dagger + c_{i+\tau\uparrow}^\dagger c_{i\downarrow}^\dagger)/\sqrt{2}. \tag{20}$$

Here, $\Delta_d(V,t')$ indicates the optimized Δ_d [Eq. (5)] for V and t' with the other model parameters fixed. $P_d(\mathbf{r})(V,t')$ indicates the average with respect to the optimized trial state for the model

IOP Conf. Series: Journal of Physics: Conf. Series **1293** (2019) 012029 doi:10.1088/1742-6596/1293/1/012029

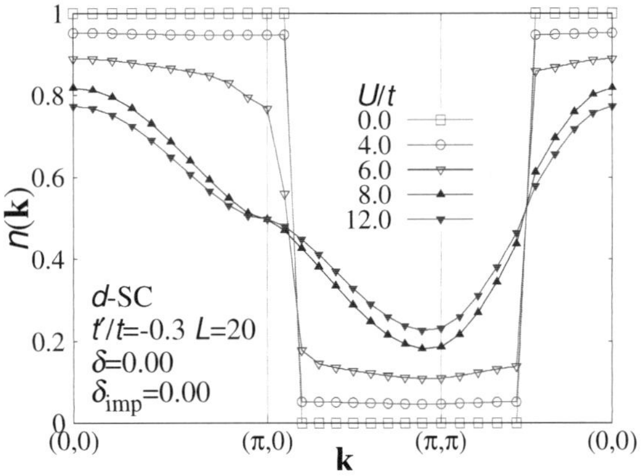

Figure 2. Evolution of momentum distribution function of d-SC state as U/t increases at half filling in uniform cases ($V = 0$) along path of $\mathbf{k}$: $(0,0)$-$(\pi,0)$-(π,π)-$(0,0)$. The values in metallic, d-SC, and Mott-insulating regimes are indicated in red colors, brown, and blue, respectively. x

parameters V and t' with the other model parameters fixed. In Fig. 1, $\mathcal{E}(V,t')$ is shown for $t' = \pm 0.3$. The symmetric relation Eq. (16) is basically satisfied, indicating the VMC calculations are correct.

We start with the half-filled case. In Fig. 1, we find that the behavior of $\mathcal{E}(V,t')$ exhibits abrupt changes at $V \left[\equiv V_U^{(\pm)}\right] \sim \pm 12t$, indicating some transitions or crossovers occur. To study this anomaly, we consider the momentum distribution function

$$n(\mathbf{k}) = \frac{1}{2}\sum_{\sigma}\langle c^{\dagger}_{\mathbf{k}\sigma}c_{\mathbf{k}\sigma}\rangle. \tag{21}$$

For recalling the behavior of $n(\mathbf{k})$, we show in Fig. 2, U/t dependence of $n(\mathbf{k})$ at half filling in the uniform cases ($V = 0$). For $V = 0$ and $\delta = 0$, it is known that the d-SC state (unless an AF order is assumed) evolves as U/t increases from metallic (normal) or weak d-SC ($U \lesssim 5t$) to d-SC ($5t \lesssim U < U_{\rm c} \sim 6.5t$) and finally to Mott insulating ($U > U_{\rm c}$) [13]. In Fig. 2, $n(\mathbf{k})$ exhibits discontinuities (Fermi surfaces, FSs) both near the antinode $(\pi,0)$ and near $(\pi/2,\pi/2)$ in the nodal direction for the metallic state. In d-SC states, the FS near $(\pi,0)$ disappears but the FS near $(\pi/2,\pi/2)$ remains, corresponding to opening a d-wave gap. In Mott insulating states, all FSs disappear, corresponding to opening of a full gap.

In Fig. 3(a), the evolution of $n(\mathbf{k})$ as V/t changes is shown for $U/t = 12$ and $t'/t = -0.3$. For $V_U^{(-)} < V < V_U^{(+)}$ (indicated with greenish colors), $n(\mathbf{k})$ exhibits no discontinuity, indicating that the Mott-insulating state for $V = 0$ continues within this range of V. For $V < V_U^{(-)}$ (warm colors), a FS appears near $(\pi/2,\pi/2)$ but $n(\mathbf{k})$ is continuous near $(\pi,0)$, indicating that d-SC is realized. Thus, an insulator-to-SC (Mott) transition is considered to occur at $V = V_U^{(-)}$. On the other hand for $V > V_U^{(+)}$ (cold colors), it seems that FSs appear both near $(\pi,0)$ and $(\pi/2,\pi/2)$, indicating that the state becomes metallic (or weak d-SC). In this case, an insulator-to-metal (Mott) transition is considered to arise at $V = V_U^{(+)}$.

It is important to check the order parameter of d-SC (P_d), and the d-wave pairing-gap parameter (Δ_d), which are shown in Fig. 4. As expected from $n(\mathbf{k})$, d-SC coherence vanishes ($P_d = 0$) in the range of $V_U^{(-)} < V < V_U^{(+)}$, but becomes finite for $V < V_U^{(-)}$. For $V > V_U^{(+)}$, P_d is small or almost vanishes, indicating that the state is weak d-SC or metallic. Thus, the interpretation of the behavior of $n(\mathbf{k})$ is confirmed. Note in Fig. 4 that Δ_d is robust in the insulating regime $|V| \lesssim |V_U^{(\pm)}|$, whereas the pairing gap becomes small in the conducting regime $|V| \gtrsim |V_U^{(\pm)}|$. This behavior of Δ_d will be understood in the same manner as in the uniform

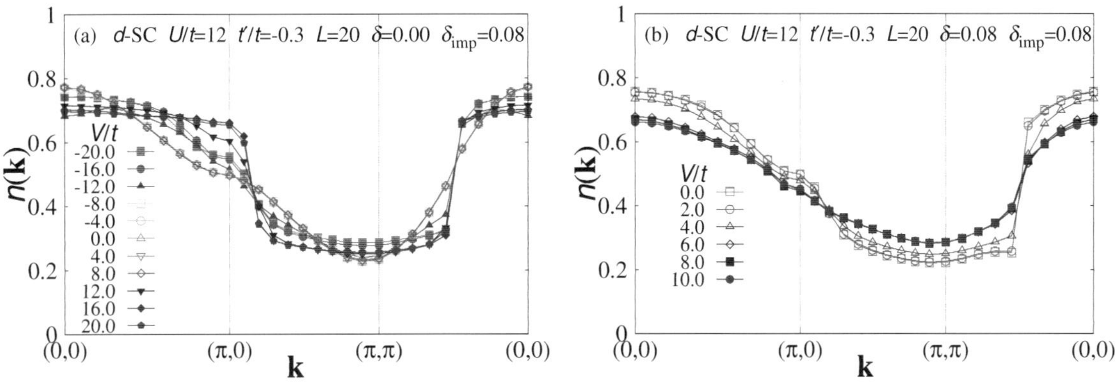

Figure 3. (a) The momentum distribution function at half filling ($\delta = 0$) for $t'/t = -0.3$ and $U/t = 12$ is shown along the same path as in Fig. 2 for various values of V/t. (b) The same quantity as in (a) but at $\delta = 0.08$ and for $V \geq 0$.

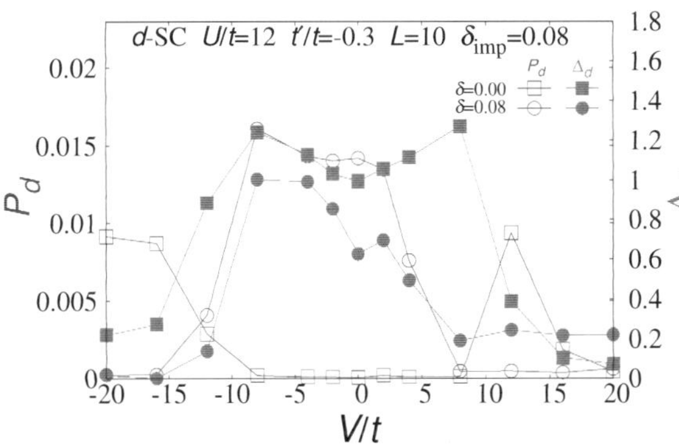

Figure 4. The d-SC correlation function P_d (d-SC order parameter, $P_d \equiv P_d(\mathbf{r}_\infty)$ with $\mathbf{r}_\infty = (L/2, L/2)$,left axis) and the optimized d-wave pairing gap parameter Δ_d (right axis) are shown for $U/t = 12$ and $t'/t = -0.3$ as functions of V/t. The results for $\delta = 0$ and 0.08 are simultaneously plotted.

cases [14, 7]: The origin of d-wave pairing is the AF spin correlation, which becomes largest in carrier-less cases and decreases as the carrier density increases. The d-SC coherence (P_d) is suppressed as the carrier density decreases because charge fluctuation is suppressed by Mott physics.

Therefore, we next consider the electron densities at the impurity and host sites. In Fig. 5(a), $n_{\rm imp}$ and $n_{\rm hst}$ are plotted as functions of V/t. In the insulating regime $|V| \lesssim |V_U^{(\pm)}|$, electron density both in the impurity and host sites are almost unity, namely, a "half-filled" state is preserved in this range of V. On the other hand in the conductive regime $|V| \gtrsim |V_U^{(\pm)}|$, the impurity sites are occupied by holons (empty sites) for $V > 0$ or doublons (doubly occupied sites) for $V < 0$. Correspondingly, the charge carriers (doublons for $V > 0$ or holons for $V < 0$) appears in the host sites and enable currents to flow. Thus, the Mott transitions arising at $V = V_U^{(\pm)}$ are filling-control-type. Because the basic mechanism of the Mott transition is the same as in the case of AF and normal states, for detailed explanation, refer to Ref. [6].

We can further discuss the reason of asymmetry in P_d with respect to $V = 0$ in Fig. 4 on the basis of the sign of charge carriers. Because the charge carriers for $V \lesssim V_U^{(-)}$ are holons, the mechanism of SC is parallel to the hole-doped case of $t'/t = -0.3$, in which d-SC is robust up to

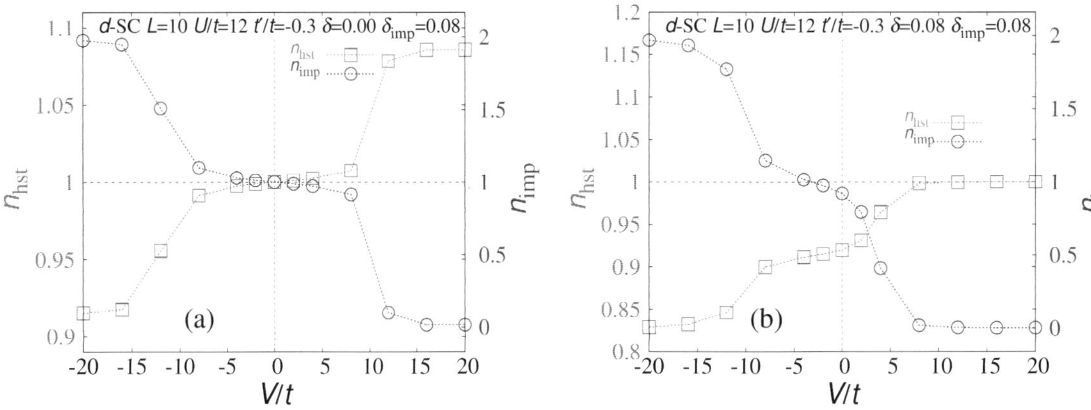

Figure 5. (a) V/t dependence of electron densities in impurity sites ($n_{\rm imp}$, right axis) and in host sites ($n_{\rm hst}$, left axis) [Eq. (15)] at half filling ($\delta = 0$). (b) The same quantities as in (a) are shown for $\delta = 0.08$.

high doping rate [14, 7, 15]. On the other hand, the charge carriers for $V \gtrsim V_U^{(+)}$ are doublons (electrons), the mechanism corresponds to the electron-doped case of $t'/t = +0.3$, in which d-SC is weak and nearly vanishes. Thus, the robustness of d-SC becomes markedly different whether the impurity potential is attractive or repulsive even at half filling, in contrast to the AF order.

Next, we take up a doped case in which the impurity density is equivalent to the doping rate ($\delta_{\rm imp} = \delta$). This case is experimentally special, but theoretically simple to consider the metal-to-insulator (Mott) transition arising at small V/t. It is known for homogeneous systems ($V = 0$) that the d-SC state is conductive (d-SC or metallic) for any U/t and δ (> 0) [14]. Below, we mainly consider the effect of repulsive V. The behavior of total energy abruptly changes at $V\,[\equiv V_{\rm M}] \sim 4t$ (not shown). We again expect that some transition occurs at $V = V_{\rm M}$. In Fig. 3(b), $n(\mathbf{k})$ is shown for various values of V/t for $\delta = 0.08$. For $V < V_{\rm M} \sim 4t$, $n(k)$ has a jump at a point near $(\pi/2, \pi/2)$ but is smooth around $(\pi, 0)$, showing a feature characteristic of d-SC. On the other hand for $V > V_{\rm M}$, $n(k)$ is always smooth on the path. It means that a d-SC-to-insulator (Mott) transition occurs at $V = V_{\rm M}$. We corroborate this point by referring to P_d in Fig. 4, where P_d is finite for $V < V_{\rm M}$ but abruptly vanishes and remains null for $V > V_{\rm M}$. In Fig. 5(b), we find $n_{\rm imp}$ and $n_{\rm hst}$ becomes 0 and 1, respectively for $V > V_{\rm M}$. The half-filled state is realized in the host sites. Thus, this transition is again a filling-control-type Mott transition. The mechanism is parallel to the corresponding transition in the AF state; see Fig. 3 in Ref [5] for details. Incidentally, in Fig. 4, Δ_d is found damped together with P_d for $V > V_{\rm M}$, suggesting that the AF correlation becomes weak in this insulating state.

Finally, let us look at the charge-density structure factor defined as

$$N(\mathbf{q}) = \frac{1}{N_{\rm s}} \sum_{i,j} e^{i\mathbf{q}\cdot(\mathbf{r}_i - \mathbf{r}_j)} \langle n_i n_j \rangle - n^2. \tag{22}$$

Fig. 6 shows the evolution of $N(\mathbf{q})$ as V/t changes for the doped case. It shows that the charge distribution hardly alters from the uniform case for $V < V_{\rm M}$, probably owing to the screening effect. On the other hand, $N(\mathbf{q})$ exhibits scattered pattern depending upon the impurity configuration. This scattered property in the large-V/t phase does not depend on whether the large-V/t phase is conducting or insulating. In this case, the phase is insulating,

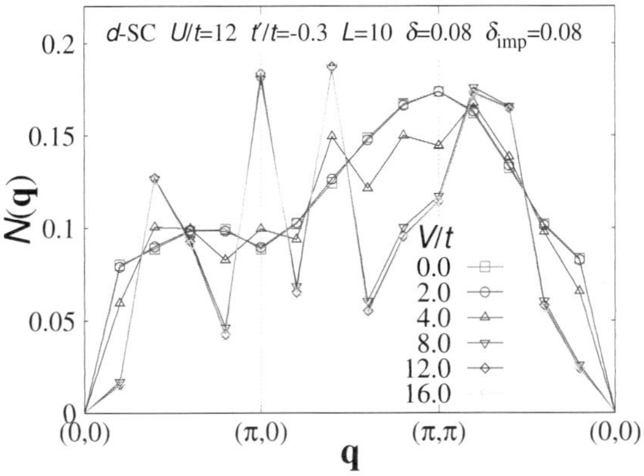

Figure 6. The charge-density structure factor $N(\mathbf{q})$ along the path of $\mathbf{q}$: $(0,0)$-$(\pi,0)$-(π,π)-$(0,0)$ is shown for several positive values of V/t. We use a certain impurity configuration for all values of V/t. As a result, the scattered forms for $V > V_{\mathrm{M}}$ becomes similar. If we use distinct impurity configurations, the patterns become random. It follows that the impurity configurations regulate the charge distribution for $V > V_{\mathrm{M}}$.

but the scattered property in $N(\mathbf{q})$ appears in the conductive states at half filling (not shown). See Fig. 2(b) in Ref. [6] for the AF state.

4. Summary and discussions

We studied the effect of point-type impurity potential on a superconducting state on a two-dimensional Hubbard model, using a variational Monte Carlo method. We found that filling-control-type insulator-to-conductor (Mott) transitions occur at $V = V_U^{(\pm)} \sim \pm U$ at half filling $(U/t = 12)$. The state for $V < V_U^{(-)}$ is d-wave superconducting and the charge carriers are holes, whereas the state for $V > V_U^{(+)}$ is metallic (or weak d-SC) and the carriers are electrons. This large asymmetry in the d-SC state is in contrast to the AF state. At a partially filling, we find a d-SC-to-insulator (Mott) transition at $V = V_{\mathrm{M}} \sim 4t$, which is somewhat larger than that for the AF state ($V_{\mathrm{M}} \sim 0 - 2t$). In this report, we took up a specific case of $\delta_{\mathrm{imp}} = \delta$ as a doped example. We found that the behavior considerably differs in other conditions. We will publish this point and other development elsewhere in near future. We acknowledge the support by Grants-in-Aid from the Ministry of Education, Culture, Sports, Science and Technology, Japan.

References

[1] For a recent review, a special issue 2012 *J. Phys. Soc. Jpn.* **81** Issue 1
[2] For instance, Pezzoli M E and Becca F 2010 *Phys. Rev. B* **81** 075106; Farhoodfar A, Gooding R J and Atkinson W A 2011 *Phys. Rev. B* **84** 205125
[3] For instance, Fukushima N, Chou C-P and Lee T K 2009 *Phys. Rev. B* **79** 184510; Tamaki H and Miyake K 2012 *J. Phys. Soc. Jpn.* **81** 124712
[4] Yokoyama H, Sato R and Kobayashi K 2017 *J. Phys. Conf. Ser.* **871** 012032
[5] Yokoyama H, Sato R and Kobayashi K 2018 *J. Phys. Conf. Ser.* **1054** 012014
[6] Yokoyama H, Sato R and Kobayashi K 2019 To be published in *J. Phys. Conf. Ser.*
[7] Sato R and Yokoyama H 2016 *J. Phys. Soc. Jpn.* **85** 124707
[8] Rice T M and Brinkman W F 1972 *Phys. Rev. B* **5** 4350
[9] de Gennes P G 1966 *Superconductivity of Metals and Alloys* (Perseus, Massachusetts) Sec. 5.1
[10] Liang S -D and Lee T K 2002 *Phys. Rev. B* **65** 214529
[11] Yanagisawa T 2003 *J. Phys. A: Math. Gen.* **36** 9337
[12] Sorella S 2001 *Phys. Rev. B* **64** 024512
[13] Yokoyama H, Ogata M and Tanaka Y 2006 *J. Phys. Soc. Jpn.* **75** 114706
[14] Yokoyama H, Ogata M, Tanaka Y, Kobayashi K and Tsuchiura H 2013 *J. Phys. Soc. Jpn.* **82** 014707
[15] Sato R and Yokoyama H 2018 *J. Phys. Soc. Jpn.* **87** 114003

IOP Conf. Series: Journal of Physics: Conf. Series **1293** (2019) 012030 doi:10.1088/1742-6596/1293/1/012030

Improvement of anisotropy of superconducting properties in Y-rich $YBa_2Cu_3O_y$ film in magnetic fields

M. Shiomi[1], Y. Ichino[1], Y. Tsuchiya[1], A. Ichinose[2] and Y. Yoshida[1]

[1] Department of Electrical Engineering, Nagoya University, Furo-cho, Chikusa-ku, Nagoya 464-8601 Japan

[2] Electric Power Engineering Research Laboratory, Central Research Institute of Electric Power Industry, 2-6-1 Nagasaka, Yokosuka, Kanagawa 240-0196, Japan

E-mail : shiomi.motoki@k.mbox.nagoya-u.ac.jp

Abstract. In order to put the $REBa_2Cu_3O_y$ superconductor into practical use, it is necessary to more isotropically improve the critical current density (J_c) with respect to an applied magnetic field angle. Therefore, artificial pinning centers (APCs) which are like a particle shape should be introduced into the superconductor. In this study, nano-particule APCs were introduced by fabricating $Y_{1+x}Ba_2Cu_3O_y$ ($x = 0 \sim 1.0$) thin films with excess Y composition, and effect of x, substrate temperature (T_ss) and pulsed laser repetition frequency (f_Ls) on the isotropy of J_c in the magnetic field was evaluated. As a result, these conditions proved to have a strong influence on isotropic improvement of J_c in the magnetic field.

1. Introduction

Since a $REBa_2Cu_3O_y$ (REBCO, RE=Rare Earth) superconductor has high critical temperature (T_c) which is over 90 K, cooling cost can be suppressed and it is expected to put into a practical use such as superconducting magnet coil. However, when applying REBCO wire as a coil, the magnetic field in the coil is not applied to the wire in a certain direction but is applied from various directions [1]. From this requirement, the REBCO superconductor having a large anisotropy for magnetic field angular dependence of J_c ($J_c(\theta)$) is required isotropic improvement of J_c (θ). As a solution, isotropic improvement of J_c (θ) has been reported by introducing artificial pinning centers (APCs) in nano-particle shape such as RE_2O_3 or RE_2BaCuO_y (RE211) in a superconductor [2, 3]. As a method of introducing them, it is reported tuning a composition of RE in a target when preparing REBCO wires by pulsed laser deposition (PLD) method [4]. Additionally, substrate temperature (T_ss) and pulsed laser repetition frequency (f_Ls) are one of the important deposition parameters when we fabricate the films by PLD method.

In this study, Our goal is to achieve a REBCO film that has isotropic enhancement of $J_c(\theta)$ at sub-cooling temperature of liquid nitrogen (65 K). We fabricated $Y_{1+x}Ba_2Cu_3O_y$ (Y_{1+x}BCO; $x = 0 \sim 1.0$) films by PLD method. We aim to control the size and number density of nanoparticles not only x but also deposition conditions, and also examine what kind of modulations the microstructure of the film might have on these conditions by $J_c(\theta)$ in magnetic field and so on.

2. Experimental details

We fabricated Y_{1+x} BCO thin films on IBAD-MgO metal substrate by PLD method using a KrF excimer laser (λ = 248 nm). The oxygen partial pressure, the energy density irradiated on the target, the film

IOP Conf. Series: Journal of Physics: Conf. Series **1293** (2019) 012030 doi:10.1088/1742-6596/1293/1/012030

thickness were 800 mTorr, 1.4 J/cm^2, 300 nm, respectively. T_s and f_L was changed from 820 to 860°C and, from 5 to 20 Hz. We observed microstructure of a film by Transmission Electron Microscopy (TEM), and evaluated critical current density in magnetic field by DC four-probes method using Physical Property Measurement System (PPMS).

Table 1. T_c and $J_c^{s.f.}$ of Y_{1+x}BCO thin films with different excess Y of x.

x	T_c [K]	$J_c^{s.f.}$ [MA/cm^2]
0	90.6	2.47
0.2	90.4	3.27
0.4	88.9	2.87
0.8	89.1	1.00
1.0	89.1	0.83

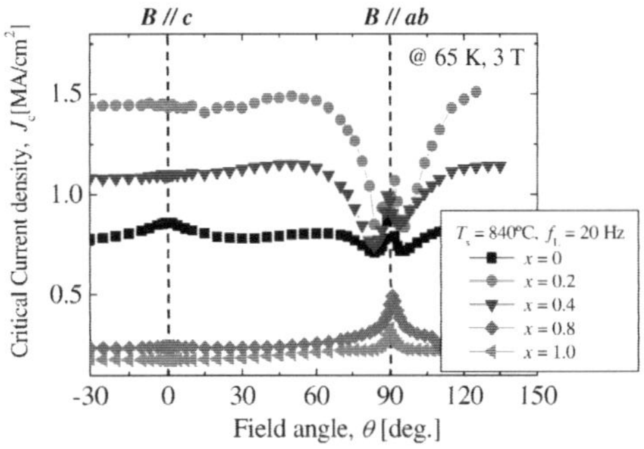

Figure 1. Magnetic Field angular dependence of J_c on Y_{1+x}BCO(x = 0 - 1.0).

3. Results and discussion

3.1. Dependence of excess Y on superconducting properties in magnetic field and microstructure

We fabricated Y_{1+x}BCO(x = 0 ~ 1.0) films at T_s = 840°C and f_L = 20 Hz. Table 1 shows T_c and critical current density in self-field ($J_c^{s.f.}$). From this result, T_c decreased by about 1.5 K when x is 0.4 or more, $J_c^{s.f.}$ became the maximum value at x = 0.2, and declined with increasing x larger than 0.2. Figure 1 indicates x dependence of $J_c(\theta)$ at 65 K and 3 T. It is the same tendency with $J_c^{s.f.}$ and became the maximum at x = 0.2 at $J_c(\theta = 0^\circ)$. Thus, x = 0.2 film contains the nanoparticles which has optimum size and number density to improve isotropic J_c.

Figure 2 shows TEM cross sectional image and diameter histogram of nanoparticles of x = 0.2 film. We measured the diameter of the nanoparticles from Fig2 (a). One can see that many nanoparticles which are Y-rich composition. The particle sizes are distributed in the range from 3 ~ 12 nm and average diameter (d_{av}) is 8.03 nm. It is considered that these nano-particles trap the quantized flux at 65 K. From these results, we found $Y_{1.2}$BCO(x = 0.2) thin film exhibited most isotropic $J_c(\theta)$.

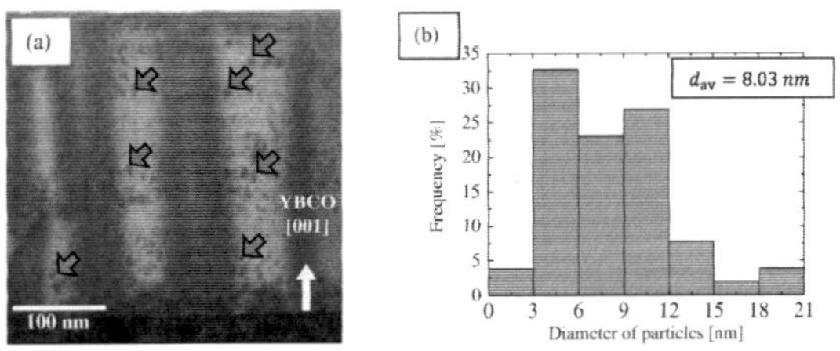

Figure 2. (a) Cross-sectional transmission electron microscopic image of the thin film with x = 0.2 and (b) diameter histogram of nanoparticles.

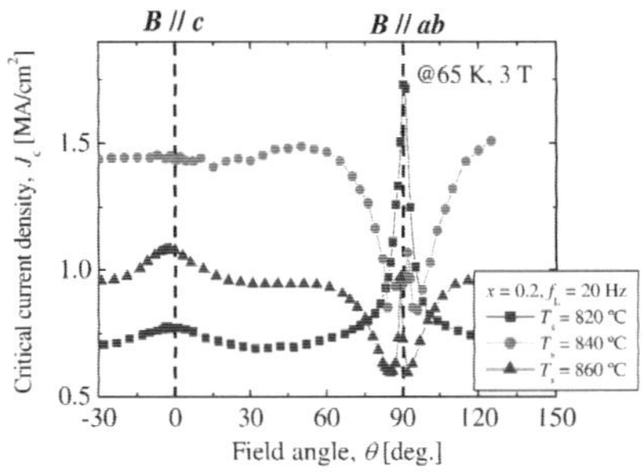

Figure 3. Magnetic Field angular dependence of J_c of $Y_{1.2}$BCO film in T_s=820 ~ 860°C.

3.2. T_s Dependence of superconducting properties in magnetic field

We evaluated $J_c(\theta)$ in magnetic field the $Y_{1.2}$BCO thin films prepared at various T_s. Figure 3 shows $J_c(\theta)$ at 65 K in 3 T of $Y_{1.2}$BCO thin films fabricated at T_s = 820 ~ 860°C and f_L = 20 Hz. This figure indicates

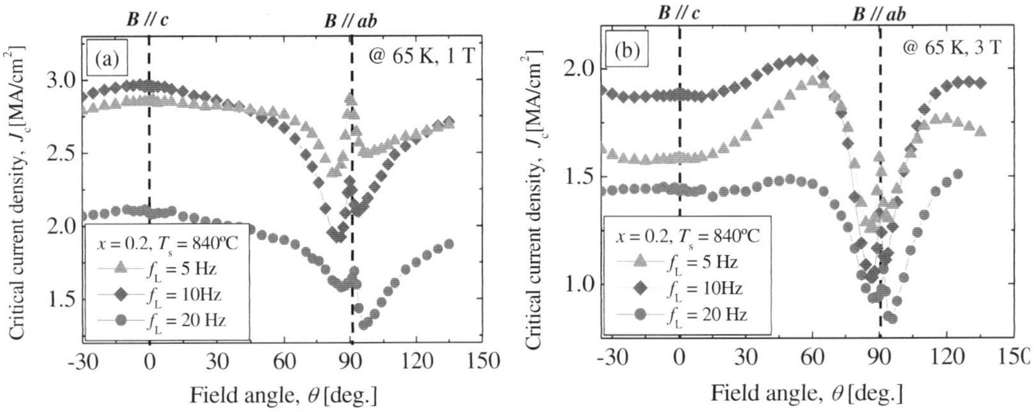

Figure 4. Magnetic Field angular dependence of J_c of $Y_{1.2}$BCO film at f_L=5 ~ 20 Hz at (a) 65 K and 1 T (b) 65 K and 3 T.

$J_c(\theta = 0^\circ)$ of $T_s = 820$°C is lower than $T_s = 840$°C. The reason is considered to be a decrease the diffusion distance of the excess Y on the substrate surface during the film growth, so that the nanoparticles become small, for such occasions they does not contribute as APC due to smaller diameter than magnetic flux diameter [5]. On the other hand, at $T_s = 860$°C, the $J_c(\theta = 0^\circ)$ decreased as compared with $T_s = 840$°C. We think that diameter of nanoparticles becomes larger due to high T_s, so that number density of nanoparticles could decrease.

3.3. Influence of f_L on the isotropy of J_c in the magnetic field
The influence of the laser repetition frequency f_L on $J_c(\theta)$ in magnetic field was evaluated. Figure 4 shows $J_c(\theta)$ at 65 K, 1 and 3 T of $Y_{1.2}$BCO thin film fabricated at $f_L = 5 \sim 20$ Hz and $T_s = 840$°C. As compared with $f_L = 20$ Hz, $f_L = 5$ and 10 Hz improved $J_c(\theta)$ for all magnetic field applied angles, and we confirmed the J_c dome around *B*//*c* appears as increasing with f_L at 1 T, while the dimple appears by decreasing with f_L at 3 T around *B*//*c*. Matsui *et al.* reported that such shapes appeared by introducing small-size and high-number density defects induced by light ions irradiation to a YBCO thin film[6]. Therefore, it is shown that higher density nanoparticle are introduced into the film due to the lowing f_L. Additionally, J_cs of *B*//*ab* for $f_L = 5$ and 10 Hz were higher than that of the f_L= 20 Hz sample. Thus, there is a possibility that *ab* correlated APCs are introduced by low f_L.

4. Summary
In this study, we fabricated Y_{1+x}BCO thin films with the *x* range from 0 to 1.0 for isotropic improvement of $J_c(\theta)$ and evaluated effects of T_ss and f_Ls. As a result, for *x*, T_c, $J_c^{s.f.}$ and $J_c(\theta)$ showed the largest at $x = 0.2$. T_s dependence of $J_c(\theta)$ at $Y_{1.2}$BCO thin film was evaluated, and we confirmed $J_c(\theta)$ curves changed by T_s. It indicates that diameter of nanoparticles changes with T_s. We finally demonstrated about $J_c(\theta)$ of $Y_{1.2}$BCO thin films which were fabricated in $f_L = 5 \sim 20$ Hz, so we achieved isotropic improvement of $J_c(\theta)$. This fact that density of nanoparticles became higher by lowering deposition rate. In addition, we suggested that ab-correlated APC which enhanced J_c of *B*//*ab* were introduced by low deposition rate.
These results proved that the deposition conditions have an influence on the isotropy of $J_c(\theta)$ and also change the nanoparticle size and density.

Acknowledgement
This work was partly supported Grants- in-Aid for Scientific Research (15K14301, 15K14302, 16K20898) from JST-ALCA. We receive the cooperation of Dr. Izumi, Dr. Ibi and Dr. Machi about IBAD-MgO tape.

References

[1] Higashikawa K, Kiss T, Inoue M, Awaji S, Watanabe K, Fukushima H, Yamada Y, Shiohara Y 2009, *Physica C*, **469**, 1776.

[2] Mele P, Matsumoto K, Horide T, Ichinose A, Mukaida M, Yoshida Y and Horii S 2007, *Supercond.Sci.Technol.*, **20,** 616.

[3] Haugan T, Barnes P N, Wheeler R, Meisenkothen F and Sumption M 2004, *Nature*, **430**, 867

[4] Wang H, Serquis A, Maiorov B, Civale L, JiaQ L, Arendt P N, Foltyn S R, MacManus-Driscoll J L, Zhang X 2006, *J.Appl.phys.*, **100**, 053904.

[5] Ozaki T, Yoshida Y, Ichino Y, Takai Y, Ichinose A, Matsumoto K, Horii S, Mukaida M and Takano Y 2010, *J.Japan Inst.Metals*, **74**, 422.

[6] Matsui H, Ootsuka T, Ogiso H, Yamasaki H, Sohma M, Yamaguchi I, Kumagai T and Manabe T 2016, *Supercond.Sci.Technol.*, **29**, 065002.

IOP Conf. Series: Journal of Physics: Conf. Series **1293** (2019) 012031 doi:10.1088/1742-6596/1293/1/012031

High-Yield Fabrication of Ag Films by Reel-to-Reel Pulsed Laser Deposition System

Jin Matsuzaka, Yuji Tsuchiya, Yusuke Ichino, Yutaka Yoshida

Abstract. Ag films are deposited on REBCO tapes as a stabilizing layer by a sputtering method in the most case. The sputtering deposition has a disadvantage of material cost for wasted Ag due to a low deposition yield, contrary to the pulsed laser deposition (PLD) method. Therefore, we developed Reel-to-Reel PLD system to fabricate Ag films on tapes with the high deposition yield. As results, the PLD method has a deposition rate up to 1 nm/s as the sputtering method. The electrical resistivity of the Ag film by the PLD method is 4.7 times higher than the sputtering method. The deposition yield of the PLD method is 13 times higher than the sputtering method. Finally, the Ag film was deposited on moving tapes by using the developed Reel-to-Reel PLD system with a uniform thickness.

1. Introduction

The high temperature superconducting REBCO tapes consist of four layers such as a metal substrate, a buffering layer, a superconducting layer, and a stabilizing layer [1][2]. The stabilizing layer protects the superconducting layer from moisture and corrosion. Furthermore, the layer locally bypasses current and diffuses heat when dissipation appears in the superconducting layer. Bilayer of Ag and Cu are commonly used as the stabilizing layer of the REBCO tapes [3][4][5][6]. Ag is the optimum material for stabilizing layer because Ag has low electrical resistivity, high thermal conductivity [7], and is chemically stable with the REBCO layer. On the other hand, Cu is much cheaper than Ag and has modest properties as the stabilizing layer. However, it is fatal that Cu degrades REBCO when Cu is deposited directly on REBCO. Therefore, the combination of Ag and Cu is commonly chosen.

In most cases, the stabilization layer is fabricated by a sputtering method or e-beam evaporation [8][9][10]. The sputtering method has a large and uniform deposition area but has a disadvantage of a low deposition yield [11][12][13][14]. Due to the wasted material, the material cost increases when an expensive material such as Ag is used. On the other hand, the pulsed laser deposition (PLD) method has a small deposition area and a high deposition yield [15][16][17]. If the PLD method is adopted to fabricate the Ag stabilizing layer, a series of operations to fabricate the superconducting layer and the stabilizing layer can be performed by one apparatus since the superconducting layer is produced by

Reel-to-Reel PLD system [18][19][20]. In this study, we fabricated Ag thin films on metal tapes by Reel-to-Reel PLD system to improve the deposition yield compared with the sputtering method.

2. Experimental methods

We fabricated Ag thin films on Hastelloy© tapes and glass substrates with 10 mm width by the PLD method and the sputtering method.

First, the PLD method is described as follows. A KrF excimer laser Coherent, COMPex201 (wavelength: 248 nm), was used for the PLD method. The purity of Ag target used for the PLD is 99.98% (Nilaco). The deposition parameters were as follows: a laser energy of 90 mJ, a repetition frequency of 100 Hz, a back pressure of 10^{-3} Pa, a target diameter of 25 mm, a distance between target and substrate of 40-80 mm, a laser spot size of 1.8-2.2 mm^2, and an Ar pressure of 0-10 Pa. The distance between target and substrate was changed by adjusting the position of substrate holder, and the laser spot size was changed by adjusting the position of the focusing lens. First, Ag films were deposited on the fixed substrates to optimize the deposition condition. Then, an Ag film was deposited on a moving substrate to demonstrate uniform deposition on tapes.

As a comparison of the PLD method, a DC magnetron sputtering apparatus ACS-40000-C2 was used for a sputtering method. The purity of Ag target used for the sputtering is 99.99% (FURUYA METAL). In the method, Ag films were deposited on fixed substrates at a distance between the target and the substrate of 150 mm and at an electric power of 120 W. The target diameter of the sputtering method was 50 mm.

We analyze both the fabrication processes of the Ag films in terms of deposition rate, surface, crystalline, electrical properties, and yield. The thicknesses of the Ag films deposited by both methods were measured with an optical interferometry microscope Nikon, ECLIPSE LV150N. The deposition rate was calculated by dividing the film thickness by the deposition time. The surface of the Ag films on Hastelloy© tapes were observed using scanning electron microscope Hitachi High-Tech Science, S-3400N. The crystallinity of the Ag films on Hastelloy© tapes were analyzed using a multifunctional X-ray diffractometer Rigaku, ATX-G. The electric properties were measured by a four-probe method using Physical Property Measurement System Quantum Design, MODEL6000. In this measurement, the Ag films with a thickness of 2 μm on glass substrates instead of the Hastelloy© tapes were used to measure the accurate resistivity.

Finally, the yields of both the method was calculated from the weight gain of Ag deposited on the substrate and the weight loss of Ag target before and after deposition. The weight loss of the Ag target was measured using a balance METTLER TOLEDO, PB303-S. In the PLD method, the weight gain of Ag deposited on substrate was estimated by summarizing of the distributed thickness of Ag along a longitudinal direction and by assuming the typical density of Ag. At this time, the film thickness along the transversal direction was assumed to be uniform. In the sputtering method, the weight gain of Ag film was estimated from the thickness of Ag films assuming that the uniform distribution of thickness throughout the deposition area of 78.5 cm^2.

3. Results and discussion

First, we discuss the experiments of the PLD method with the fixed substrate. Fig. 1 shows the distance between target and substrate dependence of deposition rate. The dotted line is a guide to the eye. The blue and the green lines assume the case for the ablated particle without and with the directivity, respectively. The blue line and the green line are based on the deposition rate at the distance between target and substrate of 20 mm. Both the lines are assumed that all particle fly directivity. However, the particle does not fly directivity and diffuse in the plane direction in reality, so these two lines do not match the experimental values. According to Fig. 1, the deposition rate was 0.028 nm/s and 0.445 nm/s when the distance between the target substrate was 80 mm and 20 mm, respectively. The deposition rate rapidly increased at the closer the distance. Theoretically, the density of the material evaporated from the target increases as decreasing the vertical distance from the target [15]. Therefore, the increasing at the deposition rate is due to vertical direction dependence of ablated Ag particle density. Ag particles evaporated from the target have directivity in the vertical direction, but some of the Ag particles diffuse in the plane direction since the plasma and the gas diffuse from a high pressure to a low pressure. Therefore, the particle density in the vertical direction decreases as going further away from target. The number of target particles deposited on the substrate and deposition rate decreases as increasing the distance between target and substrate. The particles diffuse in the vertical direction and there are few particles over an interface at which the pressure is sufficiently low, and plasma is transferred to the gas, so the deposition rate is about 0 and lower blue line. In Fig. 1, experimental values intersect with blue line at a distance between target and substrate of 80 mm. Fig. 2 shows the laser spot size dependence of deposition rate. As shown in Fig. 2, the deposition rate was 0.167 nm/s and 0.389 nm/s when the laser spot size was 1.8 mm^2 and 2.2 mm^2, respectively. The deposition rate increased for the lager laser spot size. In the PLD method, the plume shape change from spherical to linear as enlarging the laser spot size [15]. Therefore, the deposition area converges since becoming narrower the plume. The laser energy density decreases as enlarging the laser spot size since the laser energy is constant. As decreasing the laser energy density, the number of particles ablated from the target decreases and the particle energy also decreases. When it becomes lower than a certain energy, ablation disappears [21]. In our experiment, the target ablated even for the largest laser spot size. In Fig.2, the deposition rate depends mainly on plume shape since the deposition rate increases as enlarging the laser spot size. These results suggest that the smaller distance between target and substrate and the larger laser spot size increases the deposition rate of Ag in the PLD method. The deposition rate of PLD method is 3.67 times higher than that of the sputtering method since the deposition rate of sputtering method was 0.192 nm/s. For all the following results, Ag was deposited at the maximum deposition rate at distance between target of 20 mm and substrate and laser spot size of 1.8 mm^2.

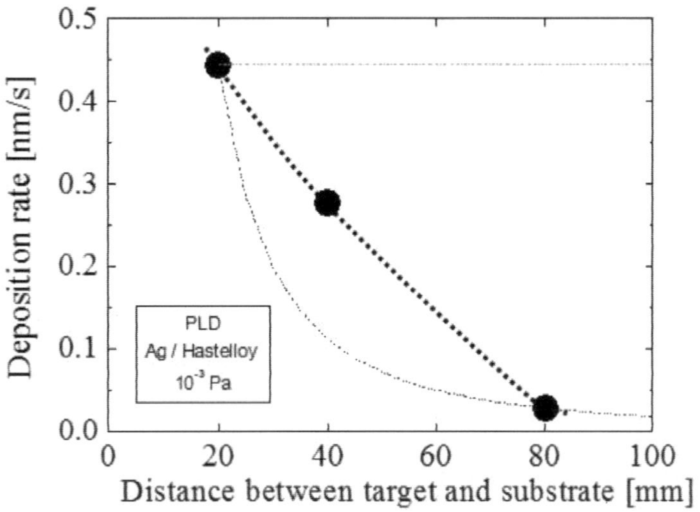

Fig. 1 Distance between target and substrate dependence of deposition rate of Ag film by PLD method.

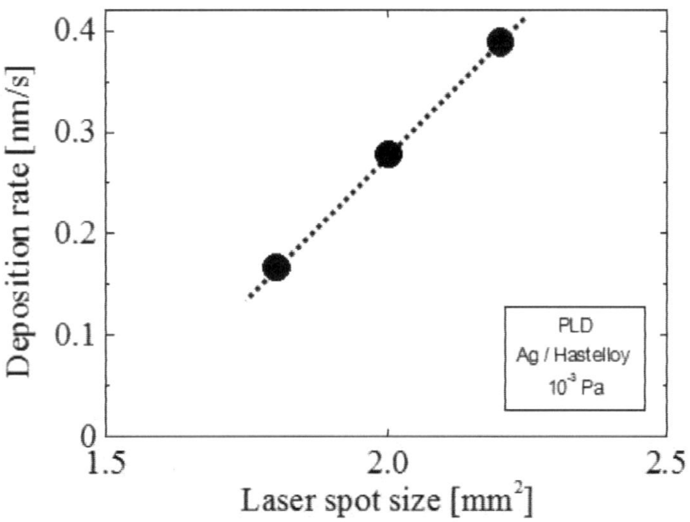

Fig. 2 Laser spot size dependence of deposition rete of Ag film for the PLD method.

Fig. 3 is scanning electron microscope (SEM) images of Ag films deposited on Hastelloy substrates by using PLD and sputtering. In the PLD film, many droplets appear with sizes of 0.34-5.72 μm^2 and at the number density of 0.021 /μm^2. For the stabilizing layer, the size and the number density of the Ag droplets should be as small as possible because the overlapping of the droplets makes voids in the Ag layer which is fatal for the protection of the REBCO layer. As shown in Fig. 3, the number density times the size is 0.0071-0.12 smaller than 1. Therefore, the Ag layer is supposed to work as a stabilizing layer. The average surface roughness Ra was 2.44 nm and 1.89 nm in the PLD and the sputtered film, respectively. Since the droplets are large, Ra of the PLD film is about 1.3 times larger than that of the sputtering film.

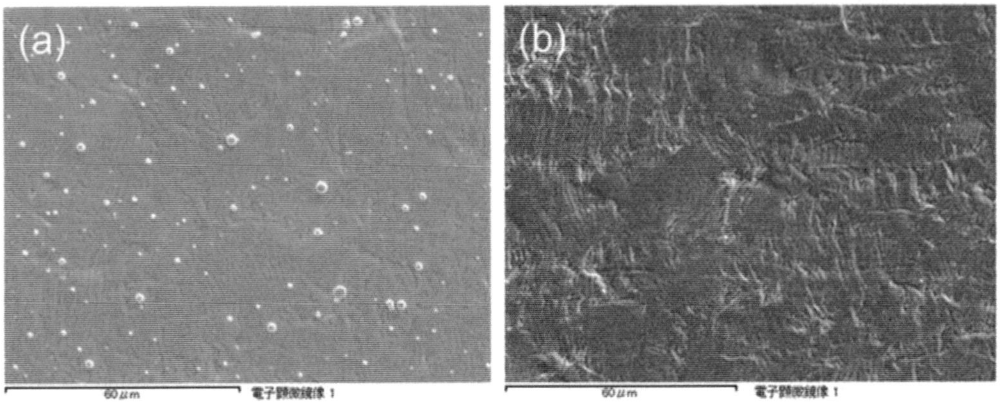

Fig. 3 SEM images of Ag thin film deposited on Hastelloy by (a)PLD and (b)sputtering.

We analyse a θ-2θ x-ray diffraction pattern of the Ag films for both the methods. However, no diffraction peak was found. This result showed that these films consisted of amorphous phase of Ag.

Next, we deposited Ag film by the PLD method on the fixed Hastelloy substrates with different Ar pressure and with keeping all the other conditions same by aiming at the higher yield. Fig. 4 shows Ar pressure dependent thickness distribution of the Ag films. At Ar pressure of 5 Pa, largest thickness of 4.2 μm was obtained at the center of the films which is coaxially with the plume during the deposition. The distribution of film thickness is similar for the deposition condition at a back pressure and the Ar pressure of 10 Pa; the center of the film has a thickness of 2.5 μm. All the profiles have Gaussian-like distribution. FWHM of the central peak of the thickness is 24, 25, and 22 mm at the Ar pressure of ~ Pa, 5Pa, 10 Pa, respectively. During the Ag deposition at the higher Ar pressure, the ablated target particles converged near the center of the plume and the mean free path of the particles becomes short [22]. According to our results, Ar pressure of 5 Pa is the optimal condition for the high yield. When it is more than that, the yield decreases since the mean free path becomes shorter than the distance of 20 mm between target and substrate, and below that the yield decreases due to diffusion of target particles.

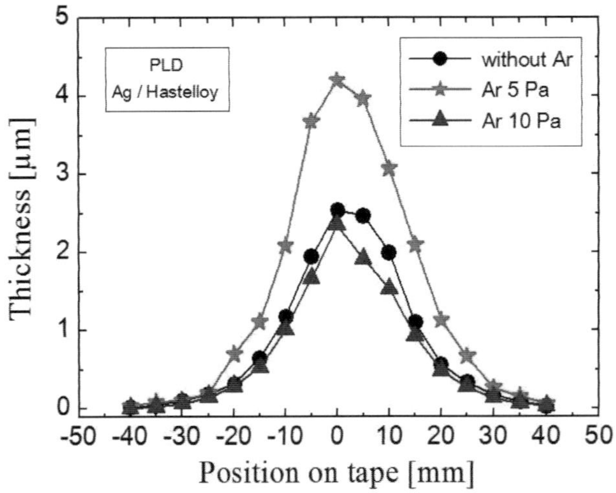

Fig. 4 Ar pressure dependence of Ag film thickness distribution.

Fig. 5 shows Ar pressure dependence of the yield of Ag films. The yield of the Ag film with Ar pressure of 5 Pa is 31% and the highest.

Table 1 shows a comparison of the deposition parameters of the PLD method at the highest yield and the sputtering method. The resistivity for both the methods were shown in Tab. 1. As a result, the Ag film fabricated by the PLD has 4.7 times higher resistivity compared with than the sputtering film. The difference in the resistivity is considered to be due to droplets in the PLD film as shown in Fig. 3. The droplets make voids which disturb the current. Therefore, the electrical resistivity of the PLD film increases. To obtain the same performance as the stabilization layer of the Ag layer, the PLD method requires 4.7 times thicker Ag than the sputtering method. On the other hand, the yield of PLD method is about 13 times higher than the yield of sputtering method. Overall, the PLD method is 2.8 times more efficient than sputtering method. As a conclusion of the comparison for both the fabrication method of the Ag film, the PLD method is superior to the sputtering method in terms of material cost reduction.

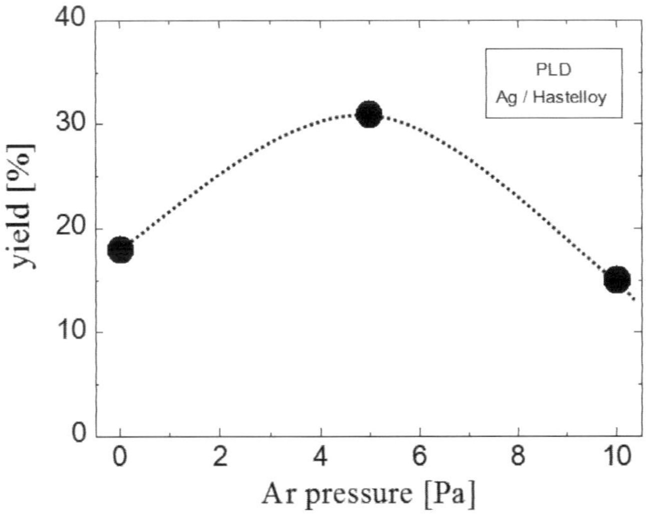

Fig. 5 Ar pressure dependence of the yield of Ag films.

Table 1 Deposition parameters of PLD and Sputtering.

	PLD	Sputtering
Deposition time [min]	60	60
Distance between target and substrate [mm]	20	150
Temperature [°C]	25	25
Max thickness [μm]	4.2	0.9
Target diameter [mm]	25	50
Electrical resistivity [Ωm]	8.78×10^{-8}	1.86×10^{-8}
Yield [%]	31	2.4

Next, we discuss the experiments of the PLD method with moving the substrate. Fig. 6 shows Ag film thickness distribution by moving PLD method using Reel-to-Reel system at tape speed of 0.36 m/h without Ar. The film thickness by moving system is controlled by tape speed. The stabilizing layer is deposited at one time with slow speed or deposited at many times with fast speed since Ag films do not change the properties when the films are laminated. In Fig. 6, the film thickness in the longitudinal direction was uniform as 300 nm. Therefore, it was suggested that the stabilizing layer of long tape is deposited by the Reel-to-Reel PLD system.

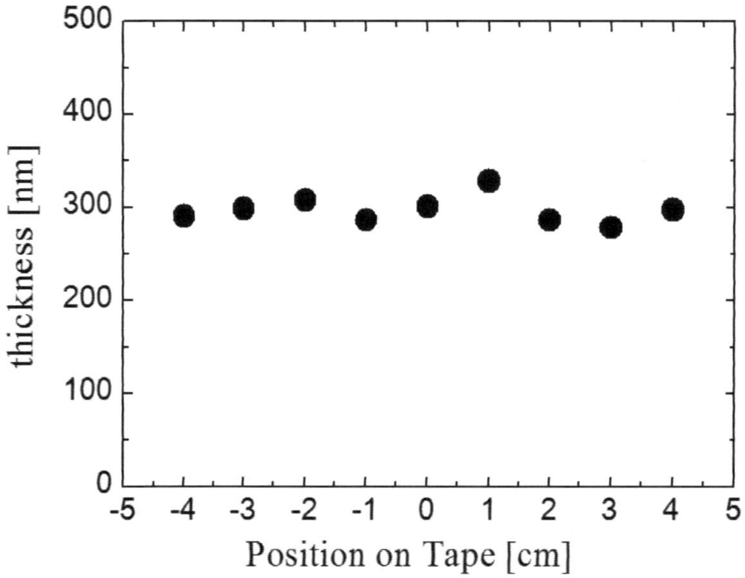

Fig. 6 Ag film thickness distribution by moving PLD method using Reel-to-Reel system.

4. Summary

We fabricated Ag thin films by the reel-to-reel PLD system to improve the deposition yield compared the sputtering method. The deposition rate of the PLD method increased by closing the distance between target and substrate and enlarging the laser spot size, and it is higher than sputtering method. The droplets in Ag were observed only in the films by the PLD method. The Ag films consisted of amorphous phase of Ag. In the PLD method, the yield of Ag was the highest at Ar pressure of 5 Pa, and it is about 13 times higher than that of the sputtering method. On the other hand, the resistivity of the PLD film is 4.7 times higher than that of the sputtering film. Overall, the PLD method is 2.8 times more efficient than the sputtering method in terms of material cost reduction. The uniform Ag film was deposited on moving tapes by using the Reel-to-Reel PLD system.

Acknowledgments

This work was partly supported by JST-ALCA.

References

[1] Shiohara Y, Yoshizumi M, Izumi T and Yamada Y 2007 *J. Japan Inst. Metals* **71** 11 pp 993-998

[2] Shiohara Y, Nakaoka K, Izumi T and Kato T 2016 *J. Japan Inst. Met.* Matter **80** 7 pp 406-419

[3] Barth C, Mondonico G and Senatore C 2015 *Supercond. Sci. Technol.* **28** 045011

[4] Shin H S and Dedicatoria M J 2013 *Supercond. Sci. Technol.* **26** 055005

[5] Shin H S, Kim K H, Dizon J R C, Kim T Y, Ko R K and Oh S S 2005 *Supercond. Sci. Technol.* **18** S364-S368

[6] Kato J Y, Sakai N, Tajima S, Miyata S, Konishi M, Yamada Y, Chikumoto N, Nakao K, Izumi T and Shiohara Y 2006 *Physica C* **445-448** 686-688

[7] Matula R A 1979 *J. Phys. Chem. Ref. Date* **8** 4

[8] Ekin J W, Larson T M, Bergren N F, Nelson A J, Swartzlander A B, Kazmerski L L, Panson A J and Blankenship B A 1988 *Appl. Phys. Lett.* **52** 21 pp 1819-1821

[9] Ekin J W, Panson A W and Blankenship B A 1988 *Appl. Phys. Lett.* **52** 4 pp 331-333

[10] Balachandran U 2002 *PRACTICAL SUPERCONDUCTOR DEVELOPMENT FOR ELECTRICAL POWER APPLICATIONS (ARGONNE NATIONAL LABORATORY)*

[11] Sigmund P 1969 *PHYSICAL REVIEW* **184** 2 pp 383-416

[12] Eckstein W 2011 *IPP-Report*

[13] Zalm P C 1983 *J. Appl. Phys.* **54** 5 pp 2660-2666

[14] Andersen H H and Bay H L 1973 *Radiation Effects* **19** pp 139-146

[15] Singh R K and Narayan J 1990 *PHYSICAL REVIEW B* **41** 13 pp 8843-8859

[16] Singh R K, Holland O W and Narayan J 1990 *J. Appl. Phys.* **68** 1 pp 233-247

[17] Nakata Y, Kaibara H, Okada T and Maeda M 1996 *J. Appl. Phys.* **80** 4 pp 2458-2466

[18] Prusseit W, Hoffmann C, Nemetschek R, Sigl G, Handke J, Lumkemann A and Kinder H 2006 *IEEE TRANSACTIONS APPLIED SUPERCONDUCTIVITY* **16** 2 pp 996-998

[19] Igarashi M, Kakimoto K, Hanyu S, Tashita C, Hayashida T, Hanada Y, Fujita S, Morita K, Nakamura N, Sutoh Y, Kutami H, Iijima Y and Saitoh T 2010 *Journal of Physics: Conference Series* **234** 022016

[20] Iijima Y, Kakimoto K, Igarashi M, Fujita S, Hirata W, Muto S, Yoshida T, Adachi Y, Daibo M, Naoe K, Fukuzaki T, Higashikawa K, Kiss T and Awaji S 2017 *IEEE Transactions on Applied Superconductivity* **27** 4

[21] Murakami K 2003 *J. Plasma Fusion Res.* **79** 10 pp 1035-1042

[22] Chen T, Li X M, Zhang S and Zeng H R 2004 *Journal of Crystal Growth* **270** pp 553-559

Effects of $Sm_{1+x}Ba_{2-x}Cu_3O_y$ Films with Non-Stoichiometric Composition Fabricated by Combinatorial Pulsed Laser Deposition method on the Superconducting Properties

G. Murase, Y. Tsuchiya, Y. Ichino and Y. Yoshida

Department of Electrical Engineering, Nagoya Unviersity, Furo-cho, Chikusa-ku, Nagoya 464-8601 Japan

E-mail: murase.gouki@h.mbox.nagoya-u.ac.jp

Abstract. $Sm_{1+x}Ba_{2-x}Cu_3O_y$ (Sm-rich) had higher critical current density (J_c) in magnetic fields and higher irreversible fields than stoichiometric $SmBa_2Cu_3O_y$ (SmBCO) thin films. In this study, to clarify the features of artificial pinning centers (APCs) doped Sm-rich films, we prepared BaHfO3 (BHO)-doped $Sm_{1+x}Ba_{2-x}Cu_3O_y$ films with $x = 0.00 \sim 0.20$ using combinatorial pulsed laser deposition (C-PLD) method on a metallic substrate and we evaluate surface morphrogies, crystal orientations and superconducting properties of the films. As a result, when x increased, the terrace width also increased. From the measurement of J_c in magnetic fields, the BHO-doped Sm-rich films showed J_c plateaus around $1 \sim 3$ T. From the angulaer dependence of J_c, the BHO-doped Sm-rich films showed J_c peak where B is parallel to c-axis. From several measurements, we discuss effects of $Sm_{1+x}Ba_{2-x}Cu_3O_y$ films with and without BHO on the superconducting properties.

1. Introduction

There are many equipments using superconductivity such as superconducting magnetic energy strage (SMES), superconducting maglev and magnetic resonance imaging (MRI). $REBa_2Cu_3O_y$ films have high critical current density in magnetic fields. Therefore they are expected to be applied for various equipments operated in magnetic fields. However, their J_c decreased in high magnetic fields. To improve J_c in magnetic fields, doping of artificial pinning centers (APCs) into REBCO films is effective[1,2]. $BaMO_3$ (M = metal) nanorod is one of the most effective APCs.
Previously, our group reported that $Sm_{1+x}Ba_{2-x}Cu_3O_y$ (Sm-rich) thin films to be changed x had higher J_c in magnetic fields and higher irreversible fields than those of stoichiometric $SmBa_2Cu_3O_y$ (SmBCO) thin [3-5]. We think that the Sm-rich films have many dislocations and low-T_c phases due to Sm/Ba substitution and they act as APCs. On the other hand, Ichinose *et al* suggesuted that Sm-rich films which contained APCs such as $BaSnO_3$ (BSO) had lower superconducting properties in magnetic fields than those of stoichiometric SmBCO films with BSO[6]. They inferred that superconducting phases need more Ba atoms than BSO and then lack of Ba atoms for formation of BSO nanorods.
It has not been discussed in details about properties of Sm-rich films contained both low-T_c phases and BMO. In this study, to clarify the features of APC-doped Sm-rich films with various x, we fabricated $Sm_{1+x}Ba_{2-x}Cu_3O_y$ films using combinatorial pulsed laser deposition (C-PLD) method. The C-PLD method enables to prepare a film with continuously changing composition across a single substrate [7].

Using the C-PLD method, high-speed research of new materials and optimizing experiment conditions were performed in previous reports [7-9]. In this study, we evaluated surface morphologies and superconducting properties of the APC-doped Sm-rich films fabricated by the C-PLD method to clarify effects of x and BMO doping on the crystal growth and the superconducting properties.

2. Experiment details

We fabricated Sm-rich films and BHO-doped Sm-rich films using the C-PLD method[7-9] with $x = 0.00 \sim 0.20$ on CeO_2-buffered IBAD-MgO substrates. To control film compositions, we used $SmBa_2Cu_3O_y$ target and $Sm_{1.2}Ba_{1.8}Cu_3O_y$ target. A KrF excimer laser (λ = 248 nm) was used as a high energy output source. The repetition frequency of the pulsed laser was 20 Hz , the laser energy density was 1.0 J/cm^2, the O_2 partial pressure was 400 mTorr, the distance between the substrate and the target was 60 mm and the substrate temperature was 860°C. BHO was doped by surface modified target technique [10]. After the deposition, oxygen was introduced into the chamber and the substrate was rapidly cooled down to room temperature. In addition, to introduce enough oxygen, post annealing in an oxygen atmosphere was performed before measurements of transport properties. The annealing temperature was 350°C which is the optimal condition for pure SmBCO films.
The crystal orientations were determined by X-ray diffraction (XRD) method. The surface morphologies of the films were observed by atomic force microscopy (AFM). Film thickness and BHO content were estimated by inductively coupled plasma optical emission spectroscopy (ICP-OES). From ICP-OES, BHO content was 4.1vol.% and film thickness was 200 nm. Superconducting properties were measured by the DC four-probes method using physical properties measurement system.

3. Results and discussions

Figure 1 shows AFM images of Sm-rich films and BHO-doped Sm-rich films. Figure 2 shows terrace width depending on x from the AFM images of Figure 1. From the AFM images, step edges of the Sm-

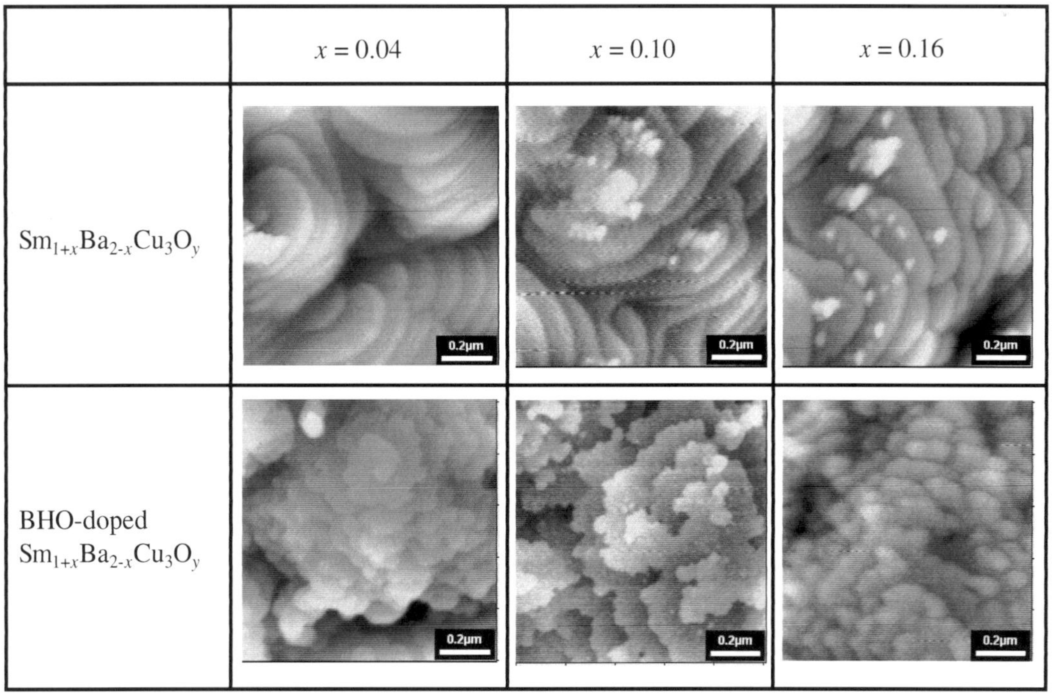

Figure 1 AFM images of the $Sm_{1+x}Ba_{2-x}Cu_3O_y$ films and the BHO-doped $Sm_{1+x}Ba_{2-x}Cu_3O_y$ films.

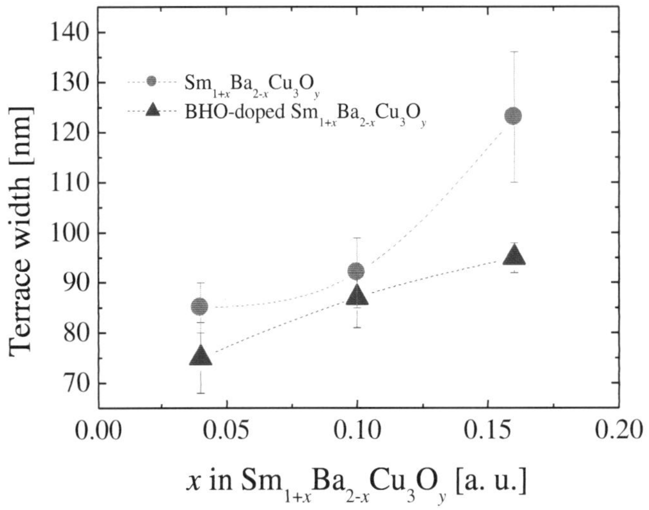

Figure 2 Terrace width depending on x.

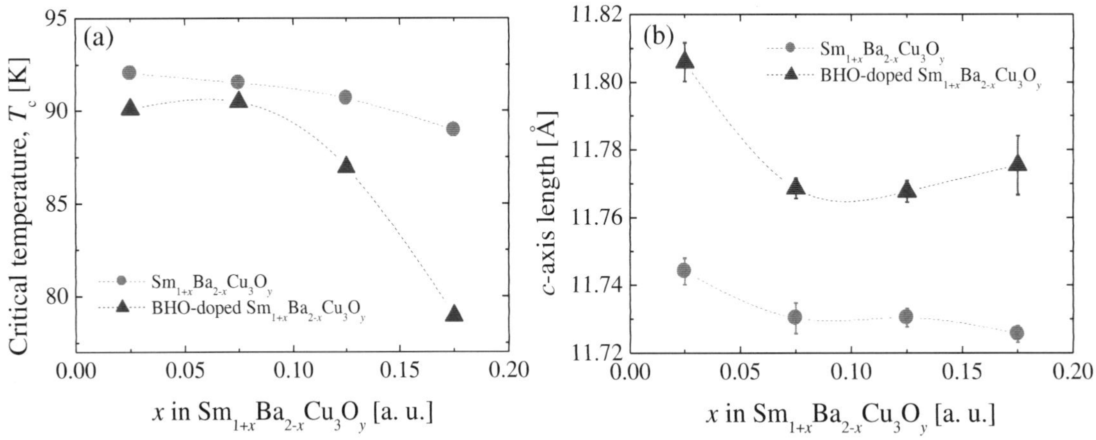

Figure 3 (a) Critical temperature (T_c) and (b) c-axis length depending on x.

rich films were smooth, but ones of the BHO-doped Sm-rich films are petaloid shape. These indicate that BHO forms nanorods [11]. In addition, precipitates due to Sm/Ba substituition were observed as x increased. From Figure 2, terrace width was gradually increased as x increased in both Sm-rich films and BHO-doped Sm-rich films. We reported the same tendency and the growth mechanism of Sm-rich films with the wide terrace [12]. In this report, we assumed that the surface diffusion distance (λ_s) of SmBCO particles increased as x increased. When x increased, the diameter of SmBCO grains increased and the density of nucleations decreased. In this case, collision frequency of SmBCO particles migrating on the terrace became lower compared to stoichiometric SmBCO. Therefore, the particles on the terrace could be moved in a longer migration range and then step edges grew faster as x increased.

Figure 3 shows critical temperature (T_c) and c-axis length depending on x. In the non-doped Sm-rich films, as x increased, T_c gradually declined from 92.0 K down to 88.9 K and c-axis length decreased monotonically. This feature is identical to $Sm_{1+x}Ba_{2-x}Cu_3O_y$ melt grown bulks [13]. On the other hand, the BHO-doped Sm-rich films showed the highest T_c = 90.4 K at x = 0.075 and T_c drastically declined over x of 0.075. Compared to the non-doped Sm-rich films, c-axis length of the BHO-doped Sm-rich films was extended indepedent on x. This is because there is lattice mismatch between SmBCO and BHO nanorods.

Figure 4 shows critical current density (J_c) in magnetic fields ($B//c$) of the films at 77 K. J_c at the self field ($J_c^{s.f.}$) of the non-doped and BHO-doped Sm-rich films decreased as x increases. The $J_c^{s.f.}$ are affected by T_c as shown in figure 3(a). Similaly, $J_c^{s.f.}$ of the BHO-doped films were lower than that of the non-doped films because T_c was lower than the non-doped films in all of x. In magnetic fields, J_c of the non-doped Sm-rich films drastically decreased regardless of x. In contrast, although the BHO-doped Sm-rich films had lower $J_c^{s.f.}$, there are plateaus of J_c around 1 ~ 3 T. The end of J_c plateau was changed with x as seen in inset of Figure 4. In brief, the end of plateau moved to the high magnetic field as x increased. This change of plateau end indicates that the number density of BHO would increase as x increased. As I mentioned about step growth mechanism from Figure 1, there is some possibility that λ_s of SmBCO depending on x affects growth of BHO nanorods. We need more detailed examinations such as an microstructure observation of the films. At the plateau area, J_c of the BHO-doped Sm-rich films was higher than that of the non-doped Sm-rich films. This suggests that BHO nanorods act as APCs.
Figure 5 shows angular dependence of J_c of the films at 77 K and 1 T. From this figure, when applied angle of the magnetic field is parallel to c-axis ($B//c$), J_c of the BHO-doped Sm-rich films was higher than that of the non-doped Sm-rich films at x = 0.025, 0.075 and J_c peak was seen regardless of x. Therefore, it is considered that BHO forms nanorods. In addition, sharpness of J_c peak has changed with x. This suggests that the number of density and the diameter of BHO nanorods would be changed depending on x . We need further examinations at higher magnetic fields or at different temperature. From the above, the BHO-doped Sm-rich films keep J_c from declining in magnetic fields especially at $B//c$, so BHO was probably introduced into the films as nanorods.

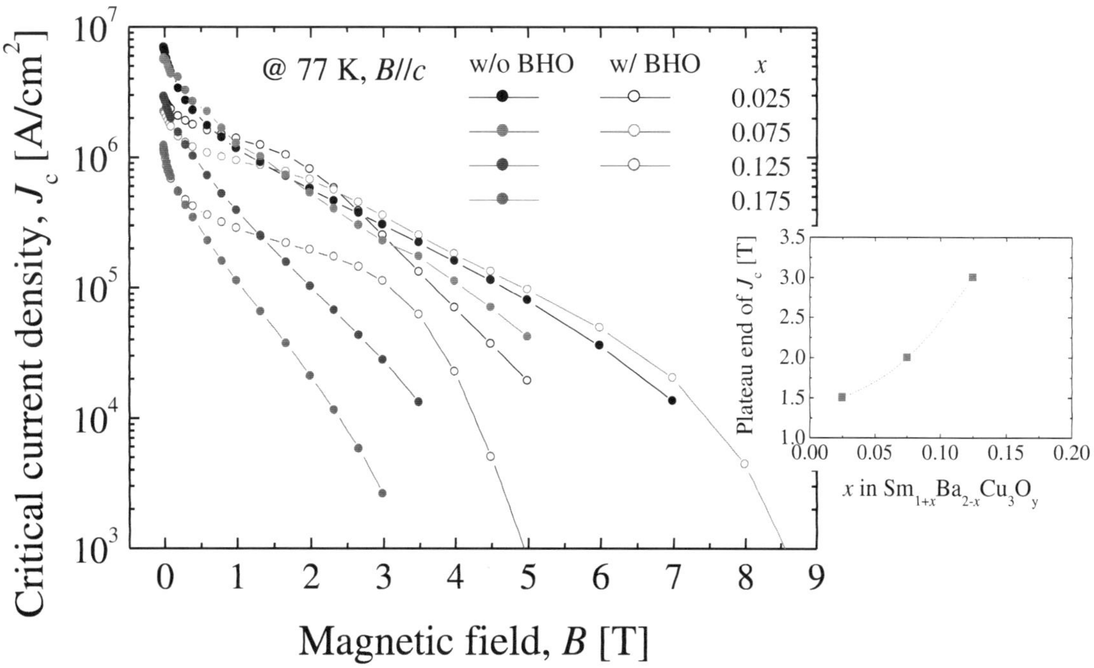

Figure 4 Critical current density (J_c) in magnetic fields (closed symbol stands the non-doped Sm-rich films, closed symbol stands the BHO-doped Sm-rich films) and inset figure represents plateau end of J_c depending on x.

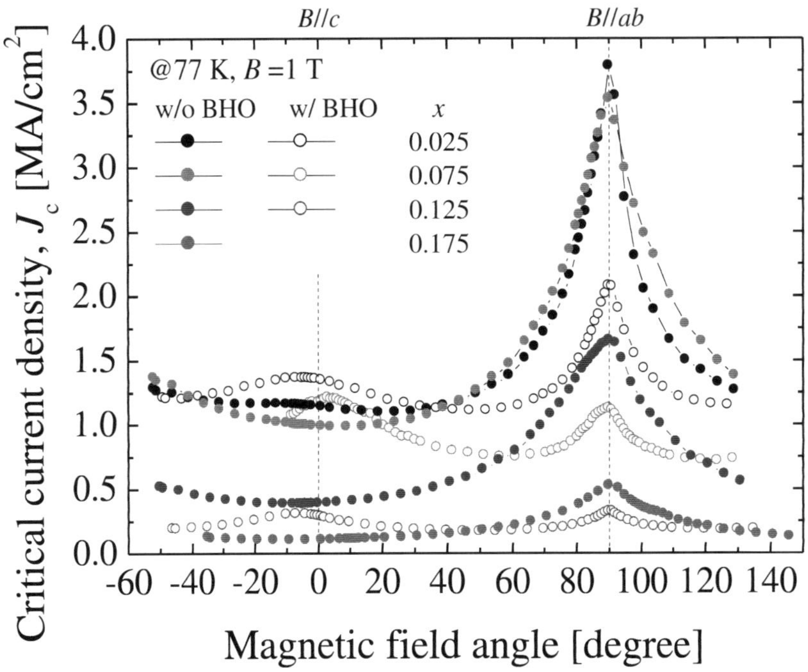

Figure 5 Angular dependence of J_c at 77 K, 1 T (closed symbol stands the non-doped Sm-rich films, closed symbol stands the BHO-doped Sm-rich films).

4. Summary

In this study, we fabricated the non-doped Sm-rich films and the BHO-doped Sm-rich films using the C-PLD method and investigated effects of x and BHO doping on the crystal growth and the superconducting properties. Compared to the non-doped Sm-rich films, in the BHO-doped Sm-rich films, c-axis length was extended and T_c was decreased. Terrace width widened as x increased. The BHO-doped Sm-rich films had higher J_c in magnetic fields than the non-doped films. This indicates that BHO acts as an APC. The BHO-doped Sm-rich films showed J_c plateau. End of J_c plateau moved to high magnetic field as x increased. In addition, from angular dependence of J_c measurement, J_c peak at $B//c$ was seen regardless of x. There is a possibility that BHO form nanorods with different number density depending on x because the end of J_c plateau and angular dependence of J_c are changed depending on x. We need more detailed investigations such as microstructure observation of the films.

Acknowledgement
This work was partly supported by a Grant-in-Aid for Scientific Research (16H04512 and 16K20898), and JST-ALCA. The metallic substrates were provided by Dr. Izumi, Dr. Ibi and Dr. Machi of AIST.

References
[1] MacManus-Doriscoll J L, Foltyn Q X, Wang H, Serquis A, Civale L, Maiorov B, Hawley M E, Maley M P and Peterson D E 2004 *Nature Materials* **3** 439
[2] Haugan T, Barns P N, Wheeler R, Meisenkothen F and Sumption M 2004 *Nature Materials* **430** 867
[3] Yoshida Y, Matsumoto K, Miura M, Ichino Y, Takai Y, Ichinose A, Horii S and Mukaida M 2005 *Japanese Journal of Applied Physics* **44** L546
[4] Yoshida Y, Matsumoto K, Miura M, Ichino Y, Takai Y, Ichinose A, Mukaida M and Horii S

2005 *Physica C* **426** 1043
[5] Miura M, Yoshida Y, Ichino Y, Ozaki T, Takai Y, Matsumoto K, Ichinose A, Horii S and Mukaida M 2006 *Japanese Journal of Applied Physics* **45** L701
[6] Ichinose A, Shun M, Sugihara K, Tsuruta A, Ichino Y, Yoshida Y, Awaji S and Matsumoto K 2016 *J. Japan Met. Mater.* **80** 420
[7] Koinuma H and Takeuchi I 2004 *Nature Materials* **3** 429
[8] Ichino Y, Yoshida Y, Yoshimura T, Ono I, Takai Y, Yoshizumi M, Izumi T and Shiohara Y 2011 *IEEE Transaction on Applied Superconductivity* **21** 2949
[9] Yoshimura T, Ichino Y, Yoshida Y, Takai Y, Kita R, Suzuki K and Takeuchi T 2011 *Physica C* **471** 947
[10] Mele P, Matsumoto K, Horide T, Ichinose A, Mukaida M, Yoshida Y, Horii S and Kita R 2008 *Nature Materials.* **3** 439
[11] Ozaki T, Yoshida Y, Ichino Y, Harada T, Takai Y, Matsumoto K, Ichinose A, Horii S, Mukaida M and Kita R 2009 *Physica C* **469** 1388
[12] Sudoh K, Yoshida Y, Ichino Y and Takai Y 2002 *Physica C* **378** 1222
[13] Sano M, Hayakawa Y and Kumagawa M 1996 *Supercond. Sci. Technol.* **9** 478

IOP Conf. Series: Journal of Physics: Conf. Series **1293** (2019) 012033 doi:10.1088/1742-6596/1293/1/012033

Evaluation of superconducting properties for $YBa_2Cu_3O_y$ coated conductors fabricated by self-heating technique in Pulsed Laser Deposition method

W. Sato, Y. Tsuchiya, Y. Ichino and Y. Yoshida

Department of Electrical Engineering, Nagoya University, Furo-cho, Chikusa-ku, Nagoya 464-8601 Japan

E-mail: sato.wataru@e.mbox.nagoya-u.ac.jp

Abstract. We fabricated $YBa_2Cu_3O_y$ coated conductors with 0.4 - 4.0 μm in the thickness on IBAD-MgO template deposited on Hastelloy metal substrates which were heated by a self-heating (S-H) technique in pulsed laser deposition method. The self-heating (S-H) technique is a method to heat a metal substrate by the Joule heat. In addition, preset substrate temperature (T_s) was gradually changed during the film deposition. It is possible to suppress *a*-axis grains even in a thick film by the S-H technique. Critical current density (J_c) and critical current (I_c) reached 2.0 MA/cm^2 and 800 A/cm-width in a film with the thickness of 4.0 μm at 77 K and self-field.

1. Introduction

$REBa_2Cu_3O_y$ (REBCO) coated conductors (CCs) are used for superconducting magnetic energy storage (SMES), magnetic resonance imaging (MRI), high temperature superconducting cables and generators. They would greatly benefit from further increase of critical currents (I_c). In order to improve the I_c, it is essential to increase the film thickness while maintaining the high critical current density (J_c). However, it has been reported that *a*-axis oriented grains are generated as the film thickness increases and J_c decreases [1]. The occurrence of *a*-axis grains is due to decrease of substrate surface temperature during the deposition. Thus, many research groups have studied the method of heating substrate. Some typical heating methods are a cold wall and a hot wall technique. Figure 1 shows the cold wall technique. A substrate is heated by using a heater [2]. It is difficult to keep the surface temperature constant during the deposition because of the indirect heating from the underside of the tape. Although this technique has such characteristics, we could fabricate $Sm_{1+x}Ba_{2-x}Cu_3O_y$ multilayer thick films with insertion of Sm-Rich Phase [3]. The hot wall technique can easily keep the surface temperature constant due to the direct radiant heating of the surface and thermal insulation effectiveness of the closed box structure.

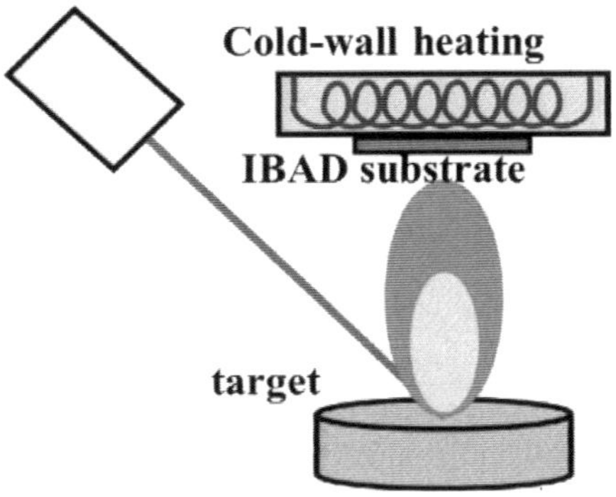

Figure 1. Schematic drawing of cold-wall technique. The substrate is heated by using a heater.

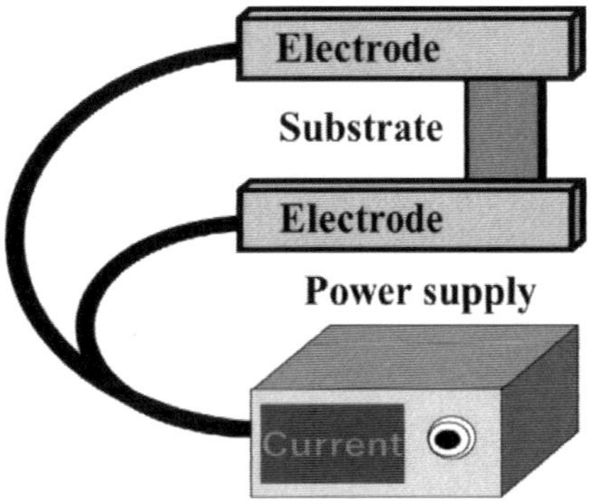

Figure 2. Schematic drawing of self-heating technique. When the metal substrate is energized, Joule heat is generated by the resistance of the substrate, so that the substrate is heated.

Fujikura Ltd. reported that $GdBa_2Cu_3O_y$ CCs with the thickness of about 6 μm could be fabricated using the hot wall technique [4]. On the other hand, in Figure 2, a self-heating (S-H) technique is a technique to heat the substrate by the Joule heat after applying a heating current through the Hastelloy metal substrates [5]. This technique provides rapid thermal response compared with the conventional heating technique that heats the substrate with a heater [6]. Ruipeng *et al.* and Selvamanickam *et al.* reported that successful fabrication of multilayer $Gd_{0.5}Y_{0.5}Ba_2Cu_3O_y$ thick CCs and Zr-doped $(Gd,Y)Ba_2Cu_3O_y$ thick film using this technique [7, 8]. Furthermore, when the substrate surface temperature decreases, the substrate surface temperature can be kept constant by momentarily raising the preset substrate temperature. Therefore, it is expected to suppress *a*-axis grains even in a thick film by the S-H technique. In this study, we evaluated the superconducting properties for $YBa_2Cu_3O_y$ (YBCO) CCs fabricated by pulsed laser deposition (PLD) method with the S-H technique in order to suppress the generation of *a*-axis oriented grains.

2. Experimental details

We fabricated YBCO films on biaxially textured oxide buffer layer of IBAD-MgO substrates which were heated by S-H technique in pulsed laser deposition (PLD) method using a Nd:YAG pulsed laser (wavelength : λ = 266 nm, repetition frequency : f = 10 Hz). Preliminary experiment, a relationship between a surface temperature of the buffer layer and a heating current was measured and the temperature controlled by the heating current was stand as T_s. The T_s was controlled according to the following equation (1) during the film deposition.

$$T_s(t) = T_{s\infty} - (T_{s\infty} - T_{s0})\exp(-t/\tau) \tag{1}$$

In equation (1), T_{s0} is initial T_s at $t = 0$, $T_{s\infty}$ is final T_s at $t = \infty$, and τ is the time constant. In this study, we fabricated YBCO films of which thicknesses were from about 0.4 to 4.0 μm was prepared under the conditions of T_{s0} = 850°C, $T_{s\infty}$ = 1045°C, τ = 1250 – ∞ sec., oxygen partial pressure P_{O2} = 400 mTorr, and laser energy density D = 1.25 J/cm^2. During the deposition, a back side temperature ($T_{b.s.}$) of REBCO deposited side were measured using pyrometer (emissivity : ε = 0.30) every 5 minutes. In other words, $T_{b.s.}$ is a temperature of hastelloy. In order to reduce oxygen deficiency, post annealing in an oxygen atmosphere of about 1 atm was performed before a measurement of transport properties.

Crystal orientation of the films were checked by X-ray diffraction (XRD) method, and inductively coupled plasma (ICP) emission spectroscopy was used for evaluation of film thickness. J_c were measured by the DC four-probes method.

3. Results and discussion

3.1. Thickness dependence of $T_{b.s.}$ and τ

Figure 3 shows time dependence of $T_s(t)$ calculated by equation (1) when τ was changed. It is confirmed that $T_s(t)$ was increased by decreasing τ. Figure 4 shows thickness dependence of $T_{b.s.}$ when τ was changed. The value on the horizontal axis was calculated from the deposition rate and the deposition time. Because $T_{b.s.}$ was observed using pyrometer, it is possibility that the emission of the plume was measured, but we confirmed that $T_{b.s.}$ did not change with and without the plume. In the case of $\tau = \infty$ corresponding to constant T_s during the deposition, it is found that $T_{b.s.}$ is decreased as the film thickness increase. As the film thickness increases, the surface roughness of the YBCO film increases. Therefore, it is considered that $T_{b.s.}$ decreased due to change of the surface emissivity by surface roughening [9]. On the other hand, it was observed that $T_{b.s.}$ was maintained when $\tau = 1250$ and 1500. Decrease of surface temperature due to the surface roughening is compensated by $T_s(t)$ change.

3.2. Thickness dependence of XRD peak intensity ratio and τ

The occurrence of *a*-axis grains is due to decrease of surface temperature during the deposition. In this study, it is expected that generation of *a*-axis grains was suppressed by $T_s(t)$, because $T_{b.s.}$ was maintained as mentioned above. Figure 5 shows thickness dependence of XRD intensity ratio of $I_{(200)} / (I_{(200)} + I_{(005)})$ depending on τ. It is observed that the *a*-axis oriented grains increased as the film thickness increases at $\tau = \infty$, which corresponds to a conventional cold-wall heating technique. Remarkably, by increasing $T_s(t)$ during the film deposition, we enabled to suppress *a*-axis oriented grain formation during the thickening.

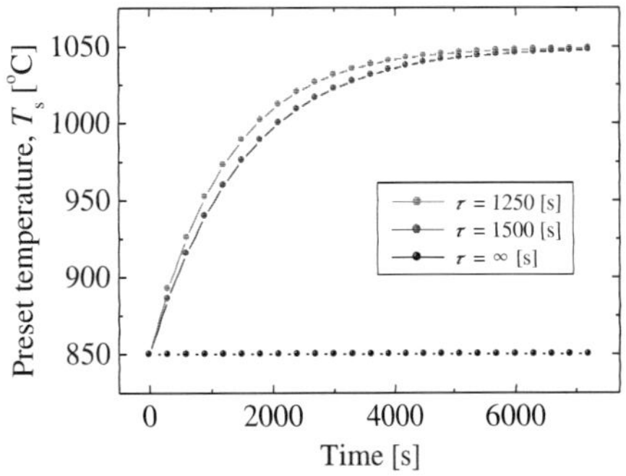

Figure 3. Time dependence of $T_s(t)$ calculated from equation (1) when τ was changed.

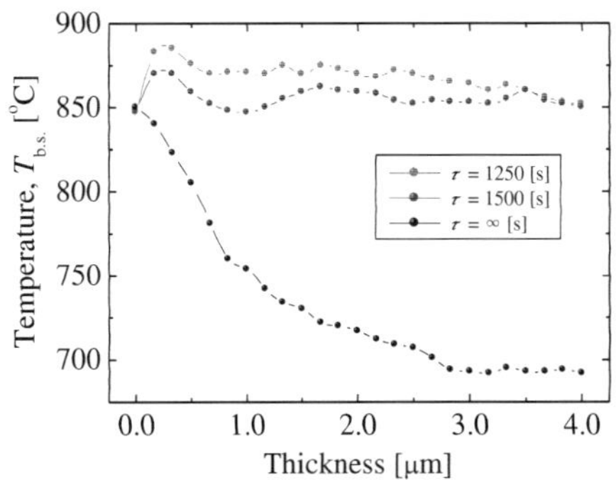

Figure 4. Thickness dependence of $T_{b.s.}$ when τ was changed.

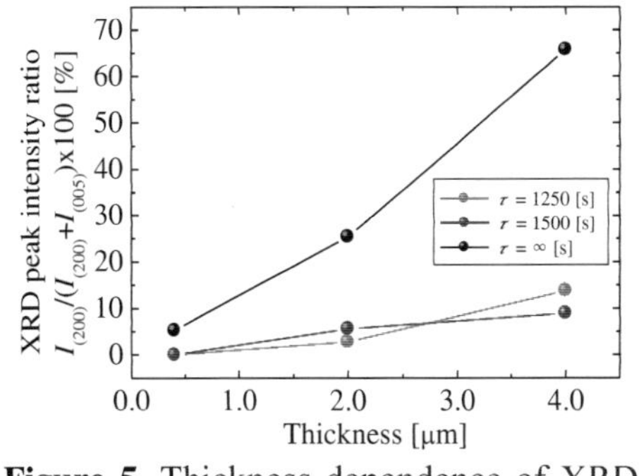

Figure 5. Thickness dependence of XRD intensity ratio of $I_{(200)} / (I_{(200)} + I_{(005)})$ depending on τ.

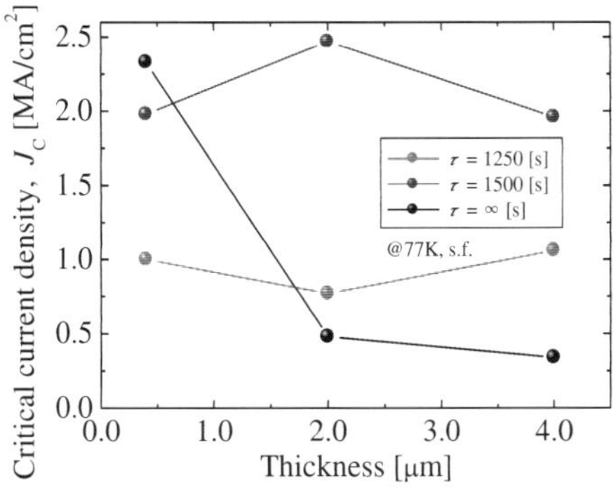

Figure 6. Thickness dependence of J_c depending on τ.

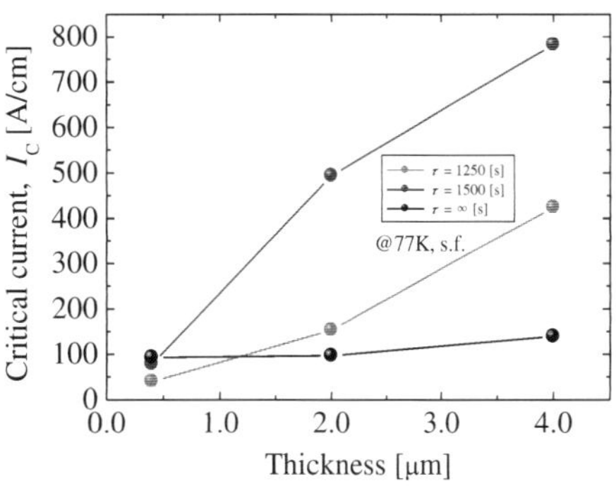

Figure 7. Thickness dependence of I_c depending on τ.

3.3. Thickness dependence of J_c, I_c and τ

It has been reported that J_c decreases as the film thickness increases due to *a*-axis oriented grains [1]. Therefore, it is expected that J_c is maintained in this study because *a*-axis oriented grains were suppressed by changing $T_s(t)$. Figure 6 and Figure 7 show thickness dependence of J_c and I_c when τ is changed. It was found that J_c was maintained at $\tau = 1250$ and 1500. So, I_c tends to increase linearly and maximum I_c reached about 800 A/cm-width in the film at $\tau = 1500$ in 4 μm thick. The J_c reached 2.0 MA/cm^2 (77 K, 0 T). On the other hand, it is confirmed that J_c decreased as the film thickness increases at $\tau = \infty$. Thus, I_c tends to be constant. These facts mean that J_c cannot be maintained as thickness increases by using only S-H technique. It is expected to improve the property of thick YBCO CCs using S-H system and gradual increase of $T_s(t)$.

4. Summary

We evaluated the superconducting properties for YBCO CCs fabricated by PLD method with S-H technique in order to suppress the generation of *a*-axis oriented grains. In addition, T_s was gradually changed during the film deposition. It is observed that $T_{b.s.}$ decreased, the *a*-axis oriented grain mixture ratio increased and J_c decreased as the film thickness increases when $\tau = \infty$, which corresponds to a conventional cold-wall technique. On the other hand, it was confirmed that *a*-axis grains were suppressed and J_c was maintained by finite τ. The J_c reached 2.0 MA/cm^2 (77 K, 0 T) at $\tau = 1500$ sec. in 4 μm thick YBCO CCs. It is expected to improve the property of thick YBCO CCs by using S-H system and gradual increase of $T_s(t)$.

Acknowledgement

This work was partly supported by JSPS (16K20898 and 16H04512), JST-ALCA project. The metal substrates were provided from AIST.

References

[1] Ibi A, Fukushima H, Kuriki R, Miyata S, Takahashi K, Kobayashi H, Konishi M, Watanabe T, Yamada Y and Shiohara Y 2006 *Physica C.* **445** 525
[2] Sekiguchi A 2016 *J. Vac. Soc. Jpn.* **7** 171
[3] Ozaki T, Yoshida Y, Miura M, Ichino Y, Takai Y, Matsumoto K, Ichinose A, Horii S and Mukaida M 2007 *IEEE Trans. Appl. Supercond.* **17** 3270
[4] Kakimoto K *et al.* 2010 *Supercond. Sci. Technol.* **23** 014016
[5] Zhai H Y, Christen H M, Martin P M, Zhang L and D. Lowndes D H 2003 *IEEE Trans. Appl. Supercond.* **13** 2622
[6] Majkic G, Galstyan E, and Selvamanickam V 2015 *IEEE Trans. Appl. Supercond.* **25** 6605304
[7] Zhao R, Zhang F, Liu Q, Xia Y, Lu Y, Cai C, Xiong J, Tao B and Li Y 2017 *Supercond. Sci. Technol.* **30** 025023

[8] Majkic G, Pratap R, Xu A, Galstyan E and Selvamanickam V 2018 *Sci. Rep.* **8** 1
[9] Sievers S, Mattheis F, Krebs H U, and Freyhardt H C 1996 *Japan. J. Appl. Phys.* **35** L74

IOP Conf. Series: Journal of Physics: Conf. Series **1293** (2019) 012034 doi:10.1088/1742-6596/1293/1/012034

Development of multi-filamentated long EuBCO coated conductors with BHO doping by plane-plume PLD method

A Ibi[1], T Machi[1], K Nakaoka[1], M Sato[1], T Izumi[1], K Higashikawa[2], T Kiss[2]

1. National Institute of Advanced Industrial Science and Technology (AIST) 1-2-1 Namiki, Tsukuba, Ibaraki 305-8564 Japan
2. Dept. of Electrical Engineering, Kyushu University 744, motooka, nishi-ku, Fukuoka-city 819-0395 Japan

akira-ibi@aist.go.jp

Abstract. Long $REBa_2Cu_3O_X$ (REBCO, RE: rare earth element) coated conductors with $BaMO_3$ (BMO, M: metal) doping have been expected for the industrial and commercial applications. It is known that critical current (I_c) of REBCO coated conductors (with or without artificial pinning centers doping) in magnetic field at high temperature is superior to the other superconducting wires. Especially, we have found that $EuBa_2Cu_3O_x$ (EuBCO) coated conductors with $BaHfO_3$ (BHO) doping show higher critical current density (J_c) and I_c at high temperature in self and magnetic fields than those of $YBa_2Cu_3O_x$ and $GdBa_2Cu_3O_x$ coated conductors with BHO doping. However, in order to realize EuBCO coated conductors with BHO doping for various applications such as an armature coil, the much higher uniformity of not only longitudinal and but transversal I_c distributions of long REBCO coated conductors with high in-field performance for filamentary structure is required. The plane-plume PLD (pulsed laser deposition) method is performed by shortening the target-substrate distance and increasing the number of plumes with scan of X-Y axes directions to increase the deposition rate and averaging the deposition conditions which should relate to obtain the uniformity I_c distributions. We fabricated the EuBCO coated conductors with BHO doping (of 0.6 m length and 10 mm in width) by the combination of the ion-beam assisted deposition (IBAD) and the plane-plume PLD method and then we processed the multi-filamentary structure EuBCO coated conductors with BHO doping (of 8 filaments of 440 μm, 0.6 m in length and 5 mm in width) by using the excimer laser scribing to reduce the AC loss and to control the shielding current. I_c distributions in a width direction of this EuBCO coated conductor with BHO doping was improved as an uniformity of filament-I_c of 4.3 %.

Recently, we fabricated the 16 filamentary and 3 m long EuBCO coated conductor with BHO doping by IBAD / plane-plume PLD method. We confirmed that the hysteresis loss was reduce to 1/16 by a 16 filamentary EuBCO coated conductor with BHO doping comparing with EuBCO coated conductor with BHO doping without filamentation.

1. Introduction

We have been investigating the $REBa_2Cu_3O_x$ (REBCO, RE: rare earth element) coated conductors with $BaMO_3$ (BMO, M: metal) doping [1-3]. In our previous work, we found that $EuBa_2Cu_3O_{7-x}$ (EuBCO) coated conductor with $BaHfO_3$ (BHO) doping by IBAD (ion-beam assisted deposition) [4] and PLD (pulsed laser deposition) methods show higher critical current (I_c) and critical currents density (J_c) at high temperatures in self and magnetic fields than those of other REBCO coated conductors with BMO doping, such as $YBa_2Cu_3O_x$ and $GdBa_2Cu_3O_x$ coated conductors with BHO doping. For example, the long EuBCO coated conductor with BHO doping showed high I_c(min) value of 141.2 A/cm-w in the all angles of applied magnetic field of 3 T at 77 K [5].

Recently, the REBCO coated conductors with BMO doping have been expected for the industrial and commercial applications at high temperatures in magnetic fields. However, the much higher uniformity of not only longitudinal and but transversal I_c distributions of long REBCO coated conductors with BMO doping with high in-field performance is required for higher performance in these applications. On the other hand, the multi-filamentated long REBCO coated conductors with BMO doping are effective for the control of the shielding current of the coil and alternating-current loss (AC loss). However, the almost previous work of filamentation for coated conductors has been attempted to pure REBCO coated conductors. Therefore, the multi-filamentary structure was also required in the BHO doped EuBCO coated conductors for the wider industrial and commercial applications.

We have tried to develop the multi-filamentary structure in long EuBCO coated conductors with BHO doping by the plane-plume PLD method and the excimer laser scribing [6].

2. Experimental procedure

The plane-plume PLD method is performed shortening the target-substrate distance (about 70 mm) to increasing the number of plumes (18 plumes) by scan in X-Y axes directions to increase the deposition rate and averaging the deposition conditions which should relate to obtain the uniformity I_c distributions [7-9]. The EuBCO layer with BHO doping were deposited by the plane-plume PLD method with vapor-liquid-solid (VLS) mode on 10 mm wide CeO_2 / $LaMnO_3$ / IBAD-MgO / Y_2O_3 / $Gd_2Zr_2O_7$ / Hatelloy C-276 substrate [10]. The value of in-plane and out-of-plane texturing degrees ($\Delta\phi$ and $\Delta\omega$) of CeO_2 cap layer were about 2 to 3 and 1 to 2 degrees, respectively. A 200 W industrial XeCl excimer laser with a wavelength of 308 nm was used at a pulse repetition rate of 300 Hz and a pulse energy of 600 to 700 mJ. The optical system is synchronized to the laser pulse and scanning of the laser beam to on a bulk EuBCO target containing BHO was controlled as well. The laser repetition rate at 300 Hz is divided into 18 plumes which are almost the same in size and shape. Typically, the pulse repetition rate of 300 Hz and a pulse energy of 600 to 700 mJ lead the deposition rate of about 24 nm/s for a EuBCO layer with BHO doping. Commercially available targets of sintered off-stoichiometric $EuBa_{1.8}Cu_{3.0}O_x$ containing 3.5 mol% doped BHO with a diameter of 6-inch were used for deposition of the EuBCO layer with BHO doping. The deposition temperature was about 930 ° C for a substrate tape with a transferring speed of 60 m/h. The oxygen pressure was maintained to be at about 800 mTorr with a flow of 10 sccm oxygen. The T-S distance (distance between the target and the substrate) was set about 70 mm although the distance in the conventional PLD method is about 100 mm. The thickness of EuBCO layer with BHO nanorods was about 700 nm for one time deposition with the substrate tape transferring speed of 60 m/h as a typical value. I_c value was measured by the conventional four-terminal method with the criterion of 1 μV/cm. The multi-filamentary structure was fabricated by using the KrF excimer laser with the wavelength of 248 nm irradiation with Reel-to-Reel system.

3. Results and discussions

In our previous work, figure 1 shows the longitudinal I_c distributions at 77 K and self-field of (a) a 93 m long EuBCO coated conductor with BHO doping by the conventional PLD method with deposition rate of 4.2 nm/s and (b) a 105 m long EuBCO coated conductor with BHO doping by the in-plume (plane-plume) PLD method with deposition rate of 24 nm/s [9]. The EuBCO layers with BHO doping

IOP Conf. Series: Journal of Physics: Conf. Series **1293** (2019) 012034 doi:10.1088/1742-6596/1293/1/012034

were about (a) 3.6 μm and (b) 0.7 μm in thickness, respectively. The transferring substrate through the inside of multi-plume with scan of X-Y (6-3) axes directions (plane-plume) is less affected by the change of plume conditions such as tilt and swing of the plume. Therefore, we thought that the longitudinal I_c distributions shows uniform since few changes in compositional shift of EuBCO layer from target, deposition rate of EuBCO layer (with change optimized deposition temperature), and EuBCO layer thickness.

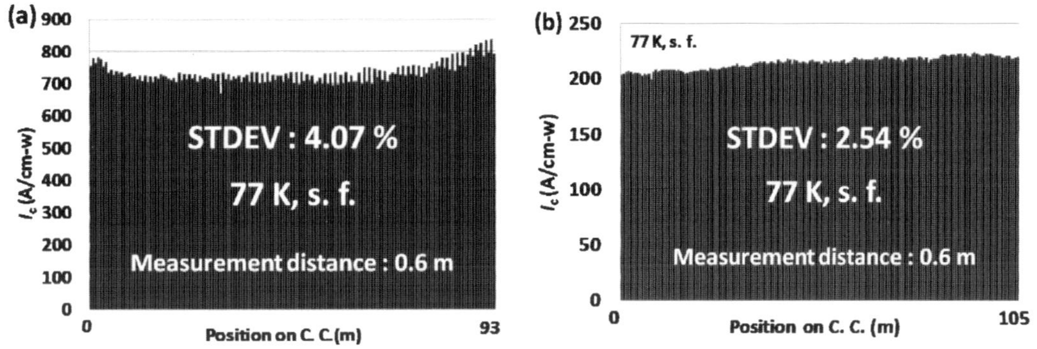

Fig. 1 Longitudinal I_c distributions at 77 K and self-field of (a) a 93 m long EuBCO coated conductor with BHO doping by conventional PLD method and (b) a 105 m long EuBCO coated conductor with BHO doping by plane-plume PLD method.

Figure 2 indicates two dimensional J_c distributions of the (a) 93.7 m long EuBCO coated conductor with BHO doping in 10 mm width by conventional PLD method and (b) 20.1 m long EuBCO coated conductor with BHO doping in 10 mm width by plane-plume PLD method by SHPM (Scanning Hall Probe Microscopy) with Reel-to-Reel system [11]. The amplitude of the sheet current density corresponded to that of critical current density almost in all area of the coated conductor except for the region where current direction changed [11]. The uniformity of in-plane J_c distribution of 20 m long EuBCO coated conductor with BHO doping was improved by the plane-plume PLD method.

In our previous work, figure 3 shows the I_c distributions of (a) 0.6 m long EuBCO coated conductor with BHO doping of 8 filaments of 440 μm in width by the conventional (out-of-plume) PLD method and (b) 0.6 m long EuBCO coated conductor with BHO doping of 8 filaments of 581 μm in width by the plane-plume PLD method [9]. We measured each I_c value of 8 filaments of 440 or 581 μm in width and slot of 20 μm in width for 0.6 m long EuBCO coated conductor with BHO doping. The variation of each I_c value were (a) 10.4 and (b) 4.3 %, respectively ($\{(I_{c\,ave.}) / (I_{c\,min.})\} / I_{c\,ave.} \times$ 100 (%)). I_c distribution in a width direction of the EuBCO coated conductor with BHO doping by the plane-plume PLD method was much improved from that of the conventional PLD method.

We fabricated a 100 m long EuBCO coated conductor with BHO doping by plane-plume PLD method and then a 3 m long sample was cut out from this 100 m long EuBCO coated conductor with BHO doping. Furthermore, we processed 16 filamentary structures to this 3 m long EuBCO coated conductor with BHO doping by excimer laser scribing method. Figure 4 (a) indicates photograph of the EuBCO coated conductor with BHO doping of 16 filaments in 10 mm width. We confirmed that the hysteresis loss was reduced to 1/16 by a 16 filamentary and 3 m long EuBCO coated conductor with BHO doping comparing with EuBCO coated conductor with BHO doping without filamentation (figire 4 (b)). From these results, J_c distribution in a width direction of these EuBCO coated conductors with BHO doping by the plane-plume PLD method is also reasonably uniform even in the long coated conductors.

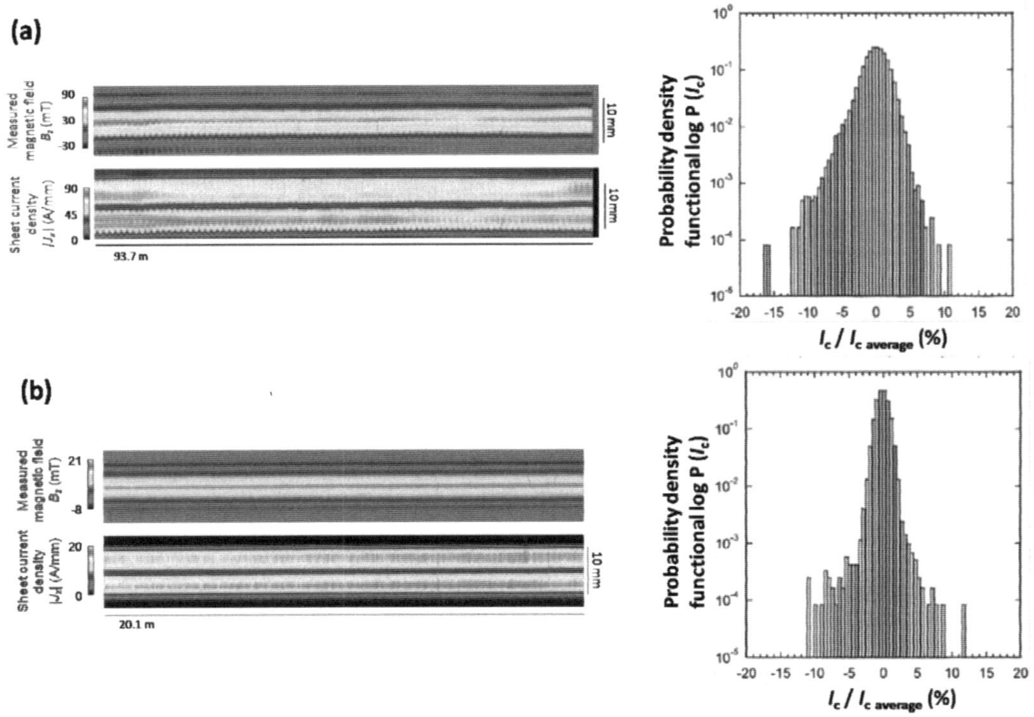

Fig. 2 Two dimensional J_c distributions of the (a) 93.7 m long EuBCO coated conductor with BHO doping in 10 mm width by conventional PLD method and (b) 20.1 m long EuBCO coated conductor with BHO doping in 10 mm width by plane-plume PLD method by SHPM with Reel-to-Reel system.

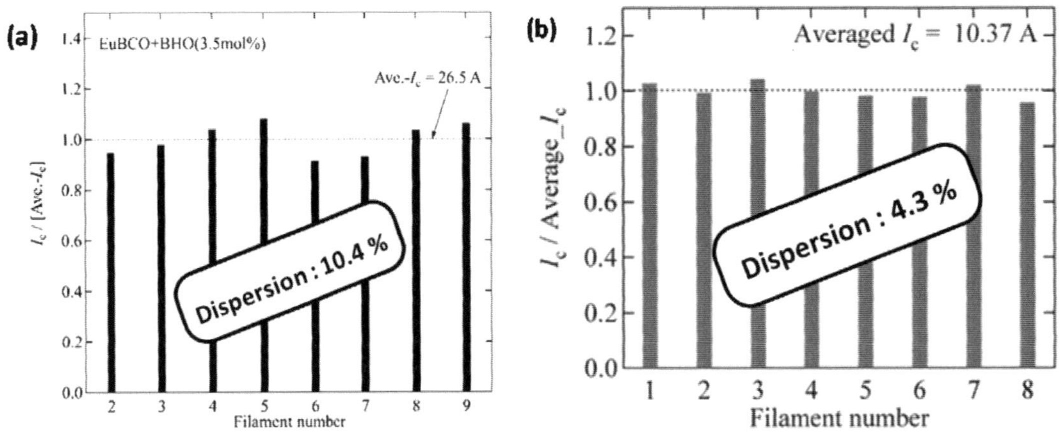

Fig. 3 Normalized I_c distributions of (a) 0.6 m long EuBCO coated conductor with BHO doping of 8 filaments of 440 μm in width by conventional PLD method and (b) 0.6 m long EuBCO coated conductor with BHO doping of 8 filaments of 581 μm in width by plane-plume PLD method.

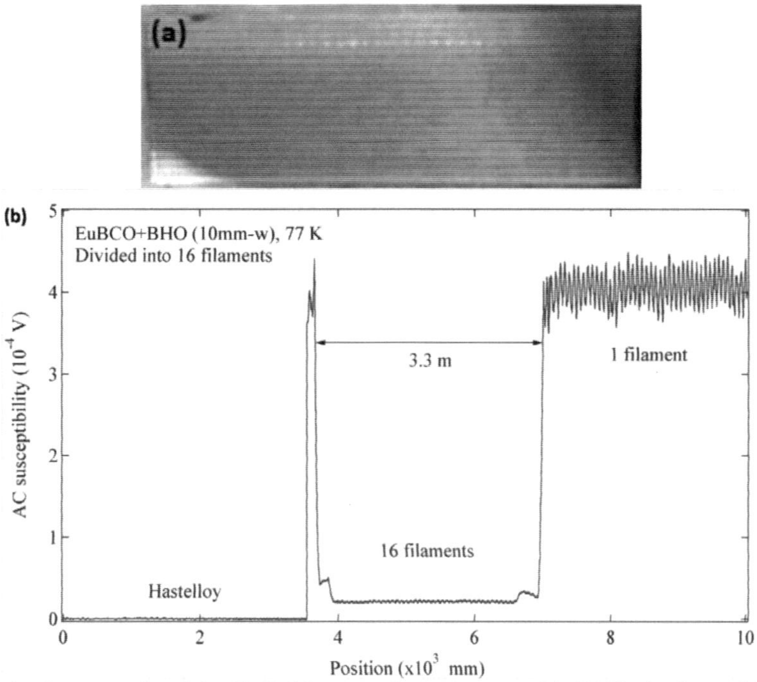

Fig. 4 A photograph of the EuBCO coated conductor with BHO doping of 16 filaments in 10 mm width and (b) hysteresis loss was reduced to 1/16 by a 16 filamentary and 3 m long EuBCO coated conductor with BHO doping comparing with EuBCO coated conductor with BHO doping without filamentation.

4. Conclusion

We have been investigating the long REBCO coated conductors with BMO doping by IBAD / conventional (out-of-plume) PLD and plane-plume PLD method. The plane-plume PLD is effective to obtain high uniformity of longitudinal and transversal J_c in long EuBCO coated conductor with BHO doping with high deposition rate and high in-field performance because of the transferring substrate through the inside of multi-plume with scan of X-Y axes direction (plane-plume) which less affected by change of plume condition such as tilt and swing of plume.

We evaluated two dimensional J_c distributions of the 20 m long and 10 mm wide EuBCO coated conductor with BHO doping by SHPM with Reel-to-Reel system. As a result, the uniformity of in-plane J_c distribution of 20 m long EuBCO coated conductor with BHO doping was improved by the plane-plume PLD method.

Moreover, we fabricated 16 filamentary and 3 m long EuBCO coated conductors with BHO doping in 10 mm width by IBAD / plane-plume PLD and excimer laser scribing method. We confirmed that the hysteresis loss was reduced to 1/16 by a 16 filamentary and 3 m long EuBCO coated conductor with BHO doping comparing with EuBCO coated conductor with BHO doping without filamentation.

Acknowledgements

This work is supported by the New Energy and Industrial Technology Development Organization (NEDO), Advanced Medical Services from the Japan Agency for Medical Reserch and development (AMED and Ministry of Economy, Trade and Industry (METI).

References

[1] Yamada Y, Takahashi K, Kobayashi H, Konishi M, Watanabe T, Ibi A, Muroga T, Miyata S 2005 *Appl. Phys. Lett.* **87** 132502
[2] Ibi A, Miyata S, Kuriki R, Kobayashi H, Fukushima H, Kinoshita A, Kato T, Hirayama T, Yamada Y, Shiohara Y 2008 *Physica C* **468** 1514-1517
[3] Tobita H, Notoh K, Higashikawa K, Inoue M, Kiss T, Kato T, Hirayama T, Yoshizumi M, Izumi T, Shiohara Y 2012 *Supercond. Sci. Technol.* **25** 062002
[4] Iijima Y, Tanabe N, Kohno O, Ikeno Y 1992 *Appl. Phys. Lett.* **60** 769
[5] Yoshida T, Ibi A, Takahashi T, Yoshizumi M, Izumi T, Shiohara Y 2014 *Physica C: Superconductivity* **504** 42-46
[6] Katayama K, Machi T, Nakamura T, Takagi Y, Nakaoka K, Yoshizumi M, Izumi T, Shiohara Y 2014 *Physics Procedia* **58** 142
[7] Chikumoto N, Lee S, Nakao K, Tanabe K 2009 *Physica C* **469** 1303
[8] Lee S, Chikumoto N, Yokoyama T, Machi T, Nakao K, Tanabe K 2009 *IEEE Trans. Appl. Supercond.* **19** 3192
[9] Ibi A, Machi T, Nakaoka K, Izumi T 2018 *Journal of Physics: Conf. Series* 1054 012032
[10] Ibi A, Yoshida T, Izumi T, Shiohara Y, Yokoe D, Kato T, Hirayama T 2016 *Physics Procedia* **81** 97-100
[11] Higashikawa K, Inoue M, Kawaguchi T, Shiohara K, Imamura K, Kiss T 2011 *Physica C* **471** 1036

IOP Conf. Series: Journal of Physics: Conf. Series **1293** (2019) 012035 doi:10.1088/1742-6596/1293/1/012035

Optimization of interim heat-treatment condition on TFA-MOD process for fabrication of $Y_{0.77}Gd_{0.23}Ba_2Cu_3O_y$ coated conductors with $BaHfO_3$

K Nakaoka[1], R Yoshida[2], M Sato[1], A Ibi[1], T Machi[1] and T. Izumi[1]

[1] National Institute of Advanced Industrial Science and Technology, 1-2-1 Namiki, Tsukuba, Ibaraki, Japan
[2] Nanostructures Research Lab., Japan Fine Ceramics Center, 2-4-1 Matsuno, Atsuta, Nagoya, Japan

E-mail: koichi.nakaoka@aist.go.jp

Abstract. We have optimized the interim heat-treatment (IHT) time in trifluoroacetate based metal organic deposition process on fabrication of the $BaHfO_3$ (BHO) doped $Y_{0.77}Gd_{0.23}Ba_2Cu_3O_y$ (YGdBCO) coated conductor (CC) in order to further improve the in-field critical current density (J_c) performance. The in-field J_c values of 0.6 μm-thick BHO doped YGdBCO CCs increased with increasing the IHT time within 90 min. The highest critical current density (J_c) value of 0.86 MA/cm^2 at 77 K in 3 T ($B//c$) was obtained for 0.6 μm-thick BHO doped YGdBCO CC with IHT time of 90 min. CCs with IHT time shorter than 90 min had relative low in-field J_c performance but high self-field J_c performance. It can be caused by the insufficient transformation to BaF_2 in IHT which is a reason for the relatively large BHO particles. In the case of applying IHT for longer time than 90 min, coarsening of CuO particles in precursor films was suggested and degradation of self- and in-field J_c performances were observed.

1. Introduction

The metal organic deposition (MOD) process using metal trifluoroacetates (TFA) has been recognized as a process for cost-effective and suitable for large-scale production to fabricate $REBa_2Cu_3O_y$ (RE: rare-earth element, REBCO) coated conductors (CCs) with high-performance in a self-field. However, it had been lower comparing with the CCs by the vapor process in a magnetic-field. On the other hand, low cost CCs with high in-field superconducting performance have been required for electric power applications such as magnetic resonance imaging, motors, generators, and accelerators.

The standard heat-treatment procedure of the conventional TFA-MOD process consists of calcination and crystallization steps. The calcination step is performed at low-temperature for thermal decomposition of a starting solution to an amorphous precursor phase. The amorphous precursor films are crystallized to REBCO phase at high-temperature during the crystallization step. In order to improve the in-field critical current density (J_c) performance of $BaMO_3$ (M: metal element, BMO) doped REBCO CCs, we have modified the heat-treatment procedure of TFA-MOD process and proposed new heat-treatment step, which was called as "interim heat-treatment (IHT)", introduced between the calcination and the crystallization steps [1-3]. The IHT gives improving homogeneous dispersion of the finer BMO particles and improving the in-field J_c performance of BMO doped REBCO CCs. The fundamental analysis of IHT was previously reported [3]. We found that the IHT is the process to transfer from BaF_2–$(Y,Gd)F_3$ solid solution (SS) to BaF_2 and Y_2O_3 which is the necessary condition of BMO nucleation. However, the IHT conditions (temperature, holding time and atmosphere etc.) have been not optimized.

In this paper, we have optimized the IHT conditions, particularly IHT time, for further improvement of in-field J_c performance of $BaHfO_3$ (BHO) doped $Y_{0.77}Gd_{0.23}Ba_2Cu_3O_y$ (YGdBCO) CCs.

2. Experimental

The starting solution was obtained by dissolving and mixing yttrium and gadolinium propionates, Ba-trifluoroacetate, copper 2-ethylhexanoate, and tetrakis(2,2,6,6-tetramethyl-3,5-heptanedionato) hafnium with the ratio of Y : Gd : Ba : Cu : Hf equal to 0.77 : 0.23 : 1.60 : 3.00 : 0.10 in 1-methoxy-2-propanol [4].

The starting solution was controlled with the total metal (excluding hafnium) concentration of 0.45 mol/L. The starting solution was deposited by spin-coating onto CeO_2 (pulsed laser deposition) buffered $LaMnO_3$ (radio frequency sputter deposition)/MgO (ion beam-assisted deposition)/Y_2O_3 (ion beam sputter deposition)/$Gd_2Zr_2O_7$ (ion beam sputter deposition)/Hastelloy™ substrates (33 mm in length and 10 mm in width) at 4800 rpm for 20 s. This spin-coated sample was placed on an alumina plate which was inserted in an electric tube furnace and then subjected to a heat-treatment consisting of a series of calcination, IHT, and crystallization steps. The calcination was performed by heating up to 500°C at the heating rate of 10°C/min in O_2 atmosphere at atmospheric pressure. To prepare thicker precursor films, we repeated the spin-coating and the calcination 20 times. Subsequently, the thick precursor films were applied to the IHT and crystallization steps for conversion to the superconducting phase. The heating profile for the crystallization step with IHT step is shown in figure 1. The IHT and crystallization processes were applied under a mixed gas atmosphere of humid Ar and O_2 (total pressure: 30 kPa, partial pressure of H_2O: 1.7 kPa, partial pressure of O_2: 0.04 kPa). Holding time in the IHT step was varied 0 to 240 min and holding time in the crystallization step was 80 min. Consequently, 0.6 μm-thick BHO doped YGdBCO films were obtained. The thickness of the films was determined by a cross-sectional scanning electron microscopic observation. A quenched sample during the IHT step was prepared by removing from the electric tube furnace and immediately dropping onto a thick copper block at room temperature. This quenched sample preparation was done within 10 sec. The BHO doped YGdBCO films were oxygen annealed at 500°C under oxygen pressure of about 200 kPa for 1 h.

X-ray diffraction (XRD) analyses were carried out with Cu Kα radiation. The critical current was measured at 77 K under self- and magnetic field by the direct-current four-probe transport technique with a criterion of 1 μV/cm.

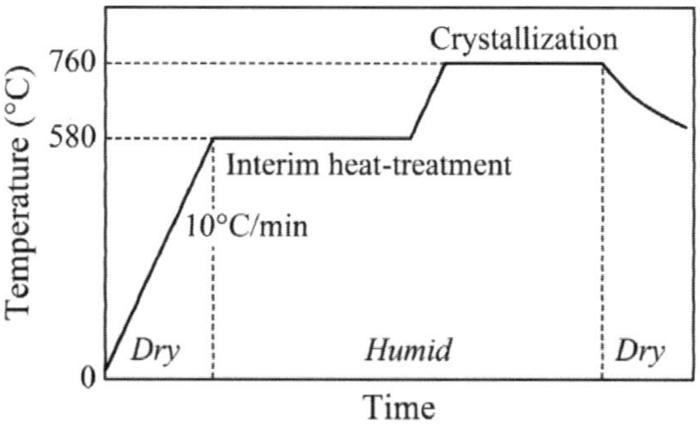

Figure 1. Heating profile of crystallization step with interim heat-treatment.

3. Results and Discussion

XRD analyses of quenched samples after holding for 0, 30, 60, 90, 120, 180, and 240 min at 580°C during the IHT step were performed to investigate the microstructure evolution in precursor films. Figure 2 shows the XRD patterns of quenched samples scanned in ranges of 2θ from 20 to 32° and from 34 to 54°. Four broad diffraction peaks excluding the peaks from the CeO_2 buffered $LaMnO_3$/MgO/Y_2O_3/$Gd_2Zr_2O_7$/Hastelloy™ substrate (2θ ~ 22.4 and 46.0°: $LaMnO_3$, 2θ ~ 29.7°: $Gd_2Zr_2O_7$, 2θ ~ 43.4°: Hastelloy™) were observed at 2θ ~ 25.6, 29.8, 42.6 and 50.5° for quenched sample with IHT time of 0 min. These four diffraction peaks correspond to a BaF_2–$(Y,Gd)F_3$ solid solution (SS) phase [3]. BaF_2 and $(Y,Gd)F_3$ are present as a SS phase in the precursor films as

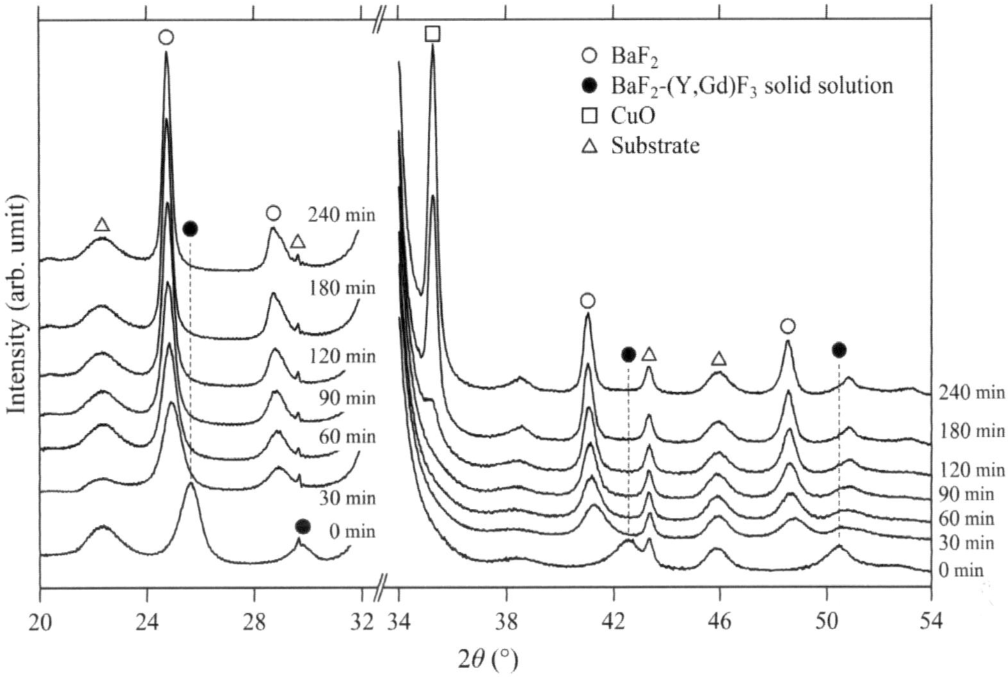

Figure 2. X-ray diffraction patterns of quenched samples after holding for 0, 30, 60, 90, 120, 180, and 240 min at 580°C during the interim heat-treatment step.

prepared. As the IHT time increased, all four diffraction peaks moved to lower 2θ angles and shifted to $2\theta \sim 24.8$, 28.8, 41.1 and 48.7° for the quenched sample with IHT time of 240 min. The diffraction peaks of $2\theta \sim 24.8$, 28.8, 41.1 and 48.7° are indexed to the (111), (200), (220) and (311) planes of pure BaF_2 phase, respectively, which indicates that the BaF_2–(Y,Gd)F_3 SS to pure BaF_2 phase transformation occurred during the IHT. Therefore, the basic reaction is proposed as follows [3];

$$2(Y,Gd)F_3 \text{ (in } BaF_2\text{–}(Y,Gd)F_3 \text{ SS)} + 3BaF_2 \text{ (in } BaF_2\text{–}(Y,Gd)F_3 \text{ SS)} + 3H_2O \text{ (g)} \rightarrow (Y,Gd)_2O_3 + 3BaF_2 + 6HF \text{ (g)}. \quad (1)$$

Additionally, new XRD diffraction peak of CuO($\bar{1}11$) around $2\theta = 35°$ was confirmed in quenched samples with long IHT time of over 90 min.

The IHT time dependence of the diffraction peak position of BaF_2–(Y,Gd)F_3 SS ($2\theta \sim 25.6°$) in quenched samples is shown in figure 3. It can clearly be seen from this figure that the peak of BaF_2–(Y,Gd)F_3 SS moved to the position of pure BaF_2 ($2\theta \sim 24.8°$) with increasing the IHT time. Furthermore, transformation reaction from BaF_2–(Y,Gd)F_3 SS phase to pure BaF_2 phase was found to achieved at IHT time of around 60 min for the samples having a final thickness of about 0.6 μm. Figure 4 shows the relationship between the XRD peak intensities of ($\bar{1}11$) plane of CuO phase in quenched samples and IHT time. The diffraction peak of CuO($\bar{1}11$) plane was observed in the quenched samples with IHT time of more than 90 min. As for the diffraction peak of CuO($\bar{1}11$) plane observed at IHT time

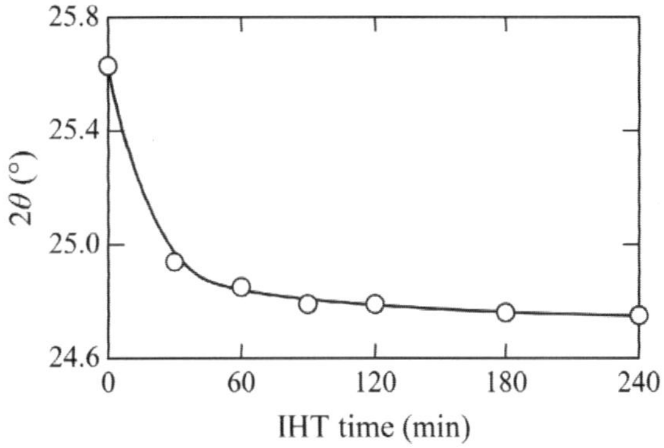

Figure 3. The IHT time dependence of the diffraction peak position of BaF_2–(Y,Gd)F_3 SS ($2\theta \sim 25.6°$) in quenched samples.

of longer than 90 min, the diffraction peak intensity increased with increasing IHT time, which might be due to coarsening of CuO particles.

The influence of IHT time on J_c performance of BHO doped YGdBCO CCs was investigated. Figure 5 shows the IHT time dependence of the J_c values at (a) 77 K in self-field and (b) 77 K in 3 T (*B*//*c*) for BHO doped YGdBCO CCs. High J_c values of above about 4.5 MA/cm^2 were obtained at 77 K in self-field (a) for the CCs with IHT time of 90 min or less. On the other hand, the J_c values decreased with increasing of the IHT time longer than 90 min. The J_c values at 77 K in 3 T (*B*//*c*) (b) increased with increasing IHT time until 90 mm. The increment of in-field J_c performance is believed from the results of the fine dispersion of BHO particles in the CCs due to the progress of IHT reaction (reaction 1). The BHO particles can be nucleated at low-temperature because the necessary condition of transformation from BaF_2–(Y,Gd)F_3 SS to BaF_2 is almost completed during the IHT step. The low-temperature nucleation and growth might affect to the suppression of growth of the BHO particles due to low diffusivity. The highest J_c value of 0.86 MA/cm^2 at 77 K in 3 T (*B*//*c*) was obtained for CC with IHT time of 90 min. The J_c values decease with increasing the IHT time over 90 min, as well as behaviour of J_c performance at 77 K in self-field (a). The degradation of J_c performance of CCs with IHT time of more than 90 min might be related to the coarsening of CuO particles in the precursor films. Although the mechanism has not yet been elucidated, it is known that there is compositional segregation within the thickness of once layer in multi-layered precursor films [5-7]. Rupich *et al* reported that a copper-rich region is presented between the individual layers in the multi-layered precursor film and the epitaxial growth of YBCO is partially disrupted as it crosses the copper-rich interface, which leads to the J_c degradation of the YBCO film [6, 7]. In our previously study, Y, Gd, Ba elements segregate at the lower part of each layer, and in contrast, Cu element at the upper part [5]. Therefore, this degradation of J_c performance of CCs with long IHT time might be due to formation of blocking layers of coarsened CuO, which inhibits the epitaxial growth of REBCO phase during the crystallization step, in the precursor film by coarsening of CuO particles.

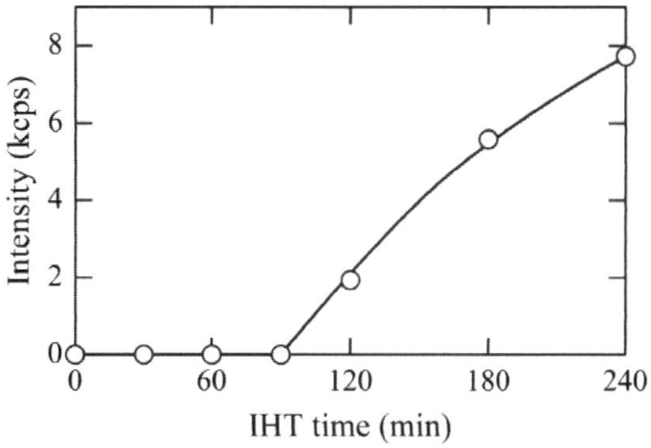

Figure 4. Relationship between the XRD peak intensities of ($\bar{1}11$) plane of CuO phase in quenched samples and IHT time.

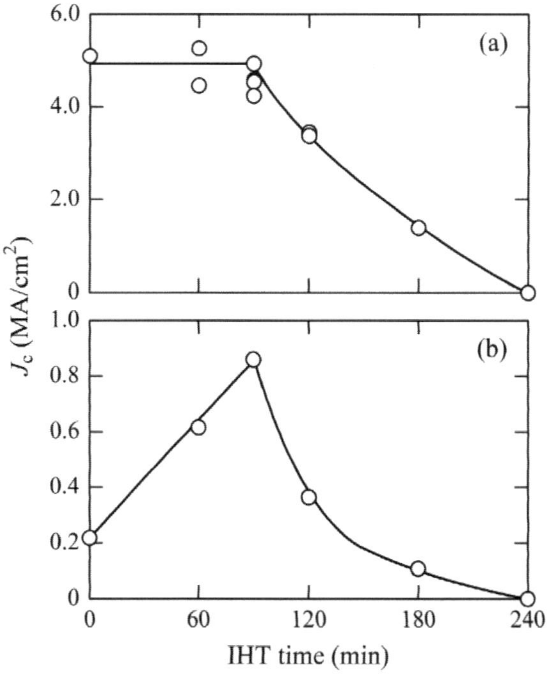

Figure 5. IHT time dependence of the J_c values at (a) 77 K in self-field and (b) 77 K in 3 T (*B*//*c*) for BHO doped YGdBCO CCs.

4. Conclusion

We have optimized the IHT time on fabrication of the 0.6 μm-thick BHO doped YGdBCO CCs for further improving the in-field J_c performance. The in-field J_c value of CCs was improved with increasing the IHT time within 90 min. The highest J_c value of 0.86 MA/cm^2 at 77 K in 3 T (*B*//*c*) was obtained for CC with IHT time of 90 min. In the case of applying IHT for longer time than 90 min, the degradation of J_c performance of CCs was confirmed. Furthermore, coarsening of CuO particles in the precursor films was observed. The degradation of J_c performance of CCs with long IHT times might be due to formation of blocking CuO layers, which inhibits the epitaxial growth of REBCO phase during the crystallization step, in the precursor film by coarsening of CuO particles. In the optimization of IHT time, we found the existence of lower and upper limits of IHT time, which are determined by the required time for complete transformation to pure BaF_2 phase (reaction 1) and the formation of the blocking layer to the epitaxial growth of REBCO phase due to coarsening of CuO, respectively.

Acknowledgments
This study was supported by the Ministry of Economy, Trade and Industry (METI), and the New Energy and Industrial Technology Development Organization (NEDO).

References

[1] Kimura K, Hironaga R, Takahashi Y, Nakanishi T, Koizumi T, Hasegawa T, Higashikawa K, Inoue M, Kiss T, Kato T, Nakamura T, Yoshizumi M, Izumi T and Shiohara Y 2013 *IEEE Trans. on Appl. Supercond.* **23** 6601704
[2] Kimura K, Hironaga R, Nakamura T, Takahashi Y, Koizumi T, Hasegawa T, Higashikawa K, Inoue M, Kiss T, Yoshizumi M, Izumi T and Shiohara Y 2015 *IEEE Trans. on Appl. Supercond.* **25** 6604204
[3] Nakaoka K, Kimura K, Kato T, Yoshida R, Usui Y, Nakamura M, Izumi T and Shiohara Y 2016 *IEEE Trans. on Appl. Supercond.* **26** 8000304
[4] Nakaoka K, Matsuda J, Kitoh Y, Goto T, Yamada Y, Izumi T and Shiohara Y 2007 *Physica C* **469** 519
[5] Nakaoka K, Yoshida R, Kimura K, Kato T, Izumi T and Shiohara Y 2017 *Supercond. Sci. Technol.* **30** 055008
[6] Rupich M W, Li X, Thieme C, Sathyamurthy S, Fleshler S, Tucker D, Thompson E, Schreiber J, Lynch J, Buczek D, DeMoranville K, Inch J, Cedrone P and Slack J 2010 *Supercond. Sci. Technol.* **23** 014015
[7] Rupich M W, Miller D, Feenstra R and Holesinger T G 2009 *Wire Development Group Presentation DOE Peer Review 2009* (Alexandria, VA, Aug. 2009)

Superconducting properties of ($Y_{1-x}Eu_x$)$Ba_2Cu_3O_y$-coated conductors prepared via trifluoroacetate–metalorganic deposition

M Sato[1], K Nakaoka[1], A Ibi[1], T Machi[1], T Izumi[1]

[1.] National Institute of Advanced Industrial Science and Technology (AIST), 1-2-1 Namiki, Tsukuba, Ibaraki 305-8564, Japan

michio.sato@aist.go.jp

Abstract. In order to investigate the effect of the partial substitution of Y by Eu on the superconducting properties of $REBa_2Cu_3O_y$ (RE: rare earth) films, we fabricated trifluoroacetete–metalorganic deposited $Y_{1-x}Eu_xBa_2Cu_3O_y$ (x=0.1, 0.23 0.3, 0.4, 0.5) films on metallic substrates with optimal crystallization conditions. The optimized crystallization conditions for fabricating $Y_{1-x}Eu_xBa_2Cu_3O_y$ coated conductors with high x composition are relatively high crystallization temperature and low oxygen pressure comparing with $YBa_2Cu_3O_y$. The films with high x exhibit a high critical temperature, irreversible fields, and high critical current density in the high-magnetic region at 77 K. However, the superiority on the critical current density of the films with high x in self and low fields at 77 K is not necessary conformed. This behavior might be explained by the influence of the melt-growth mode in the film with high x.

1. Introduction

$REBa_2Cu_3O_y$ [REBCO; rare-earth (RE) elements=Y, Gd, Eu etc.] films deposited on flexible and lightweight metallic substrates, manufactured by the ion-beam assisted deposition (IBAD) or on the rolling-assisted biaxial textured substrates, often referred to as coated conductors (CCs), hold promise for various industrial applications such as in generators, transmission, cables, and superconducting magnetic energy storage. REBCO films can be manufactured with several techniques, including the pulsed laser deposition (PLD) [1, 2], the metalorganic chemical vapor deposition (MOCVD) [3], the reactive co-evaporation (RCE) [4], and the metalorganic deposition (MOD) [5, 6]. Among these deposition techniques, the trifluoroacetate (TFA)-based MOD (TFA-MOD) is notably economical because it does not require a high-vacuum system. Although $YBa_2Cu_3O_y$ (YBCO) CCs fabricated by TFA-MOD possess good superconductivity, having a high critical temperature (T_c) and critical current density (J_c) in an external magnetic field, significant improvements to the in-field J_c are still required.

Much work in this direction has focused on partially substituting or displacing the Y in these films with other RE elements. Partial substitution of Y by Sm has been shown to drastically increase the in-field J_c in PLD-$Y_{2/3}Sm_{1/3}Ba_2Cu_3O_y$, which resulted in an in-field J_c of 0.06 MA/cm^2 at 75.5 K and a magnetic field of 5 T applied parallel to the c-axis of the films which related with the crystallographic c-axis [7]. This value is three times higher than that of YBCO. Goswami et al. reported that PLD-$Y_{1-x}Eu_xBa_2Cu_3O_y$ film has higher in-field J_c than unsubstituted YBCO because the Eu-substituted material has dense *c*-axis-correlated defects that result in an improvement of flux-pinning properties in weak magnetic fields [8]. In the case of fabrication with TFA-MOD, $Y_{0.77}RE_{0.23}Ba_2Cu_3O_y$ (RE=Sm or Gd) CCs have high in-field J_c compared with YBCO CCs [9]. Although SmBCO in bulk material has a higher maximum T_c than GdBCO at ~94 K [10], the in-field J_c properties of $Y_{0.77}Gd_{0.23}Ba_2Cu_3O_y$ are superior to those of $Y_{0.77}Sm_{0.23}Ba_2Cu_3O_y$ in the coated conductors. It might be caused by the difficulty to control of the substitution of Ba by Sm sites [11]. According to the above-mentioned knowledge, Eu was selected as a substituting element for Y because of its high T_c but little substitution into Ba site [12-14]. In this work, in order to investigate the effect of the partial substitution of Y by Eu on superconducting properties, we used the TFA-MOD process to fabricate $Y_{1-x}Eu_xBa_2Cu_3O_y$ (x=0.1, 0.23 0.3, 0.4, 0.5) CCs with optimal crystallization conditions.

2. Experimental

$Y_{1-x}Eu_xBa_2Cu_3O_y$ ($Y_{1-x}Eu_xBCO$) films were grown from organic metal solutions including Y-propionate, Eu-propionate, Ba-trifluoroacetate, and Cu-2-ethylhecanoate [15, 16] with a molar ratio of Y:Eu:Ba:Cu =1-*x*:*x*:1.5:3 onto the following substrates: laminated structure of PLD-CeO_2, Y_2O_3, $LaMnO_3$, IBAD-MgO, $Gai_2Zr_2O_7$, and Hastelloy C276. The solution, with a metal-ion concentration of 1.2 mol/L, was deposited on a CeO_2-buffered substrate by the spin-coating technique. The coated films were then decomposed onto an amorphous precursor film by heating them to 500°C in humid flowing oxygen gas with a water-vapor partial pressure [$P(H_2O)$] of 3.2 kPa and a water inlet temperature of 400°C. This coating/calcination decomposition process was repeated to obtain a final thickness of about 0.5 μm. The precursor films were then heated to their maximum temperature (T_{max}) in a humid mixed argon and oxygen atmosphere with $P(H_2O)$ of 3.2 kP. The optimized conditions had been investigated for each composition with T_{max} of 700°C–820°C and oxygen partial pressure [$P(O_2)$] ranging 0.01–0.05 kPa. Here, the optimized condition for composited CCs was meant for the highest self-field J_c at liquid nitrogen temperature. The crystalized films were then annealed at 500°C under 200 kPa for 60 min in O_2 followed by furnace cooling to room temperature.

The microstructure of the crystallized $Y_{1-x}Eu_xBCO$ films was observed using a scanning electron microscope (SEM). The self-field J_c values were measured in 1-cm-wide films by the standard four-probe DC transport method under liquid-nitrogen temperature with the criterion in voltage of 1μV/cm. For resistivity and in-field J_c measurements, the films were patterned into ~35μm-wide strips by the laser etching. These measurements were performed with the standard four-probe method using a Quantum Design Physical Properties Measurement System that can generate a magnetic field of up to 9 T. T_c and irreversibility field (B_{irr}) were determined using a 0.01-ρ_N sample, where ρ_N is the normal-state resistivity.

3. Results and Discussion

To find optimized conditions in crystallization step for $Y_{1-x}Eu_xBCO$ CCs, we examined the properties of the REBCO films fabricated under crystallization conditions of T_{max} of 700°C–820°C and $P(O_2)$ of 0.01–0.05 kPa. It was found that as x increases in $Y_{1-x}Eu_xBCO$ CCs, suitable T_{max} increases and $P(O_2)$ decreases. This can be explained by the high peritectic temperature of EuBCO than that of YBCO [17]. To clarify the effect of substitution by Eu in the TFA-MOD REBCO CC, we compared $Y_{0.77}Eu_{0.23}BCO$ CCs with $Y_{0.77}Gd_{0.23}BCO$ CCs formed under each optimal crystallization conditions. As a result, $Y_{0.77}Eu_{0.23}BCO$ revealed a higher T_c (=91.1 K) than $Y_{0.77}Gd_{0.23}BCO$ (T_c=90 K) even with the same substitution quantity (x=0.23).

We also measured the self-field transport properties with different quantities of substituted Eu. Figure 1 plots the substitution quantity (x) dependence of T_c and self-field J_c ($J_c^{s.f.}$) at 77 K for $Y_{1-x}Eu_xBCO$ CCs. The T_c values of $Y_{1-x}Eu_xBCO$ CCs increase from 88.5 K to 91.9 K as x increases, indicating that Eu substitution amount can make T_c higher. The improvement in T_c after partial substitution of Y by Eu is attributable to the higher maximum T_c in EuBCO than that in YBCO [12]. It also suggests that the YBCO and EuBCO phases are uniformly distributed in the $Y_{1-x}Eu_xBCO$ matrix. Moreover, we can see

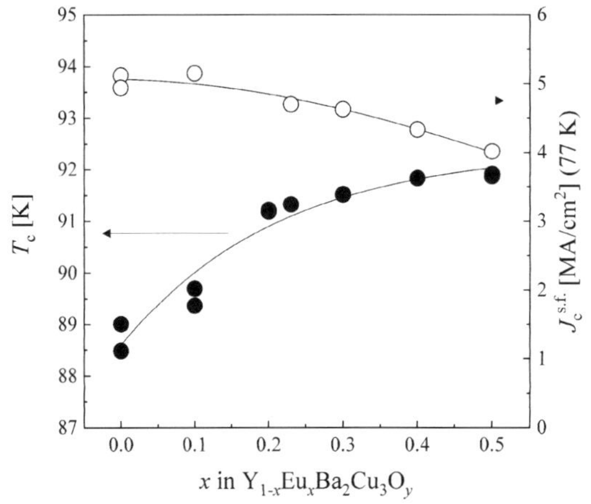

Figure 1. The dependence of T_c (left axis) and $J_c^{s.f.}$ (right axis) on quantity of substituted Eu (x) at 77 K for $Y_{1-x}Eu_xBCO$ CCs.

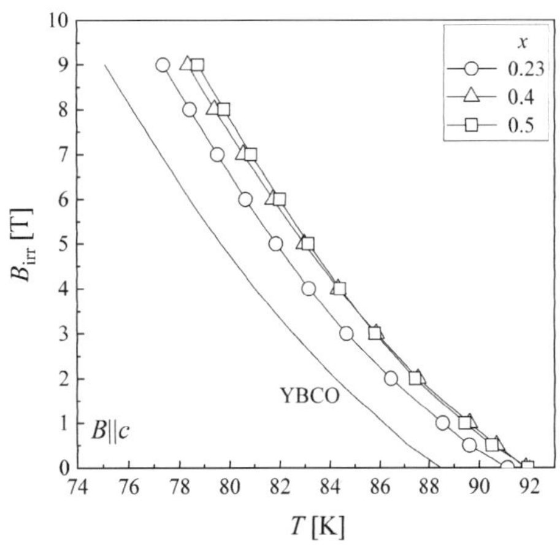

Figure 2. The dependence of B_{irr} on temperature at $B||c$ for in $Y_{1-x}Eu_xBCO$ CCs with x=0, 0.23, 0.4, and 0.5.

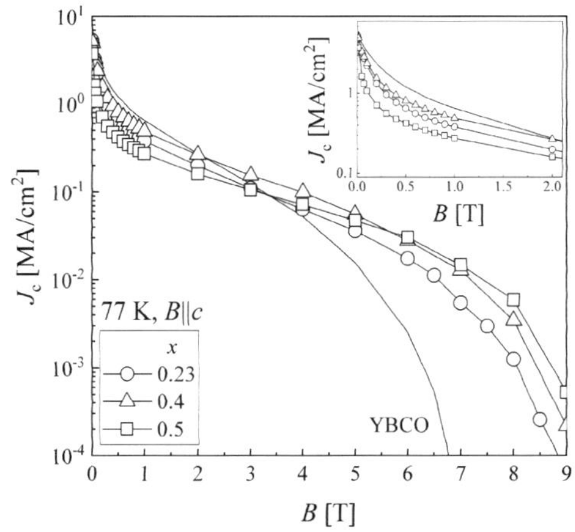

Figure 3. The dependence of J_c at 77 K and $B||c$ on applied magnetic field for $Y_{1-x}Eu_xBCO$ CCs with x=0, 0.23, 0.4, and 0.5. The inset shows the magnetic-field dependence (B < 2 T) of J_c for $Y_{1-x}Eu_xBCO$ CCs with x=0, 0.23, 0.4, and 0.5.

that the T_c values of $Y_{1-x}Eu_xBCO$ CCs do not increase linearly with Eu substitution amounts, which may imply the necessity of further optimization in the condition of the oxygen-annealing. On the other hand, the $J_c^{s.f.}$ in $Y_{1-x}Eu_xBCO$ CCs decreases slightly with increasing x. For small amounts of substitution, such as $x \leq 0.1$, the $J_c^{s.f.}$ values were in the range 4.9–5.2 MA/cm^2 at 77 K. For high substitutions amounts, such as x=0.5, the $J_c^{s.f.}$ values drop by about 10%. In general, J_c depends strongly on T_c. Therefore, the degradation of $J_c^{s.f.}$ for high x was unexpected because T_c had been improved.

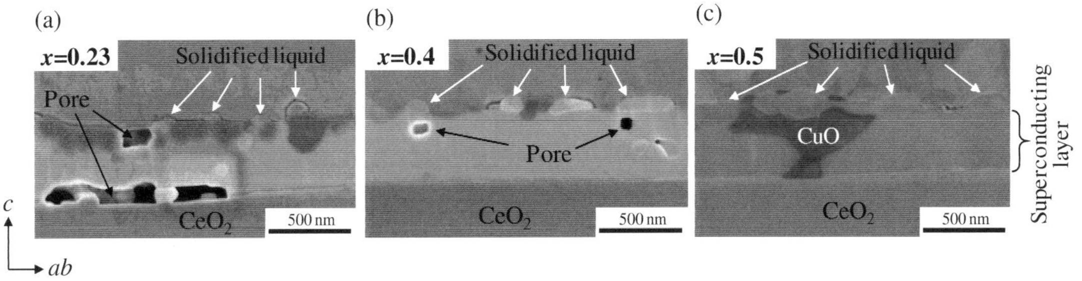

Figure 4. Cross-sectional SEM images of $Y_{1-x}Eu_xBCO$ CCs with x=(a) 0.23, (b) 0.4, and (c) 0.5.

In-field transport properties under the same conditions were also measured. Figure 2 shows the temperature dependence of B_{irr} at $B||c$ in $Y_{1-x}Eu_xBCO$ CCs with x=0, 0.23, 0.4, and 0.5. As for $B||c$, the CCs with high substitutions of Eu exhibit higher B_{irr} than those with smaller amounts of Eu, indicating that the differences in B_{irr} values are related to differences in T_c. Figure 3 showed the J_c values at 77 K and $B||c$ against the external magnetic field for $Y_{1-x}Eu_xBCO$ CCs with x=0, 0.23, 0.4, and 0.5. The inset of Figure 3 shows the magnetic field ($B < 2$ T) dependence of J_c for $Y_{1-x}Eu_xBCO$ CCs with x=0, 0.23, 0.4, and 0.5. $Y_{1-x}Eu_xBCO$ CCs have lower J_c than unsubstituted YBCO CCs (those with x=0) when $B <$ 2 T. Additionally, the J_c of $Y_{1-x}Eu_xBCO$ CCs with high x under a low magnetic field does not necessarily improve, as shown the inset of Figure 3. On the other hand, $Y_{1-x}Eu_xBCO$ CCs with high x have high J_c under a high magnetic field than the CCs with small x. The enhancement of J_c in a high magnetic field for the CCs substituted with large amounts of Eu should be a result of the improvement of B_{irr} and T_c.

To understand the self-field and low-field J_c behaviors in the Eu-substituted CCs, we observed the microstructure using SEM as shown in Figure 4. The size and the number of pores in the matrix decreases with increasing Eu substitution amount, whereas the amount of solidified liquid phase increases on the surface of the superconducting layers (see white arrow in Figure 4) accordingly. The difference in microstructures and the behaviors of electro-magnetic properties in the CCs with different substitution quantities could be explained as follows: In the case of small x at low T_{max} and high $P(O_2)$ in optimized crystallization conditions, pores result from the reactions of the non-superconducting phase entrapped in the matrix during the epitaxial growth, since this solid-state reaction causes volume shrinkage [18]. With high x at high T_{max} and low $P(O_2)$ in optimized crystallization conditions, the solidified liquid phase might be ejected because of the dominance of the melting layer at the growth front of the superconducting layer. It results in higher crystallinity and the suppression of pore formation since the intermediate phase in front of the growing interface is easier to dissolve in the liquid phase, which lead to the reduction of the trapped particles. However, the segregation of the residual phase at the grain boundary in the melt-growth mode can be assumed because the high purify grain can be grown. The segregation at the grain boundaries might degrade the grain-to-grain connectivity in the superconducting layer although the potential of the grain itself is higher due to its high crystallinity. Moreover, the agglomeration and/or coarsening CuO grains (see in Figure 4 (b)) might be cause by the influence of the melt-growth mode. Thus, the influence of the melt-growth mode might make the self-field and low-field J_c in CCs lower with high amounts of Eu substitution although T_c, B_{irr} and J_c in the high magnetic field regions are higher.

4. Conclusion

In order to investigate the effect of the partial substitution of Y by Eu on superconducting properties in the REBCO coated conductors, we fabricated $Y_{1-x}Eu_xBa_2Cu_3O_y$ (x=0.1, 0.23 0.3, 0.4, 0.5) films on metallic substrates by the TFA-MOD method under optimal crystallization conditions for each composition. As x increases in $Y_{1-x}Eu_xBCO$ CCs, suitable T_{max} increases and $P(O_2)$ decreases. The T_c values of $Y_{1-x}Eu_xBCO$ CCs increase with increasing x, which shows the highest T_c=91.9 K in this study.

However, the $J_c^{s.f.}$ in $Y_{1-x}Eu_xBCO$ CCs decreases slightly with increasing x. The CCs with high substitutions of Eu show higher B_{irr} than those with smaller amounts of Eu due to the improvement of T_c. On the other hand, $Y_{1-x}Eu_xBCO$ CCs with high x have high J_c under high magnetic field regions than the CCs with small x, although the improvement of J_c in low magnetic field for the CCs with high x does not confirmed. From microstructures, melt-growth mode in $Y_{1-x}Eu_xBCO$ CCs with high x conditions appears. The residual segregation at the grain boundary which can be assumed in the melt-growth mode might degrade the grain-to-grain connectivity in the superconducting layer. It seems to be a reason for lower J_c in low magnetic field including self-field.

Acknowledgments
This work was supported by the New Energy and Industrial Technology Development Organization (NEDO).

References
[1] Wu D X, Foltyn R S, Arendt N P, Blumenthal R W, Campbell H I, Cotton D J, Coulter Y J, Hults L W, Maley P M, Safar F H and Smith L J 1995 *Appl. Phys. Lett.* **67** 2397
[2] Ibi A, Fukushima H, Yamada Y, Miyata S, Kuriki T, Takahashi K and Y. Shiohara 2006 *Supercond. Sci. Technol.* **19** 1229
[3] Selvamaninkma V, Xu A, Liu Y, Khatri D N, Lei C, Chen Y, Galstyan E and Majkic G 2014 *Supercond. Sci. Technol.* **27** 055010
[4] Matias V, Rowley J E, Coulter Y, Maiorov B, Holesinger T, Yung C, Glyantsev V and Moeckly B 2010 *Supercond. Sci. Technol.* **23** 014018
[5] Gupta A, Jagannathan R, Cooper I E, Giess A E, Landman J I and Hussey W B 1988 *Appl. Phys. Lett.* **52** 2077
[6] Obradors X, Puig T, Pomar A, Sandiumenge F, Pinolñol S, Mestres N, Castaño O, Coll M, Cavallaro A, Palau A, González C J, Gutiérrez J, Romà N, Ricart S, Moretó M J, Rossell D M and Tendeloo V G 2004 *Supercond. Sci Technol.* **17** 1055
[7] MacManus-Driscoll L J, Foltyn R S, Maiorov B, Jia X Q, Wang H, Serquis A, Civale L, Lin Y, Hawley E M, Maley P M and Peterson D E 2005 *Appl. Phys. Lett.* **86** 032505
[8] Goswami R, Haugan J T, Barnes N P, Spanos G and Holtz L R 2010 *Physica C* **470** 318-322
[9] Miura M, Kato T, Yoshizumi M, Yamada Y, Izumi T, Hirayama T and Shiohara Y 2009 *Appl. Phys. Express* **2** 023002
[10] Yal X, Izumi T and Shiohara Y 2003 *Supercond. Sci. Technol.* **16** 4
[11] MacManus-Driscoll L J 1997 *Adv. Mater.* **9** 457
[12] William M V G and Tallon L J 1996 *Physica C* **258** 41-46
[13] Haugan J T, Campbell A T, Pierce A N, Locke F M, Maartense I and Barnes N P 2008 *Supercond. Sci. Technol.* **21** 025014
[14] Zhang S, Xu S, Fan Z, Jiang P, Han Z, Yang G and Che Y 2018 *Supercond. Sci. Technol.* **31** 125006
[15] Nakaoka K, Matsuda J, Kitho Y, Goto T, Yamada Y, Izumi T and Shiohara Y 2007 *Physica C* **463-465** 519-522
[16] Nakaoka K, Yoshizumi M, Usui Y, Izumi T and Shiohara Y 2014 *Physics Procedia* **58** 134-137
[17] Murakami M, Sakai N, Higuchi T and Yoo I S 1996 *Supercond. Sci. Technol.* **9** 1015
[18] Matuda S J, Tokunaga Y, Teranichi R, fuji H, Kaneko A, Asada S, Honjo T, Yajoma A, Iijima Y, Saitoh T, Izumi T and Shiohara Y 2004 *Physica C* **412-414** 890

IOP Conf. Series: Journal of Physics: Conf. Series **1293** (2019) 012037 doi:10.1088/1742-6596/1293/1/012037

Dependences of AC Loss in Stacked $REBa_2Cu_3O_y$ Superconducting Tapes on the Interval among Tapes under Perpendicular Magnetic Field

H Sasa[1], G Kawasaki[1], S Miura[1], M Iwakuma[1], T Izumi[2], T Machi[2] and A Ibi[2]

[1] Research Institute of Superconductor Science and Systems, Kyushu University, Fukuoka 819-0395, Japan

[2] National Institute of Advanced Industrial Science and Technology, Ibaraki 305-8564, Japan

E-mail: sasa@sc.kyushu-u.ac.jp

Abstract. In this study, we focused on the dependence of the AC loss in $REBa_2Cu_3O_y$ (REBCO, RE = rare earth, Y, Eu, Gd) superconducting tapes stacked perpendicular to the tape face on the interval between the superconducting layers. The AC loss of the 2-layer stacked tapes with various interval in perpendicular field was investigated by pickup-coil method. The AC loss for external applied magnetic field, B_m, below the penetration field, B_p, increases with increasing the interval. In the case that the ratio of the distance between the neighboring superconducting layer, D, to the width of the tape, w, (D/w) is smaller than 0.1, the AC loss of 2-layer tapes decreased to 1/2 of that of 1-layer tape. On the other hand, in the case of D/w being larger than 0.1, the AC loss of 2-layer tapes was larger than 1/2 and it approached the AC loss of 1-layer tape with increasing D/w. From the result obtained in this study, it is predicted that, for B_m below B_p, the AC loss of stacked REBCO tapes with large D/w, such as multifilamentary tapes, is larger than the AC loss of stacked non-scribed REBCO tapes. However, for B_m above B_p: a large AC loss region, it is known that the AC loss of the multifilamentary tapes decreases compared with that of non-scribed tapes. Next step, AC loss of the multifilamentary tapes for B_m below B_p will be investigated experimentally and theoretically.

1. Introduction

$REBa_2Cu_3O_y$ (REBCO, RE = rare earth, Y, Eu, Gd) is expected as a new generation material of superconducting tapes because it can be used even at high temperatures and high magnetic fields due to its excellent critical current densities and high critical temperatures. For AC applications of REBCO superconducting tapes, the AC loss generated in the tapes is one of the serious issues because it may cause a severe temperature rise of superconducting windings and result in thermal runaway. Therefore, the AC loss reduction and understanding its property in detail are absolutely essential to the AC applications. The AC loss of REBCO tapes depends on various operating conditions such as the magnetic field amplitude, B_m, the operating temperature, T [1], the number of the stacked tapes, n [2][3], the number of the filaments of the tape [4], the angle between the field and the tape face [5] and so on. The goal of our study is to establish a method which can easily estimate the amount of the AC loss for such various conditions from less experimental results.

In practical coil-based applications, the REBCO tapes are usually wound into a multilayer coil. In the case of the stacked superconducting strips, whole of them can be considered as one-body when the ratio of the distance between the neighboring strips, D, to their width, w, (D/w) is smaller than 0.1 [6]. Recently, we found that the AC loss of the REBCO tapes stacked into n-layer with $D/w < 0.1$ decreases to $1/n$ of that of 1-layer tape [2][3]. However, the REBCO tapes are preferred to be divided into a multi-filamentary structure by laser scribing technique in order to reduce the AC loss for B_m above B_p [7]. Here, B_p is the penetration field when the magnetic flux reaches the center of the filament. The D/w of each filament of the scribed tape should exceed 0.1. Therefore, the dependence of the AC loss on D/w which is larger than 0.1 should be investigated. The objective of this paper is to clarify the D/w dependence of the AC loss of stacked REBCO tapes with changing D in B_m below B_p. This study will be a help to establish AC loss estimation method for the condition of large D/w (> 0.1) such as the stacked multifilamentary REBCO tapes. The AC losses of the 2-layer stacked non-scribed tapes with various D/w under perpendicular magnetic field were measured by pickup-coil method.

2. Experimental setup

2.1. Parameters of REBCO superconducting tapes

Parameters of REBCO superconducting tapes are listed in table 1. The sample tapes were provided by the National Institute of Advanced Industrial Science and Technology. The REBCO superconducting tapes were fabricated by the combination of an ion beam assisted deposition technique and a pulsed laser deposition method. To improve the flux pinning property of the tape, $BaHfO_3$ was doped to the $EuBa_2Cu_3O_y$ superconducting matrix as an artificial pinning center [8]. The REBCO tapes were stacked into two layers with the various interval perpendicularly to the tape face as shown in figure 1. The intervals were determined in order for their ratio to be approximately 1:2:4:8. The tapes were covered with kapton tapes to insulate each other. The two tapes used in this study were cut off from one long tape. In this study, D indicates the distance between the neighboring superconducting layers as shown in figure 1.

Table 1. Parameters of the EuBCO tapes.

Dimensions of the tape	5 mm in width, 106 μm in thickness
Stabilizing layer	Ag (5 μm)
Superconducting layer	$EuBa_2Cu_3O_y + BaHfO_3$ [3.5 mol%] (0.7 μm)
Buffer layer	$CeO_2 + LaMnO_3 + MgO + Y_2O_3 + Gd_2Zr_2O_7$ (0.7 μm)
Substrate	Hastelloy (100 μm)
I_C of the tape at 77 K, self-field	200 A
Number of stacked tapes	1, 2
Interval between the neighboring a superconducting layer (corresponding D/w)	0.29, 0.65, 1.15, 2.37 mm (0.058, 0.130, 0.230, 0.474)

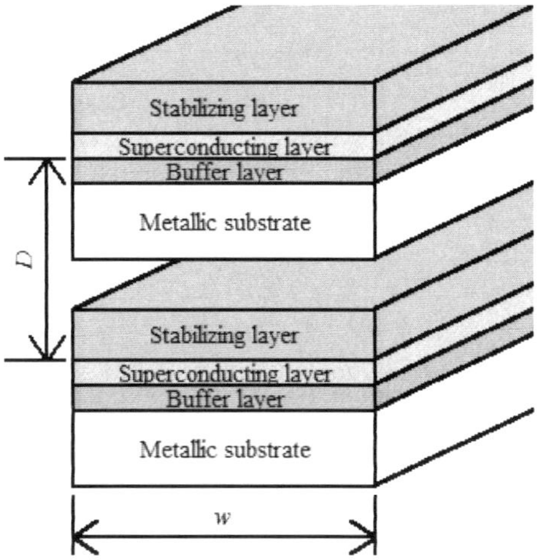

Figure 1. Schematic illustration of stacked REBCO tapes. The thickness of each layer and the width of the tapes are not on scale.

2.2. Experimental equipment

In this study, the magnetization of the REBCO superconducting tapes was measured by using a saddle-shaped pick-up coil [9]. From the measured magnetization and following equation, the AC loss of the tapes was calculated.

$$W = \int BdM = -\oint MdB \tag{1}$$

This equation says that the AC loss per unit volume per cycle, W, can be evaluated by integrating the magnetization, M, by the applied field, B, and it corresponds to the inner area of the measured M-B curve. The external magnetic field was applied perpendicularly to the tape face by a NbTi superconducting magnet cooled by liquid helium. The experimental conditions in this study are listed in table 2. It was confirmed that the AC loss of REBCO tapes has no frequency dependence [10].

Table 2. Experimental conditions.

Temperature	25-77 [K]
Field amplitude	0.004-4.3 [T]
Field frequency	0.01-0.2 [Hz]
Field angle	Perpendicular to the tape face
Method	Saddle-shaped pick-up coil method

3. Results and discussion

Figure 2 shows the observed B_m dependences of the AC loss of a REBCO tape and the 2-layer REBCO tapes with various D/w for B_m below B_p at 50 K. The inset shows the observed AC loss curves in a broad range. The case of 1-layer tape can be considered as the case of 2-layer tapes separated infinitely. The influence due to the magnetic field occurred by another tape is infinitely small in the case of the infinitely separated two tapes. Therefore, the 1-layer case is considered as the case of the 2-layer tapes with D/w being infinity in this study.

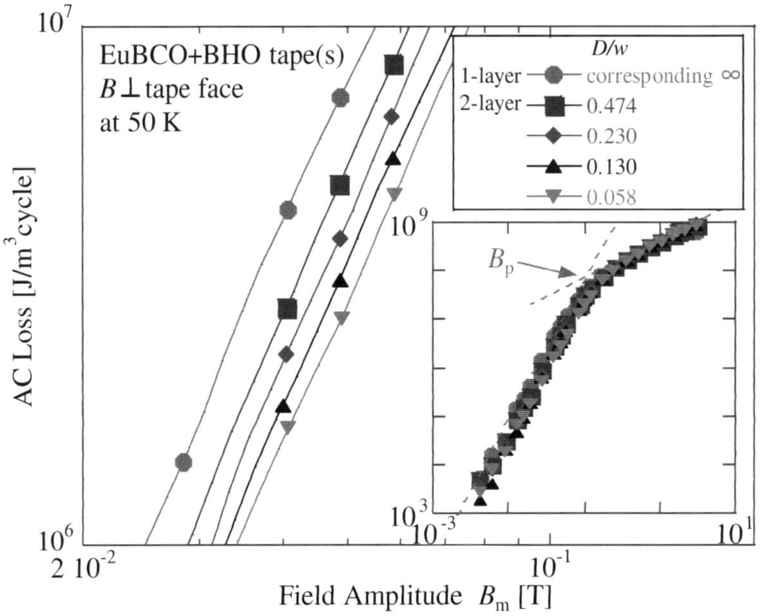

Figure 2. Observed B_m dependences of the AC loss of 1-layer tape and of 2-layer tapes with D/w = 0.474, 0.230, 0.130 and 0.058 at 50 K.

The B_p corresponds to the breaking point of the AC loss curve and can be determined from the intersection point of two approximate lines for above and below B_p as shown in the inset in figure 2. The B_p values are 0.1-0.17 T at 50 K. The AC loss for B_m above B_p has no interval dependence. In B_m above B_p, B_m is much larger than the diamagnetic field and the difference of the magnetization due to the difference of the interval is very small. Therefore, the AC loss has no interval dependence. On the other hand, the AC loss for B_m below B_p increases as increasing D/w. In the case of the narrow interval, the demagnetization field is mutually intensified by the magnetic field created by the shielding current in another tape, and the penetration depth is smaller than the wider interval case. Therefore, the AC loss increases with increasing D/w. Similar property could be also observed at other temperatures. Such observed property that the AC loss for B_m below B_p increases with increasing the interval agrees with the result obtained from the numerical calculation [11].

The AC losses for B_m below 1/2 of B_p at 50 K normalized by the following equation are shown in figure 3. In equation (2), the units of the AC loss and the field amplitude B_m are [J/m³cycle] and [T], respectively.

$$6.81\times10^{10} B_m^{3.03} \tag{2}$$

The power of equation (2) was determined from the average value of the power of the approximation curve of the AC loss curves below B_p for each D/w. The coefficient of equation (2) was determined in order for the average value of the normalized AC loss of 1-layer tape to be one.

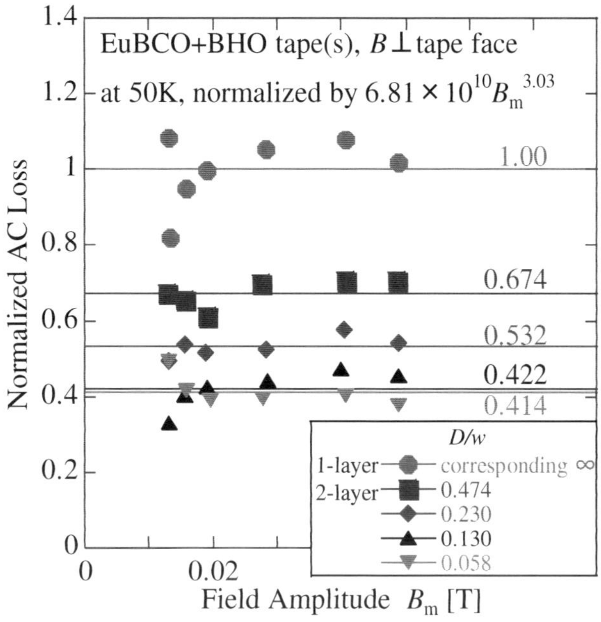

Figure 3. B_m dependences of normalized AC loss of 1-layer tape and 2-layer tapes with different interval at 50 K. The horizontal lines indicate the average values of the normalized AC losses for each *D/w*.

The horizontal lines in figure 3 and the values written above or below the lines correspond to the average values of the normalized AC losses for each *D/w*. In figure 3, the AC losses for B_m below 10^{-2} T are omitted because they include large errors compared with those for B_m above 10^{-2} T due to the small magnetization.

The average values of normalized AC losses obtained in the same way are plotted in figure 4. Note that the horizontal axis of figure 4 is *w/D*. For comparison, the theoretical prediction which is for the case that the number of the stacked tapes is infinity and expressed as following equation [6][12] is also shown as dashed line in figure 4.

$$W = \frac{1}{wt}\frac{\mu_0 I_C^2}{\pi}\frac{1}{6}\left(\frac{H_0}{I_C/2\pi w}\right)^4 \left\{\frac{\tanh(\pi w/2D)}{\pi w/2D}\right\}^2 \tag{3}$$

Here, t, I_C and H_0 are the thickness of the superconducting layer of the tapes, the critical current of the tapes and the field amplitude, respectively.

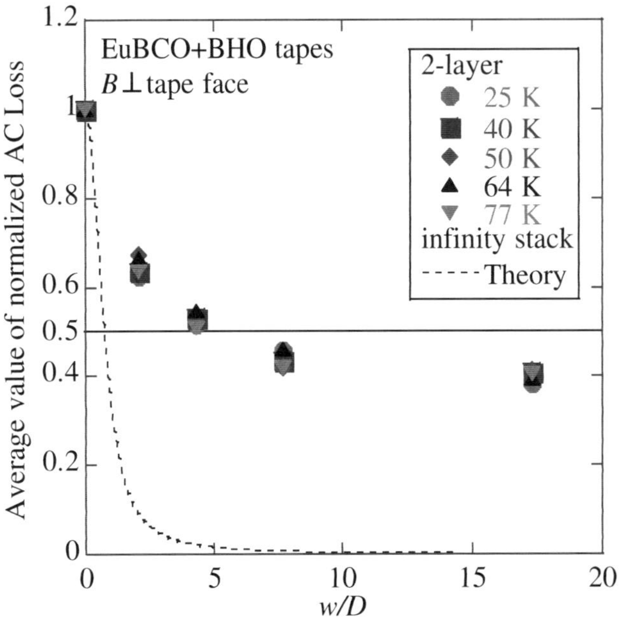

Figure 4. The interval dependence of the average values of the normalized AC losses of 2-layer tapes for $B_m < B_p$ at each temperature.

The observed interval dependence of the AC loss for B_m below B_p has no temperature dependence. As mentioned earlier, it has been already cleared that the AC loss of the n-layer tapes with $D/w < 0.1$ (i.e. $w/D > 10$) decreases to $1/n$ of that of 1-layer tape. Therefore, for $w/D > 10$, the theoretical prediction curve shown in figure 4 is almost zero. In the case of experimental results of 2-layer tapes, the normalized AC loss converges almost 1/2 for $w/D > 10$. The reason why the AC loss of 2-layer tapes with $D/w < 0.1$ decreases to 1/2 of that of a single tape is explained in Ref. [3] in detail. Brief explanation is as follows. In the case of stack with $D/w < 0.1$, the stacked two tapes can be regarded as twice the thickness of a single tape [6]. The magnetization of the 2-layer stacked tapes is 1/2 of that of a single tape from the viewpoint of the demagnetization coefficient. Therefore, the AC loss decreases to 1/2. The reason for the normalized AC loss slightly below 1/2 might be non-uniform superconducting-property distribution due to its local inhomogeneity. The experimental result and the theoretical prediction for infinity stack have the similar property, which the normalized AC loss for $w/D > 10$ approaches each convergence value and that for $w/D < 10$ increases sharply with decreasing w/D. It is predicted that the experimental result approaches the theoretical curve as increasing the number of the stacked tapes.

4. Conclusion

In this study, the dependence of the AC loss of 2-layer stacked EuBCO superconducting tapes on the interval between the neighboring superconducting layer was investigated experimentally. The AC loss for B_m above B_p had no interval dependence, whereas the AC loss for B_m below B_p increased with increasing the interval. Such observed property agreed with the numerical analysis. Furthermore, in comparison with the theoretical prediction for infinite stack, the observed results showed similar interval dependence of the AC loss. This study will be helpful to understand the AC loss property of the stacked multifilamentary REBCO tapes. Next step, the influence of the magnetic field by the filaments located side by side should be considered.

Acknowledgement
This research was partially supported by the Japan Science and Technology Agency (JST): Advanced Low Carbon Technology Research and Development Program (ALCA) and the Japan Society for the Promotion of Science (JSPS): Grant-in-Aid-for Scientific Research (JP18H03783 and JP17H06931).

References

[1] Iwakuma M, Nigo M, Inoue D, Miyamoto N, Funaki K, Iijima Y, Saitoh T, Yamada Y, Izumi T and Shiohara Y 2006 *Supercond. Sci. Technol.* **19** 350-8
[2] Sasa H, Ito T, Iwakuma M, Miura S, Izumi T, Machi T and Ibi A 2018 *J. Phys.: Conf. Ser.* **1054** 012037
[3] Sasa H, Kawasaki G, Miura S, Iwakuma M, Izumi T, Machi T and Ibi A *J. Appl. Phys.* (*Submitted*)
[4] Ito T, Iwakuma M, Miura S, Izumi T, Adachi K, Machi T and Ibi A 2018 *IEEE Trans. Appl. Supercond.* **28** 8200505
[5] Iwakuma M, Nigo M, Inoue D, Kiss T, Funaki K, Iijima Y, Saitoh T, Yamada Y and Shiohara Y 2005 *IEEE Trans. Appl. Supercond.* **15** 1562-5
[6] Mawatari Y 1996 *Phys. Rev.* B **54** 13215-21
[7] Katayama K, Machi T, Nakamura T, Takagi Y, Nakaoka K, Yoshizumi M, Izumi T and Shiohara Y 2014 *Physics Procedia* **58** 142-5
[8] Tobita H, Notoh K, Higashikawa K, Inoue M, Kiss T, Kato T, Hirayama T, Yoshizumi M, Izumi T and Shiohara Y 2012 *Supercond. Sci. Technol.* **25** 062002
[9] Iwakuma M, Nanri M, Fukui M, Fukuda Y, Kajikawa K and Funaki K 2003 *Supercond. Sci. Technol.* **16** 545-56
[10] Iwakuma M, Hayashi H, Okamoto H, Tomioka A, Konno M, Saito T, Iijima Y, Suzuki Y, Yoshida S, Yamada Y, Izumi T and Shiohara Y 2009 *Physica* C **469** 1726-32
[11] Pardo E, Sanchez A and Navau C 2003 *Phys. Rev.* B **67** 104517
[12] Mawatari Y 1997 *IEEE Trans. Appl. Supercond.* **7** 1216-9

IOP Conf. Series: Journal of Physics: Conf. Series **1293** (2019) 012038 doi:10.1088/1742-6596/1293/1/012038

Fabrication of a Compact High-field Magnet by Coated Conductor Stacks

T Hashimoto[1], S Pyon[1], Y Iijima[2], S Sugiura[3], S Uji[3], T Terashima[3], and T Tamegai[1]

[1] Department of Applied Physics, The University of Tokyo, 7-3-1 Hongo, Bunkyo-ku, Tokyo 113-8656, Japan

[2] Fujikura Ltd., 1-5-1 Kiba, Koto-ku, Tokyo 135-8512, Japan

[3] Reserch Center for Functional Materials Quantum Transport Properties Group, National Institute for Materials Science, 3-13 Sakura, Tsukuba, Ibaraki 305-0003, Japan

E-mail: th_shin236@yahoo.co.jp

Abstract. A bulk magnet was fabricated by stacking coated conductors with enhanced J_c by introducing nanorods. The fabricated CC stack consists of 100 pieces of CCs, and it was magnetized by field-cooling method in an applied field of 14.5 T down to 4.2 K. By reducing the magnetic field down to zero at a ramp of rate 1 T/min, we have succeeded in trapping 12.6 T at the center of the CC stack. We also confirmed that the CC stack can sustain the high trapped field for more than 20 min, which demonstrates the practical usefulness of such a staked CC magnet.

1. Introduction

Coated conductors (CCs) are widely used in applications such as carrying large current without energy dissipation and fabricating high-field magnets for various purposes. They can also be used as a trapped-field magnet by stacking short segments of CCs. There are a lot of studies of a trapped-field magnet using bulk superconductors by field-cooling mehod. Although there are problems of mechanical strength and thermomagnetic instability when magnetizing a magnetic field in a bulk superconductor [1, 2], it is possible to obtain a high and stable magnetic field by stacking CCs [3, 4]. There have been several studies on trapped-field magnet using bulk materials and stacks of CCs [5, 6, 7]. We have fabricated a CC stack with two stacks of 130 pieces of GdBCO CCs and trapped 7.92 T at the center of the stacks [8]. We have also fabricated a CC stack with two stacks of 80 pieces of irradiated GdBCO CCs and trapped 9 T at the center of the stacks, and also reported that flux jump occurred when trying to trap a large magnetic field quickly [9]. It is also reported that the record-high trapped magnetic field of 17.6 T has been achieved by using two kinds of CC stacks [10].

In the present study, we aimed at enhancing the trapped magnetic field by fabricating the bulk magnet using EuBCO CCs with better $J_c - H$ characteristics. For that purpose, $BaHfO_3$ nanorods are introduced into EuBCO CCs. Two stacks of each 50 pieces of EuBCO CCs were placed next to each other and miniature Hall probes for measuring the trapped field were placed at the center of the stacks. They were cooled down to 4.2 K in a magnetic field of 14.5 T, and the field was reduced to zero at 1 T/min. In this condition, we have succeeded in trapping 12.6 T at the center of the stacks.

2. Experimental Methods

In the present study, we used EuBCO CCs with $BaHfO_3$ nanorods by using hot-wall pulsed-laser deposition process. 2.5 μm thick EuBCO CC was deposited on 50 μm hasteroy tape, and ~3 μm Ag layer was deposited for protection. The bulk magnet consists of two piles of each 50 CC stack, and local magnetic field at the center of the two stacks was measured with a miniature GaAs Hall probes (HG-0711) made by Asahi Kasei Microdevices as shown in Fig. 1(a). The dimensions of the Hall probe used as a magnetic sensor was 1.2 × 0.5 mm^2 and the thickness was 0.3 mm. It has been confirmed that ion-implanted GaAs Hall probes can be operated even at temperatures much lower than the specified temperature range and linearity with respect to the magnetic field is good enough due to their low mobilities [11]. Two Hall probes were soldered on a thin printed circuit board to be arranged at an interval of 1.8 mm. Therefore, it was possible to measure the local field not only at the center but also at the position 1.8 mm from the center. Since the thickness of the CC is 60 μm, the total height of the bulk magnet is only 6.5 mm including the part of Hall probes. In addition, in order to investigate the stability of the temperature, the temperature change was also measured using a Cernox sensor (CX-1050, Lakeshore [12]). For trapping magnetic field, we used a 15 T superconducting magnet at national institute for materials science (NIMS). The CC was placed in variable-temperature insert and cooled via liquid helium at 4.2 K and 1 atm helium gas at other temperatures. First, a magnetic field of 14.5 T was applied to the CC stacks at 100 K above T_c(~93 K), and the CC stacks was cooled to the target temperature. Finally, the external field was ramped down to zero slowly, leaving a trapped field in the stack.

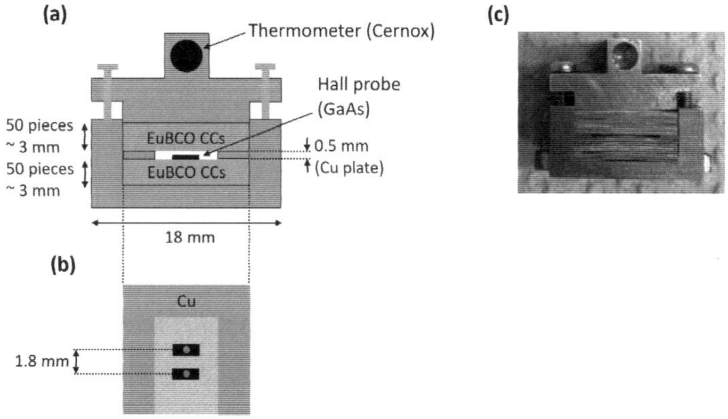

Figure 1. (a) Schematic vertical cross section of a bulk magnet consisting of double stack of GdBCO CCs and (b) arrangements of Hall probes between the two CC stacks. (c) A photo of the bulk magnet and its casing.

3. Results and Discussion

It is important to have good high-field performance to trap large magnetic field. Fig. 2(a) shows the external magnetic field dependence of J_c of EuBCO CCs at seraval temperatures. $J_c - H$ characteristics of GdBCO CC used in our previous study [9] is also shown in Fig. 2(b). It demonstrates that J_c of the EuBCO CC is improved especially at the high magnetic field compared with GdBCO CC.

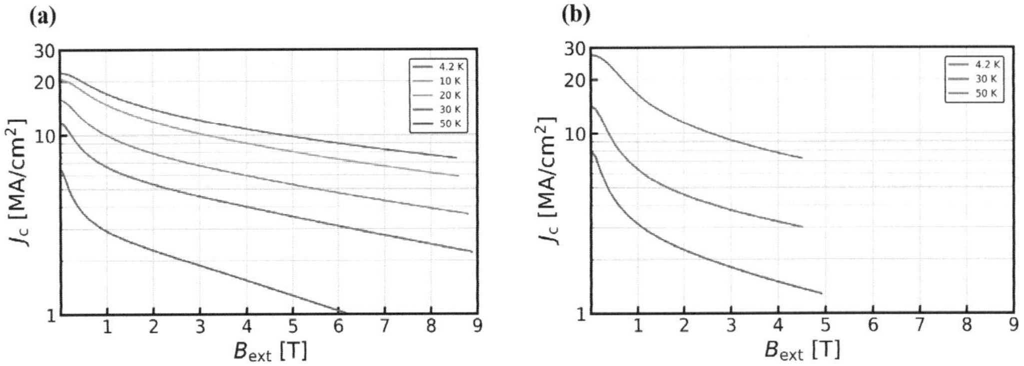

Figure 2. Magnetic field dependence of J_c at various temperatures of the (a) EuBCO and (b) GdBCO CCs.

Magnetization processes at the center and 1.8 mm away from the center as functions of applied field at 4.2 K are shown in Fig. 3. It is clear that we have succeeded in trapping 12.6 T in the center of CCs. This trapped field value is much larger than 9 T trapped by 80 stacks of CCs reported in our previous study [9]. Although the trapped field is lower than the latest report by Cambridge group [10], the height of our CC stacks is 1/6 of their stacks, and time to trap field is 37 times faster. It should be noted that the initial rapid temperature change above 12 T is due to the fact that the experiment was started from a slightly lower temperature. A part of temperature increase below 12 T ($\sim$0.035 K) is due to the magnetoresistance of Cernox sensor, and the rest of the increase ($\sim$0.03 K) is induced by the flux motion during the magnetization process.

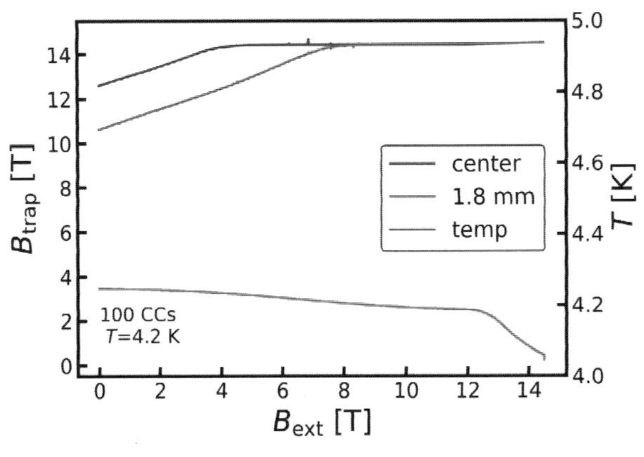

Figure 3. Local magnetic field in the magnetization process with the field-ramp rate of 1 T/min at two locations of the CC stack magnet at 4.2 K. Temperature of the stack is also plotted.

The field trapped in the center of the CC stacks at different temperatures are shown in Fig. 5(a). The ramp rates used for applied field were 1 T/min at 77 K, 30 K, 20 K and 4.2 K. This means that it took only 15 min to trap high magnetic field at 4.2 K. At 10 K, when the ramp rate was 1 T/min, flux jumps were observed and the trapped field rapidly decreased due to low heat capacity and insufficient cooling by helium gas as is shown in Fig. 4(a). Therefore, the ramp rate was reduced to 0.3 T/min at 10 K (Fig. 4(b)). It clearly demonstrates that as the temperature is decreased, the trapped field increases. However, the enhancement of the trapped field becomes weaker below 10 K. This behavior is similar to $J_c - T$ dependence shown in Fig. 5(b).

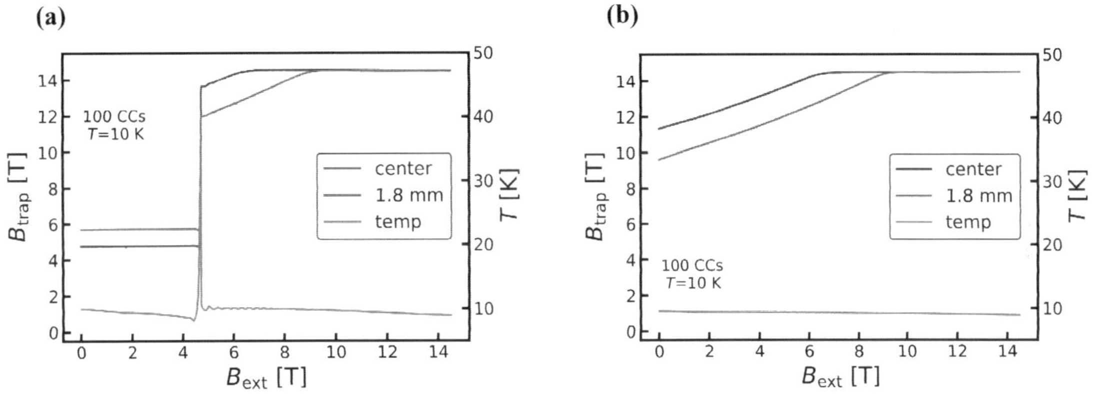

Figure 4. Local magnetic field in the magnetization process with field-ramp rates of (a) 1 T/min and (b) 0.3 T/min at two locations of the CC stack magnet at 10 K. Temperature of the stack is also plotted.

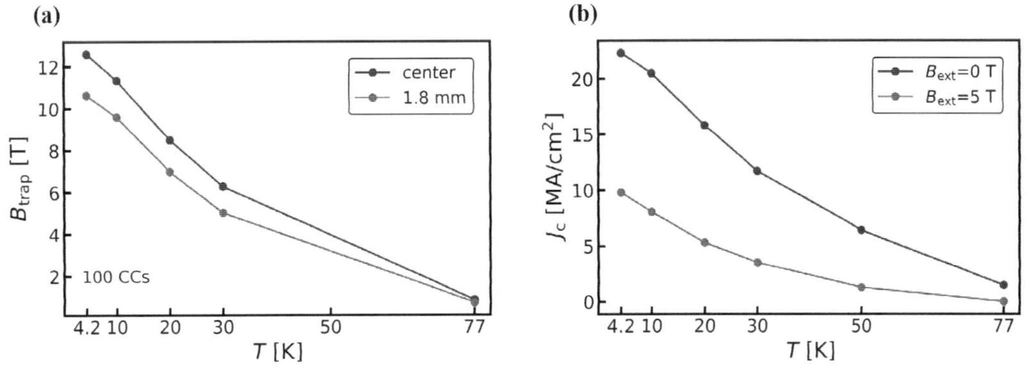

Figure 5. Temperature dependence of (a) the trapped magnetic field in the stack and (b) J_c of the EuBCO CC.

In the real application of the stack magnet of CCs, it is important to sustain a large magnetic field for a long time. In order to check the stability of the trapped field, flux creep was measured for the trapped field up to 20 min after field cooling as shown in Fig. 6.

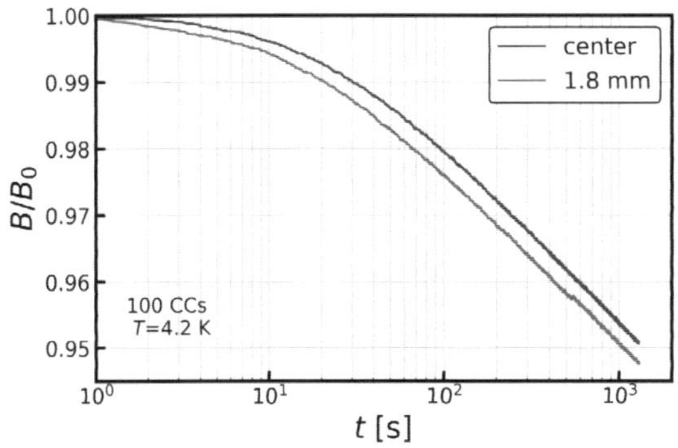

Figure 6. Flux creep measurements for trapped field at different positions in the first 20 min after magnetization. The field-ramp rate of the magnetization process is 1 T/min. B_0 is the trapped field 3 s after magnetization.

It can be seen that the normalized relaxation rate (dln M/dln t) is 0.012 even after 20 min. It is necessary for the application to examine how much the trapped field is maintained after magnetizing the CC stacks. When one attempts to use CC stacks as a high-field permanent magnet at higher temperature, a similar confirmation of long-term stability of the trapped field is important.

4. Summary
We have fabricated a bulk magnet by stacking 100 pieces of EuBCO coated conductors with $BaHfO_3$ nanorods. A large magnetic field of 12.6 T was successfully trapped in this CC stack after field-cooling and reducing the field at a rate of 1 T/min. The trapped field was very stable, and more than 95 % of the field was maintained at the center of the stack even 20 min after the magnetization.

Acknowledgements
We thank Prof. Z. X. Shi at Southeast University for helping the evaluation of J_c of the EuBCO coated conductor.

References
[1] Tomita M and Murakami M 2000 *Supercond. Sci. Technol.* **13** 722
[2] Tomita M and Murakami M 2003 *Nature* **106** 104
[3] Weijers H W, Markiewicz W D, Voran A J, Gundlach S R, Sheppard W R, Jarvis B, Johnson Z L, Noyes P D, Lu J, Kandel H, Bai H, Gavrilin A V, Viouchkov Y L, Larbalestier D C and Abraimov D V 2014 *IEEE Trans. Appl. Supercond.* **24** 4301805
[4] Durrell J, Dennis A, Jaroszynski J, Ainslie M, Palmer K, Shi Y, Campbell A, Hull J, Strasik M, Hellstrom E and Cardwell D 2014 *Supercond. Sci. Technol.* **27** 082001
[5] Hahn S, Beom Kim S, Cheol Ahn M, Voccio J, Bascun J and Iwasa Y 2010 *IEEE Trans. Appl. Supercond.* **20** 1037
[6] Patel A, Filar K, Nizhankovskii V I, Hopkins S C and Glowacki B A 2013 *Appl. Phys.*

Lett. **102** 102601
[7] Patel A, Hopkins S C and Growacki B A 2013 *Supercond. Sci. Technol.* **26** 032001
[8] Tamegai T, Hirai T, Sun Y and Pyon S 2016 *Physica C* **530** 20
[9] Hashimoto T, Pyon S and Tamegai T 2018 *J. Phys.: Conf. Ser.* **1054** 012050
[10] Patel A, Baskys A, Mitchell-Williams T, McCaul A, Coniglio W, Hnisch J, Lao M and Glowacki B A 2018 *Supercond. Sci. Technol.* **31** 102789
[11] Tamegai T, Iye Y, Oguro I and Kishio K 1993 *Physica C* **213** 33
[12] https://www.lakeshore.com/Pages/Home.aspx

IOP Conf. Series: Journal of Physics: Conf. Series **1293** (2019) 012039 doi:10.1088/1742-6596/1293/1/012039

A Study on different Narrowing Processes in the Manufacturing of 1mm HTS Tapes

L P Yao[1], F Gu[1], M Y Yin[2], Y Q Li[1], M Song[3], L Li[3], Z Y Li[1]

[1] Department of Electrical Engineering, Shanghai Jiao Tong University, No.800 Dongchuan Road, Minhang District, Shanghai, 200240, China
[2] Sino-Korean School of Multimedia Design, Shanghai University of Engineering Science, Shanghai,200240, China
[3]Guangdong Power Grid Corporation, Guangzhou, 510623, China

E-mail: lizhuyong@sjtu.edu.cn

Abstract—This paper provides a reference for manufacturing the narrow HTS tapes. The narrow tapes has many advantages such as smaller AC (Alternating Current) loss and suppressed screen current compared with normal tapes. Therefore, it's meaningful to manufacture the narrow tapes. Traditionally, the way to manufacture the 1 mm HTS tapes is to use the mechanic narrowing process In this paper, we proposed two narrowing process to manufacture the 1 mm HTS (high temperature superconducting) tapes, one is using the cutting machine, the other is using the laser. The critical current tests and microscope observation tests were carried out in order to make a comparison with these two methods. Results show that the laser narrowing process has a smaller critical current loss but has dross, and the mechanical narrowing process shows a larger critical current loss but clearer section stratification.

1. Introduction

Since the discovery of the second generation (2G) high temperature superconducting (HTS) tapes in the early 1990s, it has been rapidly developed and used in various fields, such as superconducting machines, superconducting transformers, magnetic levitation, etc[1]. With the development of the 2G HTS tapes [2], the narrowing and packaging process[3] of the HTS tapes has been widely concerned, because of the smaller AC loss and more suppressed screen current[4], and making good quality tapes is always a concern for future use. Moreover, the narrowing process has a great influence on the quality of the tapes, in this paper, we studied and tested two narrowing processes for comparison.

Traditionally, the commercial HTS tapes have width of 2 mm-12 mm. In this paper, a 1 mm-width HTS tape is manufactured and tested, and a novel narrowing process for such width of the HTS tape is proposed by the laser. Two narrowing processes are tested in this paper in order to compare and verify the feasibility of these two methods. One is cut mechanically by a cutting machine and the other one is by laser. In the experiments, a 4 mm 2G HTS tape is incised into four 1 mm wide HTS tapes. The used

IOP Conf. Series: Journal of Physics: Conf. Series **1293** (2019) 012039 doi:10.1088/1742-6596/1293/1/012039

HTS tapes are the second generation HTS wires, i.e., REBCO tapes, manufactured by SuNAM Company, the specific parameters of the tapes is listed in Table I. Different sections of the 1 mm HTS tapes which is cut by different method is observed and compared using the microscope. Moreover, current-voltage tests are carried on these 1 mm HTS tapes and the critical currents are studied and compared.

Table I. Specific parameters of REBCO tapes.

Parameters	Value
Manufacturer	SuNAM
Type	REBCO
Insulation	No
Width and Thickness	4.1 mm × 0.142mm
Material of the metal substrate	Non-magnetic stainless steel
Thickness of superconducting layer	~1.3 μm
Thickness of silver layers	~1 μm
Thickness of substrate layer (including buffer)	~108 μm
Thickness of copper plating layer, each side	~20 μm

2. Narrowing process

Fig. 1 shows the two different narrowing equipments, Fig.1(a) is the mechanical narrowing machine and Fig.1(b) is the laser one. We used the rolling shearing machine to narrow the HTS tapes, and two cutting tools are used in this rolling shearing machine for manufacturing the 1mm HTS tapes. The tool bits used in the machine narrowing process has 4×1mm knife edge. Mechanical narrowing process is using the cutting machine (XTF 80 made by our laboratory) to incise the 4 mm wide HTS tape into narrower HTS tapes. The mechanical cutting machine we used is the commercial one frequently used to fabricate flat copper wires for conventional transformers [5]. After aligning the equipment and the tapes, select an appropriate cutter, keep prosper cutting strength and cutting speed, then incise the 4 mm wide HTS tapes into 1 mm wide tapes. In the mechanical narrowing process, the cutting speed is 3m/s, and the human factors such as the alignment of the blade position and artificial tensile may have an impact on the HTS tapes.

(a) (b)

Fig. 1. The narrowing equipment. (a) mechanical narrowing machine; (b) laser narrowing machine;

Laser narrowing machine is using the laser beam to heat the HTS tapes. In this study, the machine we used is in this paper is a UV (ultraviolet) laser marking machine modeled SWLASER-F3W made by Shanghai SW Laser Technology CO., LTD. In the laser narrowing process, the laser narrowing machine we used has the following conditions: the laser oscillation method we used is a solid-state laser and the Q-switch frequency range of the laser is 20-200kHz. Moreover, the laser UV-wavelength is 355 nm, and no more assist gas is used in the process. The maximum processing speed of the laser narrowing machine is 100 m/h. Furthermore, the laser narrowing machine has a closed loop control with tension and speed, and the processing jigs is absent but limit screws are present.

Adjust the position of the laser lens in order to make the reflect laser focuses on the tapes first during the laser narrowing process. Then, fix the position of the strip and adjust the parameters and marking position of the laser marking. Finally, open the air extracting device to absorb the air between the cutting platform and the HTS tape, and run the device. In this study, the cutting speed is set at 50 m/s for the accordance with that of the mechanical narrowing machine, and the frequency is 40 kHz with the pulse width is 25 μs.

In order to test the different narrowing methods on manufacturing 1mm-width HTS tapes, the critical current tests and the microscope observation tests of the HTS tapes are performed. In this paper, the brief analysis and comparison of these two narrowing methods for manufacturing the 1mm HTS tapes is discussed based on these two tests (i.e. critical current test and microscope observation tests).

3. Results and discussions

3.1 The critical current tests

The critical current tests are performed on the 4mm-width HTS tapes firstly, then blow-dry these tapes quickly. After that, using these two narrowing processes to incise the 4mm HTS tapes into 1mm wide. Then perform the same critical current tests on these 1mm HTS tapes. we used transport measurement method, i.e., four-probe method in the critical current tests. The HTS tapes we used are 1m long, and the voltage terminals of the HTS tape is approximately 50 cm long. According to the 1 μV/cm threshold, the critical voltage is 50 μV. The ramping rates of the applied current are the same, i.e., 1 A/s. All of the critical current tests were performed in a liquid nitrogen (LN2) bath at 77 K. The results are showed in table II.

Table II. Critical currents of the test tapes.

		Ic of the 1mm HTS tapes				Sum of Ic of 4 1mm HTS tapes	Ic of the 4.1mm HTS tapes	Ic loss percent
		1	2	3	4			
Mechanical narrowing	1	40	40	44	30	154	202	23.8%
	2	40	40	35	36	151	200	24.5%
	3	40	46	37	47	170	204	16.7%
Laser narrowing	1	28	53	64	62	207	212	2.4%
	2	40	40	50	61	191	204	6.4%
	3	51	40	44	60	195	204	4.4%

As it showed, the critical current of the 4mm HTS tapes are similar which is around 200A. The sum of the critical current of the four 1mm HTS tapes in the same narrowing process is also similar, which is 155 A for mechanical narrowing process and 195 A for laser narrowing process. The Ic(critical current) loss in the mechanical narrowing process is around 20%, while 5% of that in the laser narrowing process. From the results, the critical current is almost no loss during the laser narrowing process.

3.2 The microscope observation tests

In order to have a deep study on the critical current influence of the narrowing process on the HTS tapes, some microscope observation tests are performed on these cutting sections of the 1mm tapes along the narrowing direction. Fig. 2 and fig. 3 show the photos of the sections of the mechanical narrowing process and laser narrowing process, respectively. The figure (a) and (b) shows the sections of the tapes under an optical microscope and an electron microscope, respectively. And figure (c), (d) and (e) is the elemental analysis using the EDS (Energy Disperse Spectroscopy) which is component analysis in the red box in figure (b). As the figure shows, (c), (d) and (e) represent the distribution of copper, silver and REBCO on the cutting surface of the tapes, respectively.

As the fig. 2 (a) and fig. 3 (a) show, the layer construction of the superconductors of the mechanical narrowing process is much clearer than in laser cutting, and the surface is much more unsmooth. As shown in the figures, the cutting sections of the tapes are damaged due to the slitting methods. Referring to the papers [6]-[7], the distributions of critical current densities in narrowing tapes are uneven, this contributes to the I_c suppresses due to mechanical and laser narrowing.

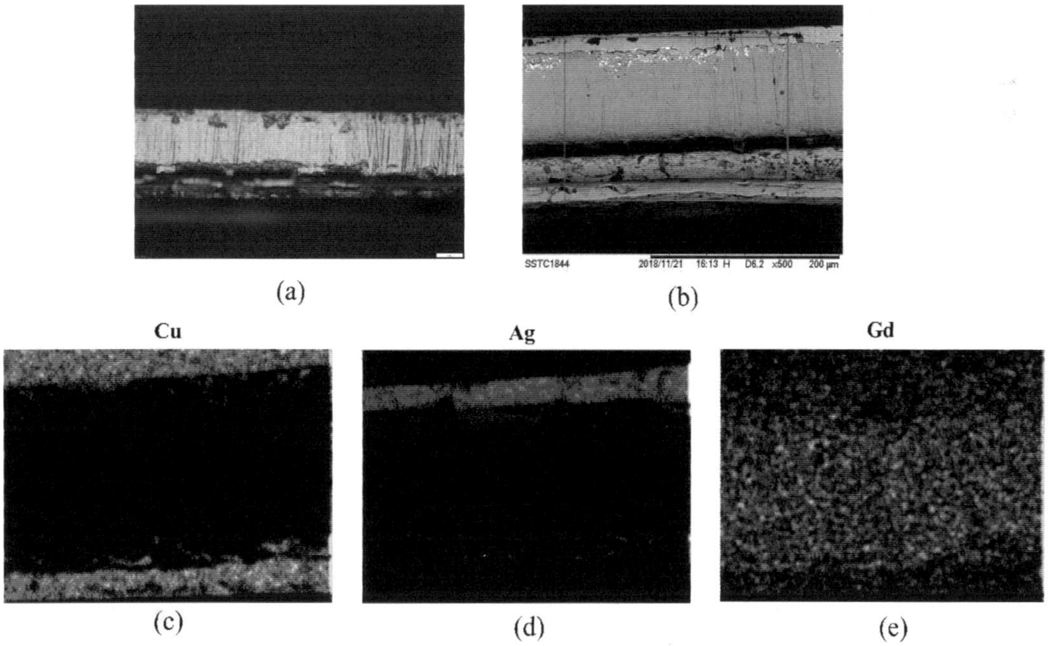

Fig. 2. The microscope sections of the mechanical narrowing process. (a) optical microscope; (b) electron microscope; (c) Cu in EDS (Energy Disperse Spectroscopy); (d) Ag in EDS; (e) GdBCO in EDS.

As the elemental analysis diagrams in (b), (c) and (e) show, the diffusion range of the copper and silver layer of the mechanical narrowing process is less related as they permeate to other layers in the laser narrowing process. Since the melting point of the silver in the HTS tapes is the lowest and it has a high fluidity after liquefaction, the distribution of silver is wide in the laser narrowing process. Moreover, the silver filaments can be seen figure 3 (a).

In addition, the GdBCO disperses to all parts of the sections both in these two narrowing processes. Therefore, compared with the mechanical narrowing process, using the laser to cut may "burn" the surface of the tapes, and the material in all layers will mix together. The main reason for the above results is that the formation mechanism is different. Mechanical cutting is due to the pressure, while the laser cutting is due to excessive energy.

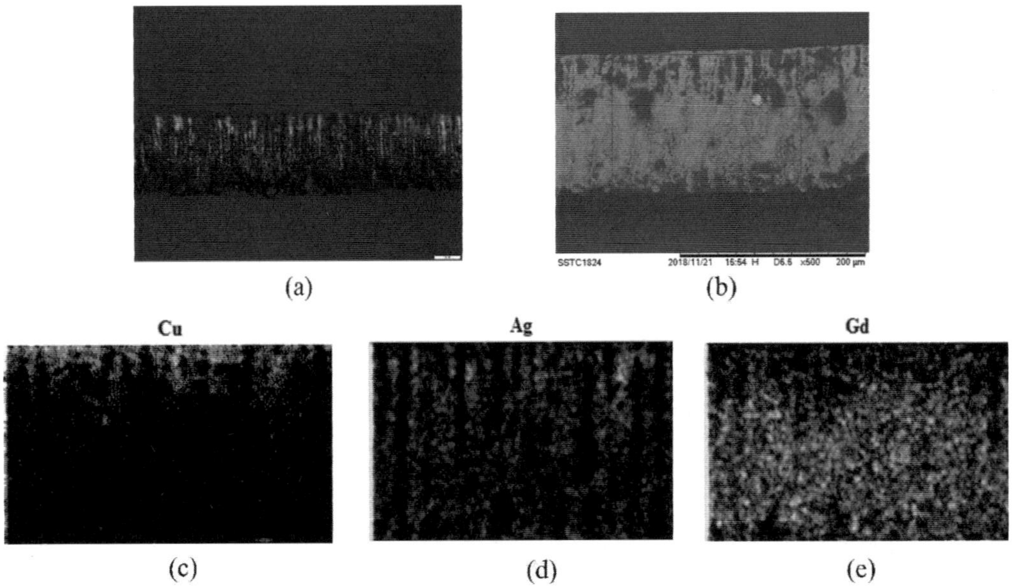

Fig. 3. The microscope sections of the laser narrowing process. (a) optical microscope; (b) electron microscope; (c) Cu in EDS; (d) Ag in EDS; (e) GdBCO in EDS.

Fig. 4. The microscope sections of the mechanical narrowing process. (a) No wrapped; (b) Half rapped; (c) All wrapped.

Moreover, in the mechanical narrowing process, the quality of the narrowing process has a great influence on the cutting sections. According to the proportion of the area covered by the copper layer to the total section, the cutting sections in the mechanical narrowing process is divided into three categories: no wrapped, half wrapped and all wrapped, which are showed in fig. 4. The copper layer wrapped part of the other layers downward along the cutting direction. The coverage area of the copper layer is uncertain which is related to many human factors such as the cutting speed, cutting strength and etc. In the mechanical narrowing process, the interface defects are easy to occur on the cutting sections, i.e., cracks. In contrast, the laser narrowing process is more stable for the cutting sections are similar.

4. Conclusion

In this paper, two narrowing methods used to manufacture the 1 mm-width HTS tapes were experimentally investigated and compared. In the experiment, the 1 mm HTS tapes were manufactured by the mechanical narrowing process and laser narrowing process, respectively. And the critical current tests and microscope observations are performed on these tapes. The microscope tests including using an optical microscope and an electron microscope, and an elemental analysis using the EDS is also performed. The following are the conclusions were derived:

1) In the critical current tests, the 1mm tapes manufactured by the mechanical narrowing process have larger critical current loss than that in the laser narrowing process, which shows the mechanical narrowing process caused greater loss.
2) In the microscope tests, the stratification of the HTS tapes is much clear in the mechanical narrowing process, while it seems the laser cutting is more scattered.
3) The mechanical narrowing process has many advantages, it will not affect by the thickness of the tapes, the process is simple, and the section looks better. But it's relative to the human factors so it's not very stable, and the Ic loss is a little large.
4) Laser narrowing process can be used in narrowing long HTS tapes, and it has high precision and low critical current loss. However, it has dross and high narrowing costs.

The results demonstrate that the mechanical narrowing process and laser narrowing process cut both ways and are both feasible for manufacturing the 1 mm tapes. This paper could provide a possible reference for manufacturing the narrow HTS tapes. Process improvements for manufacturing narrow tapes are included in subsequent studies.

Acknowledgement
This work was supported by the National Key Research and Development Program 2017YFB0902301.

References

[1] Young Jin Hwang, Jae Young Jang, Seunghyun Song, and SangGap Lee 2016 *IEEE Trans. Appl. Supercond.* **27(4)** 4603405
[2] Drew W. Hazelton, Venkat Selvamanickam, Jason M. Duval, David C. Larbalestier, William Denis Markiewicz, Hubertus W. Weijers, and Ronald L. Holtz 2009 *IEEE Trans. Appl. Supercond.* **19(3)** 2218-2222
[3] Y. Wang, S. Baasansuren, C. Xue, and T. Hasegawa 2016 *IEEE Trans. Appl. Supercond.* **26(4)**

4804406
[4] Z. Y. Li, J. Li, Y. Wang, Z. Yao, Z. R. Kang, B. Yuan, Z. D. Yang, Z. Jin, and Z. Hong 2016 *IEEE Trans. Appl. Supercond.* **26(4)** 8201104
[5] Z. Y. Li , Y. Q. Li , M. Y. Wang, D. M. Xi, J. W. Zhang , Y. H. Ma, Z. Hong, Z. Jin, and K. Ryu 2018 *IEEE Trans. Appl. Supercond.* **28(3)** 4800405
[6] M.Yagia, S.Mukoyamaa, H.Hiranoa, N.Amemiyab, A.Ishiyamac, S.Nagayad, N.Kashimad, Y.Shioharae 2007 Physica C **463-465** 1154-1158
[7] K. Higashikawa et al. 2013 *IEEE Trans. Appl. Supercond.* **23(3)** 6602704

Study of the superconducting layer microstructure and $(Nb,Ti,Ta)_3Sn$ bronze strands properties

I M Abdyukhanov, V I Pantsyrny, A G Silaev, A S Tsapleva, N V Konovalova, M V Alekseev, E A Dergunova, K A Mareev, M N Nasibulin, V A Drobyshev, M V Kravtsova, P A Lykianov and M V Krylova

SC A A Bochvar High-Technology Research Institute of Inorganic Materials, Moscow, 5A Rogova Street, Russia

mavalekseev@bochvar.ru

Abstract. Superconducting materials with high current carrying capacity operating in magnetic fields up to 18 T at 4.2 K are required for promising applications in magnetic systems of high-energy physics devices, NMR-spectroscopes and others. The paper presents the study results of the Nb_3Sn strands manufactured by the bronze route differed in the filament composition. The filaments were made from the alloys Nb-3.5 wt. %Ta and Nb-7.0wt.%Ta and the matrix was produced from the Cu-15.5wt.%Sn-0.25wt.%Ti alloy. To form the superconducting Nb_3Sn phase a series of reaction heat treatments of conductor samples were carried out by two-stage regimes with the temperature of the second stage varying from 650 to 720 °C. The microstructure of the Nb_3Sn layer was studied by scanning electron microscopy and their electro-physical characteristics were also measured. It was shown that after reaction at 575 °C, 100 h + 650 °C, 200 h samples with different Ta content had almost the same current-carrying capacity in the field range 12-18 T at 4.2 K. The critical temperature of superconductors decreased from 17.5 to 16.5 K with an increase of the second stage of reaction temperature from 650 to 720 °C.

1. Introduction

Superconducting materials with high current carrying capacity in magnetic fields up to 18 T at 4.2 K are required to create magnetic systems of high-energy physics devices, such as accelerators of elementary particles or NMR spectroscopes. Nb_3Sn superconductors manufactured by the bronze route are widely used to operate in magnetic fields up to 12 T. Ti or Ta doping are widely used to increase current-carrying capacity of Nb_3Sn superconductors in higher fields, wherein the alloying element may be introduced into Cu-Sn matrix and into starting filament material.

It is shown in paper [1], that a small amount of alloying additives of transition elements significantly increases the value of the upper critical field (B_{C2}) of Nb_3Sn superconductors. Thus, the maximum value of B_{C2} achieved as a result of conductor doping was 27 T at 4.2 K (with the addition of ~1.5at.%Ti or ~ 4at.%Ta) compared to 23,5 T for undoped one. Other studies [2-5] were aimed at determination an optimal Ti content in the conductor also doped of 7.5 wt. %Ta. The authors of [2] showed that superconductors with adding of 7.5wt.%Ta in Nb filaments and 0.25wt.%Ti in bronze matrix had a maximum critical current density (Jc) and B_{C2}.

It was noted that the value of B_{C2} and the current carrying capacity of superconductors decreased with an increase of the Ti content up to 1wt.% in filaments or at simultaneous use of 0.75wt.%Ti both in filaments and in matrix. The critical temperature (Tc) of the samples was also reduced by increasing the Ti content in the superconductor filaments up to 2wt.% at 2.2 K, the Ta content was 7.5wt.%. Joint doping of 0.25wt.%Ti and 7.5wt.%Ta also reduced the Tc of 0.7 K compared to the only Ta doped superconductor. It was shown in [2] that the optimal regime of reaction heat treatment of conductors with joint doping of 0.25wt.%Ti and 7.5wt.%Ta was 600 °C, 100 h+670 °C, 150 h. In work [4] it was shown that at joint alloying of 1wt.%Ti+7.5wt.%Ta the value of B_{C2} was lower than at doping of 0.25wt.%Ti+7.5wt.%Ta. Jc of such conductors reached a maximum during reaction at 670 °C, 100 h. The value of Tc also reached a maximum in this case. B_{C2} reached 27.5 T in wire doped with 7.5wt.%Ta by reducing Ti content up to 0.18wt.% [5]. It was also found that the distribution of alloying elements varies in the width of the layer. Minimal Ti content and the maximum content of Ta were in the center of it, but at the periphery the situation was reversed: the Ta content decreased and the mass fraction of Ti increased. The authors of this work also pointed to the decrease of T_c with increase of mass fraction of alloying elements and they associated it with the presence of inhomogeneity in the superconducting layer and the formation of zones with too high or insufficient content of alloying elements.

Thus, the analysis of previous works showed that the optimal content of Ti in the bronze matrix was 0.25 wt. %. However, data on the effect of Ta content in filaments on the electrophysical properties of the wire was not enough.

The aim of this work was to study the effect of different Ta content in filaments on the Nb_3Sn layer microstructure in wire doped of Ti and Ta simultaneously, and the dependence of the microstructure on the reaction heat treatment regime, as well as to determine the electrophysical properties of these superconductors.

2. Experimental

2.1. Strands manufacture and reaction heat treatment

Nb_3Sn superconductor samples of 1 mm in diameter with 68406 Nb filaments differed by the Ta content (T3 and T7) were manufactured by the bronze route. Figure 1 shows the structure of sample T3 cross section with Nb-Ta filaments uniformly distributed in the bronze matrix doped with 0.25wt.%Ti. Copper stabilization is separated from the superconducting core by Nb diffusion barrier with Ta inserts.

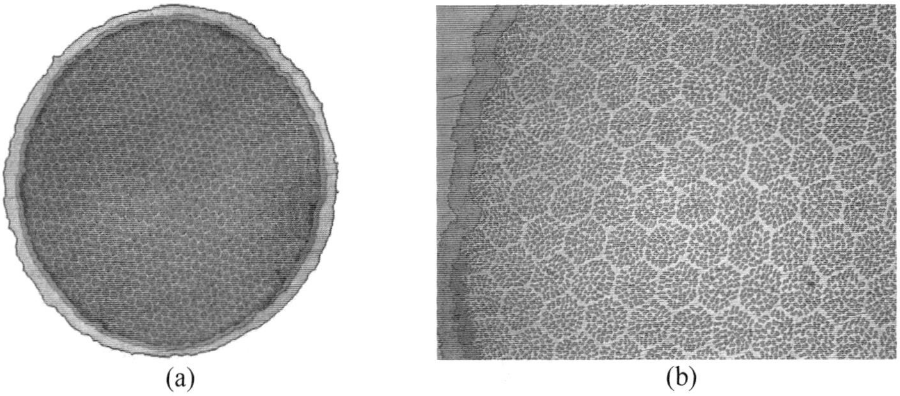

(a) (b)

Figure 1. Optical micrograph of the wire cross section at ø 1mm (a) and fragment (b).

Table 1 shows the characteristics of the two experimental strands investigated.

The reaction heat treatment for the superconducting phase formation was performed under vacuum (<10^{-5} Torr) by a series of two-stage treatments: 575°C/100h+(650;680;700;720)°C/100h and 575°C/100 h+650 °C/200h.

Table 1. Nb_3Sn strands 1 mm in dia parameters.

Type of sample	T3	T7
Bronze composition, wt. %	Cu-15.5Sn-0.25Ti	Cu-15.5Sn-0.25Ti
Filament composition, wt. %	Nb-3.5Ta	Nb-7.0Ta
Number of doubled filaments in group	39	39
Number of filaments	68406	68406
Volume fraction of Cu in strand, %	12	12
Volume fraction of Nb in superconducting core (calculated), %	32	32

2.2. Investigation of the superconducting phase microstructure and critical characteristics measurements

The fractured cross sections of reacted samples T3 and T7 were observed in scanning electron microscope (SEM) for microstructure investigation of the superconducting phase. The quantitative analysis of the grain structure morphology and average grain size was carried out by the method of random secants from SEM micrographs.

For superconducting phase composition study the micro-x-ray spectral analysis was carried out on polished conductor cross sections.

The critical current (Ic) of the conductor samples was determined on the base of transport measurements (by the criterion of 0.1 μV/cm) at 4.2 K in a magnetic field of 12-18 T in SC"VNIINM" and in the High Magnetic Field Laboratory (Grenoble, France). Measurements were performed on the ITER-VAMAS type holder 32 mm in diameter.

Critical temperature (T_C) was determined by the volt - temperature characteristics (VTC) of the samples in accordance with the recommendations of the standard IEC-61788-10 using device SR-SC-05 of Sniper-Researches production. VTC were measured on samples of 60-70 mm in length which were soldered to the current and potential contacts on the substrate of the probe equipped with a temperature sensor Cernox CX-1030-SD-4B. The total temperature measurement error (taking into account the sensor error and temperature fluctuations) for the range of 20-40 K was ~30 mK.

3. Results and discussion

The microstructure of the superconducting layer of conductors T3 and T7 samples after reaction heat treatment by different regimes (Table 2) has been studied in this work. As one can see in figure 2, the Nb_3Sn layer structure of studied samples is characterized by the presence of the equiaxed and columnar grain zones and the residual Nb zone.

Table 2. Results of quantitative analysis of Nb_3Sn grain structure morphology and average grain size of conductor after different reaction heat treatment regimes.

Regime of reaction heat treatment	T3			T7		
	Fraction of equiaxed grains, %	Fraction of columnar grains, %	Average grain size, nm	Fraction of equiaxed grains, %	Fraction of columnar grains, %	Average grain size, nm
575°C-100h+650°C-200h	44.4	49.5	84±10	54.5	40.6	84±10
575°C-100h+650°C-100h	55.0	36.5	71±8	55.6	39.1	75±9
575°C-100h+680°C-100h	59.0	35.9	96±3	57.3	36.6	77±4
575°C-100h+700°C-100h	65.4	33.9	107±7	75.6	22.9	93±11
575°C-100h+720°C-100h	72.7	26.2	111±14	80.1	19.4	107±13

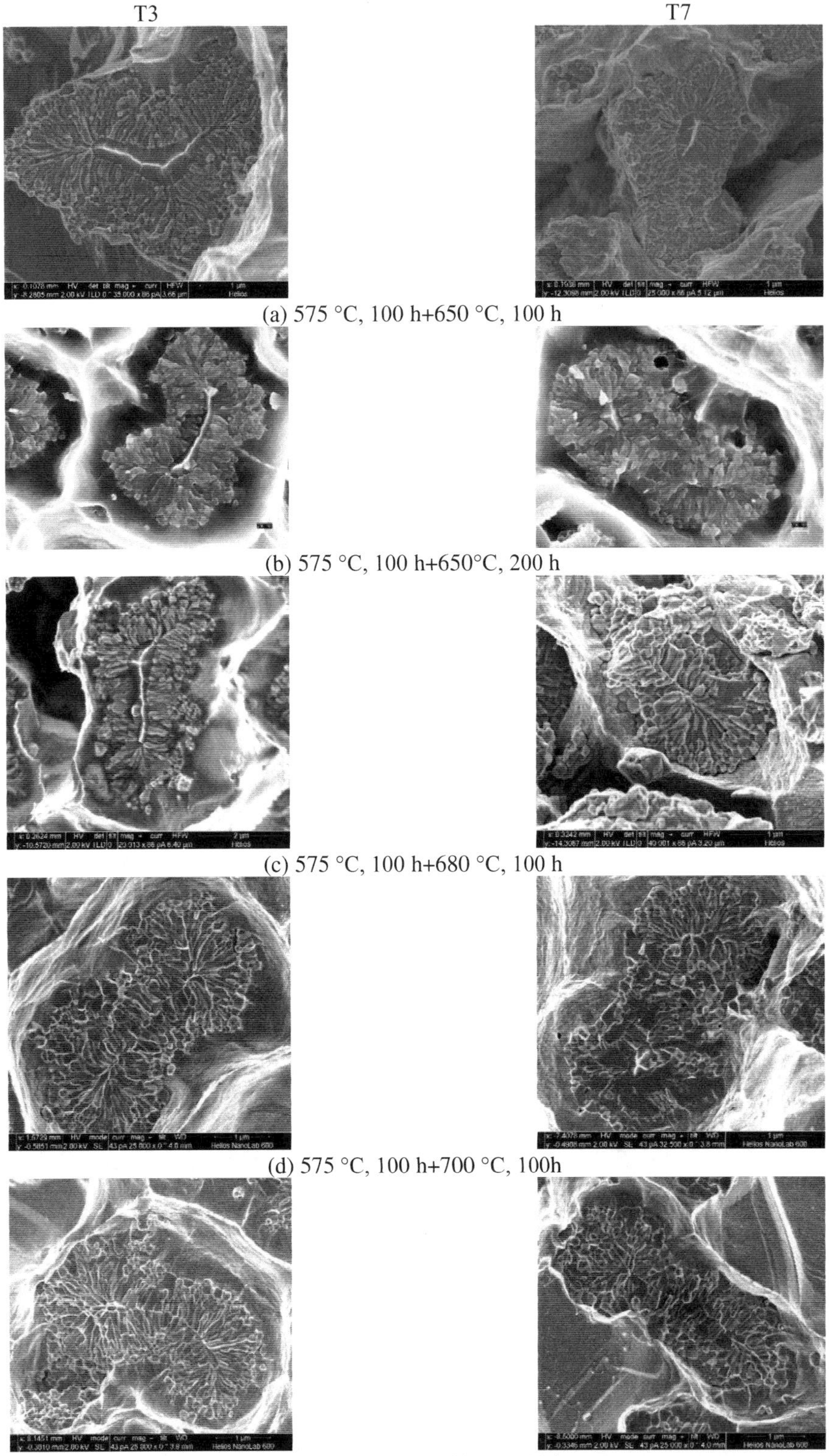

Figure 2. Microstructure of the Nb_3Sn superconducting phase of strands samples with different Ta content in filaments (T3 and T7) after reaction by different regimes.

It was established that the residual fraction of Nb in both samples decreased after increase of the second stage temperature from 650 to 720 °C, but the fraction of equiaxed grains increased (table 2). At the same time there is a noticeable decrease of the fraction of columnar grains: in the T7 sample from 39.1 to 19.4 % and in the T3 sample from 36.5 to 26.2 %.

Herewith in the T3 strand the columnar grains fraction is higher or at the same level in comparison with T7 after reaction by all regimes (table 2). The increase of the superconducting phase average grain size of both samples was observed at the increase in the temperature of the second stage treatment. The maximum difference in grain size was found after reaction in temperature range 680-700°C. Despite the different content of Ta in the filaments a significant difference had not been found in the fractions of columnar and equiaxed grains in the studied samples.

The results of X-ray analysis of the superconducting layer composition after different heat treatment regimes are presented in table 3.

Table 3. Chemical composition of the Nb_3Sn superconducting layer of the studied samples after reaction heat treatment by different regimes.

Regime of reaction heat treatment	T3	T7
	Sn content in superconducting layer, at.%	
575 °C, 100 h + 650 °C, 100 h	20.2±2.8	20.3±2.1
575 °C, 100 h + 680 °C, 100 h	20.2±2.6	20.9±2.3
575 °C, 100 h + 700 °C, 100 h	22.5±1.1	21.9±2.0
575 °C, 100 h + 720 °C, 100 h	22.1±2.8	21.2±2.2

Despite the different Ta content in the filaments of T3 and T7 samples no difference in the amount of Sn in the superconducting layer was found. It was noted that the Sn content in the superconducting phase of both samples was slightly increased from 20.2-20.3 up to 21.2-22.1 at.% after increase of reaction temperature up to 720 °C.

The dependence of Jc on the magnetic field of T3 and T7 samples is shown in figure 3. As one can see, both samples of 1.0 mm in diameter had almost the same value of Jc in the field range from 12 to 18 T at 4.2 K after heat treatment at 575°C, 100 h + 650 °C, 200 h. Analysis of the results showed that the increase in Ta content from 3.5 to 7.0 wt. % had no noticeable effect on the current carrying capacity of these conductors.

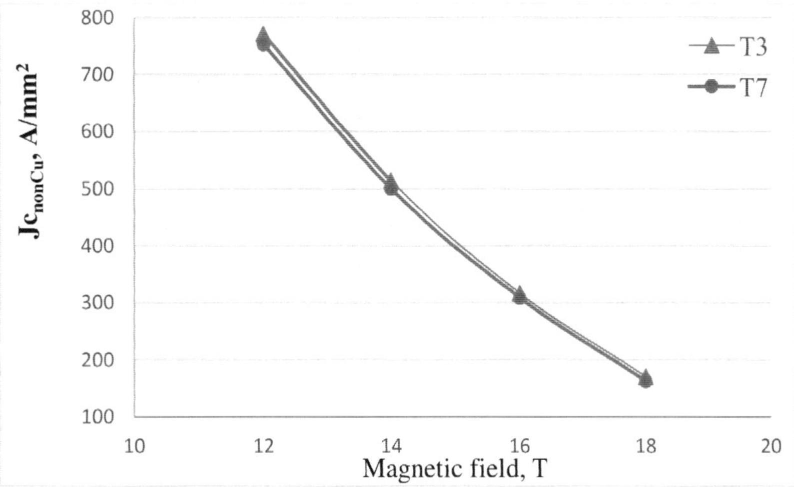

Figure 3. Jc_{nonCu} versus magnetic field (at 4.2 K) for the T3 and T7 strands 1 mm in diameter.

The Tc dependence on the temperature of the second stage of reaction heat treatment of the studied samples is shown in figure 4.

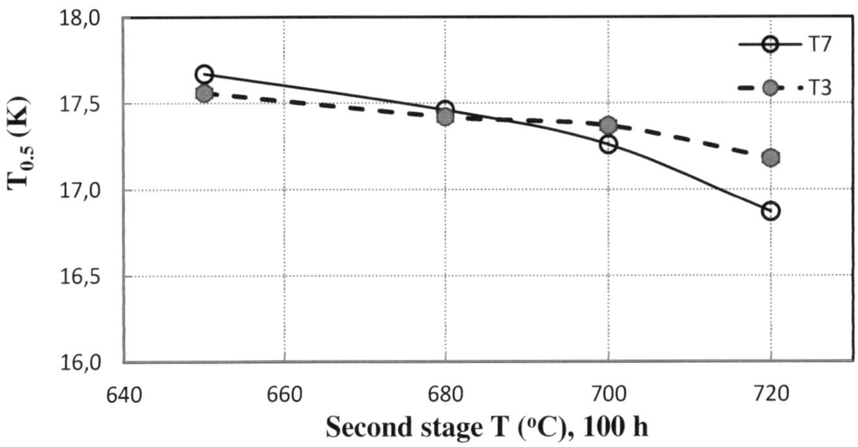

Figure 4. Tc dependence on the reaction heat treatment second stage temperature of the T3 and T7 samples.

Analysis of the results showed that the increase of the heat treatment second stage temperature from 650 to 720°C led to the decrease of the strands T3 and T7 Tc value by 0.38 and 0.8 K (from 17.6 and 17.7 K to 17.2 and 16.9 K), respectively. This may be due to the heterogeneity of the alloying elements distribution in the superconducting phase. It is known [1] that a small amount of Ti additive (~1 at.%) or Ta (~3 at.%) provides a slight increase of the Tc value compared to pure Nb_3Sn. However, the increase of alloying elements amount can significantly reduce the value of Tc. The maximum values of Tc 17.6-17.7 K were achieved on the samples after the reaction heat treatment at 575 °C, 100 h + 650 °C,100 h.

4. Conclusion

The microstructure and electrophysical characteristics of the bronze route Nb_3Sn strands with filaments containing 3.5 or 7.0 wt.% Ta in the matrix doped by 0.25wt.%Ti have been studied. Analysis of the superconducting layer microstructure in strands with filaments differed by Ta content showed that the increase of the second stage temperature from 650 to 720 °C led to columnar grains and residual niobium fraction decrease and equiaxed grains fracture increase. The average grain size increased from 71 to 111 nm in the strands of both types.

It was found that the value of Tc of the investigated strands decreased with the increase of the heat treatment second stage temperature. At the same time Tc value is reduced for the strands with 3.5 and 7.0 wt.% Ta content by 0.38 and 0.8 K, respectively, that could be connected with the greater heterogeneity of the superconducting layer composition of the conductor with a higher content of Ta.

It was shown that the current carrying capacity of these superconductors in the range of magnetic fields 12 - 18 T practically did not depend on the amount of Ta in the filaments. It was found that the increase in Ta content from 3.5 to 7 wt.% did not lead to a significant change in the microstructure of the superconducting layer and an increase of Jc in the magnetic fields of more than 12 T, which indicated the possibility of reducing content of Ta down to 3.5 wt.%.

Acknowledgments

The authors are grateful to the High Magnetic Field Laboratory (Grenoble, France) for the providing of cryogenic magnetic facilities and to Dr. Xavier Chaud for all assistance in the measurements. Much gratitude is expressed to the National Research Center "Kurchatov Institute" (Moscow, Russia) for the providing of the technical opportunity to conduct electron microscopic studies.

References

[1] Suenaga M, Welch D O, Sabatini R L, Kammerer O F and Okuda S 1986 Superconducting critical temperatures, critical magnetic fields, lattice parameters, and chemical compositions of pure and alloyed Nb_3Sn produced by the bronze process *J. Appl. Phys.* **59** 840–53

[2] Abächerli V, Uglietti D, Lezza P, Seeber B, Flükiger R, Cantoni M and Buffat P A 2005 The influence of Ti doping methods on the high field performance of $(Nb; Ta; Ti)_3Sn$ multifilamentary wires using Osprey bronze *IEEE Trans. Appl. Supercond.* **15** 3482–85

[3] Uglietti D, Abächerli V, Buta F, Seeber B and Flükiger R 2006 The effect of Ti addition on critical parameters and transport properties of quaternary $(Nb; Ta; Ti)_3Sn$ wires *Supercond. Sci. Technol.* **19** 1185–90

[4] Abächerli V, Buta F, Uglietti D, Senatore C, Seeber B and Flükiger R 2007 Investigation on the Effect of Ta Additions on Jc and n of $(Nb; Ti)_3Sn$ Bronze Processed Multifilamentary Wires at High Magnetic Fields *IEEE Trans. Appl. Supercond.* **17** 2564–67

[5] Senatore C, Abächerli V, Cantoniand M and Flükiger R 2007 Distribution of Tc from calorimetry and the determination of Sn gradients inbronze route Nb_3Sn wires with an internal and external Ti source *Supercond. Sci. Technol.* **20(9)** 217–22

IOP Conf. Series: Journal of Physics: Conf. Series **1293** (2019) 012041 doi:10.1088/1742-6596/1293/1/012041

Development of monitor system for parallel-type superconducting level sensor for liquid hydrogen

Naoki Tanaka[1], Kazuhiro Kajikawa[1], Hidetoshi Oguro[2], Makoto Sugino[3], Tsutomu Nakanishi[3] and Itsuo Aoki[3]

[1] Department of Electrical and Electronic Engineering, Kyushu University, Fukuoka 819-0395, Japan
[2] Department of Materials Science, Tokai University, Hiratsuka 259-1292, Japan
[3] Jecc Torisha Co. Ltd., Kawagoe 350-0833, Japan

tanaka@sc.kyushu-u.ac.jp

Abstract. We develop a monitor system to control a parallel-type superconducting (SC) level sensor and to indicate the level of liquid hydrogen in a container. The parallel-type level sensor is composed of an MgB_2 wire and non-SC wire, which are vertically located in parallel and electrically connected in series. We assemble the monitor unit using a power supply, central processing unit (CPU), analog-to-digital (AD)/digital-to-analog (DA) converter, touch panel, voltage dividers and isolators. The terminal voltages of the MgB_2 wire and non-SC wire with a constant current produced by the power supply are stored in the memory of CPU through the isolators, voltage dividers and AD converter. The liquid level is calculated in the CPU according to a pre-written program code. The obtained result is displayed on the touch panel and also output externally via the DA converter and isolator. Furthermore, the normal operation of the developed monitor system is validated using dummy voltages from an oscillator.

1. Introduction

Hydrogen that produces only water after its combustion is expected as a clean energy, and the liquefied hydrogen might be used for mass storage and transport in near future. At that time, it is necessary to understand the accurate amount of liquid hydrogen in the container. Superconducting (SC) level sensors for liquid hydrogen using MgB_2 wires, which have a critical temperature of about 39 K [1] and which becomes SC state in the liquid hydrogen with the boiling temperature of about 20 K at atmospheric pressure, have been studied [2–4]. Haberstroh et al. have investigated an SC level sensor for liquid hydrogen using the MgB_2 wire reinforced by stainless-steel (SS) sheath [2]. Takeda et al. have studied an external-heating-type level sensor in which a heater wire is wholly wound on the MgB_2 wire to improve the sensitivity [3]. Our group has proposed a parallel-type SC level sensor composed of MgB_2 wire A and non-SC wire B [4]. This sensor has an advantage that it is hardly affected by various conditions of gas such as temperature, pressure and object, and therefore it is expected that the reproducibility of sensor output is very good. Our group has also developed a monitor system to control the parallel-type SC level sensor and to indicate liquid level in the container. This monitor unit was composed of a power supply, central processing unit (CPU), touch panel and so on. Although the installed monitor system has been working properly for more than six years [5], it was expensive because a commercial programmable logic controller (PLC) was used there.

In this study, another type of monitor system for the parallel-type SC level sensor is developed to reduce the costs of components. Moreover, it is confirmed using dummy voltages from an oscillator whether the fabricated monitor system works properly or not.

2. Principle of parallel-type SC level sensor

Figure 1 shows the principle diagram of the parallel-type SC level sensor [4]. The level sensor is composed of two thin wires, SC wire A and non-SC wire B, and they are vertically located in parallel and electrically connected in series in the container. When an optimal current I is applied to these wires, the upper part of the SC wire A in gas generates a voltage drop V_G, whereas its lower part in liquid has no voltage drop due to the SC state. Thus, the terminal voltage of the SC wire A, V_A, is equal to V_G. In the case of the non-SC wire B, the upper part in gas also generates a voltage drop V_G if the temperature dependence of resistivity is identical to that for the SC wire A. On the other hand, the lower part of the non-SC wire B generates a voltage drop V_L depending on liquid temperature T_b. So the terminal voltage of the non-SC wire B, V_B, is given by $V_B = V_G + V_L$. When the difference voltage, $V_P = V_B - V_A$, is observed, at that time, the voltage drop of the non-SC wire B in liquid, V_L, is finally obtained. Therefore, this sensor has an advantage that it is hardly affected by the conditions of gas, so that the reproducibility of sensor output is expected to be very good.

The MgB_2 wire fabricated by the powder-in-tube (PIT) method [6] has been used as the SC wire A up to now. On the other hand, the non-SC wire B might be prepared from some candidates such as a wire just after drawing process for the SC wire A (as-drawn wire), a wire made of only sheath material (solid wire) and so on. Figure 2 shows the temperature dependence of electrical resistivity for the SC wire A with SS sheath and its as-drawn wire for numerical simulation [7]. The diameters and lengths of these wires are assumed to be 0.1 mm and 1 m, respectively. The temperature profiles in these wires are numerically calculated by using a one-dimensional heat balance equation. It has been well known that these temperature profiles are determined by only the balance between local heating and cooling, and not affected by the thermal conduction of the wires [8]. Figure 3 shows the numerical results of output voltages for the SC wire A and as-drawn wire B, V_A and V_B, during slowly filling or spontaneously evaporating the liquid hydrogen [7]. It is found that the output voltage V_A increases with rising up or falling down of the liquid level, whereas the output voltage V_B doesn't vary so much. Contrary to the ideal principle explained in the antecedent paragraph, the voltage drops V_G for the SC wire A and non-SC wire B in gas are different from each other due to the discrepancy between their dependence of

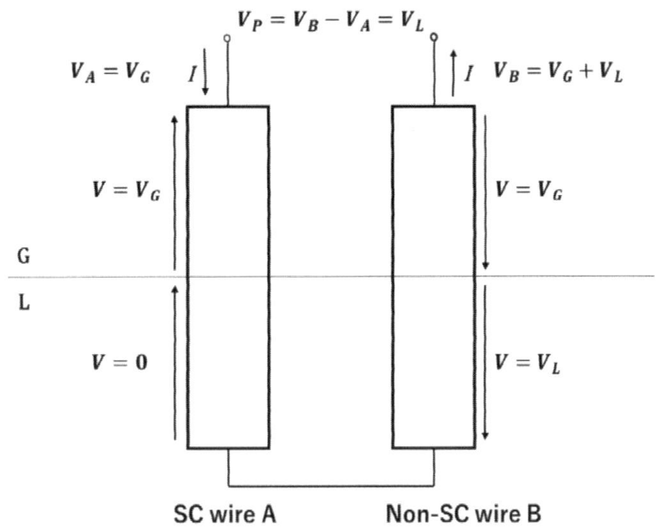

Figure 1. Principle diagram of parallel-type SC level sensor [4].

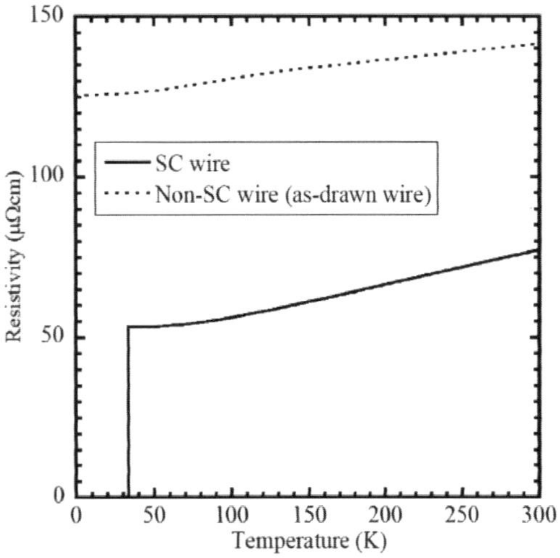

Figure 2. Example of temperature dependence of electrical resistivity for SC wire A and its as-drawn wire B used for numerical simulation [7].

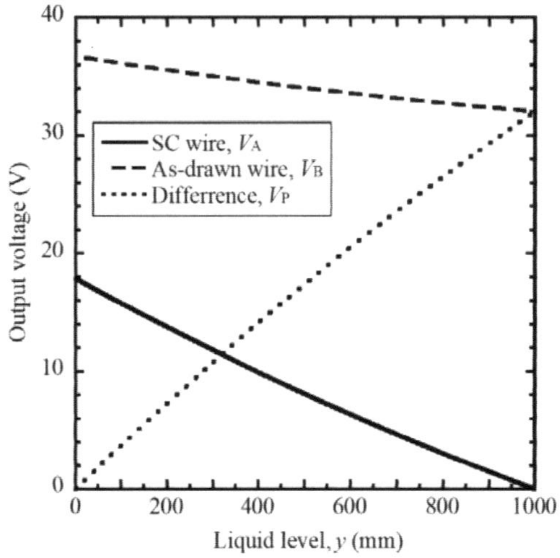

Figure 3. Output voltages from SC wire A and its as-drawn wire B [7]. The difference in voltage between them is also estimated.

resistivity on temperature as shown in figure 2. In order to realize the difference voltage V_P equal to zero when the liquid level is at the bottom, therefore, V_P plotted by a dotted line in figure 3 is estimated as [7]

$$V_P(y) = V_B(y) - \alpha V_A(y), \quad (1)$$

where y is the vertical position of liquid level, and the adjustment parameter α is given by

$$\alpha = \frac{V_{B1}}{V_{A1}} \quad (2)$$

with $V_{A1} = V_A(0)$ and $V_{B1} = V_B(0)$ when the origin of the liquid level y is set at the bottom of the sensing wire. It can be seen that the difference voltage V_P has an almost linear relationship with the liquid level y, and thus the calibration method of linear approximation using only five parameters, V_{A1}, V_{B1}, V_{A2}, V_{B2} and L, is carried out in this study, where $V_{A2} = V_A(L/2)$ and $V_{B2} = V_B(L/2)$ with the effective length L of sensing wire. By using the temporary parameter c expressed by

$$c = \frac{L/2}{V_{B2} - \alpha V_{A2}}, \quad (3)$$

the liquid level y can be finally estimated as

$$y = c(V_B - \alpha V_A). \quad (4)$$

3. Development of monitor system

A monitor system adaptable to the parallel-type SC level sensor is constructed. Figure 4 shows the circuit diagram of the monitor system. The monitor system is composed of a power adapter, CPU, analog-to-digital (AD)/digital-to-analog (DA) converter, touch panel, power supply for level sensor, voltage dividers and isolators. Figure 5 shows the flowchart of monitor system operation. The output voltages, V_A and V_B, are stored in the memory of CPU through the isolators, voltage dividers and AD converter. After that, the liquid level is calculated in the CPU by following a pre-installed program code. The calculation result is displayed on the touch panel and also output externally via the DA converter and isolator. Moreover, the touch panel can display not only the liquid level in the units of both millimeters and percent, but also output voltages, V_A and V_B, individually. Furthermore, the start or stop operation of level sensor can be controlled on the touch panel.

The Raspberry Pi 3 Model B is used as the CPU [9]. The Raspberry Pi is a series of small single-board computers developed in United Kingdom to promote teaching of basic computer science in

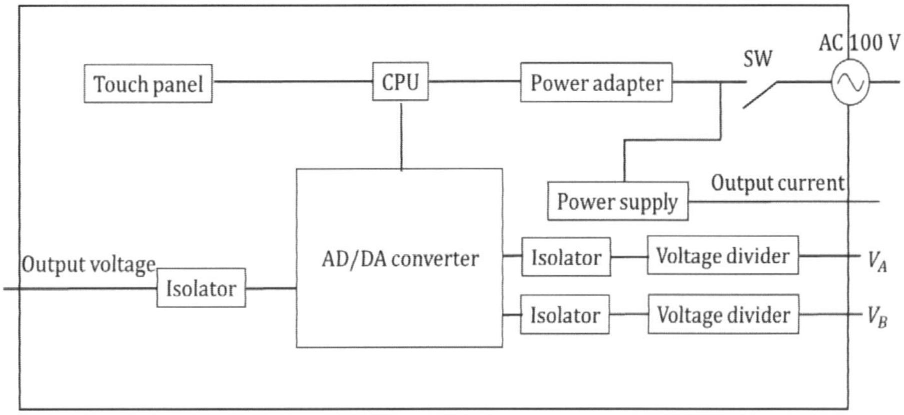

Figure 4. Circuit diagram of monitor system.

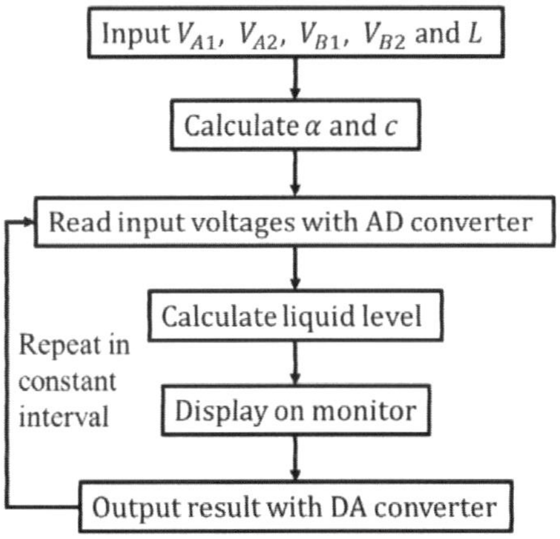

Figure 5. Flowchart of monitor system operation.

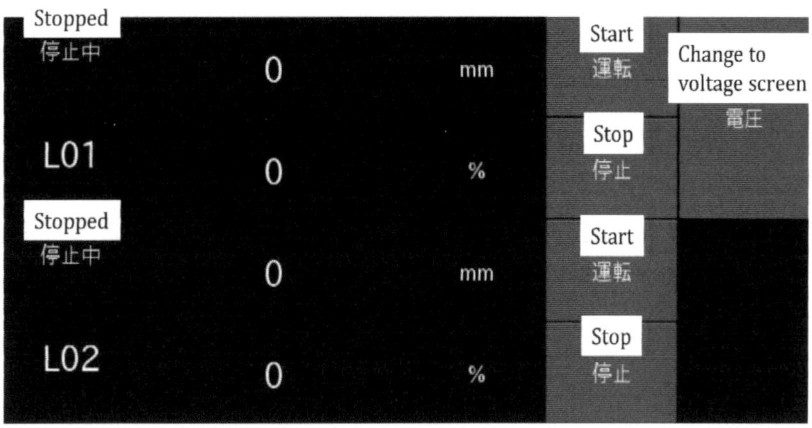

Figure 6. Example of display screen.

schools, and popular among users such as robotics. The monitor system using the Raspberry Pi is much cheaper than that developed previously using the commercial PLC [5]. However, there is no AD/DA function in the Raspberry Pi itself, and thus the High-Precision AD/DA Board for the Raspberry Pi is used [10]. This AD/DA board is composed of the AD converter with 8 channels and 24-bit resolution, and the DA converter with 2 channels and 16-bit resolution, which have enough performance for the monitor in this study. Figure 6 shows one of display screens for the developed monitor system. An application code adaptable to touch panel operation is developed by using the Kivy that is one of open source libraries of the Python [11]. The 7-inch touch screen monitor for the Raspberry Pi is used for display, and the level sensor can be controlled on this touch screen.

4. Preliminary confirmation test of monitor operation

In order to confirm whether the developed monitor system works properly or not, the preliminary test is carried out. The setting parameters for the test are listed in table 1. The time-varying voltages are input to the monitor so as to simulate rising up and falling down of the liquid level. The program code, in which the liquid level of 0-100% corresponds proportionally to the output voltage of 0-10 V, is prepared in advance. Figure 7 shows the result of preliminary confirmation test, in which V_A is changed from 0 V

Table 1. Setting parameters for preliminary confirmation test of monitor system operation.

Symbols	Values
V_{A1}	4 V
V_{A2}	2 V
V_{B1}	3 V
V_{B2}	3 V
L	180 mm

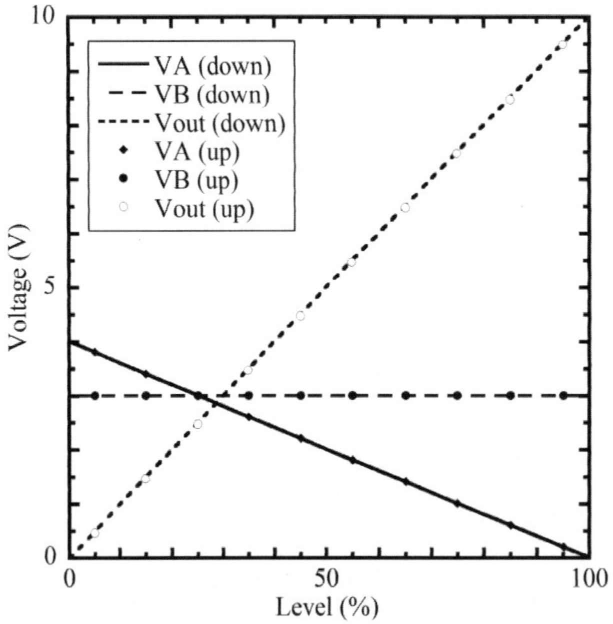

Figure 7. Result of preliminary confirmation test of monitor system operation.

to 4 V and V_B is kept constant at 3 V. The reproducibility is good for both rising up and falling down of the liquid level, and the output voltage is displayed on screen correctly.

5. Conclusions

The monitor system for the parallel-type SC level sensor for liquid hydrogen was developed. The monitor unit was composed of the power adapter, CPU, AD/DA converter, touch panel, power supply for the level sensor, voltage dividers and isolators. The liquid level was calculated in the CPU by storing the output voltages from the level sensor in its memory, and then the calculation result was displayed on the touch panel screen and also output as the analog voltage. The cost of the monitor unit was reduced very much by using the Raspberry Pi, which is the small single-board computer, as the CPU. Furthermore, the application code to control the level sensor via the touch panel was developed by using the Kivy. Finally, the preliminary confirmation test of monitor system operation was carried out, and then the good reproducibility was obtained.

References

[1] Nagamatsu J, Nakagawa N, Muranaka T, Zenitani Y and Akimitsu J 2001 *Nature* **410** 63–4
[2] Haberstroh Ch and Zick G 2006 *Adv. Cryo. Eng.* **51** 679–84
[3] Takeda M, Matsuno Y, Kodama I and Kumakura H 2008 *Adv. Cryo. Eng.* **53** 933–9
[4] Kajikawa K, Inoue T, Watanabe K, Kanazawa M, Yamada Y, Kobayashi H, Taguchi H and Aoki I 2012 *Physics Procedia* **36** 1396–401
[5] Shirai Y, Shiotsu M, Tatsumoto H, Kobayashi H, Naruo Y, Nonaka S and Inatani Y 2016 *Physics Procedia* **81** 158–61
[6] Yamada Y, Kanazawa M, Ohno T, Tachikawa K, Kajikawa K, Inoue T, Aoki I, Matsumoto A and Kumakura H 2012 *IEEE Trans. Appl. Supercond.* **22** 6200304
[7] Koshio J, Kajikawa K, Yamada Y, Makino M and Aoki I 2017 *TEION KOGAKU (J. Cryo. Soc. Jpn.)* **52** 355–62
[8] Tomachi K et al. 2009 *TEION KOGAKU (J. Cryo. Soc. Jpn.)* **44** 366-71
[9] https://www.raspberrypi.org/
[10] https://waveshare.com/wiki/High-Precision_AD/DA_Board
[11] https://kivy.org/#home

Recent Progress of Iron-Based Superconducting Round Wires

***Sunseng Pyon[1], Tsuyoshi Tamegai[1], Katsutoshi Takano[2], Hideki Kajitani[2], Norikiyo Koizumi[2] and Satoshi Awaji[3]**

[1]Department of Applied Physics, The University of Tokyo, Hongo, Bunkyo-ku, Tokyo 113-8656, Japan

[2]Naka Fusion Institute, National Institutes for Quantum and Radiological Science and Technology (QST), 801-1 Mukoyama, Naka-shi, Ibaraki 311-0193, Japan

[3]High Field Laboratory for Superconducting Materials, Institute for Materials Research, Tohoku University, Sendai 980-8577, Japan

E-mail: pyon@ap.t.u-tokyo.ac.jp

Abstract. Recent progress of round wires using iron-based superconductors is reviewed. Both transport and magnetic J_c have been enhanced by several techniques such as purification of polycrystalline powders, high-pressure sintering (HIP), and control of drawing and sintering conditions. The present record of transport J_c of round wires is realized when the wire with 122-type compounds is processed at 175 MPa with a value at 4.2 K under 100 kOe being 38 kA/cm^2 using (Ba,K)Fe_2As_2. We also discuss HIP round wires of (Sr,K)Fe_2As_2 and recently fabricated $CaKFe_4As_4$. Details of the optimization of round wires to achieve large J_c values are described.

1. Introduction

Iron-based superconductors (IBSs) are one of promising candidates for future high-field applications of superconductors because they have high critical temperature, T_c, large upper critical field, H_{c2}, and relatively low anisotropy compared with cuprate superconductors [1,2]. Among IBSs, 122-type compounds of (AE,K)Fe_2As_2 (AE = Ba, Sr) have been studied as promising materials for future application, because high critical current density, J_c, is realized in superconducting wires and tapes using (AE,K)Fe_2As_2 [3-17]. For fabrication of superconducting wires and tapes, application of pressure is the most effective method to eliminate weak links between superconducting wires. For tapes, uniaxial press is effective for enhancement of J_c in (Ba,K)Fe_2As_2 or (Sr,K)Fe_2As_2 pressed tapes [3-8]. Recently, J_c in press tape under high magnetic field of 100 kOe achieved 1.5 x 10^5 A/cm^2, which is higher than the practical level of 10^5 A/cm^2 [8]. In the case of round wires, hot isostatic pressing (HIP) technique is used for the fabrication [9-17]. As shown in figure 1, highest J_c in the HIP wire achieved approximately 2 x 10^5 A/cm^2 under self-field and 3.8 x 10^4 A/cm^2 under 100 kOe, respectively. Although J_c in the HIP round wires is still lower than that in pressed tapes, they are more promising for high-field applications because its round shape is much more advantageous and convenient than the textured tapes for a wide range of applications.

In this report, recent progress of round wires using 122-type IBSs, mainly (Ba,K)Fe_2As_2, is reviewed. Both transport and magnetic J_c have been enhanced by several techniques such as purification of polycrystalline powders, HIP process, and control of drawing and sintering conditions. The present record of transport J_c of round wires is realized when the wire is processed at 175 MPa with a value at 4.2 K under 100 kOe being 38 kA/cm^2. We have also fabricated HIP round wires of (Sr,K)Fe_2As_2, and 1144-type IBSs of $CaKFe_4As_4$ for the first time. Details of the optimization of round wires to achieve large J_c values and the dependence on raw materials are discussed below.

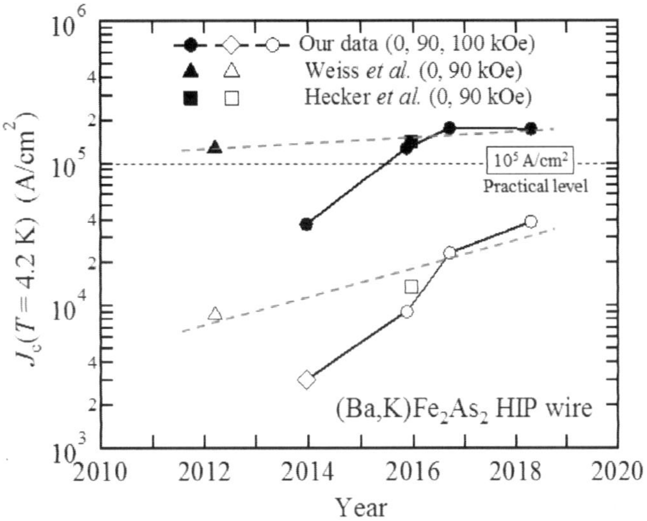

Figure 1. Evolution of transport J_c at 4.2 K under several magnetic field of the HIP round wires fabricated by our group and other groups [9,10,12,14,15,17].

2. Experiments

All superconducting wires were fabricated by *ex-situ* powder-in-tube (PIT) method. Polycrystalline powders of $Ba_{0.6}K_{0.4}Fe_2As_2$, $Sr_{0.6}K_{0.4}Fe_2As_2$ and $CaKFe_4As_4$ were prepared by the solid-state reaction. Details of powder preparation are described in previous papers [10-12,15,18]. Fabrication process of wires is schematically shown in figure 2. Ground powder was filled in a silver tube with an outer diameter of 4.5 mm and an inner diameter 3 mm. Here, three different fabrication processes were employed. First, Ag tubes filled with powder were cold drawn into a round shape with a dimeter ~1.2 mm using dies with circular holes ("drawn"). Second, they were fabricated using only a groove roller with square grooves ("rolled"). Third, they were swaged using a rotary swaging machine ("swaged"). Obtained three kinds of wires were cut into short pieces. They are put into a 1/8 inch copper tube and redrawn into a square shape with a groove roller down to a diagonal dimension of ~1.2 mm. After the drawing process, both ends of the wire were sealed by using an arc furnace. The sealed wires were sintered using the HIP technique. All wires were heated up to 400-700°C in argon atmosphere, and sintered under different pressures of 0.1-170 MPa and for 0.5-12 h. HIP processes under the highest pressure 170 MPa were performed at National Institutes for Quantum and Radiological Science and Technology (QST), and others were performed at the University of Tokyo. The critical current measurements in high magnetic fields were carried out by using the 15T-SM at High Field Laboratory for Superconducting Materials, IMR, Tohoku University. Current–voltage (*I*–*V*) characteristics up to 140 kOe were measured by the four-probe method with solder for contacts. Bulk magnetization was measured to evaluate T_c and magnetic J_c by a superconducting quantum interference device (SQUID) magnetometer (MPMS-5XL, Quantum Design). Magnetic J_c is evaluated using the extended Bean model. Vickers hardness, *Hv*, was measured on the polished surface of the wire core. For MO imaging, the HIP wire and tape were cut and the transverse cross sections were polished with lapping films. An iron–garnet indicator film was placed in direct contact with the sample and the whole assembly was attached to the cold finger of a He-flow cryostat (Microstat-HR, Oxford Instruments). MO images were acquired by using a cooled-CCD camera with 12-bit resolution (ORCA-ER, Hamamatsu). The phase identification and the evaluation of texturing of the core of the HIP wire were carried out by powder XRD with Cu-Kα radiation (Smartlab, Rigaku).

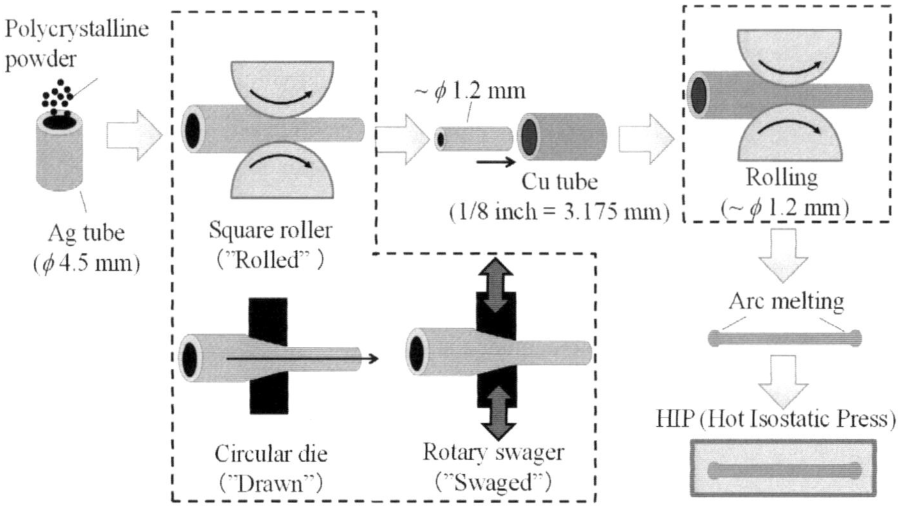

Figure 2. The fabrication process of PIT-HIP round wires.

3. Results and discussion

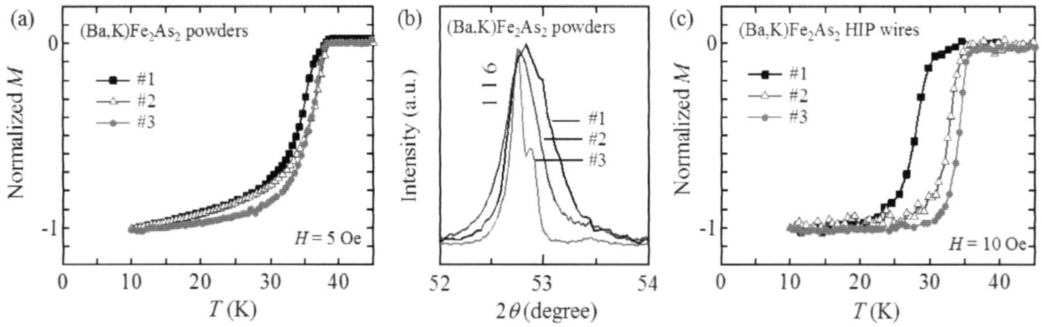

Figure 3. (a) Magnetization as a function of temperature at 5 Oe for (Ba,K)Fe_2As_2 powders. (b) (116) peaks in x-ray diffraction patterns of (Ba,K)Fe_2As_2 powders. (c) Magnetization as a function of temperature at 10 Oe for (Ba,K)Fe_2As_2 HIP wires #1, #2, and #3 are correspond to the data of ref. [10], ref. [12], and ref. [15], respectively.

Figure 3(a) shows temperature dependence of magnetization of (Ba,K)Fe_2As_2 powders. Powders of #1, #2, and #3 correspond to the data of ref. [10], ref. [12], and ref. [15], respectively. The quality of powder #2 were improved compared to that of powder #1 by changing the synthesis conditions, such as nominal compositions, using elements or precursors as raw materials, heating temperature and time, and so on [10, 12]. Synthesis processes were further improved by adopting the ball milling process to mix raw materials and sintering by packing raw materials in Nb tube, and powder #3 was obtained [15]. As a result, magnetization of the newest powder of #3 shows sharper drop at T_c. Improvements of synthesis processes are also confirmed from x-ray diffraction measurements. No impurity peaks are identified in #3. As shown in figure 3(b), it is obvious that the newest powder #3 shows narrow peak profile and clear separation of $K_{\alpha 1}$ and $K_{\alpha 2}$ peaks is observed. This narrow diffraction peak also guarantees homogeneity of grains in the wire core. Clear difference of T_c in HIP wires using these polycrystalline powders are shown in figure 3(c). It should be noted that wire fabrication conditions such as sintering pressure and temperature of the HIP wires #2 and #3 are almost the same. Nevertheless, T_c and J_c in the HIP wire #3 are larger than those in the wire #2. These results clearly indicate that purification of polycrystalline powders is effective for enhancement of J_c.

In the process of detailed evaluation of the properties of the wire, such as T_c and J_c at each step of the wire fabrication, we unexpectedly found degradation of polycrystalline powders in the wire core, including its superconducting properties, during the drawing process [12]. This is clearly seen in figure 4(a) as a decrease of shielding volume fraction and slight suppression of T_c before sintering as the diameter of the wire becomes smaller. We have also found that degradation of superconducting properties of the core of the HIP wire by the drawing process can be recovered by the following heat treatment. As shown in figure 4(b) the reduced T_c by the repeated drawing processes is partly recovered through the HIP treatment. As the sintering temperature increases, the T_c of the sintered wire increases. The highest T_c of the HIP wire is ~35 K, which is close to T_c = 38 K of the as-grown powder. It suggests that sintering at higher temperatures is very effective for the recovery of T_c and shielding volume fraction. It should be noted that the sintering temperature is limited by the melting temperature of Ag/Cu alloy of ~770 °C, in the case of present HIP method using Ag and Cu sheaths. The sintering-temperature dependence of magnetic J_c is similar to that of T_c as shown in figure 4(c). The dashed line indicates the melting point of Ag/Cu alloy of ~770 °C. Similar to the case of T_c, J_c in the HIP wire sintered at 713 °C has the highest self-field J_c. These results clearly indicate that a higher sintering temperature is favorable for increasing J_c. Sintering-temperature dependence of both T_c and J_c implies that damaged polycrystalline powders in the core of the HIP wire during the drawing process are recovered by high-temperature sintering.

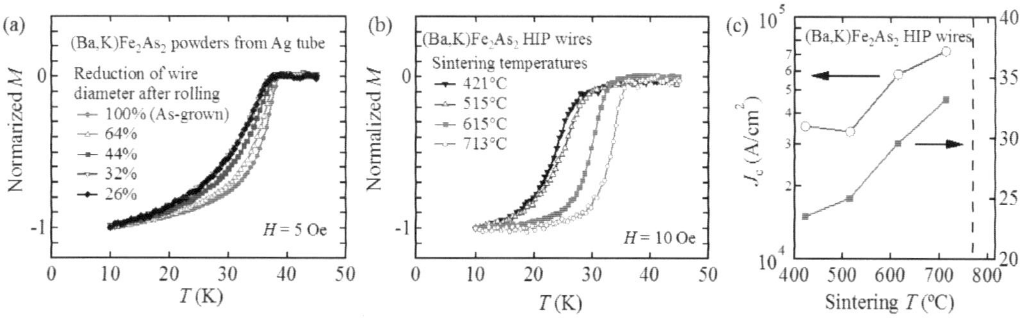

Figure 4. (a) Normalized magnetization as a function of temperature of (Ba,K)Fe_2As_2 powders which were extracted Ag sheathed wires of various rolling steps. Reduction of wire diameter after rolling is shown in the figure. (b) Normalized magnetization as a function of temperature of (Ba,K)Fe_2As_2 HIP wires which were sintered in argon atmosphere under 9 MPa pressure for 12 h at various temperatures. (c) Sintering temperature dependence of J_c and T_c (midpoint) in HIP wires estimated from the magnetization measurements [12].

Next, we focus on the effects of sintering pressure on physical properties in the HIP wire. To improve weak links between grains, higher density is advantageous. Sintering at higher pressure should be effective for that. Superconducting properties in the HIP wires sintered at 700°C under several pressures were evaluated from magnetization measurements as shown in figures 5(a) and (b). In HIP wire sintered under the highest pressure of 175 MPa, highest T_c and J_c are realized. These results are summarized in figures 5(c) and (d). Clearly, T_c and J_c are correlated to the sintering pressure. Furthermore *H*v also shows positive correlation with the sintering pressure shown in figure 5(c). This suggests the improvement of weak-links in grains and promotion of the reaction of raw materials by

IOP Conf. Series: Journal of Physics: Conf. Series **1293** (2019) 012042 doi:10.1088/1742-6596/1293/1/012042

densification. It is clearly confirmed systematically in IBS round wire that high core density plays a key role in achieving the practical level of J_c, as discussed in IBS pressed tape [3].

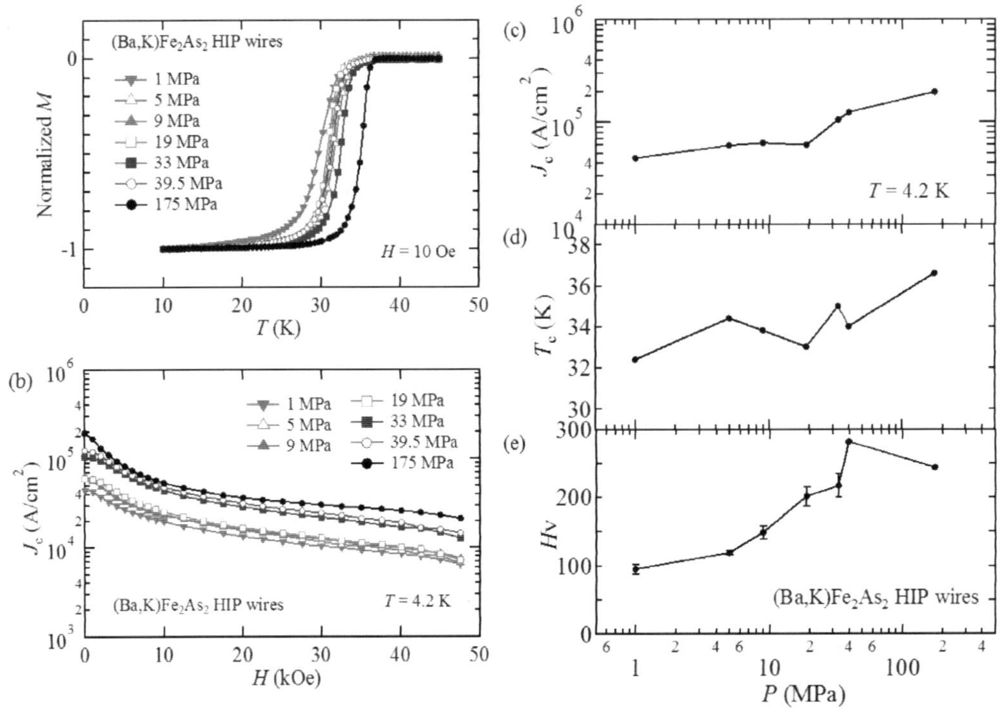

Figure 5. (a) Normalized magnetization as a function of temperature of (Ba,K)Fe_2As_2 "swaged" HIP wires which were sintered at in argon atmosphere under several pressures at 700°C. (b) Magnetic J_c as a function of magnetic field of these wires. Sintering pressure dependence of (c) magnetic J_c at 4.2 K under self-filed, (b) T_c, and (c) Vickers hardness (Hv) in these wires.

Recently, we found that wire-drawing process strongly affect their J_c. Magnetic field dependence of transport J_c in "drawn", "swaged", and "rolled" HIP wires are shown in figure 6(a). Their raw material of polycrystalline powders and HIP sintering conditions are the same. Nevertheless, J_c at high field of 100 kOe in "drawn", "swaged", and "rolled" HIP wires are 38 kA/cm^2 [17], 32 kA/cm^2, 20 kA/cm^2 [15], respectively. Let us discuss the reason why we obtained significantly different values. First possible reason is the difference of degree of degradation of superconducting core caused in the course of different wire fabrication processes. Figure 6(b) shows the temperature dependence of magnetization in these HIP wires. The T_c of the "drawn" and "swaged" HIP wires are approximately 1 K higher compared with that in "rolled" HIP wire. Higher T_c suggests more homogeneous carrier distribution. Considering the fact that the same powder was used for them, wire fabrication with more homogeneous force using dies is effective to suppress the degradation of the core during the drawing process and help recovery in the sintering process as described in figure 4 [12]. Second and most possible reason is the difference of degree of texturing of the superconducting core caused by different wire fabrication processes. Texturing in the core of the HIP wire was analysed using x-ray diffraction. HIP wires were cut and polished to prepare two kinds of surfaces, which are perpendicular (transverse) and parallel (longitudinal) to the current direction, as shown in figures 7(a) and (b), respectively. XRD intensities for (002) and (103) peaks for two kinds of surfaces are shown and compared in figures 7(a) and (b). It should be noted that the intensity in the left part of figure 7(a) is multiplied approximately by 2.2 to normalize the incident x-ray intensity for the comparison with the right part of figure 7(a). Obviously, the ratio of peak intensity of (002) and that of (003), r, shows clear difference.

Smaller r is observed when the x-ray is reflected by the surface perpendicular to the current as shown in figure 7. In the case of pressed tapes, where grains are well-textured along the c-axis, the value of r becomes much larger than the unity [3]. Reduction of the intensity of the (002) peak shown in figure 7(a) implies that the fraction of grains with their c-axis perpendicular to the current is reduced. 122-type materials are layered compounds and they usually show plate-like shapes. It is suggested that these grains in the core are textured during the drawing and groove-rolling processes, although the degree of texturing is much weaker compared with tapes. During fabrication processes, plate-like grains could be concentrically textured, as shown by the yellow broken lines in inset of figure 7(c). To determine the degree of texturing, we made a series of XRD measurements as the wire core is successively thinned down. The prepared surface is parametrized by α, as described in inset of figure 7(c). It is expected that the intensity of the (00l) peaks is enhanced near the surface, while it is suppressed near the center. We repeated the polishing of the HIP wire, and the XRD measurements, alternately. Estimated r values are plotted as a function of α in figure 7(c). The general trend of r as a function of α is similar to the expected one. This result also suggests that grains in the core are textured during the wire-fabrication process. Furthermore, compared with the "drawn" HIP wire, the value of r in the "rolled" HIP wire is smaller. This implies that the degree of texturing is enhanced by the drawing process using dies compared with the groove-rolling process. The weaker texturing in the groove-rolled wire may be due to the less symmetric shape of the core.

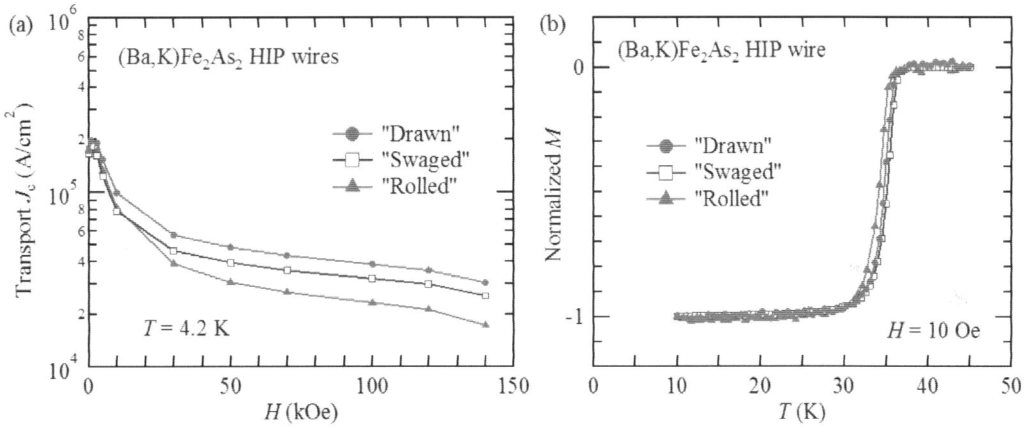

Figure 6. (a) Transport J_c as a function of magnetic field in "drawn" (ref. [17]), "swaged", and "rolled"(ref. [15]) $(Ba,K)Fe_2As_2$ HIP wires. (b) Normalized magnetization as a function of temperature in these wires.

Finally, we discuss several IBSs for raw materials of PIT wires. Both the transport and magnetic J_c for the $(Ba,K)Fe_2As_2$, $(Sr,K)Fe_2As_2$, and $CaKFe_4As_4$ HIP wires from our previous studies are shown in figure 8(a) [11,17,18]. $(Ba,K)Fe_2As_2$ HIP wire has the highest transport J_c among all IBS round wires, at present. From the optical images of the transverse cross sections of the HIP wires in figures 8(b)-(d), copper, silver, and superconducting core regions are clearly identified and no voids are observed. Figures 8(e)-(g) show MO images of the transverse cross section of the core of the HIP wires correspond to figures 8 (b)-(d) in the remanent state at 5 K, respectively. The smooth variation of the trapped magnetic field in the HIP wires at 5 K indicates the presence of a uniform bulk current flowing in the wires core across grains. However, J_c at magnetic field of 10 kOe in $(Sr,K)Fe_2As_2$ and $CaKFe_4As_4$ HIP wires are significantly lower than that in $(Ba,K)Fe_2As_2$ HIP wire. For $(Sr,K)Fe_2As_2$ and $CaKFe_4As_4$ HIP wires, the condition of polycrystalline-powder synthesis and wire fabrication processes are still not improved and impurities in the core of the wire still remain [18]. They still remain as candidate of raw materials for high J_c HIP wire. Actually, the tape of $(Sr,K)Fe_2As_2$ reaches significantly high J_c [4,5]. The value of J_c in single crystal $CaKFe_4As_4$ is comparable to that in single crystal of (Ba,K)Fe_2As_2 [19], and further enhancement of J_c in both wires and tapes using $CaKFe_4As_4$ is expected [18,20]. Further improvements of wire fabrication process are demanded for them.

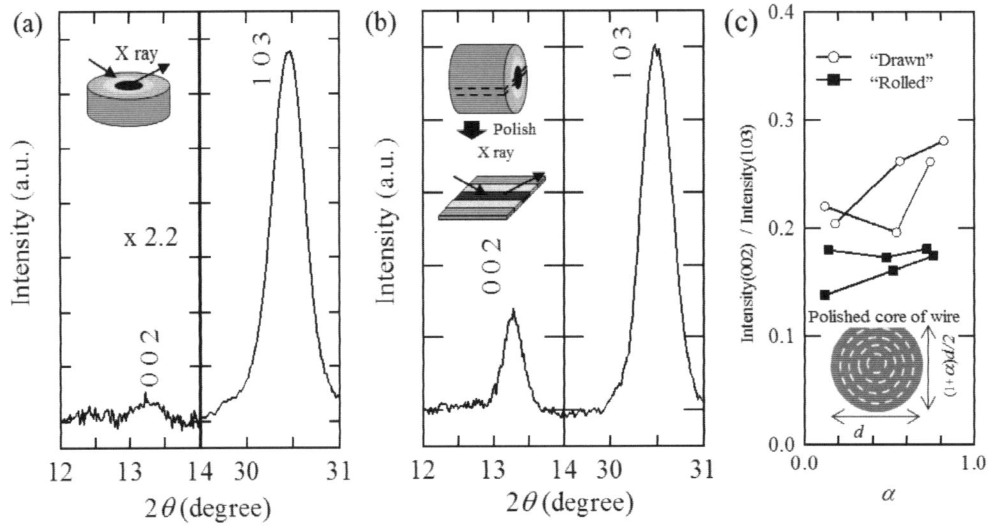

Figure 7. (a) (b) Powder XRD data of (002) and (103) peaks for two different cross sections of the HIP wires. Insets show the schematics of each cross section. Intensity in the left part of figure (a) is multiplied by 2.2 to normalize the incident x-ray intensity for the comparison with the right part of figure (b). Inset of (c) shows the schematic of the polished core of wire. d is the diameter of the core and α is the parameter that characterizes the exposed surface of the core. Yellow dotted lines indicate the possible texturing of grains in the core. (c) α dependence of the ratio of intensities of powder XRD peaks of (002) and (103), r, of the HIP wire [17].

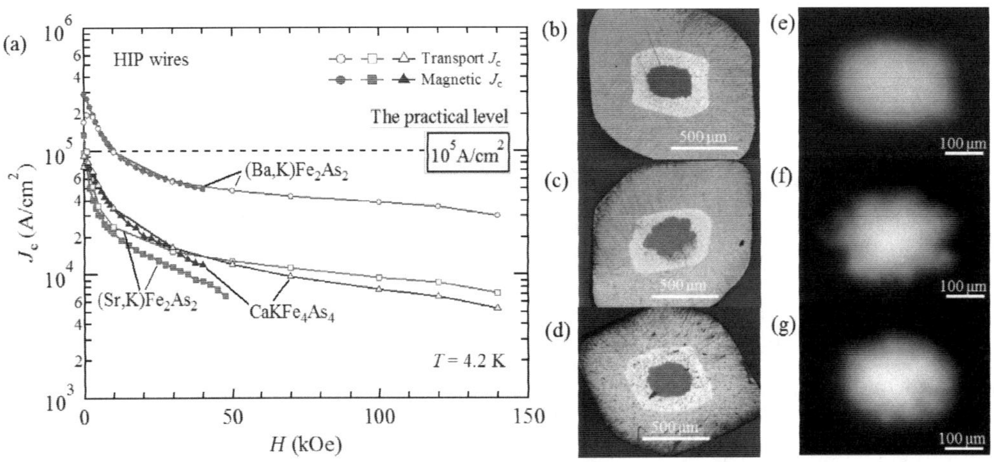

Figure 8. Transport J_c as function of magnetic field in the HIP wires using (Ba,K)Fe_2As_2 [17], (Sr,K)Fe_2As_2 [11] and $CaKFe_4As_4$ [18]. Optical micrograph of the HIP wires of (b) (Ba,K)Fe_2As_2, (c) (Sr,K)Fe_2As_2, and (d) $CaKFe_4As_4$. Corresponding MO images of the HIP wires in the remanent state at 5 K of (e) (Ba,K)Fe_2As_2, (f) (Sr,K)Fe_2As_2, and (g) $CaKFe_4As_4$.

4. Summary

Recent progress of IBS round wires is reviewed. Transport J_c has been enhanced by successive introductions of several techniques such as purification of polycrystalline powders, high-pressure sintering, control of drawing and sintering condition, the improvement of links in grains and promotion of the reaction of raw materials by densification, and enhancing the texturing inside the wire core by introduction of dies or rotary swager in the drawing process. The present records of transport J_c of round wires are realized for that processed at 175 MPa with values at 4.2 K under self-field and 100 kOe being 2.0 x 10^5 A/cm^2 and 3.8 x 10^4 A/cm^2, respectively, using (Ba,K)Fe_2As_2. (Sr,K)Fe_2As_2 and $CaKFe_4As_4$ are also possible candidates of raw materials for high-J_c round wires. We expect that further enhancement of J_c in these HIP wires can be realized by further purification of raw materials, enhancing the degree of texturing and further densification of the core using higher pressure.

Acknowledgments

This work was partially supported by a Grant-in-Aid for Scientific Research (A) (17H01141), a Grant-in-Aid for Young Scientists (B) (No.16K17745), and the Japan-China Bilateral Joint Research Project by the Japan Society for the Promotion of Science (JSPS).

References

[1] Ma Y 2012 *Supercond. Sci. Technol.* **25** 113001
[2] Hosono H, Yamamoto A, Hiramatsu H and Ma Y 2018 *Mater. Today* **21**, 278
[3] Gao Z, Togano K, Matsumoto A and Kumakura H 2014 *Sci. Rep.* **4** 4065
[4] Lin H, Yao C, Zhang X, Zhang H, Wang D, Zhang Q, Ma Y W, Awaji S and Watanabe K 2014 *Sci. Rep.* **4** 4465
[5] Lin H, Yao C, Zhang X, Dong C, Zhang H, Wang D, Zhang Q, Ma Y, Awaji S, Watanabe K, Tian H and Li J 2014 *Sci. Rep.* **4** 6944
[6] Zhang X, Yao C, Lin H, Cai Y, Chen Z, Li J, Dong C, Zhang Q, Wang D, Ma Y, Oguro H, Awaji S and Watanabe K 2014 *Appl. Phys. Lett.* **104** 202601
[7] Gao Z, Togano K, Matsumoto A and Kumakura H 2015 *Supercond. Sci. Technol.* **28** 012001
[8] Huang H, Yao C, Dong C, Zhang X, Wang D, Cheng Z, Li J, Awaji S, Wen H and Ma Y 2018 *Supercond. Sci. Technol.* **31** 015017
[9] Weiss J D, Tarantini C, Jiang J, Kametani F, Polyanskii A A, Larbalestier D C and Hellstrom E E 2012 *Nat. Mat.* **11** 682
[10] Pyon S, Taen T, Ohtake F, Tsuchiya Y, Inoue H, Akiyama H, Kajitani H, Koizumi N, Okayasu S and Tamegai T 2013 *Appl. Phys. Express* **6** 123101
[11] Pyon S, Tsuchiya Y, Inoue H, Kajitani H, Koizumi N, Awaji S, Watanabe K and Tamegai T 2014 *Supercond. Sci. Technol.* **27** 095002
[12] Pyon S, Yamasaki Y, Kajitani H, Koizumi N, Tsuchiya Y, Awaji S, Watanabe K and Tamegai T 2015 *Supercond. Sci. Technol.* **28** 125014
[13] Tamegai T, Pyon S, Tsuchiya Y, Inoue H, Koizumi N and Kajitani H 2015 *IEEE Trans. Appl. Supercond.* **25** 7300504
[14] Hecher J, Baumgartner T, Weiss J D, Tarantini C, Yamamoto A, Jiang J, Hellstrom E E, Larbalestier D C and Eisterer M 2016 *Supercond. Sci. Technol.* **29** 025004
[15] Pyon S, Suwa T, Park A, Kajitani H, Koizumi N, Tsuchiya Y, Awaji S, Watanabe K and Tamegai T 2016 *Supercond. Sci. Technol.* **29** 115002
[16] Tamegai T, Suwa T, Pyon S, Kajitani H, Takano K, Koizumi N, Awaji S, and Watanabe K 2017 *IOP Conf. Ser.: Mater. Sci. Eng.* **279** 012028
[17] Pyon S, Suwa T, Tamegai T, Takano K, Kajitani H, Koizumi N, Awaji S, Zhou N and Shi Z 2018 *Supercond. Sci. Technol.* **31** 055016
[18] Pyon S, Miyawaki D, Veshchunov I, Tamegai T, Takano K, Kajitani H, Koizumi N and Awaji S 2018 *Appl. Phys. Express* **11** 123101
[19] Tamegai T, Takahashi A, Pyon S, Ito N, Ishida S, Iyo A, Eisaki H, Imai M, Abe H, Terashima T, Ooi S and Ichinose A, 2018 *12th Int. Conf. Materials and Mechanisms of Superconductivity and High Temperature Superconductors (M2S2018)*

IOP Conf. Series: Journal of Physics: Conf. Series **1293** (2019) 012042 doi:10.1088/1742-6596/1293/1/012042

[20] Cheng Z, Dong C, Huang H, Liu S, Zhu Y, Wang D, Vlasko-Vlasov V, Welp U, Kwok W-K and Ma Y 2019 *Supercond. Sci. Technol.* **32** 015008

IOP Conf. Series: Journal of Physics: Conf. Series **1293** (2019) 012043 doi:10.1088/1742-6596/1293/1/012043

Fabrication of (Ba,Na)Fe_2As_2 round wires using HIP process

D Miyawaki[1], S Pyon[1], T Tamegai[1], S Awaji[2], K Takano[3], H Kajitani[3], and N Koizumi[3]

[1] Department of Applied Physics, The University of Tokyo, Hongo, Bunkyo-ku, Tokyo 113-8656, Japan

[2] High Field Laboratory for Superconducting Materials, Institute for Materials Research, Tohoku University, Sendai 980-8577, Japan

[3] Naka Fusion Institute, National Institutes for Quantum and Radiological Science and Technology (QST), Naka, Ibaraki 311-0193, Japan

Email: miyadai883@gmail.com

Abstract. We have fabricated round wires of $Ba_{0.6}Na_{0.4}Fe_2As_2$ using powder-in-tube technique and hot isostatic press (HIP) treatment. Transport J_c was measured at 4.2 K of two $Ba_{0.6}Na_{0.4}Fe_2As_2$ round wires with Ag/Cu double sheath which were processed at 700 °C for 4 h at 175 MPa. Transport J_c of one of the wires reached 76 kA/cm^2 and 24 kA/cm^2 at self-field and 100 kOe, respectively, while J_c of another wire is slightly larger at low fields. It means that J_c of $Ba_{0.6}Na_{0.4}Fe_2As_2$ wire is resilient against large magnetic field, which demonstrates very promising characteristics of this material. Detailed characterizations of wires using X-ray diffraction and compositional mapping by EDX are also presented. We also explore the better condition for synthesis of $Ba_{0.6}Na_{0.4}Fe_2As_2$ by changing the reaction temperature.

1. Introduction

The discovery of iron-based superconductors (IBSs) in 2008 has prompted a great interest in their application potentiality [1]. Soon after the discovery, several types of high temperature superconductors in iron pnictides were discovered. IBSs are high-temperature superconductors that have high critical temperature T_c, high critical magnetic field H_{c2}, small anisotropy γ, and large critical current density J_c under high magnetic fields. Among them, superconducting wires and tapes have been fabricated mainly by using (AE,K)Fe_2As_2 materials (AE = Ba or Sr), so called '122-type' [2-8], because of their high H_{c2} [9, 10] and low γ(~2) [10, 11], as well as their high T_c (~36-38 K) [12, 13]. They are expected to be put into practical applications such as wires for high-field magnets. In particular, superconducting tapes and wires using K-doped 122 type ((Ba,K)Fe_2As_2 and (Sr,K)Fe_2As_2) have been extensively studied. Recently, we have reported fabrication and characterization of (Sr,Na)Fe_2As_2 tapes, and demonstrated that J_c in this system is 47 kA/cm^2 even at 100 kOe at 4.2 K [14]. However, it became clear that segregation of Na in the wire core causes a problem in this system. Reduction of Na content is one way to achieve better performance. However, since antiferromagnetic phase is extended to x = 0.5 in $Sr_{1-x}Na_xFe_2As_2$ [15], pursuing underdoped sample is not a good option. So, we chose (Ba,Na)Fe_2As_2 as an alternative, where antiferromagnetic phase region is much narrower [16].

2. Experiments

First, we synthesized polycrystalline samples of $Ba_{0.6}Na_{0.4}Fe_2As_2$. Synthesis processes were conducted in an argon filled glove box. Starting materials (BaAs, NaAs, Fe_2As) were ground into powder with an agate mortar and pestle. Starting materials were synthesized by a method similar to that of (Sr,Na)Fe_2As_2 [17]. Polycrystalline samples were synthesized at 770 °C for 15 h in a stainless steel pipe, where pellets of the starting materials were placed. Then, the reacted materials were ground into powder, pelletized, and reacted at 770 °C for 15 h again. The prepared polycrystalline materials were ground into powder. The ground powder was tightly packed into a silver tube with outer and inner diameters of 4.5 and 3.0 mm, respectively. Some of tubes were groove-rolled into a wire with 1.3 x 1.3 mm^2 cross section, and others were swaged using a rotary swager into the diameter of 1.3 mm. After cutting them into short pieces, one of the pieces was put into 1/8 inch copper tube and redrawn into a square shape with a groove roller down to a diagonal dimension of 1.2 mm. After these processes, both ends of the wire were sealed using an arc furnace. The sealed wires were sintered using the hot isostatic press (HIP) technique. Wires were heated for 4 h at 700 °C in an argon atmosphere under different pressures up to 175 MPa. HIP processes were performed at National Institutes for Quantum and Radiological Science and Technology. We also synthesized polycrystalline samples of $Ba_{0.6}Na_{0.4}Fe_2As_2$ at different reaction temperatures (800 °C, 820 °C, and 850 °C).

Magnetic measurements were conducted using a commercial SQUID magnetometer (MPMS-XL5, Quantum Design). Transport J_c was measured at 4.2 K by the standard four-probe method in magnetic fields up to 140 kOe. Powder X-ray diffraction (XRD) measurements were conducted using Cu-Kα radiation (Smartlab, Rigaku). SEM images of the core were taken using SEM-EDX (S-4300, Hitachi High-Technologies equipped with EMAX x-act, HORIBA).

3. Results and discussion

3.1. Transport J_c measurements at 4.2 K in magnetic fields up to 140 kOe

Figure 1 compares transport J_c at 4.2 K of two $Ba_{0.6}Na_{0.4}Fe_2As_2$ round wires with Ag/Cu double sheath which were processed at 700 °C for 4 h at 175 MPa. In the drawing process, wire #1 was swaged, and wire #2 was rolled. Transport J_c of wire #1 is 76 kA/cm^2 and 24 kA/cm^2 at self-field and 100 kOe, respectively. On the other hand, J_c at 100 kOe of wire #2 is lower than that of #1, while self-field J_c is slightly larger. Even at a high magnetic field of 100 kOe, J_c of wire #1 is reduced to only one third of the self-field value.

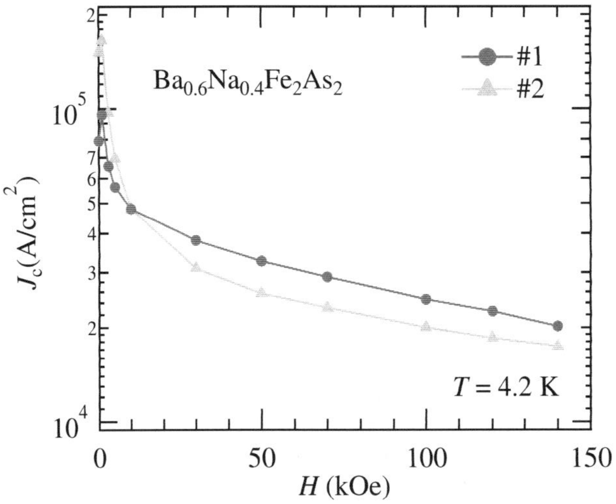

Figure 1. Magnetic field dependence of transport J_c at 4.2 K of two $Ba_{0.6}Na_{0.4}Fe_2As_2$ HIP round wires processed at 700 °C and 175 MPa for 4 h.

3.2. Magnetic susceptibility measurements

Figure 2 shows *M-T* measurements of $Ba_{0.6}Na_{0.4}Fe_2As_2$ polycrystalline powders sintered once and twice at 770 °C, and the HIP wire. Comparing polycrystals sintered once and twice, the transition of twice sintered became a little sharper. Although T_c of the HIP wire became lower than that of polycrystals, the transition became much sharper.

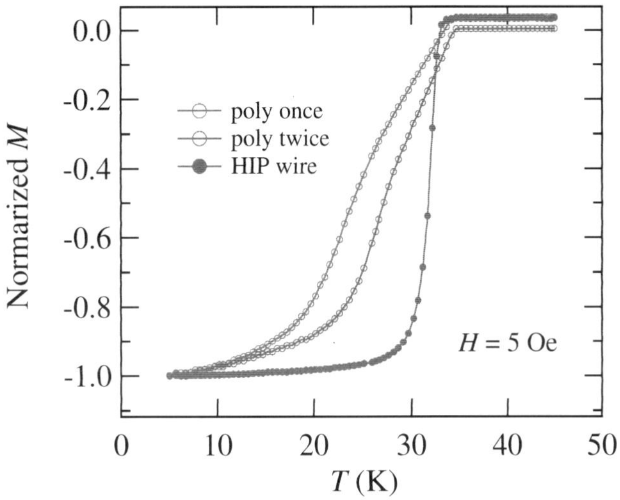

Figure 2. Normalized magnetization at H = 5 Oe as a function of temperature of $Ba_{0.6}Na_{0.4}Fe_2As_2$ polycrystalline powders sintered once and twice at 770 °C, and the HIP wire.

Figure 3 shows normalized magnetization as a function of temperature of $Ba_{0.6}Na_{0.4}Fe_2As_2$ polycrystalline powders sintered at different temperatures. By increasing the sintering temperature from 770 °C, the transition became sharper as in samples sintered at 800 °C and 820 °C. However, when the sintering temperature reached 850 °C, T_c dropped markedly and the transition became broader.

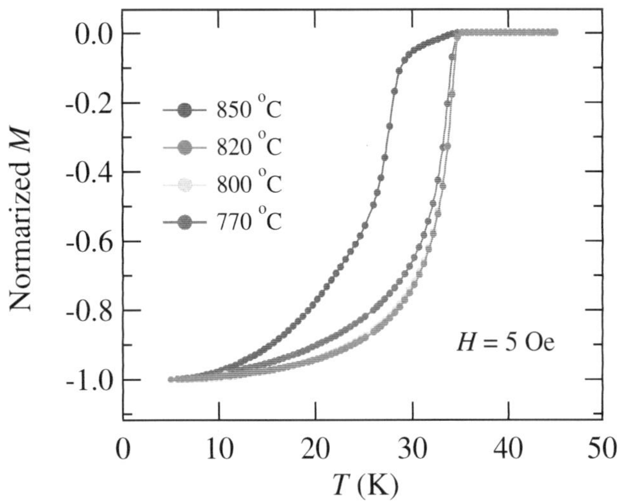

Figure 3. Normalized magnetization at H = 5 Oe as a function of temperature of $Ba_{0.6}Na_{0.4}Fe_2As_2$ polycrystalline powders sintered twice at different temperatures (770, 800, 820, and 850 °C).

3.3. X-ray diffraction analyses and SEM images

Figure 4 shows X-ray diffraction patterns of the HIP wire and polycrystalline powders (sintered at 770 °C) of $Ba_{0.6}Na_{0.4}Fe_2As_2$. Compared to the polycrystalline powders, peaks of the HIP wire are sharper. Although peaks of impurities (Fe_2As and FeAs) can be seen in powder sample, the amount of impurities in the HIP wire is less. SEM images in Fig. 5 shows Fe_2As impurities exist in the core of the HIP wire, and some parts are Ba rich. In addition, Na-rich region can also be identified as in the case of $(Sr,Na)Fe_2As_2$ [14].

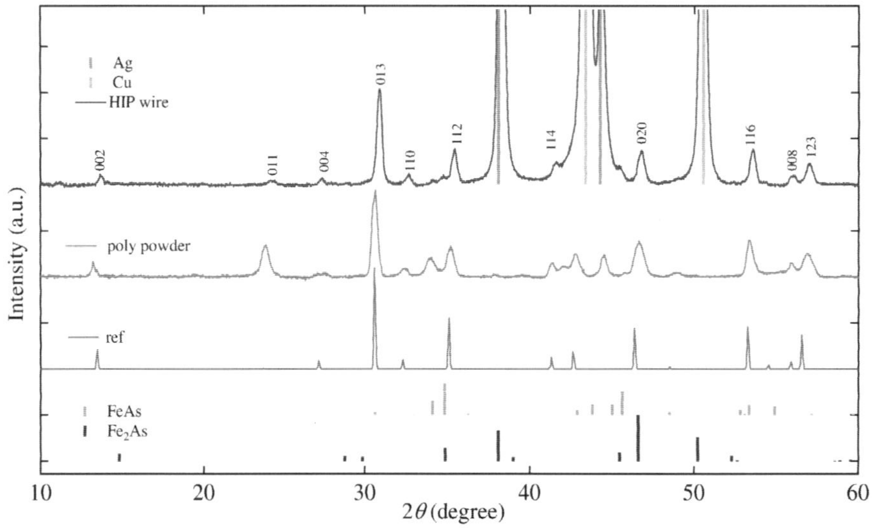

Figure 4. X-ray diffraction patterns of the HIP wire and polycrystalline powders of $Ba_{0.6}Na_{0.4}Fe_2As_2$ prepared at 770 °C. Reference data is calculated as explained in ref. [18].

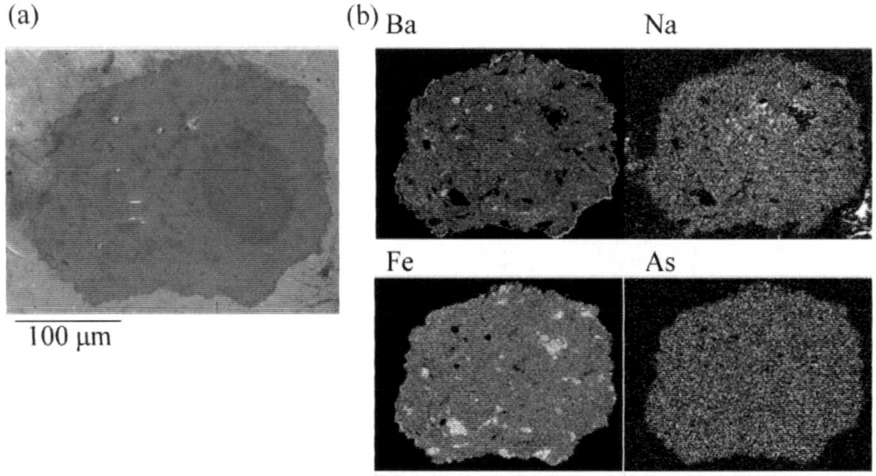

Figure 5. (a) SEM image of the cross section of the HIP wire. (b) Compositional mappings of the HIP wire by EDX.

In Figure 6, polycrystals sintered at 770 °C had most impurities of FeAs and Fe_2As. Polycrystals sintered at 800 and 820 °C had less impurities, and polycrystals sintered at 850 °C had least impurities. As the sintering temperature increases, the amount of impurities became less.

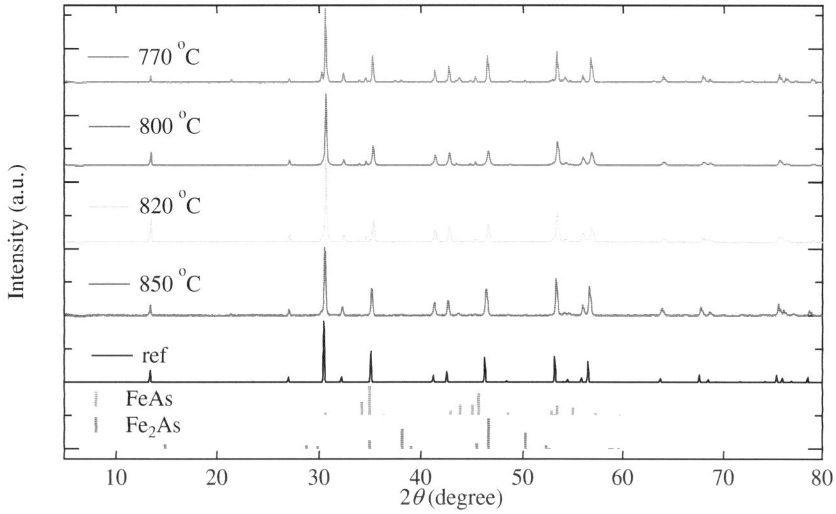

Figure 6. X-ray diffraction patterns of polycrystalline powders of $Ba_{0.6}Na_{0.4}Fe_2As_2$ sintered at 770, 800, 820, and 850 °C. Reference data is calculated as explained in ref. [18].

Although there are still impurities in the HIP wire using polycrystalline sample sintered at 770 °C, J_c was as large as that in (Ba,K)Fe_2As_2. By comparing data in Fig. 6 and Fig. 3, it is clear that the polycrystals sintered at 800 and 820 °C had less impurities and sharper transition. Although T_c of the polycrystals sintered at 850 °C is lower, the content of impurities is minimum. Compositional mapping of the sample sintered at 850 °C by EDX shows that the average content of Na is 0.3 as compared with the initial value of 0.4. We speculate that this reduction of average Na content is the origin of the suppression of T_c in sample sintered at 850 °C. We thus conclude that the optimal temperature for the preparation of $Ba_{0.6}Na_{0.4}Fe_2As_2$ is between 800 °C and 820 °C. J_c of HIP wires using polycrystals sintered at the optimal temperature is expected to be high.

4. Summary

We fabricated round wires of $Ba_{0.6}Na_{0.4}Fe_2As_2$ using powder-in-tube technique and HIP treatment. The maximum transport J_c at 4.2 K and 100 kOe of $Ba_{0.6}Na_{0.4}Fe_2As_2$ wire sintered at 770 °C reaches 24 kA/cm^2. It is remarkable that J_c in this wire is reduced to only one third of the self-field value at 100 kOe. Although there are still impurities in the HIP wire using polycrystalline powder sample sintered at 770 °C and it is not optimal, J_c is comparable to that in (Ba,K)Fe_2As_2 wire. From *M-T* and XRD measurements, we propose that the optimal temperature for the preparation of $Ba_{0.6}Na_{0.4}Fe_2As_2$ is between 800 °C and 820 °C.

References

[1] Kamihara Y, Watanabe T, Hirano M and Hosono H 2008 *J. Am. Chem. Soc.* **130** 3296
[2] Ma Y W, Gao Z S, Qi Y P, Zhang X P, Wang L, Zhang Z Y and Wang D L 2009 *Physica* C **469** 651
[3] Togano K, Matsumoto A and Kumakura H 2011 *Appl. Phys. Express* **4** 043101
[4] Pyon S, Tsuchiya Y, Inoue H, Kajitani H, Koizumi N, Awaji S, Watanabe K and Tamegai T 2014 *Supercond. Sci. Technol.* **27** 095002
[5] Pyon S, Yamasaki Y, Kajitani H, Koizumi N, Tsuchiya Y, Awaji S, Watanabe K and Tamegai T 2015 *Supercond. Sci. Technol.* **28** 125014
[6] Weiss J D, Tarantini C, Jiang J, Kametani F, Polyanskii A A, Larbalestier D C and Hellstrom E 2012 *Nat. Mater.* **11** 682
[7] Pyon S, Suwa T, Park A, Kajitani H, Koizumi N, Tsuchiya Y, Awaji S, Watanabe K and Tamegai T

2016 *Supercond. Sci. Technol.* **29** 115002

[8] Tamegai T, Suwa T, Pyon S, Kajitani H, Takano K, Koizumi N, Awaji S and Watanabe K 2017 *IOP Conf. Ser. Mater. Sci. Eng.* **279** 012028

[9] Putti M *et al.* 2010 *Supercond. Sci. Technol.* **23** 034003

[10] Wang Z S, Luo H Q, Ren C and Wen H H 2008 *Phys. Rev.* B **78** 140501R

[11] Yuan H Q, Singleton J, Balakirev F F, Baily S A, Chen G F, Luo J L and Wang N L 2009 *Nature* **457** 565

[12] Rotter M M, Tegel M, Johrendt D, Schellenberg I, Hermes W and Pöttgen R 2008 *Phys. Rev.* B **78** 020503

[13] Sasmal K, Lv B, Lorenz B, Guloy A M, Chen F, Xue Y Y and Chu C W 2008 *Phys. Rev. Lett.* **101** 107007

[14] Suwa T, Pyon S, Tamegai T, and Awaji S 2018 *Appl. Phys. Express* **11** 063101

[15] Taddei K *et al.* 2016 *Phys. Rev.* B **93** 134510

[16] Avci S *et al.* 2013 *Phys. Rev.* B **88** 094510

[17] Shinohara N *et al.* 2015 *Supercond. Sci. Technol.* **28** 062001

[18] The reference data are calculated using software *Crystal Diffract, Hulinks.* The lattice constants of $Ba_{0.6}Na_{0.4}Fe_2As_2$ are taken from ref. [19]. Since there are no available data of atomic coordinates of $Ba_{0.6}N_{0.4}Fe_2As_2$, the atomic coordinates of $Ba_{0.6}K_{0.4}Fe_2As_2$ [20] are used instead.

[19] Cortes-Gil R, Parker D R, Pitcher M J, Hadermann J, and Clarke S J 2010 *Chem. mater* **22** 4304

[20] Rotter M, Tegel M, and Johrendt D 2008 *PRL* **101** 107006

IOP Conf. Series: Journal of Physics: Conf. Series **1293** (2019) 012044 doi:10.1088/1742-6596/1293/1/012044

Generation of Uniform Magnetic Field between Face-to-Face HTS Bulk Magnets

T Oka[1], K Higa[2], S Tsunoda[2], J Ogawa[2], S Fukui[2], N Inoue[1], M Miryala[1], M Murakami[1], K Yokoyama[3] and T Nakamura[4]

[1] Shibaura Institute of technology, 3-7-5 Toyosu, Koto-Ward, Tokyo 135-8548, Japan
[2] Niigata University, 8050 Ikarashi-Ninocho, Nishi-Ward, Niigata 950-2181, Japan
[3] Ashikaga University, 268-1 Ohmae-cho, Ashikaga, Tochigi 326-8558, Japan
[4] RIKEN, 1-7-22, Suehiro, Tsurumi-Ward, Yokohama, Kanagawa 230-0045, Japan

E-mail: okat@sic.shibaura-it.ac.jp

Abstract. Aiming to develop the compact nuclear magnetic resonance devices, the authors have been developing the uniform magnetic fields in the space between the face-to-face magnetic poles which contain HTS bulk magnets. The authors deformed the original convex shape of the magnetic field distribution to concave by attaching a ferromagnetic iron plate of 20 mm and 100 mm in diameter on one of the pole surfaces. The magnetic field distribution precisely measured along x-axis changed from concave to convex with increasing distance from the magnetic pole surface, which suggested the presence of magnetically flat region. From the results of mapping on the field uniformity data, we have succeeded in obtaining the field uniformity under 500ppm in the space of 4 x 4 x 0.8 mm^3 when the magnetic poles emitting concave and convex field distributions were settled face-to-face with various gaps 30-70 mm. This performance exceeded the target value of 1,500ppm which is necessary to detect the NMR signals.

1. Introduction

In general, since nuclear magnetic resonance (NMR) magnets require us extremely-high uniform magnetic field spaces, it has been suspected to be impossible to form such homogeneous field distributions with use of the high temperature bulk magnets (abbreviated as bulk magnets) which originally exhibit the conical-shaped magnetic field with steep field gradient [1]. Only as for the magnetic field intensity, the bulk magnets would be feasible candidates as intense field generators for NMR/MRI (Magnetic Resonance Imaging) devises when we would refer to the performances reported as 16 T, 17.24 T, and 17.6 T in the past papers [2-4].

According to the reports [5, 6], the field uniformity less than 1,500ppm should be required to detect the NMR signals. Then, we changed the shape of magnetic field distribution from convex to concave by attaching an iron plate on one of the pole surfaces. As shown in Figure 1, the magnetic field distribution exhibits convex shape up to 1.47 T. It was deformed to be concave, exhibiting 0.80 T at the center. The magnetic poles were activated by the pulsed field magnetizing (PFM) technique beforehand by feeding the current to a pulse coil.

As shown in Figure 2, the magnetic poles were settled face-to-face with various gap spaces, and the magnetic-field uniformity was estimated there. When we combined the concave and convex field distributions to compensate the uneven field distributions, the best uniformity reached 358ppm in the 30 mm gap, which exceeded the target value of 1,500ppm [7]. In addition, the numerical simulation

IOP Conf. Series: Journal of Physics: Conf. Series **1293** (2019) 012044 doi:10.1088/1742-6596/1293/1/012044

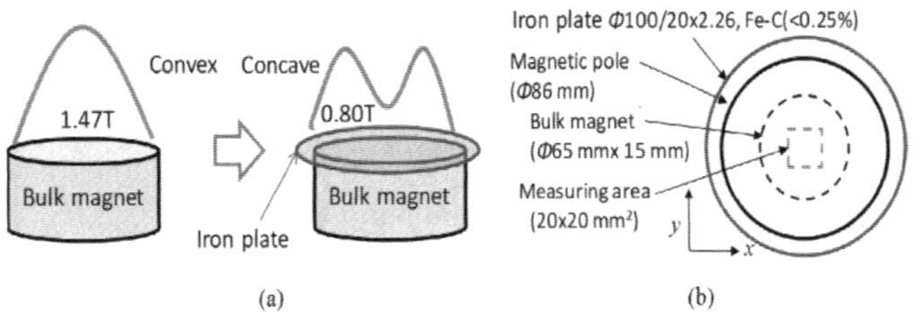

Figure 1. Change of magnetic field distribution by attaching an iron plate (a), and the top view (b).

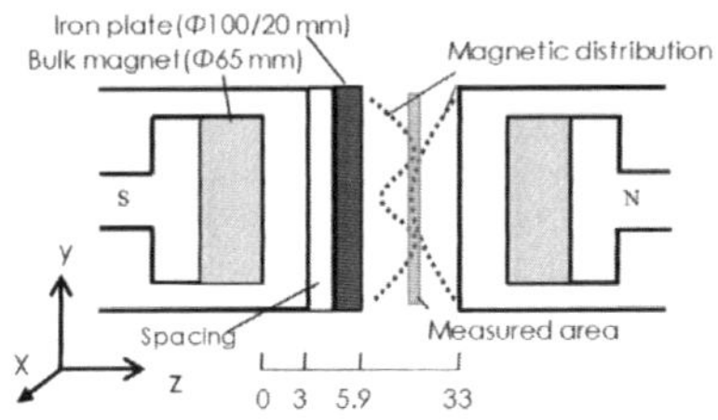

Figure 2. Schematic structure of face-to-face arranged magnetic poles.

showed us the best uniformity of 30ppm at 1.1 T, which also suggested us the feasibility of novel compact NMR device [8]. In the study, the authors investigated the magnetic field distribution more precisely than ever by evaluating the uniformity on x-y plane to estimate the deformation by iron plates.

2. Experimental procedure

2.1. Magnetic pole structures and arrangements

We employed the single and the face-to-face bulk magnet systems, which were kept at 31 K by GM-cryocooler in the series of experiments. The actual instances of system configuration are shown elsewhere [8-10]. Figure 1 shows how to change the magnetic field distribution from the original convex to concave shapes by attaching an iron plate of 100 mm in diameter. The top view along z-axis indicates the measuring area by scanning Hall probe, the location of bulk magnet in the magnetic pole chamber, and an iron plate on the pole surface. In the experiment, a couple of iron plates of 20 mm and 100 mm in diameter and 2.26 mm in thickness were attached on one magnetic pole in turn.

Figure 2 shows a schematic structure of the face-to-face arranged magnetic poles with and without an iron plate. The magnetic fields of convex and concave shapes were combined with various gaps of 30-70 mm. In the figure, the scale indicates the position of the x-y plane on which the field intensity data were measured. Note that the distance from the bulk magnet surface to the iron plate surface is 5.9 mm, which is defined as the null point.

The bulk magnet was activated by the pulsed field magnetizing method beforehand, emitting magnetic flux density of 1.47 T on the left-hand side pole and 1.8 T on the other side pole.

2.2. Evaluation of magnetic field uniformity

Figure 3a shows an evaluation method for magnetic field uniformity. A facet of 4 mm x 4 mm area is derived from the magnetic field distribution map, and the uniformity is calculated by the formula (1) by 21 data points except those from four corners of the region. As shown in Figure 3b, the uniformity

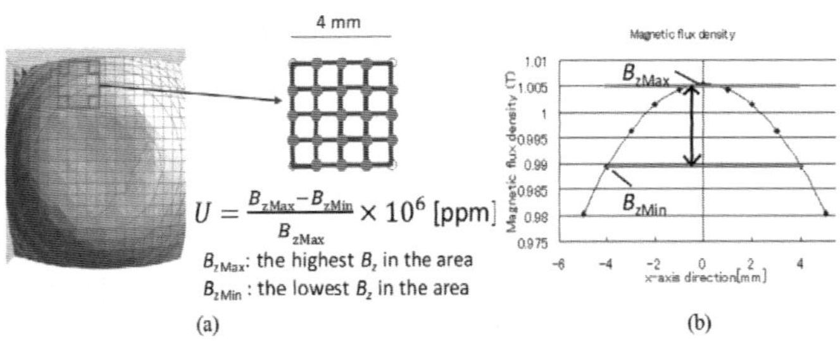

Figure 3. Evaluation for magnetic field uniformity.

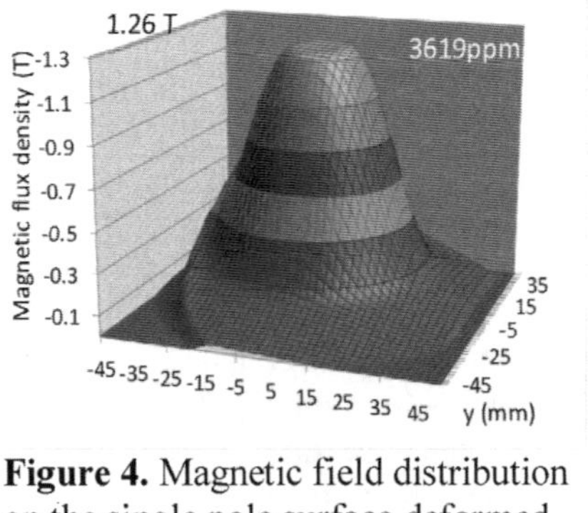

Figure 4. Magnetic field distribution on the single pole surface deformed by attaching a 20-mm iron plate.

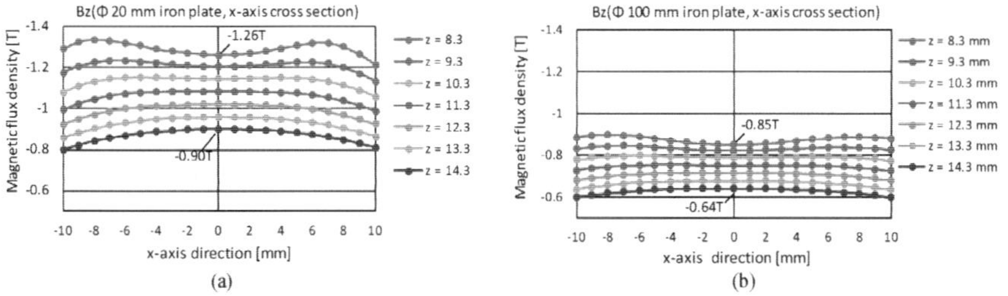

Figure 5. Position dependence of magnetic field distribution along x-axis with 20 mm (a) and 100 mm (b) iron plates in diameter.

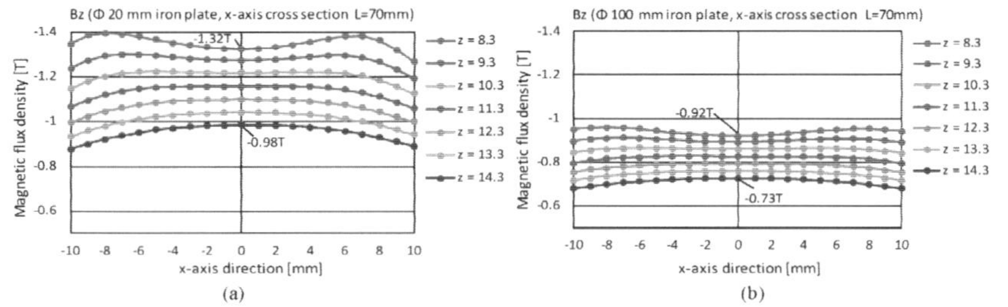

Figure 6. Magnetic field distribution along x-axis with 20 mm and 100 mm iron plates attached on the left-hand side of face-to-face settled poles with a gap 70 mm.

data were also derived from the averaged 5-point data in the 4-mm span of x- or y-axes cross section as in the manner in the past papers [8-10], where B_{zMax} and B_{zMin} mean the highest and lowest B_z values in the area, respectively.

$$U = \frac{B_{zMax} - B_{zMin}}{B_{zMax}} \times 10^6 \text{ [ppm]} \qquad (1)$$

3. Results and Discussion

An iron plate was attached on the left pole of face-to-face system. In Figure 4, the magnetic field distribution gives us a concave shape on the iron plate surface due to its shielding effect. The iron plate was composed of Fe-C (<0.25%) with the dimensions of 20 mm in diameter and 2.26 mm in thickness. The original intensity 1.47 T changed to 1.26 T at the center of the bulk surface. As for the single pole, the uniformity was evaluated as 3,619ppm (parts per million) at the center point.

Figure 5 shows the unidirectional magnetic flux distribution measured along x-axis as a function of distance from the iron surface. The negative values mean S pole. The distribution changed from convex shape to concave with increasing distance. The magnetic field intensity changed from 1.26 T at 8.3 mm to 0.90 T at 14.3 mm. One may expect the magnetically-flat region at around 11.3 mm from the null point. When we replaced the 20-mm iron plate to 100-mm in diameter, the magnetic field intensity degraded due to enhanced shielding effect in comparison with that of 20 mm in diameter.

When we refer to the face-to-face system with a gap of 70 mm between the poles, as shown in Figure 6b, the magnetic field intensity measured at 8.3 mm increased to 1.32 T at the center point. As shown in Figure 6b, the intensity degraded to 0.92 T at 8.3 mm when the iron plate size was changed from 20 mm to 100 mm.

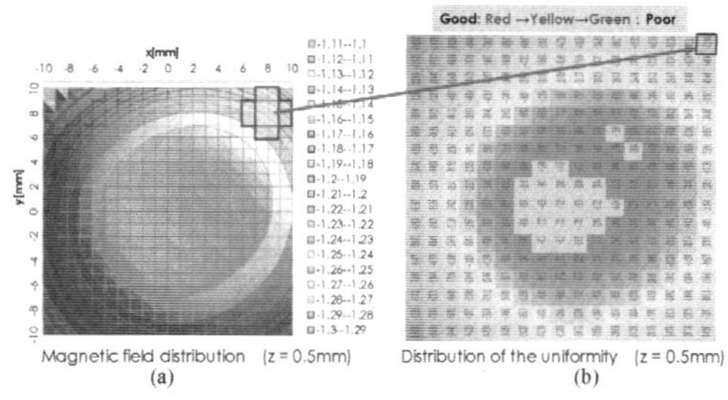

Figure 7. Mapping of uniformity data.

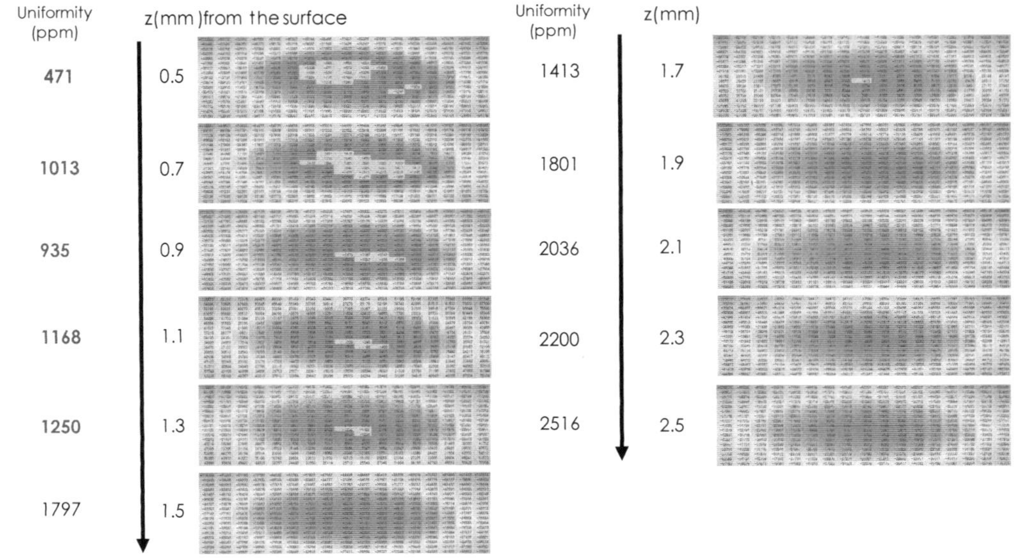

Figure 8. Thickness of uniform area along z-axis.

Figure 7 shows the mapping of trapped fields and uniformity data. The facets indicate the uniformity which were calculated by averaging surrounding 21-point uniformity data. The centered position of pink region corresponds to the area less than 1,500ppm. The best uniformity reached 463ppm at the maximum field of 1.25 T. Figure 8 shows the mapping of magnetic field data measured in 1 mm pitch on x-y plane. The sheets show the uniformity along z-axis of 0.5-2.5mm from the pole surfaces. As for the region which exhibited the uniformity less than 1,500ppm, we obtained the uniform area in the range of 0.5-1.3 mm along z-axis from the surface. Then, we have succeeded in obtaining the uniform field area less than 1,500ppm in the space of 4 mm x 4 mm x 0.8 mm.

Figure 9a shows the position dependence of uniformity data vs position along z-axis, which were obtained in x-y planes, in the cross section of x-axis, and y-axis, respectively. We obtained the best uniformity data 878ppm at z=10.9 mm in the plane, 706ppm in x-axis, and 788ppm in y-axis directions. As shown in Figure 9b, we attempted to evaluate the reproducibility of the measurement. One sees slight deviation in the repeatedly-measured data, which clearly exhibited however the presence of uniform region less than 1,500ppm in the range of 0.8 mm, from 0.5 mm to 1.3 mm of z-axis from the pole surface. As this uniformity is sufficient to detect NMR signals.

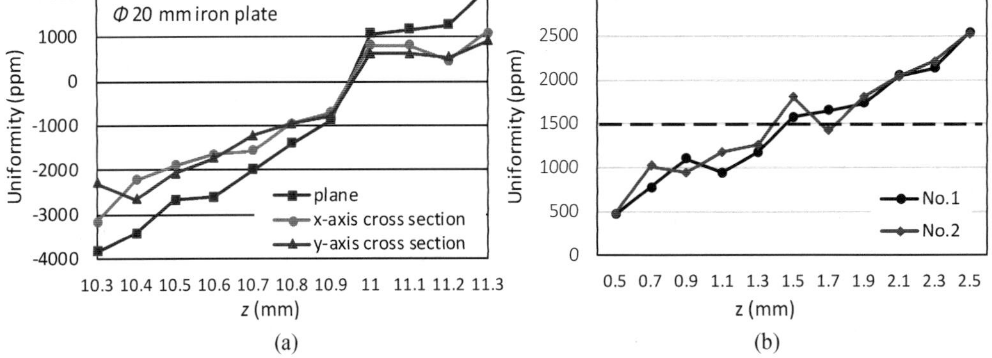

Figure 9. Position dependence of uniformity measured along z-axis, obtained in the x-y plane, x- and y-axis cross sections, and uniformity along x- and y-axes cross sectional distribution measured in the gap of magnetic poles (a), while (b) shows the reproducibility of magnetic field measurement.

4. Conclusion

We discussed about our recent attempts to develop the uniform magnetic field space for NMR applications. Attaching iron plate to one magnetic pole was found to be sufficiently effective in forming magnetically uniform field space, and we have succeeded in obtaining the field space with the best uniformity less than 500ppm at 0.5 mm position, and less than 1,500ppm in the range 0.5-1.3 mm. This implies that the application to the novel compact NMR/MRI devices must be feasible with use of HTS bulk magnets and PFM activation.

References

[1] Oka T, Yokoyama K, Itoh Y, Yanagi Y, Yoshikawa M, Ikuta H, Mizutani U, Okada H and Noto K 2003 A 3 T magnetic field generator using melt-processed bulk superconductors as trapped field magnets and its applications *Physica C* **392–396** 709

[2] Gruss S, Fucks G, Krabbes G, Verges P, Schaetzle P, Mueller K-H, Fink J and Schultz L 2001 Trapped field beyond 14 Tesla in Bulk $YBa_2Cu_3O_{7-d}$ *IEEE Trans. Appl. Supercond.* **11** 3720

[3] Tomita M and Murakami M 2003 High-temperature superconductor bulk magnets that can trap magnetic fields of over 17 Tesla at 29 K *Nature* **421** 517

[4] Durrell J, Dennis A, Jaroszynski J, Ainslie M, Palmer K, Shi Y-H, Campbell A, Hull J, Strasik M, Hellstrom E and Cardwell D 2014 A trapped field of 17.6 T in melt-processed bulk Gd-Ba-Cu-O reinforced with shrink-fit steel *Supercond. Sci. Technol.* **27** 082001

[5] Nakamura T, Itoh Y, Yoshikawa M, Oka T and Uzawa J 2007 Development of a Superconducting Magnet for Nuclear Magnetic Resonance Using Bulk High-Temperature Superconducting Materials *Concepts in Magnetic Resonance Part B (Magnetic Resonance Engineering)* **31B(2)** 65

[6] Ogawa K, Nakamura T, Terada Y, Kose K and Haishi T 2011 Development of a magnetic resonance microscope using a high T_c bulk superconducting magnet *Appl. Phys. Lett.* **98** 234101-1

[7] Oka T, Hirayama E, Takahashi Y, Kanai T, Ogawa J, Fukui S, Sato T, Yokoyama K and Nakamura T 2015 Attempt to generate a strong and uniform magnetic field by face-to-face HTS bulk elements in a magnet system *Physics Procedia* **67** 1005

[8] Oka T, Takahashi Y, Yaginuma S, Ogawa J, Fukui S, Sato T, Yokoyama K and Nakamura T 2016 Uniform magnetic field between face-to-face settled HTS bulk magnets combining concave and convex magnetic field distributions *Physics Procedia* **81** 45

[9] Oka T, Ogawa J, Fukui S, Sato T, Nakano T, Yokoyama K, Nakamura T, Fujishiro H and Noto K 2018 A Review Article: Compact Magnetic Field Generators Containing HTS Bulk Magnets Cooled by Refrigerators and Their Feasible Applications *High-Temperature Superconductors: Occurrence, Synthesis and Applications* ed M Miryala and M Koblischka (Nova Science Publishers) chapter 13 pp 291-306

[10] Oka T, Takeda A, Shimizu K, Higa K, Tsunoda S. Ogawa J, Fukui S, Sato T, Yokoyama K and Nakamura T 2018 Feasible Applications of Bulk HTS Magnet and Magnetic Field-Capturing Characteristics in Their Activation Process *IEEE Trans. Appl. Supercond.* **29** 1

IOP Conf. Series: Journal of Physics: Conf. Series **1293** (2019) 012045 doi:10.1088/1742-6596/1293/1/012045

Fracture strength properties of (Gd,Y)BaCuO large single-grain bulk at liquid nitrogen temperature

A Murakami[1] and A Iwamoto[2]

[1] National Institute of Technology, Ichinoseki College, Takanashi, Hagisho, Ichinoseki, Iwate 021-8511, Japan
[2] National Institute for Fusion Science, 322-6 Oroshi-cho, Toki, Gifu 509-5292, Japan

E-mail: amura@ichinoseki.ac.jp

Abstract. In order to investigate the fracture strength properties of a (Gd,Y)BaCuO large single-grain bulk with 90 mm in diameter, bending tests were carried out at liquid nitrogen temperature 77 K for the specimens cut from the bulk sample. The large single-grain bulk was fabricated by RE compositional gradient technique to overcome the problem of undesirable nucleation apart from the seed crystal. One precursor that consisted of some regions with different Gd and Y contents was used for the fabrication. Bending tests for the plain and the V-notched specimens were carried out to evaluate the fracture strength and fracture toughness, respectively. The fracture strength values of the specimens that included the interface between two regions with different Gd and Y contents were not lower than those of the specimens that did not include the interface, which is similar to the bending test results at room temperature. The fracture strength values at 77 K were higher than those at room temperature.

1. Introduction

It is well-known that REBaCuO superconducting bulk materials, where RE denotes yttrium or lanthanoid element, have to be a single-grain for their excellent superconducting properties. Thus, REBaCuO bulk materials are fabricated by melt-processing using a seed crystal. Enlargement of REBaCuO bulk size is effective in improving the performances for some superconducting devices. However, it is difficult to obtain large single-grain bulk piece without undesirable nucleation apart from the seed crystal. In order to overcome the problem of undesirable nucleation, RE compositional gradient technique has been developed [1,2]. RE compositional gradient technique is based on the difference in the peritectic decomposition temperature of REBaCuO. One precursor that consists of some regions with different RE contents is used for the fabrication. It has been reported that the trapped magnetic field profile of a large single-grain bulk fabricated by RE compositional gradient technique had single peak [2], which is similar to the trapped magnetic field profiles of smaller REBaCuO single-grain bulks fabricated by using a precursor with no compositional gradient.

Since superconducting bulk materials are subjected to electro-magnetic force and thermal stress in the superconducting devices [3,4], evaluations of the fracture strength properties are indispensable for their practical applications. The present authors previously evaluated the fracture strength of a

(Gd,Y)BaCuO large single-grain bulk fabricated by RE compositional gradient technique through bending tests at room temperature; the fracture strength of the large single-grain bulk was comparable to that of a smaller GdBaCuO bulk fabricated by using a precursor with no compositional gradient [5].

In this study, the fracture strength properties were evaluated at liquid nitrogen temperature 77 K through bending tests for the specimens cut from a (Gd,Y)BaCuO large single-grain bulk sample. It is considered that the evaluations at 77 K are informative for the development of superconducting devices equipped with REBaCuO single-grain bulks.

2. Experimental procedure

Bending test specimens were cut from a (Gd,Y)BaCuO single-grain bulk sample with 90 mm in diameter fabricated by Nippon Steel & Sumitomo Metal Corporation using RE compositional gradient technique. As schematically shown in Figure 1, one precursor with 25 mol% RE_2BaCuO_5 (RE211) was prepared such that the Y content increased with increase of the distance from the centre of it. 1 wt% Ce was added to the precursor to inhibit coarsening of RE211 particles [6,7]. 16.7 wt% Ag was also added to the precursor. Ag addition is effective in improving the fracture strength properties [8-10]. The precursor was heated in air up to 1443 K, kept at that temperature for 1 h and then cooled down to 1310 K. After that, one (Nd,Sm)BaCuO seed crystal was placed on the top of it. Oxygen annealing was conducted at 673 K for 100 h for the melt-grown bulk sample.

Bending test specimens with the dimensions of 2.8 x 2.1 x 24 mm^3 were cut from the bulk sample as schematically shown in Figure 1. Specimens that did not include the interface between two regions with different Gd and Y contents are denoted as Specimens A. On the other hand, specimens that included the interface are denoted as Specimens B. Plain specimens and V-notched specimens were used for the evaluations of the fracture strength and fracture toughness, respectively, as schematically shown in Figure 2 (a) and (b). While Specimens A1, A3, A5, B1 and B2 were used for the fracture strength evaluation, Specimens A2, A4 and A6 were used for the fracture toughness evaluation. V-notch was introduced by using razor with diamond paste. The notch root radius was about 20 μm. 4-

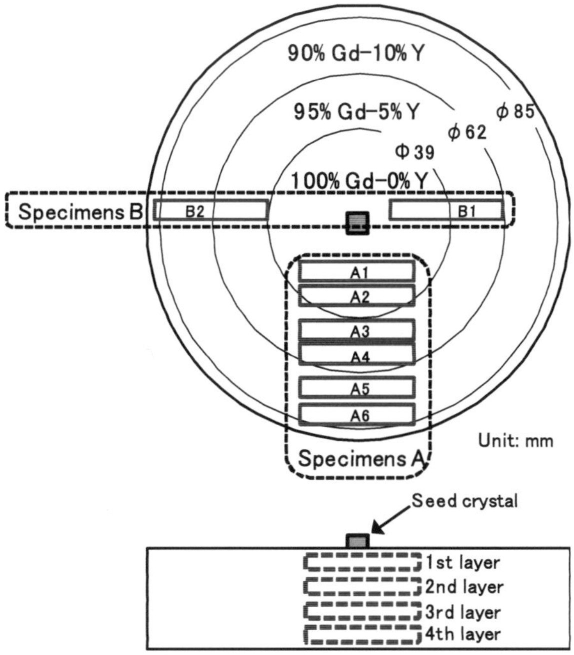

Figure 1. Schematic illustration of cutting out of bending test specimens.

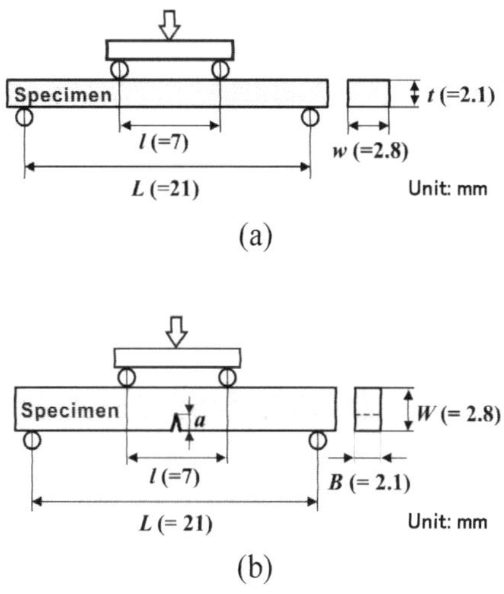

Figure 2. Schematic illustration of 4-point bending tests for (a) fracture strength and (b) fracture toughness evaluations, respectively.

point bending load was applied at 77 K under the crosshead speed of 0.2 mm/min. Before the loading, the specimen was immersed into the liquid nitrogen bath, together with the 4-point bending jig. Fracture strength σ was calculated by the following equation.

$$\sigma = \frac{3P(L-l)}{2wt^2} \quad (1)$$

Where P is the maximum load, L is outer supporting span, l is inner loading span, w and t are width and thickness of the plain specimens, respectively. Fracture toughness K_{IC} was calculated by the following equations [11].

$$K_{IC} = \sigma\sqrt{\pi a}F(a/W) \quad (2)$$

$$\sigma = \frac{3P(L-l)}{2BW^2} \quad (3)$$

$$F(a/W) = \sqrt{\frac{2W}{\pi a}\tan(\frac{\pi a}{2W})}\,\frac{0.923+0.199(1-\sin\frac{\pi a}{2W})^4}{\cos\frac{\pi a}{2W}} \quad (4)$$

Where σ is the fracture strength, P is the maximum load, a is V-notch depth, L is outer supporting span, l is inner loading span, W and B are height and thickness of the V-notched specimens, respectively. Fracture surfaces of the V-notched specimens were observed by using digital microscope KEYENCE VHX-500 with 3D observation mode. The V-notch depth was measured on the fracture surfaces by using image analysis software.

3. Fracture strength at 77 K

Figure 3 presents the fracture strength values at 77 K of Specimens A1, A3, A5, B1 and B2 cut from 1st-4th layers of the bulk sample (see Figure 1). Open and solid symbols represent the data of Specimens A and B, respectively. Since it has been reported that the fracture strength of YBaCuO bulk was superior to those of other REBaCuO bulks [12], it was expected that the fracture strength increased with increase of the distance from the centre of the bulk sample. However, such an increase is not observed for the bending test results. The fracture strength values of Specimens B2 are slightly lower than those of Specimens B1. However, the distributions of pores and RE211 particles of Specimens B1 and B2 were similar to each other.

Figure 4 presents comparison of the fracture strength at 77 K between Specimens A and B. The average fracture strength values at room temperature reported elsewhere [5] are also shown for reference. The fracture strength values of Specimens B are not lower than those of Specimens A. Thus, it is demonstrated that the interfaces between two regions with different Gd and Y contents do not cause the deterioration of the fracture strength for the whole bulk material. The average fracture strength values at 77 K are higher than those at room temperature. It is deduced that such an improvement of the fracture strength is mainly attributable to the decrease of the inter-atomic distance by cooling. It has been reported that the fracture strength values of conventional GdBaCuO single-grain bulks with 30-40 mm in diameter fabricated by using a precursor with no compositional gradient were 97 MPa at 77 K [13], which was obtained under the same testing conditions as those of the bending test carried out in this study, and 69-115 MPa at room temperature [10]. The fracture strength value commonly depends on the testing conditions, such as the dimensions of specimen and loading method. Thus, it is considered that the fracture strength of the (Gd,Y)BaCuO large single-grain bulk fabricated by RE compositional technique is comparable to that of the smaller GdBaCuO single-grain bulk. It has been reported that the fracture strength values at 77 K of a (Gd,Dy)BaCuO bulk fabricated by RE compositional gradient technique were around 90 MPa [14]. Thus, it is also considered that the fracture strength values of the (Gd,Y)BaCuO bulk tested in this study are comparable to those of the (Gd,Dy)BaCuO bulk.

Figure 5 presents Weibull plots of the fracture strength data at 77 K. F, σ and m represent cumulative fracture probability, fracture strength and Weibull coefficient, respectively. Weibull coefficient values of the fracture strength at room temperature [5] are also shown for reference. Weibull coefficient value is obtained through linear fitting of the Weibull plot data, and it is commonly used for the evaluation of scatter of the data; larger Weibull coefficient value means smaller scatter of the data. Weibull coefficient values of Specimens A are smaller than those of Specimens B. It is deduced that the smaller Weibull coefficient values of Specimens A are associated with the large distance between Specimens A1 and A5. There is no significant difference in the Weibull coefficient value between 77 K and room temperature. Meanwhile, it seems that fitted curve with curvature changing in the middle, which consists of two straight lines, can be drawn for the Weibull plot data, so there is a possibility that this bulk has two fracture modes.

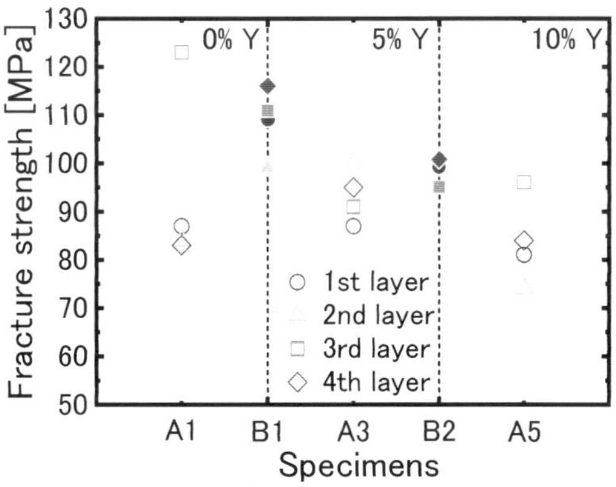

Figure 3. Fracture strength values at 77 K for Specimens A1, A3, A5, B1 and B2 cut from 1st-4th layers of the bulk sample (see Figure 1).

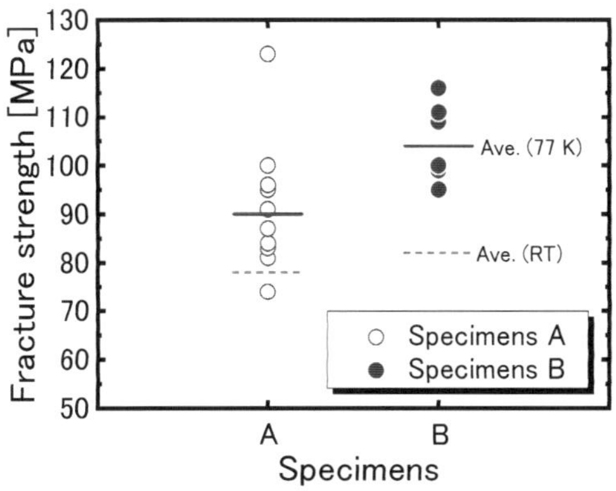

Figure 4. Fracture strength of Specimens A and B at 77 K. The average fracture strength values at room temperature reported elsewhere [5] are also shown for reference.

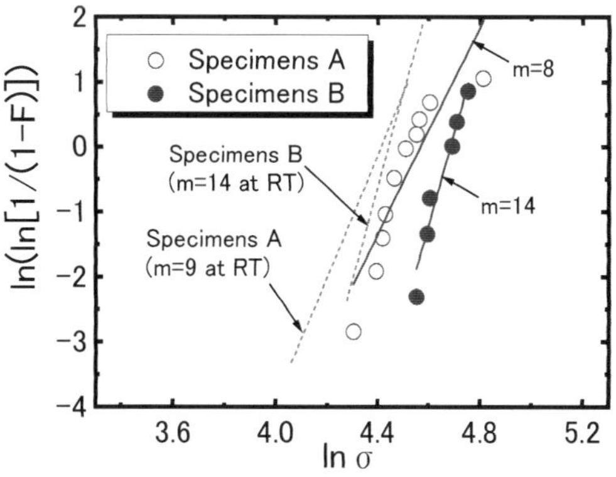

Figure 5. Weibull plots of the fracture strength data at 77 K. F, σ and m represent cumulative fracture probability, fracture strength and Weibull coefficient, respectively. Weibull coefficient values of fracture strength at room temperature [5] are also shown for reference.

4. Fracture toughness at 77 K

Figure 6 presents the fracture surface of a V-notched specimen. In order to evaluate the fracture toughness, V-notch depth was measured on the fracture surfaces. Figure 7 presents the fracture toughness values at 77 K of Specimens A2, A4 and A6. Since it is impossible to adjust the V-notch to the interface between two regions with different Gd and Y contents, the fracture toughness tests of Specimens B were not carried out. It has been reported that the fracture toughness values at 77 K of a (Gd,Dy)BaCuO large single-grain bulk fabricated by RE compositional gradient technique were around 1.0-1.8 MPa $m^{1/2}$ [15]. Fracture toughness values of the (Gd,Y)BaCuO bulk tested in this study are comparable to the values of the (Gd,Dy)BaCuO. While the fracture strength value does not increase with increase of the distance from the seed crystal as mentioned above, the fracture toughness value slightly increases. It is considered that Y addition is effective in improving the crack propagation resistance.

Figure 6. Fracture surface of V-notched specimen.

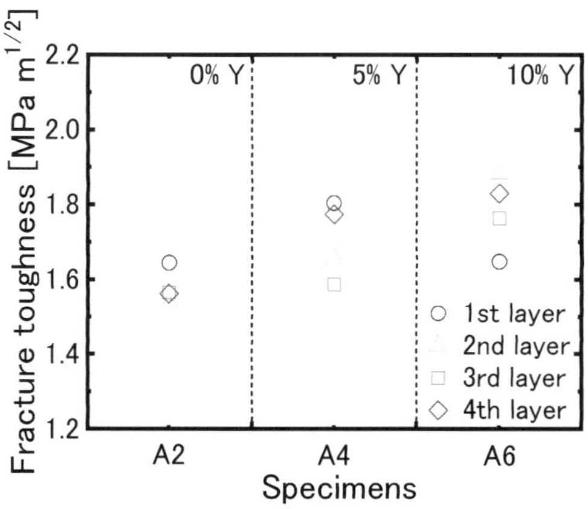

Figure 7. Fracture toughness of Specimens A2, A4 and A6 evaluated through bending tests at 77 K for V-notched specimens.

5. Summary

In order to investigate the fracture strength properties of a (Gd,Y)BaCuO large single-grain bulk with 90 mm in diameter fabricated by RE compositional gradient technique, bending tests were carried out at 77 K for the plain and the V-notched specimens cut from the bulk sample. The fracture strength values at 77 K of the specimens that included the interface between two regions with different Gd and Y contents were not lower than those of the specimens that did not include the interface, which is similar to the bending test results at room temperature. The fracture strength at 77 K evaluated in this study was superior to that at room temperature evaluated in the previous study. The fracture strength of the large single-grain bulk was comparable to those of smaller single-grain bulks fabricated by using a precursor with no compositional gradient. While the fracture strength values did not increase with increase of the distance from the seed crystal, the fracture toughness values slightly increased.

Acknowledgements

The authors thank students of National Institute of Technology, Ichinoseki College for their relevant technical support. This work was supported in part by KAKENHI (17K06825) and NIFS Collaboration Research Program (NIFS17KECA050).

IOP Conf. Series: Journal of Physics: Conf. Series **1293** (2019) 012045 doi:10.1088/1742-6596/1293/1/012045

References

[1] Morita M, Teshima H and Hirano H 2006 *Nippon Steel Tech. Rep.* **93** 18

[2] Teshima H, Morita M, Okajima N, Terao Y, Sekino M, Ohsaki H and Murakami A 2013 *Proc. ICEC 24 - ICMC 2012* 879

[3] Ren Y, Weinstein R, Liu J, Sawh R P and Foster C 1995 *Phys. C, Supercond.* **251** 15

[4] Miyamoto T, Nagashima K, Sakai N and Murakami M 2000 *Supercond. Sci. Technol.* **13** 816

[5] Murakami A and Iwamoto A *IEEE Trans. Appl. Supercond.* Accepted

[6] Muralidhar M, Jirsa M, Iida K and Murakami M 2001 *Phys. C* **357-360** 665

[7] Chen S K, Zhou L, Wang K G, Wu X Z, Zhang P X and Feng Y, 2002 *Phys. C* **366** 190

[8] Matsui M, Sakai N and Murakami M 2002 *Supercond. Sci. Technol.* **15** 1092

[9] Fujimoto H 2003 *Supercond. Sci. Technol.* **16** 1115

[10] Nariki S, Sakai N, Tomita M and Murakami M 2002 *Phys. C* **378-381** 779

[11] Tada H, Paris P C and Irwin G 1985 *The Stress Analysis of Crack Handbook.* Del Res. Corp.

[12] Katagiri K, Murakami A and Kasaba K 2008 *Cryogenics* **48** 87

[13] Fujimoto H and Murakami A 2011 *IEEE Trans. Appl. Supercond.* **21** 2718

[14] Murakami A, Teshima H, Morita M, Kudo T and Iwamoto A 2014 *J. Phys.: Conf. Ser.* **507** 012034

[15] Murakami A, Fujimoto H and Iwamoto A 2013 *IEEE Trans. Appl. Supercond.* **23** 6800505

IOP Conf. Series: Journal of Physics: Conf. Series **1293** (2019) 012046 doi:10.1088/1742-6596/1293/1/012046

Improvement of the trapped field of REBCO bulk activated by pulsed field magnetization with a large soft-iron yoke

K Yokoyama[1] and T Oka[2]

[1] Division of Electrical and Electronic Engineering, Ashikaga University, 268-1 Omae-cho, Ashikaga, Tochigi 326-8558, Japan
[2] Materials Science and Engineering, Shibaura Institute of Technology 3-7-5 Toyosu, Koto-Ward, Tokyo, 135-8548 Japan

Abstract. We study to increase the trapped magnetic field of a REBCO bulk magnet excited by pulsed field magnetization (PFM). In PFM, a soft-iron yoke is used to expose the bulk to the magnetic field for a long time. Although a small yoke 40 mm in diameter was used in our previous study, it was changed to a large yoke 64 mm in diameter in this paper. Accordingly, the pulse width was extended by 13%. When a GdBCO bulk 60 mm in diameter and 20 mm thick was magnetized by varying an amplitude of the applied field and temperature in a bulk magnet system using a two-stage GM-type refrigerator with a cooling capacity of 12 W at 20 K, the maximum trapped flux density reached 4.07 T on the center of the bulk surface and 2.56 T on the pole surface at 20 K. Moreover, the total magnetic flux was 3.28 mWb, which was about 13% larger than that of a small yoke. This result suggests that this method is useful in practice because the trapped field can be enhanced only by changing the soft-iron yoke.

1. Introduction

The trapped magnetic field of a REBCO bulk increases due to improvement of the mechanical strength and critical current density [1-4]. Recently, a roadmap of HTS bulk for scientific and industrial applications was published [5], and it is expected that research on its material, magnetization, and practical use has been accelerated. When a bulk is applied as a strong magnetic field generator, magnetization is a very important problem. Pulsed field magnetization (PFM) has some advantages, such as a high versatility of exciting equipment and short operating time, while an amplitude of the trapped field is low as compared with field cooling method. Then, various methods to improve the trapped field performance are considered, and they can be broadly divided into two approaches: one is the improvement of the magnetizing method, and the other is the modification of the exciting equipment. In the former, several advanced PFM methods such as IMRA, MPSC, and MMPSC were proposed [6-8], and the high trapped field of 5.2 T was achieved by the MMPSC method [8]. In the latter, a split coil was used, and the pulse width was extended by changing the inductance of the exciting coil and the capacitance of the condenser of the current source [9-12]. However, the cost of the above modifications is significant.

In our magnetizing system, a bulk is put between two soft-iron yokes to expose the bulk to the magnetic field for a long time. Moreover, they have a role in protecting the refrigerator from the impact caused by repulsive force when applying a strong magnetic field because the attractive force operates on yokes in the direction of pulling toward the center of the coil, although a repulsive force operates on the bulk in the direction in which the sample gets away from the coil. Since we used a small bulk 36 mm in diameter when we started this study, a small soft-iron yoke 40 mm in diameter was applied. In this paper, the yoke was changed to a large one 64 mm in diameter because a large bulk 60 mm in diameter was applied presently. By using a large yoke, we expected that the pulse width would be extended because the magnetic flux hold in a soft-iron yoke was increased. This paper evaluates the influence of the yoke size on a magnetizing characteristic quantitatively by applying a single pulsed field with varying an amplitude of the magnetic field and temperature to a GdBCO bulk 60 mm in diameter using the two different yokes above in a bulk magnet system using a two-stage GM-type refrigerator with cooling ability of 12 W at 20 K.

2. Experimental

Figure 1 shows a schematic of a bulk magnet system when a magnetizing coil (ID of 93 mm, OD of 124 mm, and height of 48 mm) was attached to the magnetic pole. A bulk was put between two soft-iron yokes; one was arranged under the sample (referred to as the "lower-yoke"). Another was placed on the upper side of the sample (referred to as the "upper-yoke"), which was attached to a magnetizing coil, and, thus, it was removed together with the coil after magnetization. The lower yoke was made of permendur, and the dimensions were 64 mm in diameter and 20 mm thick. The upper yoke was made of pure iron, and two different sizes were used; one was 40 mm in diameter and 47 mm thick, and the other was 64 mm in diameter and 47 mm thick. Figure 2 shows the time changes of the applied field normalized by the maximum flux density when the large yoke (L-yoko) and the small yoke (S-yoke) were used. The pulse width of the L-yoke was 1.3 times longer than that of the S-yoke as an intended aim.

A high-performance GdBCO bulk 60 mm in diameter and 20 mm thick that was reinforced mechanically with a SUS316L steel ring 2 mm thick (Nippon Steel & Sumitomo Metal Corporation) was inserted in a sample holder, and it was fixed to the sample stage with screws. A transverse-type Hall sensor (BHT-921, F.W.BELL) was attached on the center of sample surface with Kapton tape for measuring the time change of the magnetic flux density during magnetization, and a Cernox temperature sensor (CX-1030-SD-HT-1.4L/J, Lake Shore Cryotronics) was also adhered near the Hall sensor with GE7031 varnish for measuring the time change of temperature. After a chamber was covered and evacuated by a vacuum pump, the bulk was cooled and the temperature was adjusted to 20, 30, 40, and 50 K by a temperature controller. A single pulsed field was applied with varying an amplitude from 3.9 to 6.2 T for each temperature, where the rising time and the fall time of the pulsed field were 10 ms and 100 ms, respectively. During magnetization, the magnetic flux density and temperature were monitored in intervals of 100 μs and 100 ms, respectively. After magnetization, the three-dimensional Hall sensor (BH-703, F.W.BELL) was scanned on the magnetic pole surface with a 2 mm pitch for measuring the distribution of the trapped flux density; moreover, the total magnetic flux was calculated by integrating the measured flux density over the whole region.

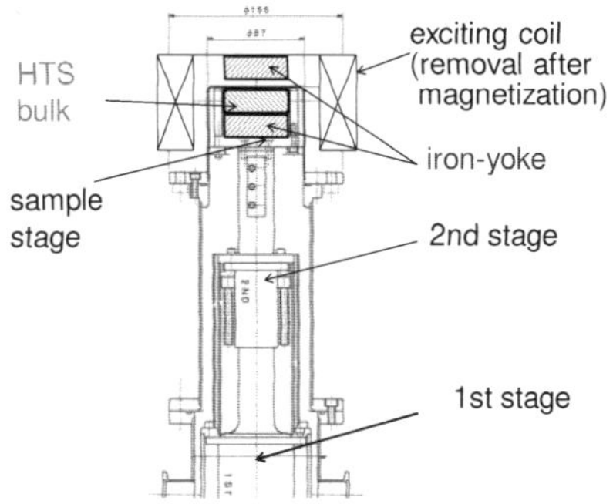

Figure 1. Schematic of experimental setting for PFM. An HTS bulk was put between two soft-iron yokes.

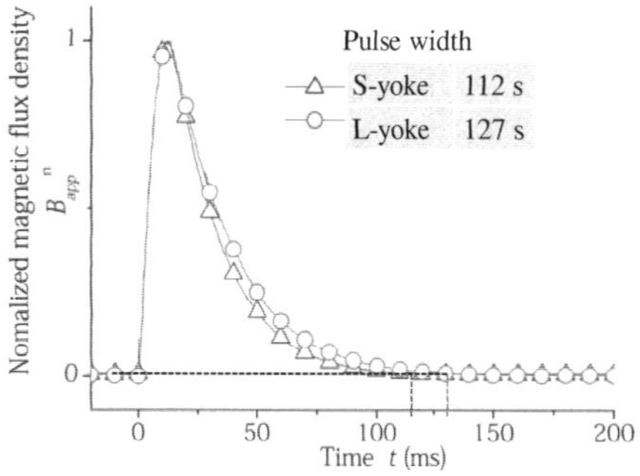

Figure 2. Time changes of normalized applied field of the large yoke (L-yoke) and the small yoke (S-yoke).

3. Results and discussion

Figure 3 shows the time responses of the magnetic flux density, B_z, on the center of the bulk surface for each applied field in (a) the L-yoke and (b) the S-yoke. In the L-yoke at 20, 30, and 40 K, a trapped field increased as an increasing an applied field, B_{app}, increased from 3.9 T to 5.4 T; moreover, the decrease in B_z after reaching a peak was slight, indicating that the flux flow hardly happened.

Therefore, a high trapped field was achieved for B_{app}=4.6 and 5.4 T. However, flux jump occurred for B_{app}=6.2 T; accordingly, the B_z value decreased greatly. The B_z value increased gradually after reaching the minimum value. This means that magnetic flux gathers from a peripheral part in the center by the flux flow. In the S-yoke, on the other hand, flux jump occurred with a lower applied field, B_{app}=5.4 T at 30 and 40 K and B_{app}=4.6 T at 50 K, as compared with the L-yoke. These results suggest that it is hard for flux jump to occur in a high applied field due to the extension of the pulse width by using a large yoke.

Aiming to find a suitable applied field for a high field trapping, an applied field was increased by 0.1 T from 5.4 to 6.2 T at 20 K. When a magnetic field of 5.8 T was applied, a trapped field of 4.07 T was achieved.

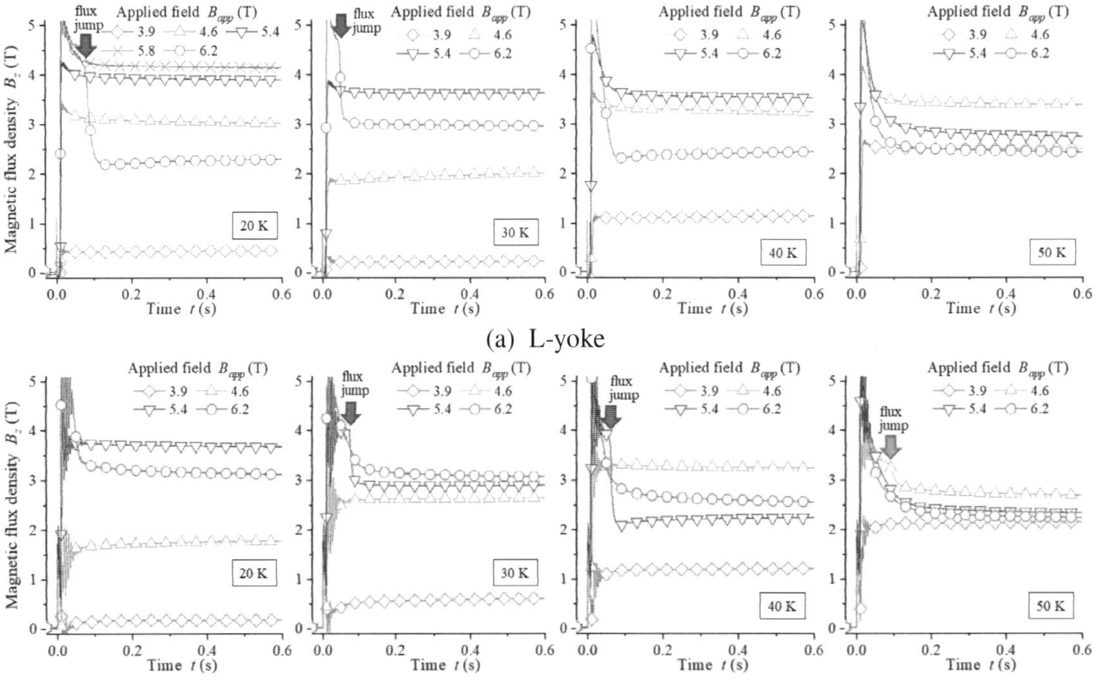

(b) S-yoke

Figure 3. Time responses of magnetic flux density on the center of the bulk surface at 20-50 K in (a) the L-yoke and (b) the S-yoke.

Figure 4 illustrates the magnetic field distributions on the pole surface for B_{app}=3.9-5.4 T at 20-50 K in the L-yoke and the S-yoke. For B_{app}=3.9 T in the L-yoke, the trapped flux density on the center of the sample was low, as compared with the peripheral part at 20 and 30 K, indicating that the magnetic flux did not sufficiently reach the center of the sample, due to a strong magnetic shield. On the other hand, the magnetic flux reached the center due to a low magnetic shield, and the distributions were concentrically shaped at 40 and 50 K. For B_{app}=4.6 and 5.4 T, the trapped magnetic flux increased with increasing the applied field at 20 and 30 K, while it decreased with increasing the applied field. An area of high flux density of more than 2.0 T appeared for B_{app}=4.6 and 5.4 T at 20, 30, and 40 K, and that of B_{app}=5.4 T at 20 K was the widest. In the S-yoke, on the other hand, the tendency of field trapping was the same as that of the L-yoke; however, the amplitude of the trapped field was lower than that of the L-yoke.

Figure 5 shows the magnetic flux density, B_z, on the center of the bulk surface (open symbols) and on the pole surface (closed symbols) as a function of the applied field, B_{app}, in the L-yoke (red) and the S-yoke (blue) at 20, 30, 40, and 50 K. Although both tendencies of variation were the same, the B_z value of the L-yoke was bigger than that of the S-yoke in almost all cases. The maximum trapped flux density on the bulk surface was 4.07 T for B_{app}=5.8 T at 20 K, and the maximum flux density on the pole surface was 2.35 T for B_{app}=5.4 T at 20 K.

Figure 6 compares the total magnetic flux, Φ, that is calculated using the distribution data from figure 4 between the L-yoke (red circle) and the S-yoke (blue triangle) at 20-50 K. The variation of Φ did not always correspond to the variation of B_z in figure 5 because the magnetic flux decreased in the peripheral part due to the flux flow, even though the maximum trapped field at the center was high.

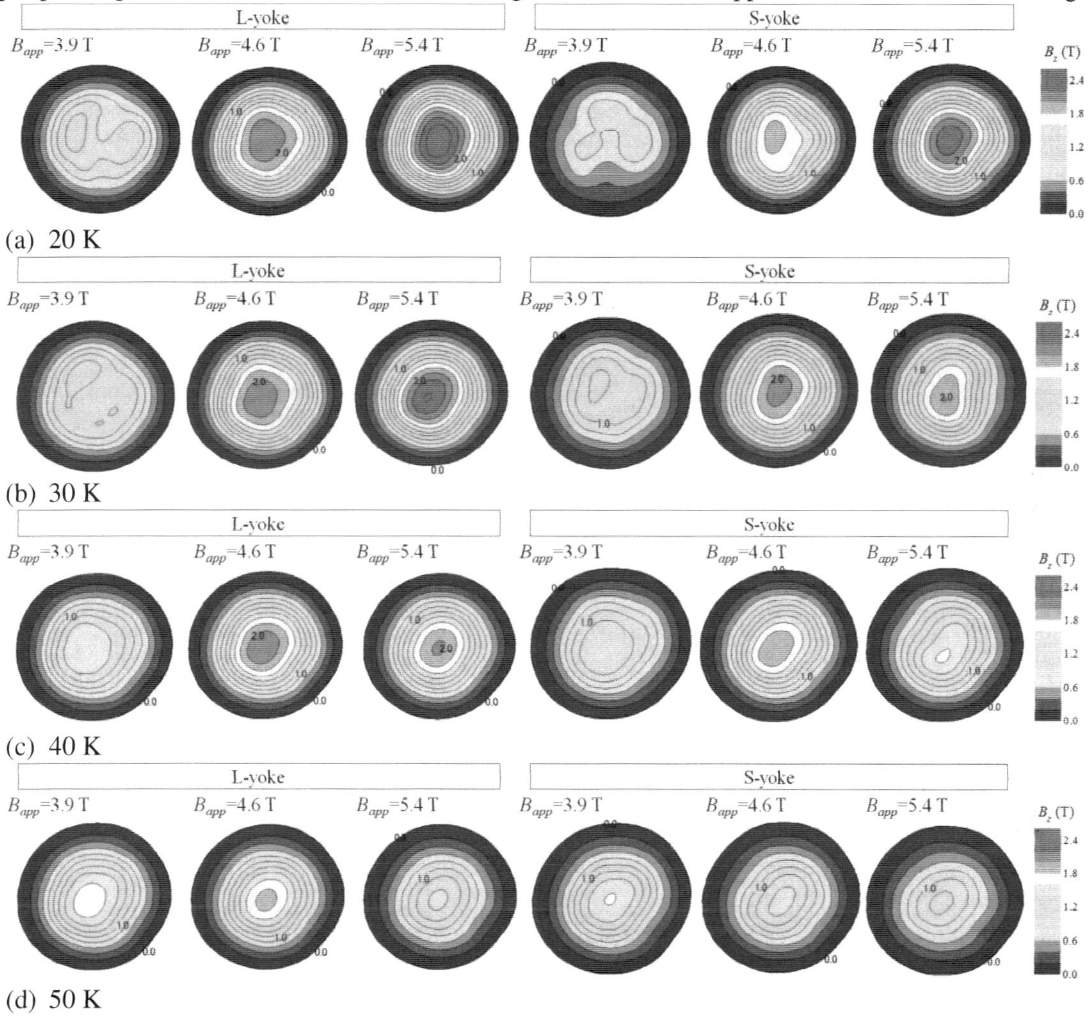

Figure 4. Trapped field distributions on the magnetic pole for B_{app}=3.9-5.4 T at 20-50 K.

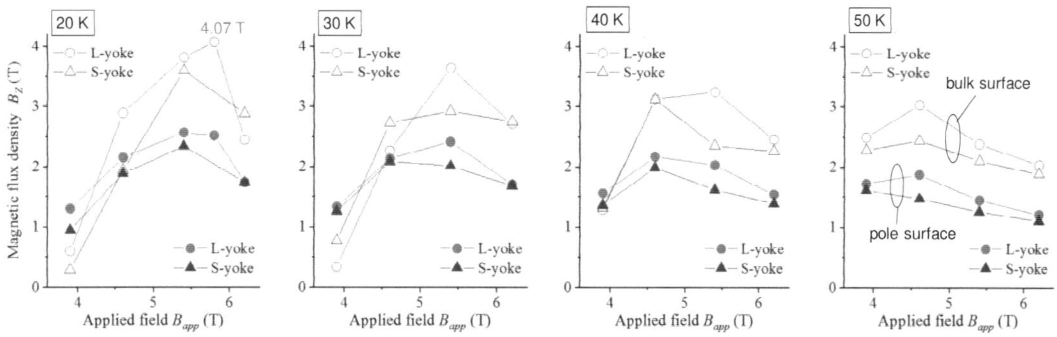

Figure 5. Magnetic flux densities on the center of the bulk surface (open symbol) and on the magnetic pole surface (solid symbol) as a function of the applied field for the L-yoke (red) and the S-yoke (blue).

Fortunately, the total magnetic flux can be increased by applying a low magnetic field to the bulk several times.

Figure 7 compares the maximum temperature between the L-yoke (red circle) and the S-yoke (blue triangle) at 20-50 K. The temperature rise of the L-yoke was larger than that of the S-yoke at all applied fields and all temperatures. Although the PFM of a long magnetic pulse with a slow rising time had the effect of suppressing heat generation, the rising time of the L-yoke was the same as that of the S-yoke in this method; moreover, the time when the sample was exposed to the magnetic field was extended. Therefore, the large amount of magnetic flux penetrated the bulk, and, consequently, the temperature increased. Although the trapped field of the bulk at a critical state decreased because J_c decreased with increasing temperature, a bulk magnetized by PFM was not in a critical state; thus, there was a margin of the trappable magnetic field. Therefore, the temperature rise did not influence the trapped field property in this experiment.

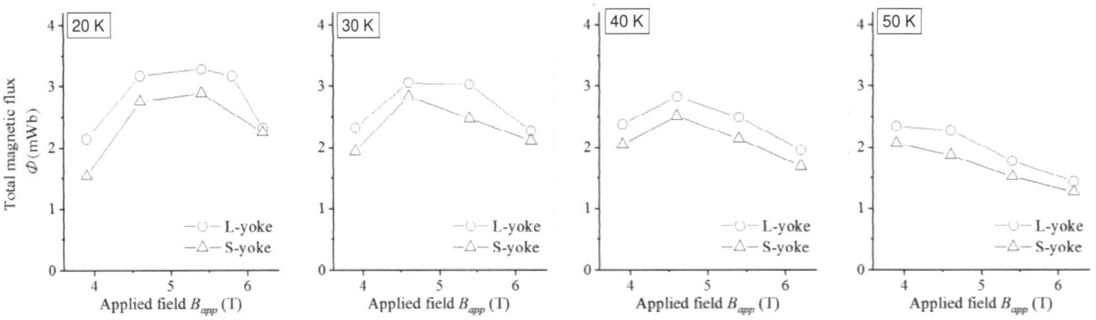

Figure 6. Comparison of the total magnetic flux between the L-yoke (red circle) and the S-yoke (blue triangle).

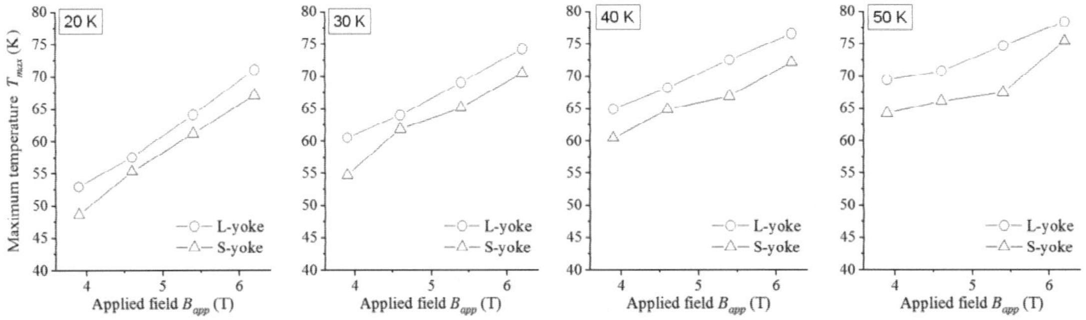

Figure 7. Comparison of the maximum temperature between the L-yoke (red circle) and the S-yoke (blue triangle).

4. Conclusion

We investigated the influence of the size of the soft-iron yoke to the trapped field performance. The yoke was applied to expose the bulk to the magnetic field for a long time. Although a small yoke 40 mm in diameter was used in the previous experiment, a large yoke 64 mm in diameter was fabricated in this study; consequently, the pulse width was extended by 13%. A single pulsed field was applied to a GdBCO bulk 60 mm in diameter and 20 mm thick with varying the amplitude of the pulsed field and temperature, and the magnetic flux density and temperature on the bulk surface and the magnetic field distribution on the pole surface were evaluated. The maximum trapped flux density of 4.07 T was achieved on the center of the bulk surface for B_{app}=5.8 T at 20 K. Moreover, the total magnetic flux was 3.28 mWb, and the maximum flux density on the pole surface was 2.56 T, which were improved by approximately 13% and 10%, respectively, as compared with those of the S-yoke. At present, the same examination was performed on a desktop-type bulk magnet system using a Stirling refrigerator. If the effectiveness of this method can be confirmed, it will be useful practically because the trapped field can be enhanced only by changing the yoke. We will also consider industrial applications of the bulk magnet system.

References

[1] J H Durrell, A R Dennis, J Jaroszynski, M D Ainslie, K G B Palmer, Y H Shi, A M Campbell, J Hull, M Strasik, E E Hell-strom and D A Cardwell 2014 *Supercond. Sci. Technol.* **27** 082001
[2] D Zhao, M Izumi, T Fujimoto, Y Zhang, W L Zhou and K Xu 2015 *IEEE Trans. Appl. Supercond.* **25** 6800204
[3] S Nariki, H Teshim and M Morita 2016 *IEEE Trans. Appl. Supercond.* **26** 7200404
[4] Y Setoyama, J Shimoyama, H Ogino and K Kishio 2016 *IEEE Trans. Appl. Supercond.* **26** 7200304
[5] J H Durrell, M D Ainslie, D Zhou, P Vanderbemden, T Brad-shaw, S Speller, M Filipenko and D A Cardwell 2018 *Supercond. Sci. Technol.* **31** 103501
[6] H Ikuta, T Hosokawa, H Ishihara, M Yoshikawa, Y Yanagi, Y Itoh, T Oka and U Mizutani 2001 *IEEE Trans. Appl. Supercond.* **11** 1 p3716
[7] M Sander, U Shutter, R Koch and M Läser 2000 *Supercond. Sci. Technol.* **13** 6 p841
[8] H Fujishiro, T Tateiwa, A Fujiwara, T Oka and H Hayashi 2006 *Physica C* **445-448** p334
[9] M D Ainslie, H Fujishiro, H Mochizuki, K Takahashi, Y-H Shi, D K Namburi, J Zou, D Zhou, A R Dennis and D A Cardwell 2016 *Supercond. Sci. Technol.* **29** 074003
[10] K Takahashi, M D Ainslie, H. Fujishiro, T Naito 2017 *IEEE Trans. Appl. Supercond.* **27** 6800205
[11] H Fujishiro and T Naito 2010 *Supercond. Sci.Technol.* **23** 105021
[12] Z Huang, H S Ruiz, Y Zhai, J Geng, B Shen, T A Coombs 2016 *IEEE Trans. Appl. Supercond.* **26** 5202105
[13] K Yokoyama, T Oka and K Noto 2010 *IEEE Trans. Appl. Supercond.* **20** p973
[14] K Yokoyama, E Kulawansha, Z Yuanding and T Oka 2016 *IEEE Trans. Appl. Supercond.* **26** 4 4601204
[15] K Yokoyama, A Katsuki, A Miura and T Oka 2018 *IEEE Trans. Appl. Supercond.* **28** 6800404

Processing and Characterization of Charcoal Added Bulk MgB_2 Superconductor

J Longji Dadiel[1,*], M Muralidhar[1,*], Sai Srikanth Arvapalli[1], S Pavan Kumar Naik[1,2] and M Murakami[1]

[1]Superconducting Material Laboratory, Graduate School of Science and Engineering, Shibaura Institute of Technology, 3-7-5 Toyosu, Koto-ku, Tokyo 135-8546, Japan
[2]Electronics and Photonics Research Institute, National Institute of Advanced Industrial Science and Technology (AIST), 1-1-1 Central 2, Umezono, Tsukuba, Ibaraki, 305-8568, Japan

[*]miryala1@shibaura-it.ac.jp / mg17009@shibaura-it.ac.jp

Abstract: In this study, we focus on methods to further improve the flux pinning and critical current density (J_c) of disk shaped MgB_2 bulk superconductors by a comparative study with locally obtained charcoal at optimized processing conditions. The doping of charcoal was varied at 0, 0.2, 0.4 and 0.6 wt%. Bulk MgB_2 samples were produced by in-situ solid state reaction in Ar gas and ambient pressures at 775 °C, employing high purity of Mg metal and nano B powder mixed in a fixed ratio of Mg:B = 1:2 and optimal conditions. The phase has been investigated by X-ray diffractometer (XRD). The effects of the dopants on superconducting properties especially the field dependence of J_c have been determined by superconducting quantum interference device (SQUID) and critical temperature (T_c). Measurements of the magnetic susceptibility showed sharp superconducting transitions ($\Delta T_c \sim$ 1 K). J_c at self-field and 20 K was determined to be ~280 and 467 kA/cm^2 for the 0 and 0.4 wt% charcoal added samples respectively. The structural and superconducting results were correlated with the addition of charcoal in the final product.

Keywords: MgB_2; charcoal; X-ray diffraction; critical current density, flux pinning, scanning electron microscope

1.Introduction

The discovery of superconductivity in MgB_2 is one of the most intriguing achievements owing to its relatively high critical temperature (T_c of 39 K) compared to the conventional low-temperature superconductors like Nb_3Sn, NbTi. MgB_2 offers the possibility of wide engineering applications in the temperature range 20-30 K, where Nb_3Sn and Ni-Ti alloy are not applicable due to their low T_c. Further, the field trapping capability for the practical applications of bulk MgB_2 magnets is similar to melt-textured $REBa_2Cu_3O_y$, which can be utilized as a trapped field magnet in nuclear magnetic resonance (NMR) and magnetic resonance imaging (MRI) [1-6]. Considerable efforts have been made to improve the critical current density (J_c), upper critical field (H_{c2}), and irreversibility field (H_{irr}) of this superconductor [7]. Also, significant progress has been made concerning the development of processing techniques, improving the flux pining, and large size MgB_2 bulk growth [1-5]. To utilize this material for industrial applications, including superconducting super-magnets, a further improvement of J_c is crucial, especially in sintered bulks [6, 8]. Doping of elements or compounds have been found to be effective for improving the superconducting properties of bulk, tapes, wires, and films, especially under magnetic fields [9]. Several approaches have been developed for incorporating the effective flux pinning centers in bulk MgB_2 material in-order to improve the superconducting performance. In high magnetic fields MgB_2 exhibits a decrease in J_c, due to its poor grain connectivity and lack of strong pinning centers. Earlier studies indicated that an inclusion of nano-dopants enables control of MgB_2 microstructure by introducing nano-pinning centers into the superconducting matrix. Among several substitutions, carbon doping is found to be effective [10].

In our previous work, we investigated the effect of nanoscopic diamond particles addition on bulk MgB_2 material by means of X-ray diffraction (XRD), superconducting transition temperature and critical current density at 20K [11]. The optimal amount nanoscale diamond addition to MgB_2 makes it possible to form a high density of nano-inclusions in MgB_2 matrix and improved the J_c dramatically. However, to reduce the cost of production one must look for the naturally available other form of carbon sources. To improve the pinning performance of bulk MgB_2 materials, we tried to utilize other form of carbon sources such as activated charcoal in combination with optimum sintering temperature set in earlier studies [6]. Due to its low cost, we have selected charcoal as a carbon source for a carbon-doped MgB_2, studied the variation of T_c, triggering of different phases and J_c. The best sample indicated a high critical current density around 467 $kAcm^{-2}$ at 20 K and the superconducting transition temperature around 37.5 K.

2.Experimental details

The MgB_2 samples with the varying contents of charcoal were synthesized by an *in situ* liquid-solid reaction. The starting powder materials were Mg powder (99.9% purity, 325 mesh), nano sized amorphous boron powder (98.5% purity, ~250 nm), and locally obtained charcoal powder from charcoal market, Plateau raiders, Jos-Nigeria as dopant. The charcoal powder was collected and pulverized at the mineral processing department of the National Metallurgical Development Centre Nigeria. Ball milling was carried out employing BICO made (395-50 model, 220 V,3ph, 60 cycles) with ball (3/4 inch - 1½ inch) to powder ratio of 10:1. The precursors were pressed into pellets of 10 mm in diameter and 7 mm thickness by use of a uniaxial press. The samples were sealed in titanium foils and then sintered at optimal temperature of 775 °C for 3 h in a tube furnace under pure argon atmosphere prior to cooling to room temperature [6]. The crystal structures and constituent phases of the fabricated MgB_2 samples were investigated using a high-resolution X-ray powder diffractometer (Rigaku/ SmartLab), using Cu-K_α radiation (λ = 1.5418 Å). The XRD patterns were collected with a step size of 0.02° over a 2θ range from 20° to 90°. The microstructural properties were investigated employing field emission scanning electron microscopy (FESEM, Jeol JSM-7100F model). Temperature and field dependent characterization of the samples was conducted by subjecting smaller specimens of dimension ~2.0×2.0×1.0 mm^3 from the sintered pellets using a SQUID magnetometer (MPMS5). The J_c values were estimated from the magnetization hysteresis loops (M-H loops) swept from 0 to 5 T at a constant temperature of 20 K based on the modified Bean critical state model:

$$J_c = \frac{20\Delta m}{a^2 d\left(b-\frac{a}{3}\right)} \ (\mathrm{A/cm^2})$$

where d is the sample thickness, a and b are the cross-sectional dimensions (in mm) with $b \geq a$, and Δm is the difference of the magnetic moments (in emu) during increasing and decreasing field in M-H loop [12]. The critical temperatures were also determined by magnetic susceptibility measurements from the onset of diamagnetism.

3.Results and discussion

Further, the charcoal powder was sieved and ball milled for 2 hours down to from <40 to 160 nm particle size as analyzed using the zeta potential particle analyzer. In figure 1, the obtained size distribution histogram of the charcoal powders is presented which shows that the particles are distributed majorly in the size range of 60-100 nm. The Mg and B powders were mixed in the nominal ratio of 1:2 while varying the compositions of the charcoal for each sample produced at 0, 0.2, 0.4 and 0.6 wt% under argon atmosphere in a glove box.

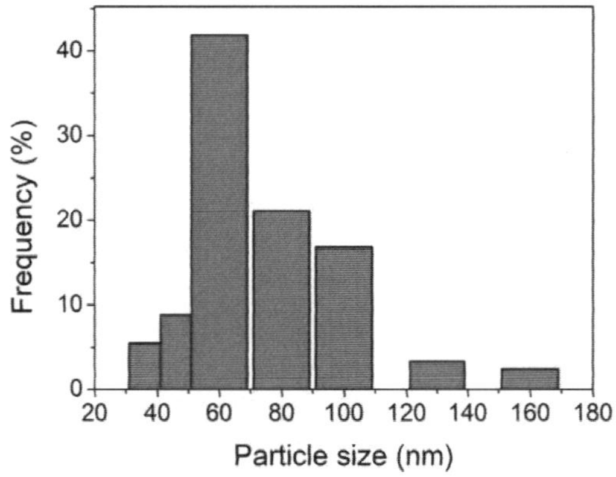

Figure 1. Charcoal particles size histograms as obtained from the Zeta potential measurement after sieving and ball milling.

Figure 2 shows the XRD patterns of all synthesized MgB_2 samples, which gives the information on the crystallographic phases at different levels of charcoal doping. All the major Bragg peaks of MgB_2 hexagonal structure can be identified, indicating that samples mainly consist of the desired MgB_2 phase with minimal amount of MgO reflecting in all samples as impurity. The formation of MgO could not be completely prevented due to high reactivity of magnesium with oxygen, we were able to observe only MgO impurity by XRD, similar to our reports [13]. The lattice constants were evaluated from the XRD results which indicate little strains in direction of higher angles as shown in Table 1. The slight variation in the lattice constants is an indication that the amount of C-doping from the charcoal was effective in causing a small lattice distortions indicating minute C substitution in samples. Whereas, the c lattice constant is ~3.525 Å and is almost invariant. These results are similar to previous reports available with lowest C-doping in MgB_2 samples [14-15]. The effective C substitution (C_{eff}) in MgB_2 samples was determined using the relation described in the previous literature [16]. It was found that the C_{eff} values are ~0.007 which confirms to the chemical formula of the ~$MgB_{1.993}C_{0.007}$.

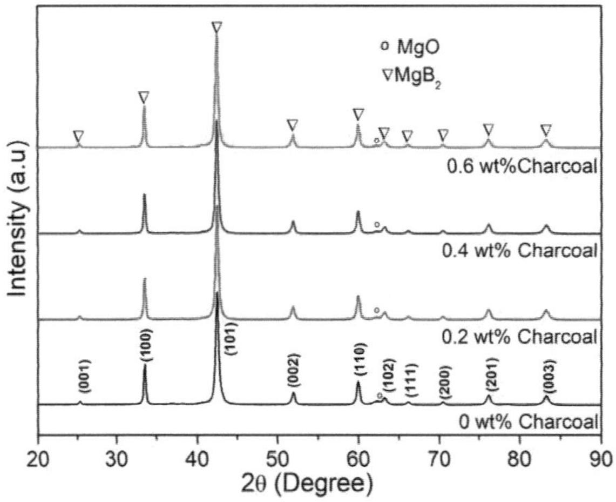

Figure 2. (a) X-ray diffraction patterns of charcoal doped MgB_2 bulk samples produced using the mixture of the nano boron powders and different charcoal contents in sintering process

No particular shift in Bragg peaks was observed in XRD which reveals that considerable amount of substitution has not happened and not to mention the added charcoal was very minute. However, as usual slight MgO impurity phase formation can be seen.

Table 1. Starting compositions, lattice constants, onset and offset T_c*'s* compared with the changing transitions

Sample	Nominal ratio	***Lattice parameters***		***Superconducting transition (K)***		
		a (Å)	***c (Å)***	$T_{c,onset}$	$T_{c,zero}$	ΔT_c
Reference	MgB_2 + 0 wt% Charc	3.0864	3.5253	38.3	37.9	0.4
1	MgB_2 + 0.2 wt% Charc	3.0836	3.5253	37.9	37.2	0.7
2	MgB_2 + 0.4 wt% Charc	3.0834	3.5248	38.0	37.2	0.8
3	MgB_2 + 0.6 wt% Charc	3.0834	3.5248	37.9	36.5	1.4

The DC magnetic susceptibility curves were measured on the MgB_2 samples with a varying content of the charcoal powder in a magnetic field of 1 mT. All samples show sharp superconducting transitions (with transition width ΔT_c of ~ 1 K). There was a slight change in the onset of the critical temperatures ($T_{c,onset}$) with inclusion of charcoal (Fig 3, left). Here we obtained the T_c onset around 37.9 K for 0.6 wt% of charcoal as compared to high T_c onset of 38.3 K for pure sample. More details regarding T_c can be found in Table.1. This could indicate that we have to find an optimum between possible superconductivity degradation and pinning enhancement [11]. The temperature dependence of the shown magnetization curves indicated that the onset of T_c for the charcoal added samples is slightly lower than that of the pure sample. In a typical carbon doped MgB_2, the T_c is known to be dependent on the amount of carbon substitution into the boron sites which means that the more carbon substitution corresponds to a lower T_c [17]. However, our results showed very little T_c degradation which might be because of MgO and possibly the lower content of charcoal addition and small amount of effective C substitution as supported from the XRD results.

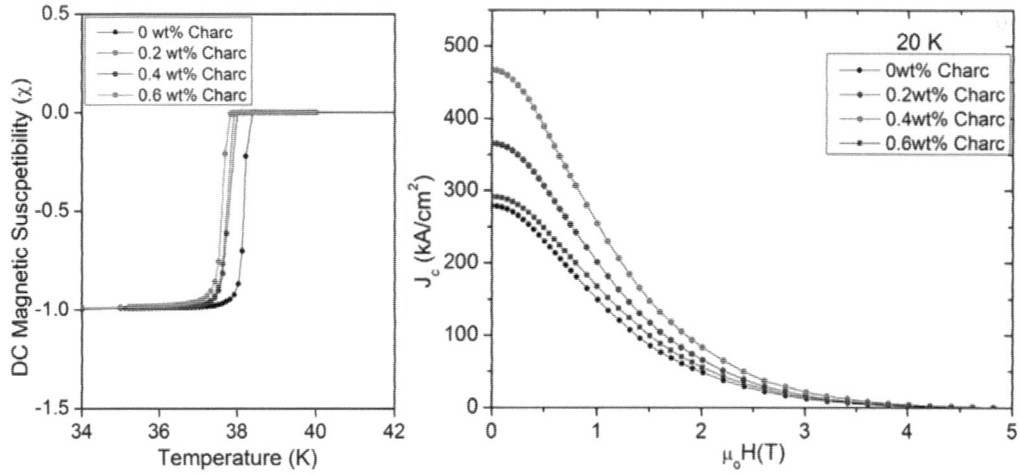

Figure 3. Superconducting transition in the bulk MgB_2-charcoal produced using the mixture of the amorphous boron powders in sintering process (left). The field dependence of the critical current densities for charcoal added MgB_2 at T = 20 K (right). All samples are sintered at 775 °C for 3h in argon atmosphere.

The variation of the J_c at 20 K as the function of the applied magnetic field was analyzed to investigate the effect of charcoal addition. Figure 3 (right) shows the J_c curves at 20 K as a function of applied magnetic field for the bulk MgB_2 material with varying content of charcoal powder. The self-field J_c values of 0, 0.2, 0.4 and 0.6 wt% charcoal added MgB_2 samples are 280, 366, 467 and 341 kA/cm^2 respectively. The self-field critical current density around 467 kA/cm^2 was observed at 20 K for the best sample as compared to the J_c of 280 kA/cm^2 for the pure sample. MgB_2 sample with 0.4 wt% of charcoal powder showed higher critical current density than the 0 wt% charcoal added MgB_2 as well as the other variables indicating the optimal composition among the present series. If the J_c values are compared at 1 T applied field for 0, 0.2, 0.4 and 0.6 wt% charcoal added MgB_2 samples are 164, 201, 280 and 183 kA/cm^2 respectively. We can say that optimization was effective for the 0.4wt% charcoal added sample. To understand the mechanism for the improved superconducting performance we tried to test the microstructure. Similarly, high J_c results were obtained previously with carbon coated boron which has high carbon content where the substitution can be seen clearly [18].

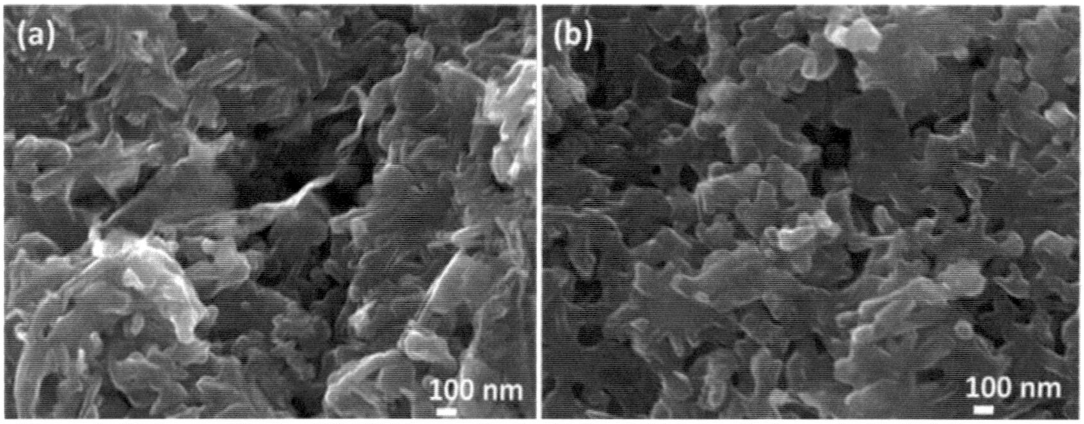

Figure 4. High magnification FE-SEM images for the (a) 0 and (b) 0.4 wt% charcoal added MgB_2 samples

Figure 4 shows the higher magnification FE-SEM images recorded on the fractured surface of 0 and 0.4 wt% charcoal added MgB_2 samples. From the keen observation of micro-structural features, it is evident that the 0.4 wt% charcoal added sample contain fine sized grains compared to former sample. The elementary pinning force in MgB_2 is dependent on its nano-structural control. J_c at low field is simply determined by the grain connectivity, grain sizes and pinning force of the grain boundaries [19]. The results of the microstructural analysis show the presence of refined grains in the charcoal added sample and we emphasize that these grains aid in enhancing the J_c. These refined grains are possibly created by nano-sized charcoal particles and create strong pinning effects which are responsible for the improved flux pinning. The superconducting J_c is highest in the 0.4 wt% charcoal added sample and we think that this may be because of the optimal distribution of refined grains.

4. Conclusions

MgB_2 bulk samples were successfully synthesized by a solid state sintering process and various amounts of charcoal addition. XRD studies indicated that all samples were single phase MgB_2 along with a small quantity of MgO. All the charcoal added samples showed a sharp superconducting transition around 38 K. The occurrence of superconductivity is slightly affected by the charcoal addition compared to pure MgB_2 materials. The effective grain refinement supported superior J_c performance in 0.4 wt% charcoal added sample. Among all additions, 0.4 wt% charcoal was found to be optimal for superior field dependency of J_c value of 467 kA/cm^2 at 20 K in self-field and 280 kA/cm^2 at 1 T. This work shows that addition of charcoal particles is a promising way to improve the superconducting performance of the MgB_2 bulk materials along with reducing the economic cost via cheap and abundant charcoal additive.

Acknowledgments
The paper was supported by Grant-in-Aid FD research budget code: 112261, Shibaura Institute of Technology (SIT). Longji would also like to thank the Japan International Cooperation Agency (JICA) for financial support through the ABE Initiative program. Thanks to Prof. Tetsuo Oka for continued support and advice.

REFERENCES

1. Vinod K, Abhilash Kumar R G and Syamaprasad U 2007 *Supercond. Sci. Technol.* **20** R1.
2. Perini E and Ginuchi G 2009 *Supercond. Sci. Technol.* **22** 045021.
3. Kajikawa K and Nakamura T 2009 *IEEE Trans. Appl. Supercond.* **19** 1669.
4. Gunchi G, Ripamonti G, Cavallin T and Bassani E 2014 *Cryogenics* **46** 23.
5. Kim J H, Oh S, Heo Y-U, Hata S, Kumakura H, Matsumoto A, Mitsuhara M, Choi S, Shimada, Maeda Y M, MacManus-Driscoll J L and Dou S X, 2012 *NPG Asia Materials*, **4** doi:10.1038/am.2012.3.
6. Muralidhar M, Inoue K, Koblischka MR, Tomita M and Murakami M 2014 *J. Alloys Compd.* **608** 102.
7. Fuchs G, Müller K-H, Handstein A, Nenkov K, Narozhnyi V.N, Eckert D, Wolf M and Schultz L 2011 *Solid State Communications* **118** 497.
8. Muralidhar M, Nozaki K, Kobayashi H, Zeng X L, Koblischka-Veneva A, Koblischka MR, Inoue K and Murakami M 2015 *J. Alloys Compd.* **649** 833.
9. Anurag Gupta and A V Narlikar 2009 *Supercond. Sci. Technol.* **22** 125029.
10. Muralidhar M, Higuchi M, Jirsa M, Diko P, Kobal I and Murakami M 2017 *IEEE Transactions of Applied Superconductivity,* **27** 6201104.
11. Longji Dadiel J, Muralidhar M and Murakami M 2018 *J. Phys. Conf. Ser.* **1054** 012052.
12. Chen D and Goldfarb R B 1989 *J. Appl. Phys.* **66** 2489.
13. Muralidhar M, Kenta N, Koblischka MR and Murakami M 2015 *Physica Status Solidi a*, **212** 2141.
14. Lee S, Masui T, Yamamoto A, Uchiyama H and Tajima S 2003 *Physica C* **397** 7.
15. Yamamoto A, Shimoyama J, Ueda S, Iwayama I, Horii S, Kishio K 2005 *Supercond. Sci. Technol.* **18** 1323.
16. Avdeeva M, Jorgensena J D, Ribeirob R A, BudÕkob S L and Canfield P C 2003 *Physica C* **387** 301.
17. Wilke R H T, Bud'ko S L, Canfield P C and Finnemore D K, 2004 *Phys. Rev. Lett.* **92** 217003.
18. Muralidhar M, Higuchi M, Diko P, Jirsa M and Murakami M 2017 *J. Phys. Conf. Ser.* **871** 012056.
19. Hampshire D P and Jones H 1987 *J. Phys. C* **21** 419.

IOP Conf. Series: Journal of Physics: Conf. Series **1293** (2019) 012048 doi:10.1088/1742-6596/1293/1/012048

Linewidth dependence of NbN-based microwave kinetic inductance detectors

S Negishi[1], H Mikami[1], S Hashimoto[1], K Nakajima[2], H Terai[3], S Tanaka[1] and S Ariyoshi[1]

[1] Toyohashi University of Technology, 1-1 Hibarigaoka, Tempaku, Toyohashi, Aichi 441-8580, Japan
[2] Yamagata University, 4-3-16 Jonan, Yonezawa, Yamagata 992-8510, Japan
[3] National Institute of Information and Communications Technology, 588-2 Iwaoka, Nishi-ku, Kobe, Hyogo 651-2492, Japan

E-mail address: ariyoshi@ens.tut.ac.jp

Abstract. We analysed, fabricated, and characterised niobium nitride (NbN)-based microwave kinetic inductance detectors (MKIDs) with a rewound strip structure (spiral-MKIDs). To control resonance characteristics precisely, the linewidth (w) dependence of rewound spiral resonators varying between w = 10, 20, and 40 μm was analysed using an electromagnetic field simulator; the resonance frequencies (f_0's) were obtained as 4.42, 4.58, and 5.15 GHz respectively, proving that the inductance decreases as the w increases. To verify the simulation results experimentally, NbN-based MKID arrays with different linewidths were fabricated on 10×10 mm^2 sapphire substrates, and cooled down to 3 K. The microwave characteristics with a vector network analyser, and f_0's were found to be 4.11, 4.23, and 4.85 GHz for the MKIDs with w = 10, 20, and 40 μm, respectively. These values are in agreement with the simulation results within 7.6% accuracy. The other resonance characteristics such as the scattering-matrix element (S_{21}) and loaded quality factor (Q_L) showed the similar w dependence with the simulation. Spiral-MKIDs with narrower w is expected to have better detector performance owing to shallower S_{21} and higher Q_L.

1. Introduction

Terahertz waves—frequency range of 0.1–10 THz—exhibit the transparency of radio waves and straightness of light waves simultaneously. Terahertz wave detection has been used for non-invasive and non-destructive imaging inside an object application. Recent advances in terahertz wave technology are expected to be applied in various fields, such as medical science, security, and agriculture [1]. The microwave kinetic inductance detector (MKID) [2] detects terahertz waves when the kinetic inductance of a superconducting film changes due to the absorption of a terahertz photon. The MKIDs consist of multiple resonators coupled with a microwave through-line, where the resonance frequency shift of each resonator is detected by the frequency-multiplexed readout scheme. All of the components in MKIDs are made up of a single-layer film, which enables relatively easy fabrication of a large-scale array. Many other groups [3] have been developing MKIDs to achieve ultimate sensitivity performance mainly used for astronomical observations. In these detectors, low transition temperature (T_c) materials, such as Al [4] or TiN [5], are commonly employed. However, their operations require expensive and/or large cooling systems employing a dilution refrigerator (0.1 K) or closed-cycle ^{3}He sorption pump (0.3 K).

On the other hand, NbN material can provide relatively high sensitivity performance at 3 K; this temperature becomes easily achievable using a compact ^{4}He refrigerator.

In this study, we describe NbN-based MKIDs with rewound spiral resonators (spiral-MKIDs) operated at 3 K. The linewidth dependence of rewound spiral resonators is analysed using an electromagnetic field simulator. We then explain the fabrication of the spiral-MKIDs and characterise the electrical performance of the detector from the following perspectives: resonance frequency, scattering-matrix element, and loaded quality factor.

2. Detector design and analysis

Electromagnetic field analysis is useful for effective optimisation of the MKID design. To control resonance characteristics precisely, the linewidth dependence of a rewound spiral resonator [6–8] was analysed via a simulator, *Sonnet Lite*. Figure 1 shows a resonator designed from the analysis model [9]. The total length of the spiral is approximately 12.5 mm corresponding to the half wavelength of the resonance frequency of around 5 GHz on the sapphire substrate. Coplanar waveguide (CPW) through-line is employed for the excitation and readout. The line and gap widths of the through-line are 40, and 10 μm respectively, which corresponded to a specific impedance of 50 Ω.

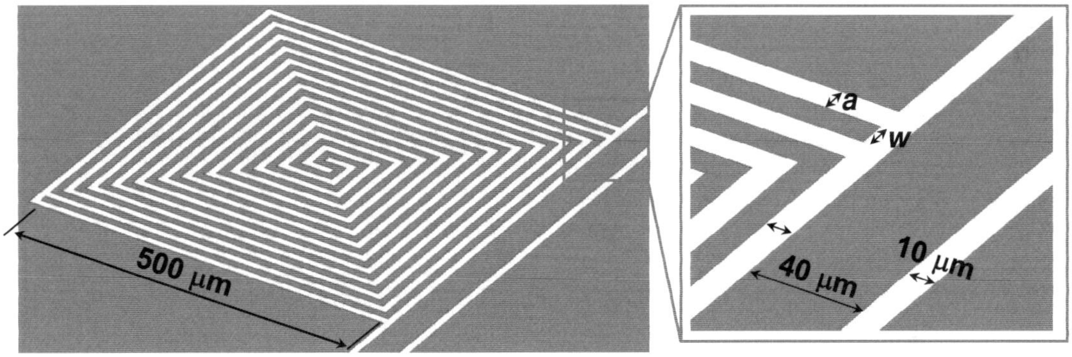

Figure 1. Analysis model of one-pixel spiral-MKID (bird's-eye view).
Dark and light regions are the NbN film and sapphire substrate, respectively.

The analysis was conducted under the following conditions: frequency range of 1–10 GHz; relative permittivity of the substrate, 9.8; substrate thickness, 0.5 mm; thin film (100 nm) AC resistance, $R_{rf} = 5.88 \times 10^{-26}$ Ω Hz^{-2}/sq; and sheet inductance, L_S = 1.838 pH/sq, where the R_{rf} and L_S were recalculated from previously reported values obtained at thickness of 150 nm thickness [10]. The analytical model is defined to be surrounded by air at the upper space and silicon at the lower space of the substrate. Resonance characteristics were analysed on the three resonator models with a fixed line space, a = 10 μm and linewidth (w) varying to 10, 20, and 40 μm, giving the resonator size of 500×500 μm^2, 650×650 μm^2, and 800×800 μm^2, respectively. Figure 2 (a) shows the simulation results of microwave resonance characteristics. The resonance frequencies (f_0's) obtained were 4.42, 4.58, and 5.15 GHz respectively, proving that the inductance decreases as the resonator linewidth increases. The f_0 simulated for the MKID with w = 40 μm was blue-shifted by 0.73 GHz compared with that for 10 μm. The scattering-matrix element (S_{21}) and loaded quality factor (Q_L) were also analysed for each w and shown in Figure 2 (b). Q_L was obtained from the f_0 and its resonance dip width (Δf), which is defined as $Q_L = f_0 / \Delta f$. S_{21} values were found to be -30.1 dB for w = 10 μm, -39.5 dB for w = 20 μm, and -45.7 dB for w = 40 μm at the resonance dip; thus, narrower the w is the shallower the S_{21}. On the contrary, the narrower the w higher the Q_L with individual Q_L values as 4020 for w = 10 μm, 2230 for w = 20 μm, and 970 for w = 40 μm. The simulation results revealed that spiral-MKIDs with narrower linewidths are expected to have a better detector performance owing to lager S_{21} and higher Q_L.

IOP Conf. Series: Journal of Physics: Conf. Series **1293** (2019) 012048 doi:10.1088/1742-6596/1293/1/012048

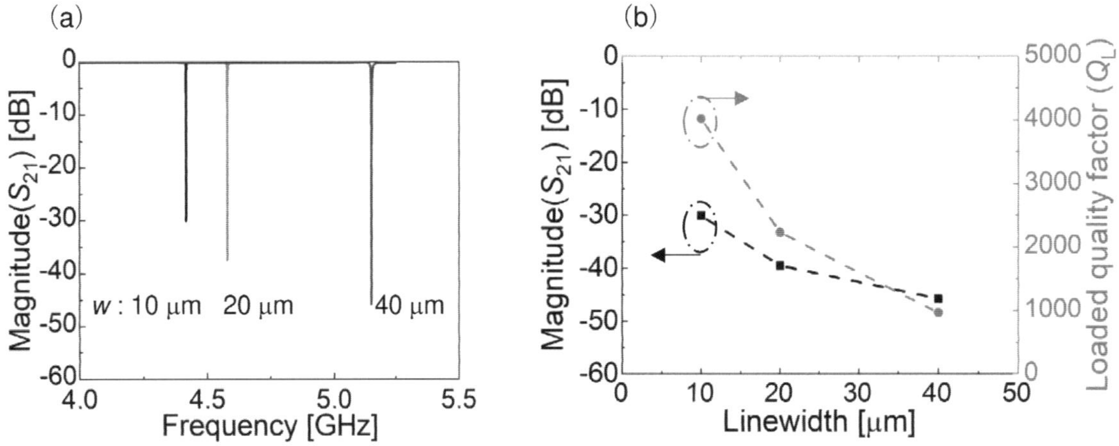

Figure 2. (a) Simulated resonant curve of the NbN-based MKIDs with different linewidths, (b) Dependence of S_{21} and Q_L on linewidth of the MKIDs.

3. MKID fabrication

To verify the simulation results experimentally, we deposited a 100 nm thick NbN film on a sapphire substrate using DC magnetron sputtering. Spiral-MKIDs with three different linewidths (w = 10, 20, and 40 μm) were then fabricated using a standard photolithography and Ar-ion etching. Figure 3 shows a 25-pixel prototype array on a 10×10 mm^2 sapphire substrate and an enlarged image of a single pixel. Both sides of the through-line and edges of the resonators are grounded at the same plane. The resonator length of each detector pixel differs by 30 μm and is connected to a single signal readout line by capacitive coupling, as this configuration is capable of the frequency-domain multiplexing.

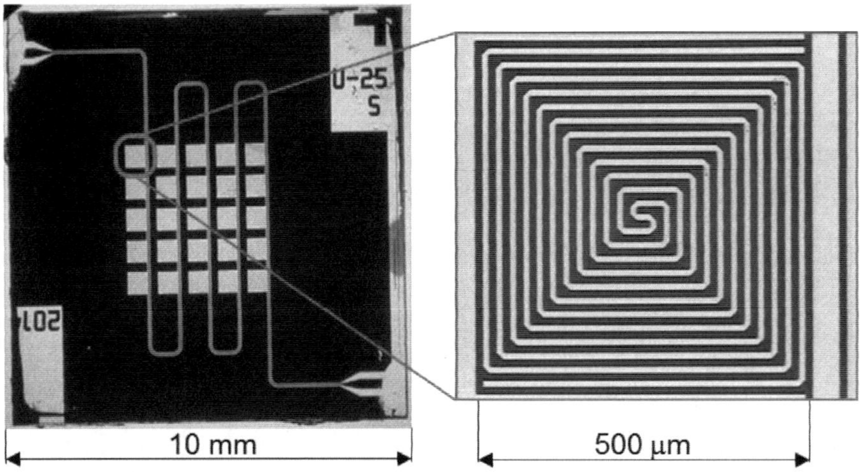

Figure 3. NbN-based MKID array fabricated on a sapphire substrate (left). Enlarged image of a single pixel (right). In the right image, yellow and gray regions are the NbN film and sapphire substrate, respectively.

4. Electrical evaluation

The fabricated MKID chip was mounted on a cryostat based on a ^{4}He pulse tube refrigerator and cooled down to 3 K. Microwave resonance characteristics of each detector chip were evaluated using a vector network analyser. The output power of the network analyser was set to -40 dBm. The transmission signal was measured through a cryogenic high-electron-mobility transistor amplifier and ambient temperature amplifier with a gains of 30 dB and 28 dB, respectively, to achieve low-noise measurements.

Figure 4 (a) shows the microwave resonance characteristics of the 25-pixel MKID array with w = 10 μm measured at 3 K without calibration. The half-wavelength resonance dips were clearly observed in the frequency range from 4.11–4.40 GHz; the frequency interval between adjacent pixels was 13 MHz with a standard deviation of 3 MHz. For w = 20 μm (Fig.4 (b)) and 40 μm (Fig.4 (c)), their resonance dips were shifted to the higher frequency range such as 4.23–4.62 and 4.85–5.48 GHz, respectively. This frequency shift can be explained from the general behavior that the inductance decreases with increasing linewidth (w), and the resonance frequency is inversely proportional to the square root of the inductance (*i.e.* $f = 1/2\pi\sqrt{LC}$). Figure 4 (d) shows the linewidth dependence of S_{21} and Q_L extracted from the resonance data. The linewidth difference is observed not to affect S_{21} experimental results. The Q_L becomes higher at a narrower linewidth. This is because the narrower linewidth which weakens the coupling strength between the detector edge and readout through-line. On the other hand, the Q_L of 1240

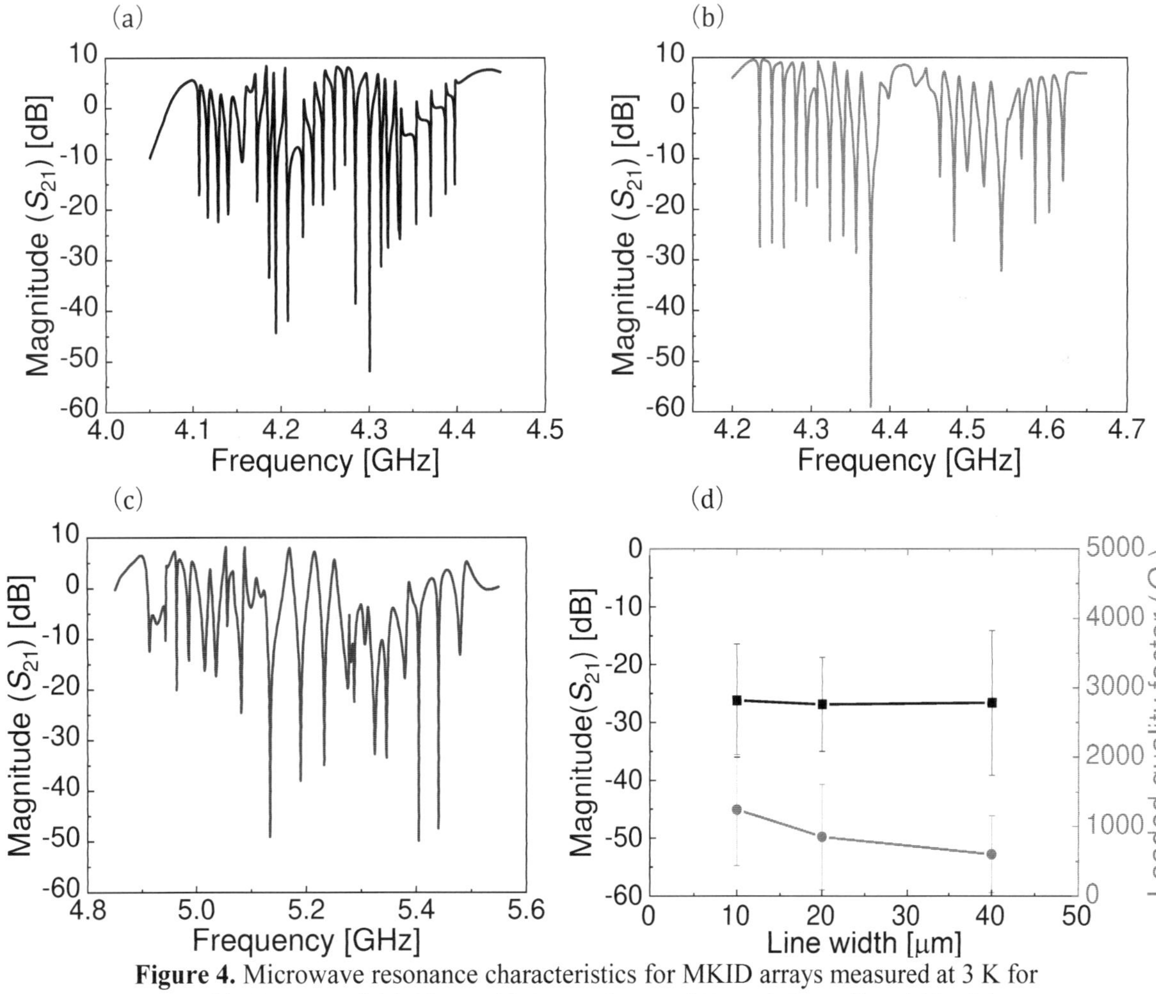

Figure 4. Microwave resonance characteristics for MKID arrays measured at 3 K for (a) w = 10 μm, (b) 20 μm, and (c) 40 μm, (d) Dependence of S_{21} and Q_L on linewidth of the resonators.

for 10 μm, for example, is lower than previously reported results—2000 (at 150 nm, 4 K) [10] and 16000 (at 140 nm, 3.2 K) [9].

Table 1 summaries the comparison of simulation and experimental results. The values of f_0 match the simulated results within 7.6 % accuracy the maximum, and the frequency shift (0.74 GHz) between $w = 10$ and 40 μm is also in good agreement with the simulated result (0.73 GHz). The values of S_{21} and Q_L is calculated by averaging all of the resonance characteristics. However, the discrepancy for S_{21} and Q_L is relatively large compared with that of f_0. This is probably caused by the NbN film quality such as transition temperature and residual resistance, which strongly affects the resonance shape. Deposition of the higher quality film as well as optimisation of the fabrication process are expected to improve the detector performance such as noise equivalent power and response speed.

Table 1. Comparison of simulation and experimental results for each linewidth

	The first resonance frequency f_0 [GHz]	Average scattering-matrix element S_{21} [dB]	Average loaded quality factor Q_L [-]
	$w = 10$ μm		
Simulation	4.42	-30.1	4020
Experiment	4.11	-26.2	1240
(Standard deviation)		(9.8)	(800)
(Error)	(7.0%)	(14.9%)	(69.2%)
	$w = 20$ μm		
Simulation	4.58	-39.5	2230
Experiment	4.23	-26.9	850
(standard deviation)		(8.1)	(760)
(Error)	(7.6%)	(31.9%)	(61.9%)
	$w = 40$ μm		
Simulation	5.15	-45.7	970
Experiment	4.85	-26.6	600
(standard deviation)		(12.5)	(550)
(Error)	(5.8%)	(41.8%)	(38.1%)

5. Conclusion

To predict the microwave resonance characteristics of spiral-MKIDs prior to the fabrication is one of the important issues for optimising the detector performance effectively. For this purpose, we analysed, fabricated and evaluated NbN-based MKIDs with three different linewidths (10, 20, and 40 μm). The simulated resonance frequency was in good agreement with the experimental results within 7.6% maximum accuracy. In terms of the scattering-matrix element (S_{21}) and loaded quality factor (Q_L), the experimental results had similar linewidth dependence as that obtained from the simulation. Thus, spiral-MKIDs with narrower linewidth are expected to realise higher sensitivity performance owing to shallower S_{21} and higher Q_L. To further confirm our conclusions drawn here, optical measurements such as noise equivalent power and response speed are some of the factors that need to be considered for future studies on these systems.

Acknowledgments

The authors would like to thank Professor Takashi Noguchi (National Astronomical Observatory of Japan) for his valuable suggestions for the development of the detector. This study was partly supported by the Grant-in-Aid for Scientific Research (B) (No. 17H02809) from the Ministry of Education, Culture, Sports, Science and Technology of Japan.

IOP Conf. Series: Journal of Physics: Conf. Series **1293** (2019) 012048 doi:10.1088/1742-6596/1293/1/012048

References

[1] Lee Y S 2009 *Principles of Terahertz Science and Technology* (New York: Springer).
[2] Day P K, LeDuc H G, Mazin B A, Vayonakis A and Zmuidzinas J 2003 *Nature* **425** 817
[3] Monfardini A *et al* 2010 *Astron. Astrophy.* **521**, A29
[4] Baselmans J, Yates S J C, Barends R, Lankwarden Y J Y, Gao J, Hoevers H and Klapwijk T M 2008 *J. Low Temp. Phys.* **151** 524–29
[5] Leduc H G *et al* 2010 *Appl. Phys. Lett.* **97** 102509
[6] Ma Z and Sakurai E 2005 *IEICE Trans. Electron.* **E88-C** No.2 216
[7] Ma Z and Kobayashi Y 2005 *IEICE Trans. Electron.* **E88-C** No.7 1406
[8] Hayashi K *et al* 2014 *J. Low Temp. Phys.* **507** 042015
[9] Ariyoshi S *et al* 2013 *Appl. Phys. Exp.* **6** 064103
[10] Hayashi K, Saito A, Sawada T, Ogawa Y, Nakajima K, Yamada H, Ariyoshi S and Ohshima S 2013 *Phys. Procedia* **45** 213–16

IOP Conf. Series: Journal of Physics: Conf. Series **1293** (2019) 012049 doi:10.1088/1742-6596/1293/1/012049

Plug-in Wire for 200-pixel Superconducting Tunnel Junction X-ray Detector Array for Helium-3 Cryostat

***S Shiki[1], G Fujii[1], M Ukibe[1]**

[1]Department of Nanoelectronics, National Institute of Advanced Industrial Science and Technology, 1-1 Umezono, Tsukuba, Ibaraki 305-8568, Japan

E-mail: s-shiki@aist.go.jp

Abstract. A plug-in wire system is designed for a 200-pixel superconducting tunnel junction (STJ) X-ray detector array for a helium-3 cryostat. The STJ array and the room-temperature electronics are electrically connected by wire modules that have connectors at both ends. Two types of wire module are used to balance the electric conductance and thermal conductance under various temperature conditions. One type uses a copper-based miniature coaxial cable harness, and is utilized to electrically connect the room-temperature electronics to the 3 K stage. The other type uses woven wire looms of twisted-pairs of NbTi wire, and is used to electrically connect the 3 K stage to the helium-3 stage. With the installation of all wire modules, the temperature of the cold stage is less than 310 mK, and the holding time is more than 80 hours. The plug-in wire system is suitable for large-scale STJ X-ray detector arrays.

1. Introduction

Superconducting tunnel junction (STJ) X-ray detectors are promising for materials analysis because of their high energy resolution and high counting rate capability. 100-pixel STJ arrays have been applied in fluorescence yield X-ray absorption spectroscopy [1,2] and elemental analysis of scanning electron microscopy [3]. However, to realize a more practical analytical instrument based on STJ X-ray detectors, it is necessary to overcome the following two technical difficulties. One is achieving throughput equal to that of a silicon drift detector by making a large pixel array. The other is developing a scalable wiring system for connecting the STJ array detectors in the cold stage of a helium-3 cryostat and the room-temperature (RT) readout electronics. Large-format STJ arrays (1000-pixel array with an area of 10 mm^2) have been already developed [4]. However, realizing the scalable wiring system is still difficult because in a conventional wiring system for a cryostat, all wires are usually soldered by hand. Therefore, we have developed a plug-in wire system in which all electrical wires are attached to connectors at both ends. These connectors are used to electrically connect the STJ array to the RT electronics. In this paper, we show the performance of a helium-3 cryostat with the plug-in wire system designed for a 200-pixel STJ array.

2. Cryostat configuration

The plug-in wire system is installed in a cryostat consisting of a ^{3}He sorption cooler and a pulse tube refrigerator supplied by Niki Glass Company. The cryostat can achieve a temperature of below 1 K, which is the operation temperature of a Nb STJ (Fig. 1).

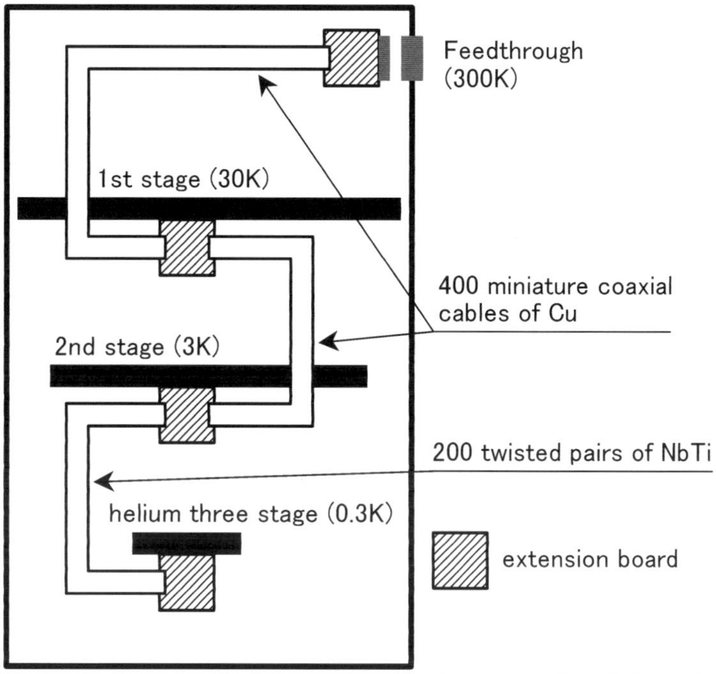

Fig. 1 Schematic diagram of proposed plug-in wire system for a helium-3 cryostat.

In the cryostat, three different temperature conditions are used to achieve a cold temperature environment for efficient STJ detector operation. The temperatures in these three conditions are 30 K, 3 K, and 300 mK, respectively. The 30 and 3 K conditions are maintained by the pulse tube refrigerator and the 300 mK condition is maintained by the ^{3}He sorption cooler. There are three stages for the three temperature conditions for utilizing the different temperature environments (30 K, 3 K, and 300 mK) for experiments, called the first stage, second stage, and ^{3}He stage, respectively. All stages are under the vacuum conditions for thermal insulation from the RT environment. With this configuration, our cryostat could hold a temperature of about 300 mK for 130 hours before the installation of the plug-in wire system.

3. Plug-in wire system

To achieve a long cooling time, the thermal flow from RT to the ^{3}He stage via wires utilized for electrical connections is normally minimized by using wires made of high-resistivity materials, which makes the total resistance between the STJ detectors and RT electronics large. To reduce the noise of readout electronics, it is necessary to keep the resistance between detectors and readout electronics as small as possible. To meet these contradictory requirements, the plug-in wire system is divided into three portions based on the temperature conditions, which are RT to the first stage (RT-1S), first stage to second stage (1S-2S), and second stage to ^{3}He stage (2S-^{3}He), respectively. In each portion, the electrical connection is realized using a wire harness module with 50 lines for 25 STJ pixels. The number of wire harnesses is determined by the pixel number of STJ arrays. Two lines are required for one STJ pixel because a common ground is not used in our cryostat. Therefore, eight wire harnesses are necessary in a portion for the 200-pixel array.

Electrical connections between wire harnesses in different portions are established by four extension boards, at RT and in the first stage, second stage, and ^{3}He stage, respectively, without soldering by hand. The extension board at RT has a 50-pin D-Sub connector and a 50-pin surface-mount connector (Cabline-SS). The other three extension boards have two 50-pin surface-mount

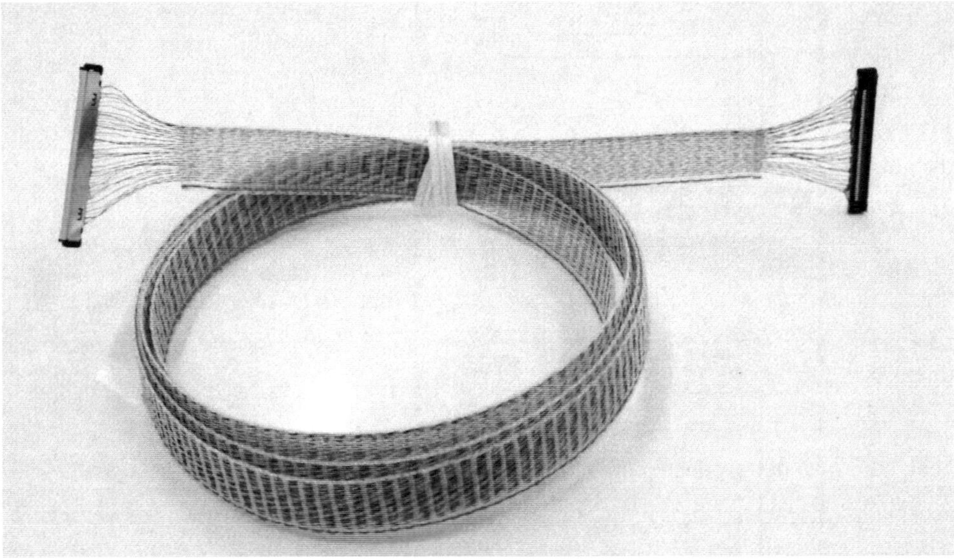

Fig. 2 Photograph of 25-pair woven wire loom made of NbTi.

connectors. In addition, the 50-pin D-Sub connector on the extension board at RT is linked to a hermetically sealed male/male 50-pin D-Sub connector of a vacuum feedthrough.

A wire harness consisting of 50 AWG42 coaxial cables and 50-pin connectors is used in the RT-1S and 1S-2S portions. The coaxial cable (resistance: 7 Ω/m) is made of tin-doped copper with an unknown dopant concentration. Because in the RT-1S and 1S-2S portions, the reduction of the thermal flow from RT is the most important. If conventional cables made of pure Cu (resistivity: 5-6 Ω/m) are used, the heat load to the second stage will be approximately 0.5 W, which exceeds the cooling capability of the pulse tube refrigerator.

In the remaining portion (2S-^{3}He), a wire harness consisting of woven wire looms of 25 twisted pairs of AWG42 NbTi wires is used because NbTi becomes a superconductor at the temperature in this portion and exhibits low thermal conductivity in addition to zero resistance (Fig. 2). In this portion, low thermal conduction and high electrical conduction are realized simultaneously.

The thermal flow through the above wire system was estimated as follows. The thermal flow can be determined using the total cross section, wire length, and wire thermal conductivity [5-6]. The total length of all wires in the harness is 1.0 m. The thermal conductivity of AWG42 coaxial cable at above 8 K has not been reported in the literature. However, we assumed that the thermal flow in the coaxial cable is almost equal to that in BeCu with a diameter of 0.33 mm because the thermal conductance of the coaxial cable is approximately half of that of a 0.86-mm semirigid coaxial cable of BeCu at temperatures below 8 K [7]. The estimated thermal loads for the first, second, and ^{3}He stages are 3 W, 30 mW, and 1.4 μW, respectively. With these thermal loads, it should be possible to obtain a long holding time at 300 mK in our cryostat. The temperature and holding time were measured to investigate the effect of the plug-in wire system on cooling performance.

4. Cooling performance

The temperature of the ^{3}He stage was measured to investigate the effect of the plug-in wire system on cooling performance (Fig. 3). The temperature of the cold stage was less than 310 mK, and the holding time was more than 80 hours. The holding time decreased by approximately 50 hours due to the installation of the plug-in wire system. The main reason for the holding time reduction is thermal load through the coaxial cables between RT and the second stage. For 200 pixels, the cooling performance is sufficient for experiments. However, since the thermal load increases with increasing number of pixels, the thermal load for a 1000-pixel STJ array would be much larger than the cooling capability of the ^{3}He sorption cooler. Therefore, the heat load through the wire harnesses should be reduced by using conductor materials with low thermal conductivity, such as phosphor bronze, constantan, manganin, or stainless steel, instead of tin-doped copper.

The ^{3}He cryostat was combined with fluorescence yield X-ray absorption spectroscopy [8] on beamline BL-11A in the Photon Factory at the High Energy Accelerator Research Organization (KEK) to show the reliability of the plug-in wire system. A cold finger with a length of approximately 70 cm was attached to the ^{3}He stage to realize high sensitivity to fluorescent X-rays from a sample. Figure 4 shows the temperature curve of the ^{3}He stage. The temperature of the ^{3}He stage is 317 mK, with a holding time of 35 hours. Some peaks appear in the temperature curve. The peaks occurred when the experimental configuration was changed. The holding time decreased by 45 hours due to the installation of the cold finger. The main reason for the reduction in holding time is the heat capacity of the cold finger and the thermal load through the support of the cold finger. The cooling performance of the ^{3}He cryostat is sufficient for fluorescence yield X-ray absorption spectroscopy experiments with a synchrotron source, for which a holding time of more than one day is required.

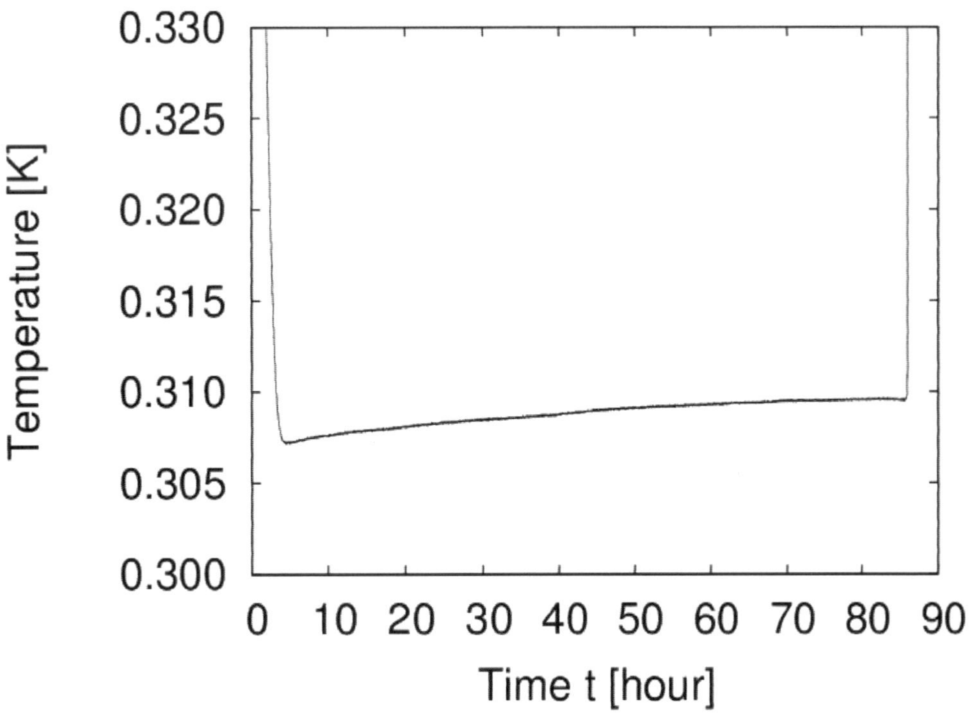

Fig. 3 Temperature curve of a helium-3 cryostat with proposed plug-in wire system.

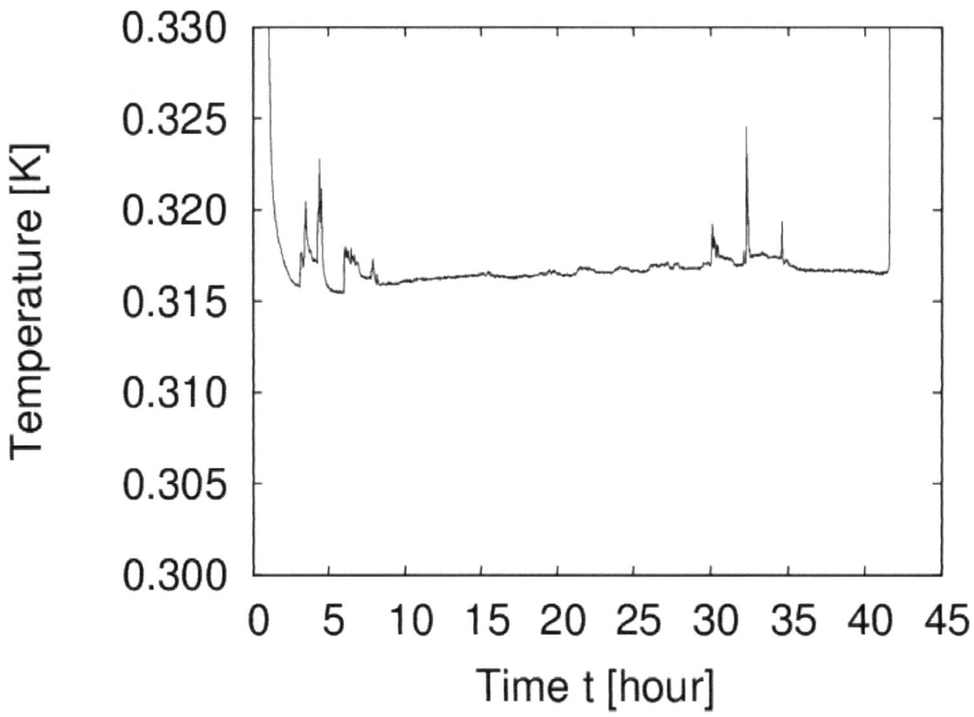

Fig. 4 Temperature curve of a helium-3 cryostat with proposed plug-in wire system combined with synchrotron beamline BL-11A at the Photon Factory at KEK.

5. Conclusion

A scalable plug-in wire system was developed for large-format STJ arrays. The proposed plug-in wires for a 200-pixel STJ array were installed in a cryostat with a ^{3}He sorption cooler. The temperature of the cold stage was less than 310 mK, with a holding time of 80 hours, using the plug-in wire system.

Acknowledgement

This work was carried out as part of Toyota's collaborative research program. This work was performed under the approval of the Photon Factory Program Advisory Committee (Proposal No. 2017G625).

References

[1] Friedrich S, Harris J, Warburton W K, Carpenter H M, Hall J A, Cantor R 2014 *J. Low Temp. Phys.* **176** 553–559
[2] Shiki S, Ukibe M, Kitajima Y, Ohkubo M 2012 *J. Low Temp. Phys.* **167** 748–753
[3] Fujii G, Ukibe M, Shiki S, Ohkubo M 2017 *X-Ray Spectrometry* **46** 325–329
[4] Fujii G, Ukibe M, Shiki S, Ohkubo M 2017 *J. Low Temp. Phys.* **184** 194–199
[5] Garwin R L 1956 *Rev. Sci. Instruments* **27** 826–828
[6] Olson J R 1997 *Cryogenics* **33** 729–731
[7] Kushino A, Kasai S, Ukibe M, Ohkubo M 2018 *J. Low Temp Phys* **193** 611–617
[8] Shiki S, Zen N, Ukibe M, Ohkubo M 2009 *AIP Conference Proceedings* **1185** 409–412

IOP Conf. Series: Journal of Physics: Conf. Series **1293** (2019) 012050 doi:10.1088/1742-6596/1293/1/012050

Ginzburg-Landau theory for the operation principle of superconducting delay-line inductance detectors

T Koyama[1] **and T Ishida**[1,2]
[1]Division of Quantum and Radiation Engineering, Osaka Prefecture University, Sakai, Osaka 599-8570, Japan
[2]NanoSquare Research Institute, Osaka Prefecture University, Sakai, Osaka 599-8531, Japan
E-mail: tkoyama67@nifty.com

Abstract. Transmission of voltage pulses generated by local heating in a kinetic inductance detector made of a superconducting stripline is theoretically investigated on the basis of the TDGL equation. The effect of thermal diffusion is also incorporated into the electrodynamics of this system. We derive an equation that can describe the generation and transmission of voltage pulses in this detector. A pair of voltage pulses with opposite polarities are created when a spatiotemporal variation in the superconducting order parameter occurs in a small region of this superconducting stripline. We clarify the characteristic feature in the shape of the voltage pulse.

1. Introduction

Radiation detectors using superconducting circuits such as transition edge sensor (TES)[1, 2], superconducting nanowire single-photon detector (SNSPD)[3, 4] and microwave kinetic inductance detector (MKID)[5] have been widely utilized in various fields using radiations. Recently, a new type of neutron detector called *Current-Biased Kinetic Inductance Detector* (CB-KID), in which high spatial- and energy-resolved imaging is possible, has been intensively developed by Ishida's group[6, 7, 8, 9, 10, 11, 12]. This neutron detector consists of two orthogonal meandering superconducting nanowires deposited on a basal plane composed of insulating and superconducting layer stacks. The S-I-S structure in this detector functions as an efficient microwave delay-line. The superconducting nanowires in this detector are further covered with a ^{10}B-layer, which absorbs neutrons well. A neutron captured by a boron nucleus in the ^{10}B-layer causes a nuclear reaction (n+^{10}B$\rightarrow$ ^{7}Li +^{4}He) and then generates heat. As a result, the superconducting nanowires contacted with the ^{10}B-layer are locally heated, that is, a hot spot is created in the nanowires. As discussed by Ishida's group, a local heating of the current-biased superconducting nanowire causes variation of the kinetic inductance and creates voltage pulses traveling along the nanowire. Hence, one can read out the position and time of the neutron capture events from the arrival times of the pulses at the two nanowire's ends.

In our previous paper[13] we formulated an electrodynamic theory for the operation principle of CB-KID. In this theory the Maxwell-London equation with a non-uniform London penetration depth is utilized to derive an equation that can describe the electromagnetic propagation modes in the detector having an S-I-S waveguide structure. It was shown that a transient hot spot appearing in the nanowire excites a pair of voltage pulses with opposite polarities and these pulses propagate with a Swihart velocity[14] toward the two nanowire's ends.

In this paper we develop the previous theory for the operation principle of CB-KID on the basis of the time-dependent Ginzburg-Landau (TDGL) equation instead of the London equation. To incorporate the effect of heat flow from a hot spot the thermal diffusion equation in the superconducting state is also utilized. We derive an integral equation, which is similar to the one given in the previous paper, for the voltage propagation modes in this detector. The shape of the voltage pulses generated by a hot spot can be obtained by solving this equation. It is shown that the voltage pulse excited by a transient hot spot has a time width nearly equal to the life time of the hot spot.

2. Dynamics of the order parameter in a superconducting stripline

2.1. TDGL equation

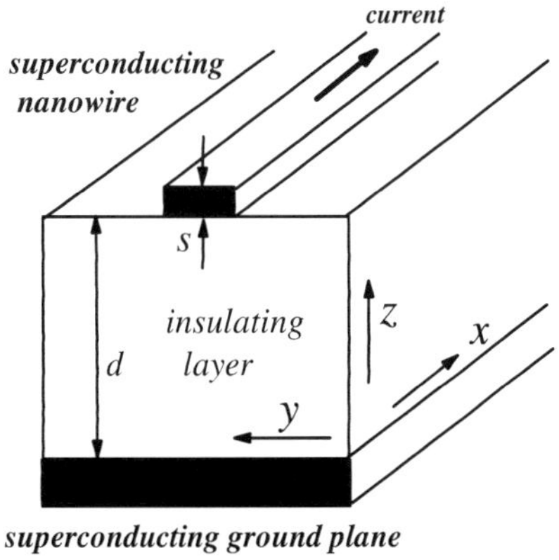

Figure 1. Schematical view of the stripline in CB-KID. A bias current I flows in the superconducting nanowire. Voltage signals can propagate in the insulating layer along the stripline.

Consider a long superconducting stripline with rectangular cross section of width L and thickness s ($L \gg s$), which is deposited on a basal plane made of a stack of insulating and superconducting layers and forms a structure which can be regarded as an S-I-S waveguide (Fig.1). In the following we use the coordinates, x, y and z, as follows, $x \parallel$ the stripline, and $y, z \parallel$ the two sides of the rectangular cross section, and $-\infty < x < +\infty$, $-L/2 < y < L/2$ and $d < z < d+s$ with d being the thickness of the insulating basal plane. Let us assume that the superconducting order parameter $\Psi(\mathbf{r},t)$ of the stripline obeys the TDGL equation, which is given in the absence of an external magnetic field as

$$\gamma\partial_t\Psi(\mathbf{r},t) = -\Big[-\frac{\hbar^2}{2m^*}\nabla^2\Psi(\mathbf{r},t) + \alpha(T)\Psi(\mathbf{r},t) + \beta|\Psi(\mathbf{r},t)|^2\Psi(\mathbf{r},t)\Big]. \tag{1}$$

Here, the parameter γ is the quasi-particle damping constant. Suppose that a high-energy particle, e.g. ^{4}He (α-particle), penetrates into the stripline and stops at a site inside it. In this event the temperature of the region around the site rises, i.e., $T \to T(\mathbf{r},t) = T + T_1(\mathbf{r},t)$, that is, one can expect that a hot spot is created inside the stripline by absorbing a high-energy particle. In this paper we study the case that an increase in temperature is not large in this incident, i.e., $T_1(\mathbf{r},t) \ll T_c$. To incorporate the effect of a hot spot into the dynamics of the order parameter we use the TDGL equation in which the parameter $\alpha(T)$ is extended as

$$\alpha(T) \to \alpha(T + T_1(\mathbf{r},t)) \simeq \alpha(T) + \alpha_1 T_1(\mathbf{r},t) + \mathcal{O}(T_1(\mathbf{r},t)^2). \tag{2}$$

Furthermore, we express the order parameter as

$$\Psi(\mathbf{r},t)/\Psi_0 = 1 + \psi(\mathbf{r},t), \tag{3}$$

where $\Psi_0 = \sqrt{|\alpha(T)|/\beta}$ is the order parameter in the state without a hot spot and $\psi(\mathbf{r},t)$ represents the deviation from it. Substituting Eqs.(2) and (3) into Eq.(1), one finds the equation for $\psi(\mathbf{r},t)$ in the linear approximation as

$$\left[D^{-1}\partial_t - \nabla^2 + \xi^{-2}\right]\psi(\mathbf{r},t) = -\eta T_1(\mathbf{r},t), \tag{4}$$

where the parameters, D, ξ and η are defined as

$$D^{-1} = 2m^*\gamma/\hbar, \quad \xi = \hbar/\sqrt{4m^*|\alpha(T)|}, \quad \eta = 2m^*\alpha_1/\hbar^2. \tag{5}$$

Consider a case where the size of a hot spot is of the order of 1μm, which is much larger than the coherence length, $\xi \sim 0.1\mu$m. In this case, since one can expect $|\nabla\psi| \ll \xi^{-1}\psi$, Eq.(4) is approximated as

$$\left[D^{-1}\partial_t + \xi^{-2}\right]\psi(\mathbf{r},t) \simeq -\eta T_1(\mathbf{r},t). \tag{6}$$

This equation is easily solved as

$$\psi(\mathbf{r},t) = -D\eta\int_0^t d\tau e^{-D(t-\tau)/\xi^2}T_1(\mathbf{r},\tau), \tag{7}$$

when a hot spot appears at $t=0$. We assume that the relaxation time of $T_1(\mathbf{r},t)$ is much longer than the quasi-electron relaxation time $\xi^2/D \equiv \tau_{qp}$. In this case Eq.(7) can be expanded as

$$\psi(\mathbf{r},t) = -D\eta\Big\{-\frac{\xi^2}{D}[e^{-t/\tau_{qp}} - 1]T_1(\mathbf{r},t)$$

$$-(\frac{\xi^2}{D})^2[1 - (1 + \frac{Dt}{\xi^2})e^{-t/\tau_{qp}}]\partial_t T_1(\mathbf{r},t) + \cdots\Big\}, \tag{8}$$

Then, in the lowest order approximation one finds the solution,

$$\psi(\mathbf{r},t) \simeq -\eta\xi^2[1 - e^{-t/\tau_{qp}}]T_1(\mathbf{r},t), \tag{9}$$

which indicates that the variation of the order parameter due to a hot spot is proportional to that of the local temperature. Using Eq.(9), one can derive the equation for the superconducting current flowing along the stripline in the presence of a hot spot as follows,

$$j_x(\mathbf{r},t) = -\frac{e^{*2}\Psi_0^2}{m^*c}(1+\psi(\mathbf{r},t))^2[A_x(\mathbf{r},t) - \frac{\hbar c}{e^*}\nabla_x\varphi(\mathbf{r},t)]$$

$$\simeq -\frac{c}{4\pi\lambda_L^2}[1 - 2\eta\xi^2(1 - e^{-t/\tau_{qp}})T_1(\mathbf{r},t)]A_x^{\varphi}(\mathbf{r},t), \tag{10}$$

up to the first order in $T_1(\mathbf{r},t)$, where $A_x^{\varphi} = A_x - (\hbar c/e^*)\nabla_x\varphi$ with φ being the phase of the order parameter and $\lambda_L = \sqrt{m^*c^2/4\pi e^{*2}\Psi_0^2}$ is understood to be the London penetration depth. Note that Eq.(10) corresponds to the generalized London equation given in our previous paper[13].

2.2. *Thermal diffusion*

Let us now construct a theory for the dynamics of the superconducting order parameter, taking account of the effect of thermal diffusion in the presence of a hot spot. For this we investigate the spatiotemporal variation of the temperature in the stripline having a hot spot, making use of the phenomenological argument given in [15, 16], which is based on the TDGL theory.

Suppose that a hot spot appears at a certain time in a finite region V inside the stripline and it decays through the thermal diffusion along the stripline and also into the basal plane. Consider the free energy of the superconducting electrons in this region. In the presence of a hot spot it depends on time and can be expressed as $F = \int_V d\mathbf{r}(f_n(\mathbf{r},t)+f_s(\mathbf{r},t))$, where $f_n(\mathbf{r},t)$ and $f_s(\mathbf{r},t)$ represent the free energy densities, respectively, in the normal and superconducting states. Note that $\int_V d\mathbf{r} f_s(\mathbf{r},t) < 0$ in the superconducting state, while $\int_V d\mathbf{r} f_s(\mathbf{r},t) = 0$ in the normal state. The heat diffusion equation is derived from the relation, $dF/dt = \int_V d\mathbf{r} q(\mathbf{r},t)$ in the presence of a heat source, where $q(\mathbf{r},t)$ in the present case is understood to be the heat quantity supplied by a hot spot. Let us now calculate the time derivative dF/dt. For the normal component one can utilize the conventional relation in the phenomenological heat transport theory,

$$\frac{d}{dt}\int_V d\mathbf{r} f_n(\mathbf{r},t) = \int_V d\mathbf{r}\Big\{C_V\partial_t T(\mathbf{r},t) + \nabla\cdot\mathbf{j}_n^Q(\mathbf{r},t)\Big\}, \tag{11}$$

where C_V and $j_n^Q(\mathbf{r},t) = -\kappa_n\nabla T(\mathbf{r},t)$ are, respectively, the specific heat and the heat current in the normal state with κ_n being the normal state thermal conductivity. For the superconducting component $f_s(\mathbf{r},t)$ we use the GL free energy and assume that the time dependence of the order parameter obeys the TDGL equation given in Eq.(1). Then, after some calculations one finds the relation,

$$\frac{d}{dt}\int_V f_s(\mathbf{r},t) = \int_V d\mathbf{r}\Big\{\partial_t\alpha(T(\mathbf{r},t))|\Psi(\mathbf{r},t)|^2 - 2\gamma|\partial_t\Psi(\mathbf{r},t)|^2 + \nabla\cdot\mathbf{j}_s^Q(\mathbf{r},t)\Big\}, \tag{12}$$

where $\mathbf{j}_s(\mathbf{r},t)$ is the heat current coming from the relaxation of the order parameter,

$$\mathbf{j}_s^Q(\mathbf{r},t) = \frac{-\hbar}{2m^*}(\nabla\Psi^*(\mathbf{r},t)\cdot\partial_t\Psi(\mathbf{r},t) + \partial_t\Psi^*(\mathbf{r},t)\cdot\nabla\Psi(\mathbf{r},t)). \tag{13}$$

From Eqs.(11) and (12) it follows the thermal diffusion equation in the superconducting state,

$$C_V\partial_t T(\mathbf{r},t) + \partial_t\alpha(T(\mathbf{r},t))|\Psi(\mathbf{r},t)|^2 - 2\gamma|\partial_t\Psi(\mathbf{r},t)|^2 + \nabla\cdot[\mathbf{j}_n^Q(\mathbf{r},t) + \mathbf{j}_s^Q(\mathbf{r},t)] = q(\mathbf{r},t). \tag{14}$$

Furthermore, since $T_1(\mathbf{r},t) \ll T_c$, one can linearize Eq.(14) with respect to $T_1(\mathbf{r},t)$, noticing the fact that $|\partial_t\Psi|^2 \sim \mathcal{O}(T_1^2)$ and $\mathbf{j}_s^Q \sim \mathcal{O}(T_1^2)$, as follows,

$$C_S\partial_t T_1(\mathbf{r},t) - \kappa_n\nabla^2 T_1(\mathbf{r},t) = q(\mathbf{r},t), \tag{15}$$

with $C_S = C_V + |\Psi_0|^2\alpha_1$. Let us solve Eq.(15) in a stripline system such as CB-KID. Note that the stripline in CB-KID is very thin, $s \sim 40$nm, so that we neglect the temperature variation along the z-direction inside the stripline, i.e., $T_1(\mathbf{r},t) \simeq T_1(x,y,t)$. Furthermore, we introduce the approximation that the heat flow to the basal plane per unit time is proportional to the temperature difference between the strip line and the basal plane, i.e., $\partial_z j_n^Q \simeq \zeta(T(x,y,t)-T) = \zeta T_1(x,y,t)$, assuming that the temperature of the basal plane is kept T. Then, on the basis of the above argument we modify Eq.(15) for the present stripline system as follows,

$$C_S\partial_t T_1(\mathbf{r},t) - \kappa_n(\partial_x^2+\partial_y^2)T_1(\mathbf{r},t) + \zeta T_1(\mathbf{r},t) = q_0\delta(x)\delta(y-y_0)\delta(t), \tag{16}$$

in the case where a hot spot appears at $t = 0$ and its center is located at $\mathbf{r} = (0, y_0)$. Note that the width $L \sim 1\mu$m of the stripline in CB-KID is not large. Accordingly, we solve Eq.(16) for the temperature averaged along the y-direction, i.e., $\tilde{T}_1(x,t) = L^{-1}\int_{-L/2}^{L/2} dy T_1(x,y,t)$, under the boundary condition $\partial_y T_1|_{y=\pm L/2} = 0$. The equation for $\tilde{T}_1(x,t)$ is easily obtained as

$$C_S \partial_t \tilde{T}_1(x,t) - \kappa_n \partial_x^2 \tilde{T}_1(x,t) + \zeta \tilde{T}_1(x,t) = q_0 \delta(x)\delta(t), \tag{17}$$

and its solution is given as

$$\hat{T}_1(x,t) = \frac{q_0 x_h}{C_S}\sqrt{\frac{\tau_z}{t}}\exp\Big[-\frac{t}{\tau_z} - \frac{1}{2}(\frac{\tau_z}{t})(\frac{x}{x_h})^2\Big]\Theta(t), \tag{18}$$

where $\tau_z = C_S/\zeta$, $x_h = \sqrt{2\kappa_n/\zeta}$ and $\Theta(t)$ is the Heaviside step function. Thus, one understands that the time dependence of the order parameter is given as

$$\psi(x,t) = -W[1 - e^{-t/\tau_{qp}}]\sqrt{\frac{\tau_z}{t}}\exp\Big[-\frac{t}{\tau_z} - \frac{1}{2}(\frac{\tau_z}{t})(\frac{x}{x_h})^2\Big]\Theta(t), \tag{19}$$

with $W = q_0 \eta \xi^2 x_h / C_S$. Eq.(19) indicates that the size of the hot spot is x_h and its relaxation time is equal to τ_z.

3. Voltage signals propagating along the stripline

In this section the electrodynamic response of a superconducting stripline to a transient hot spot is investigated on the basis of the TDGL theory developed in Sec.2. In [13] we derived the equation for the superconducting phase difference in a thin superconducting stripline forming a S-I-S structure such as CB-KID, which is given as

$$\frac{\lambda_L + d}{d}\frac{\epsilon}{c^2}\partial_t^2\theta(x,t) - \partial_x^2\theta(x,t) = \frac{e^*}{\hbar c}A_x^{\varphi}(x,d,t), \tag{20}$$

where $\theta(x,t)$ is the superconducting phase difference between the stripline and the superconducting basal plane and ϵ is the dielectric constant of the insulating layer with a thickness d (see Fig.1). The vector potential $A_x^{\varphi}(x,d,t)$ in Eq.(20) can be eliminated by using the solution of the Maxwell equation in the region inside the stripline as shown in [13]. When the GL theory is used for the superconducting current flowing in the stripline, the Maxwell equation to determine $A_x^{\varphi}(x,d,t)$ is obtained as

$$\partial_z B_y(x,z,t) = \lambda_L^{-2}(1 + \psi(x,z,t))^2 A_x^{\varphi}(x,z,t). \tag{21}$$

In [13] we present a way to solve Eq.(21) under the suitable boundary condition at the interface between the superconducting stripline and the insulatng basal plane. Then, following the procedure given in [13], one can derive the equation with a source term from Eq.(20) under the assumption $\psi(x,z,t) \ll 1$ as

$$v^{-2}\partial_t^2\theta(x,t) - \partial_x^2\theta(x,t) = F(x,t), \tag{22}$$

where v is the velocity defined as $v = d^{1/2}[d + \lambda_L(1 + \coth(s/\lambda_L)]^{-1/2} c\epsilon^{-1/2}$ and the source term $F(x,t)$, which stems from the appearance of a transient hot spot, is given as

$$\begin{aligned} F(x,t) &= (2\lambda_L/d)\coth(s/\lambda_L)(\epsilon/c^2)\partial_t^2\theta(x,t)\cdot\psi(x,d,t) \\ &+2\lambda_L\coth(s/\lambda_L)[d + \lambda_L(1 + \coth(s/\lambda_L)]^{-1}\partial_x\theta(x,t)\cdot\partial_x\psi(x,d,t). \end{aligned} \tag{23}$$

Note that $F(x,t) = 0$ for $t < 0$, when the hot spot appears at $t = 0$. Eq.(22) can be transformed into the integral form as

$$\theta(x,t) = \theta_0(x,t) + \int_{-\infty}^{\infty} dx' \int_{-\infty}^{\infty} dt' G^R(x - x', t - t')F(x',t'), \tag{24}$$

where $\theta_0(x,t)$ is a solution of the homogeneous equation, $[v^{-2}\partial_t^2 - \partial_x^2]\theta_0(x,t) = 0$, and $G^R(x,t)$ is the retarded Green function satisfying the equation, $[v^{-2}\partial_t^2 - \partial_x^2]G^R(x,t) = \delta(x)\delta(t)$, i.e.,

$$G^R(x,t) = \Theta(t)\int_{-\infty}^{\infty}\frac{dq}{2\pi}\frac{iv}{2q}[e^{iq(x-vt)} - e^{i(x+vt)}]. \tag{25}$$

It is noted that Eq.(24) is an integral equation, since $F(x,t)$ contains $\theta(x,t)$. In the following we study the case in which a constant bias current I flows in the stripline and a hot spot appears at $(x,t) = (0,0)$. In this case the phase difference $\theta_0(x,t)$ is understood to come from the magnetic flux induced by the bias current. We express the magnetic flux B_0 induced by the bias current as $B_0 = LI$, introducing the inductance L phenomenologically. Then, using the Josephson relation, one can derive the phase difference, $\theta_0(x,t) \equiv \theta_0(x)$, as

$$\theta_0(x) = (e^*/\hbar c)(d + \lambda_L[1 + \coth(s/\lambda_L)])LIx + \text{const.} \tag{26}$$

Let us now solve Eq.(24). The equation can be solved systematically by means of the iteration method. We seek the lowest order solution, which is the first order with respect to $\psi(x,t)$. The source term $F(x,t)$ in Eq.(23) in this case is approximated as

$$F(x,t) \simeq 2(e^*/\hbar c)\lambda_L\coth(s/\lambda_L)LI\partial_x\psi(x,t) \equiv (2e^*K/\hbar)LI\partial_x\psi(x,t), \tag{27}$$

with $K = (\lambda_L/c)\coth(s/\lambda_L)$, which is obtained by substituting $\theta_0(x)$ into $\theta(x,t)$ in Eq.(23). Furthermore, using the Josephson relation for the voltage difference, $V(x,t) = -\int_{-\infty}^{d} dzE_x = -(\hbar/e^*)\partial_t\theta(x,t)$, one can obtain the lowest order solution for the voltage difference from Eqs.(24) and (27) as follows,

$$V(x,t) = -2KLI\int_{-\infty}^{\infty}dx'\int_{-\infty}^{\infty}dt'\partial_t G^R(x-x',t-t')\partial_{x'}\psi(x',d,t'). \tag{28}$$

The time derivative of the retarded Green function in Eq.(28) is derived from Eq.(25) as

$$\partial_t G^R(x,t) = \frac{v^2}{2}[\delta(x-vt) + \delta(x+vt)]\Theta(t). \tag{29}$$

Then, Eq.(28) yields the solution,

$$V(x,t) = -KLv^2\int_{-\infty}^{\infty}dx'\Big\{\Theta(x-x')\partial_{x'}\psi(x',t')|_{t'=(x'-z^{(+)})/v}$$
$$+\Theta(x'-x)\partial_{x'}\psi(x',t')|_{t'=(z^{(-)}-x')/v}\Big\}\cdot I, \tag{30}$$

with $z^{(\pm)} = x \mp vt$. Note that Eq.(30) is expressed as $V(x,t) = V^{(+)}(x,t) + V^{(-)}(x,t)$, where $V^{(+)}(x,t)$ and $V^{(-)}(x,t)$ are the voltage signals propagating in the $+x$- and $-x$-directions, respectively. It is easy to see that in the asymptotic region, i.e., $x \to \pm\infty$ and $|z^{(\pm)}| < \infty$, the voltage signals (30) depend only on the variables $z^{(\pm)}$, i.e.,

$$\lim_{x\to+\infty} V^{(+)}(x,t) = \hat{V}^{(+)}(z^{(+)}), \qquad \lim_{x\to-\infty} V^{(-)}(x,t) = \hat{V}^{(-)}(z^{(-)}), \tag{31}$$

that is, the two voltage pulses propagate in the opposite directions with a constant speed v. When Eq.(19) is utilized for ψ, the voltage pulses, $\hat{V}^{(\pm)}(z^{(\pm)})$, are given as

$$\hat{V}^{(+)}(z^{(+)}) = -KLWv^2\int_0^{\infty}dr(1 - e^{-r/v\tau_{qp}})(\frac{v\tau_z}{r})^{3/2}(\frac{r + z^{(+)}}{x_h})$$

IOP Conf. Series: Journal of Physics: Conf. Series **1293** (2019) 012050 doi:10.1088/1742-6596/1293/1/012050

$$\times \exp\Big[-\frac{r}{v\tau_z} - \frac{1}{2}(\frac{v\tau_z}{r})(\frac{r+z^{(+)}}{x_h})^2\Big]\cdot I \equiv \tilde{L}^{(+)}(z^{(+)})I, \tag{32}$$

and

$$\hat{V}^{(-)}(z^{(-)}) = KLWv^2 \int_0^\infty dr(1-e^{-r/v\tau_{qp}})(\frac{v\tau_z}{r})^{3/2}(\frac{r-z^{(-)}}{x_h})$$

$$\times \exp\Big[-\frac{r}{v\tau_z} - \frac{1}{2}(\frac{v\tau_z}{r})(\frac{r-z^{(-)}}{x_h})^2\Big]\cdot I \equiv \tilde{L}^{(-)}(z^{(-)})I, \tag{33}$$

where $\tilde{L}^{(\pm))}(z^{(\pm)})$ is understood to be the time derivative of the kinetic inductance of the stripline. It is also noted that the voltage pulses $V^{(\pm)}$ have opposite polarities since $V^{(+)}(z) = -V^{(-)}(-z)$. Let us now calculate the voltage pulse $V^{(-)}(z^{(-)})$ given in Eq.(33). The integral in Eq.(33) can be evaluated as a function of the two dimensionless parameters, $v\tau_{qp}/x_h$ and $v\tau_z/x_h$. Suppose that the parameters, τ_{qp}, τ_z, v and x_h, take values in the region of $\tau_{qp} \sim 10^{-12}$sec, $\tau_z \sim 10^{-11}$sec, $v \sim 10^9$cm$\cdot$ sec^{-1} and $x_h \sim 10^{-3}$cm, i.e., $v\tau_{qp}/x_h \sim 1$ and $v\tau_z/x_h \sim 10$, because the parameter values of the CB-KID made of Nb nanowires are expected to be in this region. Fig.2 shows the $z^{(-)}$ dependence of the normalized voltage pulse, $V_N(z^{(-)}) \equiv \hat{V}(z^{(-)})/KLWIv^2x_h$ for $v\tau_{qp}/x_h = 5.0$ and three different values of $v\tau_z/x_h$. As seen in this figure, the voltage pulse steeply rises first and then gradually attenuates. Such a feature in the shape of the voltage pulse has been observed in several kinetic inductance detectors[12, 17]. It is also noticed that the width of the voltage pulse increases with τ_z, the relaxation time of the hot spot. From this result one may conclude that the voltage pulses generated by a transient hot spot in the kinetic inductance stripline have a time width of the order of the life time of the hot spot.

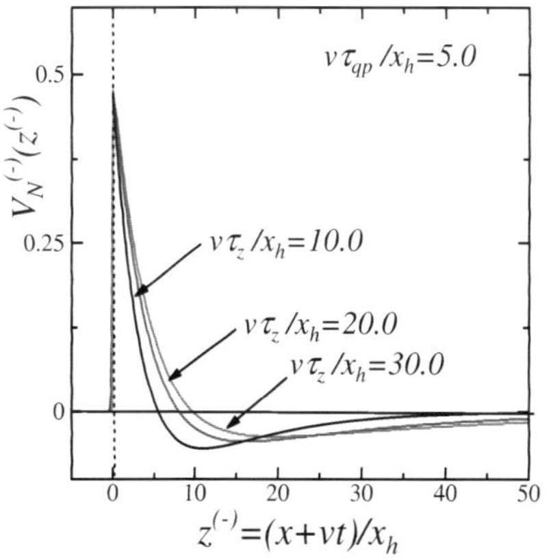

Figure 2. Shape of the normalized voltage pulse propagating along the stripline.

Let us next investigate the temperature dependence of the voltage pulse. We assume that the temperature dependence appears through the temperature variation of the London penetration depth only. In this case the velocity v and the coefficients, K and W, which are included in the prefactor in Eq.(33), depend on temperature. In Fig.3 we plot the voltage pulses as a function of time for two different temperatures, $T/T_c = 0$ and 0.8. It is seen that the amplitude of the voltage pulse increases with temperature. This result indicates a possibility that the detector shows higher sensitivity at high temperatures. To obtain the voltage pulses on a more reliable footing in the high temperature region near T_c full numerical calculations will be required for the coupled equations (1) and (14), because the present theory is valid in the region $|\psi(\mathbf{r},t)| \ll \Psi_0$, which is well satisfied in the low temperature region, $T \ll T_c$.

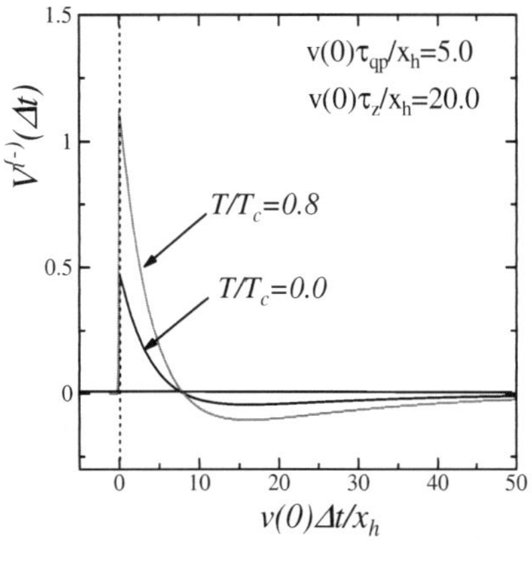

Figure 3. Voltage pulses at $T/T_c = 0$ and 0.8. $v(0)$ is the velocity at $T = 0$ and Δt is the time measured from $t = -x/v(T)$. For the temperature dependence of the London penetration depth we use the two-fluid model, i.e., $\lambda_L(T) = \lambda_L(0)/\sqrt{1-(T/T_c)^4}$.

4. Summary

In summary we have constructed an electrodynamic theory for the operation principle of a kinetic inductance detector having a S-I-S waveguide structure on the basis of the TDGL theory. The heat flow effect from a hot spot is also incorporated into our theory, using the phenomenological thermal diffusion equation in the superconducting state. We have clarified the feature of the voltage pulses generated by local heating in this detector.

Acknowledgments

The authors thank M. Machida for helpful discussions. This work was partially supported by Grant-in Aid for Scientific Research (S) (No. 23226019) and (A) (No. 16H02450) from JSPS, Japan.

References

[1] Irwin K D, Hilton G C, Martinis J M, and Cabrera B 1995 *Rev. Sci. Instrum.* **66** 5311
[2] Chervenak A, Irvin K D, Grossman E N, Martinis J M, Reintsema C D, and Huber M E 1999 *Appl. Phys. Lett.* **74** 4043
[3] Goltsman J A, Okunev O, Chuikova G, Lipatov A, Semenov A, Smirnov K, Voronov B, Dzardanov C A, Williams C, and Sobolewski R 2001 *Appl. Phys. Lett.* **79** 705
[4] Miki S, Yamashita T, Fujiwara M, Sasaki M, and Wang Z 2010 *Optics Letters* **35** 2133
[5] Mazin B A, Day P K, Zmuidzinas J and Leduc H G 2002 *AIP Conf. Proc.* **I** 309
[6] Ishida T, Yagi I, Yoshioka N, Huy H T, Yotsuya T, Satoh K, Uno M, Shimakage H, Miki S, Wang Z, Kato M, Machida M and Hojou H 2008 *J. Low Temp. Phys.* **151** 1074
[7] Shibata H 2014 *Appl. Phys. Express* **7** 103101
[8] Shishido H, Yoshida T and Ishida T 2015 *Appl. Phys. Express* **8** 113101
[9] Miyajima S, Shishido H, Narukami Y, Yoshioka N, Fujimaki A, Hidaka M, Oikawa K, Harada M, Oku T, Arai M and Ishida T 2017 *Nucl. Instr. Meth.* A **842** 71
[10] Narukami Y, Miyajima S, Fujimaki A, Hidaka M and Ishida T 2015 *IEEE. Trans. Appl. Supercond.* **25** 2400904
[11] Narukami Y, Oikawa K, Harada M, Oku T, Arai M, Hidaka M, Fujimaki A and Ishida T 2015 *Appl. Phys. Lett.* **107** 23260
[12] Shishido H, Miki Y, Yamaguchi H, Iizawa Y, Dang V, Kojima K, Koyama T, Oikawa K, Harada M, Miyajima S and Hidaka M 2018 *Phys. Rev. Appl.* **10** 044044
[13] Koyama T and Ishida T 2018 *J. Phys. Conf. Ser.* **1054** 012055
[14] Swihart J 1961 *J. Appl. Phys.* **32** 461
[15] Machida M, Koyama T, Kato M and Ishida T 2004 *Nucl. Instr. and Meth.* **A529** 409
[16] Shapiro I, Pechenik E and Shapiro B Ya 2001 *Phys. Rev. B* **63** 184520
[17] Zhao Q Y, Zhu D, Calandri N, Dane A E. McCaughan, Bellei F, Wang H Z, Santavicca D and Berggren K K 2017 *Nature Photonics* **11** 247

IOP Conf. Series: Journal of Physics: Conf. Series **1293** (2019) 012051 doi:10.1088/1742-6596/1293/1/012051

Temperature dependent characteristics of neutron signals from a current-biased Nb nanowire detector with ^{10}B converter

The Dang Vu[1,*], Yuki Iizawa[2], Kazuma Nishimura[2], Hiroaki Shishido[2,3], Kenji M Kojima[4,7,†], Kenichi Oikawa[1], Masahide Harada[1], Shigeyuki Miyajima[2,5], Mutsuo Hidaka[6], Takayuki Oku[1], Kazuhiko Soyama[1], Kazuya Aizawa[1], Tomio Koyama[7], and Takekazu Ishida[3,7]

[1]Materials and Life Science Division, J-PARC Center, Japan Atomic Energy Agency, Tokai, Ibaraki 319-1195, Japan

[2]Department of Physics and Electronics, Osaka Prefecture University, Sakai, Osaka 599-8531, Japan

[3]NanoSquare Research Institute, Osaka Prefecture University, Sakai, Osaka 599-8531, Japan

[4]Muon Science Laboratory and Condensed Matter Research Center, Institute of Materials Structure Science, KEK, Tsukuba, Ibaraki 305-0801, Japan

[5]Advanced ICT Research Institute, National Institute of Information and Communications Technology, Kobe, Hyogo 651-2492, Japan

[6]National Institute of Advanced Industrial Science and Technology, Tsukuba, Ibaraki 305-8568, Japan

[7]Division of Quantum and Radiation Engineering, Osaka Prefecture University, Sakai, Osaka 599-8570, Japan

[†]Present address: Centre for Molecular and Materials Science, TRIUMF and Stewart Blusson Quantum Matter Institute, University of British Columbia, Vancouver, BC, V6T 2A3 and V6T 1Z4, Canada

*vu.thedang@jaea.go.jp

Abstract. We are developing a new type of the neutron imager based on a superconducting neutron detector. We previously succeeded in constructing and demonstrating neutron detection capability of a superconducting current-biased kinetic inductance detector (CB-KID). In order to improve the spatial resolution and detection efficiency, the characteristics of a superconducting neutron detector have been studied systematically in the present work. As an extension of studying the characteristics of neutron detector, we investigated temperature dependence of neutron signal such as propagation velocity and the signal amplitude as a function of time of flight (ToF) with temperature. We consider that it is important to understand the temperature dependence of the signal to improve the spatial resolution and detection efficiency of a superconducting neutron detector.

1. Introduction

Neutron imaging has been recognized as a powerful technique in conducting non-destructive analysis in the wide areas such as science, medical field, and industrial technology to study the properties of materials, product components, and various systems without causing damage. Since the neutron beam can penetrate deeply into most materials except hydrogen, lithium and boron. Therefore, neutron imaging is in many instances suitable than X-ray or gamma-ray imaging because some matters are more visible with neutron beams whenever it is almost impossible to see with X-ray imaging techniques. Neutron imaging in non-destructive transmission imaging, tomography, dark-field image and visualization of magnetic-field distribution has been applied to various materials [1], intermetallic alloys [2], fuel cells [3], lithium ion batteries [4], cultural heritages [5], and others. The study of neutron imaging was started in the late nineteen thirties and the neutron radiographs of reasonable quality were first taken by Thewlis in UK in 1955 [6]. In recent decades, neutron-imaging technology has been developed rather rapidly to achieve high resolution, energy-dispersive imaging and high-speed readout system. The neutron imagers are known as a photographic-film detector [7], a scintillator-and-storage-phosphor detector [5], a gas neutron detector [8,9], a semiconductor (solid-state) detector [10].

Our work aims at obtaining a neutron imager with high-spatial resolution, high sensitivity, high temporal resolution, and high-speed measurement instrumentation by constructing a new delay-line current-biased kinetic inductance detector (CB-KID) system. The CB-KID system consists of an orthogonal X- and Y-meanderlines and an enriched ^{10}B neutron absorption layer superimposed on top of CB-KID [11–13]. Koyama and Ishida [14] proposed a theory of the signal propagation along a superconducting waveguide having an S-I-S structure within the framework of the London-Maxwell theory to explain operating mechanism of the CB-KID. They predicted the propagation velocity as a function of temperature, of which we would like to verify the validity of the theoretical prediction of the propagation velocity. We also intend to study the temperature-dependent characteristics of neutron signals, i.e., the signal height and the signal width as a function of detector temperature. We design the superconducting neutron detector with eight layers by a computer-aided-design software (Layout Editor) to fabricate the superconductive device at the Clean Room for Analog-Digital-Superconductivity (CRAVITY) of the National Institute of Advanced Industrial Science and Technology (AIST). We use a Gifford-McMahon (GM) cryocooler to cool down a neutron detector. The temperature of detector is controlled by a temperature controller (Cryocon Inc.) using a LabVIEW program.

2. Operation principle of neutron detector

A neutron detector is constructed by a pair of X- and Y-CB-KIDs [11–13], orthogonal to each other, which consist of superconducting-nanowire meanderlines (Fig.1(a)) and a ^{10}B layer working as a neutron particle converter (Fig.1(b)). A pulsed voltage signal is generated by a local transient change in n_s induced by the nuclear reaction between a ^{10}B nucleus and an incident neutron through a change in local kinetic inductance L_k in a superconducting wire. The kinetic inductance L_k is expressed by

$$L_k = m_s l / n_s q_s^2 S \quad (1)$$

where m_s is the effective mass of the Cooper pair, q_s is the electric charge of the Cooper pair, n_s is the Cooper pair density, l is the length of stripline, and S is the cross-sectional area of superconducting nanowire. A nuclear reaction mainly produces a 1.47-MeV ^{4}He particle and a 0.88-MeV ^{7}Li particle, of which one particle reaches the Nb nanowire to create a hot spot in a stripline of the detector, where a rapid reduction of n_s occurs locally at a tiny spot of stripline with a length Δl ($\ll l$) to yield a transient change in L_k. According to equation (2), a pair of pulsed-voltage signals are generated in the stripline biased by the DC current while the sign (polarity) of the signal is dependent on the current direction and the propagation direction. The pulsed-voltage signal (Fig. 1(c)) propagates as an electromagnetic wave pulse along the stripline (meanderline) toward one of the end electrodes with an opposite polarity and the electromagnetic wave is always confined in between the superconducting meanderline and superconducting ground plane as a Swihart wave [14,15]. After the output signals from CB-KID were amplified by ultra-low noise amplifiers, the position (x,y) of neutron nuclear event was evaluated with the aid of a time-to-digital converter (TDC) of the Kalliope-DC readout circuit. By combining two X and Y meanderlines, orthogonal to each other, we can detect a correct position of neutron beam arrival by using equation (3) (see below).

Phenomenological, the voltage V across the hot spot (Fig. 1(c)) is expressed as

$$V = I_b\left(\frac{dL_k}{dt} + \frac{dL_m}{dt}\right) + (L_k + L_m)\frac{dI_b}{dt} \simeq I_b\frac{dL_k}{dt} \tag{2}$$

where the magnetic inductance L_m is kept constant with time and $dI_b/dt \simeq 0$ as far as the detector remains in the superconducting state. As shown in Fig. 1 (b), the hot spots are created by a charged particle emitted from the neutron reaction in both the X meanderline and the Y meanderline. A pair of a positive signal and a negative signal start to propagate toward the opposite electrodes. The negative signal is inverted by using a differential amplifier. As shown in Fig. 1 (c), we can measure the differences in the timestamps Δt_x and Δt_y of pulsed signals arrived at anode electrode and cathode electrode of X, Y meanderline. Therefore, the position of a nuclear event (x, y) is calculated as

$$\begin{cases} x = \mathrm{ceil}\left(\frac{\Delta t_x v_x}{2h_x}\right) p_x \\ y = \mathrm{ceil}\left(\frac{\Delta t_y v_y}{2h_y}\right) p_y \end{cases} \tag{3}$$

where the origin of a coordinate is at the center of a detector, p_x, p_y are the repetition pitches for the X, Y meanderline; v_x, v_y are velocities of neutron pulsed in x, y meanderline; h_x, h_y are the lengths of a single-microstrip segment in X, Y meanderline. The ceil is a function to give an integer of the operant. In our neutron detector, we design $p_x = p_y = p$ and $h_x = h_y = h$ to obtain a square sensitive area of the detector. This also simplifies the imaging process in building neutron image.

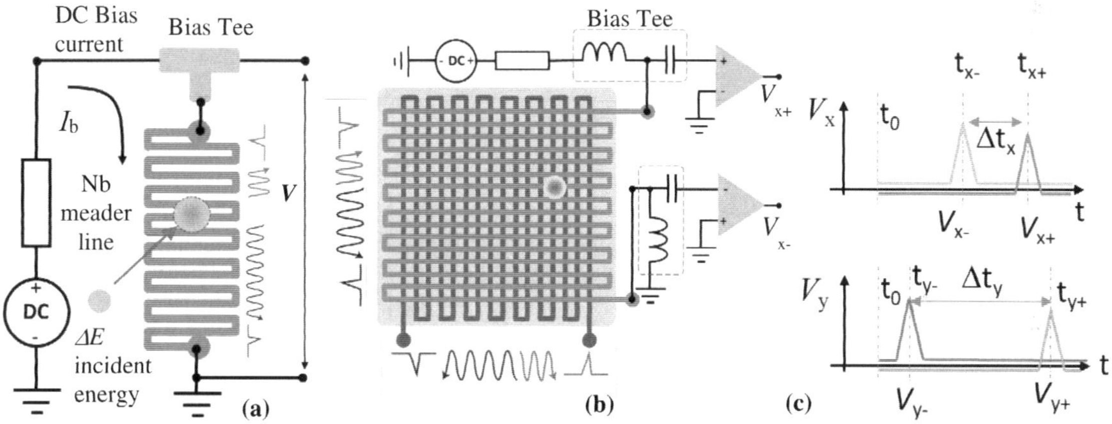

Fig 1 (a). A structure of single CB-KID with DC bias current; (b). The operating principle of delay-line CB-KID with two orthogonal meanderlines; (c) The schematic wave forms of the pulsed signals amplified by a non-inverted amplifier and an inverted amplifier to produce output signals with positive polarity. The event time t_0 should be the same in both the X and Y meanderlines.

3. Neutron detector and experimental details

3.1. Neutron detector

Fig. 2 (a) shows the schematic view of our delay-line CB-KID system while Fig. 2 (b) is a cross sectional image of the X-Y CB-KID system. Our superconducting neutron detector is designed by a computer-aided design (CAD) software (Layout Editor) so as to meet the requirement of AIST standard process. Fig. 2 (c) is a schematic drawing of a corner of our neutron detector CB-KID. The neutron detector consists of two superconducting nano-wires (X and Y) in a form of meanderlines. The neutron absorption layer consists of an assembly of very fine enriched ^{10}B particles, which were painted uniformly on top of the meanderlines in a form of ethanol solution of ^{10}B fine particles. Using this method, the thickness of the ^{10}B layer was thick enough more than 10μm compared to the ranges of ^4He and ^7Li particles. In fabrication, the turning points of X and Y meanderlines are rounded to ensure smooth propagations of the electromagnetic waves along the whole meanderline. In addition, the meanderline width is kept constant even at a turning point. This is to reduce the refection of the pulsed signals by minimizing a change in the matching impedance along the whole meanderline. Both the *X* and *Y* superconducting Nb meanderlines consist of 10000 microstripline segments of 0.9-μm width, and 0.6-μm spacing of segment-length 15 mm with a sensitive area as 15x15 mm^2. The repetition period (pitch) is 1.5 μm while the total length of the meanderline extends to a scale of 151 m. Our detector consists of seven layers deposited on a thermally oxidized Si substrate as shown schematically in cross section [16] of Fig. 2 (b). We expect that the neutron detector would have the specification of high resolution of 15x15μm^2 and megapixel imaging.

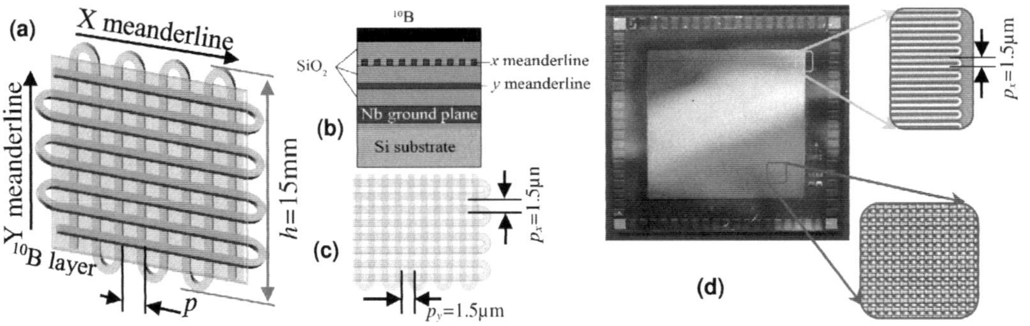

Fig. 2. (a) Schematic image of the CB-KID system. An *X* CB-KID meander is orthogonal to a *Y* CB-KID meander. The ^{10}B conversion layer is superimposed on the CB-KID system. (b) Schematic cross sectional image of the *X Y* CB-KID system (c) Layout editor design image has the repetition pitch of *x* and *y* by 1.5μm. (d) Photograph of a neutron detector fabricated by us and photographs taken by a laser microscope.

3.2. Experimental details

The superconducting neutron detector is cooled down to 4K by using a Gifford–McMahon (GM) refrigerator to conduct a long-time measurement at a stable cryogenic temperature. A Cernox thermometer and a heater are installed in the neighbourhood of the neutron detector. The temperature of the CB-KID sensor is controlled by a temperature controller (Cryocon Model 42) using a LabVIEW program. Two DC adjustable voltage sources are used to feed a DC bias current to X and Y meanderlines. The DC currents of the X and Y detectors are adjusted to obtain proper amplitudes of pulsed voltages of neutron events, which are suitable to be measured by a high-speed digital oscilloscope and a time-to-digital converter (TDC) of the Kalliope-DC readout circuit [17]. The voltage signal from neutron event on the meanderline is amplified by a low-noise amplifier (NF SA-430F5) to feed the TDC module of the Kalliope-DC readout circuit or a 2.5-GHz sampling digital oscilloscope (Teledyne LeCroy HDO4104-MS). A signal pulse from the CB-KID was transmitted through SMA and MMCX connectors and semi-rigid cables to reduce noises in the cryostat.

4. Experimental results

4.1. Temperature dependence of propagation velocity

In the preceding studies, we have investigated the temperature-dependent characteristics of propagation velocity of a superconducting neutron detector both theoretically and experimentally [14,18]. Since the propagation of pulsed voltage along a meanderline would be important to confirm the operating principle of a neutron detector. We conducted the signal transmission measurement as was reported earlier [18], and we extended our investigation to the pulsed voltage amplitude as a function of temperature, where we controlled the temperature of Nb-wire meanderlines from 3.8 K to 8.3 K but did not feed a DC bias current to a superconducting Nb meanderline. We inputted a test pulse to one electrode of the meanderline and detected a transmitted signal after reaching the opposite electrode. The test pulse signal was created by using a digital function generator and a passive *RC* differentiation circuit with the repetition frequency of 100 kHz. The signal width of the output pulse from the differentiation circuit was determined by a rise time, and was much sharper than the original square wave of 100ns width. The sharp pulsed signals were applied to X and Y meanderlines with the length (l_x = l_y) as 151 m, and the same signal was fed to two short superconducting lines ($l_s \approx 0$) fabricated on the same CB-KID chip to compensate a possible delay caused by inside semi-rigid wiring in the cryostat. Four outputs of superconducting lines were connected to four channels of a high-speed oscilloscope. Since the relation of l_x, $l_y >> l_s \approx 0$ is satisfied, the outputs of two short superconducting lines can be used to compensate the delay time of the cryostat wiring in evaluating a delay time in transmitting along meanderlines (see a start pulsed signal of the inset of Fig. 3 at $t = 0$). A difference between the output signal from a short superconducting line and the output signal from the meanderline, we calculated the propagation velocity of the pulsed signal along the superconducting nano-striplines. Figure 3 shows our experimental results (circle point curve) on the propagation velocity of a pulsed signal along a nanowire-stripline as a function of temperature. The propagation velocity of a pulsed signal decreases rapidly toward zero when the temperature increases very close to T_c. However, the propagation velocity is rather stable at lower temperatures. We understand that the propagation velocity depends on the structure of stripline, i.e., the thickness of insulating layer, the thickness of the nanowire, and the critical temperature T_c. The solid line of Fig. 3 is a least-squares fitting line of experimental results by the theoretical equation (4) derived by Koyama and Ishida [14] as

$$v = \frac{c}{\sqrt{\epsilon}} \sqrt{\frac{d}{d + \lambda_L(1 + \coth(s/\lambda_L))}} \tag{4}$$

where c is the speed of light in a vacuum, ϵ is the dielectric constant, d is the thickness of insulating layer, s is the thickness of the nanowire, respectively, T_c is the critical temperature and T is the temperature of superconducting nanowire, respectively. The London penetration depth can be approximated by the two fluid model as

$$\lambda_L(T) = \lambda_L(0)/\sqrt{1 - (T/T_c)^4}\ . \tag{5}$$

In the inset of Fig. 3, we found that the amplitude of pulsed voltage also depends on the temperature of the striplines. Firstly, we notice that the amplitude of a pulsed signal after transmitting the

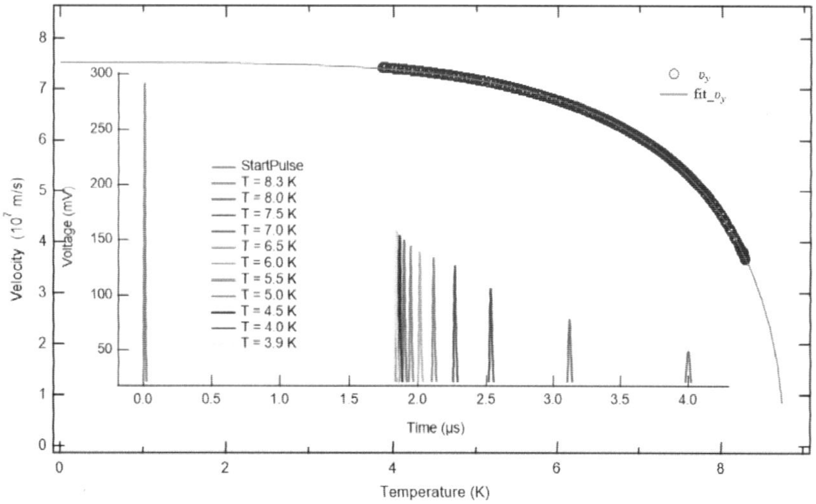

Fig. 3. Temperature dependence of propagation velocity and the transmitted signal of a pulsed voltage along the meanderline. The experimental results (open circles) on the propagation velocity of a pulse along the nano-stripline is shown as a function of temperature. The solid line is the fitting line with equation (4), where the parameters are $c = 2.99792458\times10^8$ m/s, d =300 nm, s =40 nm are fixed while others are he fitting parameters as $T_c = 8.6736 \pm 0.000341$ K, $\lambda(0) = 2.3121\times10^{-7} \pm 3.39\times 10^{-10}$ m, ϵ = 2.0397 ±0.00473. We find a good agreement between experimental observations and theoretical predictions. The inset shows a start-pulse as a function of time and the transmitted pulses from the meander.

meanderline is reduced compared with the "start pulse" at t=0 of the inset of Fig. 3. The transmitted pulsed voltage gives a smaller amplitude as the temperature increases toward the critical temperature of Nb nano-wire. Note that we did not apply the DC bias current to meanderline in conducting the velocity measurements, and the pulsed voltage was fed from outside of CB-KID as an external signal. This is because the small DC bias current does not change the propagation velocity appreciably compared to the bias-free case. Therefore, the meanderline works as a superconducting transmission line. The temperature variation of the propagation velocity is primarily due to the reduction of the Cooper pairs in the superconducting nano-wire. As seen in the inset of Fig. 3, the amplitude of the transmitted signal decreases as the temperature increases toward to the critical temperature T_c. We consider that this is caused by energy dissipation due to excess quasi particles at higher temperatures.

4.2. Temperature dependence of pulsed voltage amplitude

To study the temperature-dependent characteristics of the superconducting neutron detector, we performed the experiment with the neutron beam. The experiment was conducted at the beam line 10 (BL10) of Materials and Life Science Division (MLF) of the J-PARC Center, Japan Atomic Energy Agency (JAEA). Pulsed neutron beam appears at 25-Hz repetition rate, and the start pulse of each pulsed neutron beam is available as the MLF trigger. The BL10 is intended to provide a suitable neutron environment for testing various novel detectors devices, and the ideas for novel experimental techniques. The experimental system is described in Fig. 1 (b). To obtain systematic characteristics of neutron signal, we collected data from the neutron signal by using the time-to-digital convertor (TDC) of the Kalliope-DC readout circuit and the high speed digital oscilloscope, where the temperature was controlled from 4.0 K to 7.5 K, the meanderline was biased by a DC current $I_b = 50\mu A$ and the number of MLF triggers was 2500 to accumulate data for the measurements at each temperature. Figure 4 shows the histogram of the peak voltage with the changing of detector temperature, which was taken by the high-speed 2.5 GHz oscilloscope with the time duration from 8 ms to 13 ms in the time-of-flight (ToF) after each MLF trigger pulse. We found that the peak voltage of neutron signal changes as a function of temperature. In the value of peak signal smaller than 5 mV, the amplitude of neutron signal almost does not change appreciably. We assume that this is the contribution of the noises in signal which is not so strongly affected by a change in temperature while the amplitude of neutron signal ($V_p > 5$ mV) increases clearly when the temperature of neutron detector approaches to the critical temperature of Nb. In order to explain the experimental results, Koyama and Ishida [19] studied the Ginzburg-Landau theory for the operation principle of superconducting delay-line inductance detectors which would be explained the changing of amplitude of neutron signal with temperature.

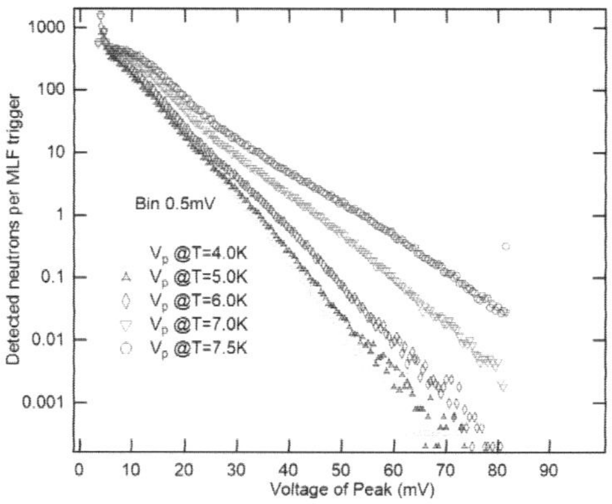

Fig. 4. The histogram of the peak voltage with changing temperature. The neutron intensity is evaluated for a given bin of 0.5 mV.

$$\hat{V}^{(\pm)}(Z^{\pm}) = KLWv^2 \int_0^{\propto} dr \left(1 - e^{-\frac{r}{v\tau_{qp}}}\right)\left(\frac{v\tau}{r}\right)^{\frac{3}{2}}\left(\frac{r - z^{(\pm)}}{x_h}\right)$$
$$\times \exp\left[-\frac{r}{v\tau_z} - \frac{1}{2}\left(\frac{r}{v\tau_z}\right)\left(\frac{r-z^{(\pm)}}{x_h}\right)^2\right] \cdot I \equiv \tilde{L}^{(\pm)}(z^{\pm})I \qquad (6)$$

where $\hat{V}^{(\pm)}(Z^{\pm})$ are the voltage signals propagating in the $+x$- and $-x$-directions, with $z(\pm) = x \mp vt$, respectively; $K = (\lambda_L/c)\coth(s/\lambda_L)$; $W = q_0\eta\xi^2 x_h/C_S$; L is inductance of meanderlines; v is the propagation velocity. The explicit expression for the voltage pulse is derived equation (6) which these expressions include the velocity v, K and W (see [19] for further details of notations).

As mentioned by Koyama and Ishida [14], the propagation velocity v is a function of temperature and K, W are also the functions of the London penetration depth. That is, these coefficients also cause the temperature dependence of the voltage pulse. We conclude that the amplitude of pulsed voltage is strongly affected by the temperature of superconducting meanderline. Detailed comparisons will be done in our future work.

In order to investigate the usefulness of the neutron detector as a function of time of flight (ToF) with changing temperature, we took the time-of-flight spectrum for each changing temperature by using the TDC module of the Kalliope-DC circuit. Note that ToF required for a 14-m flight path can be converted to neutron energy or wavelength. Fig. 5 (a) shows the signal intensity as a function of ToF with changing temperature. There is a clear difference in the event intensity as a function of time of flight (ToF) with changing temperature. The increase of signal intensity is strong when the temperature approaches to a critical temperature. In order to improve the efficiency of our neutron detector, we compared the experimental ToF spectra at $T = 4.0$ K, 5.0 K, 6.0 K, 7.0K, and 7.5 K with the predicted ToF spectrum obtained by simulations. Fig. 5 (b) shows the prediction of neutron intensity as a function of time of flight [20], of which the Monte Carlo particle and heavy ion transport code PHITS was used to deal with particle transport phenomena in a wide energy range. Detailed comparison with the prediction will be published elsewhere. The detection efficiency would be influenced by the $1/v$ law of the neutron absorption crosssection between ^{10}B and a neutron, where v is the velocity of incident neutron. In addition, the constituent materials near the CB-KID sensor also influence the shape of the neutron spectrum of Fig. 5 (a) from 5 ms to 20 ms. The important implication of Fig. 5 (a) is that we have to optimize the temperature of operating the superconducting neutron detector to enhance the detection efficiency of neutrons. Of course, the nuclear reaction rate does not depend on the detector temperature. Therefore, further theoretical studies on how the detector signal is created by the local heat dissipation on the superconducting meanderline. Our results suggest that the detector temperature is a key to enhance the detection efficiency of the CB-KID.

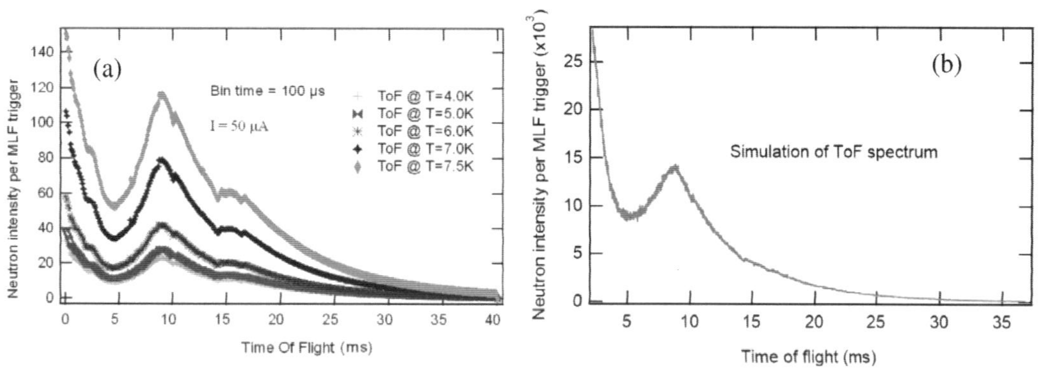

Fig 5. (a). Signal intensity as a function of ToF (time of flight) with changing temperature. (b) The simulated neutron flux as a function of ToF.

5. Summary

We systematically studied the temperature dependent characteristics of the neutron signal by using our superconducting neutron detector CB-KID. The temperature dependence of propagation velocity is in good agreement with the theoretical velocity derived by Koyama and Ishida. We observed the neutron signal of the CB-KID by using the beam line BL10 of Materials and Life Science Experimental Facility (MLF) of J-PARC. The features such as the peak voltage and the neutron intensity as a function of ToF revealed a strong temperature-dependent feature of the nanowire meanderline. The present results imply that the Ginzburg-Landau theory works well in explaining the operation principle of superconducting delay-line kinetic inductance detectors. Further understandings of the temperature-dependent features of superconducting neutron detector will be very important in order to improve the sensitivity and the resolution of our superconducting neutron detector.

Acknowledgments

Authors are wishing to acknowledge assistance and the encouragement from colleagues, the special work provided by technical staff and financial support from organizations for unnumbered assistances at MLF. This work is supported by Grant-in-Aid for Scientific Research (A) No.16H02450 from JSPS. The devices were fabricated in the clean room for analog-digital superconductivity (CRAVITY). This work is supported of MLF program (Proposal No. 2016B0012, 2017A0011, 2017B0014, 2018A0109, No. 2018P0201).

References

[1] Kardjilov N, Manke I, Hilger A, Strobl M and Banhart J 2011 Neutron imaging in materials science *Mater. Today* **14** 248–56

[2] Tremsin A S, Rakovan J, Shinohara T, Kockelmann W, Losko A S and Vogel S C 2004 Non-Destructive Study of Bulk Crystallinity and Elemental Composition of Natural Gold Single Crystal Samples by Energy- Resolved Neutron Imaging 2017 *Scientific Reports* **7** Article number: 40759

[3] Satija R, Jacobson D L, Arif M and Werner S A 2004 In situ neutron imaging technique for evaluation of water management systems in operating PEM fuel cells *J. Power Sources* **129** 238–245

[4] Siegela J B, Stefanopouloua A G, Hagansb P, Dingc Y and Gorsichc D 2013 Expansion of Lithium Ion Pouch Cell Batteries: Observations from Neutron Imaging *J. Electrochem. Soc.* **160** A1031–8

[5] Lehmann E H, Frei G, Ku G and Boillat P 2007 The micro-setup for neutron imaging : A major step forward to improve the spatial resolution *Nucl. Instruments Methods Phys. Res.* **576** 389–396

[6] Thewlis J 1956 Neutron radiography *Br. J. Appl. Phys.* **7** 345–349

[7] Pugliesi R and Andrade M L G 1997 Study of Cracking in Concrete by Neutron Radiography *Appl. Radiat. Isot* **48** 339–344

[8] Korff S A and Danforth W E 1939 Neutron Measurements with BoronTrifiuoride Counters *Phys. Rev.* **55** 980

[9] Batchelor R, Aves R and Skyrme T H R 1955 Helium - 3 Filled Proportional Counter for Neutron Spectroscopy *Rev. Sci. Instrum.* **26** 1037

[10] Kumashiro Y and Okada Y 1985 Schottky barrier diodes using thick, well-characterized boron phosphide wafers *Appl. Phys. Lett.* **47** 64

[11] Ishida T, Nishikawa M, Fujita Y, Okayasu S, Katagiri M, Satoh K, Yotsuya T, Shimakage H, Miki S, Wang Z, Machida M and Kano T 2008 Superconducting MgB_2 Thin Film Detector for Neutrons *J Low Temp Phys* **151** 1074–1079

[12] Shishido H, Miyajima S, Narukami Y, Oikawa K, Harada M, Oku T, Arai M, Hidaka M, Fujimaki A and Ishida T 2015 Neutron detection using a current biased kinetic inductance detector *Appl. Phys. Lett.* **107** 232601

[13] Yoshioka N, Yagi I, Shishido H, Yotsuya T, Miyajima S, Fujimaki A, Miki S, Wang Z and Ishida T 2013 Current-Biased Kinetic Inductance Detector Using MgB_2 Nanowires for Detecting Neutrons *IEEE Trans. Appl. Supercond.* **23** 2400604

[14] Koyama T and Ishida T 2018 Electrodynamic theory for the operation principle of a superconducting kinetic inductance stripline detector Electrodynamic theory for the operation principle of a superconducting kinetic inductance stripline detector *J. Phys. Conf. Ser.* **1054** 012055

[15] Swihart J C 1961 Field Solution for a Thin-Film Superconducting Strip Transmission Line *J. Appl. Phys.* **32** 461

[16] Shishido H, Miki Y, Yamaguchi H, Iizawa Y, Dang V T, Kojima K M, Koyama T, Oikawa K, Harada M, Miyajima S, Hidaka M, Oku T, Soyama K, Suzuki S Y and Ishida T 2018 High-Speed Neutron Imaging Using a Current-Biased Delay-Line Detector of Kinetic Inductance *Phys. Rev. Appl.* **10** 044044

[17] Kojima K M, Murakami T, Takahashi Y and Lee H 2014 New μSR spectrometer at J-PARC MUSE based on Kalliope detectors New μ SR spectrometer at J-PARC MUSE based on Kalliope detectors *J. Phys. Conf. Ser.* **551** 012063

[18] Iizawa Y, Yamaguchi H, Miki Y, Nishimura K, Shishido H, Kojima K M, Oikawa K, Harada M, Miyajima S, Hidaka M, Oku T, Soyama K, Koyama T and Ishida T 2018 Physical

characteristics of delay-line current- biased kinetic inductance detector Physical characteristics of delay-line current-biased kinetic inductance detector *J. Phys. Conf. Ser.* **1054** 012056
[19] Koyama T and Ishida T 2019 *J. Phys. Conf. Ser* to be published
[20] Harada M, Maekawa F, Oikawa K, Meigo S, Takada H and Futakawa M 2011 Application and Validation of Particle Transport Code PHITS in Design of J-PARC 1 MW Spallation Neutron Source *Prog. Nucl. Sci. Technol.* **2** 872–878

IOP Conf. Series: Journal of Physics: Conf. Series **1293** (2019) 012052 doi:10.1088/1742-6596/1293/1/012052

Development of High Throughput X-ray detectors using Superconducting Tunnel Junctions with a large area size

Y Fujisawa[1], G Fujii[2], M Ukibe[2], S Shiki[2], M Naruse[1], H Myoren[1], and T Taino[1]

[1] Graduate School of Science and Engineering Saitama University, 225 Shimo-Ohkubo, Sakura-ku, Saitama, 338-8570, Japan
[2] Department of Nanoelectronics, National Institute of Advanced Industrial Science and Technology, 1-1 Umezono, Tsukuba, Ibaraki 305-8568, Japan

taino@mail.saitama-u.ac.jp

Abstract. We have developed 100-pixel Nb/Al based superconducting tunnel junction (STJ) array detectors which was enlarged the pixel size to improve X-rays detection throughput. The two different sizes of 100 μm × 100 μm and 160 μm × 160 μm STJs which consisting of Nb (100 nm)/Al (70 nm)- AlO_X /Al (70 nm)/Nb (300 nm) were fabricated to determine the effective size of a single STJ. The leakage current, the fabrication yield and the energy resolution for Al- K_α of 100 μm-square and 160 μm-square STJs were 9.3 ± 4.6 nA and 12.3 ± 9.7 nA, 85 % and 84 %, 23.6 ± 4.6 eV and 27.5 ± 3.8 eV, respectively. Since these exhibited almost equal performance, it is found that STJs with a size of 160 μm × 160 μm can be used as a high throughput X-ray detector.

1. Introduction

Materials analysis using X-rays, particularly fluorescence analysis is utilized in various research fields. In conventional X-rays analytical instruments, semiconductor detectors are used to detect X-rays. However, by using the conventional ones, it is difficult to perform the analysis of light trace elements in advanced functional materials, which are important for realizing the low-carbon society. Because semiconductor X-ray detectors can't distinguish clearly the characteristic X-ray peaks of light elements due to the limitation of its energy resolution. On the other hand, a superconducting tunnel junction (STJ) X-ray detector can exhibit theoretically several tens of times higher energy resolution than that of semiconductor detectors. Therefore, to realize advance X-ray analytical instruments with a quite high energy resolution, we have developed an STJ array detector. Our fabricated STJ arrays consisting of 100 pixels have already been applied for an X-ray absorption fine structure measurement of nitrogen dopants in n-SiCs [1]. However, the analysis throughput of our STJ array detectors is currently less than one-hundredth of semiconductor detectors one. To overcome the above drawback, there are two methods as follows. One is to increase the pixel number in the array. In fact, we had already developed a 400-pixel STJ array detector [2]. The other is to enlarge the pixel size. In the first place the energy resolution of the total detection system using STJs, $\Delta E_{\mathrm{total}}$, is assumed to be $\Delta E_{\mathrm{total}}^2 = \Delta E_{\mathrm{int}}^2 + \Delta E_{\mathrm{ele}}^2$. Here, ΔE_{int} is the intrinsic detector energy resolution of the STJ and ΔE_{ele} is the electronic noise. Generally, the larger the area size of STJ becomes, the bigger the ΔE_{ele} becomes due to an enlargement of the capacitance of the STJs, which leads to the degradation of the $\Delta E_{\mathrm{total}}$. However, if ΔE_{ele} is sufficiently smaller than ΔE_{int}, degradation of $\Delta E_{\mathrm{total}}$ can be fairly small [3]. At this moment, the ΔE_{ele} of our measurement system has already been small, then it is possible to determine an appropriate STJ size to satisfy the energy resolution required for individual applications by evaluating the performance of the different size of the STJs. In addition, the fabrication yield is very important to realize the large detection area for array detector. This means the ratio of the number of STJs with the low subgap leakage current to the number of STJs in the whole array.

In this paper, we fabricated two types of Nb/Al STJs with different areas to determine the effective limit of the size of a single STJ. The fabricated STJs were evaluated by the current-voltage (*I*-*V*) characteristics, fabrication yield, $\Delta E_{\mathrm{total}}$ and ΔE_{ele} on the each STJ area size.

2. Fabrication and Experiment

Two types of STJs with different areas of 100 μm × 100 μm (100 μm-square) and 160 μm × 160 μm (160 μm-square) were designed. 100-pixel STJ array detectors of each size were fabricated in the Clean Room for Analog & digital superconductiVITY (CRAVITY) [4]. All STJs have a same multilayer structure with Nb (100 nm)/Al (70 nm)- AlO_X /Al (70 nm)/Nb (300 nm). The STJs were fabricated using photolithography, DC magnetron sputtering, lift-off technique, reactive ion etching, and wet etching. These processes are explained by reference [5] in detail. The multilayer was fabricated by only DC sputtering equipment without breaking vacuum and tunneling barrier was formed by oxidation with O_2 gas. The Josephson current density (J_C) of the STJ was designed to be $200\ \mathrm{A/cm^2}$. The STJ chips were fabricated on a 3-inch Si wafer. Each chip has an area of 10 mm × 10 mm. The 100-pixel STJ array was on the chip. Figure 1 (a) and (b) show a 100 μm-square and a 160 μm-square STJ in a 100-pixel array. The bias current is fed through the contact on the top electrode, which indicated with "A" in the figure 1. The almost area of the top electrode aren't covered by the SiO_2 layer except for the area underneath top electrode (Nb) of the STJ. This structure can realize a high detection efficiency for low energy X-ray photons.

The STJs were cooled at ~0.34 K in ^{3}He cryostat. To evaluate fabrication yield of two types of 100-pixel STJ array, we measured the subgap leakage current (I_{leak}) of the all STJ. The I_{leak} was defined around energy gap (Δ / e). The magnetic field was applied to suppress the Josephson current along the diagonal direction of STJ. Two kinds of 100-pixel STJ array were irradiated Al-K_α (1487 eV) X-ray to evaluate the detection performance. The fluorescent X-ray was generated using an Al

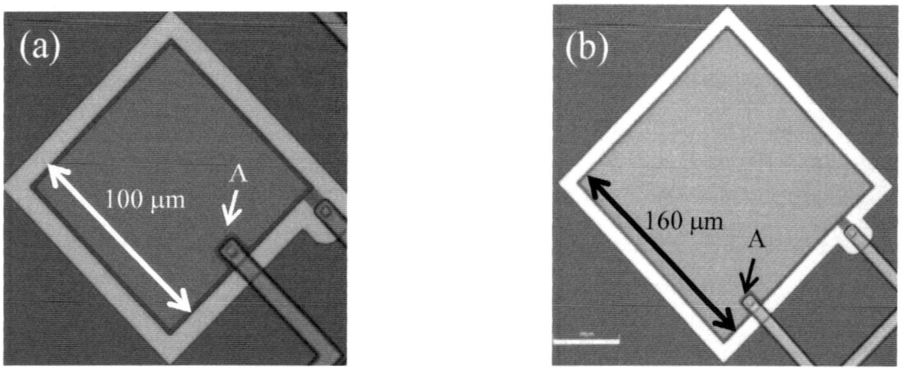

Figure 1. Microphotograph of the fabricated STJ with (a) 100 μm-square and (b) 160 μm-square

target and carbon nanostructure electron emitter. The output signals of STJs were amplified by 16-channel charge-sensitive amplifier and processed by 16-channel FPGA DSP based pulse height analyzer.

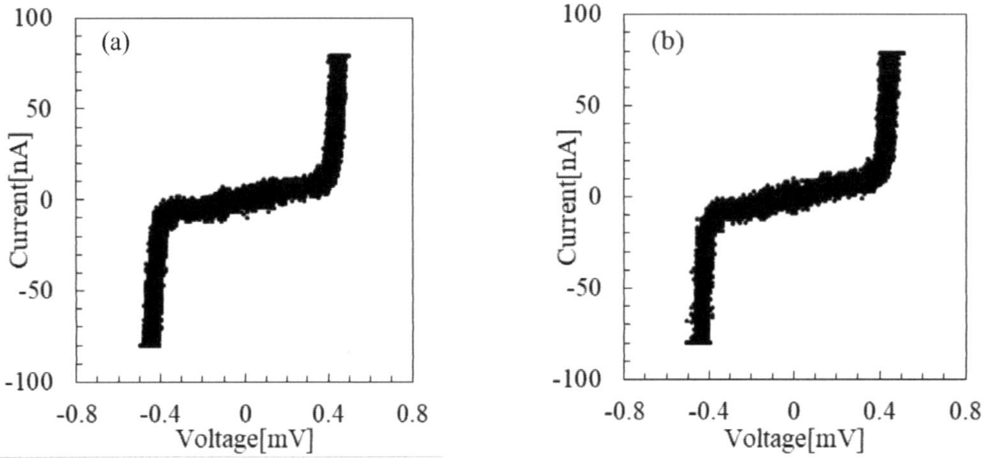

Figure 2. *I-V* characteristics of STJ with **(a)** 100 μm-square STJ and **(b)** 160 μm-square STJ

3. Results

Figure 2 shows typical *I-V* characteristics of two types of the STJs. In these figures, the I_{leak} of 100 μm and 160 μm-square STJ were about 5.3 and 6.8 nA, respectively. The fabrication yields of the 100-pixel array of 100 μm and 160 μm-square STJ were 85 and 84 %. This means the total sensitive areas of the STJ array, which is the summation of the all STJ area size with the low I_{leak}, are 0.85 mm^2 with 100 μm-STJ and 2.15 mm^2 with 160 μm-STJ. The average and standard deviations of the I_{leak} of 100 μm and 160 μm-square STJ were 9.3 ± 4.6 and 12.3 ± 9.7 nA, respectively. This uniformity of the fabricated STJ is considered to be obtained due to the high reproducibility of CRAVITY.

In the preliminary X-ray detection experiment, the constant bias current was fed to the array STJs because of excellent uniformity of the I_{leak}. Figure 3 (a) shows typical spectra obtained by two different sizes of the STJ irradiated by Al- K_α. Peaks around 1300 and 1500 eV are generated in the bottom and top electrode of the STJ by X-rays absorption, respectively. These spectra of Al- K_α (top) are normalized by the number of these peak counts. Two peaks about 1800 eV correspond to the pulser. The ΔE_{total} of 100 μm and 160 μm-square STJ were 22.4 and 28.5 eV, respectively. These results show that both types of STJs exhibit better energy resolution than that of silicon drift detector (SDD) [6]. The ΔE_{total} of 160 μm-square STJ was slight inferior to 100 μm-square STJ.

Figure 3 (b) shows the distribution of ΔE_{total} about two different sizes of STJ in the 100-pixel arrays. The best ΔE_{total} of 100 μm and 160 μm-square STJ are 17.6 eV and 20.6 eV. The average and standard deviations of ΔE_{total} of 100 μm and 160 μm-square STJ were 23.6 ± 4.6 and 27.5 ± 3.8 eV, respectively. The ΔE_{ele} of the pulser was measured to be 7.5 eV with 100 μm-square STJ and 13.0 eV with 160 μm-square STJ. By using equation 1, the average of ΔE_{int} of 100 μm and 160 μm-square STJ were calculated to be 21.5 and 23.4 eV. Since ΔE_{int} of 160 μm-square is 1.1 times larger than that of 100 μm-square, these are almost equal. Thus, this degradation of ΔE_{total} was caused by increasing of the ΔE_{ele} for enlargement STJ size. Anyway, ΔE_{total} of 160 μm-square is sufficient to distinguish K-lines of various elements.

We succeeded in the enlargement of detection area of an STJ and good energy resolution.

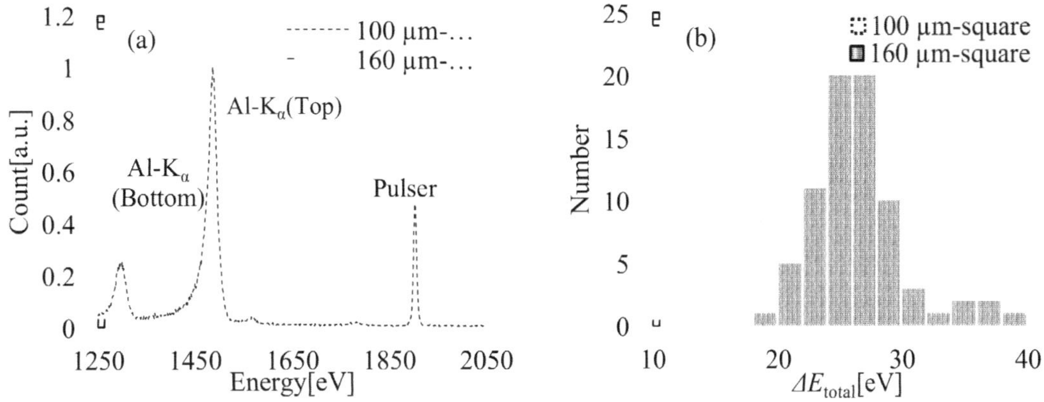

Figure 3. **(a)** Energy spectrum measured by STJ for Al-K_α line. **(b)** Histogram of ΔE_{total} measured by 100 array STJ.

4. Conclusion

The 100-pixel array of 160 μm-square Nb/Al STJs was fabricated to improve throughput of detector. The fabrication yield of the 100-pixel array was 84 %. The average and standard deviations of I_{leak} was 12.3 ± 9.7 nA. The average and standard deviations of ΔE_{total} was 27.5 ± 3.8 eV. This result shows that it is possible to operate STJ array detectors with 2.5 times larger detection area than that of the conventional STJ array consisting of 100 μm-square STJs, indicating that throughput of STJ can be improved with keeping enough high energy resolution.

Acknowledgements

A part of this work was conducted at Advanced Characterization Nanotechnology Platform of AIST, supported by "Nanotechnology Platform" of the Ministry of Education, Culture, Sports, Science and Technology (MEXT), Japan.

References

[1] Shiki S, Ukibe M, Kitajima Y and Ohkubo M 2012 *J. Low Temp. Phys.* **167** 748
[2] Fujii G, Ukibe M, Shiki S and Ohkubo M 2014 *J. Low Temp. Phys.* **176** 255
[3] Ukibe M, Kishimoto M, Katagiri M, Kurakado M and Nakazawa M 1997 *Nucl. Instrum. and Methods Phys. Res. A* **401** 299
[4] Web site: http://unit.aist.go.jp/neri/cravity/en/index.html
[5] Ukibe M, Kushino A, Chen Y. and Ohkubo M 2007 *X-ray Spectrom.* **36** 260
[6] Schlosser D M, Lechner P, Lutz G, Niculae A, Soltau H, Struder L, Eckhardt R, Hermenau K, Schaller G, Schopper F, Jaritschin O, Liebel A, Simsek A, Fiorini C and Longoni A 2010 *Nucl. Instrum. and Methods Phys. Res. A* **624** 270

IOP Conf. Series: Journal of Physics: Conf. Series **1293** (2019) 012053 doi:10.1088/1742-6596/1293/1/012053

Multipoint measurements of a Pipe Using HTS-SQUID and Magnetostriction-Based Ultrasonic Guided Wave

Y Azuma[1], T Terawaka[1], S Kubota[1], Y Yokouchi[1], Y Hatsukade[1], S Adachi[2] and K Tanabe[2]

[1]Kindai University, 1 Umenobe, Takaya, Higashi-Hiroshima, Hiroshima 739-2116, Japan

[2]SUSTERA, 2-11-19 Minowa-cho, MinatoKita-ku, Yokohama, Kanagawa 223-0051, Japan

Abstract. This paper describes study on remote inspection technology for pipes by using magnetostriction-based ultrasonic guided wave and high temperature superconductor (HTS) superconducting quantum interference device (SQUID) gradiometer. Magnetized nickel plates were adhered on an aluminium pipe sample, in order to use them as magnetostriction-based guided wave transceivers. A pair of 10mm-wide nickel plate with different angle arrangement were used to transceive uniformly distributed guided waves at all angles. A field coil was wound around one set of the nickel plates as a transmitter, while the other was used as a receiver. T (0, 1) mode guided wave was generated on the pipe by supplying a burst sine wave current of one cycle at several tens kHz to the coil. Multipoint measurements of the T (0, 1) mode guided waves around the pipe's circumference were carried out by setting the HTS-SQUID gradiometer above the receiver with lift-off of about 9 mm and rotating the pipe for 360 degrees. Signal of reflected wave from an artificial slit on the sample was well detected. We simulated the distribution of the guided wave propagating on the pipe with the slit using an ultrasonic simulator and compared the distribution with the experiment result. The guided wave signal distributions including the defect signal obtained by experiment and simulation agreed well.

1. Introduction

Recently, ultrasonic guided wave testing have been applied to long distance structures in chemical plants and electric power facilities [1-3]. Although various modes exist in the ultrasonic guided waves, T (0, 1) mode guided wave, in which group velocity is constant and mode conversion does not occur, is often used for pipes in the testing [4]. We have studied remote non-contact guided wave testing technology combining magnetostriction-based T (0, 1) mode guided waves [5] and high-temperature superconductor (HTS) SQUID gradiometer [6-8]. Typical HTS-SQUID has two to three orders of magnitude higher sensitivity than conventional semiconductor sensors and the sensitivity ranges from DC to MHz, including bandwidth of the guided wave from 10 to 100 kHz. In the magnetostriction-based guided wave testing using the conventional induction coil, since guided waves are transmitted and received by the coil wound around the entire circumference of the pipe, axial position of defect in a pipe can be detected. However, it is difficult to localize its circumferential position (see Figure 1 (a)). On the other hand, since a HTS-SQUID is used as a very high sensitive point sensor with small pickup area, by replacing the receiver coil of the conventional method with HTS-SQUID, it is possible to measure the magnetic signal distribution derived from the guided waves around the pipe circumference (see Figure 1 (b)). In this study, we applied our HTS-SQUID-based guided wave testing system to an aluminium

IOP Conf. Series: Journal of Physics: Conf. Series **1293** (2019) 012053 doi:10.1088/1742-6596/1293/1/012053

pipe with an artificial slit and performed multipoint measurements of the T (0, 1) mode guided waves by rotating the pipe for 360 degrees. We analysed signal distribution of the guided waves including reflected wave from the slit using an ultrasonic simulator and compared simulated result with the experimental results.

2. Sample and measurement system

2.1. Sample

An aluminium pipe of 3.0 mm in thickness, 90 mm in outer diameter, and 1500 mm in length was used. At positions of 300 mm and 600 mm away from the left end of the pipe, a pair of nickel plates of 0.2 mm in thickness, 10 mm in width and 283 mm in length were wound around circumference of the pipe and firmly adhered by resin bond. In order to disperse edge effect in magnetization of the nickel plates and generate the guided wave as uniform as possible on the entire circumference, the magnetized nickel plates were adhered as one set with shifting angles as illustrated in the left of Figure 2. Face-to-face parts of the nickel plates were aligned at 0 and 180 degrees. Before the adhesion, the nickel plates were magnetized using a solenoid coil, inside which the plates were exposed by a DC magnetic field of about 10 kA/m toward its longitudinal direction. The nickel plates at 300 mm away from the left end of the pipe was wound with a field coil of 30 turn using Litz wire to reduce the skin effect and used as a transmitter of the T (0, 1) mode guided waves. The other nickel plates at 600 mm was used to receive the T (0, 1) mode guided waves by converting the vibration to magnetic signals oscillating toward pipe's axial direction. Firstly, the sample without defect was measured using after-mentioned HTS-SQUID-based guided wave testing system. Secondary, a circumferential slit of 40 mm in width, 5 mm in the length was fabricated at 0 degree and 1050 mm away from the left end (see Figure 2) and measured in the same manner as the first measurement.

2.2. HTS-SQUID-based guided wave testing system

We have constructed the HTS-SQUID-based guided wave testing system for pipes based on the magnetostrictive effects and a HTS-SQUID gradiometer. Figure 2 shows the schematic diagram of the system. The function generator supplies burst wave voltage to the power amplifier, in order to feed burst wave current into the transmitter to generate magnetic field on the nickel plates of the transmitter on the pipe. The T (0, 1) mode guided wave is generated by the magnetostrictive effect in the nickel plates on the entire circumference of the pipe, and it propagates through the pipe's wall. The guided wave is converted into a magnetic signal by the inverse magnetostrictive effect of the receiver and measured by the HTS-SQUID gradiometer located above the plate. The HTS-SQUID gradiometer with ramp-edge Josephson junctions and a planar first-order differential pickup coil made by SUSTERA was used in this work [9]. The gradiometer was cooled at about 73.5 K by a coaxial pulse-tube cryocooler [10]. Flux noise level of the SQUID operated in the normal environment using a commercial SQUID electronics

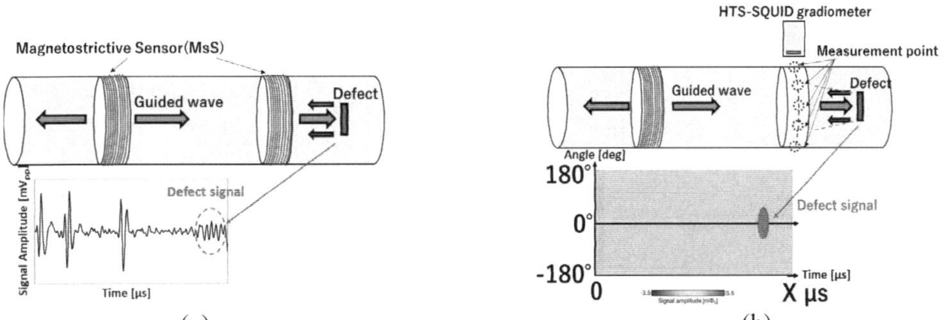

Figure 1. Schematic image of conventional magnetostriction-based T (0, 1) mode guided wave testing on a pipe. (a) Magnetostrictive sensors (MsS). (b) High temperature superconductor superconducting quantum interference device (HTS-SQUID).

is about 15 $\mu\Phi_0/Hz^{1/2}$ in a range from 10 Hz to 30 kHz. The voltage from the SQUID electronics is averaged 256 times to increase signal-to-noise ratio (SNR) and measured by a digital oscilloscope.

3. Experiments and results

3.1. Multipoint measurements around pipe's circumference

One cycle burst currents of $0.3A_{pp}$ @ 60 kHz were repeatedly applied to the coil of the transmitter to generate the T (0, 1) mode guided waves on the sample pipe. Magnetic signals due to the guided waves above the nickel plates were measured by the HTS-SQUID gradiometer, which was set to measure dB_x/dz with a lift-off distance of about 9 mm. At first, the sample without defect was measured. Based on this method, multipoint measurement was performed while rotating the pipe from -180 to +180 degrees at every 45 degrees. Next, a penetrating slit of 40 mm in width in the circumferential direction and 5 mm in length was artificially made on the pipe at 0 degree and 1050 mm from the left end of the pipe. The same multipoint measurement was performed on the defected sample pipe.

Figure 3 shows the magnetic flux signals measured at 0 degree of the pipe. In all the results, a signal was measured from around 0 μs. It is thought that the electrical signals probably originated from the supplied current to the transmitter were mixed in the measuring equipment. Magnetic signals labelled as "T1" and "T2" were measured from about 97 μs and 290 μs, respectively. These signals are the incident wave and reflected wave from the left edge, which propagated toward the both sides of the transmitter after their generation. In the case of the pipe with the slit, a signal "D1" was measured from about 390 μs, which was a reflected wave from the slit. The magnetic signals "T1", "T2", and "D1" were measured at all angles, although the waveforms of "D1" were different at angles. The experimentally obtained group velocity for the guided wave is 3102 mm/ms in aluminum.

Figure 4 (a), and (b) show contour maps of the measured signals, which was draw on a plane of the time and angle, using all measured waveforms around the entire circumference of the pipe without and with the slit. The magnetic signals of the incident waves and reflected waves from the left end of the pipe were detected from about 97 μs and 290 μs, respectively. The peak amplitudes of these waves were approximately uniform at all angles. In Figure 4 (b), the reflected wave from the slit was measured from 390-430 μs at entire angles. The distribution of the reflection wave from the slit looks approximately symmetric about 0 degree, where the centre of the slit located. Next, we analysed the distribution of the measured guided waves on the sample pipe using the 2D ultrasonic simulator SWAN21 [11].

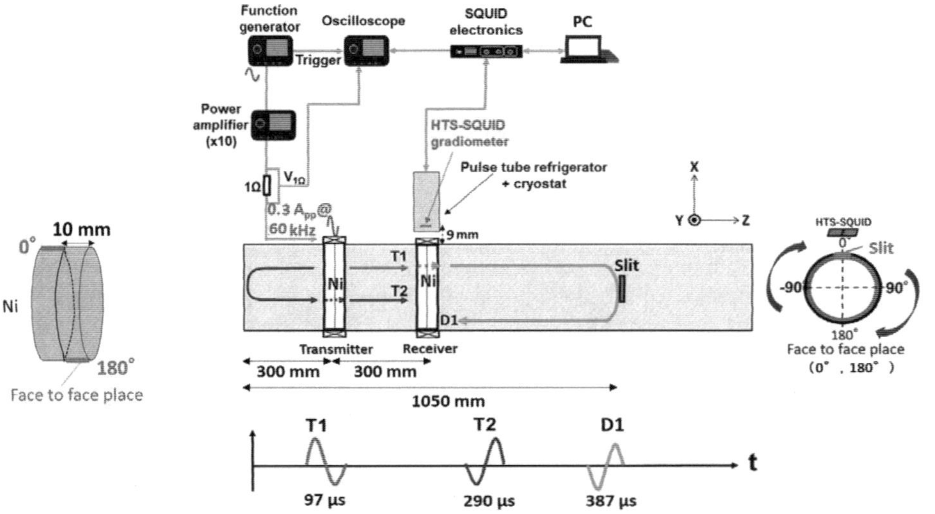

Figure 2. Schematic diagram of HTS-SQUID-based guided wave testing system. In the bottom, estimated waveform, which will be measured by the HTS-SQUID gradiometer above the receiving nickel plates on the sample pipe, is shown with measured times of T (0, 1) mode guided waves propagating at velocity of about 3100 mm/ms on aluminium.

3.2. Simulation of guided wave on plate model with slit

We performed simulation, which was aimed to emulate multipoint measurement on the pipe with the slit using ultrasonic propagation analysis simulator SWAN 21. The contour maps obtained in the experiments were projection of the three dimensional distribution of the magnetic signals above and around the pipe's circumference into a two-dimensional map. Here, we used the simulator SWAN21 to analyse ultrasonic guided wave on a two-dimensional thickless plate model. We note that the difference between the experiment and this simulation is that the edges (or boundaries) at ±180 degrees on the pipe were connected, thus the guided wave can go through the edges from +180 to -180, and vice verca in the experiment, while the edges at ±180 are not connected and the guided waves reflect at the both edges in the simulation. To emulate the T (0, 1) mode guided wave in the plate model, we used SH guided waves, which are transverse waves and have the group verocity of about 3100 mm/ms in aluminum, which is the same as the T (0, 1) mode guided waves on the pipe [3]. Figure 5 shows the schematic illustaration of the simulation model with the virtually opened sample pipe. The model in the simulation aims to emulate the experiments with the sample pipe opened at ±180 degrees (the position of the slit is 0 degree). The width and the length of the model are 283 mm and 1200 mm, corresponding to the distance from the transmitter to the right end of the pipe. The probe position was determined due to limit that the probe can only set on the boundary of the model. A measurement line for the guided wave was set at 300 mm away from the left end of the model.

Figure 6 shows the contour map obtained by the simulation. Since the simulation model is made with the length of 1200 mm, the reflected wave "T2" from the left end of the pipe does not appear. The magnetic signals as "T1" and "D1" were measured from about 104 μs and 395 μs, respectively. The time differences between the experiments and the simulation were about 7 μs and 8 μs for "T1" nad "D1", respectively. Although the generation times of the signals "T1" and "D1" obtained by the simulation slightly differs from those obtained by the experiments, the distribution of the simulated guided waves well agreed with that in experiment. The slight delay is due to the probe and receiver in the simulation do not have width, while the transmitter and the receiver in the experiment has the widths of 20 mm in total. Comparing the defect signal distributions of Figure 4 (b) and Figure 6, they resemble well especially around 400 μs, where the SNR in the experimental result is good. These results shows that the defect signal distribution can be accurately obtained by multipoint measurements using the HTS-SQUID gradiometer. The axial position and the center position of the defect should be localized from the symmetric distribution. In order to analyze defect signal distribution in more detail, it is necessary to increase the SNR of the experiment result. The SNR is partly due to the lift-off distance of about 9 mm, which is limitted due to leakage flux from the magnetized nickel plates below the gradiometer. We are trying to suppress this flux utilizing active magnetic shield technique, which is described in ref. [12].

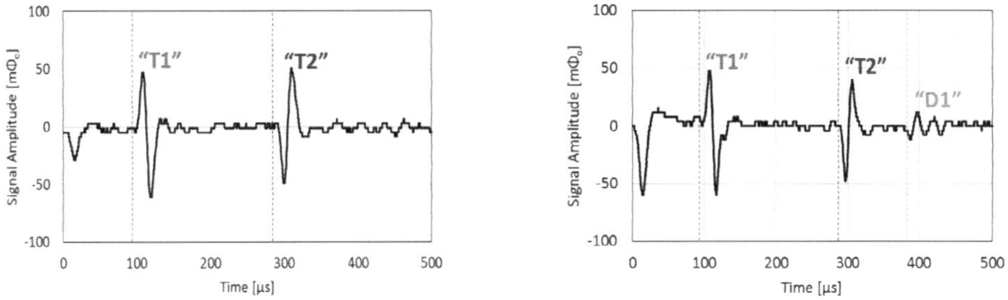

(a) (b)

Figure 3. Magnetic flux signals measured at 0 degree of the pipe.
(a) Pipe without slit. (b) Pipe with slit.

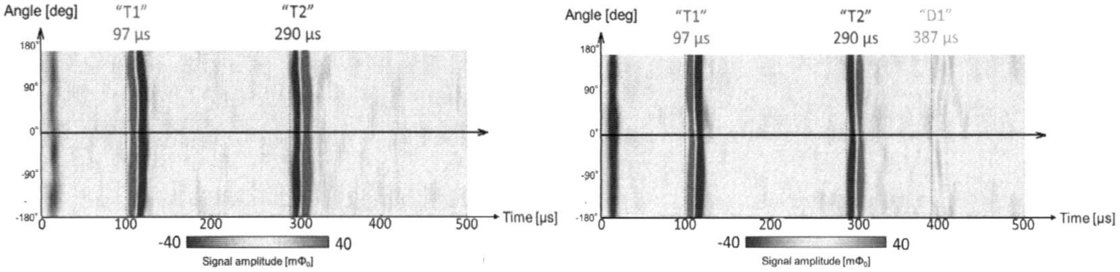

(a) (b)

Figure 4. Contour maps obtained by multipoint magnetic flux measurement. (a) Pipe without slit. (b) Pipe with slit.

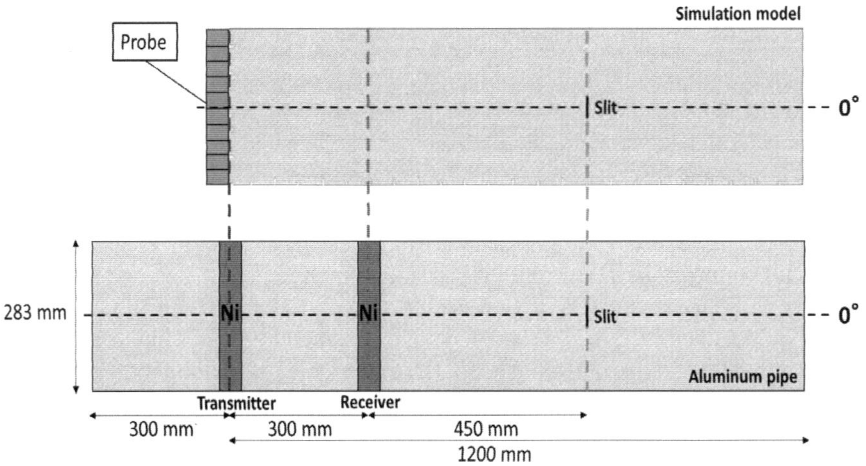

Figure 5. Simulation model (above), which emulates the experiment with the sample pip opened at ±180 degrees (the position of the slit is 0 degree) (below). Array of ultrasonic probes (transmitter) was set at the left end of the model, while measurement line of guided wave was set on the line at 300 mm away from the left end.

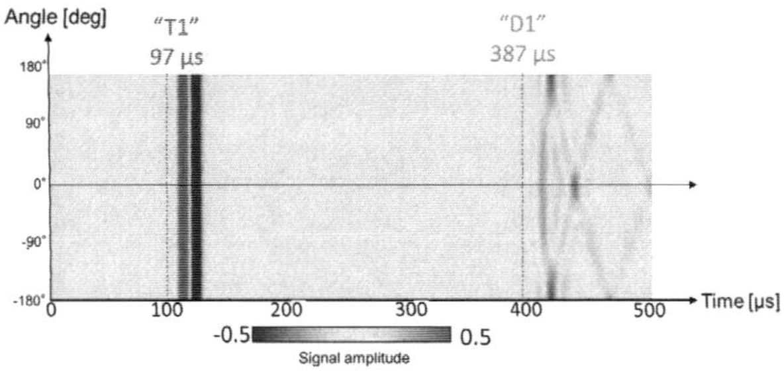

Figure 6. Contour map of SH guided waves on the model obtained by simulation.

4. Conclusion

In this study, we performed the multipoint measurements around the aluminium pipe without and with the slit using the magnetostriction-based T (0, 1) mode guided waves and HTS-SQUID-based system. The pair of 10 mm-wide nickel plates with the different angle arrangement resulted in the uniform distributions of the incident waves and reflected waves from the left pipe edge at all angles. The reflection wave due to the artificially-made slit was measured at all angles with different waveforms, and the distribution of the defect signal was axisymmetric. The incident and defect signal distributions obtained by the experiment and the 2D simulation agreed well. These result demonstrates that defect signal distribution around pipe can be accurately obtained by multipoint measurement using HTS-SQUID.

Acknowledgments

Authors would like to acknowledge great support and encouragement from Dr. Yoshikazu Tanaka of Hiroshima University for this work. This work was supported partly by Electric Technology Research Foundation of Chugoku, Japan, and is supported partly by MEXT/JSPS KAKENHI Grant Number 16K06395.

References

[1] Rose J L 1999 *Ultrasonic Waves in Solid Media* (Cambridge University Press, Cambridge)
[2] Nagai T, Hyodo M, and Takamura K 2002 *Journal of JSNDI* **51** 622 (in Japanese)
[3] Rose J L 2014 *Ultrasonic Guided Waves in Solid Media* (Cambridge University Press, New York)
[4] Tanaka Y, Ikeda T, Toiyama K, Kuwano S and Fujimoto Y, *Transactions of the Japan Society of Mechanical Engineers.* **72** 951 (in Japanese)
[5] Kwun H and Bartels K. A 1998 *Ultrasonic* **26** 171
[6] Hatsukade Y, Kobayashi T, Nakaie S, Masutani N, and Tanaka Y 2017 *IEEE Trans. Appl. Supercond.* **27** 1600104
[7] Hatsukade Y, Masutani N, Teranishi S, Masamoto K, and Kanenaga S, Adachi S, and Tanabe K 2017 *IOP Conf. Series: Journal of Physics: Conf. Series* **871** 012072
[8] Masutani N, Teranishi S, Masamoto K, and Kanenaga S, Hatsukade Y, Adachi S, and Tanabe K 2017 *IOP Conf. Series: Journal of Physics: Conf. Series* **871** 012074
[9] Tsukamoto A, Hato T, Adachi S, Motoori M, Sugisaki M, and Tanabe K 2016 *ISEC 2015 Proceedings* SQ-O06
[10] Hatsukade Y, Inaba I, Maruno Y and Tanaka S 2005 *IEEE Trans. Appl. Supercond.* **15** 723
[11] http://www.swan21.com/
[12] Hatsukade Y, Yoshida K, Kage T, Suzuki T, Tanaka S 2015 *IEEE Trans. Appl. Supercond.* **25** 1601105

IOP Conf. Series: Journal of Physics: Conf. Series **1293** (2019) 012054 doi:10.1088/1742-6596/1293/1/012054

Reduction of Environmental Magnetic Field Noise for Detection of Small Magnetic Contaminants

Takao Nishikawa and Ken Sakuta

The Graduate School of Electronic Systems Engineering, The University of Shiga Prefecture, 2500 Hassaka-cho, Hikone-City, Shiga 522-8533 Japan

oh23tnishikawa@ec.usp.ac.jp

Abstract. When performing high-sensitivity magnetic measurements via a high-sensitivity magnetic sensor such as a SQUID (Superconducting QUantum Interference Device) magnetometer, it is necessary to shield from environmental magnetic noise by employing a magnetic shield box. In general, however, the low frequency of such noise decreases the shielding performance of the magnetic shield box. Hence, the magnetic metallic contaminant signal and the low-frequency ambient magnetic field noise overlap, which makes filtering a challenging task. Therefore, we studied a system that suppresses gradient magnetic noise such as low-frequency environmental noise superimposed on a contaminant signal and commercial power supply noise localized spatially.

1. Introduction

Recently, studies have been conducted to detect magnetic contaminants in industrial products by using a highly sensitive magnetic sensor such as a SQUID magnetometer [1-2]. It is possible to detect magnetic contaminants with high accuracy by measuring the residual magnetism generated by the contaminants. In the case of high-sensitivity detection, it is necessary to reduce environmental magnetic noise in advance. Therefore, a magnetic shield box made of a ferromagnetic material such as permalloy is used. However, generally, the shielding factor of such a shield box decreases as the frequency of noise decreases. In the case of magnetic contaminant detection, slow and nonstationary changes in a magnetic field are measured, magnetic contaminant detection signal overlaps low-frequency environmental magnetic noise, and simple filter processing becomes difficult. In addition, to effectively utilize the dynamic range, it is desirable to remove the noise associated with the commercial power supply in advance. Therefore, active magnetic shields were introduced [2-3]. A noise detection sensor is installed, and a weak magnetic field is generated near the noise detection sensor by applying negative feedback via compensation coil of the output signal. In the case of magnetic contaminant detection, this technique is particularly effective because the related equipment can be miniaturized to such an extent that it can accommodate the inspection object. However, if a magnetic gradient occurs between the noise detection sensor and the signal detection sensor, the shielding effect of the signal detection sensor deteriorates. Therefore, a method of improving the shielding effect by correcting the compensation signal by extracting only the noise signal from the signal detection sensor based on the principle of phase detection was studied.

2. Active Magnet Shield System overview

2.1. Basic system configuration and basic performance

The basic configuration schematic of this system is shown in Figure 1. This system uses a general-purpose fluxgate magnetometer (Bartington Instruments) as a noise detection sensor and a SQUID magnetometer (The Jülich SQUID Company) as a signal detection sensor. The output signal of the fluxgate is amplified by a preamplifier, and a control circuit comprising multiple analog filters restricts the band to the low-frequency and commercial power supply frequency bands. The restricted signal is negative feedback to the fluxgate via the compensation coil, and it actively cancels the environmental magnetic noise around the SQUID by generating a compensation signal. Figure 2 shows the results of attenuation performance measurements of this system against pseudo-noise or ambient magnetic noise applied from the outside. Figure 2 means that when the amount of magnetic noise at the SQUID is exactly equal to the amount of magnetic noise at the fluxgate, the system can possibly reduce low frequency noise up to 10 Hz and noise at 60Hz (from the commercial power supply) by 35 dB or more at the SQUID.

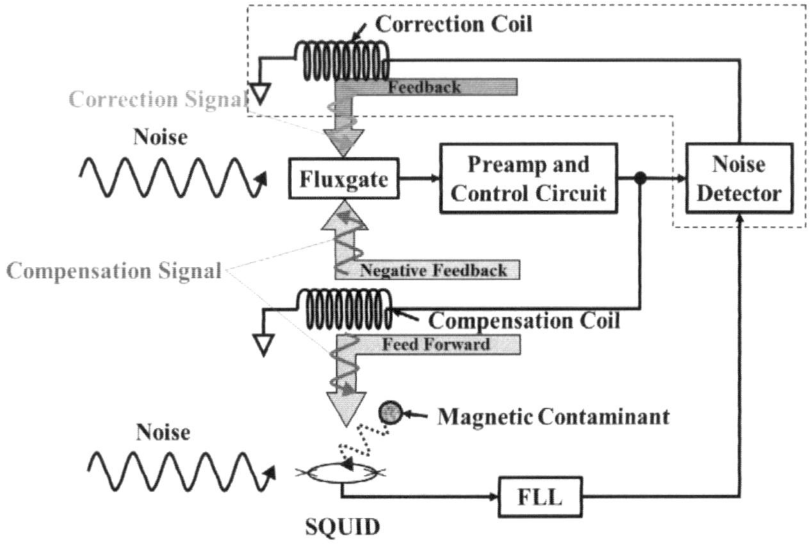

Figure 1. Schematic of active shielding system

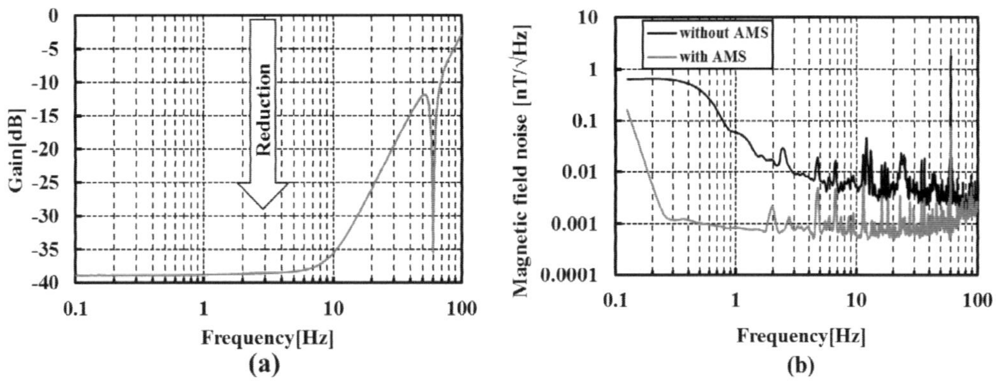

Figure 2. Frequency characteristics of shielding effect at fluxgate
(a) against applied pseudo magnetic noise and
(b) against environmental magnetic noise.

We installed the active shielding system in a magnetic contaminant detection system and conducted an experiment. The sample was pre-magnetized with a permanent magnet and passed under the SQUID via an automatic stage. In order to simulate a strict noise environment, 1 Hz and 60 Hz pseudo noise was generated from the coil at approximately 2.5 m from the SQUID. The sample was a SUJ 2 ball (0.5 mm), and the liftoff was 10 mm. As shown in Figure 3, low-frequency magnetic noise superimposed on the frequency signal with the contaminant signal was reduced by approximately 35 dB at 1 Hz and 29 dB at 60 Hz by applying the active magnetic shield, and the contaminant signal could be visualized clearly.

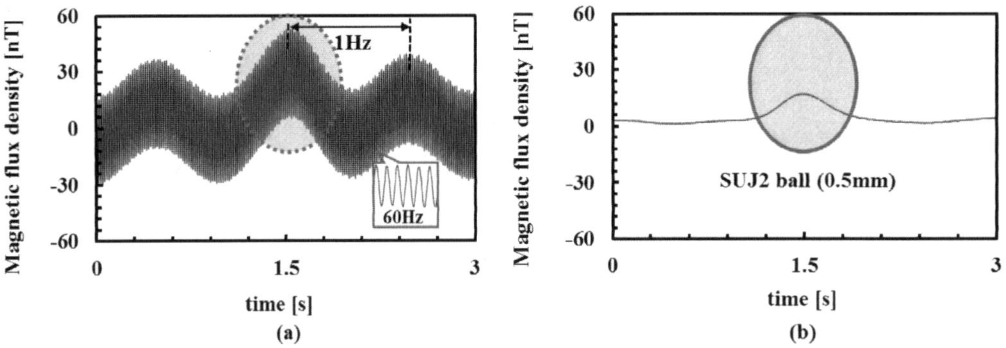

Figure 3. Waveform of contaminant detection with pseudo-noise
(a) without AMS and (b) with AMS.

2.2. System corresponding to gradient magnetic noise

As shown in Figures 2 and 3, the shielding effect against noise at 60Hz (from commercial power supply) is 35 dB at the fluxgate, whereas the shielding effect at the SQUID is 29 dB. As one of these factors, a gradient is generated in the magnetic noise generated by the two magnetometers; thus, the shielding effect at the SQUID may be deteriorated. Moreover, as the source of noise approaches the system, the gradient between the two magnetometers increases and the shielding effect deteriorates further. In this case, when the source of noise moves, it is necessary to adjust the extent of compensation sequentially, which is not practical.

Therefore, we examined a method of adjusting the compensation according to the following principle. (Figure 4)

i. By negative feedback of the magnetic noise detected at the fluxgate via the compensation coil, a magnetic field that cancels the detected magnetic noise is applied to the fluxgate and the SQUID.
ii. When there is a gradient between fluxgate and SQUID, magnetic noise remains in SQUID. Only the residual noise at the SQUID is extracted by phase detection and applied to the fluxgate as a correction magnetic field via the correction coil.
iii. The correcting magnetic field applied in step ii is detected as a noise at the fluxgate and a magnetic field that cancels the correcting magnetic field is applied to the fluxgate and the SQUID as in step i.
iv. As a result, the compensation magnetic field is corrected, and the residual noise of the SQUID is reduced.

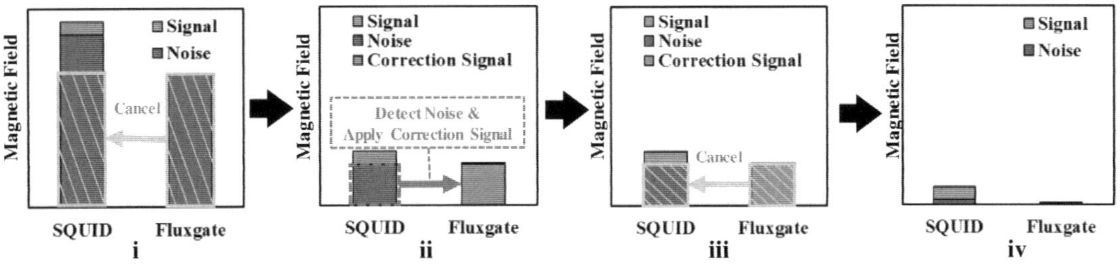

Figure 4. Principle of correction method

A schematic of the principle for detecting only the noise signal from the SQUID output signal is shown in Figure 5. The detector involves the use of two multipliers and one low-pass filter, and the principle is the same as that of phase detection. The detector operates by reproducing the original signal by multiplying the DC signal output from the LPF with the reference signal. Moreover, to prevent the amplitude of the reference signal from being dependent on the output signal, a circuit is provided for maintaining the constant amplitude of the reference signal input to the circuit. The output signal of the SQUID is used as an input signal, and the output signal of the control circuit (detected magnetic field at the fluxgate) is used as the reference signal. The output signal is converted into a magnetic field by the correction coil and applied to the fluxgate. Therefore, it is possible to extract only the noise signal from the SQUID output signal based on the noise signal detected by the fluxgate. Moreover, to prevent the amplitude of the reference signal from being dependent on the output signal, it is necessary to maintain the constant amplitude of the reference signal input to the circuit. Therefore, an amplitude adjustment circuit comprising a high-gain amplifier and a band pass filter is provided.

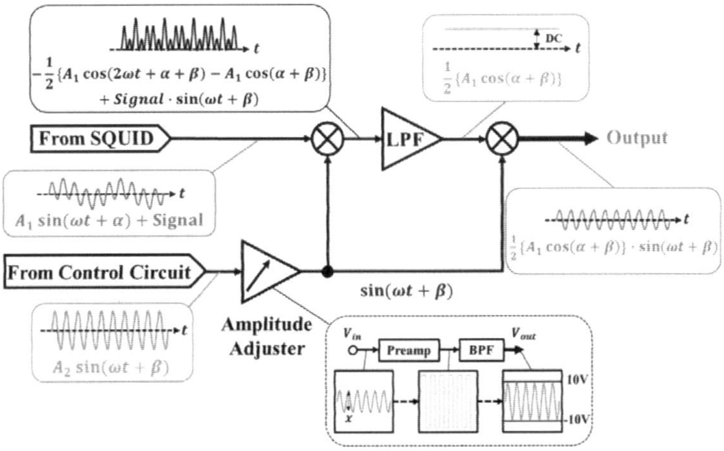

Figure 5. Principle of noise detection

3. Gradient magnetic noise reduction experiment

To confirm the effectiveness of the proposed method, experiments were conducted with the configurations shown in Figure 6. The distance between the system and the noise source coil was set to 0.5–2.5 m, and the shielding effect with or without driving of the noise detection circuit was confirmed. The magnitude of the pseudo magnetic noise was about 100 nT, and the frequency was 60 Hz. The results are shown in Figure 7. In the case of no correction, as the noise source approached the system, the shielding effect decreased greatly. This indicated that the magnetic gradient between the fluxgate and the SQUID increased. However, by correcting the compensation signal with the proposed method, the shielding effect was improved by about 10 dB. Moreover, the compensation signal became stronger and correction was confirmed. However, the slope of the shielding effect within the distance less than 1.5 m was not different between with the correction and without correction. The reason for this is

IOP Conf. Series: Journal of Physics: Conf. Series **1293** (2019) 012054 doi:10.1088/1742-6596/1293/1/012054

considered to be that eddy currents are generated on the wall of the simple magnetic shield box and the output of the noise detector is decreased due to the phase difference between the SQUID and the fluxgate. In the future, we believe that it is necessary to improve the noise detector to a two-phase detection scheme etc. that does not require phase adjustment.

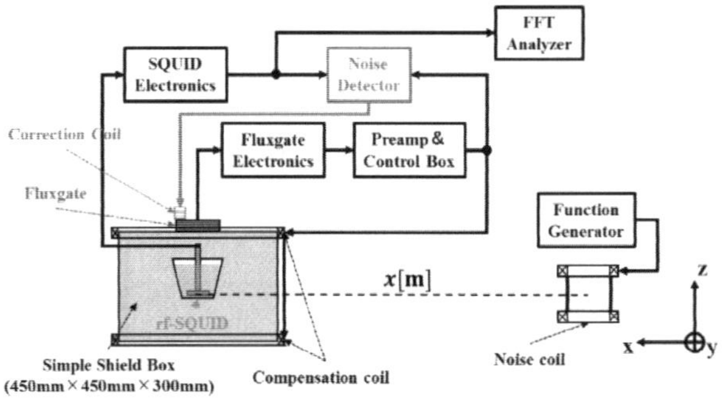

Figure 6. Gradient magnetic noise reduction experiment system

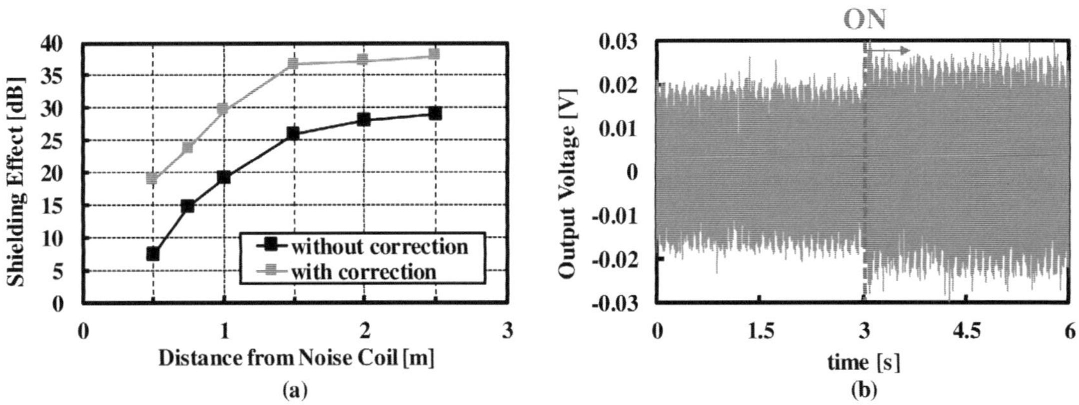

Figure 7. Effects of proposed method:
(a) Shielding effect and (b) Correction of compensation signal.

4. Conclusion

When the position of the noise source changes, a magnetic gradient is generated between the noise detection sensor and the signal detection sensor and the shielding effect deteriorates. Therefore, we devised a method of extracting only the noise signal from the signal detection sensor output signal by phase detection and applying it to the noise detection sensor to adjust the compensation signal. As a result, we were able to improve the shielding performance by approximately 10 dB.

References

[1] S Tanaka M Natsume M Uchida N Hotta T Matsuda Z A Spanut Y Hatsukade 2004 *Supercond. Sci. Technol.* 17 pp 620–623.
[2] S Tanaka H Fujita Y Hatsukade T Otani S Suzuki T Nagaishi 2007 *Supercond. Sci. Technol.* 20 S385.
[3] Y Hatsukade T inaba Y Maruno S Tanaka 2005 *IEEE TRANS. APPL. SUPERCOND* vol 15 no 2 p 723
[4] F. Shanehsazzadeh M Fardmanesh 2017 *IEEE TRANS. APPL. SUPERCOND* vol 28 no 4 pp 1-5

IOP Conf. Series: Journal of Physics: Conf. Series **1293** (2019) 012055 doi:10.1088/1742-6596/1293/1/012055

Effects and Design Considerations of Inserted Resonator to Improve Efficiency Including Rectifying Circuit for Superconducting Wireless Power Transfer System for Electric Vehicle

Yoon Do CHUNG[1], Chang Young LEE[2], Eun Young PARK[3]

[1]Dept. of electrical engineering, Suwon Science College, Kyounggi-do, 18516, Korea
[2]Korea Railroad Research Institute (KRRI), Kyounggi-do, 16105, Korea
[3]Dept. of food & nutrition, Korea Christian University, Seoul, 07661, Korea

Corresponding E-mail: ydchung@ssc.ac.kr

Abstract. The wireless power transfer (WPT) technology based on strongly resonance coupled method realizes large power charging without any wires through the air. Recently, the WPT systems have started to be applied to the wireless charging for electrical vehicles (EVs) because of their advantages compared with the wired counterparts, such as convenient, safety and fearless transmission of power. However, it has obstacles to commercialize such as delivery distance and efficiency. To solve the problems, we proposed the technical fusion using HTS antenna in the WPT system, it is called as, superconducting wireless power transfer for electric vehicle (SUWPT4EV) system. Since the superconducting wire has merits a larger current density and higher Q value than normal conducting wire, the HTS antenna enables to deliver a mass amount of electric energy in spite of a small scale antenna. Thus, SUWPT4EV system has been expected as a reasonable option to improve the transfer efficiency of large electric power. Authors already demonstrated the efficienct effects for inserted resonator using simply copper tube wires, which structure is helix type of multi layers. In this study, as a practical approach, authors evaluated the SUWPT4EV system with inserted copper type resonator, which structure is a rectangular shape of single layer to reduce installation volume. Based on the evaluation results, authors deduced the practically alternative structure for the inserted resonator to reduce the installation space as well as to improve transmission power, compared with non-resonator SUWPT4EV system under radio frequency (RF) power of 370 kHz below 500 W.

1. Introduction

Magnetically coupled resonators, each comprising a coil in series with a capacitor, were historically first used for wireless power transfer (WPT) by Nikola Tesla in the late 1800s. Nowadays, the WPT technology has been interested again because it is offers the promise of cutting the wires, allowing users to seamlessly recharge them portable devices as safely as power is transmitted through the air. On 2007 the researchers from MIT proposed resonance coils of the same resonant frequency to transfer wireless power over a distance of meters. The experimented efficiency was about 40% at a distance of 2 m with 13.56MHz power source. Such an exploring research has been promisingly emerged in a variety of applications such as consumer electronics, medical devices, and transportation charging system since there is the desire to use seamlessly recharges them in order to offer the possibility of connector-free devices [1]. As one of applications, the WPT systems have started to be applied to the charging of electrical vehicles (EVs) because of their advantages compared with the wired counterparts, such as no exposed wires, safety, convenient charging and fearless transmission of large power. However, limited distance and efficiency of the WPT system has been major obstacles in commercialization service [2],[3]. To overcome mentioned problems, we already proposed the combinations of WPT technology with HTS antenna for EV system [4], it is called as, superconducting wireless power transfer for EV (SUWPT4EV) system, which enables to transfer larger power efficiently compared with normal conducting antenna [5], [6]. Fortunately, since superconducting wire has a larger current density and higher Q value, HTS antenna can deliver a mass amount of electric energy in spite of a small scale antenna [7],[8]. On the other hand, in the superconduting equipment, a cooling vessel is a requisite subsystem, which is a main obstacle for installation space and handling to commercialize in the EV charging system. However, as even a normal conducting antenna coil for large power transfer should be installed underground due to safety and maintenance, the underground cooling vessel for HTS antenna doesn't put obstacle in the wireless EV charging system anymore.

Authors already confirmed the efficient effects of efficiency improvement for an inserted resonator, which structure is multi layers made by tube type copper wire, between HTS antenna (Tx) and copper receiver (Rx) [9]. In order to adopt the resonator practically, the installation volume of inserted resonator should be reduced. In addition, as the strongly thermal energy of inserted resonator is inevitably generated, the cooling subsystem for the resonator should be considered. In this study, as a practical approach, authors propose an inserted resonator, which structure is single layer to reduce installation volume, between antenna Tx and receiver Rx coils in order to reduce the installation space. As the inserted single layer resonator can be installed in the inside of cover at cooling vessel easily, the unavoidable thermal energy of inserted resonator can be easily cooled with HTS antenna as shown in Figure 1. In this paper, we examine the improvement effects of efficiency and delivery distance using inserted single layer resonator in the SUWPT4EV system compared with non-resonator charging system at the same experimental sequence. The performance is carried out with the RF input power of 370 KHz from 100 to 500W at 40 cm distance. Based on the evaluation results, authors propose alternative structure for the single layer inserted resonator to reduce the installation space as well as to improve transmission power.

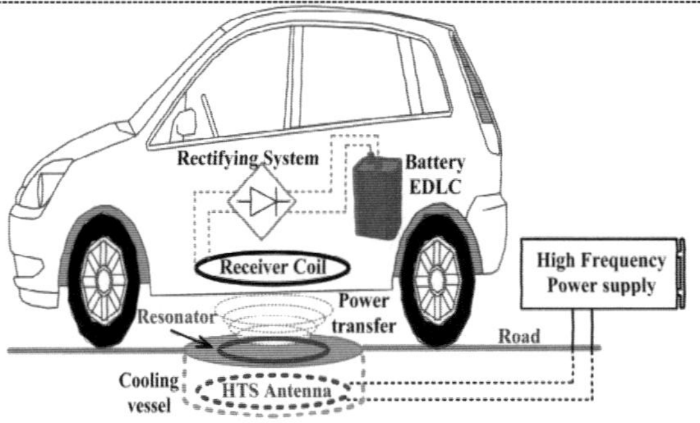

Figure 1. Illustration of superconducting wireless power transfer for electric vehicle (SUWPT4EV) with inserted resonator system.

2. Mechanism and properties

2.1. Structure and mechanism

As shown in Fig. 2, the basic principle of SUWPT4EV system with inserted resonator via coupled magnetic resonance is proposed. Such a system consists of RF power source (Vs), HTS Tx coil, normal conducting resonator Sx coil, normal conducting Rx coil, rectifying circuit and load. If the distance between Tx and Rx is broadened, the phases of voltage and current waves for each coil are proportionally shifted. That is, the resonance coupling for each coil doesn't match due to varying distance. Thus, the shifted phase of transmitting wave causes reactive power loss (thermal loss). From this point of view, the Sx coil plays a role to keep strong resonance coupling and reduce phase shift between Tx and Rx, which associates with techniques to improve efficiency and expand distance. The inductance of Tx coil, L_2 and capacitor C_2 constitute a source resonance circuit to generate an alternative non-radiative magnetic field corresponding to L_3 and C_3 resonance circuit. In theory, when the oscillation of source resonance circuit is tuned with receiver resonance circuit, the resonance current is highest and then induced magnetic field intensity of Rx coil is maximized. In the system, a variable distance of d_{12} takes a role to tune resonance coupling between Tx and Rx coils. When the inductor L and capacitor C in resonance circuit is determined, the circuit resonance frequency f could be calculated from Equation (1):

$$f = \frac{1}{2\pi\sqrt{LC}} \tag{1}$$

The transfer efficiency can be explained by quality factor Q. Higher Q indicates a lower rate of energy loss relative to the stored energy of the resonator; the oscillations die out more slowly. The magnetic resonant coupling contains creating an LC resonance, and transferring power with electromagnetic coupling. The Q value is defined by this ratio. The general definition of the quality factor is based on the ratio of apparent power to the power losses in a device. The antenna coil forms a series RL circuit and the Q factor is expressed as:

$$Q = \frac{1}{R}\sqrt{\frac{L}{C}} = \frac{\omega L}{R} = \frac{2\pi f L}{R} = 2\pi f\,\frac{Energy\ stored}{Power\ loss} \tag{2}$$

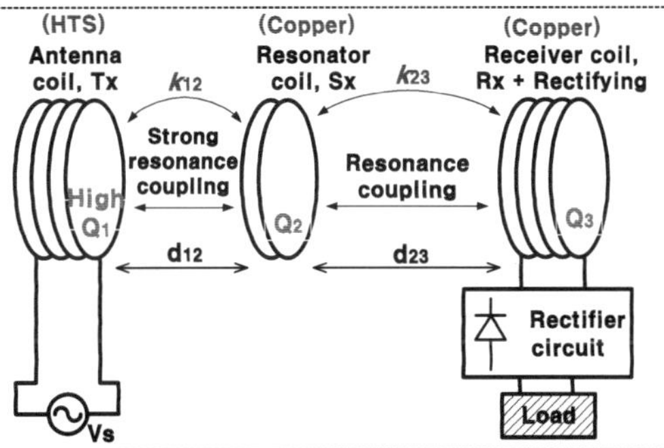

Figure 2. Schematic diagram of superconducting wireless power transfer (SUWPT) system with three-separate resonance coils.

Figure 3 shows the equivalent circuit of SUWPT4EV with inserted resonator system. The R_1, R_2, R_3, I_1, I_2, I_3 are the internal resistance and induced current of three-separate resonance coils, respectively. M_{ij} and k_{ij} means mutual inductance and coupling coefficient between coupling coils. They are calculated as [7]

$$k_{xy} = \frac{M_{xy}}{\sqrt{L_x L_y}} \tag{3}$$

IOP Conf. Series: Journal of Physics: Conf. Series **1293** (2019) 012055 doi:10.1088/1742-6596/1293/1/012055

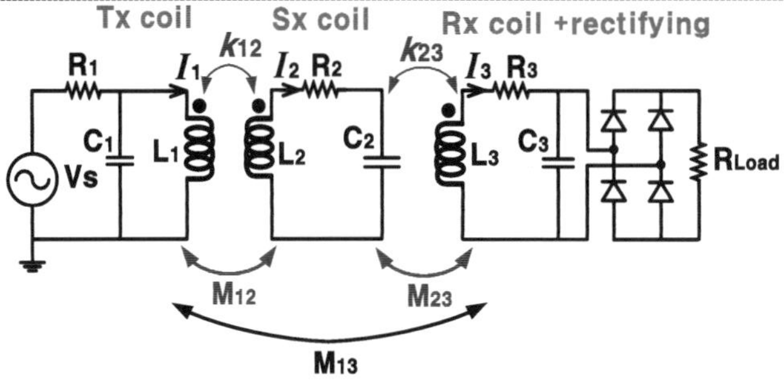

Figure 3. Equivalent circuit diagram of SUWPT4EV with inserted resonator system of Figure 2.

2.2. Thermal distribution of resonance coils

Generally, the thermal coupled loss of resonance wires (copper and superconducting coils) are caused by AC loss, which is the sum of magnetization losses and transport current losses during AC charging in an AC magnetic field. The resonance coupling coil also can cause thermal loss depended on the *LC* resonance condition. In the case of non-resonator, the thermal loss of Tx is relatively higher than Rx coil as shown Figure 4 (a) since the transferred power is relatively small. Generally, the transferred distance corresponding to induced electro-magnetic field (EMF) by Tx coil can be depended on the diameter of Tx coil. However, the diameter of Tx coil is limitation due to restricted length and impedance of Tx coil. From these reasons, the inserted resonator, which keeps strong resonance coupling, is one of reasonable options to expand the trasnferred distance, since the radius of induced EMF by Tx coil can be strongly expanded using inserted resonators. As well as, in the case of inserted Sx coil, since the radius of induced EMF can be expanded, the transferred power at the Rx coil is relatively higher compared with that under non-resonator coil. However, the thermal losses of Sx and Rx are strongly caused as shown in Figure 4 (b). Since the total impedance of Sx coil, which is only connected with capacitor, is smaller than that of Rx including load. If the load is connected with Sx coil, the transferred power of Rx coil would be relatively reduced. Also, if the induced thermal loss of Sx coil can be minimized by cooling device, the transferred power at Rx coil would be maximized. That means the inserted Sx coil plays a role to preserve the induced EMF by Tx coil and thus it delivers the EMF energy to Rx coil. In the viewpoints of efficicnecy, the design of cooling device for inserted Sx coil should be practically considered. The experimental performance of Figure 4 is carried with RF genertor 370 kHz of 200 W and it keeps reflect power below 10 W.

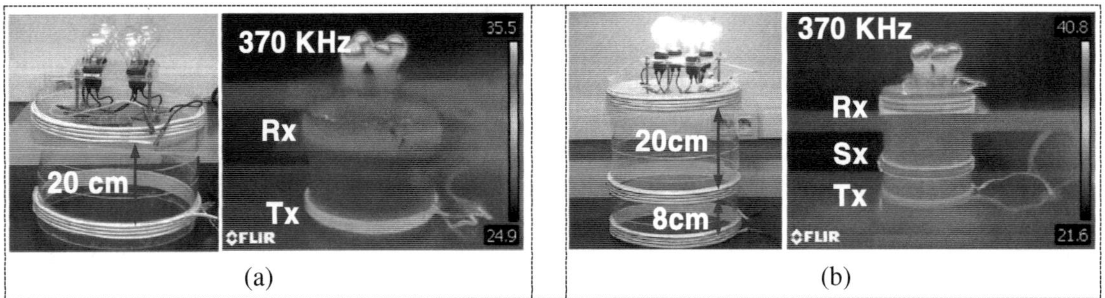

Figure 4. Photograph of experimental performance and thermal distributions with (a) non-resonator and (b) inserted resonator (Sx) between Tx and Rx coils at 200 W 370 KHz, respectively.

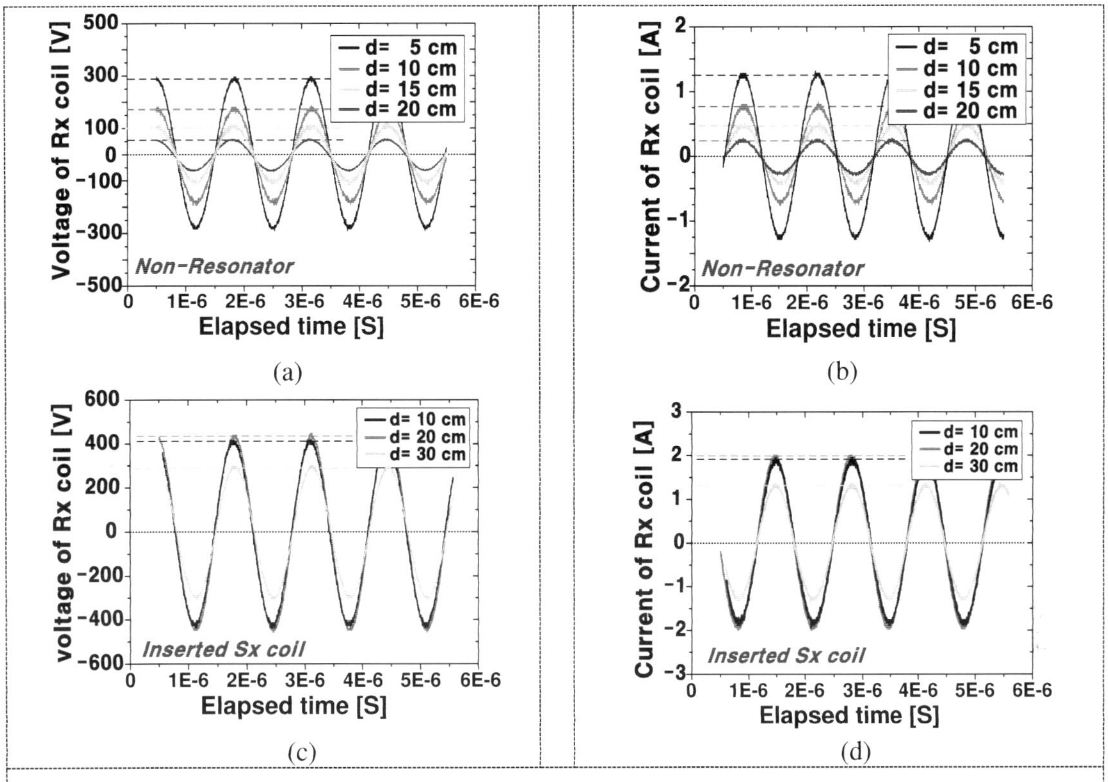

Figure 5. Measured results of distributed voltage and current waves at Rx coil (a) voltage of Rx coil without Sx coil (b) current of Rx coil without Sx coil (c) voltage of Rx coil including Sx coil (d) current of Rx coil including Sx coil in the 370 KHz with 200 W, respectively: d means interval between Tx and Rx coils.

Figure 5 shows the measured results of voltage and current waves for different intervals between Tx and Rx coils with 200W of 370 kHz RF generator under including Sx and non Sx coils, respectively. The intervals d is changed from 5 cm to 20 cm between Tx and Rx coils. The reflected power rates keep below 1 %, which is indicated in the protection device of RF generator. Definitely, the strong resonance waves with resonator Sx coil can spread far away compared with non-Sx coil under same conditions. As well as, the amplitudes of transferred voltage and current waves at Rx coil were markedly reduced depended on enhancing intervals as shown in (a) and (b). On the other hand, in the case of including Sx coil, the amplitudes of transferred voltage and current waves were slowly reduced corresponding to enhancing intervals as shown in (c) and (d). Obviously, the amplitude of transferred voltage and current waves at d= 20 cm under including Sx coil is higher over two times compared with non Sx coil. In other words, the transferred distance can be expanded by inserted resonator due to keep strong resonance coupling. The transferred power at Rx coil with inserted Sx coil is higher over 25 % compared with non-resonator Sx coil as shown in Figure 5. Even though the inserted Sx coil plays a role to keep strong electromagnetic field, as the generation of heat for inserted resonator, the cooling device for inserted Sx coil should be considered in order to keep strong resonance coupling. Thus, the Sx coil has a benefit to expand the transfer distance. However, since the thermal loss is strongly generated in the Sx and Rx coil at the large intervals between Sx and Rx coils, the cooling device for Sx and Rx coil is required.

3. Formatting the text

3.1. Structure and mechanism

In this experiment, we examined the effects of transmission efficiency and delivery distance with inserted resonator antenna including rectifying circuit of the SUWPT4EV system. The rectangular shape resonator, which is wound into one layer by helix type, was made from copper tape in order to install on the cover of cooling vessel easily. The HTS Tx and copper Rx coil are wound by helix type. The dimensional and design specifications of Tx, Sx and Rx coils and RF power are shown in Tables I and II, respectively. The rectified waves of load are measured by two pairs of fast recovery diodes. To confirm the transfer power, the multi 30 W bulbs are attached in the receiver coil as a load. The experiment is carried out below 300 W of RF input power 370 KHz for case I and II sequences. To compare transferred power ratio of cases I and II fairly, the reflect power ratio is kept below 1 %. The distances of d_1, d_2 and d_3 are fixed at 40, 30, 10 cm in the cases I and II as shown in Figure 6. Authors adopted a cooling box made from Styrofoam. In the next, we will fabricate the GFRP vessel and glass cover for practical approach. Figure 6 shows experimental setup and performance sequences without rectifying circuit at input power of 200 W under the cases I and II, respectively. The distances of d_1, d_2 and d_3 are fixed at 40, 30, 10 cm in the cases I and II as shown in Fig. 4. We adopted a cooling box made from Styrofoam. In the next, we will fabricate the GFRP vessel and glass cover for practical approach. Apparently, it is experimentally confirmed that transmission power at Rx coil with resonator is larger than that without resonator under the same condition. That means, the resonator keeps strong resonance coupling and thus the transmission distance can be expanded. The measured Q value for copper cable at 370 KHz, 300 K is 110. Also the Q value for cooled copper cable at 370 KHz, 77 K is 130 and 160. Also, at 370KHz, the Q value for HTS tape is 190 at the same length resonance coil.

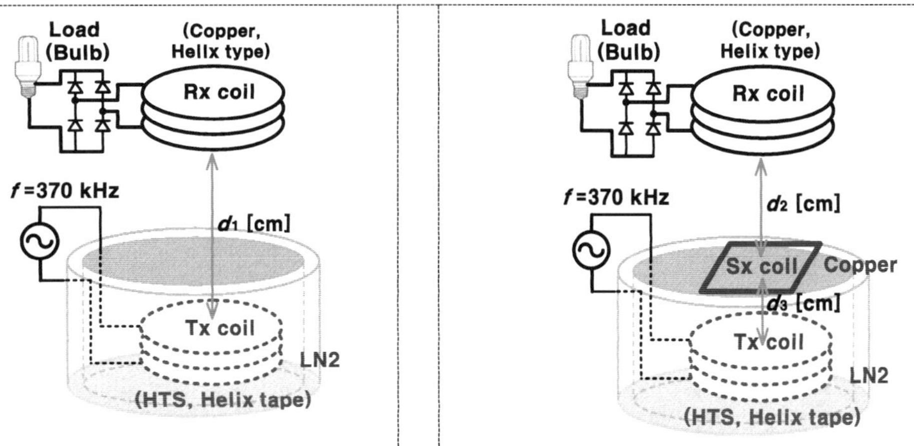

(a) Case I (b) Case II

Figure 6. Schematic illustration of different experiment sequences for transmission mode including setup under (a) non-resonator and (b) inserted resonator, respectively

Table 1. Dimensional specifications of Tx, Sx, and Rx coils

Parameters	Dimensions
HTS wire AMSC 344S (Thickness, width)	Ic = 72 A @ 77 K (0.3 mm, 4 mm)
Inductance and impedance of HTS Tx coil @300K, 370kHz	7.96 [μH], 18.42 [Ω]
Cable diamneter of of Rx coil	φ 10 mm
Inductance and impedance of Rx coil@300K, 370kHz	9.23 [μH], 21.42 [Ω]
Size of Sx coil (Thickness, width)	0.2 mm, 10 mm
Inductance and impedance of Sx coil@300K, 370kHz	10.43 [μH], 24.2 [Ω]
Load (bulb)Capacitor	30W
Capacitor of Sx and Rx coils	18[nF]

Figure 7 shows experimental setup and performance sequences without rectifying circuit at input power of 200 W under the cases I and II, respectively. Apparently, it is experimentally confirmed that transmission power at Rx coil with resonator is larger than that without resonator under the same condition. That means, the resonator keeps strong resonance coupling and thus the transmission distance can be expanded.

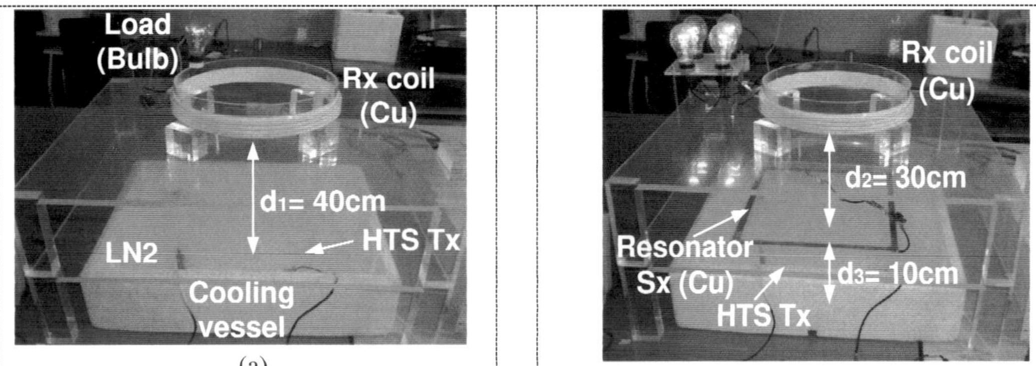

Figure 7. Photograph of experimental setup and different performances without rectifying circuit in the case I and case II of Figure 6 with input of 200 W; including HTS Tx, copper Sx, Rx and load (bulb)

Table 2. Specifications of RF power

Parameters	Dimensions
Frequency	370 kHz
Harmonics	-50dbc
Rating Power	1 KW
Output resolution	1 W
Reflect power limit	200 W

3.2. Experimental results

Figure 8 (a) shows the experimental results of current and voltage distributions for HTS Tx and copper Rx in the case I at 40 cm under input power of 200 W. As surely seen, the phases of voltage and current of Rx are shifted compared with input phases of Tx coil. The shifted angle is about 79° between Tx and Rx coils. That means it is difficult to keep resonance coupling condition, if the interval is expanded between Tx and Rx coils, even though reflected energy loss is minimized. Figure 8 (b) shows measured results of current and voltage distributions for HTS Tx and copper Rx in the case II with inserted resonator under input power of 200 W. The wave shifting between Tx and Rx coils is apparently reduced due to inserted resonance. That means the inserted resonance coil keeps strong coupling easily. Thus, the inserted resonator can minimize the reactive power loss at Rx coils.

Figure 9 (a) shows measured results of input voltage at Tx, rectified voltage and rectified current including rectifying circuit and load under Case I. The peak voltage of Tx coil, rectified voltage and rectified current are 432 V_{peak}, 384 V_{peak} and 0.18A_{peak}, respectively. The shifted phase angle is about 79°. Figure 9 (b) shows measured results of input voltage at Tx, rectified voltage and rectified current including rectifying circuit and load under Case II. The peak voltage of Tx coil, rectified voltage and rectified current are 491 V_{peak}, 442 V_{peak} and 0.21A_{peak}, respectively. Surely, the transferred voltage and rectified current are improved over 20 % and 15 %, respectively, due to inserted resonator as shown in Figures 9. As the inserted resonator can reduce reactive power loss at Tx and Rx coils, transfer ratio can be improved effectively even though the interval is expanded between Tx and Rx coils.

Figure 10 (a) shows measured results of rectified voltage distributions at load including input power of 100, 200, 300 W, respectively, under Case I. The peak values at input 100, 200, 300 W are 250, 332, 357 V, respectively. As well as, the wave shifting is enhanced depended on increasing input powers. On the other hand, rectifying waves are stably improved under inserted resonator without wave shifting as shown in Figure 10 (b). The peak values at input 100, 200, 300 W are 251, 349, 417 V, respectively.

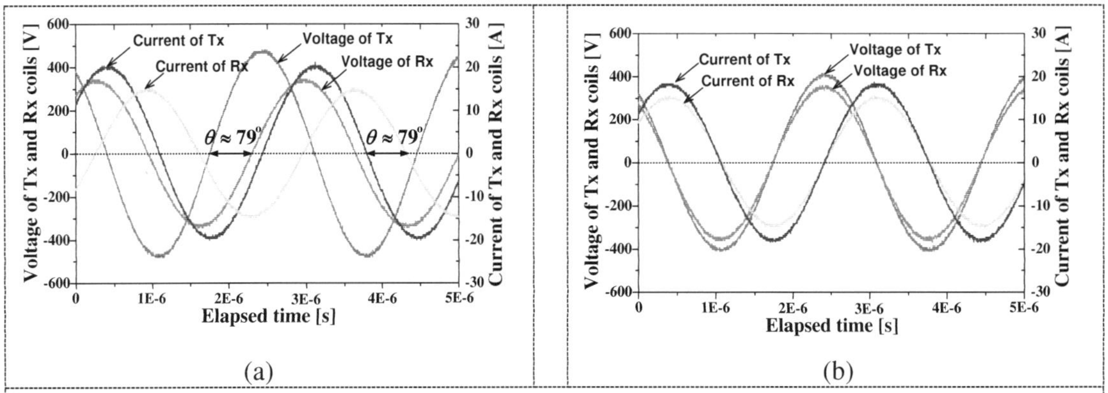

Figure 8. Measured results of voltage and current distributions at Tx and Rx coils, respectively, in the input power of 200 W under the (a) Case I and (b) Case II, respectively.

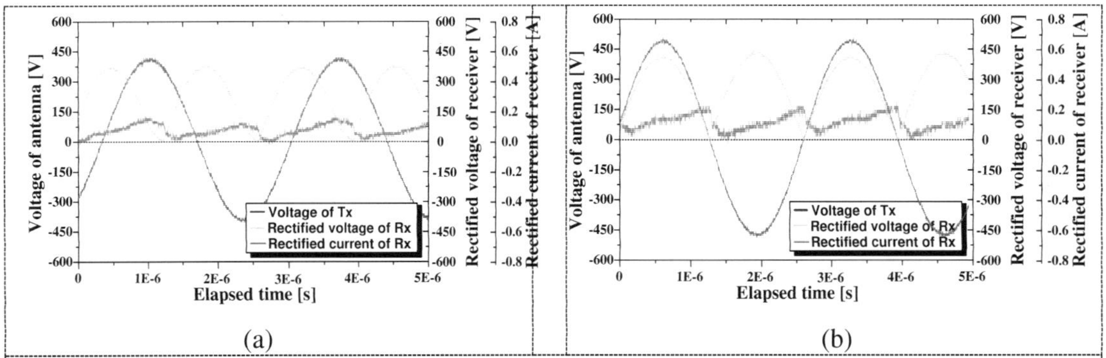

Figure 9. Measured results of input voltage at Tx coil and, rectified voltage and current of load at input power of 200 W under the (a) Case I and (b) Case II, respectively.

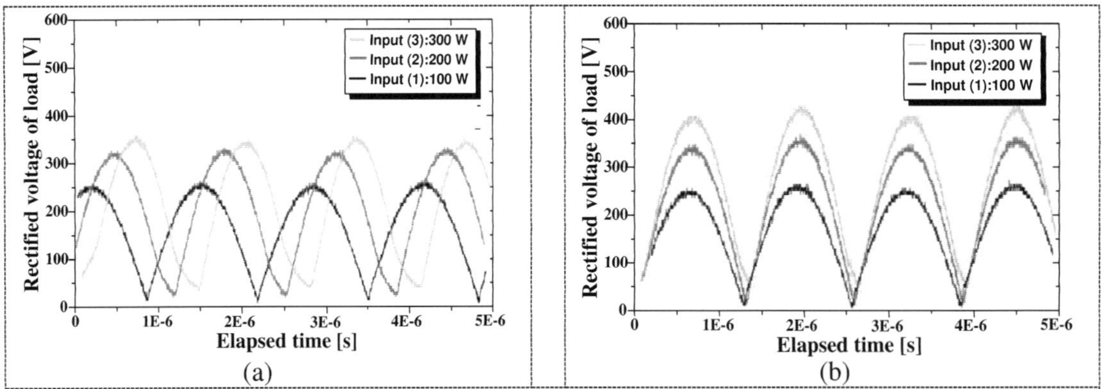

Figure 10. Measured results of rectified voltage distributions at input power of 100, 200, 300 W respectively under the (a) Case I and (b) Case II, respectively.

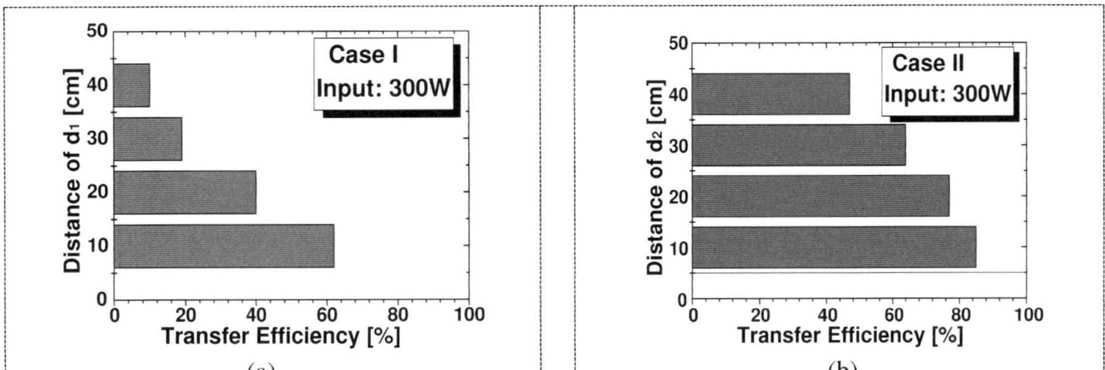

Figure 11. Measured results of transfer efficiency at Rx coil with 300 W under the (a) Case I and (b) Case II of Figure 7, respectively.

Clearly, in the case of non-inserted resonator, the wave shifting is gradually enlarged under increasing input power and it causes reactive losses, which generates thermal loss at Rx and Tx coils. On the other hand, waves shifting of transferred waves can be apparently reduced under increasing input power due to inserted resonator. That means inserted Sx coil has a role to keep stable transfer power including rectifying circuit under increasing input power. Figure. 11 shows the measured transfer efficiency for Cases I and II of Figure 7 with 300W of 370kHz, respectively. . Surely, the transfer efficiency with Sx coil at Case II is slowly decreased depended on the increasing intervals. On the other hand, the Case I is sharply decreased depended on the increasing intervals. Based on the experimental results, it is evaluated that the tape type copper Rx coil, which is installed at the outside of cover in cooling vessel, can play a role to improve the transmission efficiency easily. Thus, the copper resonator Sx coil is one of reasonable options to improve the transmitting ratio and spread the waves widely.

4. Conclusion

We point out that this study achieved wireless power transfer technology with HTS antenna coil in order to improve the delivery efficiency of the electric power for EV using inserted resonator. The properties and relations for inserted resonator between Tx and Rx coils were successfully performed. Especially, we have shown the possibility of impedance matching method for different materials (HTS and normal conductor), sizes (tape and tube types) and environments (between room temperature and low temperature) at 370 KHz resonant frequency within 40 cm. Thus, we obtained that the improved transmitting ratio of voltage and current using inserted resonator are over 25 %. That means the inserted copper resonator, which is easily installed at the cover of cooling, is one of reasonable options to improve the transmitting ratio even for superconducting antenna conditions. However, since it is confirmed that the inserted resonator can generate heating loss at Sx as well as Rx coils, the design of cooling device for inserted Sx coil should be considered to improve transfer efficiency. The economic efficiency of cooling vessel should be investigated under structural design in the next. Our research target is to evaluate such a transfer technology to EV fast and high power charging system.

Acknowledgments

This research was supported by Basic Science Research Program through the National Research Foundation of Korea (NRF) funded by the Ministry of Education (Grant No: 2017R1D1A1A09081100).

References

[1] Brown W C 1984 *IEEE Trans.* **32** 1230
[2] Sun T Xie X and Wang Z 2013 *Wireless Power Transfer for Medical Microsystems* New York: Springer 34

[3] Murakami J Sato F Watanabe T Matsuki H Kikuchi S Harakaiwa K and Satoh T 1996 *IEEE Trans. Magn.* **32** 5017
[4] Chung Y D Lee C Y Jo H C and Yim S W 2014 *Physica C* **504** 115
[5] Kim D W Chung Y D Kang H K Yoon Y S and Ko T K 2012 *IEEE Trans. Appl. Supercond.* **22** 5400604
[6] Zhang G Yu H Jing L Li J Liu Q and Feng X 2013 *IEEE Trans. Appl. Supercond.* **23** 4600505
[7] Dai H Liu Y Chen G Wu X and He T 2014 *Proc. IEEE Conf. on Computer Communications* 1105
[8] Wu K Choudhury D and Matsumoto H 2013 *Proc. IEEE* **101** 1271
[9] Chung Y D Lee C Y Kim D W Yoon Y S Jo H C Hwang Y J 2014 *IEEE Trans. Appl. Supercond.* **24** 0500205

IOP Conf. Series: Journal of Physics: Conf. Series **1293** (2019) 012056 doi:10.1088/1742-6596/1293/1/012056

Power enhancement of the high-T_c superconducting terahertz emitter with a modified device structure

H Minami[1, 2], Y Ono[1], K Murayama[1], Y Tanabe[1], K Nakamura[1], S Kusunose[1], T Kashiwagi[1, 2], M Tsujimoto[1, 2] and K Kadowaki[2, 3]

[1] Graduate School of Pure and Applied Sciences, University of Tsukuba, 1-1-1 Ten-nodai, Tsukuba, Ibaraki 305-8571, Japan
[2] Division of Materials Science, University of Tsukuba, 1-1-1 Ten-nodai, Tsukuba, Ibaraki 305-8573, Japan
[3] Algae Biomass and Energy System (ABES) Research and Development Center, University of Tsukuba, 1-1-1, Tennodai, Tsukuba Ibaraki 305-8572, Japan

E-mail: minami@bk.tsukuba.ac.jp

Abstract. Continuous and monochromatic terahertz electromagnetic waves can be generated with sizable power (~30 μW) by the mesa-shaped device made of the high-T_c superconductor $Bi_2Sr_2CaCu_2O_{8+\delta}$, by synchronizing the phase of the Josephson currents between multi-stacked intrinsic Josephson junctions. From the previous experimental results on the line width, spatial distribution of the emission, *etc*., we think that the relatively weaker power may originate from the partial failure of the phase synchronization of the Josephson currents in the mesa device. This may be improved by modifying the device structure to enhance the Josephson current density inside the mesa. In the present study, an array of four rectangle-shaped mesas with the dimensions of 80±5 μm×350 μm×3.2 μm was fabricated on a thin $Bi_2Sr_2CaCu_2O_{8+\delta}$ single crystal, where several slots were made around the mesas by photolithography and chemical etching techniques in order to weaken electrical connection between mesa and superconductor base. The best result of the emission power for the array so far obtained is ~80 μW at 0.42 THz, which is about 2.7 times bigger than the previous champion data in our group.

1. Introduction

Terahertz (THz) electromagnetic waves have attracted much attention as a new "light", because they have huge potential applications to various fields such as spectroscopic analyses and identification of materials, non-destructive sensing and imaging for security, inspection of products and medicine, and high-speed communications, quantum computation, *etc.* [1]. The time domain spectroscopy technique (THz-TDS) becomes popular and used widely in spectroscopic research fields. However, applications of monochromatic THz waves are very limited so far because of lack of high-performance emitters and detectors. Monochromatic source is indispensable in spectroscopic research fields, and particularly in high-speed communications. Development of compact high-performance solid-state emitters, as well as highly sensitive detectors, at the THz frequency region are highly challenging topic of research.

High-T_c superconductor $Bi_2Sr_2CaCu_2O_{8+\delta}$ (Bi2212) is well-known to consist of a multi-stack of equivalent atomic-scale Josephson junctions (670 layers per 1 μm thick) commonly referred to as the intrinsic Josephson junctions (IJJs) [2]. The continuous and monochromatic THz electromagnetic waves whose frequencies obey the *ac*-Josephson effect can be generated with sizable power (~30 μW), by passing *dc* current to IJJs to synchronize the phase of the Josephson currents between IJJs in the mesa-shaped device made of Bi2212 [3-5]. At present, the THz emission with frequencies beyond 2 THz can be generated, and the power of ~1 μW at 1 THz can be obtained at 40 K [6]. It is also possible to generate a sizable power (~2 μW) at 0.43 THz from the device cooled in liquid nitrogen [7,8]. As the power consumption of this device is enough low to work using a small battery, a compact cryocooler will be sufficient to cool this device. One of the main subjects is to increase the emission power of the devices to the level of 1 mW.

From the previous experimental results on the line width [9], spatial distribution [10,11] of the emission, *etc*., we think that one of the reasons of the relatively weak emission power may be due to the partial failure of the coherent phase synchronization of the Josephson current building up in the mesa device. According to the theory applied to the Josephson junction array oscillators for low-T_c superconductors [12,13], the line width is estimated to be ~10 kHz at 50 K for the present device. However, the observed line width is about 23 MHz even at the best value [9]. In this device, it is known that an extremely inhomogeneous temperature distribution referred to as the hot-spot emerges at high bias currents [14-16]. This concentration of the superconducting quasi-particles originates from the instability on their flow due to the steep negative slope of the *c*-axis resistivity. Dynamic fluctuation due to this instability may be a cause of the broad line width. Actually, the emission line width has been observed to be strikingly improved after the hot-spot formation [9]. In addition, the Josephson current density inside the mesa may not be enough high to develop the phase synchronization between IJJs completely, because of good electric connection between mesa and superconductor base. Actually, Josephson current has been observed to leak out from mesas, from the fact that the THz emission is affected by external structures [17].

The Josephson current density may be improved by adjusting the impedance matching by simply modifying the device structure. In the present study, we have tried to control the electrical connection between mesa and superconductor base, by making several slots in the superconductor base around the mesa. In addition, we have tried to synchronize simultaneously four mesas which are made on a piece of Bi2212 crystal. Best result so far obtained is the emission power of ~80 μW for the array of four mesas at 0.42 THz, which amounts to about 2.7 times more intense than the previous champion data in our group [5].

2. Experiment

2.1. Fabrication of Bi2212 mesa devices

Figure 1 (a) shows a photograph of the mesa device used for the present study. After a piece of a thin Bi2212 single crystal was glued on a sapphire substrate by thin polyimide, it was cleaved to ~6 μm thick and was immediately deposited by Ag and Au films of 10 nm, respectively. Then, four mesas with an identical rectangular shape were etched by an argon ion milling technique using metal masks twice to determine the width first, then the length, second, of the mesa [18]. One of the mesas is highlighted by a yellow rectangle in figure 1 (a). They were arrayed with an interval of 80 μm, and the dimensions of each mesa measured by an atomic force microscope were 80±5 μm in width, 350 μm in length, 3.2 μm in height, and the thickness of the underlying superconductor was ~3 μm. Finally, several slots as seen in figure 1 (a) were made in the superconductor base around each mesa by photolithography and chemical etching techniques [19].

In order to bias the mesa device, it was turned over and attached on plate electrodes of Au film with 0.5 μm thick, separated into two with 100 μm gap, on an alumina substrate, as shown in figure 1 (b). Two sets of mesas in which two mesas were connected in parallel were connected in series, as shown in the inset of figure 1 (c). Figure 1 (c) shows the temperature dependence of the mesa resistance measured by two terminal method at 100 μA. It is seen that the contact resistance between mesa and electrodes is negligibly small. The device structure is similar to the one which demonstrated the THz

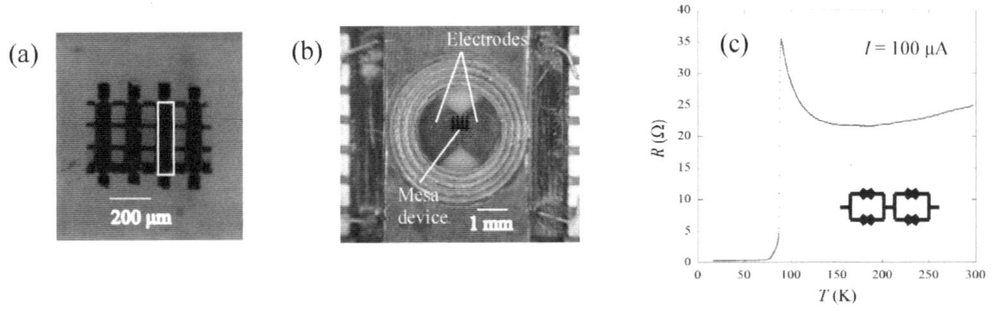

Figure 1. (a) A photograph of the mesa device fabricated on a sapphire substrate. (b) The mesa device mounted on electrode pads on an alumina substrate. (c) Temperature dependence of the resistance of the mesa device.

emission in liquid nitrogen previously [7], where the Joule heat was removed not only from the thin superconductor base but also from the top of mesa. Because of the high heat removal efficiency and small heat generation due to the low contact resistance, the THz emission can be observed up to bath temperature T_b=80 K in this present array structure.

2.2. Measurements of THz-wave emitting characteristics

The mesa array was mounted on a copper finger cooled by a He-flow cryostat (Oxford Instruments, CF1104), and the temperature dependence of the resistance and the current-voltage characteristics (*IVCs*) were measured by a conventional two-terminal method. For the detection of THz emission, the same setup as the previous experiments was used for the direct comparison of emission intensity [5]. A lock-in technique was used with an InSb hot electron bolometer (HEB) detector (QMC Instruments, QFI/2BI) with a pre-amplifier gain of 1000 and an optical chopper with a frequency of 90 Hz. The InSb HEB detector has been calibrated using black body radiation, and the optical responsivity is 3.3 mV/nW. The emission spectra between 0.1 and 2 THz were measured by a Fourier transform infrared (FTIR) spectrometer (JASCO, FARIS-1) with the resolution of 8.0 GHz, using a TGS pyroelectric detector at room temperature. The THz wave from the mesa device was collected and guided to the FTIR spectrometer by a concave mirror.

3. Results and discussion

The *IVCs* of the mesa device were measured by ramping up and down the bias current at T_b from 30 K to 80 K, and the output voltages from the lock-in amplifier proportional to that of the HEB detector were simultaneously recorded. The THz waves emitted through the sapphire substrate were measured in the direction right above the sample. Figure 2 (a) shows the *IVC* and the THz emission intensity as functions of the applied voltage and current, at T_b=72.5 K at which the emission intensity reaches at its maximum as seen in figure 2 (b). The emission intensity at its maximum is 2.7 times bigger than the previous champion data in our group [5], and the detected power is estimated to be 0.78 μW. Taking a value of 0.02 sr for the solid angle of detection and the attenuation factor of the cryostat window made of quartz glass of 0.75, the total power integrated, assuming a spatial distribution of the radiation power as seen in figure 4 in the reference [5], is estimated to be ~80 μW.

Figure 2 (c) shows the emission spectrum from the mesa device (red curve) measured at T_b=72.5 K by an FTIR spectrometer with a TGS pyroelectric detector at room temperature. Note that the ordinate is plotted by the logarithmic scale. The peak intensity at 0.42 THz exceeds more than 10^3 times in comparison with the detector noise. A broad and flat spectrum obtained from a mercury lamp (blue curve) is added for the comparison. Both measurements were performed with the same experimental set-up except for the data accumulation number as indicated in the graph. The emission frequency f of 0.42 THz agrees with the cavity resonance frequency for the long rectangular mesa, $f_c=c_0/2nw$, where

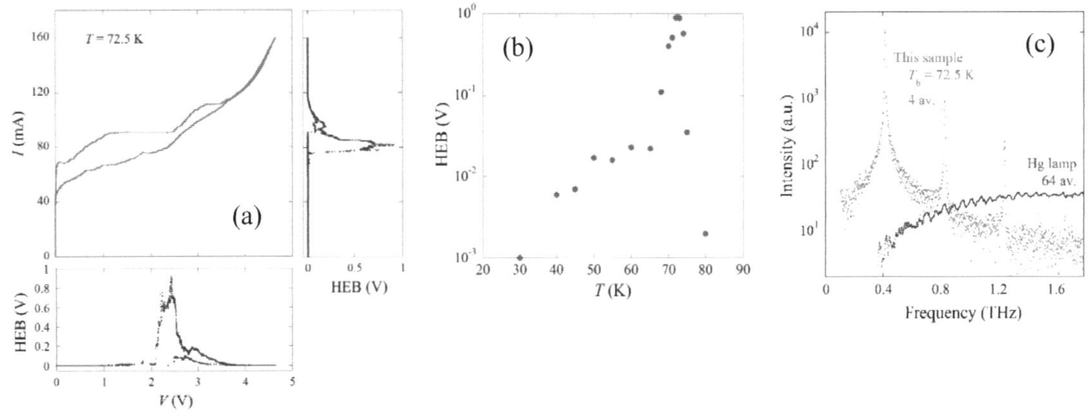

Figure 2. (a) The *IVC* of the mesa device and the effective output voltage of the HEB detector as functions of applied bias voltage and current to the mesa device at T_b=72.5 K. (b) The T_b dependence of the maximum emission intensity. (c) The THz emission spectrum from the mesa device measured at T_b=72.5 K (red curve) compared to that of a mercury lamp (blue curve).

c_0 is the speed of light in vacuum, n the refractive index of Bi2212, and w the width of the mesa [3]. Our previous investigation has shown that the harmonic peaks originate from the harmonic components of emission as well as the harmonic noise coming from the measurement apparatus. The number of active IJJs biased by voltage, N, can be estimated from the *ac*-Josephson relation, $f=(2e/h)(V/N)$, where e is the elementary charge, h the Planck's constant, V applied voltage where the THz emission occurs. The estimated N=2500~2700 is slightly larger than 2100 estimated from the total number of IJJs in a mesa. This would mean that two mesas may be electrically connected in series and partially form one set of mesas with the Josephson voltage state. Actually, as seen later in figure 3 (a), the emission peaks are simultaneously observed up to two in one spectrum.

In figure 2 (b), it is interesting to point out that a sudden increase of the emission intensity starts above 65 K. The peak emission intensity at 72.5 K reaches 40 times larger than that at 65 K. Similar increase has been observed in previous results [5], although it was not so remarkable as present observation. Since the previous device also had a slot at each corner of the rectangular mesa, this sudden enhancement of the THz emission might be the effect of slots surrounding the mesas. It is known from the previous studies that the hot-spot does not appear above about 55-60 K [15], which means that the instability on the quasi-particle flow declines at high temperature. In addition to that, because the *rf*-current density may be enhanced by the effect of slots, the stronger phase synchronization of the Josephson currents may be induced in the direction of mesa length. The exact mechanism is still not clear yet, but this phenomenon may provide us a hint to enhance further the THz emission from the IJJ device. This mesa array emits the THz wave up to T_b=80 K, just below superconducting transition temperature T_c=90 K. This is due to the high heat removal efficiency of this device and the low contact resistance between the mesa and the electrodes. There may be further improvement on these points in the future.

Figure 3 (a) presents the emission spectra at T_b=72.5 K measured at several bias currents. The FTIR spectrometer has an instrumental frequency resolution limit of 8 GHz. At higher current more than

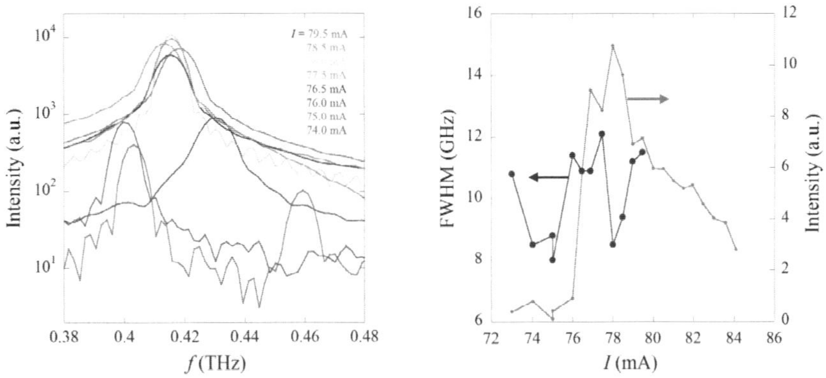

Figure 3. (a) The emission spectra at T_b=72.5 K measured at various bias currents using an FTIR spectrometer with a TGS pyroelectric detector. (b) The peak intensities and the FWHMs collected from the emission spectra are plotted as a function of the bias current.

76.5 mA, emission peaks appear around 0.42 THz, but the peak jumps to 0.432 THz at 76.0 mA. Decreasing the bias current further, the emission line splits into two at 75.0 mA, and then only one peak with lower frequency survives at 74.0 mA. It is probably because the Josephson frequency may shift away from the cavity resonance frequency f_c of one of two mesas. The emission frequency, *i.e.*, the Josephson frequency, must shift together with applied voltage in accordance with the *ac*-Josephson relation. Figure 3 (b) presents the peak intensities and the line widths by the full width at half maximum (FWHM) collected from the emission spectra data. The FWHM takes the minimum, when the spectrum splits into two peaks at 75.0 mA and the emission intensity takes its maximum at 78.0 mA. The values of FWHM are very close to the resolution limit of the FTIR spectrometer. That means that actual emission line widths must be much sharper than the resolution limit of 8 GHz. The results of figure 3 (a) and (b) suggest that the frequency pulling mechanism is working here between two mesas having slightly different f_c's and as a result two mesas are synchronously coupled at 78.0 mA.

In summary, the new experimental results of THz wave emission are reported from a mesa device made of high-T_c superconductor Bi2212. The array structure of mesas with surrounding several slots were fabricated and tested. First of all, the emission power of ~80 μW at 0.42 THz is obtained. This power is about 2.7 times bigger than the previous champion data in our group. Secondly, in the temperature dependence of the emission intensity, a striking sudden enhancement of about 40 times is observed around 72.5 K. Proper understanding of this phenomenon may provide us a clue to generate higher power emission from the IJJ device.

Acknowledgments

This work was supported by JSPS KAKENHI Grant Number 15K04688 and 15H01996, and University of Tsukuba Basic Research Support Program Type B.

References

[1] Tonouchi M 2007 *Nature Photon.* **1** 97
[2] Kleiner R, Steinmeyer F, Kunkel G and Müller P 1992 *Phys. Rev. Lett.* **68** 2394
[3] Ozyuzer L, Koshelev A E, Kurter C, Gopalsami N, Li Q, Tachiki M, Kadowaki K, Yamamoto T, Minami H, Yamaguchi H, Tachiki T, Gray K E, Kwok W-K and Welp U 2007 *Science* **318** 1291
[4] Welp U, Kadowaki K and Kleiner R 2013 *Nature Photon.* **7** 702
[5] Sekimoto S, Watanabe C, Minami H, Yamamoto T, Kashiwagi T, Klemm R A and Kadowaki K 2013 *Appl. Phys. Lett.* **103** 182601

[6] Kashiwagi T, Sakamoto K, Kubo H, Shibano Y, Enomoto T, Kitamura T, Asanuma K, Yasui T, Watanabe C, Nakade K, Saiwai Y, Katsuragawa T, Tsujimoto M, Yoshizaki R, Yamamoto T, Mianmi H, Klemm R A and Kadowaki K 2015 *Appl. Phys. Lett.* **107** 082601
[7] Mianmi H, Watanabe C, Kashiwagi T, Yamamoto T, Kadowaki K and Klemm R A 2016 *J. Phys. Condens. Matter* **28** 025701
[8] Hao L Y, Ji M, Yuan J, An D Y, Li M Y, Zhou X J, Huang Y, Sun H C, Zhu Q, Rudau F, Wieland R, Kinev N, Li J, Xu W W, Jin B B, Chen J, Hatano T, Koshelets V P, Koelle D, Kleiner R, Wang H B and Wu P H 2015 *Phys. Rev. Applied* **3** 024006
[9] Li M, Yuan J, Kinev N, Li J, Gross B, Guénon S, Ishii A, Hirata K, Hatano T, Koelle D, Kleiner R, Koshelets V P, Wang H and Wu P 2012 *Phys. Rev.* B **86** 060505(R)
[10] Minami H, Kakeya I, Yamaguchi H, Yamamoto T and Kadowaki K 2009 *Appl. Phys. Lett.* **95** 232511
[11] Kadowaki K, Tsujimoto M, Yamaki K, Yamamoto T, Kashiwagi T, Minami H, Tachiki M and Klemm R A 2010 *J. Phys. Soc. Jpn.* **79** 023703
[12] Jain A K, Likharev K K, Lukens J E and Sauvageau J E 1984 *Phys. Rep.* **109** 309
[13] Darula M, Doderer T and Beuven S 1999 *Supercond. Sci. Technol.* **12** R1
[14] Wang H B, Guénon S, Yuan J, Iishi A, Arisawa S, Hatano T, Yamashita T, Koelle D and Kleiner R 2009 *Phys. Rev. Lett.* **102** 017006
[15] Minami H, Watanabe C, Sato K, Sekimoto S, Yamamoto T, Kashiwagi T, Klemm R A and Kadowaki K 2014 *Phys. Rev.* B **89** 054503
[16] Watanabe C, Minami H, Yamamoto T, Kashiwagi T, Klemm R A and Kadowaki K 2014 *J. Phys.: Condens. Matter* **26** 172201
[17] Minami H, Koike T, Orita N, Kashiwagi T, Tsujimoto M, Yamamoto T and Kadowaki K 2012 *J. Phys.: Conference Series* **400** 022072
[18] Orita N, Minami H, Koike T, Yamamoto T and Kadowaki K 2010 *Physica* C **470** S786
Minami H, Tsujimoto M, Kashiwagi T, Yamamoto T and Kadowaki K 2012 *IEICE Trans. Electron.* **E95-C** 347
[19] Shibano Y, Kashiwagi T, Komori Y, Sakamoto K, Tanabe Y, Yamamoto T, Minami H and Kadowaki K 2019 *AIP Advances* **9** 015116

Local heating effects on the radiation intensity of high-T_c superconducting terahertz emitters

K Nakamura[1], H Minami[1, 2], R Ota[1], K Murayama[1], Y Ono[1], S Kusunose[1], T Kashiwagi[1, 2], M Tsujimoto[1, 2] and K Kadowaki[2, 3]

[1] Graduate School of Pure and Applied Sciences, University of Tsukuba, 1-1-1 Ten-nodai, Tsukuba, Ibaraki 305-8571, Japan
[2] Division of Materials Science, University of Tsukuba, 1-1-1 Ten-nodai, Tsukuba, Ibaraki 305-8573, Japan
[3] Algae Biomass and Energy System (ABES) Research and Development Center, University of Tsukuba, 1-1-1, Tennodai, Tsukuba Ibaraki 305-8572, Japan

E-mail: minami@bk.tsukuba.ac.jp

Abstract. The terahertz electromagnetic waves can be generated by the mesa device made of a piece of high-T_c superconductor $Bi_2Sr_2CaCu_2O_{8+\delta}$ single crystal, which is known to be comprised of multi-stacks of intrinsic Josephson junctions. In the case of a thick mesa containing the large number of junctions, large amount of Joule heat is generated, resulting in the formation of the inhomogeneous temperature distribution $T(r)$ in the mesa. It is known from previous studies that $T(r)$ strongly affects the emission intensity and line width. When the mesa is irradiated partly by a focused laser beam, the terahertz radiation intensity grows almost two times as compared with the standard biases without laser irradiation as was found previously. It is also found that the enhanced emission remains without losing the intensity even after switching off the laser power. This result suggests that the normal current distribution in the mesa is modified irreversibly when the device is biased under laser irradiation, according to the thermoreflectance measurement.

1. Introduction

Recently, terahertz (THz) electromagnetic waves have attracted much attention because of various potential applications such as molecular level spectroscopic identification, inspection of products, pharmaceutical studies, medicine, *etc.* [1]. However, the research field has not been progressed much because of the lack of high performance THz emitters and detectors so far available. Coherent and continuous THz emission has been discovered from the mesa device made of high transition temperature T_c superconductor $Bi_2Sr_2CaCu_2O_{8+\delta}$ (Bi2212) single crystal comprised of multi-stacks of intrinsic Josephson junctions (IJJs) [2-4]. The IJJs consist of the atomic thin insulating Bi_2O_2 layers sandwiched by superconducting CuO_2 double layers stacking along the *c*-axis direction. The device has attracted much attention to be as the new type of THz emitters because the emission mechanism is completely different from the conventional ones.

The THz emission from IJJs occurs when the following two conditions are fulfilled. One is the *ac*-Josephson effect expressed as $f = (2e/h)(V/N)$, where f is the radiation frequency, e the elementary charge, h the Planck's constant, V the voltage applied to the mesa and N the total number of the IJJ stacking in the mesa [5]. Another is the cavity resonance condition expressed as $f = c_0/n\lambda = c_0/2nw$, where c_0 is the speed of light in vacuum, n is the refractive index of Bi2212, λ is the wavelength and w is the width of the mesa [2]. The power of THz emission from IJJs mesa is at present of ~30 μW at its best [6], which is not sufficient to the practically required power of ~1 mW. Stronger emission power has been desired for various applications.

An idea to generate stronger emission power is to increase the number of Josephson junctions using thicker mesas because the emission power $P \propto N^2$. However, a thicker mesa containing the larger number of junctions produces larger amount of Joule heat, so that the temperature of the mesa in the lateral direction of the Josephson junctions increases inhomogeneously. At high bias currents, the heated area abruptly shrinks and simultaneously the normal currents concentrate in there, which is known as the formation of a hot-spot in which the temperature is extremely increased to even higher than T_c [7-10]. That shows that this system has an instability on the normal current flow. Dynamic fluctuation due to the instability may impede the development of the in-plane phase synchronization in the mesa. Actually, the emission line width is known to be strongly changed by the bias current, especially it is strikingly improved after the hot-spot was formed [11]. The emission intensity and frequency have also been observed to be affected by the formation of the hot-spot and also the size and position as well [12]. According to the computer simulation that the hot-spot position can be easily changed by the local heating, they have shown that artificial local heating by a focused laser beam makes the emission power almost double of the largest one in the standard bias process without laser beam [13].

In a theoretical model, the intense THz emission is expected from IJJs by using an external local-heating source [14]. The THz emission is investigated there by solving the sine-Gordon, Maxwell, and thermal diffusion equations simultaneously, when the top electrode of the mesa is heated locally by the external heat source. It is interesting to point out that the local temperature increase by laser heating to just below T_c is more effective for high-power emission.

In the present study, we have investigated the effect of local heating on the emission intensity by using a focused laser beam and measuring temperature distribution of the mesa using a thermoreflectance microscopy observation [15]. From these results, a clue to produce higher power emission can be found.

2. Experimental details

2.1. Fabrication of Bi2212 mesa devices

High-quality single crystals of Bi2212 were grown by a traveling solvent floating zone method [16]. Rectangular mesas were fabricated on a thin Bi2212 crystal as follows. A piece of Bi2212 single crystal was cleaved to several μm thick, deposited immediately by thin Ag and Au (~20 nm and ~10 nm, respectively) on the surface, and glued on a alumina substrate by thin polyimide. Then, mesas were fabricated by Ar ion milling technique using metal masks twice to determine the width and the length. The second milling reached down to the alumina substrate and some parts of crystal were milled out completely, as shown in figure 1 (a) and it is illustrated in figure 1 (b). In order to attach a current lead to the top of the mesa, polyimide and CaF_2 were first deposited to avoid current leakage, and then thin Ag and Au films were deposited as current leads. Figure 1 (a) shows a photograph of the mesa devices. One of the mesas is highlighted by the yellow rectangle. The stereographic view of the fabricated mesa devices is given in figure 1 (b) to show the details as described above. The dimensions of the mesa are 80 μm in width, 400 μm in length and 3.0 μm in thickness, and a thickness of the underlying superconductor is less than 3.0 μm.

2.2. Experimental setup

A schematic view of the experimental setup is shown in figure 1 (c). We used a red color laser beam with the wave length of 660 nm and with a maximum power of 80 mW. The laser spot can be focused in a circular form with a diameter of 30 μm on the sample surface by using a beam expander and a focusing lens. The sample was mounted on a Cu cold finger cooled by a He flow cryostat (Oxford Instruments, CF1104). THz radiation was detected by a modulation technique using an InSb hot-electron

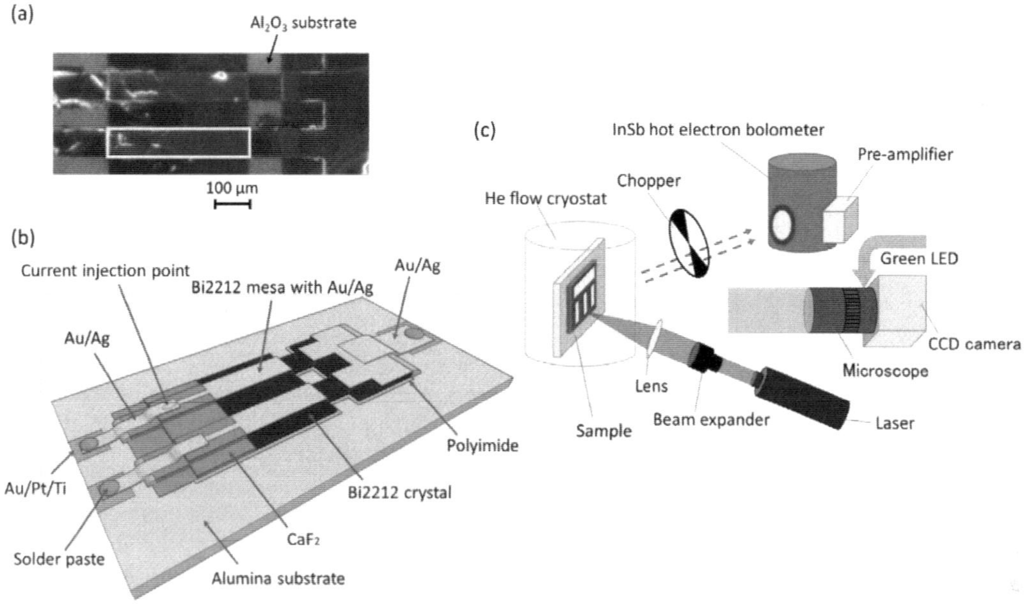

Figure 1. (a)A photograph of the fabricated array of the mesa devices. One of the mesas is highlighted by the yellow rectangle. (b)A stereographic view of the mesa devices. (c) A schematic view of the experimental setup with a laser beam. THz radiation was detected by an InSb hot electron bolometer. A microscope and a CCD camera were used for the thermoreflectance temperature distribution measurement of the surface of the mesa device and for observing the position of a laser beam.

bolometer and a lock-in amplifier. In addition, the temperature distribution of the surface of the mesa device was measured by thermoreflectance microscopy (TRM) [15]. This thermography technique is based on the experimental fact that the relative change of the reflectance at the device surface takes place with the temperature. Weak green light was irradiated on the whole surface of the device, and the reflected light was detected by a 12-bit CCD camera. The TRM image is presented as $\Delta I/I_0$ that indicates the change in reflectance from zero bias, where I_0 is the intensity of the green component in the RGB image data at zero bias and ΔI is the difference of that between at biased and zero bias conditions on the *I-V* characteristics. In order to improve the image quality, the obtained images were averaged 100 times by a computer. Although the TRM image does not tell directly the temperature distribution at the mesa surface, it provides qualitative but useful information of relative change of the surface temperature. The microscope and the camera were also used to observe the position where the laser beam was irradiated.

3. Results and discussion

The current-voltage (*I-V*) characteristics, emission intensity, and TRM measurements were all performed by continuously sweeping up and down the bias points, while keeping the power of the laser beam to the irradiation position. The laser beam was moved at nine different positions on a mesa. When the laser beam was located at some corners or at the center of a mesa for example, we did not observe significant enhancement of the emission intensity. Only when the laser beam is irradiated at a position indicated by a red dot in figure 2 (a), significant enhancement of the radiation intensity was observed. It is a corner opposite to the current injection point in the mesa. This is in contradiction to the previous result in which the enhancement of emission was observed when it was irradiated near the current injection point [13]. The reason of the contradiction is not clear at this moment.

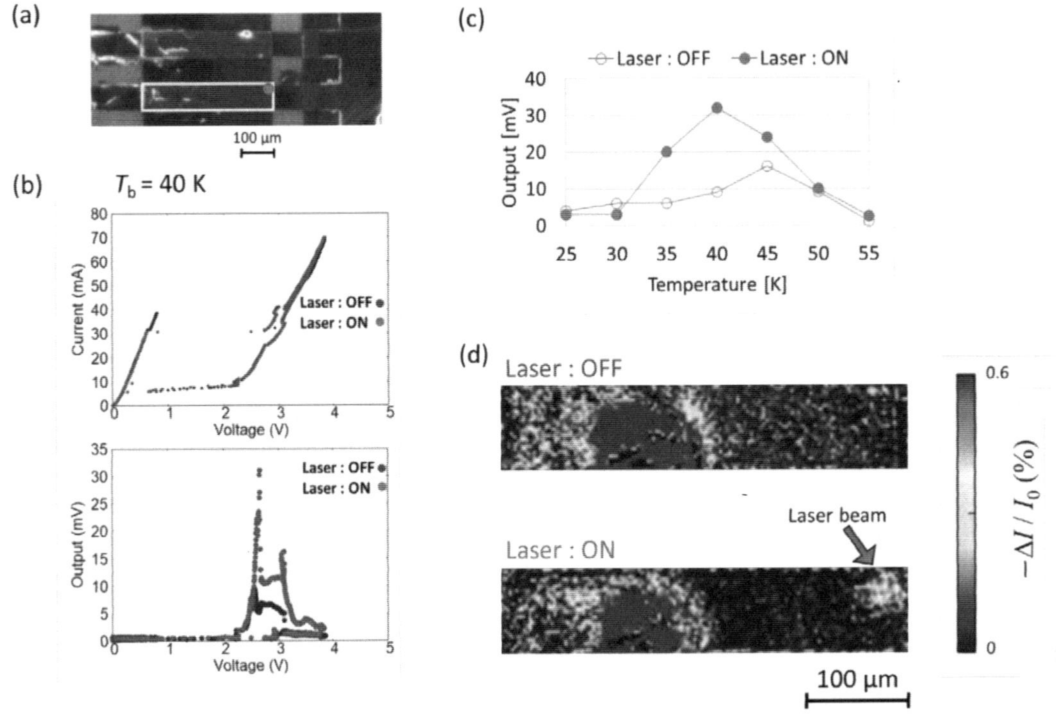

Figure 2. (a)A red dot indicates a position where a laser beam of the diameter of 30 μm was irradiated. (b)The *I-V* characteristics and the emission intensity as a function of the applied voltage at T_b=40 K when the mesa is biased with (red) and without (purple) laser irradiation at the red dot position shown in (a). (c)The maximum emission intensity as a function of T_b. The mesa is biased with (filled circles) and without (open circles) laser irradiation. (d)The TRM images of the mesa surface at T_b=40 K and V=2.9 V. The surface temperature is high at the position where the reflectance decreases.

Figure 2 (b) shows the *I-V* characteristics and the emission intensity as a function of the applied voltage at bath temperature T_b=40 K, when the mesa is biased with (red) and without (purple) laser irradiation of the power of 80 mW focused at the red dot position in figure 2 (a). As the laser is irradiated on the Ag/Au film, only a few percent of the laser power must be absorbed [8]. Although no significant difference was observed in the *I-V* characteristics, the emission intensity observed is increased more than three times under laser irradiation. Interestingly, at T_b=35 K and 40 K, the enhanced emission is observed without losing the intensity even after the laser is switched off at the fixed bias point where the emission intensity is at its maximum. In contrast to this phenomenon, in the case that the mesa is irradiated at the same bias point after biasing without irradiation, the enhancement was not observed in the intensity.

Figure 2 (c) shows the T_b dependence of the maximum emission intensity, where the mesa is biased with (filled circles) and without (open circles) laser irradiation at the power of 80 mW located at the red dot position in figure 2 (a). It is clear that at T_b=35 K, 40 K and 45 K, the emission intensity is enhanced with laser irradiation.

Figure 2 (d) shows the TRM images of the mesa when the mesa is biased at V=2.9 V at T_b=40 K. The hot-spot is clearly observed at the position slightly left from the center of the mesa. This may be due to Joule heat produced by the contact resistance at the current injection point. In the lower image, the mesa was biased while irradiating a laser beam at the red dot position in figure 2 (a), whereas it was biased without irradiation in the upper image. Each image shows the change of the reflected light intensity from zero bias current. Therefore, the heating by the laser beam itself should not be included in the lower image. A hot spot appeared at the laser irradiation point is thus due to the concentration of the normal current flow guided by local heating. Actually, the temperature slightly decreases at other parts because of the decrease of normal currents. That may be a reason why the emission is enhanced with irradiation. In the previous experiment, we have demonstrated that the position of a hot-spot can be reversibly controlled by local heating using a laser beam [13]. In the present case, however, the hot spot remains at the source position even after switching off the laser, so the emission also remain without losing the intensity.

In summary, we have observed that the emission intensity from the high-T_c superconducting terahertz emitting device is enhanced by local heating with a laser beam irradiation. The emission intensity reaches up to twice as large as the largest one obtained without laser irradiation, similar to the previous research. However, the enhanced emission is observed to remain without losing the intensity even after the laser power is switched off. This result suggests the distribution of the normal current strongly influences the emission intensity. We have reconfirmed that the terahertz emitting phenomenon of this device is deeply related to the temperature distribution, *i.e.,* normal current flow in the mesa.

Acknowledgments

This work was supported by JSPS KAKENHI Grant Number 15K04688 and 15H01996, and University of Tsukuba Basic Research Support Program Type B.

References

[1] Tonouchi M 2007 *Nature Photon.* **1** 97
[2] Ozyuzer L, Koshelev A E, Kurter C, Gopalsami N, Li Q, Tachiki M, Kadowaki K, Yamamoto T, Minami H, Yamaguchi H, Tachiki T, Gray K E, Kwok W-K and Welp U 2007 *Science* **318** 1291
[3] Welp U, Kadowaki K and Kleiner R 2013 *Nature Photon.* **7** 702
[4] Kadowaki K, Yamaguchi H, Kawamata K, Yamamoto T, Minami H, Kakeya I, Welp U, Ozyuzer L, Koshelev A, Kurter C, Gray K E and Kwok W-K 2008 *Physica* C **468** 634
[5] Josephson B D 1962 *Phys. Lett.* **1** 251
[6] Sekimoto S, Watanabe C, Minami H, Yamamoto T, Kashiwagi T, Klemm R A and Kadowaki K 2013 *Appl. Phys. Lett.* **103** 182601
[7] Skocpol W J, Beasley M R and Tinkham M 1974 *J. Appl. Phys.* **45** 4054
[8] Wang H B, Guénon S, Gross B, Yuan J, Jiang Z G, Zhong Y Y, Grünzweig M, Iishi A, Wu P H, Hatano T, Koelle D and Kleiner R 2010 *Phys. Rev. Lett.* **105** 057002
[9] Benseman T M, Koshelev A E, Vlasko-Vlasov V, Hao Y, Kwok W-K, Welp U, Keiser C, Gross B, Lange M, Kölle D, Kleiner R, Minami H, Watanabe C and Kadowaki K 2015 *Phys. Rev. Appl.* **3** 044017
[10] Minami H, Watanabe C, Sato K, Sekimoto S, Yamamoto T, Kashiwagi T, Klemm R A and Kadowaki K 2014 *Phys. Rev.* B **89** 054503
[11] Li M, Yuan J, Kinev N, Li J, Gross B, Guénon S, Ishii A, Hirata K, Hatano T, Koelle D, Kleiner R, Koshelets V P, Wang H and Wu P 2012 *Phys. Rev.* B **86** 060505(R)
[12] Watanabe C, Minami H, Yamamoto T, Kashiwagi T, Klemm R A and Kadowaki K 2014 *J. Phys.: Condens. Matter* **26** 172201
[13] Watanabe C, Minami H, Kitamura T, Asanuma K, Nakade K, Yasui T, Saiwai Y, Shibano Y, Yamamoto T, Kashiwagi T, Klemm R A and Kadowaki K 2015 *Appl. Phys. Lett.* **106** 042603
[14] Asai H and Kawabata S 2014 *Appl. Phys. Lett.* **104** 112601
[15] Kashiwagi T, Tanaka T, Watanabe C, Kubo H, Komori Y, Yuasa T, Tanabe Y, Ota R, Kuwano G, Nakamura K, Tsujimoto M, Minami H,Yamamoto T, Klemm R A and Kadowaki K 2017 *J. Appl. Phys.* **122** 233902
[16] Mochiku T and Kadowaki K 1994 *Physica* C **235** 523

IOP Conf. Series: Journal of Physics: Conf. Series **1293** (2019) 012058 doi:10.1088/1742-6596/1293/1/012058

Estimation of electricity storage capacity of compact SMESs composed of stacks of Si-wafers loaded with superconducting thin film coils in spiral trenches formed by MEMS process

Y Ichiki[1], A Ichiki[2], T Hioki[1], M Sasaki[3], J-H Noh[4], O Takai[4], H Honma[4] and T Motohiro[1,2]

[1] Graduate School of Engineering, Nagoya University, Furo-cho, Chikusa-ku, Nagoya,464-8603, Japan
[2] Institutes of Innovation for Future Society, Nagoya University, same as above
[3] Graduate School of Eng., Toyota Technological Institute, Hisakata 2-12-1, Tempaku-ku, Nagoya 468-8511, Japan
[4] Mater. & Surf. Eng. Res. Inst., Kanto-Gakuin University, 1162-2, Ogikubo, Odawara, Kanagawa 250-0042, Japan

E-mail: tomoyoshi.motohiro@chem.material.nagoya-u.ac.jp

Abstract. A method to estimate the electricity storage capacity of our compact superconducting magnetic energy storage system composed of stacks of Si-wafers loaded with superconducting thin film coils in spiral trenches formed by MEMS process was developed and tested for initial 6 designs of wafer coils. The results show energy storage density as low as around 6 Wh/ℓ. This is not caused by the limitation of the hoop stress but by the limitation of the maximum magnetic flux density. Further calculations for optimum wafer coil designs will be continued in which the effect of the maximum magnetic flux energy and the effect of the maximum hoop stress on the limit of the amount of the electrical energy storage come into a good balance.

1. Introduction

We have been developing superconducting thin film coils in a spiral trench on a Si wafers for compact SMESs as shown in figure 1. A proof of concept has been performed using NbN thin films showing energy storage of 0.01 mJ [1]. Increasing NbN thickness by mitigating film stress [2], the stored energy increased up to 0.1 mJ [3]. For further improvement, we moved on to replacement of NbN by $YBa_2Cu_3O_{7-\delta}$ (Y123) [4]. This replacement is to enable use of a cold source of liquid hydrogen: 20K or even of liquid nitrogen: 77K instead of use of a cryogenic refrigerator at 13K or lower in the case of NbN. Another benefit of this replacement is that the critical superconducting current density of Y123 can be about 2×10^{10} A/m^2 which is about 18 times larger than that of NbN [5]. Since the amount of magnetic energy storage is proportional to the square of the current, we may expect $18\times18 = 324$ times larger energy storage in the case of Y123 in comparison to the case of NbN. Here, it should be noted that the electromagnetic stress imposed on the superconducting coil becomes also 324 times larger. In a typical favourable design of the spiral coil in the case of NbN, the four units system shown in figure 1 together with a cryogenic refrigerator, gives the approximate estimate of the magnetic energy storage volume density as high as 1.8Wh/ℓ and the highest electromagnetic stress near the

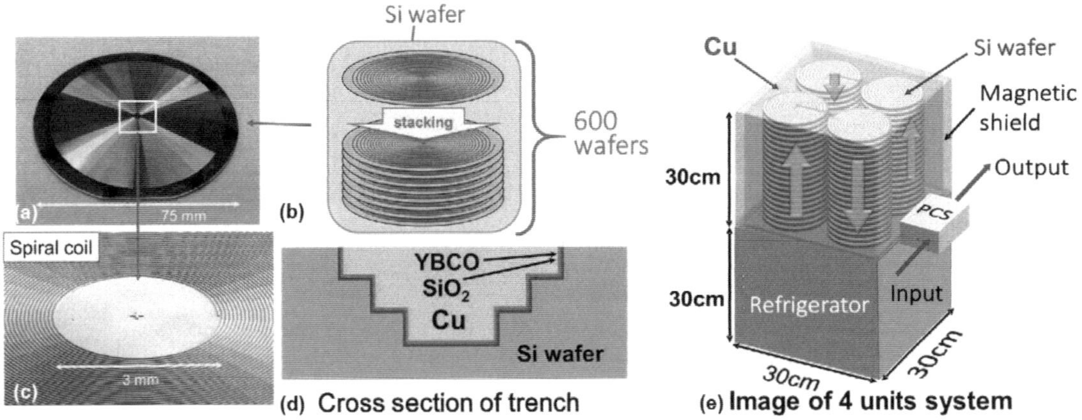

Figure 1 Superconducting coil composed of 4 coil-stack units. Each of the units is made of stacked 600 Si wafers engraved with a superconducting thin film coil in a spiral trench formed with MEMS technology.

central axis of the coil unit as high as 0.11 GPa which is well below the mechanical strength limit of a single crystalline Si: 4 GPa. However, these values are estimated to be 583 Wh/ ℓ and 35.6 GPa, respectively, by the replacement of NbN by Y123. This means the maximum superconducting current in the coil must be much reduced from the value derived from the critical current density. Then, we came on to more precise estimations of magnetic field, hoop stress and energy storage density for more concrete design of the wafer coil, and some initial results are presented here.

2. Method of estimation

2.1. Design factors of the spiral wafer coil and relationships of the related variables

Figure 2 shows a schematic illustration of about one half of the cross-section of a wafer coil design.

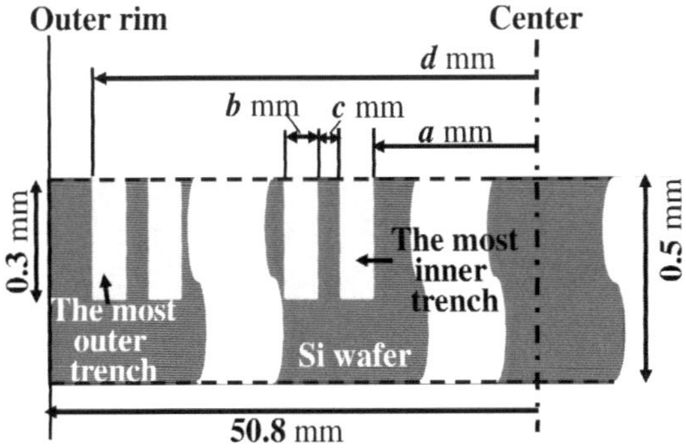

Figure 2 Schematic illustration of almost one-half of the cross-section of a wafer coil design in a Si wafer of 101.6 mm in diameter and 0.5 mm in thickness. Notations mean the inner radius of the coil: *a* mm, the trench width: *b* mm, the trench wall thickness:*c* mm and the outer radius of the coil *d* mm, The yellow rectangles indicates cross sections of trenches.

The lengths: *a*, *b*, *c* define the geometry of the coil with a spiral trenches of fixed 0.3 mm in depth. Number of turns of the spiral coil: *n* was taken as large as possible within the condition that $d = a + n\,(b+c) - c \leqq 46.5$ mm. The value: 4.3 mm was taken at the rim of the wafer coil for ease of handling. There are two types of the spiral wafer coil under the same design factors: *a*, *b*, *c*: one is left-handed and the other is right-handed. Two types of the wafer coil are to be stacked alternately by wafer-bonding technology, connecting the coils in series to form a cylindrical unit of 101.6 mm in diameter and $\ell = 0.3$ m (=0.5 mm × 600) in length, as illustrated in figure 1 (b), being composed of 600 stacked wafer coils of 0.5 mm in thickness. The total inductance *L* H was obtained using the following expression (1) for *L* for a cylindrical single-layer solenoid:

$$L=K\cdot\mu_0\cdot\pi\cdot r^2\cdot N^2/\ell. \tag{1}$$

Here, μ_0 stands for space permeability $4\pi\times10^{-7}$ H/m, r {= $(a+d)/2$} stands for a mean radius of the spiral coil, N (=600·n) stands for total number of turns of the coil, and ℓ (=0.3 m) stands for the length of the solenoid, and K stands for Nagaoka coefficient which is approximately unity in the present case: r (~33 mm) $\ll$ ℓ (=0.3 m). If s (m^2) stands for the cross-section of the superconducting film buried in the trench, ρ (m^{-3}) stands for free electron density and e stands for elementary charge: 1.60×10^{-7} C, the charge amount q (C) in the superconducting film per length dx along the trench can be expressed as follows;

$$q = e\cdot\rho\cdot s\cdot \mathrm{d}x \tag{2}$$

If v stands for the velocity of electrons, the electric current I (A) is expressed as follows;

$$I = e\cdot\rho\cdot s\cdot v \tag{3}$$

IF B (T) stands for the magnetic flux density at the trench, the Lorentz force F imposed on the current I can be expressed as follows:

$$F = q\cdot v\cdot B = I\cdot \mathrm{d}x\cdot B \tag{4}$$

Since the corresponding surface area on the trench wall is 0.0003 m $\times$dx, the hoop stress S (Pa) on the trench wall can be expressed as follows:

$$S = I\cdot \mathrm{d}x\cdot B/(0.0003\times \mathrm{d}x) = I\cdot B/0.0003 \tag{5}$$

The amount of electric energy E (J) stored in a cylindrical unit illustrated in figure 1 (b) can be expressed as follows;

$$E = L\cdot I^2/2 \tag{6}$$

Figure 3 illustrates the magnetic flux density distribution on the spiral wafer coil having a steep peak at the center of the wafer. The peak value B_{peak} may exceed the threshold value $B_{\mathrm{threshold}}$ which is

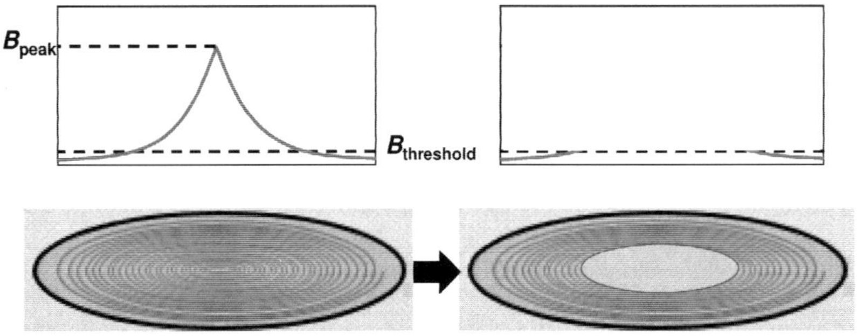

Figure 3 Schematic illustration of the magnetic flux density distribution having a steep peak B_{peak} at the center of the spiral wafer coil and required coil design to reduce the maximum magnetic flux density below the threshold value $B_{\mathrm{threshold}}$ which is determined by the maximum hoop stress or by critical magnetic flux density of the superconductor thin film.

determined by the maximum hoop stress endurable by the mechanical strength of Si wafer or determined by the critical magnetic flux density of the superconductor thin film. To reduce the maximum flux density below $B_{\mathrm{threshold}}$, optimum coil design must be explored. To do this, calculation of magnetic flux density distribution at the innermost coil at the radius a shown in figure 2 was performed changing the design parameters a, b and c in figure 2. An original calculation code (Python 3.6) based on the Biot-Savart law was developed and checked in a typical case of a circular ring current as shown in figure 4. The magnetic flux density at the center of the circular ring current can be analytically derived as $\mu_0\cdot I/(2a)$, where a is the radius of the circular current. Inserting the value for I and a, the 2.85 mT was obtained. The Python code also gave 2.85 mT, separately. Therefore, we thought it confirmed that the developed calculation code was working correctly.

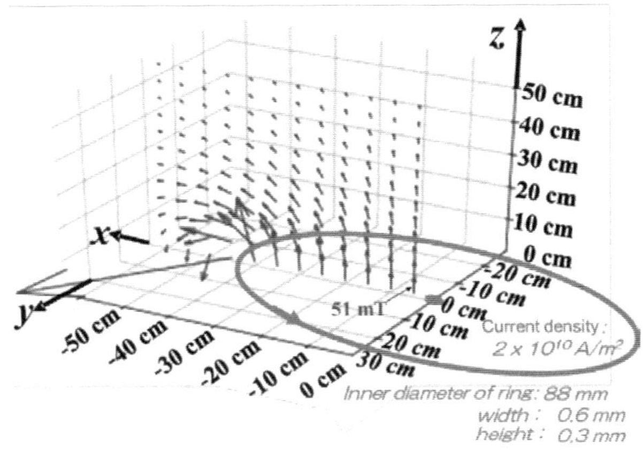

Figure 4 An example of calculation of magnetic flux density distribution indicated with blue vectors around the circular ring coil with the size and current indicated in the figure by developed Python 3.6 code.

2.2. Numerical calculation of spacial distribution of magnetic flux density

As for the calculation of the magnetic flux density distribution generated by the wafer-coil, we replaced the spiral coil by concentric circular currents with the same sizes and spacing for a calculation model. First, the magnetic flux density distribution along the line normal to the wafer coil surface on the innermost ring were calculated as shown in figure 5(a). Then, the calculated magnetic flux density distributions for 600 wafers were summed up as typically shown in figure 5 (b).

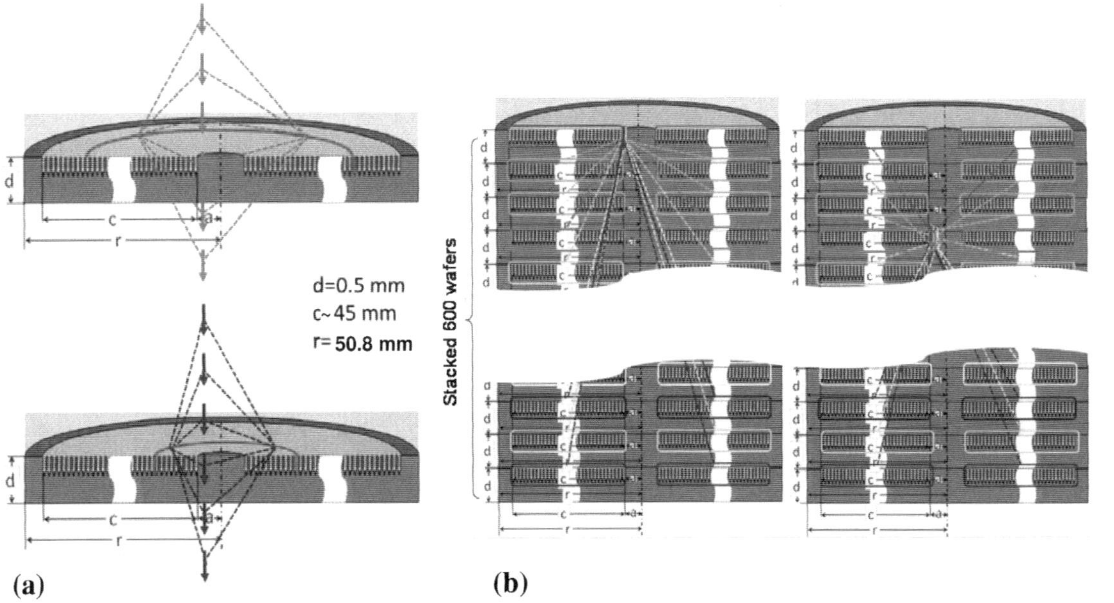

Figure 5 (a) Calculation of magnetic flux density distribution along the line normal to the one wafer coil on the innermost circular ring current, the calculated magnetic flux densities for all the concentric circular ring current were summed up. (b) Summing up of all the magnetic flux density caused by 600 stacked wafer coils.

3. Results of estimation

Figure 6 shows a typical example of the calculated result. The value of E obtained from the expression (6) was 1186 Wh/ℓ in the case of the system illustrated in figure 1 (e). Here, the critical current density of Y123 is assumed to be 2×10^{10} A/m^2. Therefore,

$$I = 2\times10^{10}\cdot s$$

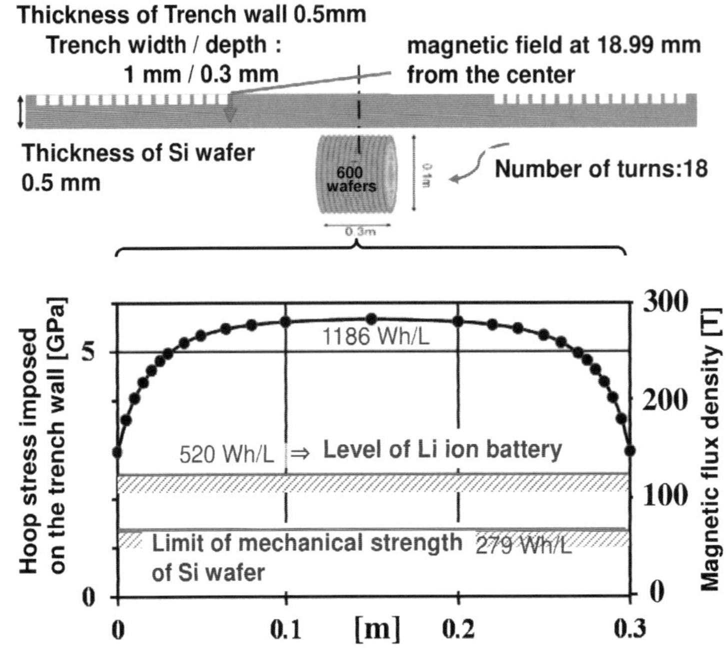

Figure 6 Example of the calculated result of the magnetic flux density for a=19 mm, b=1.0 mm, c=0.5 mm and n=18 in the notation indicated in figure 2. A schematic illustration of the wafer coil design for these a, b. c, n is given at the upper side of this figure for ease of comprehension. The plot of the calculated magnetic field is given as a function of the position along the length of 0.3 m of the cylindrical wafer coil stack unit whose schematic illustration is given just above the graph for ease of comprehension. According to the relational expression (5), the corresponding hoop stress is also given in the vertical scale on the left.

However, the hoop stress exceeded 5 GPa in the most of the region between 0 m and 0.3 m. Taking the safty factor:1/3 into account, to reduce the maximum hoop stress down to one third of the limit of the maximum stress:4 GPa for the Si wafer, that is, 1.3 GPa, E is reduced to 279 Wh/ℓ by reducing I. Here, I can be reduced by decreasing the cross-sectional area of superconductor thin film and increasing the cross-sectional area of the electroplated Cu stabilizing layer in the trench [2]. The corresponding maximum magnetic flux density is 137 T. Although the critical current density of Y123 shows a slow decrease with applied magnetic flux density, the reported data under the magnetic field does not exceed 25 T, which is far below the 137 T. To reduce this magnetic flux density down to ,for example, 20 [T], E is reduced to 5.92 Wh/ℓ. This is because E is proportional to I^2 while B is proportional to I. Table 1 summarize the calculated results for six different design factors for a,b,c

Table 1. Results of estimation for initial 6 spiral trench designs.

Design No.	a (mm)	b (mm)	c (mm)	E at 20 T	E at 1.3 GPa / B
1	13	1.5	0.5	5.33 (Wh/ℓ)	238 (Wh/ℓ) / 134 (T)
2	13	1.0	0.5	4.90 (Wh/ℓ)	281 (Wh/ℓ) / 151 (T)
3	13	0.75	0.25	5.92 (Wh/ℓ)	485 (Wh/ℓ) / 184(T)
4	19	1.0	0.5	5.92 (Wh/ℓ)	279 (Wh/ℓ) / 137 (T)
5	19	0.5	0.1	6.78 (Wh/ℓ)	781 (Wh/ℓ) / 214 (T)
6	19	0.1	0.05	6.77 (Wh/ℓ)	3082 (Wh/ℓ) / 427 (T)

including the result described above.

4. Conclusion

Method to estimate the electricity storage capacity of compact SMESs composed of stacks of Si-wafers loaded with superconducting thin film coils in spiral trenches formed by MEMS process was developed using Python 3.6 code. In initially estimated 6 designs, the calculated results show energy storage density as low as around 6 Wh/ℓ. This is because the maximum magnetic flux density limited the amount of electric energy storage rather than the maximum hoop stress. Further calculations for optimum wafer coil designs will be continued in which the effect of the maximum magnetic flux density and the effect of the maximum hoop stress on the limit of the amount of the electrical energy storage come into a good balance.

Acknowledgement

This work is partly supported by NEDO Advanced Research Program for Energy and Environmental Technology (16101979-0)

References

[1] Sugimoto N, Iguchi N, Kusano Y, Fukano T, Hioki T, Ichiki A, Bessho T and Motohiro T 2017 *Supercond. Sci. Technol.* **30** 015014
[2] Suzuki Y, Iguchi N, Adachi K, Hioki T, Ichiki A, Hsu C-W, Kumagai S, Sasaki M and Motohiro T 2017 *IOP Conf. Series: Journal of Physics: Conf. Series* **871** 012071
[3] Suzuki Y, *et al.* T 2017 *IOP Conf. Series: Journal of Physics: Conf. Series* **897**, 012019
[4] Ichiki Y, *et al.* 2018 *IOP Conf. Series: Journal of Physics: Conf. Series* **1054** 012065
[5] Sugimoto N and Motohiro T 2012 *Vacuum* **93** 13

Measurements of phase shifts in YBCO transmission lines for evaluation of kinetic inductance change

R Ishida[1], T Goto[1], H Shimakage[1], M Takeda[2]

[1] Graduate School of Science and Engineering, Ibaraki University, Hitachi, Ibaraki 316-8511, Japan
[2] Graduate School of Integrated Science and Technology, Shizuoka University, Hamamatsu, Shizuoka 432-8011, Japan

e-mail : 18nm610a@vc.ibaraki.ac.jp

Abstract. We fabricated a coplanar waveguide (CPW) using high temperature superconductor $YBa_2Cu_3O_{7-\delta}$ (YBCO) and evaluated its characteristics for the realization of a superconducting parametric amplifier. A YBCO thin film with a film thickness of 200 nm was formed by the PLD method using a YAG laser on a 10 × 10 mm MgO substrate. The YBCO thin films were processed into the CPW shape with a line width of 8 μm, a gap width of 8 μm and a line length of 5 mm by photolithography and sputter etching. Using a network analyzer, phase differences of microwave were measured at dc current bias from 0 mA to 14 mA. From the measurement results, it was found that the phase difference decreased with increasing temperature. In addition, it was found that the phase shifts greatly varied near the superconducting transition temperature at each current bias. We observed a maximum phase difference $\Delta\theta$ of -0.08 rad at a bias current of 14.0 mA, which is the closest to the transition temperature. A parametric gain was estimated to be 0.15 dB from the measured maximum phase difference.

1. Introduction

As the terahertz wave lies between the radio waves and light waves, the developments of the terahertz devices are still undeveloped. However, in the terahertz band, there are many unique spectra of many materials, which is called a fingerprint spectrum. From its characteristics, applications in various fields such as life science and security are expected [1]. At present, the use of terahertz waves has technical difficulties such as improvement of S/N ratio, and devices such as detectors and amplifiers are under development. Regarding the amplifier, HEMT (High Electron Mobility Transistor) amplifiers [2] has being been utilized in the microwave band for many years. HEMTs are the typical device as the microwave amplifier with wide band characteristic. Furthermore, they have already been put into practical use for satellite broadcasting equipment, and so on. However, its noise performance remains at several tens of times of the quantum limit. On the other hand, the operation of the superconducting parametric amplifier, which utilizes a transmission line, has been proposed recently [3-7]. In general, the superconducting transmission line has two inductances such as the kinetic inductance (L_k) and a magnetic inductance (L_m), and then a total inductance (L_t) is the sum of the two values. The L_k, which is an intrinsic inductance of the superconductor transmission line, plays a main role for the operation of the parametric amplifier. Therefore, in the superconducting parametric amplifier, the value of L_k has to be dominant over L_m. In general, the L_k has a nonlinearity with respect to applied current I as follows:

IOP Conf. Series: Journal of Physics: Conf. Series **1293** (2019) 012059 doi:10.1088/1742-6596/1293/1/012059

$$L_k(I) \approx L_k(0)\left[1+\left(\frac{I}{I^*}\right)^2\right], \tag{1}$$

where I^* is the strength of the nonlinearity [3]. In superconducting parametric amplifiers using Nb as a superconductor, it is reported that amplification of 10 dB can be obtained at 9 to 11 GHz [3]. The results showed the wideband operation, high dynamic range, and low noise performance in the microwave band. However, since the upper limit response frequency of Nb is around 700 GHz, it does not be operated in the terahertz band. Therefore, in order to realize an amplifier operation in the terahertz band, we have been studying a superconducting parametric amplifier using $YBa_2Cu_3O_7$ (YBCO) with an upper limit response frequency of around 7 THz [8]. The superconducting parametric amplifiers generally are composed of a transmission line, and are required a long and narrow signal line to increase the gain [9, 10]. In this paper, we report on the experimental procedures of the fabrication of the YBCO CPW transmission line and experimental evaluations as the parametric amplifier.

2. Fabrication of YBCO transmission lines and measurement method.

2.1. Fabrication of YBCO transmission lines

First, YBCO thin films are deposited on the MgO substrates by a pulsed laser deposition (PLD) method. After that, annealing treatment is performed in oxygen atmosphere. In the deposition process, the substrate temperature is fixed at 820 °C with an oxygen pressure of 400 mTorr. A YAG laser power density is 60 mJ/cm^2 with a repetition frequency of 10 Hz, and the distance between the target and the substrate is 3.5 cm. The annealing treatment after the deposition is carried out for 60 minutes at an oxygen pressure of 300 Torr and a temperature of 430 ° C. After that, a coplanar line with a line width of 8 μm, a gap width of 8 μm, and a line length of 5 mm is formed by photolithography and sputter-etching. The sputter-etching is performed for 70 minutes at an argon pressure of 4.0 mTorr and a supply power of 40 W. Figure 1 shows a picture of the fabricated sample. Figure 2 shows a tapered structure for preventing impedance mismatching between the coaxial cable and the coplanar line.

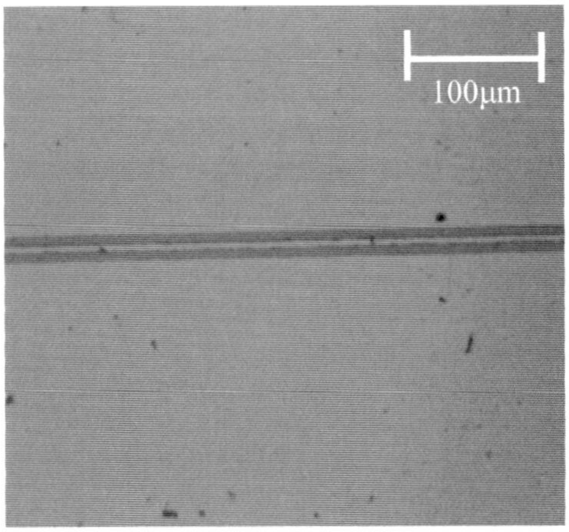

Figure 1. Microphotograph of YBCO-CPW

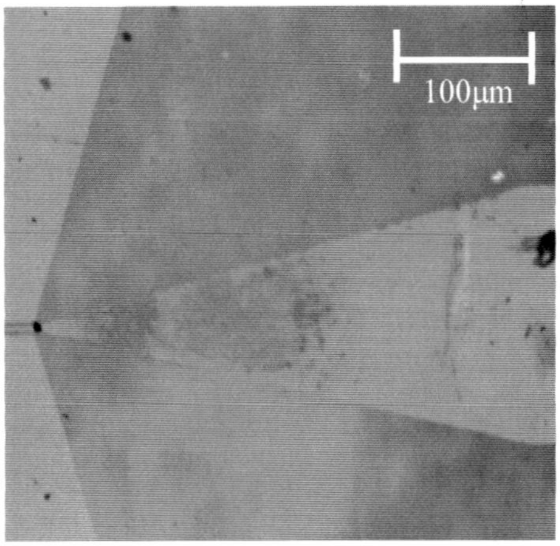

Figure 2. Microphotograph of tapered structure

2.2. Measurement method

As shown in equation (1), the L_k of the superconducting transmission line changes nonlinearly when the dc current is applied. A propagation constant consisting of inductance and capacitance also changes as follows:

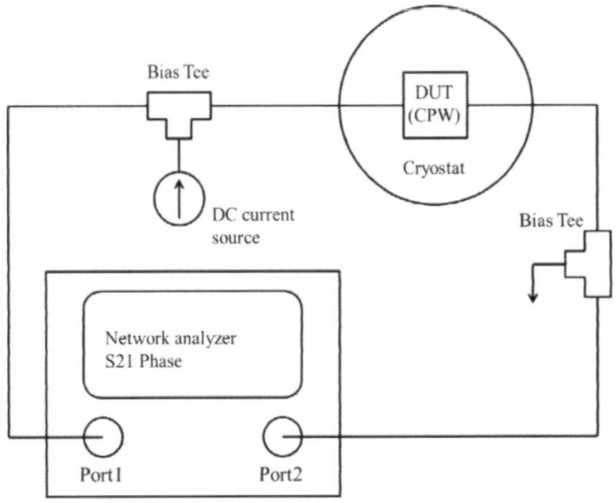

Figure 3. Measurement setup

$$\Delta\theta = [\beta(0) - \beta(I)]l = \omega\left[\sqrt{L_T(0)C} - \sqrt{L_T(I)C}\right]l. \quad (2)$$

Here, Δθ is an amount of a nonlinear phase difference and β is a propagation constant. We expected that it is possible to evaluate the changes in kinetic inductance by measuring the phase difference. Figure 3 shows a schematic configuration of the phase difference measurement. In the measurement system, the fabricated YBCO coplanar transmission line was connected to the SMA connector and cooled to a low temperature by a refrigerator. The microwave was introduced into the YBCO-CPW using a network analyzer, and dc current was supplied using a bias Tee.

3. Measurement results of phase difference on the YBCO transmission line

Using the measurement system (Fig. 3), the transmission characteristics and the phase difference of the fabricated sample were measured. In order to investigate the superconducting critical temperature of the CPW signal line, resistance-temperature (R-T) characteristics at each bias current were measured. The measurement results are shown in Figure 4. Although the measurements were made by a two-terminal measurement, the value of contact resistance was subtracted in the Figure 4. The measurement was implemented by changing the applied current from 1.0 mA to 14.0 mA. As the current value increased, a decrease of the zero-resistance temperature was observed, which reflects the critical current of the thin film. Next, the phase difference during current bias were measured with the network analyzer. The measurements were carried out at around 4 GHz and power of -10 dBm. Figure 5(a) shows temperature dependences of the phase difference when the applied current was varied from 0.0 mA to 14.0 mA. From figure 5(a), the phase difference tends to decrease as the temperature increases for all bias currents. The measurement result caused by the modulation of the kinetic inductance due to temperature change. Moreover, the phase difference greatly changes around the superconducting transition temperature for all bias currents. Figure 5(b) shows an enlarged view from 20 K to 30 K in Figure 5(a). When the temperature was about 25 K, a large phase difference with the current bias of 14.0 mA was observed.

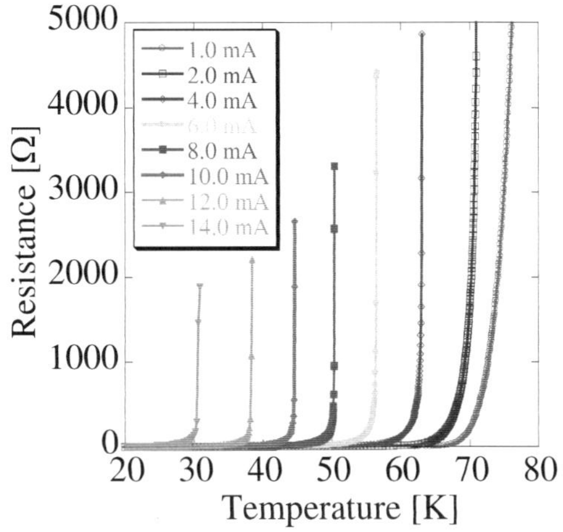

Figure 4. R-T measurement in various current biases

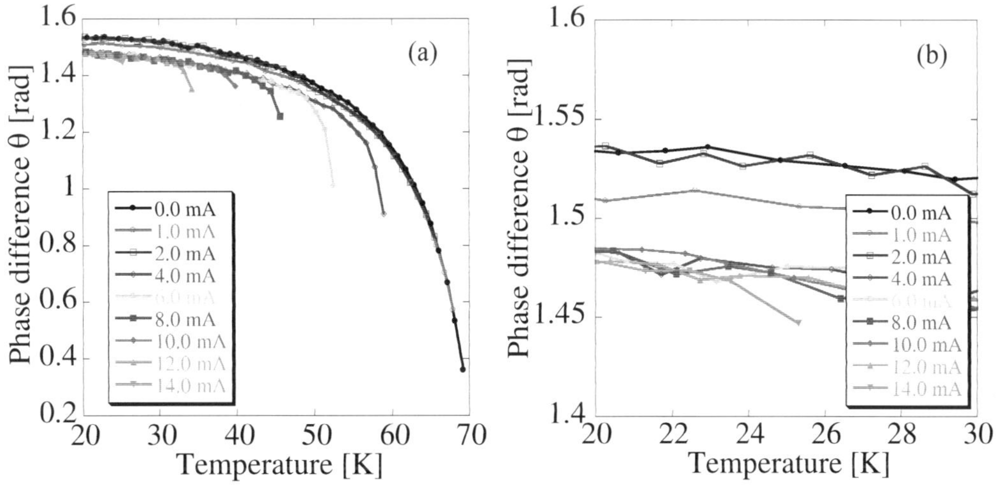

Figure 5. (a) The temperature dependences of the phase shifts when the bias currents are applied from 0.0 mA to 14.0 mA. (b) An enlarged view of Figure 3(a) from 20 K to 30 K.

Next, the bias current dependence of the phase difference amount Δθ at 25 K is shown in Figure 6. Here, the plots were normalized with the phase difference Δθ of 0 rad at bias current of 0 mA. We observed the phase difference Δθ of -0.08 rad at 14.0 mA, which was at the closest to the transition temperature. A solid line in the figure 6 is the phase difference Δθ numerically calculated from the shape of YBCO-CPW using a theoretical equation [11]. The experimental results were very close to the numerical calculations. Therefore, this phase difference Δθ is thought to be due to the kinetic inductance modulation by the current bias. Similar measurements are made at different temperatures, and amplification gains were calculated from equation [3];

$$G_s = 1 + (\Delta\theta)^2. \quad (3)$$

At 51 K, the maximum phase difference was observed. Figure 7 shows the parametric amplification gains calculated from the equation (3) with the measured phase difference. Although the highest gain was calculated of 0.15 dB as shown in Figure 7, the value was obtained just below the superconducting transition temperature. It implies that a dynamic range of the signal is thought to be very small for assuming the practical operation. Improvement of coplanar waveguide structure is essential, and the fabrication of CPW with narrower line-width and longer line-length has been being under study. Also, since the amplification gain is proportional to the length of the transmission line from the expression (2), signal gain would be the higher with the longer transmission line.

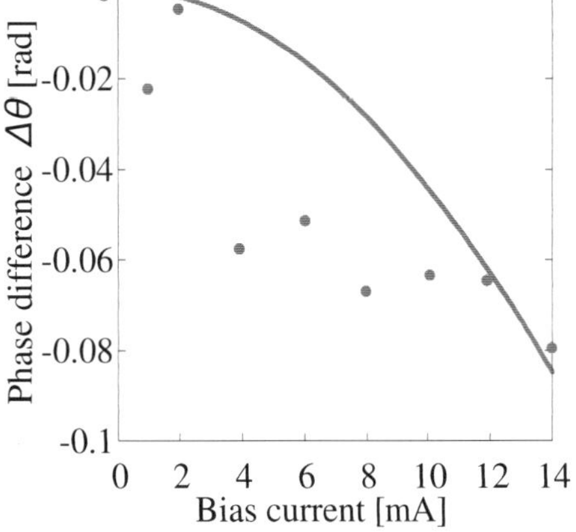

Figure 6. Bias current dependence of phase difference Δθ at 25 K. The closed circles are measured values. The solid line is numerical calculation.

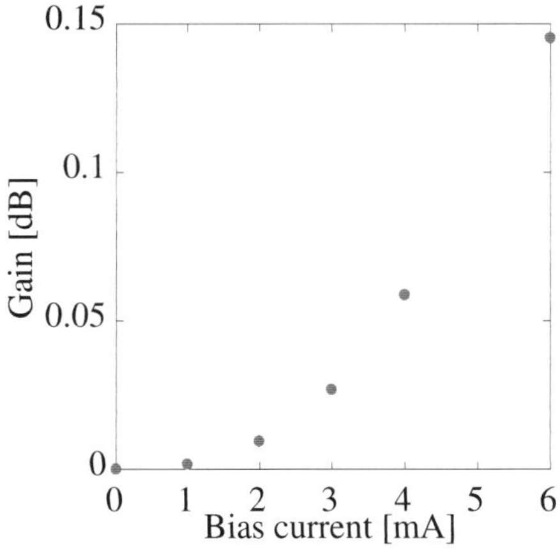

Figure 7. Parametric amplification gain calculated from phase shift Δθ at 51 K.

4. Conclusion

In this research, in order to realize the YBCO superconducting parametric amplifiers which have the potential of the terahertz band operation, the results of phase difference measurements of the YBCO CPWs were reported. The CPW with the line width of 8 μm and the line length of 5 mm was successfully fabricated, and the phase difference were observed under applying the DC bias currents. Although the phase differences were small at low temperature, the large phase differences were obtained near the critical temperature at each bias current. The results are roughly in agreement with that of numerical calculation by the theoretical kinetic inductance behavior. From the results, it implies that the phase difference obtained by the measurements are attribute to the modulation of the kinetic inductance. Moreover, using the measurement results of the phase difference at 51 K, the expected parametric gain was estimated to be 0.15 dB under bias current of 6.0 mA. However, it was impossible to obtain large phase difference especially for small bias currents. To get the large kinetic inductance modulation, the line width and the length of the CPW should be smaller and longer, respectively. In order to realize the YBCO parametric amplifier, although the required practical gain depends on each application, we are proceeding the CPW configuration improvement. Moreover, as it is important to realize a phase matching for the parametric amplifier operation, we are designing the modulation structure of the CPW.

References

[1] Federici J F, Schulkin B, Huang F, Gary D, Barat R, Oliveira F and Zimdars D 2005 *Semiconductor Science and Technology*. **20** S266
[2] Akgiray A H, Weinreb S, Leblanc R, Renvoise M, Frijlink P, Lai R and Sarkozy S 2013 *IEEE Trans. Microw*. **61** 3285-97
[3] Eom B Ho, Day P K, LeDuc H G and Zmuidzinas J 2012 *Nature Phys*. **8** 623-7
[4] Bockstiegel C, Gao J, Vissers M R, Sandberg M, Chaudhuri S, Sander A, Value L R, Irwin K D and Pappas D P 2013 *Journal of Low Temperature Physics*. **176** 476-82
[5] Chaudhuri S, Gao J and Irwin K 2015 *IEEE Transactions on Applied Superconductivity*. **25** 1500705
[6] Vissers M R, Erickson R P, Ku H S, Vale L R, Wu X, Hilton G C and Pappas D P 2016 *Applied Physics Letters*. **108** 012601

[7] Adamyan A A, de Graaf S E, Kubatkin S E and Danilov A V 2016 *Journal of Applied Physics.* **119** no. 8
[8] Goto T, Takeda M and Shimakage H 2017 The 78th JSAP Autumn Meeting, 6a-PB1-45
[9] Takeda M 2015 *IEICE Technical Report* **114** no. 406, SCE2014-50, pp. 7-12
[10] Hansryd J, Andrekson P A, Westlund M, Li J and Hedekvist P O 2002 *IEEE J. Selected Topics in Quantum Electron.* **8** 506-20

IOP Conf. Series: Journal of Physics: Conf. Series **1293** (2019) 012060 doi:10.1088/1742-6596/1293/1/012060

Design and operation of distributed double-SQUID amplifier for RSFQ circuits

Komei Higuchi, Hiroshi Shimada and Yoshinao Mizugaki
Department of Engineering Science, The University of Electro-Communications (UEC Tokyo), 1–5–1 Chofugaoka, Chofu, Tokyo, 182–8585, Japan
E-mail: higuchi@w8-7f.ee.uec.ac.jp, y.mizugaki@uec.ac.jp

Abstract. Recently, many groups have been developing on-chip Rapid Single-Flux-Quantum (RSFQ) output amplifiers which are designed to connect RSFQ circuitry to room temperature electronics. One of such amplifiers is a distributed SQUID amplifier comprising SQUIDs embedded in cascaded RSFQ RS flip-flops. In this paper, we propose a distributed double-SQUID amplifier, where a SQUID is replaced with a double-SQUID. Test circuit was fabricated using a niobium integration technology. Measurements were performed in a liquid helium bath. We confirmed that the maximum output voltage swing of a 4-stage distributed double-SQUID amplifier was 2.93 mV, which was comparable with the corresponding numerical value of 3.54 mV.

1. Introduction

Rapid Single-Flux-Quantum (RSFQ) circuitry [1] is an attractive digital technology and expected to bear the next-generation high-speed digital systems. This is because it has great advantage to low power consumption and high-speed operation. However, we have to implement superconductor–semiconductor interfaces between RSFQ circuitry and room temperature electronics, because the energy of a single flux quantum (SFQ) is relatively small. For the above reason, many groups have been developing on-chip RSFQ amplifiers which are designed to connect RSFQ circuitry to room temperature electronics.

RSFQ amplifiers can be classified according to their bias types, AC bias and DC bias. Table 1 is a list of RSFQ output amplifiers such as a Suzuki stack [2], a 4JL gate [3], a HUFFLE driver [4], a SQUID array [5] and a distributed single-SQUID amplifier (DSSA) with their bias types. Demonstrated performances reported in literature such as the maximum voltages swing (V_{max}) and maximum operation frequencies (f_{max}) are also listed with their Nb technologies.

In general, a distributed amplifier is a cascaded amplifiers coupled with signal transmission lines to obtain both wide bandwidths and high gains. For superconducting RSFQ circuitry, a 12-stage distributed amplifier was demonstrated, of which the output voltage swing was 1.75 mV for 10 Gbps data link [6]. In this paper, we refer to this distributed amplifier as a distributed single-SQUID amplifier (DSSA) for the comparison with our amplifier described below.

To improve the output voltage swing of distributed amplifiers, there are two choices available. One choice is to increase the number of stages, because the output voltage swing is expected to be proportional to the number of stages. This choice, however, requires more area occupancies and power consumption. The other choice is to increase the output voltage swing per stage if possible.

Table 1. Comparison of RSFQ output amplifiers.

	Suzuki Stack [2]	4JL [3]	HUFFLE [4]	SQUID Array [5]	DSSA [6]
Bias type	AC	AC	DC	DC	DC
V_{max} [mV]	60	2.3	1.5	3.0	1.75
f_{max} [GHz]	1.0	2.0	4.0	0.12	10
Nb tech. [kA/cm^2]	4.5	2.5	1.0	1.5	4.5
reference	[7]	[8]	[9]	[10]	[6]

In this paper, for the latter method, we propose a distributed double-SQUID amplifier (DDSA), where a SQUID in a DSSA is replaced with a double-SQUID. A double-SQUID is expected to generate output voltage swings twice as large as those of a single SQUID. Below, we present our circuit design, numerical results, and experimental results.

2. Development of a fundamental circuit stage in a distributed amplifier

Figure 1(a) shows the equivalent circuit of a fundamental circuit stage in a DSSA.

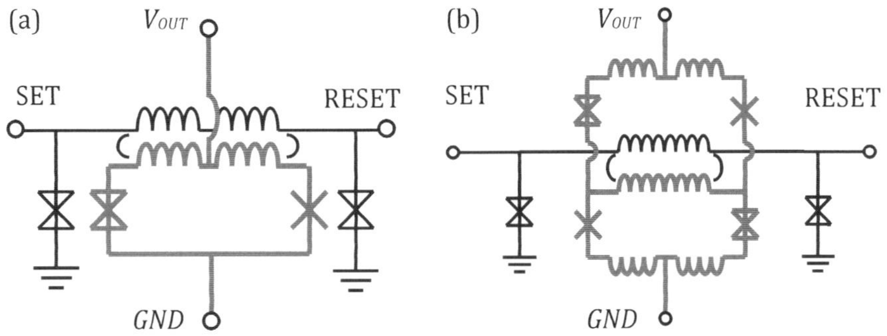

Figure 1. Equivalent circuit of a single stage in a distributed single-SQUID amplifier (DSSA) (a) and a distributed double-SQUID amplifier (DDSA) (b). "X" and hourglass symbols represent a unshunted (under-damped) junction and a critically-damped junction, respectively.

One single stage consists of two parts. One part is an SFQ storage loop with two (SET and RESET) terminals, and the other part is a SQUID coupled to the storage loop. Input SFQ pulses are transferred alternatively to either SET or RESET terminal. Alternative transmission is realized through a TFF that is presented later. When an SFQ comes from the SET terminal, it is stored in the storage loop. Then, the critical current of the SQUID is reduced and its output voltage increases. We call this state the SET state and it is a finite voltage state. Next, when an SFQ comes from the RESET terminal, it annihilates the stored SFQ. The critical current of the SQUID then returns to the original value and the output voltage decreases. We call this state the RESET state and it is the zero-voltage state. The output signal is a voltage swing between the SET and RESET state and these operations realize non-return-to-zero (NRZ) signals. In the first DSSA paper [6], the circuit layouts were designed to realize the characteristic impedance of 50 Ω, whereas termination resistors of 50 Ω was also integrated on-chip.

We propose a DDSA where SQUIDs in a DSSA are replaced with double-SQUIDs as shown in Fig. 1(b), where two stacked SQUID sharing one sensing inductor are magnetically coupled to

IOP Conf. Series: Journal of Physics: Conf. Series **1293** (2019) 012060 doi:10.1088/1742-6596/1293/1/012060

the storage loop. That is, two-fold output voltage is expected by introducing the double-SQUID structure [11].

3. Numerical comparison between DSSA and DDSA

3.1. Device parameters of a single stage in a DSSA and a DDSA
Figures 2 and 3 show the device parameters of a single stage in a DSSA and a DDSA.

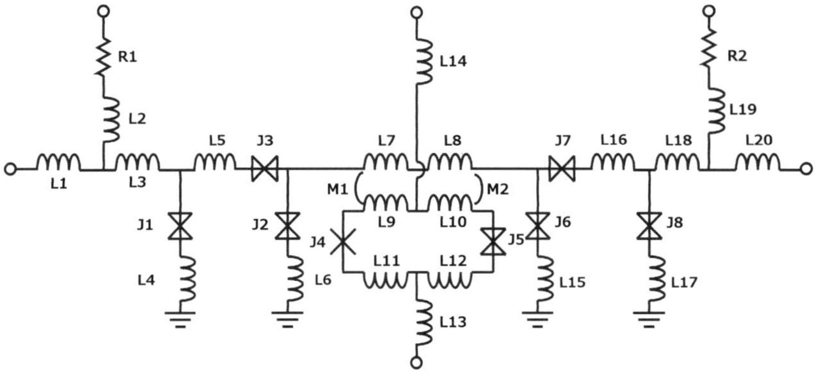

Figure 2. Equivalent circuit and parameters of a stage in a DSSA. Parameters are determined as follows using an optimization tool. The values of external shunt resistors are presented next to the critical currents of Josephson junctions. $L1 = L4 = L15 = L17 = L20 = 0.50$ pH, $L2 = L19 = 3.50$ pH, $L3 = 2.00$ pH, $L5 = 2.58$ pH, $L6 = 0.46$ pH, $L7 = L8 = 5.70$ pH, $L9 = L10 = 4.70$ pH, $L11 = L12 = 0.32$ pH, $L13 = 2.40$ pH, $L14 = 1.60$ pH, $L16 = 4.00$ pH, $L18 = 3.00$ pH, $J1 = J8 = 0.22$ mA (1.71 Ω), $J2 = J6 = 0.13$ mA (2.83 Ω), $J3 = J7 = 0.20$ mA (1.84 Ω), $J4 = 0.14$ mA, $J5 = 0.14$ mA (2.66 Ω), $R1 = 12.92$ Ω, $R2 = 15.66$ Ω, $M1 = M2 =$ 3.62pH ($k = 0.70$).

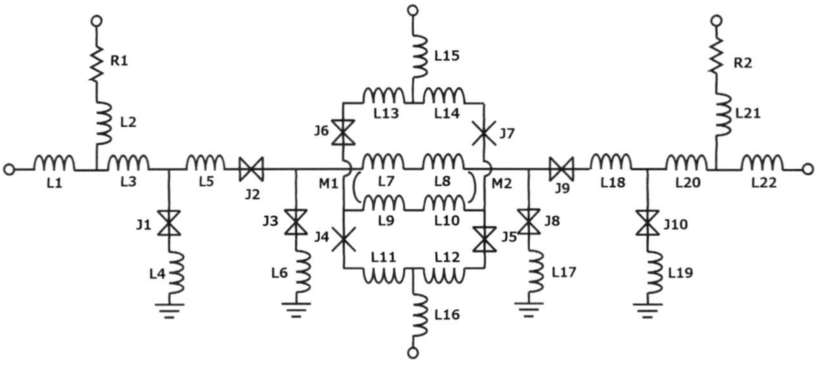

Figure 3. Equivalent circuit and parameters of a stage in a DDSA. Parameters are determined as follows using an optimization tool. The values of external shunt resistors are presented next to the critical currents of Josephson junctions. $L1 = L4 = L17 = L19 = L22 = 0.50$ pH, $L2 = L21 = 3.50$ pH, $L3 = L18 = L20 = 2.00$ pH, $L5 = 2.38$ pH, $L6 = 0.46$ pH, $L7 = L8 =$ 4.09 pH, $L9 = L10 = 4.12$ pH, $L11 = L12 = 1.80$ pH, $L13 = L14 = 1.20$ pH, $L15 = 1.60$ pH, $L16 = 2.40$ pH, $J1 = J10 = 0.22$ mA (1.71 Ω), $J2 = J9 = 0.20$ mA (1.84 Ω), $J3 = J8 = 0.13$ mA (2.83 Ω), $J4 = J7 = 0.10$ mA, $J5 = J6 = 0.10$ mA (3.73 Ω), $R1 = 17.27$ Ω, $R2 = 20.37$ Ω, $M1 = M2 = 3.20$ pH ($k = 0.73$).

We determined the parameters that realized the correct DSSA and DDSA operation by using a Josephson simulator (JSIM) [12] and an optimization tool SCOPE2 [13]. $J3$ and $J7$ in Fig. 3 as well as $J2$ and $J9$ in Fig. 3 play a role of escape junctions, which were not implemented in the original work of a DSSA [6]. Larger mutual coupling between the storage loop and SQUID was desirable. At the same time, it was important to make the LI_C products of the SQUID as small as possible for wider operating margins. Therefore, during the optimization, we kept the critical current of the SQUIDs to 0.10mA (the minimum value allowed in the fabrication process described below) and changed inductances.

3.2. Numerical results of a 4-stage DSSA and a 4-stage DDSA

Figure 4 shows numerical results of the bias current (I_{bias}) versus output voltage of a 4-stage DSSA and a 4-stage DDSA. Results for the SET state ($V_{OUT,SET}$) and RESET state ($V_{OUT,RESET}$) are plotted. The voltage of the DDSA is roughly twice of that of the DSSA. Figure 5 presents the bias current versus output voltage swing, the voltage difference calculated by subtracting $V_{OUT,RESET}$ from $V_{OUT,SET}$. (That is, $V_{swing} = V_{OUT,SET} - V_{OUT,RESET}$.)

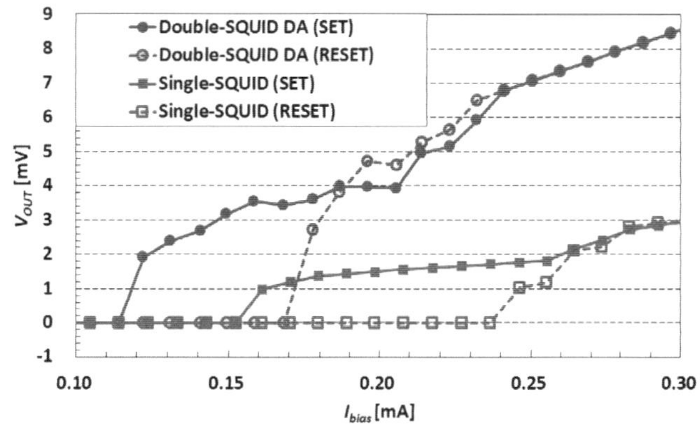

Figure 4. Numerical results of I_{bias} vs. $V_{OUT,SET}$ and $V_{OUT,RESET}$ characteristics of a 4-stage DSSA and a 4-stage DDSA.

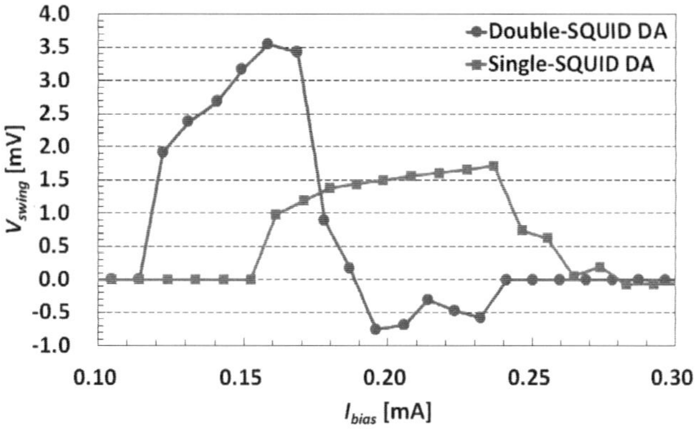

Figure 5. Numerical results of I_{bias} vs. V_{swing} characteristics of a 4-stage DSSA and a 4-stage DDSA.

The maximum output voltage swing of the 4-stage DSSA and the 4-stage DDSA are 1.71 and 3.54 mV, respectively. The V_{swing} ratio of the DDSA and DSSA is 2.07, as same as expected from the double-SQUID configuration. It is found in figure 5 that V_{swing} becomes negative for I_{bias} from 0.19 to 0.24 mA, which is because $V_{OUT,RESET}$ is greater than $V_{OUT,SET}$. (We did not pay any attention to such reversed magnitude relation during the parameter optimization.)

When we assumed that the 4-stage DDSA worked correctly until voltage swing decreased 3 dB from those of maximum value, numerical simulation indicates that the 4-stage DDSA would work for the input signal up to 9.5 Gbps. This value is 95% of that of the previous work [6] listed in Table 1. Because we focused on the voltage swing of the 4-stage DDSA, high-speed performance was put aside during the parameter optimization. Competing goals for the large voltage swing and high-speed operation will be balanced in the future.

4. Experimental characteristics of a 4-stage DDSA

4.1. Layout design and chip fabrication

We designed a single DDSA stage in an 80 × 80 μm^2 area connectable to cells in the CONNECT cell library [14]. InductEX [15] was used for extracting inductances of superconducting lines and coupling coefficients. To increase the coupling coefficient between the storage loop and double-SQUIDs, a hole in the ground plane layer was introduced, resulting in the coupling coefficient k of 0.73 in calculation.

For a test circuit, we designed a 4-stage DDSA, of which the configuration is shown in Fig. 6. The entire size of a 4-stage DDSA was approximately 680 × 1300 μm^2 including a dc/SFQ terminal.

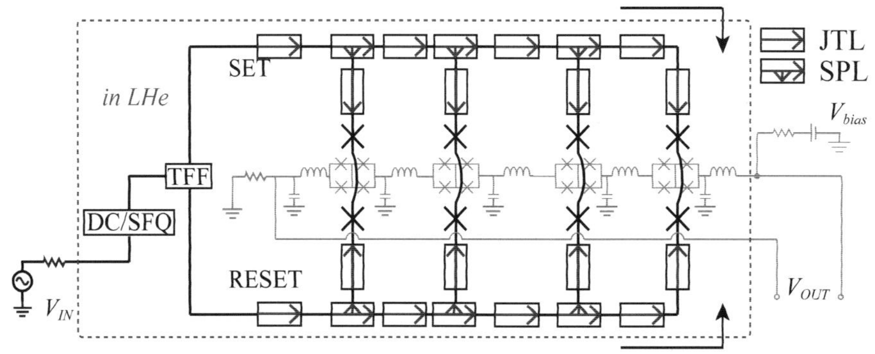

Figure 6. Configuration of the designed 4-stage DDSA with the measurement setup. The input SFQ frequency was set at 1 kHz.

Test chips were fabricated by using the 2.5 kA/cm^2 Nb/AlO$_x$/Nb integration process (the STP2 process of National Institute of Advanced Industrial Science and Technology, Japan). Figure 7 shows a micrograph of one stage in a 4-stage DDSA.

4.2. Bias current–output voltage swing characteristics

In measurements, a test chip was cooled down in a liquid helium bath. Measurement setup is illustrated in Fig. 6. The output voltage swing was acquired with an oscilloscope via a 100-fold low-noise preamplifier.

Figure 8 shows the bias current versus the output voltage swing between the SET and RESET states (I_{bias} vs. V_{swing} characteristics). The circles and triangles in Fig. 8 represent

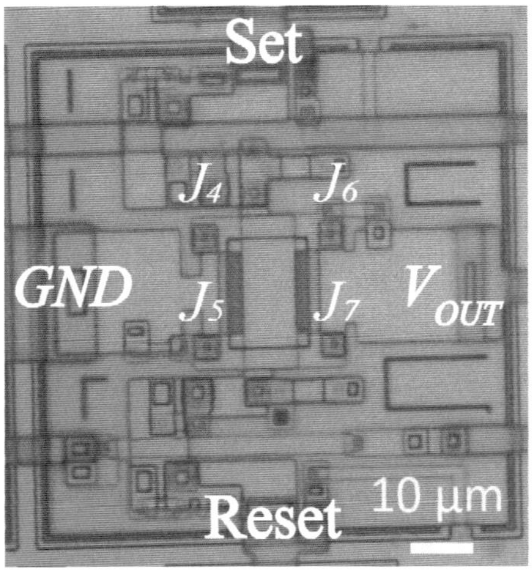

Figure 7. Photomicrograph of one stage in a 4-stage DDSA.

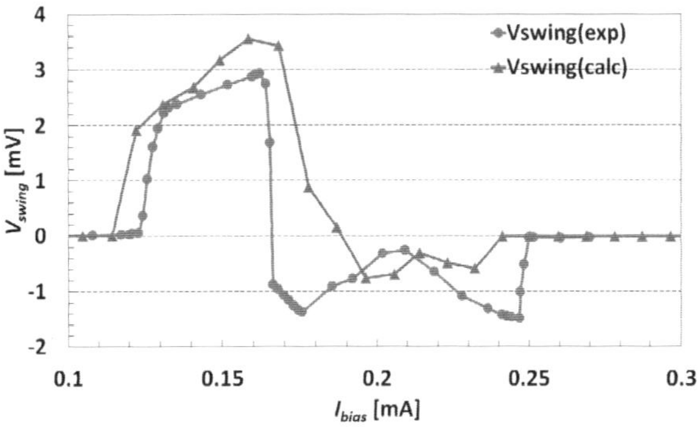

Figure 8. Experimental and numerical results of I_{bias} vs. V_{swing} characteristics of a 4-stage DDSA.

the experimental and numerical results, respectively. The maximum V_{swing} was 2.93 mV in experiments, which was comparable with the corresponding numerical value of 3.54 mV. In addition, the experimental I_{bias}–V_{swing} characteristics agreed well with numerical results, which indicated that the 4-stage DDSA worked as expected.

5. Conclusion

We designed and operated a 4-stage DDSA. A SQUID in the original DSSA was replaced with double-SQUIDs, by which two-fold output voltage was realized. One DDSA stage was designed to be 80 × 80 μm^2 and compatible with the CONNECT cell library. In experiments, the maximum output voltage swing was 2.93 mV, which was comparable with the corresponding numerical value of 3.54 mV.

Acknowledgement

The authors thank the lab members in UEC Tokyo for their fruitful discussion and technical support. This research work was partly supported by JSPS Grant-in-Aid for Scientific Research 17K04979, and by VLSI Design and Education Center (VDEC), the University of Tokyo in collaboration with Cadence Design Systems, Inc. Test circuits used in this study were fabricated using the AIST–STP2 process in the Superconducting Clean Room (CRAVITY) at the National Institute of Advanced Industrial Science and Technology (AIST). The stable supply of liquid helium from the Coordinated Center for UEC research Facilities is also acknowledged.

References

[1] Likharev K K and Semenov V K 1991 *IEEE Trans. Appl. Supercond.* **1** 3
[2] Suzuki H, Imamura T and Hasuo S 1990 *IEEE Trans. Electron Dev.* **37** 2399
[3] Nakagawa H, Sogawa E, Kosaka S, Takada S and Hayakawa H 1982 *Japan. J. Appl. Phys.* **21** L198
[4] Hebard A, Pei S, Dunkleberger L and Fulton T 1979 *IEEE Trans. Magn.* **15** 408
[5] Welty R P and Martinis J M 1991 *IEEE Trans. Magn.* **27** 2924
[6] Herr Q P 2010 *Supercond. Sci. & Technol.* **23** 022004
[7] Van Duzer T, Zheng L, Whiteley S R, Kim H, Kim J, Meng X and Ortlepp T 2013 *IEEE Trans. Appl. Supercond.* **23** 8
[8] China F, Takeuchi N, Ortlepp T, Yamanashi Y and Yoshikawa N 2018 *IEICE technical report* **117** 33
[9] Schneider D F, Lin J C, Polonsky S V, Semenov V K and Hamilton C A 1995 *IEEE Trans. Appl. Supercond.* **5** 3152
[10] Huber M E, Neil P A, Benson R G, Bums D A, Corey A, Flynn C S, Kitaygoradskaya Y, Massihzaueh O, Martinis J M and Hilton G C 2001 *IEEE Trans. Appl. Supercond.* **11** 4048
[11] Morooka T 1997 *Japan. J. Appl. Phys.* **36** L1587
[12] Fang E S and Van Duzer T 1989 *A Josephson integrated circuit simulator (JSIM) for superconductive electronics application* (Tokyo: The Japan Society of Appl. Phys.) p 407
[13] Mori N, Akahori A, Sato T, Takeuchi N, Fujimaki A and Hayakawa H 2001 *Physica C* **357–360** 1557
[14] Yorozu S, Kameda Y, Terai H, Fujimaki A, Yamada T and Tahara S 2002 *Physica C* **378–381** 1471
[15] Fourie C J and Perold W J 2005 *IEEE Trans. Appl. Supercond.* **15** 300

IOP Conf. Series: Journal of Physics: Conf. Series **1293** (2019) 012061 doi:10.1088/1742-6596/1293/1/012061

Torque Simulation on NI REBCO Pancake Coils during Quench

So Noguchi[1] and Seungyong Hahn[2]
[1] Graduate School of Information Science and Technology, Hokkaido University, Kita 14 Nishi 9, Kita-ku, Sapporo 060-0814, Japan.
[2] Department of Electrical and Computer Engineering, Seoul National University, Seoul 08826, Republic of Korea.

E-mail: **noguchi@ssi.ist.hokudai.ac.jp**

Abstract. This paper presents the unbalanced torque generated in No-Insulation (NI) $REBa_2Cu_3O_x$ (REBCO, RE = Rare Earth) pancake coils after a normal-state transition. The NI REBCO pancake coil has a high thermal stability, and it is desired to apply to ultra-high field magnets. When an NI REBCO pancake coil locally transitions into a resistive state, the operating current bypasses a normal-state-transitioned turn to adjacent turns in the coil-radial direction, and the amount of heat generation can be reduced. That is an inherent feature of NI REBCO pancake coils. However, when a radial current flows in a magnetic field, a Lorentz force is generated toward the coil-circumferential direction, i.e. torque. In case that the coil radius and the external magnetic field are very large, the extremely large torque is generated, and it could damage the magnet itself and/or the joints to terminals with a mechanical factor. Therefore, it is necessary to investigate the behavior of the torque generated in the NI REBCO pancake coil after the quench.

In this paper, we present the numerical simulation results of the behaviors of the torque generated in the NI REBCO double pancake coil after quench. Their current, thermal, and mechanical behaviors are simulated by a partial element equivalent circuit (PEEC) method and 2-D thermal finite element method (FEM). The behaviors are different depending on the coil size. In addition, the behavior is presented when the operating current is suddenly shut down after quench.

1. Introduction

$REBa_2Cu_3O_x$ (REBCO, RE = Rare Earth) pancake coils using a no-insulation (NI) winding technique [1, 2] has received attention very much. The NI winding technique greatly enhances the thermal stability of REBCO pancake coils, and it is desired to be applied to high magnetic field NMR/MRI magnet, etc. [3, 4]. The high thermal stability of NI REBCO magnets was confirmed in many experiments and numerical simulations, even though any accident caused a quench of NI REBCO magnets due to mechanical or thermal factors. When NI REBCO pancake coils locally transition into a normal state, an operating current bypasses resistive turns to adjacent turns in the radial direction. It reduces the amount of heat generation, and it brings on a high thermal stability. That is an inherent feature of NI REBCO pancake coils. However, when a radial current flows in a magnetic field, a Lorenz force is generated. The force works toward circumferential direction, i.e. torque. In the case that a large-bore coil operates at a large amount of current in a large magnetic field, a strong torque can be generated after quench, as shown in Fig. 1. In such a case, coils or joints may be damaged by unbalanced torque.

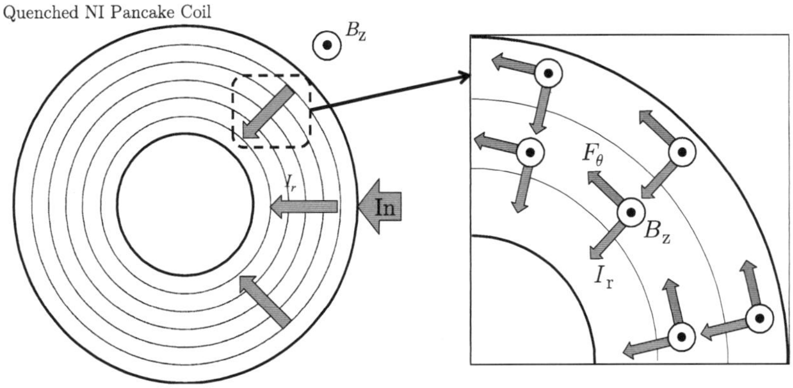

Figure 1. Schematic diagram of torque generation in quenched NI pancake coil.

So far, an overcurrent test or quench test of NI REBCO pancake coils under a magnetic field higher than 10 T has not been reported. Hence, there is no report about unbalanced torque generated in NI REBCO pancake coils. In this paper, we will discuss unbalanced torque of NI REBCO pancake coils in high magnetic field after quench.

In order to investigate the torque behavior, the numerical simulation is performed. In this paper, the current, thermal, and mechanical behaviors of the NI REBCO double pancake coil are simulated by a partial element equivalent circuit (PEEC) method and a 2-D thermal finite element method (FEM) [5, 6, 7]. The coil size affects the unbalanced torque behavior. The torque behavior is also shown when an operating current is cut off after quench.

2. Simulation Method and Models

In this paper, the current, thermal, and mechanical behaviors of NI pancake coil were simulated by a PEEC method and a 2-D thermal FEM [5, 6, 7]. Firstly, with the PEEC method, the current and electromagnetic behaviors are obtained. It derives the amount of Joule heating in the NI coil, and then the thermal behavior is simulated with the 2-D thermal FEM.

The circumferential Lorentz force generated in one partial element F_θ is calculated by the follow equation:

$$F_\theta = I_r B_z l_t \tag{1}$$

where, I_r, B_z, and l_t are the radial current flowing in a partial element, the axial magnetic field, and the length of the partial element in the radial direction, i.e. the REBCO tape thickness, respectively. The torque is the sum of the products of the radial coordinate and (1).

Table 1 lists the parameters of NI pancake coil model. The simulated model is a double pancake coil wound with SuperPower 2G HTS tape [8] using the NI winding technique, and each pancake coil has 100 turns. The equivalent circuit model of double pancake coil consists of two single pancake coils connected at innermost with a crossover turn as shown in Fig. 2. The single pancake coils are electrically and thermally insulated each other, however the heat diffuses via the crossover turn.

In the thermal simulation, an adiabatic condition is supposed. Since the magnitude of the torque varied depending on the bore diameter of coil, the simulation was performed on four different coils with radii of 50, 100, 500, and 1000 mm. Table 2 lists the simulation condition. It is supposed that the whole lower pancake coil transitions into a normal state, operating at 20 K. The magnet is operated at 500 A, and an axial external magnetic field of 10 T is applied by an outsert magnet.

Table 1. Parameters of REBCO tape and double pancake coils.

Parameters	Values
REBCO tape width (mm)	4.0
REBCO tape thickness (mm)	0.096

Copper stabilizer thickness (μm)	20.0
REBCO layer thickness (μm)	1.0
I_c @ 77 K, self-field (A)	115.0
Coil inner radius (mm)	50, 100, 500, 1000
Number of turns (each pancake coil)	100
Coil thickness (mm)	9.6
Turn-to-turn contact resistivity (μΩ•cm^2)	70.0 [9]

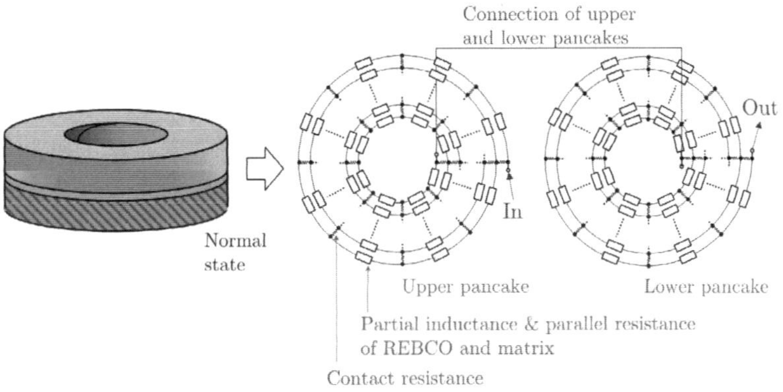

Figure 2. Schematic drawing of simulated NI double pancake coil. Equivalent circuit model of double pancake coil consists of two single pancake coils connected at innermost portion with crossover turn. It is supposed that all of lower pancake becomes normal state in simulation.

Table 2. Simulation conditions.

Parameters	Values
Simulation time step (s)	0.1
Simulation end time (s)	50
Operating current (A)	500
Axial & radial external magnetic field (T)	10.0, 0.0
Operating temperature (K)	20.0

3. Simulation Results

In this section, the simulation results of NI REBCO pancake coils after quench are presented. Since the results show different behaviors depending on the coil size, the results will be explained individually by the coil radii.

3.1. Small-Size Coils

Fig. 3 shows the total torque and the average circumferential current transition of each pancake coil at inner radii of (a) 50 and (b) 100 mm. In the graph, a counterclockwise direction is defined as positive.

In the quenched lower pancake coil, since almost of current flows toward the radial direction, the torque is generated immediately after the quench, and the strength of torque maintains afterward. The large torque appears as the coil radius is larger.

In the upper pancake coil, the torque is generated in the same direction as the lower one at the first time. This is why the current in the upper pancake coil is induced in the same radial direction as the radially-bypassing current in the lower pancake. Therefore, with the decrease of the induced current, the torque gradually decreases. The smaller the coil radius is, the faster the decay speed of torque, because the contact resistance is larger. In addition, for the 50-mm-radius pancake coils, the torque gradually increases in the counterclockwise direction after it becomes zero once.

IOP Conf. Series: Journal of Physics: Conf. Series **1293** (2019) 012061 doi:10.1088/1742-6596/1293/1/012061

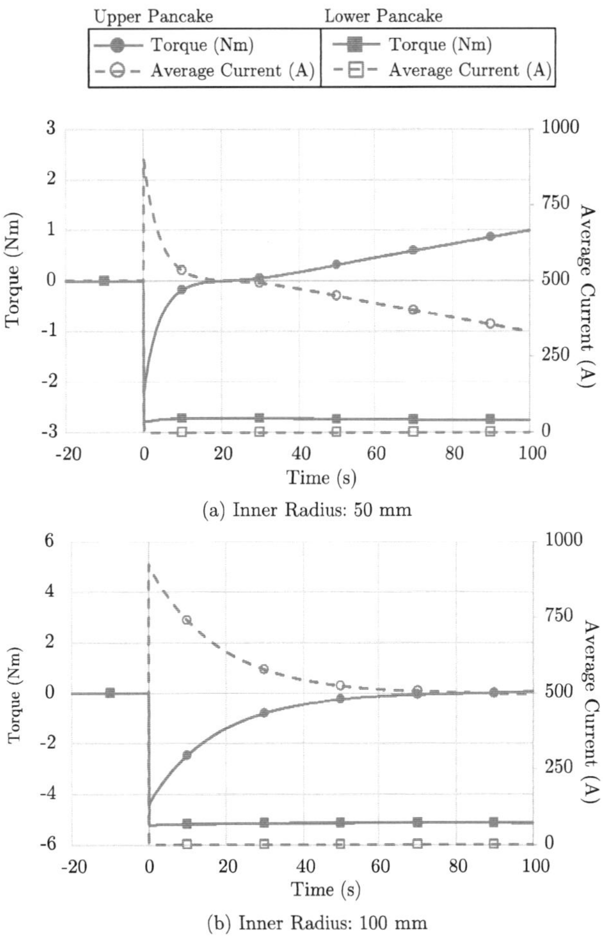

Figure 3. Torque and average circumferential current transition when inner radius is (a) 50 and (b) 100 mm. In graph, a counterclockwise direction is positive.

Fig. 4 shows the distributions of the circumferential Lorenz force in the NI pancake coils with an inner radius of 50 mm, at (a) 10 s and (b) 100 s after the appearance of normal-state transition. At 10 s, the Lorenz forces act in the same direction in both the upper and lower pancakes; however, at 100 s, the force direction oppositely changes in the inner turns of the upper pancake. This is because the normal-state transition occurs due to an induced current in the upper pancake.

Figs. 5 and 6 show the radial current distribution and the temperature distribution at 100 s, respectively. The temperature rise can be seen at the inner turns of the upper pancake. Thus, the normal-state transition occurs in the region where the radial current flows toward the inside of the coil. Thereat, the large Lorenz force is generated in the positive direction. The heat propagates from inside to outside in the upper pancake, and the torque in the positive direction increases. That is the reason of the positive torque generation. Moreover, the radial current in the lower coil concentrates into the outer current terminal, so that the Lorentz force concentrates there. Since an electric terminal is mechanically weak, a locally strong Lorentz force would deteriorate the mechanical stability.

3.2. Large-Size Coils

Fig. 7 shows the total torque and the average circumferential current as a function of time in the cases that the inner radii are (a) 500 and (b) 1000 mm, respectively, where "positive" also means the counterclockwise direction. For these large-size coils, since the torque exceeds 20 Nm, the mechanical stability of coils would decrease. Compared with small-size pancake coils presented above, the torque generated in the upper pancake decays more slowly. However, when the coil radius is 1000 mm, the sudden inversion of torque can be seen.

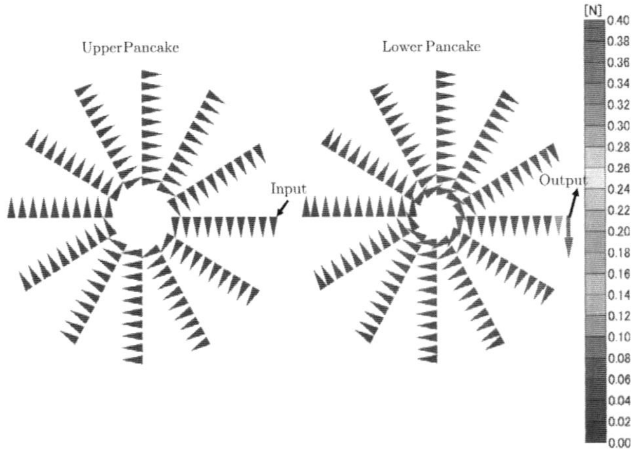

(a) At 10 s

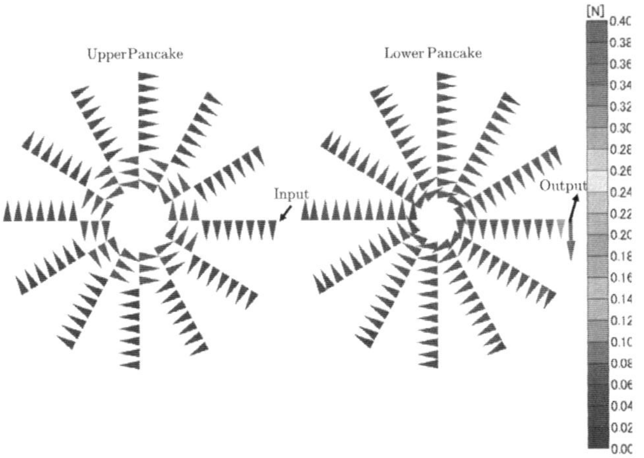

(b) At 100 s

Figure 4. Distribution of circumferential Lorentz force applied in NI pancake coils when inner radius is 50 mm at (a) 10 s and (b) 100 s after appearance of normal-state transition (not to scale).

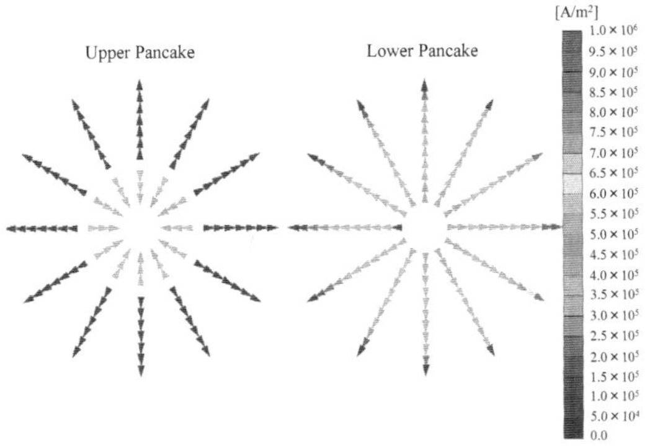

Figure 5. Distribution of radial current density when inner radius is 50 mm at 100 s after appearance of normal-state transition (not to scale).

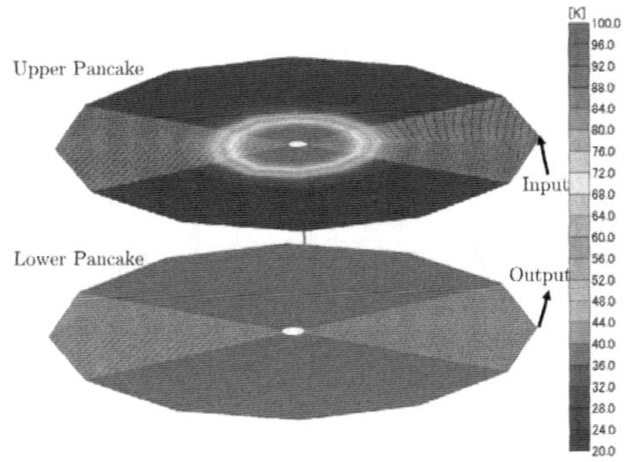

Figure 6. Distribution of temperature when inner radius is 50 mm at 10 s after appearance of normal-state transition (not to scale).

Next, Fig. 8 shows the circumferential current distributions at a coil radius of 1000 mm, on the same conditions in Fig. 4. In the upper pancake, the Lorenz force works in the negative direction at

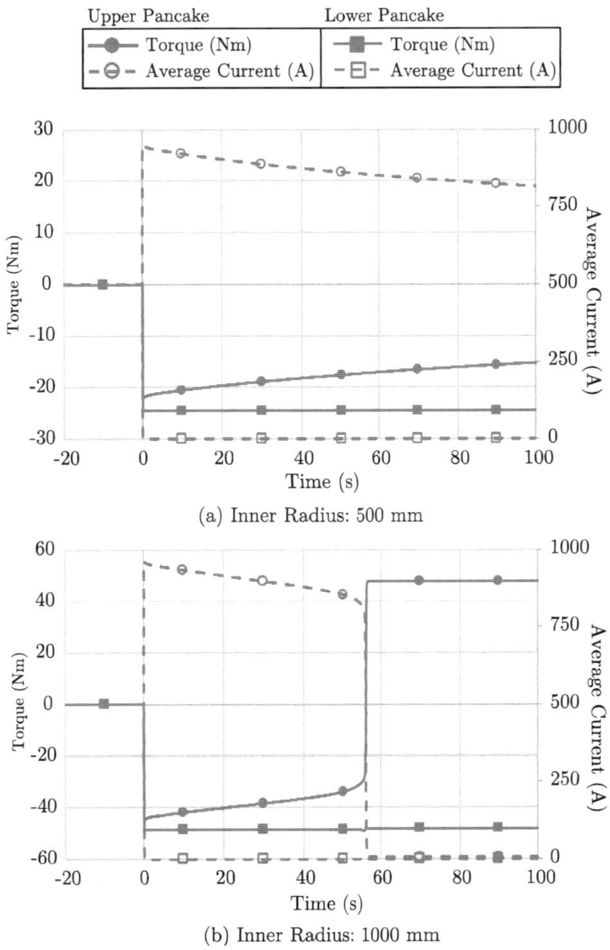

Figure 7. Torque and average circumferential current transition when inner radius is (a) 500, (b) 1000 mm. In graph, a counterclockwise direction is positive.

10 s, however it entirely changes in the positive direction at 100 s. Figs. 9 and 10 present the radial current density and temperature distributions at 100 s, respectively. When the coil radius is very large, the current straightly flows from the outer terminal to the inner crossover turn, and the Joule heat concentrates there. When a normal-state transition occurs transversely from outside to inside, the current does not circumferentially carry in the rest area of the superconducting state. As a result, the radial current toward the inside increases in the coil, and the torque direction changes immediately after a local normal-state transition.

4. Conclusion

This paper shows a risk of torque generated in high magnetic field NI REBCO pancake coils during quench. A radial current in an NI pancake coil invokes a circumferential Lorentz force. Thereby, a rotational force is generated in the NI coil. When a magnet has a large bore radius and is operated in large magnetic field, it is possible that a torque large enough to cause damage to magnet is generated.

We also performed the numerical simulation of an NI double pancake coil after a quench by PEEC and 2-D FEM. When a quench occurred, a torque was generated in coils not quenched too, due to the induced current. It gradually decreased with the decrease of induced current. The faster the decay speed of torque, the smaller the coil radius was. Here, a normal-state transition occurred due to heating by the induced current. For a small-size coil, the torque in the direction opposite to the torque initially generated was gradually generated. For a large-size coil, the sudden inversion of torque was generated.

When the current was shut off immediately, the distribution of Lorentz force changed immediately, but the total strength of them did not change at that time. Thus, it was considered that a sudden shutdown was hardly caused to break a coil.

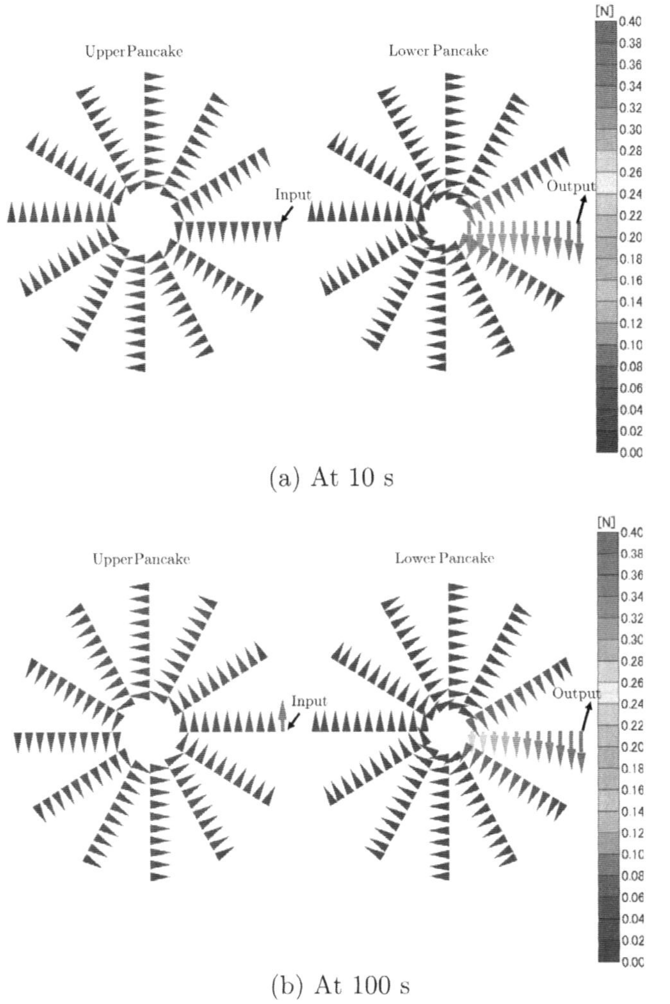

(a) At 10 s

(b) At 100 s

Figure 8. Distribution of circumferential Lorentz force applied in NI pancake coils when inner radius is 1000 mm at (a) 10 s and (b) 100 s after appearance of normal-state transition (not to scale).

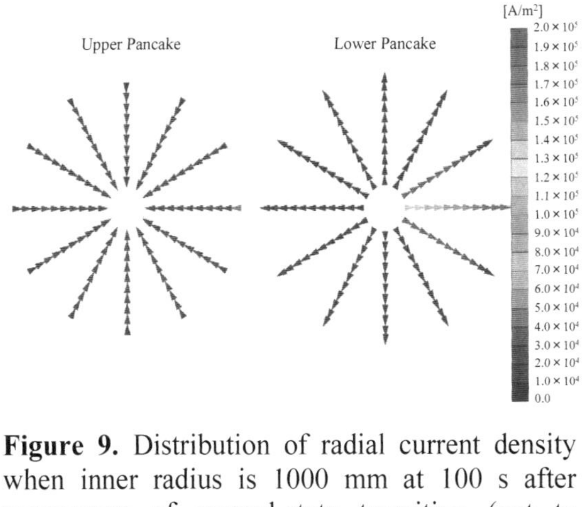

Figure 9. Distribution of radial current density when inner radius is 1000 mm at 100 s after appearance of normal-state transition (not to scale).

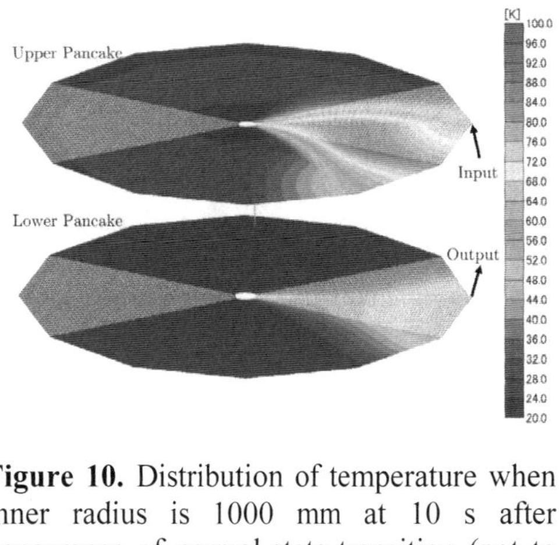

Figure 10. Distribution of temperature when inner radius is 1000 mm at 10 s after appearance of normal-state transition (not to scale). Temperature near terminal is lower than around since it is connected with copper lead which is low temperature.

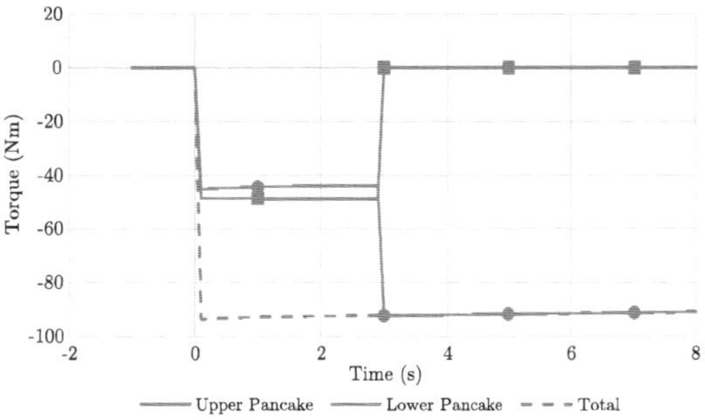

Figure 11. Transition of torque when current is cut off at 3 s after appearance of normal-state transition. Coil inner radius is 1000 mm.

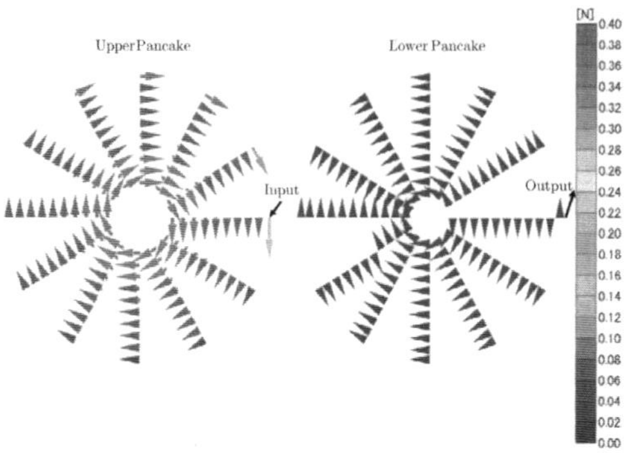

Figure 12. Distribution of circumferential Lorentz force at 3 s from current interruption. Coil inner radius is 1000 mm (not to scale).

Acknowledgments
This work was supported in part by JSPS KAKENHI grant number 15KK0192.

References

[1] Kim S B, Saito A, Kaneko T, Joo J H, Jo J M, Han Y J and Jeong H S 2012 The characteristics of the normal-zone propagation of the HTS coils with inserted Cu tape instead of electrical insulation *IEEE Trans. Appl. Supercond.* **22** 4701504

[2] Hahn S, Park D K, Bascunan J and Iwasa Y 2011 HTS pancake coils without turn-to-turn insulation *IEEE Trans. Appl. Supercond.* **21** pp 1592–1595

[3] Iwasa Y, Bascunan J, Hahn S, Voccio J, Kim Y, Lecrevisse T, Song J and Kajikawa K 2015 A high-resolution 1.3-GHz/54-mm LTS/HTS NMR magnet *IEEE Trans. Appl. Supercond.* **25** 4301205

[4] Kim K L, Song J, Choi Y H, Yang D G and Lee H 2017 Feasibility Study of a No-Insulation 1.5-T/600- mm All-REBCO Magnet for MRI Systems *IEEE Trans. Appl, Supercond.* **27** 4600204

[5] Wang T, Noguchi S, Wang X, Arakawa I, Minami K, Monma K, Ishiyama A, Hahn S and Iwasa Y 2015 Analyses of Transient Behaviors of No-Insulation REBCO Pancake Coils During Sudden Discharging and Overcurrent *IEEE Trans. Appl. Supercond.* **25** 4603409

[6] Oki T, Ikeda A, Wang T, Ishiyama A, Noguchi S, Monma K, Watanabe T and Nagaya S 2016 Evaluation on Quench Protection for No-Insulation REBCO Pancake Coil *IEEE Trans. Appl. Supercond.* **26** 4702905

[7] Noguchi S, Miyao R, Monma K, Igarahishi H, Ueda H and Ishiyama A 2017 Current Behavior Simulation in Stacked NI REBCO Pancake Coils During Local Normal- State Transition *IEEE Trans. Appl. Supercond.* **27** 4603205

[8] 2018. [Online]. Available: http://www.superpower-inc.com/content/2g-htswire

[9] Wang X, Hahn S, Kim Y, Bascunan J, Voccio J, Lee H and Iwasa Y 2013 Turn-to-turn contact characteristics for an equivalent circuit model of no-insulation ReBCO pancake coil *Supercond. Sci. Technol.* **26** 035012

Early Detection of Normal Transitions in a High Temperature Superconducting Transformer Wound with a Plurality of HTS Tapes Using the Active Power Method

N Nanato and H Aoyama

Graduate School of Natural Science and Technology, Okayama University, 3-1-1 Tsushima Naka, Kita-ku, Okayama, 700-8530, Japan

nanato@okayama-u.ac.jp

Abstract. The authors have been developing an AC power source with a high temperature superconducting (HTS) transformer which is small in size and can supply large AC current. The HTS transformer needs a detection system of normal transitions for high reliability. The authors have reported the active power method as the detection system. However, the conventional method could not detect early the normal transition in a secondary coil of the transformer which was wound with a bundle of some HTS tapes since resistive voltage in the normal area is too small due to current sharing among the HTS tapes. In this paper, the authors propose a method to detect the normal transitions in the bundle conductor by improving the active power method. The early and correct detection can be achieved by measuring each active power of each HTS tape. In order to measure the active power, current leads are attached independently to all the HTS tapes. The current leads enable to measure the resistive voltage in each HTS tape, and each active power can be measured. The proposed method works well when even one of secondary tapes transits to the normal conducting state. Experimental results show the usefulness of the proposed method.

1. Introduction

The authors have reported the active power method as the detection system of the normal transitions so far [1, 2]. The authors also have presented a small and lightweight AC current supply to grasp current conduction characteristics of an HTS sample tape [1-3]. The supply has an HTS transformer of which secondary coil is wound with a bundle of some HTS tapes [1-3]. For safe operation of the transformer, normal transitions in the transformer must be detected without any delay. However, resistive voltages due to the normal transitions become very small in the bundle wire by current sharing among the HTS tapes. Therefore the normal transitions in the secondary coil are difficult to be detected by the conventional active power method. In this paper, the authors propose an improved method to detect the normal transitions in the secondary coil. If the active power consumed in each secondary tape can be measured, the normal transitions in the secondary coil can be detected regardless of the current sharing. In order to achieve the measurement, secondary current leads are connected independently to all secondary tapes. The current leads enable to measure the voltage of each tape. The active power consumed in each tape can be detected by using the voltage of each tape. The authors show its usefulness through experiments for a hybrid HTS transformer [2] as a test sample.

2. Detection method of normal transitions in a bundle superconductor

Figure 1 shows an equivalent circuit converted to a primary side of a hybrid HTS transformer of which primary coil is a copper coil and secondary coil is wound with a bundle of HTS tapes, and Table 1 shows parameters in the figure. Number of the secondary HTS tapes is two for simplification. In the conventional active power method, the voltage signal "$v_{SC1} - v'_{SC2}$" is calculated to detect the normal transitions [2]. When only one secondary tape transits to the normal state, the current through the tape decreases by current sharing. Then the voltage signal "$v_{SC1} - v'_{SC2}$" has a much smaller resistive voltage than the inductive voltage. Also in order to detect the resistive voltage, the inductive voltages of the both two tapes must be canceled. However the both inductive voltages are difficult to be derived from the voltage signal "$v_{SC1} - v'_{SC2}$" unless the currents $\beta_1 i'_2$ and $\beta_2 i'_2$ are independently measured. Therefore the normal transitions in the secondary coil are difficult to be detected by the conventional active power method when even one of secondary tapes is in the superconducting state.

The authors propose an improved method by connecting a current lead to each secondary tape and measuring each voltage of each secondary tape. Figure 2 shows an improved circuit for the detection of the normal transitions in a bundle wire. Newly added parameters in the figure are shown in Table 2. Each secondary tape is connected to each current lead of which resistance is R'_{21} or R'_{22}. The current leads are what a current lead connecting the secondary coil and the load is divided into two, therefore $R' = R'_{21}R'_{22}/(R'_{21} + R'_{22})$.

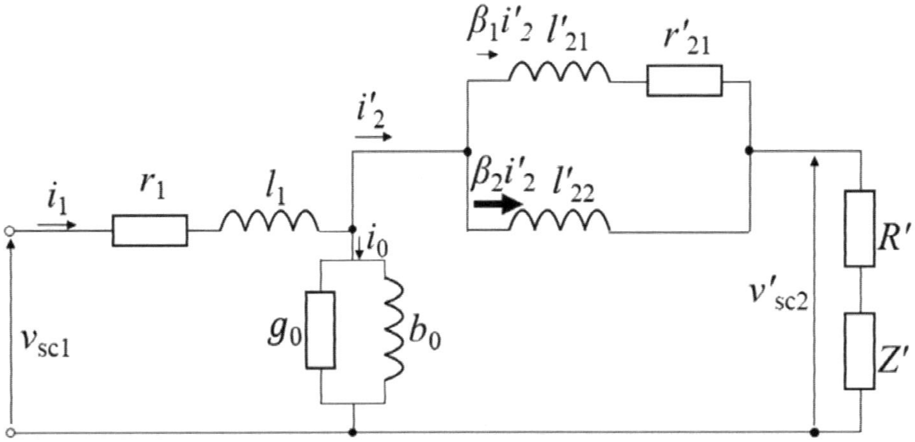

Figure 1. An equivalent circuit converted to a primary side of a hybrid HTS transformer.

Table 1. Parameters in Figure 1.

v_{SC1}	: primary voltage	v'_{SC2}	: secondary voltage
		R'	: resistance of a current lead connecting the secondary coil and the load
i_1	: primary current	Z'	: load impedance
r_1	: primary resistance	i_0	: excitation current
l_1	: primary leakage inductance	g_0	: excitation conductance
r'_{21}	: resistance of the normal area	b_0	: excitation susceptance
l'_{21} and l'_{22}	: each secondary leakage inductance for each secondary tape	$\beta_1 i'_2$ and $\beta_2 i'_2$	: each secondary current for each secondary tape, (β_1, β_2: coefficients, $\beta_1 + \beta_2 = 1,\ \beta_1 > 0, \beta_2 > 0$)

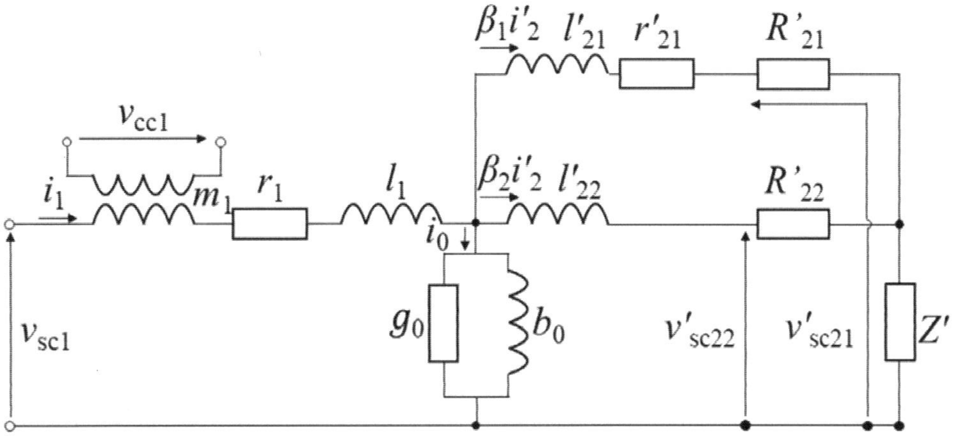

Figure 2. An improved circuit for the detection of the normal transitions in a bundle wire.

Table 2. Newly added parameters in Figure 2.

v_{CC1}	: voltage of the cancel coil	R'_{21} and R'_{22}	: each resistance of each current lead
m_1	: mutual inductance of the cancel coil	v'_{SC21} and v'_{SC22}	: each secondary voltage of each secondary tape

Following equations are obtained from Figure 2.

$$v_{SC1} - v'_{SC21} = \Delta v_1 = r_1 i_1 + l_1 \frac{di_1}{dt} + l'_{21}\beta_1 \frac{di'_2}{dt} + r'_{21}\beta_1 i'_2 \,. \tag{1}$$

$$v_{SC1} - v'_{SC22} = \Delta v_2 = r_1 i_1 + l_1 \frac{di_1}{dt} + l'_{22}\beta_2 \frac{di'_2}{dt} \,. \tag{2}$$

A following equation is derived from summation of (1) and (2).

$$\Delta v_1 + \Delta v_2 = 2r_1 i_1 + 2l_1 \frac{di_1}{dt} + (\beta_1 l'_{21} + \beta_2 l'_{22}) \frac{di'_2}{dt} + r'_{21}\beta_1 i'_2 \,. \tag{3}$$

Since $\beta_1 + \beta_2 = 1$, $l'_{21} \fallingdotseq l'_{22}$ and $i'_2 = i_1$, equation (3) is expressed as a following equation.

$$\Delta v_1 + \Delta v_2 = 2r_1 i_1 + (2l_1 + l'_{21}) \frac{di_1}{dt} + r'_{21}\beta_1 i_1 \,. \tag{4}$$

Since r_1 is a known resistance and a voltage of the cancel coil $v_{CC1} = m_1 di_1/dt$, the resistive voltage is derived as a following equation.

$$\Delta v = \Delta v_1 + \Delta v_2 - k v_{CC1} - 2r_1 i_1 = r'_{21}\beta_1 i_1, \tag{5}$$

where $k = (2l_1 + l'_{21})/m_1$.
The active power P' is calculated as follows [2].

$$P = \Delta v i_1 = r'_{21}\beta_1 i_1^2 = r_{21}\beta_1 i_2^2, \tag{6}$$

$$P' = \frac{P}{1 + sT}, \tag{7}$$

where $r'_{21} = a^2 r_{21}$, $i_2 = a i_1$ (a: turn ratio). P is an instantaneous active power in the normal area and P' is a filtered P by a low pass filter to reduce superposed noise. Equation (7) is expressed by Laplace transform and T is a time constant of the filter. P' is a parameter to detect the normal transitions.

This method is useful for 3 or more secondary tapes as well.

3. Experimental results

The authors carried out experiments on the detection of the normal transitions in the secondary coil. Configuration and specifications of the test transformer are shown in Figure 3 and Table 3, respectively. The primary and secondary coils are respectively a copper and YBCO ones. The YBCO secondary cable is a bundle of two YBCO tapes (Tape 1 and Tape 2) which are SCS4050 by Superpower. Width of the tape is 4 mm and thickness is 0.05 mm. The secondary coil was cooled at 77 K in LN_2. Amplitude of the secondary transport current was 100 Apeak and frequency 60 Hz. A heater was installed on each tape to cause the normal transitions.

v_{SC1}, v'_{SC21}, v'_{SC22}, v_{CC1} and i_1 were measured to calculate the equations from (1) to (7). The measurement and calculation were carried out by NI cRIO with FPGA chassis by National Instruments. Calculation programs of the equations from (1) to (7) were prepared by using LabVIEW software and the programs were written in the FPGA. Specifications of voltage input modules in the system were as follows: Maximum input voltage $\pm$ 10 V, Sampling rate 1 MS/s, Resolution 16-bit. Since v_{SC1} was larger than 10 V, $v_{SC1}/10$ was inputted to the module by a voltage divider.

Figure 4 shows experimental results for the normal transitions at only tape 1. The normal transitions occurred at 10 s. Temperature T_1 in the normal area of tape 1 increased gradually and temperature T_2 of tape 2 was kept at 77 K. Then P' increased gradually as well as T_1. The temperature and active power slowly increased since the current of tape 1 decreased after the normal transitions.

Figure 5 shows experimental results for the normal transitions at both two tapes. The normal transitions occurred at 11 s. Both temperatures T_1 and T_2 increased rapidly compared to Figure 4 and P' increased as well as the both temperatures. The temperatures and active power increased rapidly

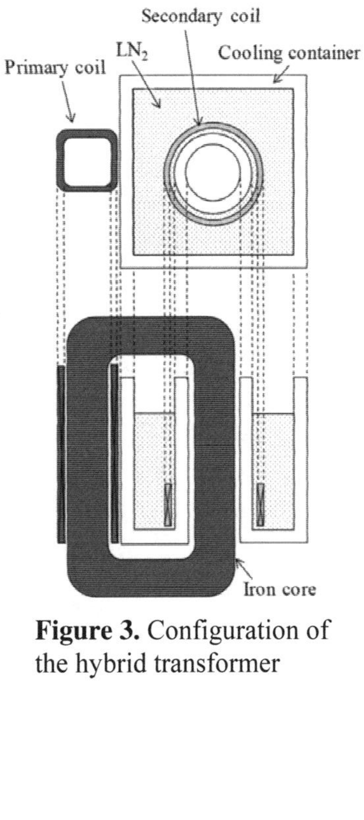

Figure 3. Configuration of the hybrid transformer

Table 3. Transformer Specifications.

	Primary coil (copper)	Secondary coil (YBCO (SCS4050))
Inner diameter	54.0 mm	89.0 mm
Outer diameter	58.0 mm	89.4 mm
Height	220.0 mm	25.0 mm
Number of turns	120	2
Resistance	392 mΩ	-
Self-inductance	395.9 mH	0.168 mH
Mutual inductance	6.55 mH	
Critical currents (Tape 1, Tape 2)	-	96, 95 A (A bundle of 2 tapes)

since the currents of tape 1 and 2 were constant before and after the normal transitions. In Figure 4, the temperature of tape 1 was around 160 K at 20 s, and P' is 1 W. In Figure 5, the temperatures of both tapes are around 200 K at 20 s, and P' is 13 W. This is because the normal area was heated by the

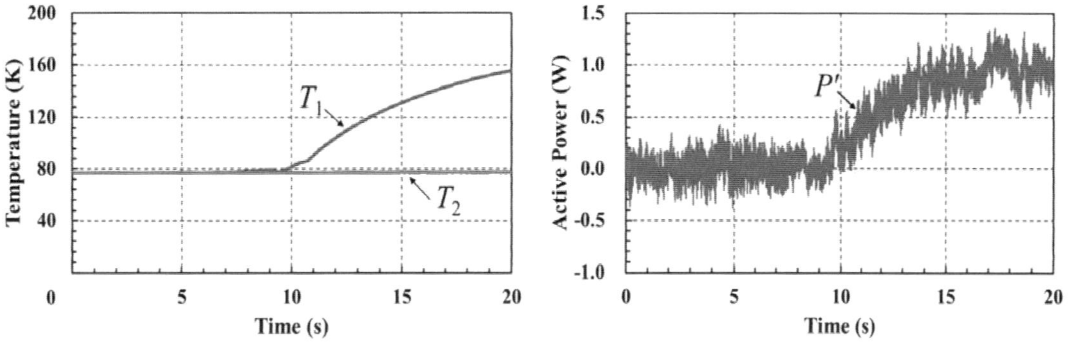

(a) Temperatures (b) Active power

Figure 4. Experimental results for the normal transitions at tape 1.

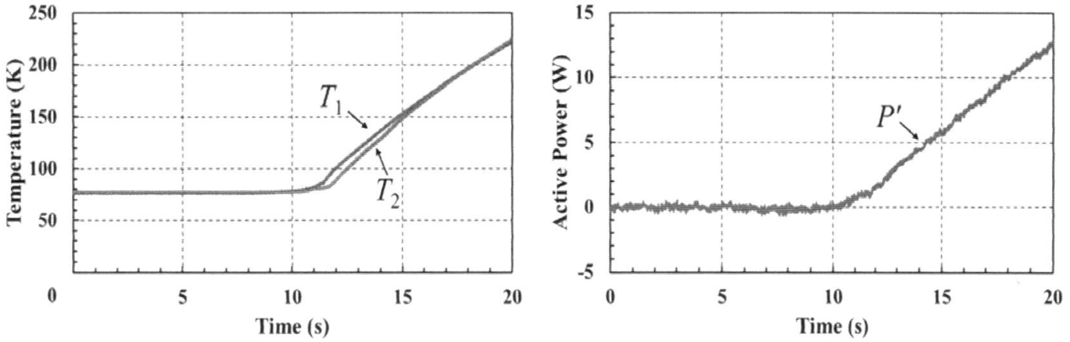

(a) Temperatures (b) Active power

Figure 5. Experimental results for the normal transitions at tapes 1 and 2.

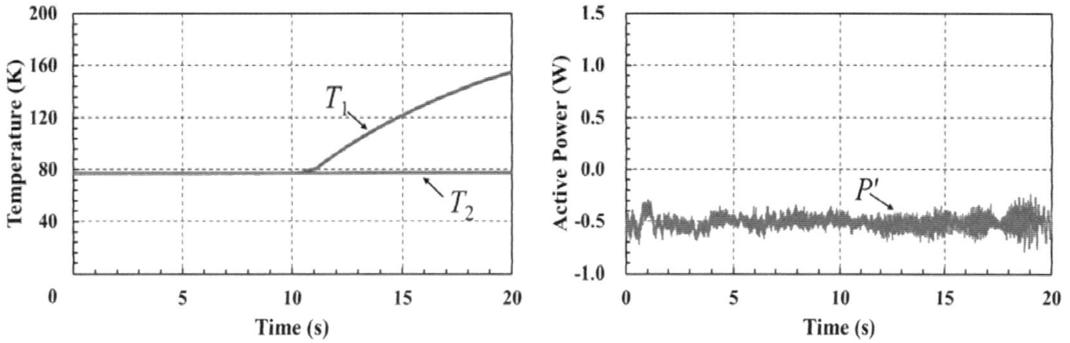

(a) Temperatures (b) Active power

Figure 6. Experimental results obtained by the conventional active power method shown in figure 1 for the normal transitions at tape 1.

mounted heater and consumed active power in the area, and amount of heat by the heater was larger than that by the consumed active power.

The results show that the proposed method works well when even one of secondary tapes transits to the normal conducting state. This means the proposed method can detect early the normal transitions. The method is suitable for a bundle secondary conductor of the HTS transformer.

Figure 6 shows experimental results obtained by the conventional active power method shown in Figure 1 for the normal transitions at only tape 1 [2]. The normal transitions occurred at 10.5 s. P' was almost -0.5 W in both the superconducting and normal state nevertheless temperature T_1 increased. As mentioned above, the results show that the normal transitions in the secondary coil are difficult to be detected by the conventional active power method. The P' was not 0 W but -0.5 W since the inductive voltages were not completely canceled.

4. Conclusion

The normal transitions in the secondary coil of the HTS transformer which is wound with a bundle of some HTS tapes are difficult to be detected by the conventional active power method. In this paper, the authors proposed the improve active power method to detect the normal transitions in the secondary coil. The method is based on measurement of each voltage of each HTS tape. The authors carried out some experiments for the test HTS transformer and showed its feasibility of the early detection of the normal transitions.

Acknowledgements

This work was supported by Electric Technology Research Foundation of Chugoku and JSPS KAKENHI Grant Number 18K04080.

References

[1] Nanato N, Tanaka S and Tenkumo S 2018 Study on a magnetic flux detection coil for detection of normal transitions in a hybrid single-phase Bi2223 superconducting transformer by the active power method, *Journal of Physics: Conference Series* **1054** 012070

[2] Nanato N, Ono T, Adachi T and Yamanishi T 2018 Protection system for normal transitions in a single-phase Bi2223 full superconducting transformer by the active power method under flowing currents of various Frequencies, *Journal of Physics: Conference Series* **1054** 012068

[3] Nanato N, Kishi N, Tanaka Y and Kondo M 2017 Basic study for a large AC current supply with a single-phase air-core Bi2223 high temperature superconducting transformer, *Journal of Physics: Conference Series* **871** 012101

IOP Conf. Series: Journal of Physics: Conf. Series **1293** (2019) 012063 doi:10.1088/1742-6596/1293/1/012063

Experimental investigation of the processes of degradation and transition to the normal state in CC-tapes under the action of current pulses

***M Osipov, S Pokrovskii, D Abin, I Anishenko and I Rudnev**

National Research Nuclear University MEPhI (Moscow Engineering Physics Institute), Kashirskoe sh. 31, 115409 Moscow, Russia

E-mail: MAOsipov1@mephi.ru

Abstract. The study of the behaviour of CC tapes under the action of nonstationary current loads was conducted: a transition to the normal state was observed, as well as processes of CC-tape degradation. The obtained data make it possible to solve the problems associated with the creation of high-speed switching devices based on high-temperature superconductors, intended for functioning in superconducting energy storage devices, energy distribution and transmission systems, current limiters, and new types of transport. In addition to superconducting switches, pulsed current loads can occur in the mentioned above systems due to short-circuit conditions or other factors, and can lead to local heating of HTSC, especially if there is an inhomogenity in the critical current in the CC-tapes. In this work we present the results of a study of non-stationary processes second generation HTS tapes heating under the impulse current loads with amplitudes up to 2.2Ic (77K, self-field). Superconducting tapes with a copper protective layer were used as the samples. The I-V characteristics of the samples were measured under the action of current pulses during cooling in liquid nitrogen. Critical parameters (amplitude, rise time of the transport current) of the current pulses, leading to the degradation of the superconductor were obtained. local current carrying characteristics of CC-tapes investigated using Hall probe magnetometry. The features of the influence of local inhomogeneities in the distribution of the critical current of the tapes on the transition of the superconductor to the normal state are revealed.

1. Introduction

An important aspect of the operation of any electrical device is its reliability. For most applications of high temperature superconductors the most unfavorable situation is the moment of transition to the normal state. At the same time, the local heat release is aggravated by the low propagation speed of the hot spot, which leads to damage to the HTS tape and, as a result, the device as a whole.

This problem is studying both theoretical and experimental research groups. In numerical calculations of this process, the final result strongly depends on the choice of the model describing the heat removal of the tape [1]. In a number of works it was shown [2–4] that there is a certain range of currents that exceed the critical value, but less than the limiting current of the thermal transition, in which a stable state is possible if the thermal balance between heat release and heat removal is observed. The article [1] shows that in the presence of a stabilizing layer, the value of the thermal transition of the HTS tape from the resistive to the normal state increases by more than 2 times compared with the corresponding current for the unstabilized tape.

The pulsed current measurement technique (PCM) has been used for a long time for characterizing superconductors at current densities near or above their critical current (Ic), in order to minimize heating effects. Such measurements are of prime importance when considering the optimization of applications such as superconducting fault current limiters, whose principle of operation is based on the superconducting to normal transition of the superconductor [5]. The results of measurements of the dependence of resistance on current during short current pulses for SuperPower tapes are given in [5, 6]. The effect of short triangular current pulses in the presence of a weak alternating magnetic field on AMSC and SuperPower tapes is described in [7]. The low speed of propagation of the normal zone leads to the occurrence of local overheatings, which destroy the tape, during inhomogeneous quenches. The literature describes ways of dealing with local overheating by increasing the thickness of the stabilizing layer of the tape [8] and by increasing the speed of propagation of the normal zone due to structural changes in the tape [9].

This paper presents experimental data on the study of the influence of stationary and non-stationary current loads with amplitudes far exceeding the critical current, and also examines the effect of initial uniformity of critical current CC-tapes on current load parameters leading to irreversible changes in the transport properties of CC-tapes.

2. Experimental methods

To study the non-stationary processes of degradation of the 2nd generation superconducting tapes, two main methods were used: measuring the current-voltage characteristics of samples in the mode of quasi-stationary flow of the transport current, as well as pulsed flow of current in the subcritical and supercritical modes, as well as local measurements of the spatial distribution of the critical current in the tapes by the method of scanning Hall magnetometry.

2.1. *Methods of stationary and non-stationary measurements of the current-voltage characteristic*

The well-known and standard approach to measuring the critical current of HTS tapes by the four-contact method was used for the I-V curve measurements. Measurement of the I-V curve in the non-stationary mode required the modification of the test bench for measuring the I-V curves — the setup for the measurement of the I-V curves was supplemented with an oscilloscope. The current source used is Agilent 6671A # J08 (basic configuration, current up to 300 A). If currents greater than 300 A are required, use an Agilent 6680A # J04 source (current up to 1000 A). If a small current step (50 mA resolution) is required, AgilentN5765A source was used. The current in the circuit was determined using the built-in shunt of the source and an Agilent 34401A voltmeter. The voltage and current were measured using a Yokogawa SL1400 oscilloscope with a 701251 module. The module has the following characteristics: the number of channels 2, the maximum measurement rate of 1 Msample/s, the measurement limits of 10mV - 200 V, the noise level of 100 μV.

Both channels of the module were used to perform the measurements. The first channel measured the voltage at the source shunt, the second channel was used to control the voltage at the potential contacts of the sample. Since high-speed ADCs do not have high sensitivity, as well as due to high noise values of the measuring module, measurements were made simultaneously by oscillograph and by more accurate voltmeters of the VAC measurement bench, which allowed us to precise the values of current and voltage on the sample when entering the quasistationary mode after a pulse increase of current.

Measurements were carried out in liquid nitrogen on a specially designed insert for I-V curves measuring. Powerful copper busbars allow a large amplitude current to pass through without significant heating due to layering and efficient cooling with liquid nitrogen. Powerful copper busbars allow a current with large amplitude to pass through without significant heating due to multilayered structure and efficient cooling with liquid nitrogen. Potential contacts are spatially separated from the current supply conductors, which also eliminates the influence of a large amount of current flowing through the buses on the voltage value on potential contacts.

2.2. *Methods of local magnetic measurements based on Hall scanning magnetometry*

For a more complete analysis of the processes of degradation of 2nd generation HTS tapes, we also used a spatial-sensitive method for studying the local current distribution in a superconductor. Based on the data of the current distribution, it can be concluded that damage occurs in the superconductor and the local regions with a lower critical current appear. In this case, one of the reasons of the appearance of damaged regions may be thermal instabilities of the superconductor when the transport current passes through the sample, especially in a non-stationary mode.

Scanning Hall magnetometry was used to study the samples. The description of the stand used in the work and the method of restoring the picture of the distribution of currents can be found in the work [10].

The stand allows to obtain primary information about the local magnetic properties of the samples. To obtain information on the current-carrying properties of superconductors, it is necessary to analyze the processes of the occurrence of currents in the sample and the restoration of the current distribution in the superconductor.

Magnetic field scanning is performed in the captured mode. It is necessary to calculate the spatial distribution of the current from the experimentally obtained surface distribution of the magnetic field of the transport current Bz (x, y). It is necessary to solve the problem of inversion of the Biot-Savart-Laplace equation for this. As a result of the calculations, two-dimensional distributions of the current components are obtained along (Jx) and across the tape (Jy). If the superconductor is defect-free, then the current flows along straight paths along the sample and at the edges of the tape the current turns, forming a closed loop. However, if regions of inhomogeneity arise, associated, for example, with defects, then the current flows around these regions, and the current component along the ribbon decreases (down to 0) and the component across the tape increases.

2.3. *Samples*

Two types of samples of CC-tapes were investigated: SuperPower tapes and American Superconductor tapes. SuperPower tapes are 4 mm wide. Substrate - Hastelloy, thickness - 50 microns on each side. The tape has a copper coating with a thickness of about 50 microns thick. The total tape thickness is about 100 microns. Tapes manufactured by American Superconductor are 4.8 mm wide. The substrate is Ni-W, with a thickness of about 50-75 microns. The tape has a copper coating with a total thickness of more than 100 microns. The total thickness of the tape is of the order of 200-250 microns, which is significantly larger than that of SuperPower tapes.

3. Results and discussion

The study of HTS tapes was carried out in three main modes: stationary mode (measurement of current-voltage characteristics, critical current determination), non-stationary mode with current pulses not exceeding 1.3 Ic samples in samples, non-stationary mode with current pulses not exceeding 2.2 Ic in samples. Also research on the local current distribution in the samples before and after the onset of degradation in the samples was conducted.

In total, 4 samples of Super Power (SP) and 4 samples of American Superconductor (AMSC) were investigated. The length of all samples was from 5 cm (6 samples) to 7 cm (2 samples).

In the stationary mode, the following characteristic results of measurements of the critical current are observed: SP samples from 97.4 to 100.5 A, AMSC samples from 110.8 to 114.1 A. Also, measurements were taken when the current flow was much higher than the critical current of the samples (up to 3 x times). All tapes showed high stability and stability of characteristics; no changes in properties in the samples were observed. The rate of current rise did not exceed 2 A / s. It should also be noted the greater homogeneity of the SP tapes (3.2–9.8%) with respect to AMSC samples (1.9–15%) prior to the study in non-stationary modes.

Studies in non-stationary regimes have revealed some general patterns of behavior of samples from both manufacturers. The occurrence of thermal instabilities during a pulses of the transport current is more critical, and after removing the current load, the degradation of the properties did not exceed 5% of the initial values. The occurrence of instability appears with a significant delay after the current pulse (from 0.2 to 6 seconds), while there is no direct correlation with the current injection modes. This may be due to the dramatically changing conditions of heat removal in liquid nitrogen. An example of this phenomenon is shown in Figure. 1. The delay in the occurrence of the peak can be explained by the low velocity of propagation of the normal zone. When a current is introduced, heat is generated, which locally overheats the superconductor. The normal zone arises and begins to grow, which leads to an increase in voltage. On the other hand, heat is removed from the superconductor to the copper layer (and further to nitrogen). After that, the superconductor is cooled and returns to the superconducting state, which leads to a decrease in voltage.

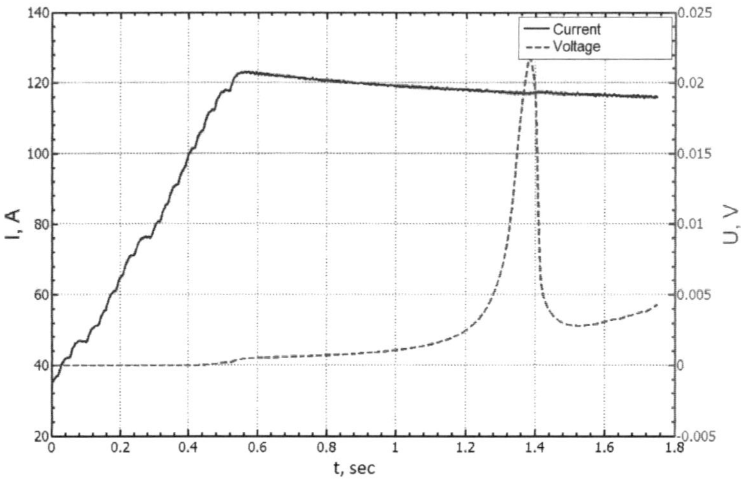

Figure 1. Dependence of the transport current and voltage on the SuperPower sample from time when current is loaded at a speed of 136.3 A / s to 115.4 A

For SP samples, an unstable state occurs when the critical current exceeds more than 1.22. In this case, there is a correlation between the local current-carrying characteristics of the tapes and the threshold current for the occurrence of instabilities. So for the initially more homogeneous tapes, the threshold current is higher. At the same time, the current rise rate is minimal for the sample with the greatest inhomogeneity.

Observed features of the behavior of the samples: the appearance of a double peak voltage on the sample after pulsed current transmission (the second peak is much smaller in amplitude, but longer than the first).

The occurrence of irreversible degradation occurs at amplitudes of 1.4 Ic and the rate of current rise is 2-3 times higher than when reversible instabilities occur. Moreover, for more homogeneous samples, the speed and amplitude are higher than for more defective ones. It should also be noted that degradation occurs locally, while a significant part of the tape does not undergo changes in superconducting characteristics. An example of the distribution of the longitudinal component of the current before and after degradation occurred is shown in Figure 2.

For AMSC samples, an unstable state occurs when the critical current exceeds more than 1.25 times. In this case, no correlation is observed between the local current-carrying characteristics of the tapes and the threshold current for the occurrence of instabilities. So for the initially more homogeneous AMSC tapes, the threshold current is lower.

Observed features of the behavior of the samples: the occurrence of a double peak voltage on the sample after pulsed current transmission (the second peak of the same amplitude) is less common than for samples SP. An example of such heterogeneity is shown in Figure 3.

For AMSC samples, an unstable state occurs when the critical current exceeds more than 1.25 times. In this case, no correlation is observed between the local current-carrying characteristics of the tapes and the threshold current for the occurrence of instabilities. So for the initially more homogeneous AMSC tapes, the threshold current is lower.

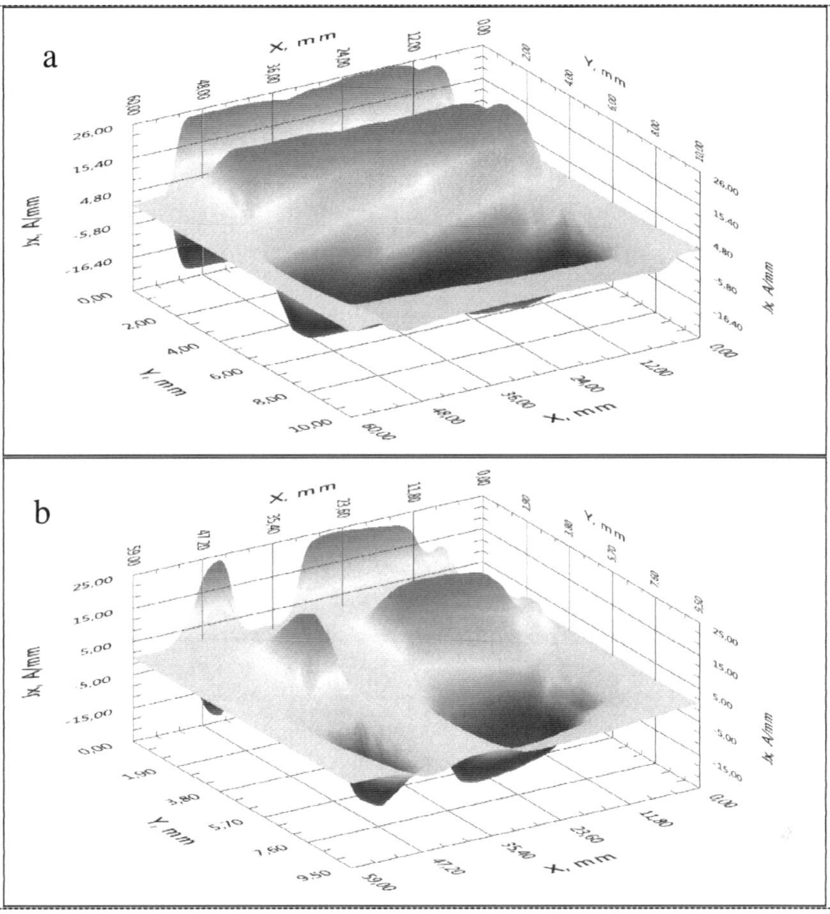

Figure 2. – The distribution of the longitudinal component of the current through the SuperPower sample 2 a) before the current load and b) after the current load

For AMSC samples, an unstable state occurs when the critical current exceeds more than 1.25 times. In this case, no correlation is observed between the local current-carrying characteristics of the tapes and the threshold current for the occurrence of instabilities. So for the initially more homogeneous AMSC tapes, the threshold current is lower.

Observed features of the behavior of the samples: the occurrence of a double peak voltage on the sample after pulsed current transmission (the second peak of the same amplitude) is less common than for samples SP. An example of such heterogeneity is shown in Figure 3.

The emergence of irreversible degradation occurs at amplitudes from 2 to 2.3 Ic and the rate of current rise is 8-10 times greater than when reversible instabilities occur. Moreover, for more homogeneous samples, the speed and amplitude are higher than for more defective ones. It should also be noted that degradation occurs locally, while a significant part of the tape does not undergo changes in superconducting characteristics. Also on sample AMSC4, irreversible deformation corresponds to a region with a significantly lower critical current.

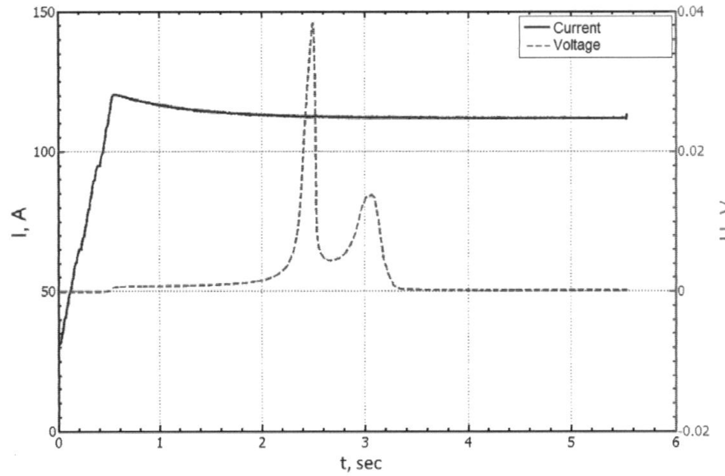

Figure 3. Dependence of the transport current and voltage on the SuperPower sample from time when current is loaded at a speed of 136.3 A / s to 115.4 A

From the comparison of samples from different manufacturers, it should be noted that AMSC samples are more resistant to the occurrence of both reversible and irreversible instabilities than SP tapes, however, the length of the degradation region of samples differs several times. Differences in the behavior of samples are mainly determined by the difference in the architecture of the tapes and the thickness of the stabilizing layers. The AMSC tapes have a significantly larger thickness of the stabilizing copper layer and substrate, which improves the removal of heat generated in the superconductor when current is ramped. This, in turn, leads to a more rapid cooling of the overheated area after the end of the current pulse. When the critical current is exceeded, the current begins to flow along the copper layer, and the thicker layer in the AMSC tapes has less resistance, and accordingly leads to less heating of the tape. All this leads to a higher stability of the AMSC tapes with respect to the current pulses

Legend used in the tables:

Ic – critical current

Iri – current of reversible instability occurrence

Vri – rate of transport current rise (occurrence of reversible instability)

Iri/Ic – current of reversible instability occurrence normalized on critical current

Vri/Ic – rate of transport current rise normalized on critical current (occurrence of reversible instability)

Iii – current of irreversible instability occurrence

Vii – rate of transport current rise (occurrence of irreversible instability)

Iii/Ic – current of irreversible instability occurrence normalized on critical current

Vii/Ic – rate of transport current rise normalized on critical current (occurrence of irreversible instability)

Table 1. Research results

Sample	Ic, A	Iri, A	Vri, A/s	Iri/I	Vri/I, 1/sec	Iii	Vii, A/s	Iii/Ic	Vii/Ic, 1/sec	Inhomo geneity, %
SP1	100,5	-	-	-	-	-	-	-	-	3.61
SP2	98,7	123,2	151.4	1.25	1.53	135.5	478.2	1.37	4.85	3.94
SP3	97,6	122,4	153.5	1.25	1.57	142	446	1.45	4.57	3.23
SP4	97,4	118,5	166.5	1.22	1.71	133	348.6	1.37	3.58	9.81
AMSC1	114.1	-	-	-	-	-	-	-	-	15.14
AMSC2	111.4	147.3	579.9	1.32	5.21	228.9	3317	2.05	29.78	9.06
AMSC3	111.6	139	387.1	1.25	3.47	260.5	3428	2.33	30.72	1.9
AMSC4	110.8	148	428	1.34	3.86	243.9	3387	2.2	30.57	15.65

4. Conclusons

Studies of the behavior of HTS tapes in nonstationary current input modes have been conducted: superconductor transition processes to a normal state and processes of degradation and of superconductor burnout at high current input rates have been observed. At current input rates above the threshold value, which is different for tapes of different architectures, thermal instability occurs, which manifests itself in the appearance of a voltage peak with a delay after the current pulse. Summarizing the obtained experimental results, we can conclude that:

1) There is no correlation between the current input modes and the delay time of the voltage peak and duration of the peak.

2) The rate of rise and amplitude of the current causing irreversible degradation is higher for more homogeneous samples than for more defective ones, for all types of samples. At the same time, the correlation between the rate of current rise and homogeneity across the tape does not have a general pattern for different types of samples.

3) The degradation of samples occurs locally and most of the sample does not undergo changes in superconducting characteristics.

Acknowledgements

The research was done under support of Russian Foundation for Basic Research (grant 17-29-10024).

References

[1] Malginov V, Malginov A, Fleishman L, and Rakitin A 2017 *Technical Physics* **62** 1516–1524
[2] Fetisov S, Vysotsky V, and Sytnikov V 2009 *IEEE Trans. Appl. Supercond.* **19** 2411–2414
[3] Fetisov S, Vysotsky V and Zubco V 2011 *IEEE Trans. Appl. Supercond.* **21** 1323–1327
[4] Zubko V, Ryabov S, Fetisov S et al. 2015 *Phys. Proc.* **67** 619–624
[5] Sirois F, Coulombe J, and Bernier A 2009 *IEEE Trans. Appl. Supercond.* **19** 3585 - 3590
[6] Sirois F, Coulombe J, Roy F and Dutoit B 2010 *Supercond. Sci. Technol.* **23** 034018
[7] Lacroix C, Sirois F, Slimani K, and Cave J 2013 *IEEE Trans. Appl. Supercond.* **23** 6601605
[8] Colangelo D and Dutoit B 2012 *Supercond. Sci. Technol.* **25** 095005
[9] Lacroix C, Sirois F and Lupien F 2017 *Supercond. Sci. Technol.* **30** 064004
[10] Pokrovskiy S, Rudnev I, Podlivaev A 2009 *J. Phys Conf. Ser.* **150** 052211

IOP Conf. Series: Journal of Physics: Conf. Series **1293** (2019) 012064 doi:10.1088/1742-6596/1293/1/012064

Numerical simulation of the fast processes in HTS tapes under the pulsed current load

I V Anischenko, S V Pokrovskii, I A Rudnev, M A Osipov and D A Abin

National Research Nuclear University MEPhI (Moscow Engineering Physics Institute), Kashirskoe sh. 31, 115409 Moscow, Russia

E-mail: mephizic@gmail.com

Abstract. This paper presents a numerical layered 2G HTS tape model with the transport current flowing through it. The simulation is performed by using FEM H-formulation method implemented in the Comsol Multyphysics software. The model provides two cooling modes: cryocooler and liquid nitrogen. LN2 cooling mode provides for a complex hysteresis character of the liquid nitrogen boiling curve and the boiling regimes changing possibility between the stationary and the bubble boiling. The delay of the voltage rising with a transport current above the critical value pulse input is shown in the liquid nitrogen cooling case. The possibility of parameters calculating for current redistribution from the HTS to the stabilizing copper layer is shown, and the stabilizing coating thickness effect on the thermal transition critical current value is studied A comparison with experimental results is carried out.

1. Introduction

In recent years, the high-T_c coated superconductors numerical simulation has become quite a popular method for the HTS behaviour predicting and optimizing the devices based on them. The increasing in number of complex devices with HTS current-carrying elements has resulted in 2-D approximations often are not sufficient. This is especially about the cases when the transport current values are close to or even higher than the HTS element critical current. In such cases the superconductor heat generation rate sharply increases. In some cases, thermal bursts are successfully discharged into the liquid refrigerant or can be compensated by cryogenic equipment. If the cooling of the superconductor is efficient and the cryogenic equipment rate is sufficient to maintain the specified thermal regime of the HTS, then there are no fundamental restrictions for use in current-carrying HTS elements of currents significantly higher than the critical one. But in order to calculate accurately such a system, it is necessary to fully simulate a superconductor and calculate the local heat generation in the 3D structure, taking into account the actual cooling and stabilization parameters of coated conductor. An even more challenging task is to simulate "hot" tapes under liquid nitrogen cooling, because for a given liquid during the rapid heat flow growth the additional overheating and boiling up may occur. In this case, the heat-removing LN2 properties in different boiling regimes differ significantly.

This paper presents the results of complex modeling of the HTS tapes with stationary current effects and non-stationary effects with current pulses. The process of transition of multilayer HTS conductors intended for use in electrical equipment from the superconducting state to the resistive and from resistive to normal when currents exceeding the critical current, as well as the stabilizing layer effect on the second-generation HTS tape is investigated. The model is universal and provides options of liquid nitrogen and liquid-free cooling. The simulation was performed using the H-formulation of finite element method (FEM) in the Comsol Multiphysics software. The choice of the formulation in the magnetic field components terms is due to the interest of further modeling of the combination of pulsed current loads in conjunction with magnetic field pulses, for the simulation of which the H-formulation has been already successfully tested.

2. Computational model

The simulated system is a layered structure of the superconducting tape containing a substrate, a superconducting layer, and also a stabilizing copper layer (figure 1.). By the others layers we neglect because of their small thickness and small influence on the thermal processes. Due to the proximity of the copper and silver specific conductance values, in the subsequent analysis it considered as a single stabilizer layer with the corresponding equivalent resistance. As characteristics of the sample, the characteristics of HTS tape "SuperOx" (Russia) were selected. The tape width was 4 mm, the superconducting layer (RE) BCO thickness - 1 μm, the copper layer thickness - 3 μm, the Hastelloy C-276 alloy substrate thickness - 0.1 mm The tape critical current was measured at the liquid nitrogen boiling point with a 1 μV / cm criterion and was ~120A, which, in terms of the cross section of the superconducting layer, gives a critical current density ~10^{10} A/m^2.

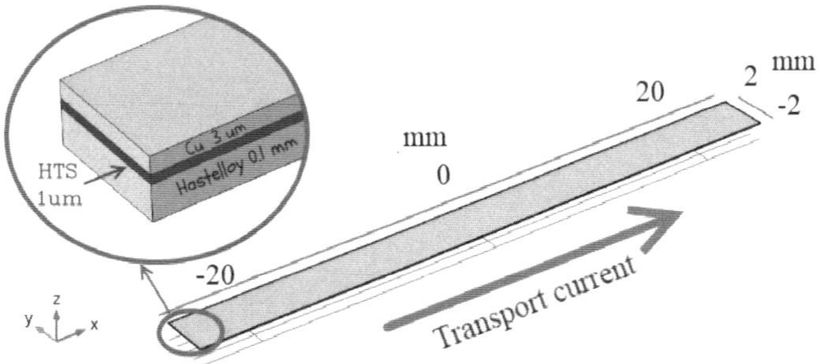

Figure 1. The illustration of the simulated experiment.

The H-formulation was implemented using the finite element method (FEM) in the *Comsol Multiphysics* numerical simulation software. The simulation is based on the Ampere and Faraday laws, and the dependent variables are the magnetic field components H_x, H_y, H_z. The nonlinear voltage dependence on current is set by the power law for current-voltage characteristic with the power of n = 25. The Pointwise Constraint boundary condition imposes the necessary restrictions on the transport current application through the end of only the superconducting layer strictly along the x axis according to a given time law. Periodic boundary conditions for dependent variables are superimposed on the ends of the tape. Such the boundary conditions formulation ensures the correct current load determination and the problem convergence.

The heat generation in the system is calculated by the formula:

$$Q = \boldsymbol{E} \cdot \boldsymbol{J} \tag{1}$$

The currents distribution $\boldsymbol{J}$ in the superconductor is determined by the current load parameters. Since the typical HTS tape self field usually do not exceed 0.1 T, it do not have a significant effect on the tape critical current and we neglect it in the calculations.

The model takes into account the tape layers real thermal characteristics. The critical current dependence on temperature is introduced in accordance with the expression (2) [1]:

$$J_{c0} = \alpha \left(1 - \left(\frac{T}{T_c}\right)^2\right)^{1.5}, \tag{2}$$

where α - fitting parameter.

The experimental critical temperature T_c is 92 K. Thermal conductivity, heat capacity and the resistance temperature dependences are entered into the model in the interpolation form for each tape layer [2].

In the model the cryofree tape cooling is performed at the lower boundary, where the cryocooler system power may vary in the range of a real cryocooler cooling system power to maintain the desired cooling temperature. Wherein the tape is placed on a copper bulk that will be considered as a cooling source. The outer boundary of the cryocooler, which is in direct contact with the HTS tape substrate, is set as a cooling source with a determined constant temperature and variable heat sink power.

The liquid nitrogen cooling realized taking into account the LN2 boiling curve [3] and takes place along the entire tape perimeter. For convective and bubble nitrogen boiling [4], appropriate heat transfer coefficients are introduced [5]. For bubble boiling, the heat transfer coefficient is calculated as:

$$\alpha_{boil} = C_h q^{0.624} \left(\rho C_p k\right)^{0.117}, \tag{3}$$

where ρ – density, C_p – heat capacity, k – thermal conductivity of liquid nitrogen, q – heat flow, C_h – coefficient [4].

The heat transfer coefficient for liquid nitrogen stationary boiling is determined by the temperature difference ΔT at the boundary between the tape and liquid nitrogen and is calculated by the formula:

$$\alpha_{conv} = C_{conv} \Delta T^{1/3}, \tag{4}$$

C_{conv} – coefficient [5].

Thus, the choice of using one or another heat transfer coefficient depends on the temperature difference ΔT and heat flux into liquid nitrogen q and is set using the conditions "*if*". Since in some liquids, including liquid nitrogen, during the rapid heat flow growth the additional overheating may occur (boiling moment delay) by several degrees ΔT_{oh} (superheating), the liquid boiling temperature may exceed usual temperature onset bubble boiling delay value ΔT_{cb} (transition from convection to boiling). Therefore, the first type of "if" condition is fulfilled when the temperature difference ΔT reaches the liquid nitrogen overheating temperature ΔT_{oh}, which in general depends on the heater surface material and the boiling liquid properties. In our case ΔT_{oh} is assumed to be constant and equal to 3K. Immediately after ΔT_{oh} is reached, a developed bubble boil begins. When the heat flux decreases, the return from bubble boiling to convection does occur not at the temperature difference ΔT_{oh}, but at the temperature difference ΔT_{cb} (the second "if" conditions type), that is, there is a boiling hysteresis. ΔT_{cb} can be found from the condition of heat transfer coefficients equality in the natural convection and bubble boiling modes. Thus, the choice of heat transfer equation coefficients occurs automatically at each time step of the solution.

In order to preserve the real geometrical dimensions of all tape layers, when building a model, mesh drawing tools for the mesh of elements, converting a square mesh into a triangular at the individual domains boundaries, and multiscale structuring were used.

3. Results and discussion

In order to evaluate the thermal processes influence the HTS tape current-voltage characteristics were built with a rather slow current flow change (0.5 A/s) and for more fast change (5 A/s) for cases of liquid nitrogen and cryocooler cooling of tape to a temperature of 77.4 K (figure 2). In the 1st case there is no cooling mode influence on critical current, therefore the direct and reverse I-V characteristics are the same. In the case of fast current increasing, the thermal processes start to effect on I-V characteristic. In both LN2 and cryocooler cooling cases the current – voltage characteristic hysteresis is observed, but for the liquid nitrogen cooling case the hysteresis is more clear. This is due to the complex nature of heat transfer to liquid nitrogen during the refrigerant heating, namely the additional overheating during the transition from convection to boiling existence, the heat transfer mechanism from convection to boiling and back repeatedly changing possibility, and the liquid nitrogen boiling curve hysteresis [3]. Indeed, for a given current input rate, the calculation showed that the condition of the liquid nitrogen boiling mode changing was performed twice — at ~ 103A transport current value (current increase) from convective to boiling and at ~ 110A (current decrease) from boiling to convective.

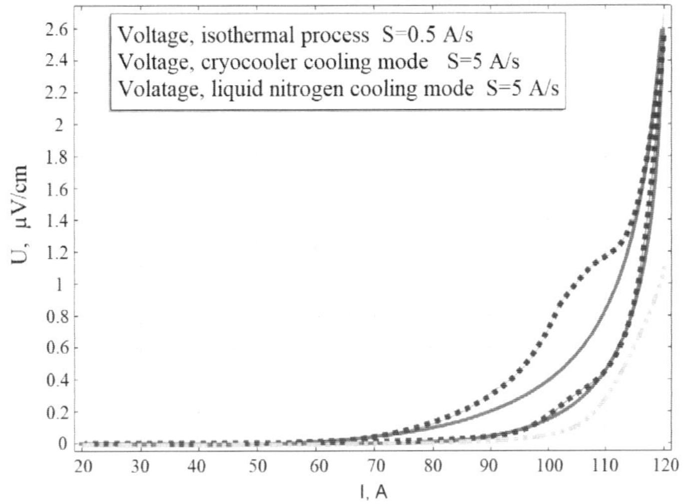

Figure 2. I-V characteristics of HTS tape (77.4 K).

The HTS tape magnetic field change dynamics during the various amplitude transport current flow is shown on the figure 3. One can see that at low currents (to $0.5J_c$) the magnetic field is zero at the HTS tape center, which indicates the currents absence in this area. Hereinafter, the J_c is the isothermal critical current value, which is 120 A. The magnetic field begins to appear in the tape center only when the current reaches more than half of critical current (0.6 J_c), and with a further current increase, only the HTS tape magnetic field amplitude changes, but the field profile shape remains unchanged. At approximately the same flowing current values (0.6 J_c ~ 72 A), a nonzero voltage on the HTS tape appears (see figure 2) and the heat generation begins to increase.

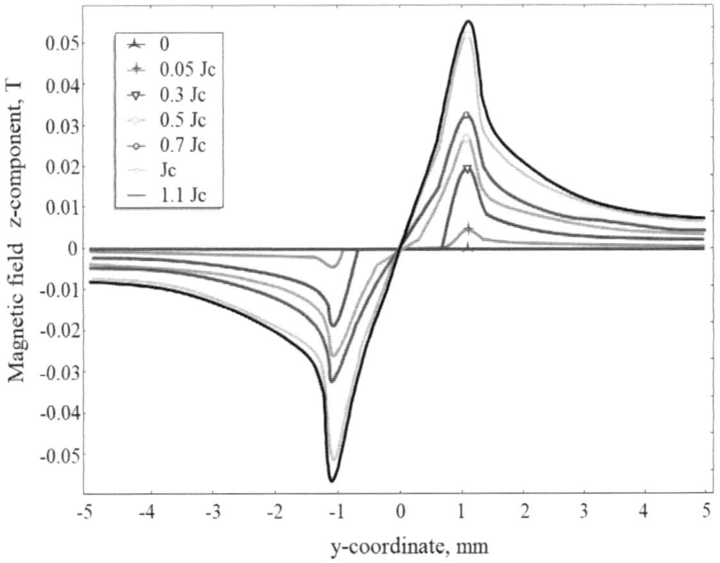

Figure 3. Intrinsic magnetic field distribution across the 2nd generation HTS tape width for various transport current amplitudes. Coordinate y=0 is the HTS tape center.

The impulse of 148 A amplitude transport current was applied to the HTS tape, cooled by liquid nitrogen, at a speed of $S = 428$ A/s, and after reaching the maximum value, the current remained constant. Figure 4 shows the HTS tape current magnitude and voltage dependence on time. The calculated and experimental results are in good agreement. It can be noted that both the experimental and calculated curves demonstrate same delay in the voltage appearance in the tape, when the current amplitude is already ~ 130% of J_c. When the superconductor critical current is exceeded, a sharp increase in heat generation leads to HTS layer overheating and the transition to the resistive state. However, this process does not appear instantly. There is a certain thermal transition critical current value J_{cT}, depending on the pulse input speed, cooling parameters and tape stabilization. This process is also influenced by the temperature delay in the transition of liquid nitrogen boiling regime from stationary to bubbly.

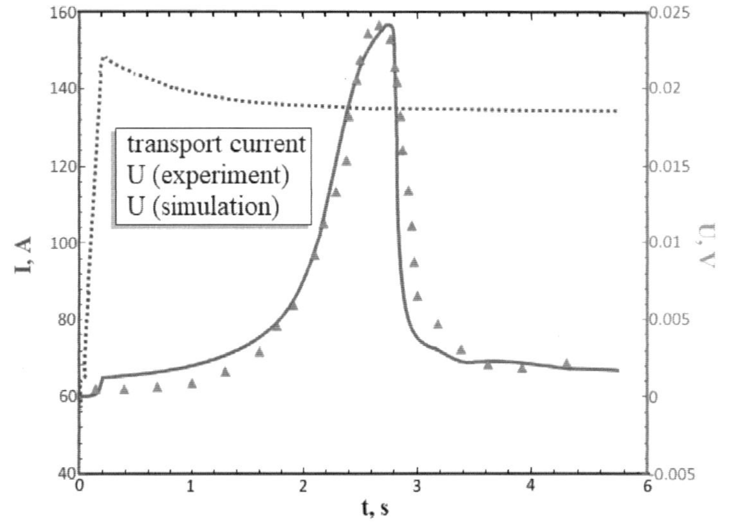

Figure 4. The dependence of the voltage of the HTS tape on time when the flowing current above the critical one.

In addition, for this current flow mode, calculations showed that with a constant transport current above the critical one, the heat flux in the superconducting layer undergoes some oscillations throughout the current load duration [6]. The heat flux increase and decrease at a constant current is the result of current redistribution between the stabilizing and superconducting HTS tape layers.

The current redistribution in the HTS coated sample and the maximum tape layers temperatures during the current load are shown on the figure 5.

While the maximum temperature gradient in the superconducting layer begins to increase, the heating in the copper layer also increases significantly. However, when the temperature gradient begins to decrease in the HTS layer, its growth continues for some time in the copper layer. This suggests a partial current flow from HTS to the copper layer. Then, the SC layer, cooled outside by liquid nitrogen, begins to return to a stable thermal state, the current is redistributed reversely, and the temperature gradient in the copper layer decreases again. It is also important to note that the temperature gradient in the Hastelloy layer follows the shape of the temperature gradient in the SC layer, but turns out to be several orders of magnitude smaller. This suggests that the substrate layer does not participate in electrodynamic processes, but only removes some of the heat from the HTS layer due to the thermal conductivity.

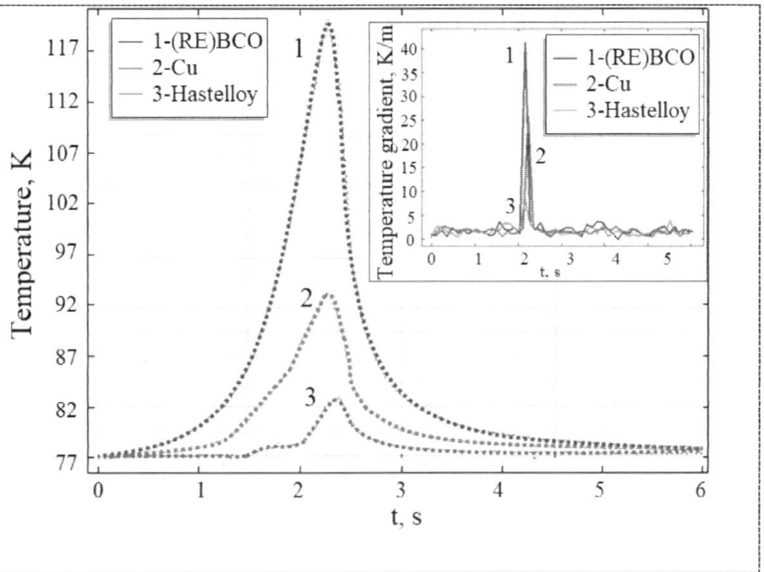

Figure 5. Dependence of the maximum temperature on time for different layers of HTS tape (insert - maximum temperature gradient on time for different layers).

The stabilizing copper layer influence on the processes occurring in the HTS tape under the overcritical current loads was also considered. It was found that, unlike a tape with a stabilizer, for a tape without copper coating, the critical current of thermal transition is determined by the instantaneous heat release value. For this reason, for HTS tapes without a stabilizing layer, the thermal transition occurs already with a small excess of the critical current (~ 105% of J_c). The critical thermal transition current for the stabilized tape was ~ 1.92 J_c *(T0)*. That is, a copper-coated tape can pass flowing currents almost 2 times the critical current at the coolant temperature. In addition, the calculation for several other copper coating layer thicknesses showed that the copper stabilizer thickness decrease leads to a decrease in J_{cT} , and an increase in it leads to an insignificant increase in J_{cT} or even a decrease. So, J_{cT} for HTS tapes with a 1 μm Cu thickness was ~ 1.42 J_c *(T0)*, for 5 μm thickness ~ 2.01 J_c, and for a 30 μm copper thickness ~ 1.64 J_c. This is the result of the fact that the stabilizer layer thickness excessive increase leads to less efficient heat transfer to liquid nitrogen. The question of the stabilizing layers thickness influence on HTS tapes characteristics for different pulsed current load modes is an interesting topic for further researches.

4. Conclusions

A numerical model for electrodynamic and thermophysical processes in 2G HTS tape under the pulse current load has been developed. As part of the thermal-coupled model cryocooler or liquid nitrogen cooling modes are possible. It has been found that there is some limiting current above the critical one, at which excessive heat release in the HTS tape does not have time to be removed by the cooling system and a superconducting transition occurs. With a fast rate of the current increasing in liquid nitrogen cooling case the local overheating of the tape HTS layer is occur and the current redistribution between the superconducting and stabilizing layers is possible.

Due to the low conductivity, the substrate does not participate in the currents redistribution, but only removes the heat from HTS layer due to own good thermal conductivity. It has been investigated that the presence of a stabilizing Cu-Ag layer with 3 μm thickness leads to the thermal transition current increasing almost 2 times more as compared with open tape. However, a too thick stabilizer layer prevents effective heat removal to liquid nitrogen and reduces the thermal transition critical current.. All this suggests the principal possibility of using stabilized HTS tapes in electrical equipment with currents substantially exceeding the critical current of HTS up to the moment when thermal bursts in the superconducting layer can be compensated by cryogenic equipment and stabilizing layers. Comparison of the calculation results with experimental data shows a good agreement.

The developed model can be successfully applied for the calculation of current-carrying elements of various designs based on second-generation HTS tape, as well as for optimizing various practical applications.

Acknowledgments
The research was done under support of Russian Foundation for Basic Research (grant 17-29-10024).

References

[1] Anishchenko I, Pokrovskii S and Rudnev I 2018 *IOP Conf. Series: Journal of Physics: Conf. Series* **945** 012015
[2] Zhang M, Matsuda K and Coombs T 2012 *Journal of Applied Physics* **112** 043912
[3] Shi M, J Ma and Wang B *Int. J. Heat Mass Transfer* 1993. **36** (18) 4461–4466.
[4] Stephan K and Abdelsalam M *Int. J. Heat Mass Transfer* 1980 **23** 73–87.M.
[5] Fetisov S, Vysotsky V and Zubko V 2011 *IEEE Trans. Appl. Supercond.* **21**(3)1323 – 1327.
[6] Deev V, Kutsenko K and Lavrukhin A 1998*Cryogenics* **38** 715–719.

IOP Conf. Series: Journal of Physics: Conf. Series **1293** (2019) 012065 doi:10.1088/1742-6596/1293/1/012065

Observation of a non-uniform current distribution in stacked high temperature superconducting tapes

T A J Meulenbroeks[1], Y Terazaki[2], S Matsunaga[3] and N Yanagi[2,3]

[1] Eindhoven University of Technology, P.O. Box 513, 5600 MB Eindhoven, The Netherlands
[2] National Institute for Fusion Science, 322-6 Oroshi-cho, Toki, Gifu 509-5292, Japan
[3] SOKENDAI (The Graduate University for Advanced Studies), 322-6 Oroshi-cho, Toki, Gifu 509-5292, Japan

E-mail: timmeulenbroeks@gmail.com

Abstract. High Temperature Superconductors (HTS) improve upon low temperature superconductors in many ways and the ability to cope with a non-uniform current distribution might be one of those improvements. To put this to the test, an experimental setup is designed to force a non-uniform current upon a stack of 5 HTS tapes, using a worst case current feeding method. The experiment can help determine the potential of this conductor design and is part of the ongoing effort to develop a non-transposed stacked HTS conductor for the nuclear fusion reactor FFHR. The results clearly show that the conductor sample is able to stably conduct a current equal to its critical current, although at an elevated electric field of roughly 5 mV/m. This means non-transposed stacked tape conductors remain stable, even if a worst case non-uniform current is constantly forced upon them. A hypothesis to explain this abnormally high electric field is formulated on the basis of the results, however additional research is needed to verify it. It states that the electric field is necessary for the tapes to share current and would mean that in a properly engineered application, these losses due to the electric field, would only occur during start-up. Overall it is clear that this experiment proves the excellent stability of non-transposed stacked HTS tapes and their ability to conduct a non-uniform current.

1. Introduction

Traditional Low Temperature Superconductors (LTS) struggle with conducting non-uniform currents and often quench prematurely when exposed to them [1]. Transposition of the conductor is necessary to avoid a non-uniform current distribution and often makes the conductor significantly more complex, especially when using High Temperature Superconductors (HTS) [2]. Besides making the conductor more complex, in some cases, transposition increases the strain in the conductor [3] [4]. This strain can reduce the performance of the superconductor [5].

Research led by the National Institute for Fusion Science (NIFS) has proven that it is possible to build and operate a coil, using a non-transposed stacked HTS tape conductor design called Stacked Tapes Assembled in Rigid Structure (STARS [6]). It is highly likely that, during the test of this coil, it had been subjected to a non-uniform current distribution but did not quench, noting that the increased thermal stability of HTS is the essential key to its success. This has been the inspiration for the main research question of this paper: Are stacked HTS tapes able to stably conduct a worst case non-uniform current? After answering the main research question, the aim is to make an assessment of the applicability of the non-transposed stacked tape conductor design.

2. Method

The goal of the experimental setup is to force a worst case non-uniform current unto a superconductor sample of 5 stacked HTS tapes. This is done by clamping the ends of the tape stack between copper connection plates and only supplying current to the top plates. It has to be noted that this is not the proper way to feed current to HTS tapes, but is used in this case, exactly because this causes a non-uniform current distribution [7]. The top plates will only be in contact with the first tape, and not with the rest of the tapes, thus the current will prefer to flow through the first superconducting layer to the other side, as all other paths offer more resistance (see Figure 1). All other paths offer more resistance because they cross resistive layers. In Figure 2 it is shown how, in a tape, the resistive layers sandwich the superconducting layer and form the resistive barrier which the current is forced to cross if it is to reach the lower tapes.

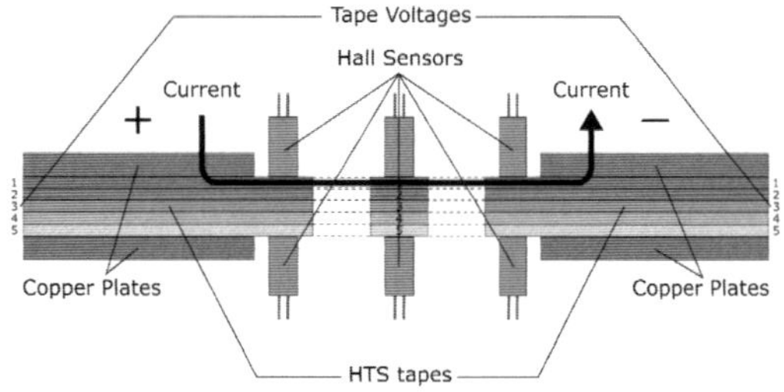

Figure 1: Simplified schematic side view of the experimental setup, showing the HTS tape stack clamped between the copper connection plates at each end. The current is only fed from the top plates, forcing the current to become non-uniform. Additionally the location of the Hall sensors and voltage taps are depicted.

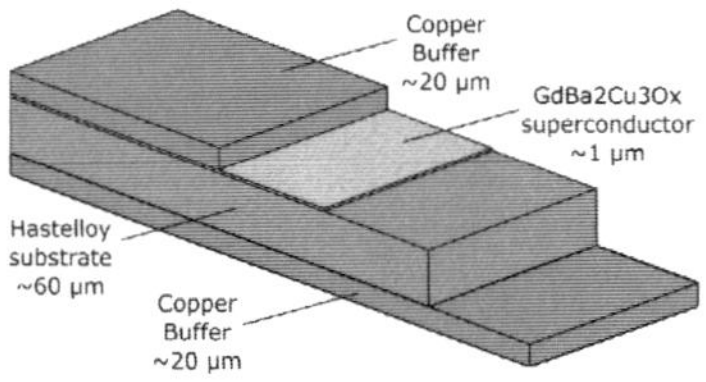

Figure 2: Simplified structure of a single HTS tape used in the experiment. The metallic layers sandwiching the superconducting layer, especially the hastelloy substrate, are the resistive barrier the current has to pass, to get to the next tape.

The tapes that make up the superconductor sample are 4mm $GdBa_2Cu_3O_x$ tapes produced by SuperOx [8]. Each tape has a minimum critical current of 120 A at 77 K and is 1.7 m long. Five of these tapes are stacked on top of each other to make up the sample, resulting in a total critical current capacity of 600 A. The tapes are stacked with their superconducting layer up, towards the current feeding connection plate so the current does not have to cross the hastelloy layer to reach the first superconducting layer. In addition to the before mentioned copper connection plates, some materials and equipment are added to the experiment to ensure realistic HTS conductor behavior. A copper jacket enveloping the tape stack is added for thermal stability and clamps are added to hold the jacket in place. Additionally the clamps apply pressure to the tape stack, reducing the contact resistance between individual tapes.

The experimental setup also needs sensors to verify whether the current distribution is indeed non-uniform and to detect any thermal runaway processes. The non-uniformity of the current distribution is measured by 3 sets of 2 Hall sensors which are depicted in Figure 1. Each set of sensors is placed perpendicular to the flat tape surface in order to measure the magnetic field generated by the current flowing through the sample. The difference between the sensors can verify whether the current distribution is asymmetric. And non-uniformity can be verified by measuring asymmetry. Two sets of Hall sensors are located at the ends of the copper jacket, as close as possible to the copper connection plates and the final sensor set is located at the middle of the jacket. All sensors are directly touching the stack of tapes via slots carved into the copper jacket to ensure the distance to the tapes is minimal and they are held in place by fiber reinforced plastic blocks to ensure minimal movement between experiments. A possible resistive transition or thermal runaway can be detected by measuring the voltage across each tape. The voltage is measured by voltage taps soldered to the top of each tape at both ends. To make the setup complete, a bakelite support plate was manufactured to minimize movement of the parts and to make insertion into the stainless steel liquid nitrogen bath more convenient. A picture of the entire setup is shown in Figure 3.

Figure 3: Picture of the experimental setup. In the bottom the copper jacket is visible. It surrounds the stack of tapes and is held together by clamps. The rectangular blocks in the middle and at the ends contain the Hall sensors. In the top the liquid nitrogen bath and the current leads are visible. The stainless steel container is roughly 2 m long.

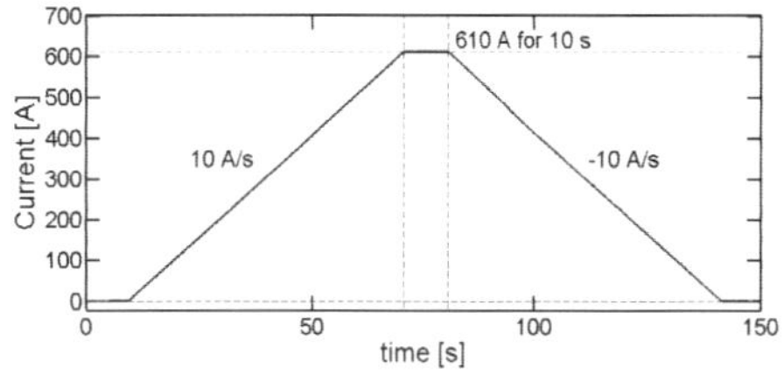

Figure 4: The structure of the current waveform applied to the sample in the 610 A test. The ramp rate and waiting period at maximum current were kept the same during all tests and were chosen as such to reduce transient effects as much as possible.

The experiment is started by lifting the experimental setup described above, into the liquid nitrogen bath and connecting a current source to the top connection plates via current leads. Next the bath is filled, cooling the setup down to 77 K. When the current waveform (Figure 4) is applied, the measurements from the voltage taps are closely monitored and if the voltage becomes unstable or in other words a thermal runaway is registered, the power supply is switched off manually. The maximum current of the waveform is increased step by step until thermal runaway has occurred.

3. Results

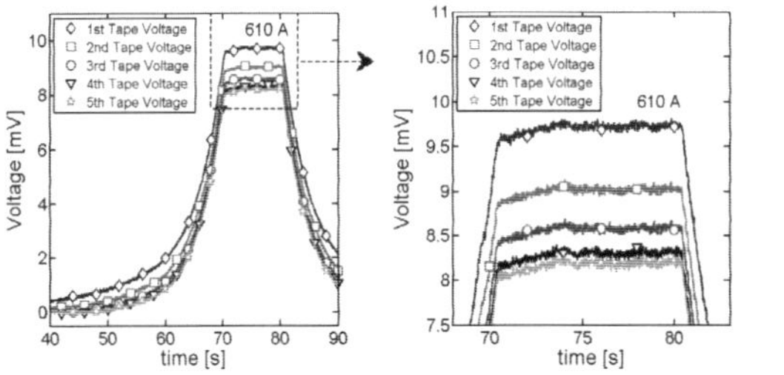

Figure 5: The tape voltage measurement that shows that a current of 610 A was stably conducted (Ic stack = 600 A), although at an elevated electric field of 5 mV/m. Distinct voltage differences between the tapes are visible

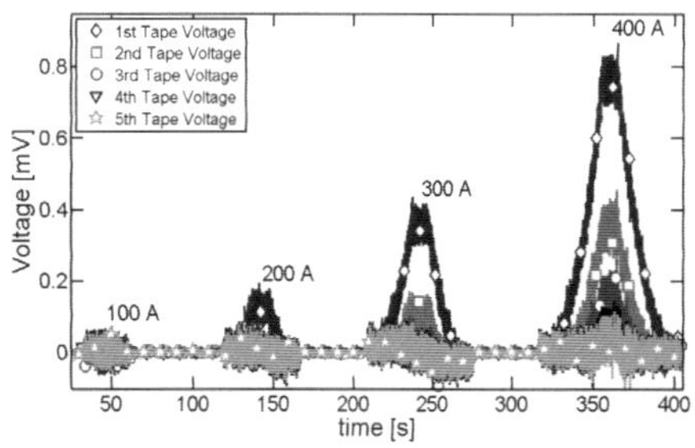

Figure 6: Development of the tape voltages while increasing the maximum current of the waveform. Note as the currents become larger, each tape voltage becomes non-zero in order from top to bottom.

Figure 5 shows that the conductor sample has stably conducted 610 A, which exceeds the minimal critical current of the sample stated in the datasheet. Though, it is important to note that even at low currents, the tape voltages exceed 10 $\mu V/m$ (see Figure 6) and at maximum current, an electric field of 5 mV/m is reached. This electric field far exceeds the threshold of 10 $\mu V/m$, commonly used for determining the critical current. Thermal runaway was verified during all tests of 620 A and more. During the constant current portion of the waveform, distinct differences between the tape voltages are observed. The differences between them become gradually smaller, the further the tape is from the current feeding connection plate.

IOP Conf. Series: Journal of Physics: Conf. Series **1293** (2019) 012065 doi:10.1088/1742-6596/1293/1/012065

Another note on Figure 6 is that as the currents become larger, each tape voltage becomes non-zero in sequence.

The results from the Hall sensors show that the current distribution at the negative and positive terminal, close to the connection plates, is non-uniform (see Figure 7). At both terminals, the measured magnetic field of the top sensor is larger than that of the bottom sensor. The result from the middle sensor shows small to no difference between the top or bottom sensors. It is likely that in the middle the current distribution is symmetric.

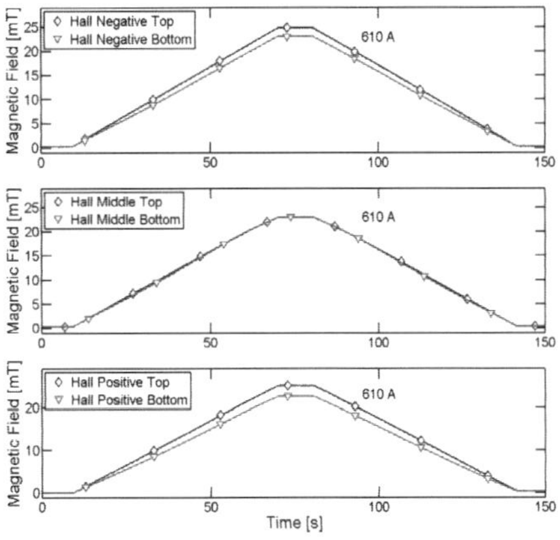

Figure 7: Results that show the current running through the sample at the positive and negative terminal is asymmetric and thus non-uniform. The result from the middle sensors imply that is likely that the current distribution is more symmetrically distributed in the middle.

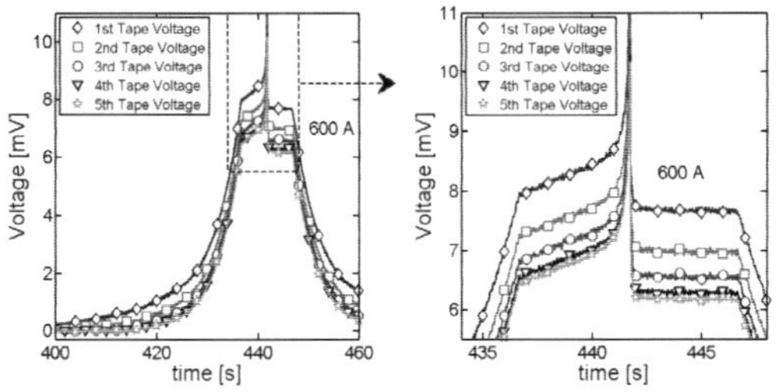

Figure 8: Graph of the recovery phenomena. Note that after the maximum current of the waveform is reached, thermal runaway of the tape voltages occurs for all tapes, followed by a sudden recovery back to a stable regime.

In Figure 8, an example of the recovery phenomena is shown. It was observed several times during tests near 600 A. After the maximum current of the waveform was reached, thermal runaway was detected, followed by an sudden crash back to stability.

4. Discussion

The fact that the sample has stably conducted a worst case non-uniform current, equal to its critical current, proves the excellent stability of non-transposed stacked HTS tapes. The tape voltages at constant current are relatively high compared to the critical electric field criterion (10 $\mu V/m$) but remain stable for 10 seconds and do not show any sign of instability/thermal runaway.

The Hall sensors confirm that the current distribution at the ends of the positive and negative terminals was asymmetric and thus non-uniform (see Figure 7), just as expected by the method. It has to be noted though, that 2D Biot-Savart analysis has shown that the uncertainty of the measurement was relatively high. The maximum deviation in magnetic field, due to the uncertainty of the sensor location was 1.4 mT. The maximum deviation in magnetic field, due to changes in the current distribution was 2.1 mT. Even though those two deviations are similar in size, there is reason to believe the measurement results from the negative and positive terminal are truthful. As pointed out in the methods, theory supports that the current distribution is non-uniform at the terminals because of the way the experiment was constructed. Additionally, it is suggested by theory, that it takes some length along the conductor to share the current between tapes [9]. Both arguments add much to the likelihood that the current distribution at the terminal was non-uniform. Remarkable is the Hall sensor results of the middle sensors. These results imply that it is likely that the current distribution naturally grows towards a symmetric distribution and will redistribute current among the tapes to do so.

Even though it is remarkable that the conductor sample is able to stably conduct a non-uniform current distribution, the electric field has to be addressed. Electric fields that exceed the standard 10 $\mu V/m$ criterion are observed at much lower currents than the critical current of the sample, some as low as 120 A. Although the diagnostic data of this experiment is not enough to explain this phenomena with absolute certainty, in the next section a hypothesis is formed based on the results observed. The hypothesis consists of two parts. The first part describes the dominant mechanism that creates the electric field at low currents (<300 A), and the second part, the dominant mechanism at high currents (>300 A). The hypothesis is built around the idea that the electric field is necessary to distribute the current between the tapes in the stack and thus the electric field is directly connected to the non-uniform current distribution.

The first part states that the electric fields are created, thus the tapes become resistive, because the critical current density is exceeded and that they are forced to share current with the tape below them. If the the total current flowing through the sample is lower than the critical current of one tape, it is likely that the superconducting layer of the first tape conducts almost all current. The paths through the others tapes are much more resistive because those paths have to cross the metallic resistive layers that sandwich the superconducting layer. The combinations of all metallic resistive layers in between two superconducting layers, for convenience sake, is called the Inter Tape Resistance Barrier (ITRB) and is shown in Figure 9. As the total sample current is increased, a point will be reached where the total sample current will exceed the critical current of the first tape and it will become resistive. As the resistance of the first tape increases, the path to the second superconducting layer becomes comparable in resistivity and the first tape will start to share current with the second tape below it. This first part of the hypothesis is supported by the results shown in Table 1 and Figure 5. Table 1 shows that during three tests, the currents at which the first and second tape become resistive, coincide closely with the critical currents of 1 and 2 tapes (120 A and 240 A respectively). Additionally the voltage differences between tapes shown in Figure 5 match with the mechanisms described in the hypothesis. The voltage differences between tapes are caused by the current flowing through the ITRB and as the current becomes smaller and smaller as it goes down the stack, the voltages differences become smaller and smaller as well.

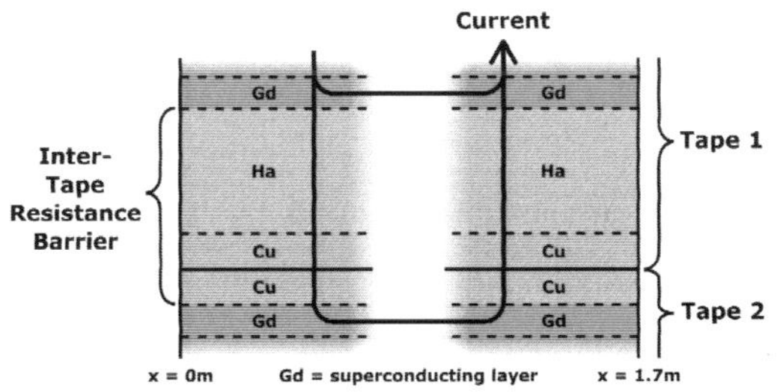

Figure 9: A simplified schematic showing the difference between two possible paths the current can take and their respective resistances. The difference in resistance being twice the Inter Tape Resistance Barrier (ITRB)

Table 1: The currents at which each tape starts its resistive transition for 3 different waveforms. Visible across all waveforms is the similarity between the current at which the first and second tape become resistive and the critical current of 1 and 2 tapes (120 A and 240 A)

Max Waveform Current	600A [A]	550A [A]	500A [A]
Tape 1	122	121	123
Tape 2	237	232	224
Tape 3	347	327	313
Tape 4	389	389	396
Tape 5	406	395	407

The second part of the hypothesis states that at some point, the increase in resistance due to the critical temperature being exceeded, becomes dominant. Note that this effect is caused by the increased resistivity described in the first part of the hypothesis. The combination of the increase in sample current and more superconducting layers becoming resistive, create local heat losses that exceed the cooling by the liquid nitrogen. This could explain the results for tape 3, 4 and 5 in Table 1 and why they occur at lower currents than their respective critical currents, 360 A, 480 A and 600 A. This could also explain the recovery phenomena. As a local hot spot boils off liquid nitrogen at a rapid rate, the nitrogen vapor is pushed outwards, creating a rapid flow that causes flash cooling, bringing the sample back to a stable resistive state. Future research might discover a way to control this phenomena and use it to suppress a thermal runaway, instead of the current approach of protection via shutdown.

If the hypothesis is assumed to be correct, this means that the method used to force the current non-uniform has caused the excessive and premature electric field and thus is the cause of the losses. In other words, to distribute the current between tapes, the electric field has to be present. If no distribution of current is necessary, the electric field would also disappear and the losses with it. This would change the way engineers approach transposition. Transposition would become an option, instead of a necessity. It would become a trade-off between additional losses in AC operation in return for a simplified conductor design and all the benefits that come with it. Moments of AC operation, like the charging and discharging of a coil would be paired with temporary increases in losses and refrigeration, up until the current has distributed itself once more. But as an upside, the manufacturing costs and mechanical performance of the coil would both improve. This opportunity is exactly why it is of paramount importance that future research is focused on verifying or disproving the hypothesis stated in this paper.

5. Conclusion

The experiment described in this paper has proven the excellent stability of non-transposed stacked HTS tapes and their ability to conduct a worst case non-uniform current. The stack of 5 HTS tapes has stably conducted a worst case non-uniform current of 610 A (Ic stack=600 A) at elevated electric fields of roughly 5 mV/m. In this paper, a hypothesis is discussed to explain this abnormally high and premature electric field. The hypothesis states that the electric field is necessary to distribute the current between tapes and that it is directly related to the non-uniform current distribution. Unfortunately, it was not possible to verify or disprove the hypothesis with the data available. If the validity of this hypothesis is confirmed, it would change the way engineers approach transposition. Instead of a necessity, it would become optional to transpose HTS cables. This opportunity is why future research should be focused on verifying or disproving the hypothesis.

Acknowledgements

Special thanks goes to FuseNet for supporting this research.

References

[1] Amemiya N 1998 Overview of Current Distribution and Re-Distribution in Superconducting Cables and Their Influence on Stability *Cryogenics* **38** 545–50
[2] Fietz W H, et al 2013 Prospects of High Temperature Superconductors for Fusion Magnets and Power Applications *Fusion Engineering and Design* **88** 440–45.
[3] Anvar V A, et al 2018 Bending of Corc® Cables and Wires: Finite Element Parametric Study and Experimental Validation *Superconductor Science and Technol.* **31** 115006
[4] Takayasu M, et al 2017 Electrical and Mechanical Characteristics of HTS Twisted Stacked-Tape Cable Conductor *IEEE Transactions on Applied Superconductivity* **27** 6900305
[5] Zhou C, et al 2016 Critical Current of Various Rebco Tapes Under Uniaxial Strain *IEEE Transactions on Applied Superconductivity* **26** 8401304
[6] Yanagi N, et al 2015 Design and Development of High-Temperature Superconducting Magnet System with Joint-Winding for the Helical Fusion Reactor *Nuclear Fusion* **55** 053021
[7] Takayasu M, et al 2014 Development of Termination Methods for 2G HTS Tape Cable Conductors *IEEE Transactions on Applied Superconductivity* **24** 6600105
[8] See http://www.superox.ru/en/
[9] Wilson M N *Superconducting Magnets* 232-35

Recent Progress of the High-Temperature Superconducting Cable Project in Japan

Tomoo Mimura[1], Takato Masuda[2], Hiroharu Yaguchi[3], Hiroyuki Fukushima[4]

[1] Tokyo Electric Power Company Holdings, Yokohama, Japan
[2] Sumitomo Electric Industries, Ltd, Osaka, Japan
[3] Mayekawa Mfg. Co., Ltd, Ibaraki Prefecture, Japan
[4] Furukawa Electric Co., Ltd, Tochigi Prefecture, Japan

Email: Mimura.Tomoo@tepco.co.jp

Abstract. We have spent considerable time studying high-temperature superconducting (HTS) cable for a long time. In Japan, a national project of the HTS cable system in a real grid operation, funded by the New Energy and Industrial Technology Development Organization (NEDO), started a decade earlier, which meant we could carry out real system operation for a year several years ago. We subsequently studied two issues on a new national project. One involved improving the performance of the cooling system, while the other involved safety verification against events possibly occurring in a system such as a ground fault or a short circuit. I will report on the outline.

1. Introduction

HTS cables have significant advantages of compactness and larger power transmission capacity and are expected to be installed in a power grid to replace existing power cables in future. In Japan, a national project funded by NEDO for HTS cables was established for about a decade, during which, despite already having a presence in the Asahi substation, actual grid operation continued for more than a year. Consequently, two further issues arose and the focus shifted to a new project several years ago. One of the issues involves recognizing the relevant aspects when troubles that can occur in the real grid also occur in the HTS cable. These include, for example, a short-circuit current flowing into an HTS cable, whereupon a ground fault occurs. It is crucial to understand the behavior of the HTS cable when the troubles occur. Other issues involve developing high-efficiency large-capacity refrigerators and confirming their reliability. In the new national project, we changed the existing Stirling type to Brayton-type refrigerators, re-energized the grid for a further year and reconfirmed efficiency, performance and reliability.

2. Outline of the research to date and the main points of this paper

The project schedule of the AC superconducting cable we are conducting is shown in figure 1. Phase 1 of the research ran up to the year 2013, while Phase 2 covered the rest of the study and in Phase 1 we connected the 240 m 66 kV superconducting cable to the real grid for the first time in Japan and reconfirmed its reliability through system operation [1], [2]. Among them, we confirmed the need to check for grid accidents, high refrigerator efficiency and problems in realizing long-term maintenance intervals. In Phase 2, to confirm these problems, we replaced the conventional Stirling refrigerator with a Brayton type developed in this research and confirmed its capacity performance and reliability [3]. The results of the short-circuit current test using a 40-m cable, its analysis and ground fault test results using a short cable were also partially reported[4], [5], [6]. The goal of the project is to create a guideline for using a superconducting cable in a real grid based on these results and summarize the high performance and reliability of the refrigerator. In particular, for the short-circuit and ground-fault currents, we examined how to handle superconducting cables after summarizing the method in the case of conventional cables. The essence of the safety guidelines of HTS cables and an evaluation of the refrigerator check test will be described below.

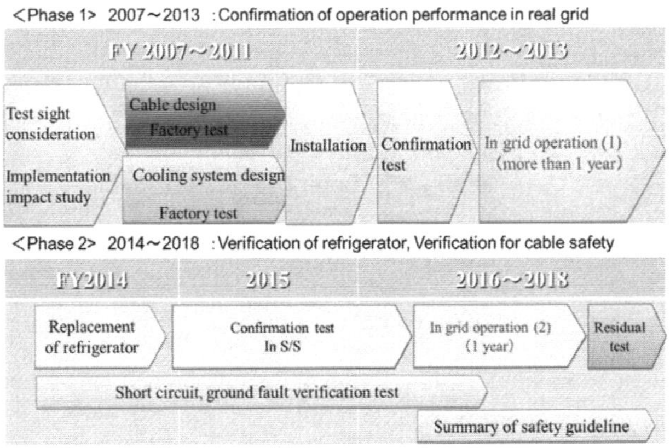

Figure 1. Schedule of National project of the HTS cable.

3. Measures against short-circuit currents and ground faults of conventional cables

The conventional cable we study is the XLPE cable, which is currently the major choice. Regarding the short-circuit current, when introducing the XLPE cable into the actual grid, the copper cross-sectional area is decided comprehensively with the short-circuit current based on the value from the user side (constant current, short-time operation current). In the case of the XLPE cable, the three allowable temperatures are set for the purpose of securing the long-term insulation performance of crosslinked polyethylene and the specific temperatures used in Japan are shown in Table 1. The key point here is the very considerable temperature difference between the constant time and the instant. The reason for the high instantaneous temperature is that it allows a temperature up to the melting point of the crosslinked polyethylene. Accordingly, it is rare for the copper cross-sectional area to be determined by the instantaneous temperature, which means the cable itself is rarely damaged due to bearing the short-circuit current.

Regarding countermeasures for ground faults, it is difficult to define general countermeasures based on the varying laws and regulations in each country. In Japan, countermeasures are taken to adopt a flame-retardant sheath or place it in a rugged container to avoid any serious impact on other equipment.

Table 1. Allowable temperatures of XLPE cables

Items	Value [degrees]
Continuous allowable temperature	90
Short-term allowable temperature	105
Instantaneous failure temperature	230

4. Important results in creating the safety guideline

4.1. About items necessary for guideline

Since few reports exist of a short-circuit current flow of HTS cable and ground faults at the pre-project stage, many tasks have been done to check what happens when each event occurs. Although we considered countermeasures from the subsequent test results, a trial and error approach was also used when finalizing countermeasures such as cases where it was difficult to put into practical use and cases where the appearance differed for long-distance cables. Accordingly, elements initially considered important were not and vice versa, making it very difficult to find general versatility to finally compile a guideline. Each mode is described below.

4.2. Short-circuit current flow
To discuss cable system safety due to the passage of a short-circuit current, we constructed a 40-m cable system. Temperature and pressure data were acquired and compared with the calculation results by a simulation code to simulate the passage of the short-circuit current via the long-distance line and the subsequent process. As described in Section 3, the current always flows after the passage of the short-circuit current, so the superconducting cable must be capable of transmitting the rated current with the temperature rising after accommodating the short-circuit current. In Japan, the maximum short-circuit condition for 66-kV class cables is 31.5 kA, 2 sec, so the calculation result under each condition using the above simulation code, with this as the upper limit, is shown in the figure 2. Although this is a design example, in the cable used for the actual grid interconnection, the upper limit temperature at which the rated current of 2kArms can flow is 92 K and the time intersecting with 92 K denotes the maximum condition under which the rated current can flow, even after a short circuit. Therefore, when a short-circuit current stricter than a certain short-circuit condition flows, the HTS cable needs to be disconnected from the grid since it cannot conduct the current constantly thereafter. Of course, since the rated current does not necessarily flow, a sequence for judging from the temperature and pressure in that situation must be provided. Also, of course, there is scope to ease the condition by the amount of copper of the former and the margin of Ic and application of the fault current limiter. However, it goes without saying that a comprehensive cost comparison is required and where the fault current limiter is not applied, there is a need to incorporate these points into the cable design.

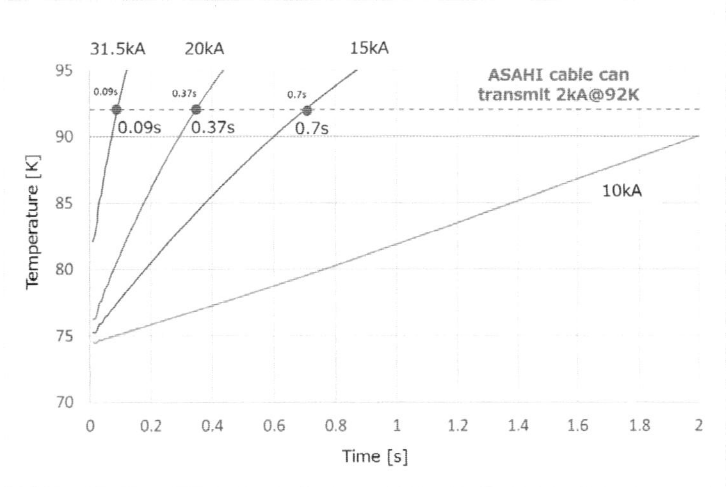

Figure 2. Condition for allowable short-circuit current

4.3. Ground Fault
As well as superconducting cables, power cable troubles mostly involve ground-fault destruction. Although the factors vary, as described in Section 3, it is important to prevent the spread of fire due to ground faults and refrain from breaking the other infrastructures with a ground-fault cable, as described in Section 3. For HTS cables, leakage of liquid nitrogen is certainly a concern during a ground fault. So

IOP Conf. Series: Journal of Physics: Conf. Series **1293** (2019) 012066 doi:10.1088/1742-6596/1293/1/012066

with prevention in mind, a ground-fault simulation test with a short cable was performed to confirm whether or not it would be possible to prevent a hole from being drilled by applying a protective layer inside the cable. Figure 3 shows the structure of the 275-kV cable and the state in which the protective layer is wound 1 cm immediately above the SUS inner pipe. Figure 4 shows the conditions of the inner and outer SUS pipes when the ground-fault current condition is 30 kA, 0.06 s. Although this condition is sufficiently feasible in our company, consequently, the direction to prevent external leakage of liquid nitrogen means an increased cable diameter, which has led to the loss of flexibility. Accordingly, we decided to rely on the leakage countermeasure described in the next section.

In addition, since the ground-fault current at the 66-kV class is standard at our level, several hundred amperes, puncturing is not done in some cases depending on circumstances. In this case, the internal pressure increased, which impacted on the pressure-resistant design of the SUS pipe itself and the range of pressure propagation in a longitudinal direction, but experiments and simulation confirmed that this level would not hinder the current design. From a user maintenance perspective, there is a need to identify the cable ground-fault position because we cannot watch the position directly, particularly in the case of installing in a duct. Also, for early restoration, there is a need to realize temperature rise and re-cooling at high speed.

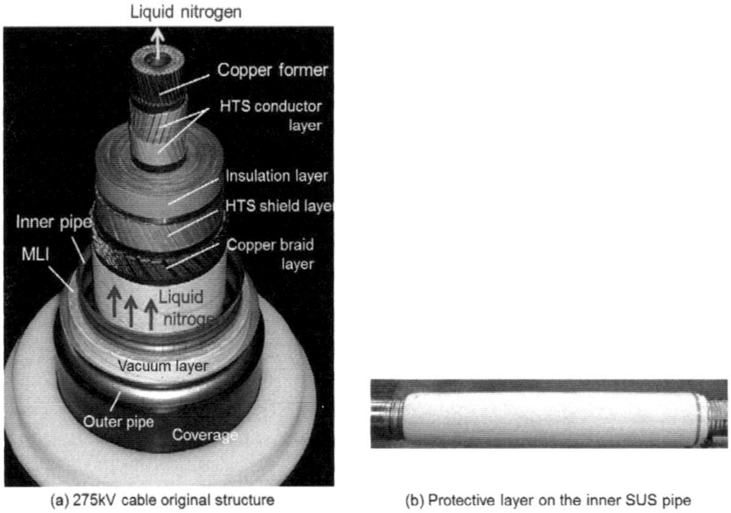

Figure 3. 275kV cable structure (a) original structure (b) additional protective layer

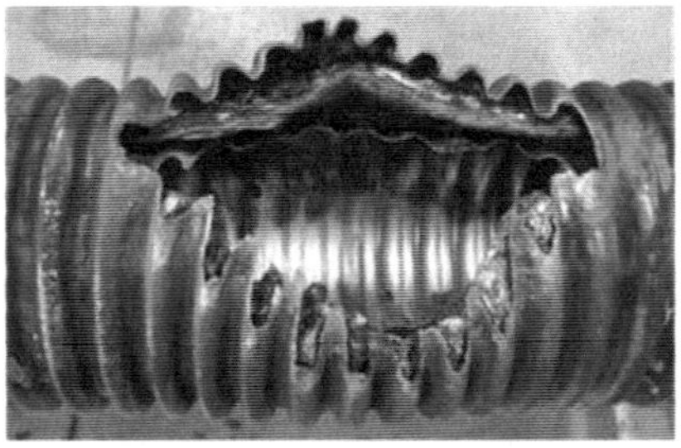

Figure 4. Broken SUS pipe by ground fault

4.4. LN2 leakage countermeasure

Initially, though external leakage via cables is assumed, connecting parts, valves, etc. slight leakage and excavation, etc., leakage at the time of a ground fault is thought to be the most severe condition because it is considered to be both quantitative and fast. Even when ground faults occur, the conditions to be considered include various factors, such as puncture size, cable height difference, leakage and the cable length. Given the difficulty in verifying these various cases this time, we built a concrete structure with a volume of about half the general manhole size used for 66 kV cables. Figure 5 shows a complete view of the manhole. A simulation test was then conducted to flow liquid nitrogen into the same. The flow rate was fixed to be close to real grid operation and basic data was acquired by measuring the temperature, pressure and oxygen concentration transition of each manhole part, the surface temperature of the installed CV cable, etc. together. Figure 6 shows the temperature transition of the thermometer installed in the concrete wall. The thermometer installed at 20, 50 mm from the floor, approached the temperature of liquid nitrogen in under 20 minutes and it was confirmed that liquid nitrogen had accumulated. Moreover, the temperature started to rise within about 30 minutes of stopping the leak. Depending on the size of the hole and the condition at the time of the ground fault, rapidly detecting the ground fault and quickly stopping LN 2 circulation allows some oxygen concentration deficiency in the manhole and is also feasible during normal cable operation. There is thought to be scope to handle the same without requiring work peculiar to superconductivity by ensuring sufficient ventilation in the manhole. The results of the surface temperature measurement of the CV cable are shown in figure 7 and the CV cable was installed at a height of 700 mm from the floor. This cable is not designed to a cold district specification, but one usually used at the temperature and usage condition of 0 ℃ or more. Following the test, the temperature dropped to about -20 ° C immediately after the leakage started and rose once the leakage had ended, but under this test condition, despite the temperature going below 0 ° C for less than 1 hour, no surface cracks were seen. We did not confirm the cable loading performance, but this will be future work.

Figure 5. Mock up for LN2 leak test in manhole

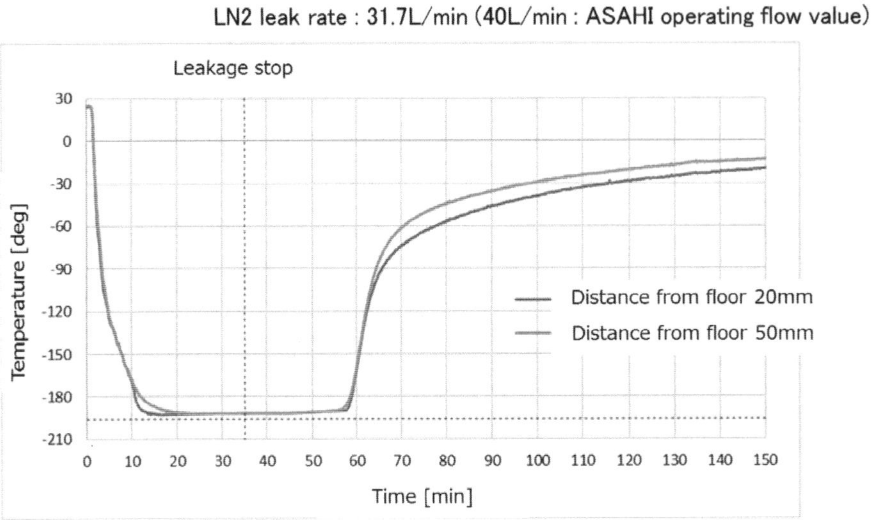

Figure 6. Examination result of LN2 leak test

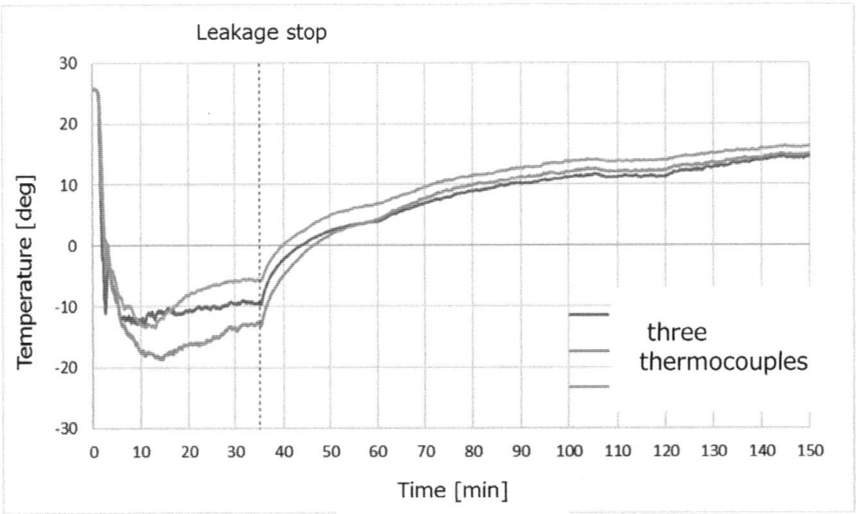

Figure 7. The surface temperature measurement results of the CV cable

4-5 Outline of the guidelines

Based on the contents and findings reported to date, we have developed safety management guidelines to manage operations. The operation management includes steady and unusual states and the contents to date correspond to an unusual state, while the contents to be included in the guideline are shown in Table 2. The steady state is summarized in the next section. Regarding short-circuit events, as described above, there are cases in which operation specific to superconductivity is required, but rough knowledge about other events was obtained. We got a prospect of real system operation by operating management based on the guidelines.

Table 2. Unusual state of safety guideline for HTS cable

Items	Contents
Short-circuit current	(1) Having a re-transmission enable/disable judgement function. (2) Allowable temperature of the cable is clear.
Ground fault	(1) Restoration work centering on oxygen luck MUST ensure safety. (2) Do not affect the soundness of other facilities. (3) Accident span can be reliably detected.
Excavation	(1) The excavation span can be identified. (2) Ensuring public safety.

5. Brayton-type Refrigerator verification result

5-1 Verification of the Brayton refrigerator

To realize a large-capacity and high-efficiency refrigerator, we developed a 5 kW @ 77 K Brayton refrigerator, confirmed it could achieve COP 0.1 at the factory test and then replaced the conventional Stirling refrigerator in the Asahi substation, which is the actual grid operation field. The system flow of the Brayton-type refrigerator and the installed equipment are shown in figure 8. Subsequently, initial cooling, including superconducting cables, was carried out and grid operation was also carried out for about one year from March 31, 2017. The results of comparing the one-year COP transition with the conventional Stirling refrigerator are shown in figure 9. Multiple Stirling refrigerators were installed to secure the refrigerator capacity, but vacuum maintenance of the on-site refrigerator, supplement of the refrigerant and overhaul etc. were required fewer than 10 times. Brayton realized operation with no maintenance at all, except for having to deliberately reduce the degree of vacuum for data confirmation. It was also confirmed that COP had twice the performance of the Stirling type. In the grid operation, since the operation was made to keep the cable operation temperature constant at 69 K, it was lower than the COP at a factory test of 77 K, but almost as expected based on temperature dependency. Although we are currently disassembling and confirming the performance deterioration at each part level, no performance degradation has occurred.

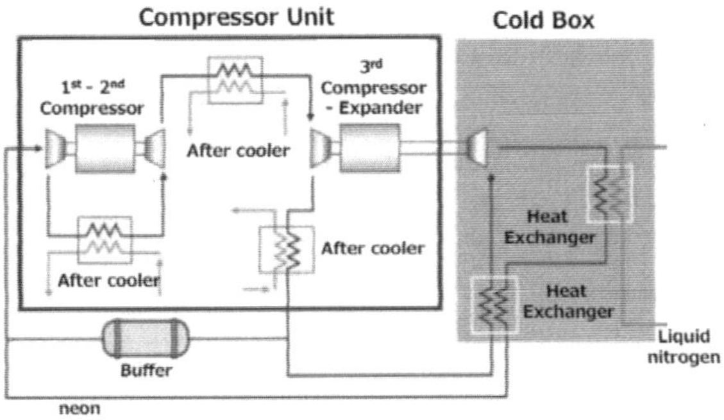

(a) Brayton refrigerator flow

(b) Refrigerator installed at Asahi S/S

Figure 8. Brayton refrigerator system flow(a) and installed at Asahi S/S(b)

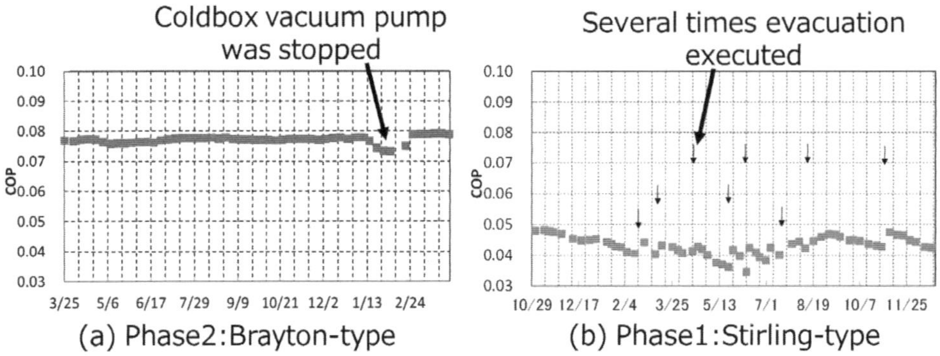

Figure 9. COP transition of refrigerator among 1 year

5-2 Steady-state of the guideline

We have already described the guidelines at the time of unusual state in the previous chapter, but outline them at the time of steady operation. For low-temperature equipment not limited to superconducting cable systems, since the information to take out each sensor from the inside of the vacuum vessel dominates, there is a need to thoroughly examine the information to be managed when designing each piece of equipment. Also crucial is knowing what threshold at which to manage the information obtained. Since superconducting cables can handle high-capacity power transmission, constant circulation of the cooling system is indispensable, so it goes without saying that improved redundancy over conventional systems is required. From this perspective, further progress in sensors and ease of exchange are also needed, which is what underpins the steady-state guidelines summarized in Table 3. Although it is not possible to uniformly describe the specific contents due to users' policy etc., I think that attention should be paid to installing detection sensors, etc., to keep the circulation system operational as much as possible.

Table 3. steady state of safety guideline for HTS cable

Items	Contents
Basically various parameters should be within the normal range	
Parameters	Temperature, Pressure, Flow rate, Vacuum level
Range	Organize thought of alarm setting

6. Conclusion

To prepare guidelines concerning the safety of superconducting cable systems, possible events in the grid were verified by actual model cables and systems. Using the results, we summarized the contents to be included in the guidelines for operating the long cable system at the time of actual application by using some simulation codes. Moreover, good results such as ensuring prescribed capacity and scope to operate with almost no maintenance, etc. were obtained by conducting a one-year grid interconnection test using a high-efficiency large-capacity Brayton refrigerator.

We are currently preparing guidelines, but expect that they will constitute reference data for designing and operating superconducting cable systems in future.

Acknowledgments
This work was supported by the New Energy and Industrial Technology Development Organization (NEDO).

References

[1] Honjo S, Mimura T, Kitoh Y, Noguchi Y, Masuda T, Yumura H, Watanabe M, Ikeuchi M, Yaguchi H, and Hara T, 2011 *IEEE Trans. Appl. Supercond.* **21**(3) 967-971
[2] Ohya M, Inagaki Y, Tatamidani K, Ito H, Saito T, et al., 2013 SEI Technical review **76**
[3] Tanaka G, Shimoda M, Yaguchi H, and Nakamura N, 2018 Asian Conference on Refrigeration and Air-conditioning **E211**
[4] Takeda N, Yasui T, Yokoo Y, Agatsuma K, Ishiyama, et al., 2017 *IEEE Trans. Appl. Supercond.* **27**(4) 5400705
[5] Masuda T, Morimura T, Nakano T, Maruyama O, Mimura T, Yasui T, Agatsuma K, and Ishiyama A, 2017 *IEEE Trans. Appl. Supercond.* **27**(4) 5401504
[6] Takagi T, Yagi M, Mukoyama S, Watanabe K, Mimura T, Maruyama O, and Nakano T, 2017 *IEEE Trans. Appl. Supercond.* **27**(4) 5401205

Low voltage and large current DC superconducting power cable designs for 10 km to 100km transmission line using experimental data of Ishikari project

Satarou Yamaguchi[1], Takeo Yamada[2], Takashi Iitsuka[2], Akio Sato[3], Toru Sawamura[4] and Victor Sytnikov[5]

[1] Chubu University, Kasugai, Aichi 487-8501, Japan
[2] JGC Corp. Yokohama, Kanagawa 220-6001, Japan
[3] JFE Steel, Chiyoda, Tokyo 100-0011, Japan
[4] SAKURA Internet Inc., Shinjuku, Tokyo 160-0023, Japan
[5] R&D Center, FGC UEC, Moscow, Russia

E-mail: yamax@isc.chubu.ac.jp

Abstract. Main merit of the DC superconducting power transmission line is low loss even in low voltage power transmission. It should be realized when the transmission length is long. Therefore, we discuss and analyse an aspect of thermo-hydrodynamic performance of Ishikari project, and show one design of 100 km transmission line. At the same time, we also discuss the cable design and the rated voltage of the cable is lower than 50 kV for 500MW transmission. This is one of the most important advantages to make the DC power grid because we can omit an ultra-high voltage substation, the primary and the secondary substations and we can connect the power stations to the distribution stations directly. Finally, we may be able to reduce the construction and operation costs of the power grid.

1. Introduction

Main merit of the DC superconducting power transmission line is low loss even in low voltage power transmission. It should be realized when the transmission length is long. Therefore, the engineering objective is how to design, construct and operate a long transmission line. It is related with the performance of thermo-hydrodynamic parameters, such as the heat leak of the cryogenic pipe, the flow rate of the liquid nitrogen (LN2), the flow resistance of the pipe, the LN2 temperature from the refrigerator, the LN2 pressure of the cryogenic pipe, the LN2 temperature profile along the transmission line, the pressure-drop of the cryogenic pipe for the circulation of the cryogen and the output pressure of the circulation pump and so on. These parameters are related with each other, and we should keep the temperature of the high temperature superconducting (HTS) cable below the design value, such as 77K. Finally, the distance between the refrigerators along the transmission line is the key to realize an actual transmission because if the refrigerator should be installed for a short distance along the transmission line, we cannot construct a long transmission line. The similar system is the natural gas pipeline, and the pump-station is installed along the pipeline. The average distance between the gas-pump stations is 5 km to 20 km inside Japan, and 100 km for the intercontinental pipeline. Therefore, we can adopt these values for the superconducting power transmission line, and this is a goal of the design and performance of the superconducting power transmission line. In order to keep the economy of the superconducting power transmission line, lower electric power loss should be realized from the lower heat leak of the cryogenic pipe and the higher COP of refrigerator system.

In order to solve the problem, we developed the radiation shield structure [1] in Ishikari project. The cross-sectional view is shown in Figure 1. There are two internal pipes. One is called as the cable pipe that the superconducting cable is set inside, and the other is called the return pipe for the circulation of the liquid nitrogen (LN2). They are made of the stainless-steel. The radiation shield is covered and connected to the return pipe thermally, and surrounded the cable pipe. Therefore, the heat flux from the outer pipe comes into the return pipe mainly, and the heat leak of the cable pipe is very small. The LN2 from the refrigerator flows into the cable pipe, then LN2 goes out from the cable pipe to the return pipe and finally goes back to the refrigerator. The cable pipe is supported by the FRP (fiber-reinforced plastic) legs from the radiation shield, Therefore, we should consider the thermal conduction of the support legs. However, the thermal conduction is very low in our design, and thus the temperature rise of the cable pipe is very low. This structure is important to make a long transmission line. If the temperature rise of the cable is high, the critical current of the HTS cable is lower along the temperature rise, and therefore we set the rated current of the HTS cable should be lower. Finally, we need many cooling stations along the cable if we keep the rated current high.

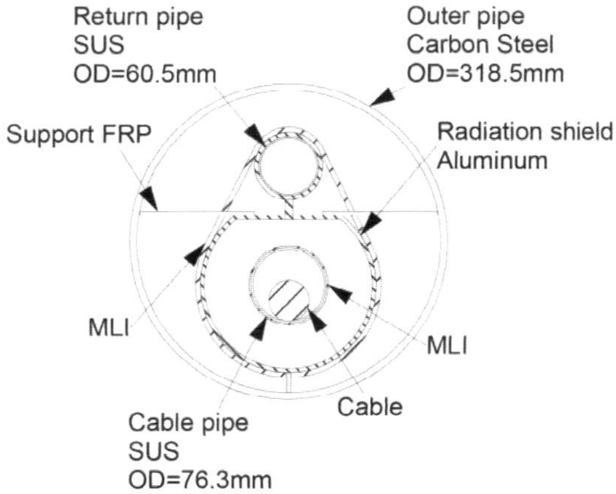

Figure 1. Cross-sectional structure of the cryogenic pipe with the radiation shield in Ishikari project.

The other aspect of the Ishikari project is to use the straight pipes in order to reduce the pressure drop of the cryogen circulations. Usually the corrugated pipe [2] is used in many superconducting cables, but we think its pressure drop is too high to construct a longer cable. This is discussed in the next section. We discuss and analyze the experimental data of Ishikari project, and we describe the thermos-hydrodynamic designs of 50 km to 100 km transmission line under the reasonable assumptions. The present power grid uses high voltage transmission lines in order to reduce the loss of electricity, and we need many substations in the grid. However, if we can use the HTS cable, the resistance of the cable is zero basically, and therefore we do not need to use high voltage grid. The voltage of the generator is ~ 20kV, and if we can connect the DC HTS cable directly after rectification, we can save the substations of the grid. Finally, we discuss the grid system using low voltage cable in the paper.

2. Experimental data from Ishikari Project and scale law for the circulation

Thermo-hydrodynamic experimental data [3] in Ishikari project are shown in the first column of the Table 1. Its length is 474 meters, and the flow rate is 30 litter/min. We can estimate the thermo-hydrodynamic data for 5 km and 10 km cables from the experimental data, such as pressure drops, the temperature rises, and the output temperature of the system, where the inlet temperature is set to be 70K. The design of the cryogenic pipe is completely the same. They are shown in the second and third columns [4].

The estimation results show that we can construct 5 km to 10km transmission line to use almost the same parts of present cryogenic system, such as the pipe, the circulation pump, but we need high refrigeration power for long transmission line. It is important that the flow rate is only 25 litter/min for the 10 km transmission line, and therefore the pressure drops of the cable and return pipes are lower than the maximum output pressure of the present circulation pump, which is 1MPa. In addition, the liquid power, that is the product of the circulation pressure and flow rate, is one of the heat-loads to the low temperature system. In case of the 10 km transmission line, the liquid power is calculated as 8.33mW/m which is much smaller than the heat leak of the cryogenic pipe. It means that we can neglect the liquid power because the total heat leak of the cryogenic pipe is ~ 1.0 W/m. Usually when the transmission line is longer, the flow rate should be high in order to keep the temperature rise of the superconducting cable for its whole length. This is necessary to keep the current carrying capacity of the superconducting cable. However, if the flow rate is high, the output pressure of the circulation pump is high. Therefore, we should consider the liquid power because it is one of the heat-loads for the cryogenic pipe, and if it is high, the economical efficiency of the superconducting cable would be worse because of high electric power consumption of refrigerator.

Table 1. The original data from Ishikari project, and the extended data for 5 km and 10 km transmission line.

Radiation Shield	Original data		
Length [m]	474	5,000	10,000
Flow Rate [L/min]	30	20	25
ΔP [kPa]_cable [K]	5.20	24.4	76.2
ΔP [kPa]_return [K]	8.50	39.8	124.5
Cable ΔT [K]	0.019	0.30	0.48
Return ΔT [K]	0.46	7.2	11.5
Outlet temp [K]	70.5	77.5	82.0
Total Pressure [kPa]	13.7	64.2	201
Heat Leak [W/m]_cable	0.034	0.034	0.034
Heat Leak [W/m]_return	0.85	0.85	0.85
Total Heat Leak [kW]	0.42	4.4	8.9

This engineering aspect is sometimes serious, and therefore we should discuss the scale law of the thermo-hydrodynamics of the cryogenic pipe. One example of the scale law for the length of cryogenic pipe is shown in Table 2.

Table 2. Thermo-hydrodynamic parameters to keep the temperature rise of the cryogenic pipe for 1 km and 10 km cables.

Distance [km]	1.0	10.0
Temperature rise [K]	3.0	3.0
Flow rate [L/min]	30	300
Output pressure [MPa]	0.1	100
Liquid power [kW]	0.05	500

We assume the parameters for 1 km transmission line in the left-hand side column of Table 2. Then, in order to keep the temperature rise of the cryogenic (cable) pipe, the flow rate, the pressure and the liquid power are estimated for the 10 km transmission line as shown in the right-hand side of Table 2. The flow rate is proportional to the length, and therefore it is 10 times higher than the value of the 1 km system because the heat leak is constant for length. Since the pressure is proportional to the square of the flow rate, it is 100 times higher for the same length, but its length is 10 times longer, therefore it is 1000 times higher for the same design of the cryogenic pipe. Since the liquid power is the product of the pressure and flow rate, the total heat load for the 1 km transmission line is 50W, and it corresponds to 50mW/m. This value is lower than usual experimental heat leaks for unit length. Therefore, it is acceptable. However, when the length is 10 km, the liquid power is 500kW, and its heat leak per length is 50W/m. We cannot accept this level of the heat load and it is higher than the heat leak of the cryogenic pipe. This kind of the analysis had been discussed in Ref. [5]. In this meaning, if we get the experimental data for 1 km transmission line shown in Table 2, we cannot extend the cable length up to 10 km.

3. Thermo-Hydrodynamic Analysis for 100km cable

It is necessary to extend the cable length longer than 10 km when we consider the present electric power grid. Therefore, after we estimated the thermo-hydrodynamic parameters for 10 km transmission, we try to extend the cable length longer using the experimental data from Ishikari project. Figure 2 shows the 50 km superconducting transmission line under the several reasonable engineering assumptions. The inlet temperature of LN2 is 70 K and the temperature rise of the cable pipe is 1 K for 50km, where the flow rate is 60 litter/min using the same heat leak for the cable pipe in Ishikari experimental data. But since the heat leak for the return pipe is high, the temperature rise is 25 K for 50km and the same flow rate.

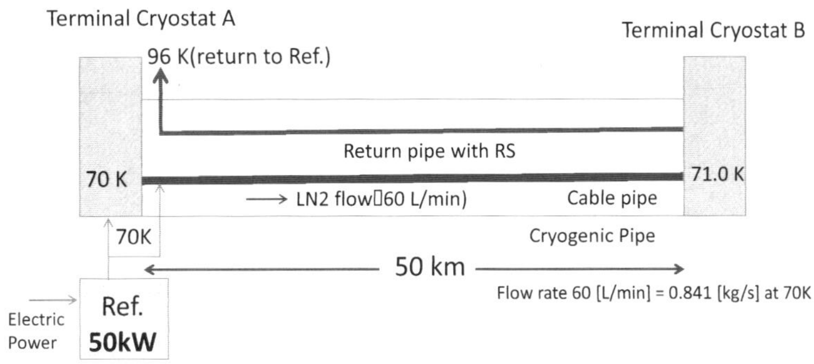

Figure 2. A design of 50 km superconducting transmission line using the experimental data from Ishikari project.

In order to reduce the pressure drops of the cable and return pipes, we will use large pipes. The size of the new large cable pipe is designed to be smaller than the radiation shield's, we can put into the cable pipe inside the radiation shield. The size of the new return pipe is also larger than the size of the pipe in Figure 1, but because it is covered by the radiation shield, the area of the radiation shield is the same area of the Ishikari design of Figure 1. Therefore, we assume that the heat leaks of the cable and return pipes are the same as the Ishikari experimental data. The new inner diameters of the cable and return pipes are 84.9 mm and 97.4 mm, individually, and we can estimate the pressure drop of the pipes, which is proportional to the fifth power of the pipe diameter using the experimental data of Table 1 and Figure 1. Finally, the pressure drops of the cable and the return pipes are 0.486 MPa and 0.495 MPa for 50 km, individually. Therefore, the liquid powers of the cable and the return pipe are 8mW/m, and lower than the heat leak, and they can be negligible like the present experimental data. The outlet temperature of the LN2 from the return pipe is 96K. The saturation pressure at that temperature is lower than 0.5 MPa. The design basis for the 50 km transmission line is almost the same as the design basis of the Ishikari project, and therefore we can use many same cryogenic parts, such as the pipe, bellows, pump and so on. This is an important to estimate the construction cost of the 50 km transmission line. Furthermore, we can use a turbo-Brayton refrigerator of 50kW at 70K in Figure 3, the COP of the refrigerator is ~ 0.12 including the water chiller and the control system of the refrigerator. It is almost 1.5 times improvement compared with the COP of the small present refrigerators.

T. Yamada et al presented the design of 100 km transmission line [6] in the annual domestic conference of Japan, 2018. They used the flush refrigerators for the return pipe in order to keep the internal pressure of the return pipe. However, if we use the flush refrigerators, the energy loss of the flushed nitrogen gas to the air is high because the temperature of the nitrogen gas is still low, and we lose a fundamental merit of the HTS cable. Figure 3 shows the different design of the 100 km transmission line without the flush refrigerators. This design does not lose the energy loss come from the liquid nitrogen loss from the flush refrigerators.

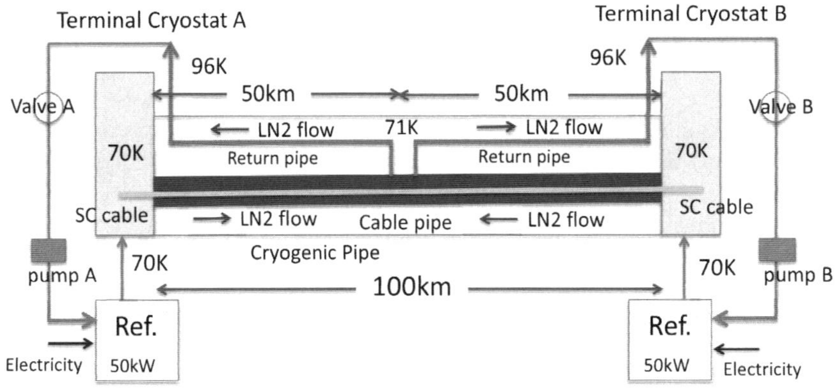

Figure 3. A design of 100 km superconducting transmission line using the experimental data from Ishikari project.

This system is combined with two 50 km systems in Figure 2. But the cable is laid through 100 km. Each return pipe is connected to the cable pipe at 50 km from each terminal cryostat. The LN2 flow is the counter flow in the cable pipe from each terminal cryostat. Two return pipes are not connected each other, but connected with the cable pipe. LN2 flows back to each terminal cryostat, and the length of the return pipe is 50 km. Therefore, we do not need to use the flush refrigerators. In order to control the flow rate of each return pipe, the valve and the circulation pump are used for each side as shown in Figure 3. Finally, the thermo-hydrodynamic parameters of the 100 km transmission line are the same as the parameters of the 50 km transmission line shown in Figure 3.

4. Electrical design of Superconducting cable

The voltage from the power station (generator) is AC 22kV, and the transformer is used to step up the voltage to AC 500kV in the ultra-voltage substation, and its high power is transmitted for a long distance in Japan. The voltage is step down from 500kV, 275kV to 66kV, 22kV in the primary and the secondary substations, finally the electric power is sent to the distribution substations, and its voltage is step down from 22kV and/or 66kV to 6.6 kV, and its power is distributed to many customers in Japan. The voltage is different from the foreign countries slightly, but the grid system is similar all over the world now.

The second merit of the superconducting power transmission line is to be able to use the low voltage system even for high electric power transmission line. The present AC power grid needs to use many step-up and step-down transformers to minimize the losses. Figure 4 shows the cross-sectional structures of the superconducting power cables, and their voltage, current and diameters.

We designed several cables to find the optimum design of the DC superconducting power transmission, and we choose that their voltages are from ±25kV to ±50kV for 500MW power transmission. The reason to choose these voltages is to pass through the ultra-high voltage substation, the primary and secondary substations of the power grid mainly. It may be effective to reduce the total cost of the grid, and we can install the HTS cables into the grid even if the cost of the HTS cable system is expensive. One is the single core cable shown in right-had side of Figure 4, and the other two are co-axial cable structures, with (center) and without (left) the earth-ground layer. The voltage of the 500

MW transmission line is 500kV in Japan, and 400kV in EU for usual AC transmission line now, and the voltage of 50kV to 25kV is not used for large electric power. But if we can use this level of the voltage, we may be able to omit an ultra-high voltage substation, the primary substation and the secondary substation from the present grid, and we can connect to the power station to the distribution substation directly. The co-axial structure is not good for an AC cable because the cable capacitance is high, and the reactive power will increase even if for a short distance, but the reactive power is almost zero because of the DC transmission. Therefore, we can use the co-axial cable for a long distance, and it should be installed underground for safety reasons mainly.

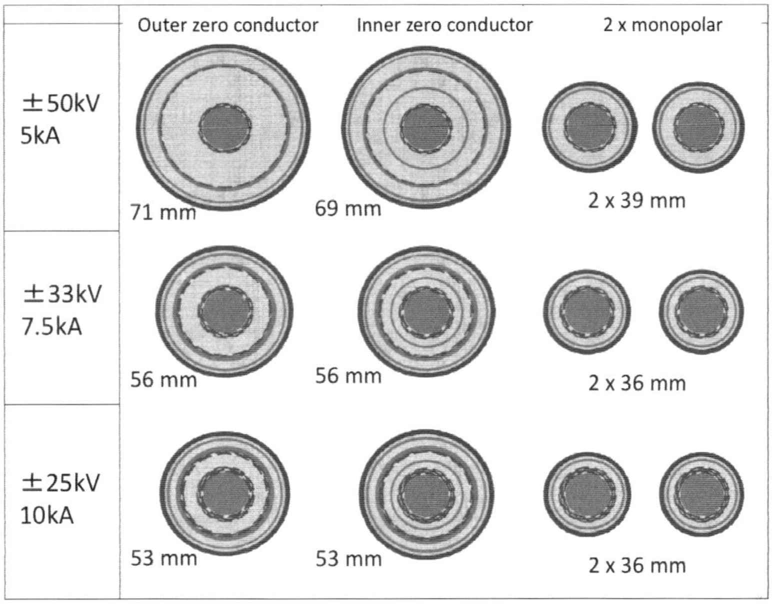

Figure 4. A design of 100 km superconducting transmission line using the experimental data from Ishikari project.

If we can omit the high voltage substations from the present grid, we may have merit to reduce the operation and the construction costs of the grid even if we should use the power converters. This cost reduction is not small. Moreover, if the voltage is ±25kV, we can avoid the corona (= electric breakdown of air at 1 atm), and the substation will be smaller. These are the second merits.

5. Summary and future plan

Before we will construct a 100 km superconducting power transmission line, we should consider to construct a 10 km-class transmission line and it should be used as the commercial system. We will be able to test and check the parameters of the superconducting power transmission system, and the technology will be completed by the construction and operations of the 10 km transmission line through these tests. All kinds of the technologies, such as the low heat leak of the cryogenic pipe, will be improved by the construction and the operation of the 10 km transmission line. The experience of the 10 km transmission line will be useful to construct the 100 km transmission line as a commercial transmission line.

On the other hand, the authors and colleagues believe that it is important to realize the 100 km transmission line significantly. Because natural-gas pipeline is one of the major energy transmission lines in the world now and its length is extended to 1000 km to connect the different countries. Many pump-stations are used to transfer the natural gas (NG) for each 100km usually. The pump stations consume an electric power to compress and transfer the NG because of its viscosity. This structure is similar to the superconducting power transmission line. When we operate the long power transmission line, the security is important and therefore the underground cable will be chosen. Low DC voltage system is also important for long operation time. After we will construct and operate the 100 km transmission line, we will be able to apply the HTS cable system to the international power grid, and connect all countries by the HTS power grid.

Acknowledgement
The authors thank to Prof. A. Iiyoshi, the Board Chairman of Chubu University, for his continuous encouragement to proceed the research, and Profs. H. Takano, N. Chikumoto, H. Watanabe and Yu. Ivanov and M. Kanda for their supports and helpful discussions in Chubu University. We also acknowledge to all members of Ishikari cool and smart energy system research association (*I-Spot*) and the study group of Superconductivity for Local, Low temp., Low cost, Liability DC superconducting power transmission line (S4L).

6. References

[1] Yamaguchi S, Kozhizuka H, Hayashi K and Sawamura T 2015 *IEEE Trans. Appl. Supercond.* **25** 5402504.
[2] Mukoyama S, Yagi M, Ichikawa M, Torii S, Takahashi T, Suzuki H, Yasuda K 2007 *IEEE Trans. Appl. Supercond.* **17** 1680.
[3] Watanabe H, Ivanov Yu, Chikumoto N, Takano H, Inoue N, Ishiyama K, Oishi Z, Koshizuka H, Watanabe M, Matsuda T, Hayashi K, Sawamura T 2017 *IEEE Trans. Appl. Supercond.* **27** 5400205.
[4] Yamaguchi S, Vyatkin V, Ivanov Yu, Watanabe H, Chikumoto N, Inoue T, Takano H 2016 *Abstract of CSSJ Conference* **93** 132 (Japanese).
[5] Yamaguchi S, Kawahara T, Hamabe M, Watanabe H, Ivanov Yu, Sun J, Iiyoshi A 2011 *Physica C*, **471** 1300.
[6] Yamada T, Iitsuka T, Sato A, Sawamura T, Yamaguchi S 2018 *Abstract of CSSJ Conference* **96** 149 (Japanese).

A Feasibility Study of Smart High-Temperature Superconducting Cable to Improve Stability of KEPCO System

Sangsoo Seo[1], Seung Ryul Lee[2], Jeonwook Cho[3]

[1] Senior Researcher, KERI (Korea Electrotechnology Research Institute), Uiwang-city, Gyeonggi-do, Korea
[2] Principal Researcher, KERI (Korea Electrotechnology Research Institute), Uiwang-city, Gyeonggi-do, Korea
[3] Principal Researcher, KERI (Korea Electrotechnology Research Institute), Changwon-city, Gyeongsangnam-do, Korea

E-mail: s_seo@keri.re.kr

Abstract. The Korean power system is a very tightly coupled island system. The installed capacity is about 100 GW and the load demand is gradually increasing. Especially in metropolitan areas where the load is concentrated, the system fault current often exceeds the breaker capacity due to the strong connection. The system operator limits the fault current by bus split operations considering the breaker capacity but this reduces system stability. Therefore, various efforts have been made using a fault current limiter to overcome these problems. A smart superconducting (SC) cable is a type of SC power transmission cable. It has not only standard structure of the existing SC power cable but also a fault current limiter. The cable has a SC characteristic in a normal state, while when a fault occurs, it can limit a fault current through generating impedance that adjusts electrical and thermal properties of SC cable and material and cross-sectional area of SC cable. In this paper, a technique is proposed to reduce system fault current while improving the system stability using a smart high-temperature SC cable. In addition, the appropriate cable capacity and locations are selected to improve the reliability of the Korean power systems using the smart SC cables. The proposed locations and capacities will be applied to the smart SC cable development project that started in May last year

1. Introduction

Due to the constant load increase, the power system is gradually operating at its stability limit. In addition, due to the development of related technologies, additional new power facilities are being installed. Especially, high temperature superconductivity (HTS) has been discovered in 1986 and drastically applied to power system facilities as an application field of the system. As an example of power system application devices, Superconducting fault current limiter (SFCL) has been developed, and a 22.9kV hybrid SFCL has been operated in Icheon substation near Seoul. The applications that are being considered are superconducting (SC) substations, energy storage systems, and superconducting (SC) cables, which can have advantages when connecting offshore wind power systems [1-5].

The Korean power system has independent systematic characteristics, which are not linked to other systems. In particular, 40% of the total load is concentrated in Seoul area. Therefore, it has a structure to supply electric power to the metropolitan area through transmission lines from the power stations in non-metropolitan area, which are constructed with lower cost than ones in urban areas. However, the power grid is tightly connected to the power supply in the dense metropolitan area. In the strongly connected power system, there is a problem that a fault current is very high when the system fault occurs. Thus, system operators have introduced various schemes into the system to limit the fault current. Typically, the bus separation and the line opening operation, the current limit reactor installation, and the system separation using BTB (Back-to-Back) HVDC (High Voltage Direct Current) have been used. Since each method has advantages and disadvantages, a smart SC cable is introduced to improve it.

Recently, there have been many attempts to apply SC cable applications to the grid [6, 7]. The SC cables have a great advantage in the operation because they have zero resistance in DC when maintaining superconductivity, however they have the disadvantage of requiring an additional cooling equipment to maintain the superconductivity. Therefore, in this paper, we will explain the merits of the SC cable with the FCL (Fault Current Limit) function that can be applied to the power system. In addition, the smart SC cable with FCL function was applied to the KEPCO system and the applicability was examined.

2. Korean power system in the metropolitan area

In the Korean power system, there is a problem that the fault current increases sharply due to the concentration of the load in the metropolitan area and the increase of the connecting lines. The system operator operates the system separately to prevent the increase of the fault current. The operation of system separation has an effect on the reduction of the fault current, but has an adverse effect on the stability of the system.

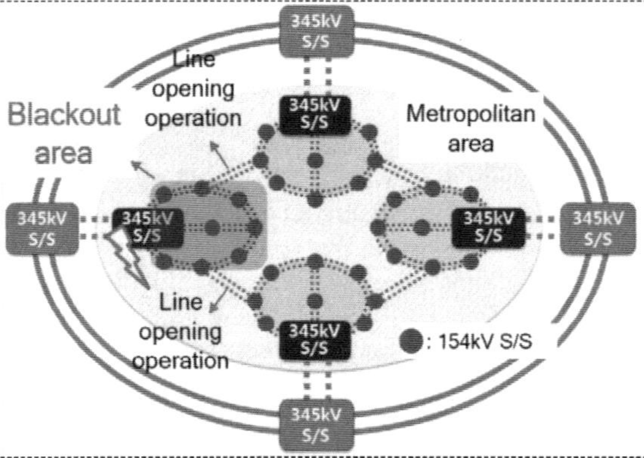

Figure 1. Korean power system configuration in metropolitan area.

The fig. 1 shows the current method of operation in the Korean metropolitan area. The metropolitan area has a loop network composed of 345kV systems. From the 345kV substation, the loads are connected by radial using 154kV networks. At this time, in order to prevent increase of fault current among the 154kV networks, each 154kV network is separately operated by the opening operation. In other words, the one 154kV network connected to the only one 345kV network. Therefore, if a fault occurs in a 345kV substation, which is a power supply line, the loads connected to the 154kV network supported by the 345kV substation cannot be supplied power because there is no bypass line due to opening operation. The blackout area in fig. 1 could be explained as an example for the disadvantage of the opening operation.

The fig. 2 shows the situation of the power grid system, when the fault occurs at Yeongseo substation (345kV). By the open operation between Shingil and Bongcheon substation, after the fault, Shingil and Yeongseo system is separated from the main system. The fault progression process is as follows. First, the insulation of the busbar breaker is broken. The system failure occurs. The protection relay in the busbar breaker is activated. All lines connected to the busbar are opened. Approximately 300,000 households of seven 154kV substations supplied from Yeongseo Substation are shut down.

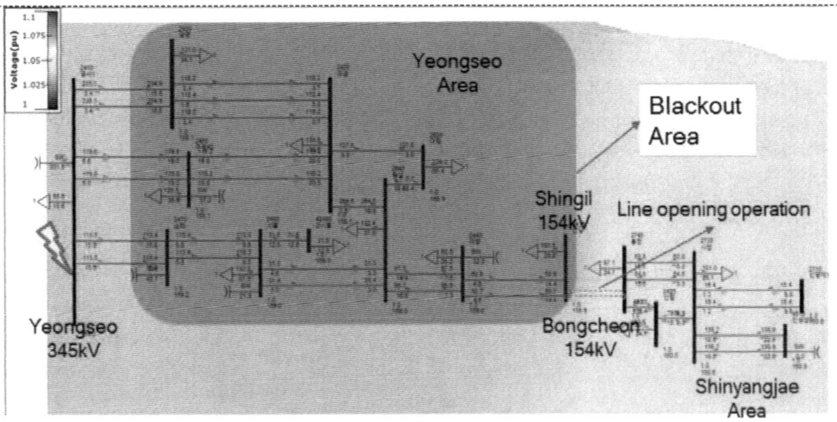

Figure 2. Yeongseo area blackout.

3. A model for smart superconducting cable

A model of the smart SC cable proposed in this paper is shown in Fig. 3. The smart SC cable with resistive type is shown in (a) and the inductive type is shown in (b). In the figures, R1 is a radius of inner nitrogen layer for cooling the superconductor. "R2-R1" is a former layer. The outer of R2 is a conducting layer using HTS tape. From the outer of the conducting layer to R3 is the insulation layer using PPLP (Polypropylene Laminated Paper). There is the external nitrogen in the shielding layer. The inductive type cable is additionally covered with a high permeability material around the resistive type cable. The material layer is generating a high current inductance when a fault current is detected. The proposed smart SC cables are testing in order to verify their characteristic.

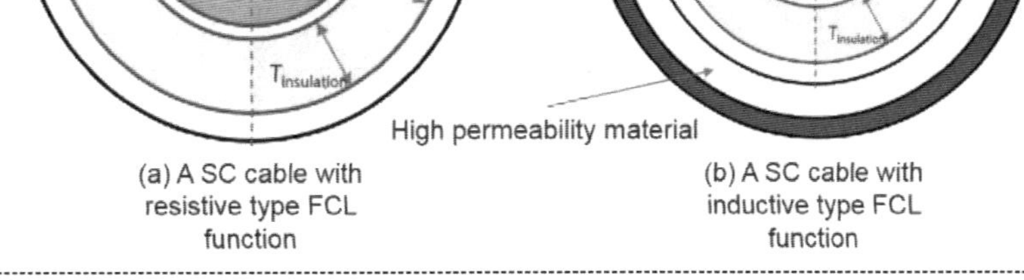

Figure 3. Smart superconducting cable with fault current limiter function.

IOP Conf. Series: Journal of Physics: Conf. Series **1293** (2019) 012068 doi:10.1088/1742-6596/1293/1/012068

4. Smart superconducting cable with fault current limit function

The smart SC cable proposed in this paper has the functions to supply the power while maintaining the superconductivity in the normal operation, to detect the fault current in short-circuit accident, and to limit the fault current by generating the high current resistance and inductance in high speed. Fig. 4 assumes that the smart SC cable is applied to the Shingil and Bongcheon substation. The Shingil-Bongcheon lines is open because of fault current problem before applying for the smart SC cable. However, in the case of an open operation, it plays a role of damaging the stability when a system failure occurs. In this paper, we propose the smart SC cable reducing the fault current. Due to the limited fault current, the Shingil-Bongcheon line will be connected. Finally, the Yeongseo area grid will be stable though the fault occurs at Yeongseo 345kV substation. In the simulation, we assumed that the resistance and inductance of the SC cables are 1 Ω / km ~ 2 Ω / km.

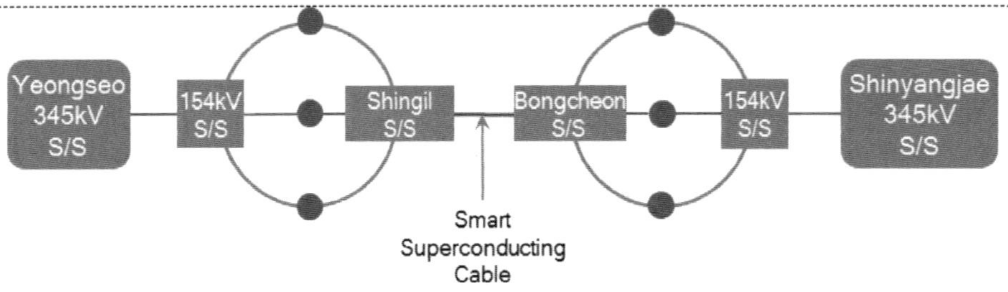

Figure 4. Smart superconducting cable with fault current limiter function.

5. Simulation results

In this paper, the feasibility of the smart SC cable with a FCL function is examined in the Korean power system in 2029, which has peak loads. The conventional line between Shingil and Bongcheon is replaced with the smart SC cable. The results of the fault current calculation are summarized in Table 1. In this paper, ASCC (Automatic Sequencing fault Current Calculation) method is used for the fault current calculation. The method is normally used in power system operators in the world. In the base-case of the conventional line connection, there are 25 buses that exceed the fault current limit. In the case of the open operation of the line, there is no bus that exceeds the fault current limit. In the case of installing a current-limiting reactor (assuming 2 Ω / km), the fault current exceeded the reference value (50kA) at four buses. In the case of applying the smart SC cable, the number of fault current exceeding buses that exceed the fault current limit is 4 to 11. Because the resistance of smart SC cable is 1 Ω / km to 2 Ω / km. This is the same result as the value of the fault current limit using the inductance. Although the smart SC cable cannot completely limit the fault current compared to the line opening, the effect of lowering the fault current is confirmed.

Fig. 4 shows the voltage on the bus when the fault occurs. The simulation scenario is as follows:

- 0.0s: start the simulation
- 0.5s: three phase fault at Yeongseo 345kV S/S
- 0.6s: the fault clear and the transformer between Yeongseo 345kV and 154kV S/S is open
- 5.0s: end the simulation

In the case that the line is completely opened, the voltage of the bus is reduced to 0 after contingency. Finally, the power failure occur on the bus. On the other hand, when the conventional line is connected and operated, it can be seen that the voltage rises to the voltage range in steady state after the contingency. However, in the case of the smart SC cable connections, the fault may not be removed because some buses exceed the fault current limit. On the other hand, the smart SC cable can be confirmed that the bus runs within the allowed voltage range after a contingency in Korea even though the bus voltage is on the lower bound of the range as shown in Fig 4. Therefore, it can be seen that the use of smart SC cable helps to improve the stability because by limiting the fault current in the event of fault and additionally enabling operation by connecting the line. However, when the fault current cannot be completely removed and the bus voltage is slightly lower after the contingency due to the resistance and the inductance of the smart SC cable, it is considered that the part needs to be improved.

Table 1. Fault current calculation results by measures for fault current limit

Measures	# of buses (50kA above)	Remarks
BaseCase (Line Connecting)	25	Yeong-Seo and Shinyangjae system (Maximum-65kA, Shinyangjae 154kV)
Line Opening	0	No bus exceeding 50kV
Fault Current Limiter with Reactor [2Ω/km]	4	Shinyangjae, Yangji, Yeoksam, Kyodae
Smart Superconducting Cable [1Ω/km]	11	Yeongseo1, Shinyangjae, Yangji, Yeoksam, Sadang, Boncheon, Kyodae, Seocho, Nonil, Shinsa, Gwacheon
Smart Superconducting Cable [2Ω/km]	4	Shinyangjae, Yangji, Yeoksam, Kyodae

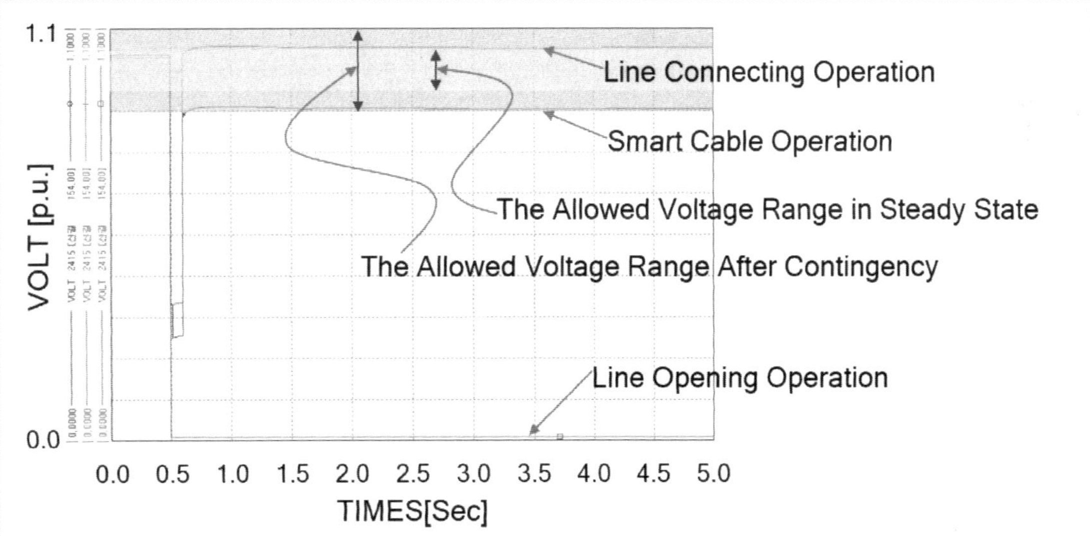

Figure 5. Shingil substation voltages

6. Conclusion

Currently, a smart SC cable with fault current limiting function is being developed in Korea. The smart SC cable was applied to the Korean power system with very robust connectivity and the feasibility study was conducted. In the metropolitan area, there are many parts that are separated and operated due to the fault current problem. If the smart SC cable is used for separated operation lines, the fault current is limited and the system stability can be improved. However, the smart SC cables have disadvantages in installation costs, and efforts to overcome them are necessary. In addition, it is possible to limit the fault current, but since there is an overload in some connected systems, the smart SC cable can be introduced considering the different system configuration in actual operation. In addition, we will review other power system application areas.

Acknowledgments

This work was supported by the Korea Institute of Energy Technology Evaluation and Planning (KETEP) and the Ministry of Trade, Industry and Energy (MOTIE) of the Republic of Korea (No. 20171220100400).
This research was supported by the KERI Primary research program of MSIT/NST (No. 18-11-N0103-08, 18-12-N0103-08).

References

[1] Hassenzahl W V 2000 Applications of superconductivity to electric power systems *IEEE Power Engineering Review* vol **20** no 5 pp 4-7
[2] Kovalsky L, Yuan X, Tekletsadik K, Keri A, Bock J and Breuer F 2006 Applications of superconducting fault current limiters in electric power transmission systems *IEEE Trans. Appl. Supercond.* vol **15** no 2 pp 2130-3
[3] Park W J, Sung B C and Park J W 2010 The Effect of SFCL on Electric Power Grid With Wind-Turbine Generation System *IEEE Trans. Appl. Supercond.* vol **20** no 3 pp 1177-81
[4] Lee B W, Park K B, Sim J, Oh I S, Lee H G, Kim H R and Hyun O B 2008 Design and Experiments Of Novel Hybrid Type Superconducting Fault Current Limiters *IEEE Trans. Appl. Supercond.* vol **18** no 2 pp 624-7
[5] Hyun O B, Sim J, Kim H R, Park K B, Yim S W and Oh I S 2009 Reliability Enhancement of the Fast Switch in a Hybrid Superconducting Fault Current Limiter by Using Power Electronic Switches *IEEE Trans. Appl. Supercond.* vol **19** no 3 pp 1843-6
[6] Sudheer T, Sarkar M, Gour A S, Rao V V and Rao B N 2017 Development and Testing of a High Temperature Superconducting (HTS) Cable for Smart Grid Applications *3rd International conference on condition assessment techniques in electrical system* pp 197-201
[7] Schmidt F and Allais A 2004 Superconducting cables for power Transmission applications – A Review *WAMS 2004 Workshop on accelerator magnet superconductor Archamps (France)* vol **38** no 17 pp 232-7

An effective design for induced current reducing of a single-phase multi-layer HTS coaxial cable with former

S Y Kang[1], S J Lee[1], D V Le[1] and M Park[1]

[1] Changwon National University, Changwon, GN, 51140, Korea

capta.paper@gmail.com

Abstract- High temperature superconducting (HTS) cables require a former layer to bypass fault currents and support superconducting tape. The former layer of the HTS cable does not conduct current in steady-state operation due to the resistance difference with the conducting layer. However, in practice, an induced current flows through the former layer due to the mutual interaction with the conducting layer. Therefore, the current distribution in the cable should be uniform and the induced current in the former layer must be minimized. This paper deals with an effective design of a single-phase multi-layer HTS coaxial (SMHTSC) cable with former layer. The optimal combination of the pitch length and the winding direction in the conducting layer and the shield layer was selected for uniform current distribution of the SMHTSC cable and minimization of induced current in the former layer. The current distribution of the SMHTSC cable was analyzed based on the numerical calculation results obtained using the impedance matching method. The results and discussions of the current distribution on the SMHTSC cable with and without former layer were compared and presented.

1. Introduction

High-temperature superconducting (HTS) cables offer a compact design and very low loss compared to conventional copper cables, and hence the HTS cables are being actively studied and developed worldwide as promising power cables for urban areas [1]-[2]. The HTS cables generally require a former located at the center of the cable, which not only acts as a support for robustly winding the superconducting tape, but also serves as a pathway to bypass the fault current. The choice of this former has a great impact on the structure and design of a cable [3]. In addition, the HTS power cables typically consist of a multi-layer structure of HTS tape wires spirally wound over coaxial former. The characteristics of multi-layers are defined by mainly two structural parameters: the pitch length, and the winding direction. The winding direction can be clockwise or anti-clockwise. The characteristics of the structural parameters play an important role in maintaining a uniform current distribution and may be influenced by the manufacturing process, so an their optimum design is required. The non-uniform current distribution consequently increases the total power loss in the cable and also reduces the cable current capacity [4]-[6].

This paper presents an effective design method of a single-phase multi-layer HTS coaxial (SMHTSC) cable with former layer. The SMHTSC cable structure is composed of a former layer, a conducting layer and a shield layer. Each layer except the former has a unique impedance that is affected by the pitch length and winding direction, and has a current distribution according to the impedance. However, when the former layer exists, the current distribution of the conducting layer is changed by the interaction with the former layer, and the induced current also flows in the former layer. Therefore, it is necessary to determine the pitch length and the winding direction so that the current distribution of the conducting layer becomes uniform and the former induced current is also minimized. The impedance matching method was used to iteratively calculate various combinations of the pitch length and the winding direction in the SMHTSC cable with and without the former layer. The pitch length and the winding direction of the SMHTSC cable, which represents the uniform current distribution in the conducting layer, were selected through the design process using the impedance matching method. By selecting the appropriate pitch length and winding direction, the induced current of the former layer was minimized to less than 27.3 A and the current distribution of the SMHTSC cable was uniformed. This study result will be useful for the development of other types of multi-layer HTS coaxial cables with former layer.

2. Numerical analysis of the SMHTSC cable

2.1. Configuration of the SMHTSC cable

Two SMHTSC cables with and without former layers are wound in the same pitch length and winding direction. Figure 1 shows the structure of a SMHTSC cable with two conducting layers and one shield layer and no former layer. Figure 2 shows the structure in which the former layer is added to the SMHTSC cable. The radius of the two conducting layers is determined by the cross-sectional area of the former layer. For the former layer, the radius is designed not to rise above a certain temperature even when the fault current is flowing. Also, in the case of shield layer, the insulation thickness is selected considering the withstand voltage, and the radius is determined accordingly. The specifications of SMHTSC cable are shown in table 1.

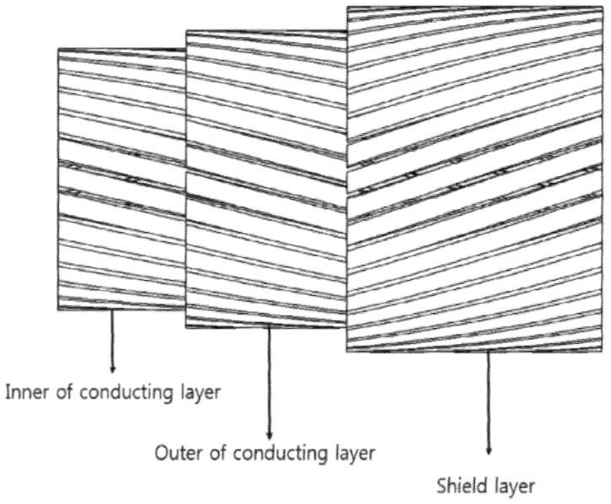

Figure 1. Structure of SMHTSC cable without former layer

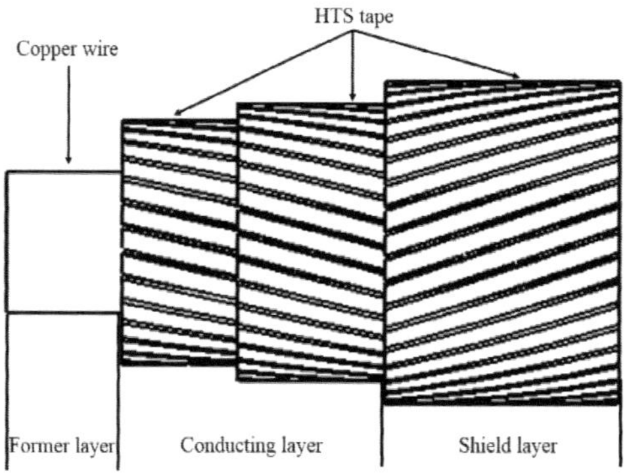

Figure 2. Structure of SMHTSC cable with former layer

Table 1. Specifications of SMHTSC cable model.

Items	Value
Voltage/capacity	23 kV/50 MVA
Rated current	1,250 A
Total cable length	1000 mm
Number of conducting layers	2
Number of shield layer	1
Radius of former layer	8 mm
Inner radius of conducting layer	9 mm
Outer radius of conducting layer	10 mm
Radius of shield layer	15 mm

2.2. Modelling of the SMHTSC cable

Figure 3 shows the equivalent circuit of the SMHTSC cable. Each layer can be represented by the resistance (R_i), the self-inductance (L_i) and the mutual inductance between the coupled two layers (M_{ij}). The self-inductance of the layer and the mutual inductance between the ith and jth layers can be calculated as shown in (1) and (2) using the magnetic field energies stored per unit volume. In addition, the inductance and mutual inductance of the former layer can be calculated as shown in (3) and (4). The impedance matrix is derived from the equivalent circuit connected to the source and the load as shown in figure 2 using (5) [7].

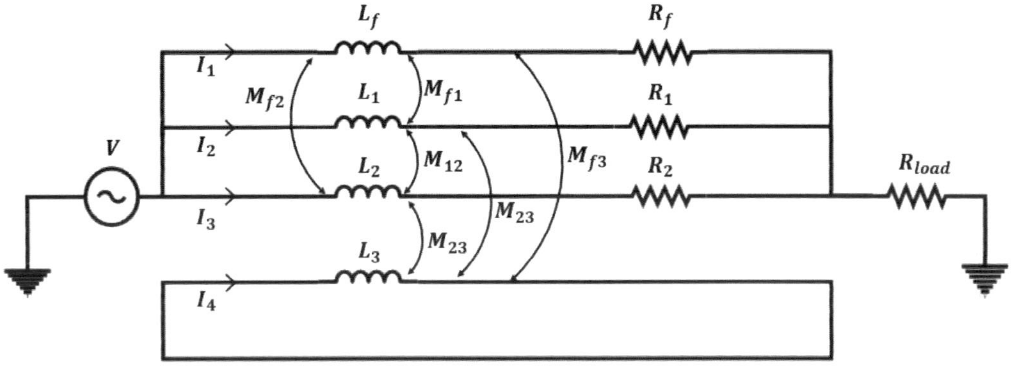

Figure 3. Circuit model of the parallel cables considering the former layer

$$L_i = \frac{u_0 \pi r_i^2}{L_{pi}^2} + \frac{u_0}{2\pi} \ln \frac{D}{r_i} \tag{1}$$

$$M_{ij} = \frac{a_i a_j \pi u_0 r_i^2}{L_{pi} L_{pj}} + \frac{u_0}{2\pi} \ln \frac{D}{r_j} \quad (r_j > r_i) \tag{2}$$

$$L_f = \frac{u_0}{8\pi} + \frac{u_0}{2\pi} \ln \frac{D}{r_f} \tag{3}$$

$$M_f = \frac{u_0 r_f^2}{3L_{pf}} + \frac{u_0}{2\pi} \ln \frac{D}{r_f} \tag{4}$$

where, u_0 is the magnetic permeability of free space, D is the outermost radius of the electrical shielding, a_i is the winding direction of clockwise, a_j is the winding direction of anti-clockwise, f is the former layer of SMHTSC cable.

$$\begin{bmatrix} V \\ V \\ V \\ 0 \end{bmatrix} = \begin{bmatrix} R_f + R_{load} & R_{load} & R_{load} & 0 \\ R_{load} & R_1 + R_{load} & R_{load} & 0 \\ R_{load} & R_{load} & R_2 + R_{load} & 0 \\ 0 & 0 & 0 & 0 \end{bmatrix} \begin{bmatrix} I_1 \\ I_2 \\ I_3 \\ I_4 \end{bmatrix} + j\omega \begin{bmatrix} L_f & M_{f1} & M_{f2} & M_{f3} \\ M_{f1} & L_1 & M_{12} & M_{13} \\ M_{f2} & M_{12} & L_2 & M_{23} \\ M_{f3} & M_{13} & M_{23} & L_3 \end{bmatrix} \begin{bmatrix} I_1 \\ I_2 \\ I_3 \\ I_4 \end{bmatrix} \tag{5}$$

where, R_1 and R_2 denote the equivalent resistances that cause AC losses, and R_{load} is the equivalent resistance of load. The SMHTSC cable was simulated under the rated current load condition, with the resistances R_1, R_2, and R_{load} of 0.015 μΩ, 0.021 μΩ and 18.4 Ω, respectively.

2.3. Analysis of the current distribution

The current distribution in the conducting and shield layers of the SMHTSC cable is analysed for various combinations of pitch length and winding direction. Table 2 shows the range of the pitch length and increase value for each layer without the former layer. The winding direction consists of clockwise and anti-clockwise directions. The pitch length and winding direction of the SMHTSC cable should be chosen to uniformly control the current distribution among each conducting layer and the shield layer. Figure 4 shows the flowchart of the impedance matching method. By using the impedance matching method, the current distribution of each layer and the induced current of the former layer are repeatedly calculated and the appropriate pitch length and winding direction can be selected.

Table 2. The range of the pitch length variation.

Items	Value
Range of pitch length	150 mm to 650 mm
Increase value of pitch length	10 mm

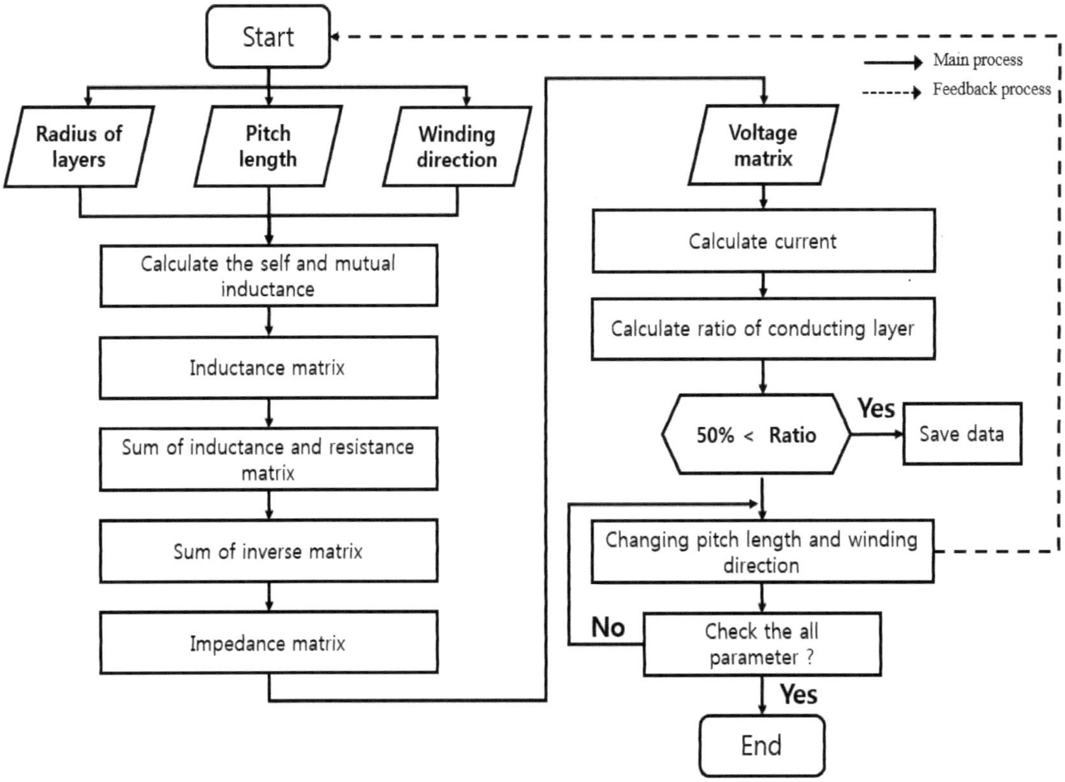

Figure 4. Flowchart of the impedance matching method

3. Calculation results and discussions

3.1. Comparison of the SMHTSC cables with and without former layer

Table 3 shows the results of repeatedly calculated current distributions for various combinations of the pitch length and winding direction in a former-less SMHTSC cable. As shown in table 2, the pitch length was changed from 150 mm to 650 mm at intervals of 10 mm, and the current ratio of the conducting layer was calculated from 70% to 90% at maximum while changing the winding direction. As a result, many current ratios were calculated, but only three cases of current ratios of 70%, 80% and 90% were considered to analyse the influence of the SMHTSC cable former layer.

Table 3. Calculation results of the current distribution in the SMHTSC cable without former layer.

Cases	Pitch length [mm]	Current of conducting and shield layers [A]	Winding direction (F= clockwise, R= anti-clockwise)	Current ratio of conducting layer [%]
Case 1	300/320/620	516.2/733.7/1157.9	R/F/R	70
Case 2	280/290/570	557.9/692.0/1156.1	F/R/F	80
Case 3	250/300/430	592.4/657.5/1145.2	R/F/R	90

The inductance value changes according to the pitch length of each layer irrespective of the presence or absence of the former layer, and the current distribution of each layer changes accordingly. Therefore, the current distribution was analysed by adding the former layer to the SMHTSC cable having the same pitch length, winding direction, and radius as in the case of no former, and the results are shown in table 4. As shown in table 4 the SMHTSC cable without the former layer had a higher current ratio of the conducting layer than with the former layer. The SMHTSC cable with the former layer has a high induced current in the former layer. Therefore, in the case of the SMHTSC cable with former layer, the current distribution and the induced current should be analysed considering the effect of the former layer and the current ratio of the conducting layer.

Table 4. Calculation results of the current distribution in the SMHTSC cable with former layer.

Cases	Pitch length [mm]	Current of conducting and shield layers [A]	Winding direction (F= clockwise, R= anti-clockwise)	Induced current of former layer [A] (Loss of former layer)	Current ratio of conducting layer [%]
Case 1	300/320/620	419.7/701.4/994.3	R/F/R	128.8 (0.2 W/m)	59
Case 2	280/290/570	455.0/497.1/1216.2	F/R/F	97.8 (0.1 W/m)	65
Case 3	250/300/430	475.9/637.0/972.9	R/F/R	136.9 (0.2 W/m)	74

3.2. An effective design of the SMHTSC cable with former

There are eight combinations of SMHTSC cables according to the winding direction. Like the pitch length, the winding direction also affects the current distribution of each conducting layer and the induced current in the former layer. In this case, the minimized induced current was the least when the winding directions of the conducting layer and the shield layer are F, R and F, respectively, as shown in figure 5. Therefore, the direction of winding of F, R, F was selected and applied to analyse the SMHTSC cable with the former layer. The calculation result of the minimum induced current according to the pitch length is shown in figure 6. As can be seen in figure 6, in the SMHTSC cable with the former layer, the induced current in the former layer can be minimized by adjusting the pitch length of each layer to an appropriate range. In particular, the pitch length change of the inner conducting layer has a greater effect on the induced current of the former layer than the other layers. Therefore, in order to effectively design a SMHTSC cable with former layer, it is necessary to select the pitch length of the inner conducting layer of 150 mm. Table 5 shows the calculated results for the pitch length range of 150 mm to 650 mm of the SMHTSC cable with former layer. The SMHTSC cable with former layer has a uniform current distribution of 98% in the conducting layer and an induced current of 27.3 A in the former layer and the copper loss of the former layer was minimized to 8.4 mW/m. It was confirmed that the current distribution of the former layer and the conducting layers could be improved through the adjustment of the pitch length and the winding direction.

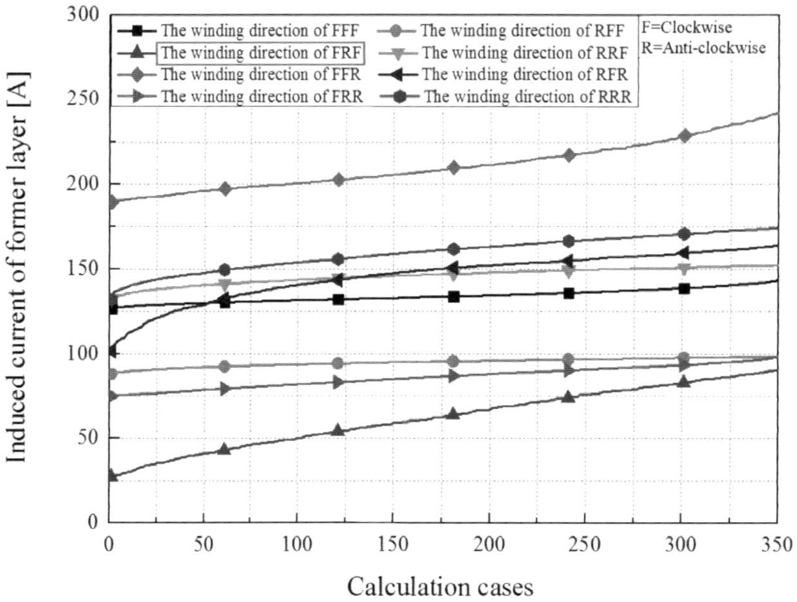

Figure 5. Calculation results of the induced current for the combinations of winding direction

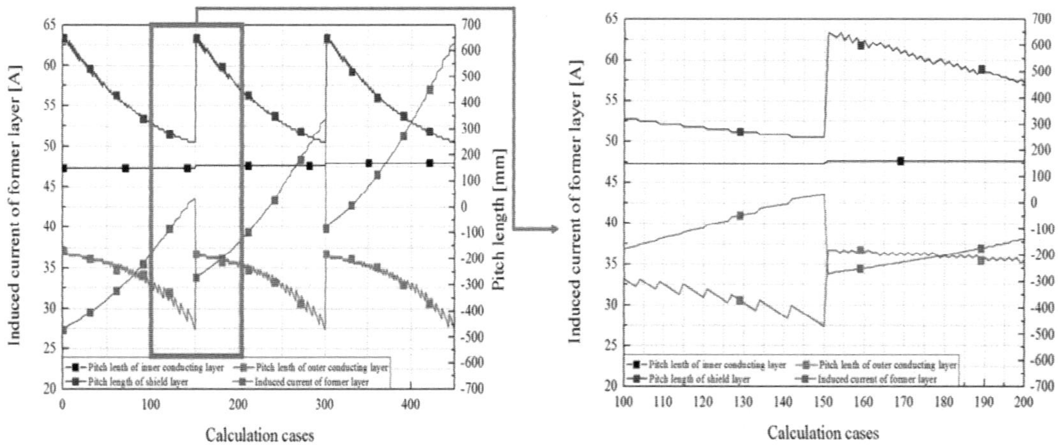

Figure 6. Calculation results of the induced current due to the pitch length combinations

Table 5. Calculation results of the effective design of the SMHTSC cable with former layer.

Items	Pitch length	Winding direction (F= clockwise, R= anti-clockwise)	Current of each layer
Former layer	-	-	27.3 A
Inner conducting layer	150 mm	F	605.8 A
Outer conducting layer	170 mm	R	616.7 A
Shield layer	650 mm	F	1237.3 A

4. Conclusions

This paper presents an effective design process for a SMHTSC cable with former layer. The influence of the former layer was analysed by comparing a SMHTSC cable with or without former layer. Both cables have the same pitch length and winding direction selected. The SMHTSC cable without the former layer showed a higher current ratio of the conducting layer than with the former layer. The SMHTSC cable with the former layer has a high induced current in the former layer. The pitch length and winding direction of each layer were chosen to reduce the induced current in the former layer. The induced current of the former layer was the smallest when the winding directions of the conducting layers and the shield layer were F, R and F, respectively. When the pitch lengths of the conducting layers and the former layer of the SMHTSC cable with the former layer were 150 mm, 170 mm and 650 mm, a 98% uniform current distribution in the conducting layers and an induced current of 27.3 A in the former layer were exhibited. The copper loss of the former layer was minimized to 8.4 mW/m. Even if the cable size and the number of layers are changed, the proposed SMHTSC cable design method can balance the current distribution of the conducting layer and minimize the induced current of the former layer. The results of this study can be used effectively to design other types of multilayer HTS coaxial cables.

Acknowledgments
This research is financially supported by Changwon National University in 2017~2018.

References

[1] Hamajima T, Yagai T, Tsuda M and Harada N 2005 Current distribution analysis in tri-axial HTS cable considering three phases *IEEE Trans. Appl. Supercond.* **15** pp 1775-78
[2] Ha S K, Kim S K, Kim J G, Park M, Yu I K, Lee S, Sim K and Kim A R 2013 Transient Characteristic Analysis of a Tri-Axial HTS Power Cable Using PSCAD/EMTDC *IEEE Trans. Appl. Supercond.* **23** 5400104
[3] Hamajima T, Yagai T and Tsuda M 2006 Analysis of Balanced Three-Phase Current Distributions in a Tri-Axial Cable *IEEE Trans. Appl. Supercond.* **16** pp 1586-89
[4] Young M A, Gouge M J, Marshall O, Demko J A, Robert C, Lue J and Fathy A E 2005 An investigation of the current distribution in triaxial cable and its operational impacts in a power system *IEEE Trans. Appl. Supercond.* **15** pp 1751–54
[5] Lee S J, Park M , Yu I K , Won Y, Kwak Y and Lee C 2018 Recent Status and Progress on HTS Cables for AC and DC Power Transmission in Korea *IEEE Trans. Appl. Supercond.* **28** 5401205
[6] Watabe A, Fukui S, Yamaguchi M, Sato T, Takao T and Tsukamoto O 2004 Numerical analysis of AC loss and current distribution characteristics of co-axial multi-layer high temperature superconducting cable *IEEE Trans. Appl. Supercond.* **14** pp 678-81
[7] Lim S H, Yim S W, Hwang S D and Han B S 2005 Influence of pitch length and winding direction on four-layer HTSC power transmission cable with a shield Layer *IEEE Trans. Appl. Supercond.* **15** pp 1735-38

IOP Conf. Series: Journal of Physics: Conf. Series **1293** (2019) 012070 doi:10.1088/1742-6596/1293/1/012070

Optimum design of cryogenic pump for circulation cooling of high temperature superconducting cables

Kenta Tadakuma[1], Kazuhiro Kajikawa[1], Yasuharu Kamioka[2], Atsushi Ishiyama[2], Shinsaku Imagawa[3], Taketsune Nakamura[4], Hirokazu Hirai[5] and Shinsuke Ozaki[5]

[1] Department of Electrical and Electronic Engineering, Kyushu University, Fukuoka 819-0395, Japan
[2] Department of Electrical Engineering and Bioscience, Waseda University, Tokyo 169-8555, Japan
[3] National Institute for Fusion Science, Toki 509-5292, Japan
[4] Department of Electrical Engineering, Kyoto University, Kyoto 615-8530, Japan
[5] Taiyo Nippon Sanso Corp., Tsukuba 300-2611, Japan

tadakuma@sc.kyushu-u.ac.jp

Abstract. Pressure drops and temperature rises caused by the cooling with liquid nitrogen are evaluated for two typical examples of HTS cables. One is the three-in-one-type AC cable located inside a corrugated pipe, and the other is the single-core-type DC cable located inside a straight pipe. The theoretical expressions of pressure drop and temperature rise indicate the contributions from three components such as the friction between the liquid nitrogen and the wetted surfaces, the ratio of the heat input to the mass flow rate of liquid nitrogen and the vertical change of position along the cooling pipe. The numerically estimated results are validated by comparing them with the test results carried out by the other groups. Furthermore, the maximum feeding distance of HTS cable can be determined by adjusting the flow rate of pump if the maximum permissible pressure drop and temperature rise are fixed in advance.

1. Introduction

High temperature superconducting (HTS) wires in the form of a tape can be cooled using liquid nitrogen with the boiling temperature of 77 K at the atmospheric pressure. A lot of research-and-development (R&D) projects for power transmission cables using the HTS wires have been in progress all over the world [1,2] because the HTS cables are expected to be small energy dissipation and low installation cost under the ground. A liquid nitrogen pump with high efficiency, enough discharge pressure and allowable maintenance interval would be required to realize a long-length HTS cable in near future, but such a pump has not been developed yet. Therefore, our group aims to develop a liquid nitrogen circulation pump composed of cryogenic magnetic bearings and HTS motor [3]. The specifications of the target pump are determined as the discharge pressure of 1 MPa using a 2-stage impeller, the flow rate of 100 L/min and the rotating speed of 5000 rpm on the basis of the specific speed in turbomachinery.

In order to design the pump, it is necessary to evaluate the pressure drop and temperature rise when the subcooled liquid nitrogen is circulated inside a cryogenic pipe with HTS cable. The pressure drop has been estimated only from the contribution of the friction between the liquid nitrogen and the wetted surfaces of pipe and cable, whereas the temperature rise has been calculated only from the heat input to the liquid nitrogen [4]. However, the simplified theoretical expressions for the pressure drop and temperature rise derived for liquid hydrogen cooling of an MgB_2 cable have included both the contributions of the friction and the heat input [5]. Therefore, it is necessary to confirm the influences of the additions on the estimated results of pressure drop and temperature rise for liquid nitrogen cooling.

In this study, we evaluate the pressure drop and temperature rise for two types of model cables when the subcooled liquid nitrogen flows through the cooling pipes. The influence of the difference in height along the cooling pipe on the pressure drop and temperature rise is also estimated. Furthermore, the maximum feeding distance of HTS cable is determined by adjusting the flow rate of liquid nitrogen if the maximum permissible pressure drop and temperature rise are fixed in advance.

2. Theoretical expressions for pressure drop and temperature rise

The pressure drop and temperature rise of the cryogenic fluid flowing through the pipe containing the HTS cable as shown in figure 1 can be expressed directly from the stationary one-dimensional balances of mass, momentum and energy as [5]

$$(1-Ma^2)\frac{1}{\rho}\frac{\mathrm{d}p}{\mathrm{d}s} = -\left\{1+\frac{(\gamma-1)Ma^2}{\alpha T}\right\}F - \frac{(\gamma-1)Ma^2}{\alpha T}\frac{W}{\dot{m}} - g\frac{\mathrm{d}z}{\mathrm{d}s}, \tag{1}$$

$$\frac{1-Ma^2}{\alpha T}C_p\frac{\mathrm{d}T}{\mathrm{d}s} = \left(\frac{1-\gamma Ma^2}{\alpha T}-1\right)F + \frac{1-\gamma Ma^2}{\alpha T}\frac{W}{\dot{m}} - g\frac{\mathrm{d}z}{\mathrm{d}s}, \tag{2}$$

where p is the pressure, T the temperature, $Ma\ (= v/c)$ the Mach number, v the flow velocity, c the speed of sound, ρ the density, $\gamma\ (= C_p/C_v)$ the specific heat ratio, C_p the specific heat at constant pressure, C_v the specific heat at constant volume, α the thermal expansivity at constant pressure, W the heat input per unit length, $\dot{m}$ the mass flow rate, and g the gravitational acceleration. The flow path s along the pipe is given by $s = \sqrt{x^2+z^2}$ if only the flow in the xz plane perpendicular to the y-axis is considered. Assuming that the symbol θ represents the angle between the horizontal direction parallel to the x-axis and the flow direction along the path s, the relationship $\mathrm{d}z/\mathrm{d}s = \sin\theta$ can be obtained [6]. Furthermore, the parameter F representing the frictional force per mass is given by [5,7]

$$F = \frac{f}{D_h}\frac{v|v|}{2}, \tag{3}$$

where f is the Darcy friction coefficient, and $D_h\ (= 4A_{\mathrm{ff}}/(P_p+P_c))$ the hydraulic diameter with the free-flow cross-sectional area A_{ff} inside the cooling pipe, and the wetted perimeters, P_p and P_c, for pipe inner surface and cable outer surface [8].

Equations (1) and (2) can be simplified as [5]

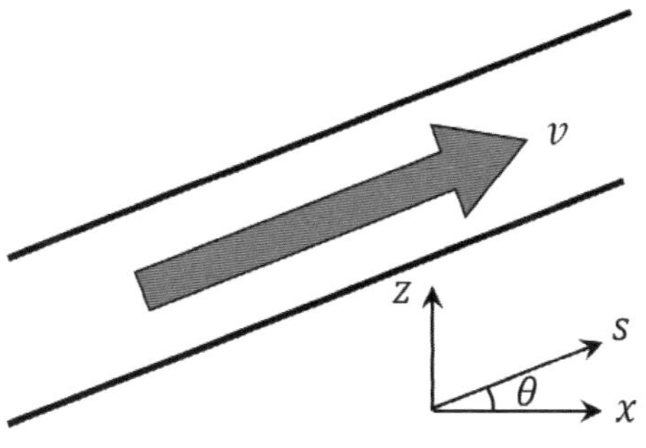

Figure 1. Tilted location of cooling pipe for HTS cable.

$$\frac{1}{\rho}\frac{\mathrm{d}p}{\mathrm{d}s} = -F - \frac{\alpha v^2}{C_p}\frac{W}{\dot{m}} - g\frac{\mathrm{d}z}{\mathrm{d}s}, \tag{4}$$

$$C_p\frac{\mathrm{d}T}{\mathrm{d}s} = (1-\alpha T)F + \frac{W}{\dot{m}} - \alpha T g\frac{\mathrm{d}z}{\mathrm{d}s}, \tag{5}$$

respectively. This is because three parameters appearing in equations (1) and (2) are extremely negligible compared with unity as $2.3 \le 10^7 Ma^2 \le 4.5$, $7.7 \le 10^7(\gamma-1)Ma^2/\alpha T \le 10.4$ and $4.4 \le 10^7\gamma Ma^2 \le 9.5$ for the subcooled liquid nitrogen with the temperature of 65-80 K by estimating them using the commercial software GASPAK [9] if the typical value for the flow velocity of $v = 0.6$ m/s is used. On the other hand, another parameter, $0.27 \le \alpha T \le 0.48$, is significant. The first, second and third terms in right-hand sides of equations (4) and (5) represent the effects of the friction, heat input and difference in height, respectively.

The equivalent friction coefficient f for the cooling pipe with HTS cable can be expressed by [10]

$$f = \frac{f_p P_p + f_c P_c}{P_p + P_c}, \tag{6}$$

where f_p and f_c are the friction coefficients of the pipe inner surface and cable outer surface, respectively. In this study, the following expressions are used as the friction coefficient f_s for the straight pipe inner surface or the HTS cable outer surface, and the friction coefficient f_r for the corrugated pipe inner surface [11]:

$$f_s = \begin{cases} 0.3164Re^{-0.25}, & Re \le 1.17\times10^5, \\ 0.0032 + 0.221Re^{-0.237}, & Re > 1.17\times10^5, \end{cases} \tag{7}$$

$$f_r = \left(1.74 - 2\log_{10}\frac{2\varepsilon}{D_i}\right)^{-2}, \quad Re > 900\left(\frac{\varepsilon}{D_i}\right), \tag{8}$$

where $Re\ (= \rho v D_h/\mu)$ is the Reynolds number, μ the viscosity, and ε the corrugation height of the corrugated pipe with the inner diameter D_i. In the model A where the corrugated pipe is used as explained in the next section, the Reynolds number Re is greater than 2970, whereas $900(\varepsilon/D_i) = 44.6$. Therefore, we can use equation (8) for the evaluation of friction coefficient. In the case of the steady flow, the product ρv is constant from the law of conservation of mass, so that the mass flow rate $\dot{m}\ (= \rho v A_{\mathrm{ff}})$ also becomes constant.

3. Evaluation of properties of liquid nitrogen circulations in model cables

In order to evaluate the pressure drop and temperature rise in the circulation cooling with subcooled liquid nitrogen, it is necessary to predetermine the structures and sizes of HTS cables and cooling pipes. The HTS cable is generally operated under the condition of AC or DC. In addition, the HTS cable for AC use has a three-in-one-type, single-core-type, or tri-axial-type structure typically. Furthermore, the use of a straight pipe as an alternative to a conventional corrugated pipe is proposed for the cooling pipe [2]. Therefore, in order to obtain a few typical sets of HTS cables and cooling pipes which cover the realistic structures as many as possible, we refer to the so-called Yokohama project [1] and Ishikari project [2] in this study. The Yokohama project is the R&D of three-in-one-type AC cable implemented under the collaboration among the Tokyo Electric Power Company, Sumitomo Electric Industries and Mayekawa Manufacturing Company. The Ishikari Project is the R&D of single-core DC cable implemented mainly by Chubu University. Figure 2 and table 1 show the structures and specifications of model cables, respectively. The model A is the three-in-one-type AC cable located inside the corrugated pipe, and the model B is the single-core-type DC cable located inside the straight pipe.

In order to validate the evaluation of pressure drop and temperature rise using equations (4) and (5), the experimental data obtained previously by the other groups are compared with the corresponding

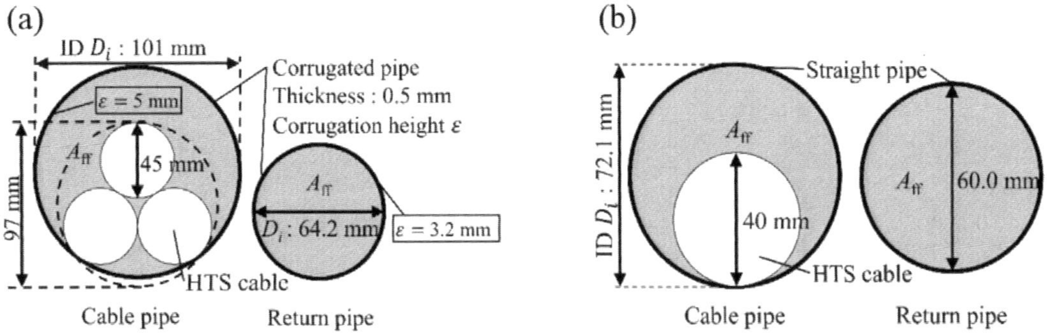

Figure 2. Cross-sectional structures of (a) three-in-one-type AC cable located inside corrugated pipe (model A) and (b) single-core-type DC cable located inside straight pipe (model B).

Table 1. Specifications of model cables.

Parameters	Model A	Model B
Reference	Yokohama project [1]	Ishikari project [2]
Free-flow cross-sectional area, A_{ff}	32.4 cm^2	28.3 cm^2
Heat inleak	1.8 W/m	1.8 W/m
AC loss for AC cable	3 W/m	-
Dielectric loss for AC cable	0.3 W/m	-
Cable inlet pressure, p_0	1.5 MPa	1.5 MPa
Cable inlet temperature, T_0	65 K	65 K
Acceptable pressure range	0.5-1.5 MPa	0.5-1.5 MPa
Acceptable temperature range	65-75 K	65-75 K

calculated results at first. In the Yokohama project [1], the pressure drop of the corrugated pipe with the three-in-one-type HTS cable has been evaluated as about 15 kPa/250 m for the flow rate of 40 L/min. Assuming the cable length of 250 m, the inlet temperature of 69 K, the inlet pressure of 0.25 MPa and the total heat input of 5 W/m, the total pressure drop can be estimated as 12.0 kPa, which is close to the above-mentioned test result. In the Ishikari project [2], on the other hand, the pressure drop for the straight pipe observed at the flow rate of 30 L/min was about 10 kPa for the cable section and the temperature rises were 0.4 K, 0.15 K and 19 mK for sections 3, 2 and 1, respectively, where the liquid nitrogen flowed through the cable pipe in the order of sections 3, 2 and 1. When the inlet temperature of 75 K, the inlet pressure of 0.25 MPa, the heat inleak of 0.818 W/m for section 3 with the length of 379 m, 0.95 W/m for section 2 with 137 m and 0.034 W/m for section 1 with 475 m are used in the calculation, the temperature rises of 369 mK, 155 mK and 21 mK are obtained for sections 3, 2 and 1, respectively, and the pressure drop is estimated as 9.7 kPa for the entire cable pipe, which are very close to the test results.

Figure 3 shows the cable length dependency of pressure drop and temperature rise in two model cables shown in figure 2. Since it is assumed that the cables are located horizontally, the effect of difference in height represented by the third terms in right-hand sides of equations (4) and (5) is not considered here. In the model A, the distance at which the pressure drop becomes equal to 1 MPa is 3.45 km together with the temperature rise of 6.47 K. In this case, the contribution of the second term in right-hand side of equation (4) is only 6.7 ppm, so that the effect of heat input can be neglected. On the other hand, the contribution of the first term in right-hand side of equation (5) is 6.0%, so that the

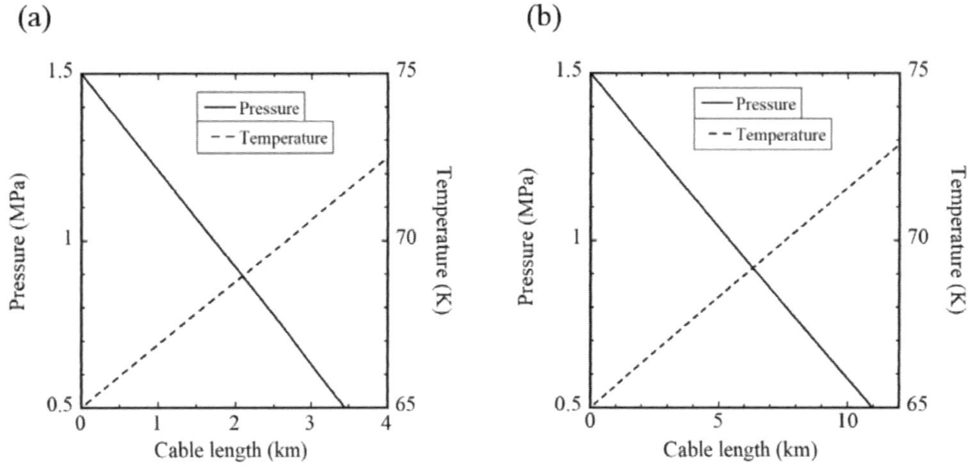

Figure 3. Profiles of pressure drop and temperature rise along HTS cables of (a) model A and (b) model B.

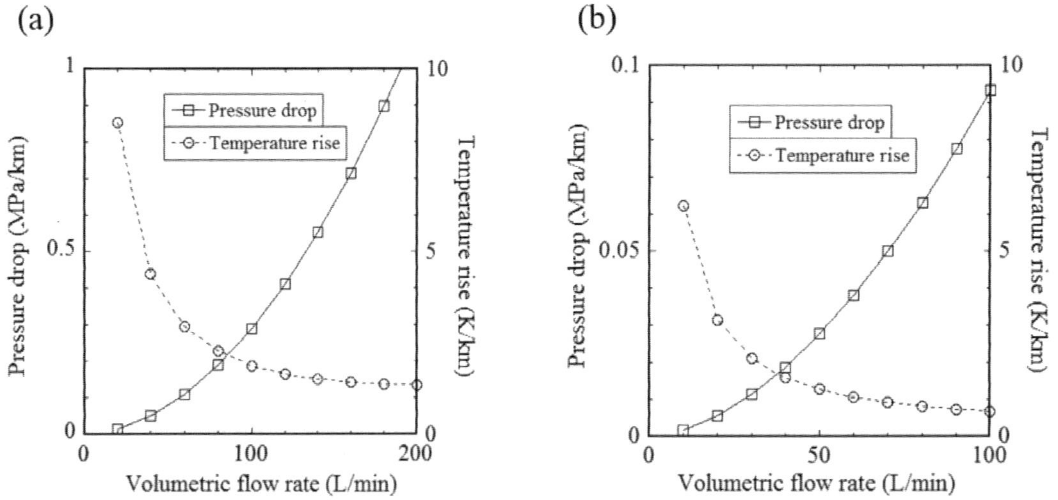

Figure 4. Flow rate dependency of pressure drop and temperature rise in (a) model A and (b) model B.

effect of friction cannot be ignored. In the case of the model B, the distance at which the pressure drop becomes equal to 1 MPa is 10.97 km together with the temperature rise of 7.20 K. In this case, the magnitude relationships between the terms in right-hand sides of equations (4) and (5) are similar to those for the model A. That is, the contributions of the second and first terms in right-hand sides of equations (4) and (5) are 9.9 ppm and 5.4%, respectively.

Figure 4 shows the flow rate dependency of pressure drop and temperature rise in the model cables. The pressure drops increase with increasing the flow rates, whereas the temperature rises have opposite dependency. This means that there are optimum conditions for flow rates, which are discussed in the next section. The pressure drops are proportional to the 1.90 and 1.76 powers of the flow rates in the models A and B, respectively.

Figure 5 shows the dependence of the difference in elevation on pressure drop and temperature rise in two model cables. When the inclination of the cable increases in the positive direction, where the

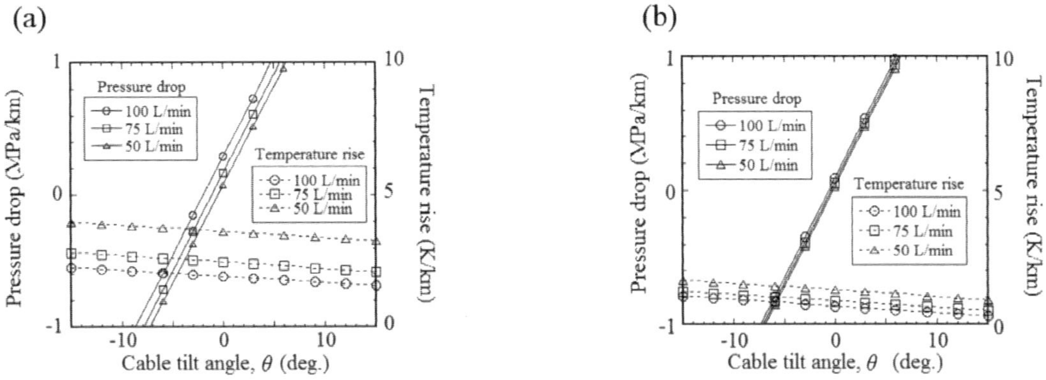

Figure 5. Dependence of difference in elevation of pressure drop and temperature rise in (a) model A and (b) model B.

liquid nitrogen flows diagonally upward, the pressure drop also increases. This is due to the own weight of liquid nitrogen. The influence of difference in height is quite significant for the pressure drop, whereas it is very small for the temperature rise.

4. Optimum design of cryogenic pump

The maximum feeding distances of HTS cables can be determined by adjusting the flow rates of pump if the maximum permissible pressure drop and temperature rise are fixed in advance, as can be found from figure 4. Therefore, we construct an optimization program code to find the flow rate at which the cable length becomes longest. In the models A and B, an optimization program code is constructed to obtain the inlet flow rate x_2 at which the cable length x_1 becomes longest within the pressure drop of 1 MPa and the temperature rise of 10 K. At that time, we use the nonlinear optimization library NLopt [12,13]. In this case, the objective function F and the inequality constraint equations are defined as

$$\begin{aligned} &\text{minimize} && F(x_1, x_2) = -x_1, \\ &\text{subject to} && p_0 - p_d \leq 1\ \text{MPa}, \\ & && T_d - T_0 \leq 10\ \text{K}, \end{aligned}$$

where p_0 and T_0 are the pressure and temperature at the cable inlet, respectively. In the case of cable pipe without return pipe, where the liquid nitrogen flow is unidirectional, p_d and T_d are considered as the pressure and temperature of the cable termination, respectively. In the case of cable pipe with return pipe, where the flow is bidirectional, p_d and T_d are considered as the pressure at the return cable end and the halfway cable termination temperature, respectively. The cable length dependency of pressure drops and temperature rises in the models A and B with optimum flow rates of liquid nitrogen is plotted in figures 6 and 7, respectively, and the optimization results are summarized in table 2. In the cases with return pipe, there are pressure drops in the return pipes, so the optimum flow rates are less than those for no return pipe. Even if the return pipes are added, the cable lengths do not become too short because the temperatures at the return pipes do not affect the temperature limit of 10 K.

5. Conclusions

The pressure drop and temperature rise were numerically evaluated when the subcooled liquid nitrogen was circulated through two types of model cables. The pressure drop and temperature rise increased and decreased with increasing the pump flow rate, respectively. When the inclination of cable increased in the positive direction, where the liquid nitrogen flowed diagonally upward, the pressure drop also increased. This was due to the own weight of liquid nitrogen. The influence of difference in height was quite significant for the pressure drop, whereas it was very small for the

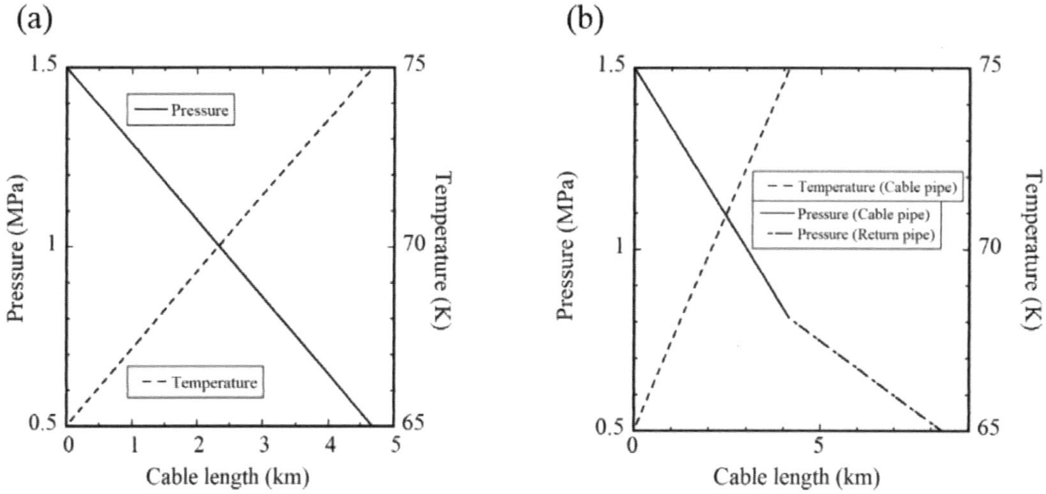

Figure 6. Profiles of pressure drop and temperature rise along model A (a) without and (b) with return pipe.

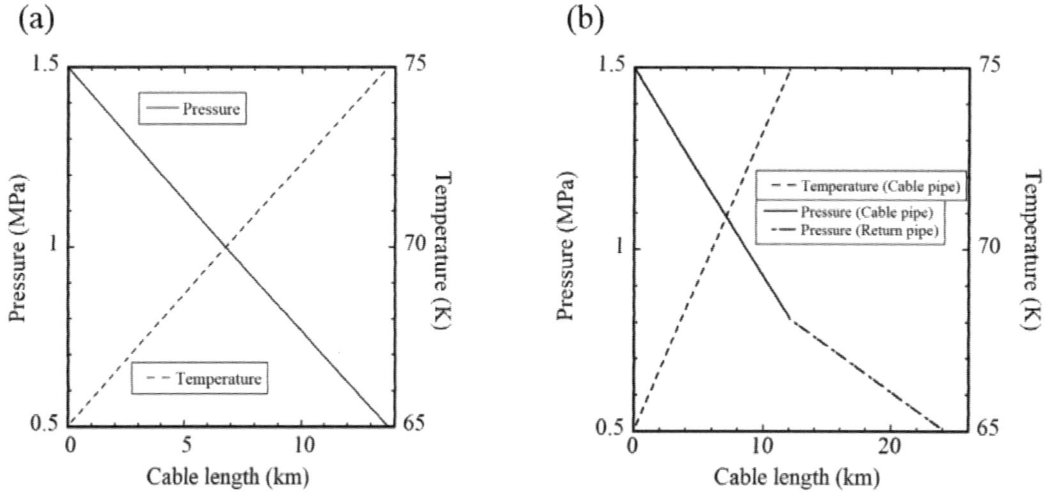

Figure 7. Profiles of pressure drop and temperature rise along model B (a) without and (b) with return pipe.

Table 2. Optimum flow rate at which the cable length becomes longest.

HTS cable	Return pipe	Optimum inlet flow rate	Cable length	Plot
Model A	None	85.2 L/min	4.68 km	Figure 6(a)
	Included	74.7 L/min	4.15 km	Figure 6(b)
Model B	None	88.3 L/min	13.74 km	Figure 7(a)
	Included	76.9 L/min	12.10 km	Figure 7(b)

temperature rise. Furthermore, we constructed the optimization program code to find the flow rate at which the cable length became longest.

Acknowledgments
This work was supported by the Advanced Low Carbon Technology Research and Development Program (ALCA) from the Japan Science and Technology Agency (JST).

References

[1] Yumura H, Ashibe Y, Ohya M, Itoh H, Watanabe M, Masuda T, Ichikawa H, Mimura T, Honjo S, Hara T, Ohno R, Shimoda M, Nakamura N, Komagome T and Yaguchi H 2013 *IEEE Trans. Appl. Supercond.* **23** 5402306
[2] Watanabe H, Ivanov Y V, Chikumoto N, Takano H, Inoue N, Yamaguchi S, Ishiyama K, Oishi Z, Koshizuka H, Watanabe M, Masuda T, Hayashi K and Sawamura T 2017 *IEEE Trans. Appl. Supercond.* **27** 5400205
[3] Kamioka Y, Kajikawa K, Hirai H, Ozaki S, Nakamura T, Imagawa S and Ishiyama A 2018 *27th Int. Cryogenic Engineering Conf. and Int. Cryogenic Materials Conf. 2018* (Oxford, UK) E-14-28
[4] Sasaki A, Ivanov Y and Yamaguchi S 2011 *Cryogenics* **51** 471–6
[5] Trevisani L, Fabbri M and Negrini F 2007 *Cryogenics* **47** 113–20
[6] Nakayama Y 2000 *Introduction to Fluid Mechanics* (Oxford: Butterworth-Heinemann) pp 56–62
[7] Bottura L 1996 *J. Comput. Phys.* **125** 26–41
[8] Barron R F 1985 *Cryogenic Systems* (2nd-ed, New York: Oxford Univ Press) pp 113–8
[9] Horizon Technologies 2011 GASPAK 3.32
[10] Ivanov Y, Watanabe H, Hamabe M, Kawahara T, Sun J and Yamaguchi S 2012 *Phys. Procedia* **36** 1372–7
[11] Nakayama Y 2000 *Introduction to Fluid Mechanics* (Oxford: Butterworth-Heinemann) pp 114–8
[12] Johnson S G 2018 *The NLopt nonlinear-optimization package* (http://ab-initio.mit.edu/nlopt, accessed on 21 September 2018)
[13] Powell M J D 1998 *Acta Numerica* **7** 287–336

IOP Conf. Series: Journal of Physics: Conf. Series **1293** (2019) 012071 doi:10.1088/1742-6596/1293/1/012071

Heat load to the cryogenic system in the 1000 m class superconducting DC power transmission system

Hirofumi Watanabe[1], Yury V Ivanov[1], Noriko Chikumoto[1], Satarou Yamaguchi[1], Kotaro Ishiyama[2], Zenji Oishi[2], Michihiko Watanabe[3] and Takato Masuda[3]

[1]Chubu University, Kasugai, Aichi 487-8501, Japan
[2]Chiyoda Corporation, Yokohama, Kanagawa, 220-8765, Japan
[3]Sumitomo Electric Industries, Ltd., Osaka, Osaka, 554-0024, Japan

E-mail: h_watanabe@isc.chubu.ac.jp

Abstract. Long-term circulation and load cycle tests of the cooling of the 1000 m class superconducting DC power transmission system in Ishikari, Japan were performed to measure the heat load delivered to and the power consumption of the cryogenic system. The average heat load and power consumption during these tests was about 2.5 and 120 kW, respectively. To increase the efficiency of superconducting power transmission systems, the selection and composition of constituent devices of the cooling system should be optimized.

1. Introduction

Superconducting power transmission presents an efficient way to send electricity long distances due to the zero resistance losses superconductivity offers. Unlike conventional power transmission systems, very little Joule heat losses are generated; however, transmission losses do occur due to the energy required to remove heat from the cryogenic system. The heat leak characteristics of cryogenic pipes at specific outer pipe temperatures have been investigated [1,2]. However, as the outer pipe temperature varies with the change of the environmental conditions such as the atmospheric temperature, weather, and direct sunlight hours, the incurred heat leak also vary. Losses occurring during circulation can also change with the change of the operation conditions of the system, such as the transmitted current, the flow rate and temperature of liquid nitrogen. Therefore, the heat load to the cryogenic system, which reflects the heat leak and losses during circulation, changes constantly during the actual operation of the superconducting power transmission system. The heat load to the cryogenic system, the cooling power of the cryocoolers, and the consumption power of the cooling system were therefore investigated using the 1000 m class superconducting DC power transmissions system in Ishikari, Japan to demonstrate the energy balance of the superconducting power transmission system [3,4,5]. The measurements were performed during long-term circulation and load cycle tests in the second cooling test of the 1000 m system [2,6]. The long-term circulation test was performed to test the stable circulation of the liquid nitrogen for about one month. This was then followed by a load cycle test, in which a current of 1 kA was turned on for 8 hours and off for 16 hours per day to simulate normal system operation for 22 days. The resulting energy balance of the 1000 m system could help improve the efficiency of superconducting power transmission systems.

2. Heat load and power consumption measurements

A piping and instrumentation diagram, including the cooling and cable systems, of the 1000 m system used here is shown in figure 1. The cooling system was situated at terminal A and consisted of two pumps (CP1 and CP2), two Turbo-Brayton cryocoolers (TB1 and TB2), two Stirling cryocoolers (ST1 and ST2), and auxiliary equipment. Two chillers (CH1 and CH2) supplied the coolant for the cryocoolers. An additional pump at terminal B (CP3) compensated for the pressure drop. The cable system consisted of the cryogenic pipe constituting a transmission line and containing two inner pipes within an outer pipe, and terminals at either end [1,2]. Of the two inner pipes, the cable pipe was used to install the cable and the return pipe was used to return the liquid nitrogen for circulation. The white and black valves shown in figure 1 represent opened and closed valves, respectively. The temperature, pressure, and liquid nitrogen level were monitored with temperature sensors "T," pressure sensors "P," and a level sensor "Lt1," respectively.

During the long-term circulation test and load cycle test, only TB1 and TB2 were used; ST1, ST2, and CP2 were cut off by the valve settings. Liquid nitrogen, which was pressurized by CP1 and cooled by TB1 and TB2, passed from the reservoir to terminal A and through the cable pipe before reaching terminal B. At terminal B, the liquid nitrogen was re-pressurized by CP3 and then passed through the return pipe before returning to the reservoir. The temperature settings of TB1 and TB2 were 70 K. Accordingly, the temperature of liquid nitrogen at the outlet of TB2 was approximately 70 K. The flow rate of liquid nitrogen was set to 40 L/min.

The heat leak was estimated for each section of the cable system (the cable pipe, the return pipe, and the terminals) by measuring the increase in temperature between the inlet and outlet of each section, the flow rate, and the specific heat capacity of the liquid nitrogen. The cryogenic pipe was installed mostly outdoors; the outer pipe temperature was measured a few meters away from the building where terminal A was installed.

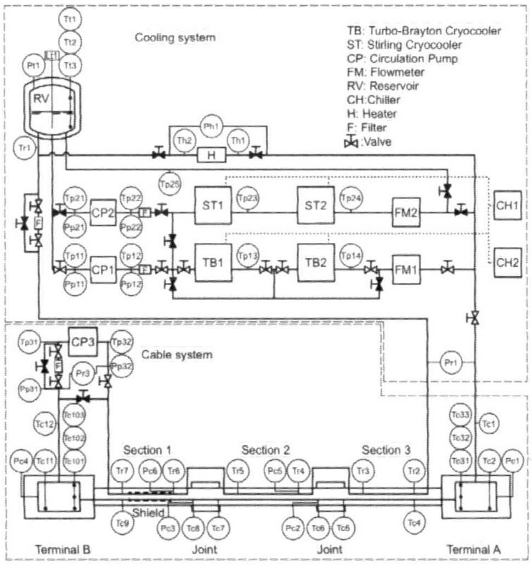

Figure 1. Piping and instrumentation diagram of the 1000 m system.

The consumed power was measured using electric power meters for TB1, TB2, CH1, and CH2 at their power inputs during the long-term circulation and load cycle tests. By a separate experiment, the consumption power of CP1, the hydraulic power, and the loss in CP1 were measured with respect to the flow rate and rotational speed of the impeller of CP1. The consumption power of CP1 was measured with an electric power meter at its power input. By referring this result with the flow rate and the rotational speeds during the long-term circulation and load cycle tests, the consumption powers of CP1 and CP3 were estimated.

3. Cooling power of the 1000 m system

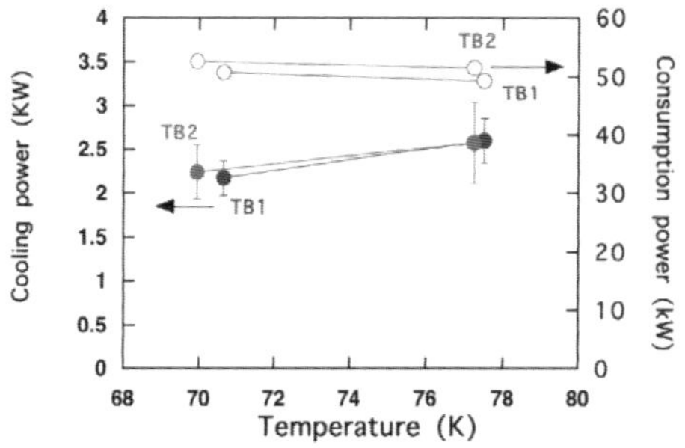

Figure 2. Cooling and power consumption of TB1 and TB2 [7].

The cryogenic system was cooled by TB1 and TB2 during the long-term circulation and load cycle tests. The nominal cooling powers of TB1 and TB2 are 2 kW at 66 K. The cooling and power consumption of TB1 and TB2 have been measured and reported previously [7], and are replicated in figure 2 for reference.

4. The heat load by the heat leak

Table 1. Heat leak of each component of the 1000 m system [2,7]. The numbers in the parentheses are the values of experimental uncertainties referred to the corresponding last digits of the results.

Outer pipe (°C)	Terminal A (kW)	Terminal B (kW)	Cable pipe (kW)	Return pipe (kW)	Total (kW)
−3.0	0.199(17)	0.233(16)	0.461(16)	0.676(16)	1.690(23)
−2.4	0.225(18)	0.248(17)	0.459(17)	0.658(18)	1.746(20)
17.4	0.226(18)	0.249(17)	0.600(17)	0.901(20)	2.091(22)

The recovery of heat leak represents a main heat load to the cryogenic system. The heat leak characteristics of each component of the 1000 m system at specific outer pipe temperatures were previously reported and are summarized in table 1 [2,7]. The heat leaks of the cable and return pipes were shown to depend on the outer pipe temperature, which increase approximately 30%–40% with an outer pipe temperature increase of 20 °C.

5. The heat load by circulation

The discharge pressure, consumption power, hydraulic power, and pump loss of CP1 were measured with respect to the flow rate and the rotational speed of its impeller and are shown in figure 3. The consumption power at the zero rotational speed was subtracted from the measured values, as the subtracted value was unrelated to the pump work. The consumed power can thus be considered the input power of the pump. The hydraulic power, which was obtained from the discharge pressure and flow rate, increased with increasing discharge pressure and flow rate. The hydraulic power, which drives the circulation, was consumed by the fluid friction and converted to the heat load to the cryogenic system. The pump loss were measured as the internal energy increase of the liquid nitrogen. A large fraction of the loss was related to the rotation of the impeller, while approximately 40 W, as shown in figure 3, was unrelated to the rotational speed and was considered as the heat leak through the inner structure of the pump. The sum of the hydraulic power and pump loss represented the total circulation loss and was nearly equal to the consumption power. Therefore, the consumption power of the pump was added as the heat load of the circulation to the cryogenic system. Since the pressure drop increased with the second powers of the flow rate, the hydraulic power and, accordingly, the input power of the pump increases with the third powers of the flow rate. As the consumption power sharply increased with the flow rate, discharge pressure, and rotational speed, increasing the efficiency of the superconducting power transmission systems, which need pumps with a very large capacity, may be difficult.

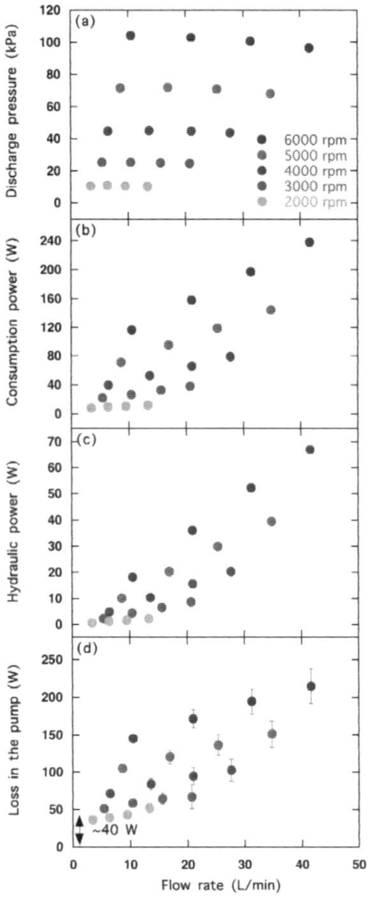

Figure 3. (a) Discharge pressure, (b) consumption power, (c) hydraulic power, and (d) loss of CP1 with respect to the flow rate and rotational speed of the impeller.

6. Heat load and power consumption measurements of 1000 m system

The total power consumption of the cooling system, representing the sum of the power consumption of TB1, TB2, CH1, and CH2, is shown in figure 4(a). The consumption of CP1 and CP3 are not shown because they were small in comparison with the others. The total heat load of the cable system is shown in figure 4(b). The heat load of the cooling system including the heat leak of the cryogenic pipe in the cooling system and the loss in CP1, which was estimated to be about 0.4 kW, is not included in the figure. To obtain the heat leak of the cryogenic pipe in the cooling system, the heat leak from Tc1 to Tr1 was subtracted from that of Tp14 to Tp11. The loss in CP1 was estimated from the data in figure 3(d). Here, the spike appearing at 2016/10/20 was caused by a temporal stop of the circulation. The pattern of the transmitted current applied to the cable during the load cycle test and the outer pipe temperatures of the cryogenic pipe are shown in figure 4(c) and 4(d), respectively. The outer pipe temperature reflects the change of the environmental conditions.

The heat loads of the cable and return pipes fluctuated and gradually decreased in phase with the outer pipe temperatures, as a large part of the cryogenic pipe is installed outdoors, so the heat leak subject to changing outer pipe temperature are accordingly subject to changing environmental conditions. The environmental conditions should therefore be considered when designing the cooling capacity and power supply systems required to operate the cooling system. On the other hand, the heat loads of the terminals were not affected by the outer pipe temperature, because they were installed in the air-conditioned buildings. However, a small fluctuation of about 40 W was seen during the load cycle test in accordance with the pattern of the transmitted current. This fluctuation was considered to be the heat load originated in the current leads by the transmitted current.

The average total heat load of the cable system was about 2.2 kW; that of the cooling system was estimated at 0.4 kW. The total heat load was thus approximately 2.6 kW. The cooling power of TB1 at 70 K was about 2.1 kW and therefore could not meet the cooling requirements of the system. As a result, TB1, at the upper stream of TB2, was in a full-load operation. The residual heat load, which exceeded the cooling power of TB1, was cooled by TB2, resulting in a partial load operation of TB2. This affected the power consumption of TB1 and TB2 and the consumption power of TB2 is an almost half of TB1. Furthermore, the power consumption of the chillers was almost 2/5 of the total power consumption. The power required for the coolant of the cryocoolers should also be considered to estimate the total efficiency of the cooling system. Partial load operation of TB2 and the use of the chillers likely reduced the efficiency of the system. Therefore, the choice of cryocooler with appropriate cooling capacity and disuse of chillers could increase the system efficiency.

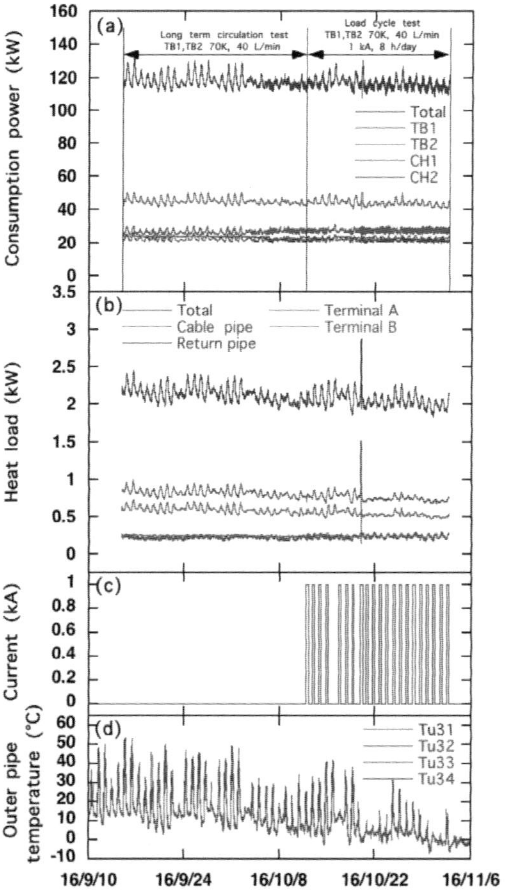

Figure 4. (a) Power consumption, (b) heat load, (c) current, and (d) outer pipe temperature during the long-term circulation and load cycle tests.

Table 2. Summary of the power consumption and heat loads for the four cases.

Case	Consumption power (kW)							HL [a] (kW)	HL/Total[a]
	TB1	TB2	CH1	CH2	CP1	CP3	Total		
1	45.3	23.6	26.7	22.2	0.23	0.10	118.1	2.2(2.6)	0.019(0.022)
2	44.3	23.1	27.4	21.5	0.23	0.10	116.7	2.1(2.5)	0.018(0.021)
3	50.7	24.3	30.4	26.5	0.23	0.10	132.3	2.4(2.8)	0.018(0.021)
4	41.4	22.0	25.3	19.6	0.23	0.10	108.6	1.9(2.3)	0.017(0.021)

[a] HL represents the heat load of the cable system. Values in parentheses include the heat load of the cooling system.

The resulting power consumption and heat load of the cable system is summarized in table 2. The heat load of the cable system includes the heat leak from Tc1 to Tr1 and the loss of CP3. The overall heat load of the system, shown in parentheses, was obtained by adding the heat load of the cooling system, 0.4 kW, to the heat load of the cable system. Case 1 and Case 2 are the averages during the long-term circulation and load cycle tests, respectively. Case 3 and Case 4 are the maximum (2016/9/16 15:05) and the minimum (2016/10/31 6:06) in the whole period, respectively. The average heat load to the cryogenic system was about 2.5 kW, while the power consumption was about 120 kW. The ratio of the total heat load to the total power consumption, which relates to the efficiency of the system, was determined to be about 0.02 in all cases.

7. Summary

The heat load to the cryogenic system and the power consumption of the cooling system were measured with the 1000 m class superconducting DC power transmission system in Ishikari, Japan. The measurements were performed during long-term circulation and load cycle tests in the second cooling test of the 1000 m system. The average heat load to the cryogenic system during these tests was about 2.5 kW, while the consumption power was about 120 kW. To increase the efficiency of superconducting power transmission systems, the efficiency of each component of the cooling system and its composition, such as the use of chillers, should be optimized.

Acknowledgement

This work was supported in part by the Japanese Ministry of Economy, Trade and Industry (METI) and New Energy and Industrial Technology Development Organization (NEDO). HW is supported by Chubu University Grant (A).

References

[1] Watanabe H, Ivanov V Y, Hamabe M, Chikumoto N, Takano H and Yamaguchi S 2015 *Physics Procedia* **67** 239-44
[2] Watanabe H, Ivanov V Y, Chikumoto N, Yamaguchi S, Ishiyama K, Oishi Z, Watanabe M and Masuda T 2018 *J. Phys. Conf. Ser.* **1054** 012076
[3] Yamaguchi S, Koshizuka H, Hayashi K and Sawamura T 2015 *IEEE Trans. Appl. Supercond.* **25** 5402504
[4] Yamaguchi S, Ivanov Y, Watanabe H, Chikumoto N, Koshiduka H, Hayashi K and Sawamura T 2016 *Physics Procedia* **81** 182-6
[5] Chikumoto N, Watanabe H, Ivanov V Y, Takano H, Yamaguchi S, Koshizuka H, Hayashi K and Sawamura T 2016 *IEEE Trans. Appl. Supercond.* **26** 5402204
[6] Chikumoto N, Watanabe H, Ivanov V Y, Takano H, Yamaguchi S, Ishiyama K, Oishi Z, Watanabe M and Masuda T 2018 *IEEE Trans. Appl. Supercond.* **28** 5401005
[7] Watanabe H *et al.* 2017 *IEEE Trans. Appl. Supercond.* **27** 5400205

Development of Single-phase Bi2223 High Temperature Superconducting Transformer with Protection System for High Frequency and Large Current Source

N Nanato, T Adachi and T Yamanishi

Graduate School of Natural Science and Technology, Okayama University,
3-1-1, Tsushima Naka, Kita-ku, Okayama, 700-8530, Japan

nanato@okayama-u.ac.jp

Abstract. The authors have been developing a large AC current source with a single-phase Bi2223 superconducting transformer. The authors aim to develop the source of supplying current with the maximum frequency of 1 kHz and the maximum amplitude of 1 kA. However, the conventional transformer has some leakage inductance resulting in large voltage drop, and therefore it was unable to transport the current with high frequency and large amplitude. In this paper, the authors propose a configuration of the transformer with less leakage inductance than inductance of the conventional one. Improved current source has output current with the maximum frequency of 1 kHz and the maximum amplitude of 900 A. In addition, a protection system for normal transition in the transformer is proposed. The authors have proposed the active power method to detect the normal transition. For high frequency and large current, however, the normal transition may be incorrectly detected by using the method due to large AC loss. The authors show a new method to avoid the incorrect detection by an empirical formula based on measurement results of the AC loss. From the experimental results of the normal transition, the authors show that the proposed method is more correct to detect the normal transition than the conventional method.

1. Introduction

The authors have developed a small and light power source that can output a current with the maximum frequency of 1 kHz and the maximum amplitude of 1 kA to measure current conduction characteristics of a high temperature superconducting (HTS) wire [1, 2]. The main part of the source is an HTS transformer. The conventional HTS transformer has some leakage inductance resulting in large voltage drop. In order to achieve the current of 1 kHz and 1 kA_{peak}, the leakage inductance must be much reduced. The main cause of the leakage inductance is gap between the primary coil and the secondary coil of the transformer. In this paper, we propose a revised structure that two primary coils with the same height as the secondary coil sandwich the secondary coil in order to reduce the gap. Achieved output current has the maximum frequency of 1 kHz and the maximum amplitude of 900 A.

Also, for the safe operation of the HTS transformer, it is essential to protect the transformer from excessive heating in the normal transition. We have proposed a protection system based on an active power method to detect the normal transition [1, 2]. However, for high frequency and large current, the normal transition may be incorrectly detected by using the method due to large AC loss. In the reference [2], we proposed a countermeasure for the correct detection, however, the calculation formula was complex and not enough for the correct detection. We propose a new method to avoid the incorrect detection based on an empirical formula to calculate the active power and show its usefulness by experimental results.

2. Development of single-phase Bi2223 HTS transformer

Figure 1 shows configuration of a proposed single-phase Bi2223 HTS transformer. In order to reduce leakage inductance, the secondary coil was sandwiched by the primary coils so as to minimize gap between the primary coils and secondary one. Height of the primary coils were the same as that of the secondary one. Specifications of the HTS transformer is shown in Table 1. Numbers of the turns were 40 and 1 for the primary coils and secondary coil, respectively. The primary rated current was 25 A_{peak} which a small power source with capacity of 3 kW $= (100\ \mathrm{V} \times 30\ \mathrm{A})$ could output. Because the secondary rated current was 1000 A_{peak}, required turn ratio was 40. The numbers of turns were the minimum to obtain the ratio. The primary coils and the secondary coil were wound with DI-BSCCO superconductor (Type H). The primary wire was a single wire because the critical current of the wire was 200 A at 77 K and self-field which was larger than the primary rated current 25 A_{peak}. The secondary wire was a bundle of 10 wires so that the critical current of the bundle became larger than the secondary rated current 1000 A_{peak}.

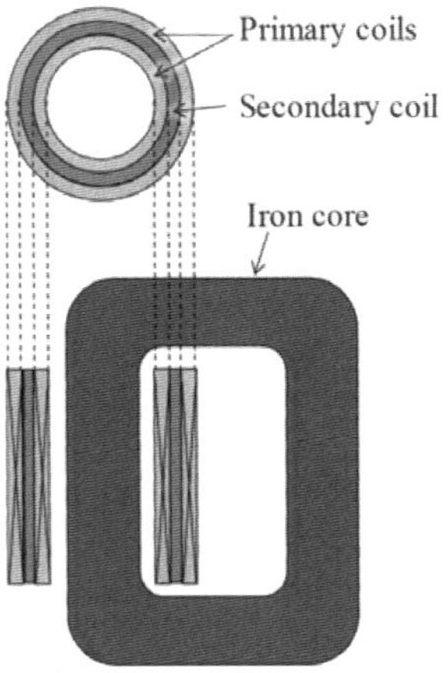

Figure 1. Configuration of the developed HTS transformer.

Table 1. Specifications of the developed transformer

Transformer			
Phase			Single
Primary rated current (A_{peak})			25
Primary rated voltage (V_{peak})			141
Secondary rated current (A_{peak})			1000
	Primary coils		Secondary coil
	Inner	Outer	
Inner diameter (mm)	89.0	89.6	89.3
Outer Diameter (mm)	89.3	89.9	89.6
Height (mm)	91.0	91.0	91.0
Number of Turns	40		1
Inductance (mH)	52.6		0.0293

Table 2 shows the leakage inductance of the developed transformer and the conventional one [2]. The leakage inductance of the HTS transformer was about 1/3 times smaller than that of the conventional one. Table 3 shows the primary currents I_1 (A_{rms}) and the primary voltages V_{SC1} (V_{rms}) for transporting the secondary current of 1 kA_{peak} at 60 Hz to the developed transformer and the conventional one. The primary currents were almost same with each other, however, the primary voltage of the developed transformer was about 1/3 times smaller than that of the conventional one due to the reduction of the leakage inductance. As shown in Figure 2, the developed transformer could output the secondary current i_2 of 900 A_{peak} at 1 kHz for the primary current i_1 of 22.5 A_{peak} and the primary voltage v_{SC1} of 141 V_{peak} (100 V_{rms}) under short-circuiting secondary coil.

Table 2. Leakage inductance

The conventional transformer (mH)	2.99
The developed transformer (mH)	0.964

Table 3. Primary currents I_1 and primary voltages V_{SC1} for transporting a secondary current of 1 kA_{peak} and 60 Hz

	I_1 (A_{rms})	V_{SC1} (V_{rms})
The conventional transformer	17.8	21.8
The developed transformer	18.1	7.98

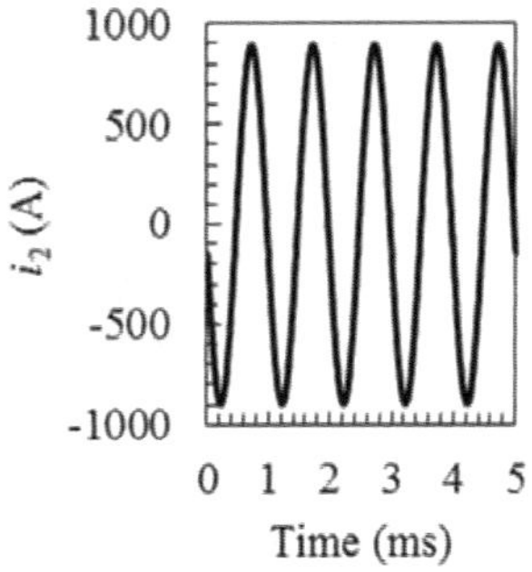

(a) Secondary current.

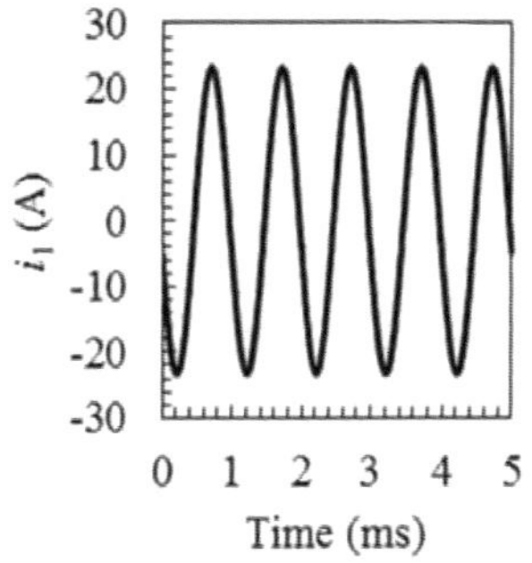

(b) Primary current.

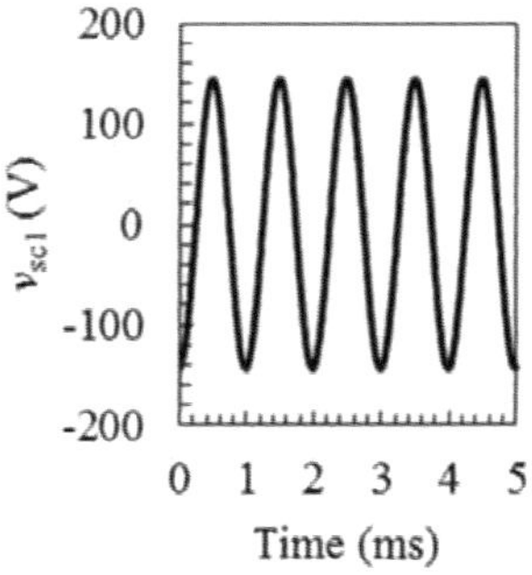

(c) Primary voltage.

Figure 2. Current transport characteristics of developed transformer.

3. Detection method of normal transition

In reference [2], the authors described that the active power method might cause incorrect detection of the normal transition due to AC loss. Also, the proposed countermeasure in [2] was complex in calculation and not enough for the correct detection. We propose a method to avoid the incorrect detection based on an empirical formula to calculate the active power.

Following equation (1) is to calculate an active power dissipated in the normal area [2].

$$P' = P_f - P_L. \tag{1}$$

P_f is the total amount of the active power dissipated in the normal area and AC loss in the transformer, and P_L is the AC loss in the transformer

The AC loss P_L is calculated by some experiments unlike the method in [2]. P_L is expressed as follows.

$$P_L = k_e I_2^2 f^2 + k_h I_2^2 f, \tag{2}$$

where I_2 is the root mean square value of the secondary current, f is the frequency and k_e and k_h are proportional coefficients. The first term on the right side shows eddy current loss and the second term shows hysteresis loss. Based on L-type equivalent circuit of the transformer, $I_2 = aI_1$ (I_1: the root mean square value of the primary current, a:turn ratio) and therefore we approximated the AC loss in the transformer by equation (2). The AC loss is calculated by equation (3).

$$P_L = \frac{1}{T}\int_0^T (v_{SC1} - av_{SC2}) \cdot i_1 dt, \tag{3}$$

where T is a periodic time of the current. By fitting (2) to (3), the coefficient k_e and k_h are determined. The calculated k_e and k_h are shown in Table 4. These are average values in a set of measurements from 60 to 1000 Hz at 900 A_{peak}.

4. Experimental results for the detection method

Figure 3 shows that P_f increases with increase of the peak current and the frequency of the secondary

Table 4. Values of the coefficient k_e and k_h

k_e	k_h
7.18×10^{-11}	4.95×10^{-8}

current due to the AC loss. Figure 4 shows that P' was suppressed to low values as compared with the P_f in Figure 3. This means that P' has little AC loss signal which causes the incorrect detection. Figure 5 shows the threshold value P_{th} and P' as function of the secondary current. P_{th} is the parameter to judge the normal transition. When P' reaches P_{th}, the transport current is cut off to protect the transformer. P_{th} must be larger than P' in the superconducting state and the normal transition must be detected so as to suppress the temperature of the normal area to less than an allowable temperature of the windings of the HTS transformer. P_{th} in Figure 5 was calculated based on the equations in [3]. The equations in [3] are based on 1 dimensional equation of heat conduction and the equation of the electric circuit. We can obtain the proper threshold to suppress the temperature of the normal area to a specified temperature by the equations. P_{th} in Figure 5 was calculated so as to suppress the maximum temperature of the normal area to 200 K which was allowable temperature of the windings [4]. The results in Figure 5 show that the proposed method can protect the HTS transformer regardless of the AC loss.

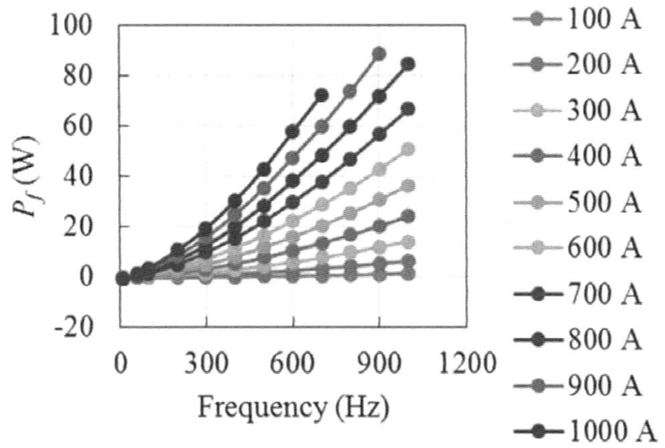

Figure 3. P_f in transporting currents with various values and frequencies.

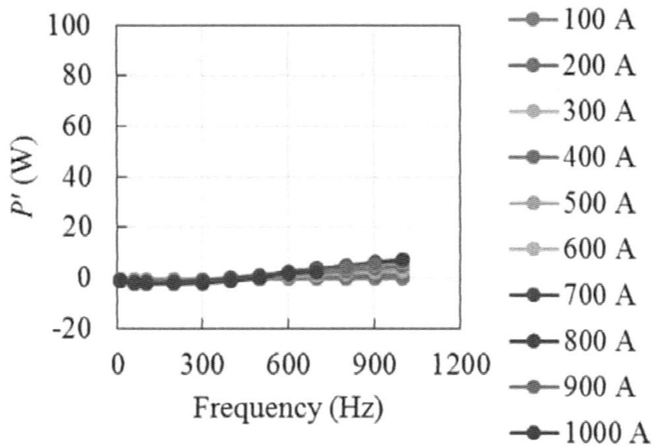

Figure 4. P' in transporting currents with various values and frequencies.

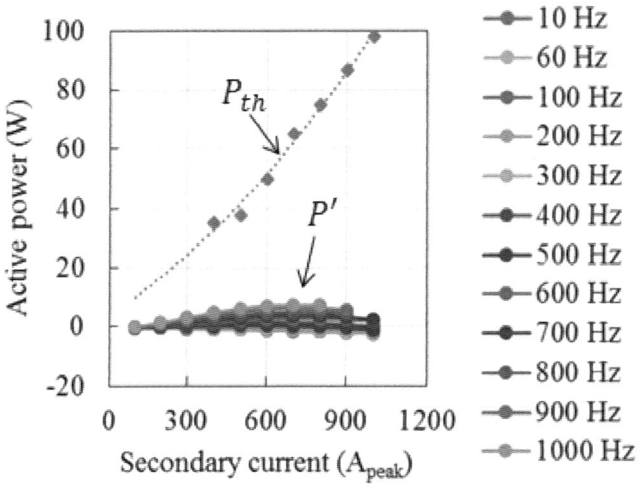

Figure 5. P_{th} and P' in transporting currents with various values and frequencies.

5. Protection tests for the Bi2223 HTS transformer

Protection tests in the normal transition were carried out by using the developed transformer. The protection circuit is shown in Figure 6 [2]. The secondary current was 850 A_{peak} at 1 kHz. The normal transition was occurred by a heater mounted on the secondary coil. When P' reached P_{th}, the thyristor switch turned off and the primary and secondary currents were shut off.

The experimental results are shown in Figure 7. Figure 7 (a), (b) and (c) show respectively P_f, P' and temperature of the normal area. P_{th} was set at 40 W for this experiment. P_f was about 120 W in the superconducting states, which was greater than 40 W until about 220 s. This means P_f causes the incorrect detection. On the other hand, P' was almost zero in the superconducting state and then it increased when the normal transition occurred at about 220 s. When P' reached 40 W, the transport currents were shut off and the temperature was suppressed less than 160 K. This means the proposed method can protect the HTS transformer correctly in the occurrence of the normal transition regardless of the AC loss.

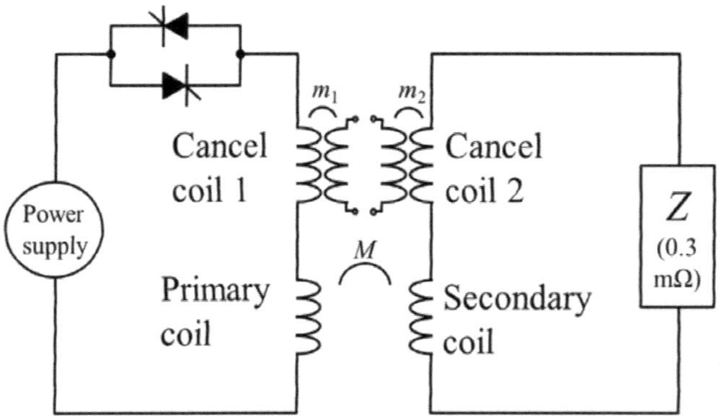

Figure 6. A protection circuit.

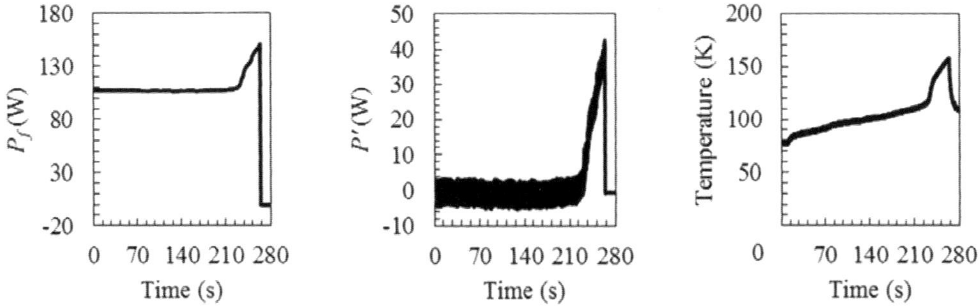

(a) P_f Signal. (b) P' Signal. (c) Temperature in the normal area.

Figure 7. Result of protection tests in the normal transition in the

6. Conclusions

The authors developed a power supply with the maximum current of 900 A_{peak} and the maximum frequency of 1 kHz by reducing leakage inductance in the HTS transformer based on structural study. For the safe operation of the transformer in the normal transition, we proposed the more correct method to detect the normal transition than the conventional method regardless of the AC loss and showed its usefulness by experimental results. We are studying a more efficient structure of the transformer to achieve the transport current of 1 kA_{peak} at 1 kHz.

Acknowledgements

This work was supported by Electric Technology Research Foundation of Chugoku and JSPS KAKENHI Grant Number 18K04080.

References

[1] Nanato N, Tanaka S and Tenkumo S 2018 Study on a magnetic flux detection coil for detection of normal transition in a hybrid single-phase Bi2223 superconducting transformer by the active power method, *Journal of Physics: Conference Series* **1054** 012070

[2] Nanato N, Ono T, Adachi T and Yamanishi T 2018 Protection system for normal transition in a single-phase Bi2223 full superconducting transformer by the active power method under flowing currents of various Frequencies, *Journal of Physics: Conference Series* **1054** 012068

[3] Nanato N, Asai W, Murase S 2012 Study on criterion for quench detection/protection of superconducting magnet based on active power method, *Physics Procedia* **27** 416

[4] Iwasa Y 2009 Case Studies in Superconducting Magnets, *Design and Operational Issues Second Edition, Springer* 505

IOP Conf. Series: Journal of Physics: Conf. Series **1293** (2019) 012073 doi:10.1088/1742-6596/1293/1/012073

Conceptual Study on lighter and more compact transmission cable systems for Hybrid Aircrafts

S Isojima[1], Y Yoshida[2], N Amemiya[3], N Sadakata[4], H Ohsaki[5] and M Okada[2]

[1] Sumitomo Electric Industries, Ltd., Japan
[2] National Institute of Advanced Industrial Science and Technology (AIST), Japan
[3] Kyoto University, Japan
[4] Fujikura Ltd., Japan
[5] Tokyo University, Japan

E-mail: isojima-shigeki@sei.co.jp

Abstract. More electrically driven aircrafts with electric propulsion systems are supposed to require 1-40 MW. In order to realize such heavily electrical aircrafts the development of lighter and more compact transmission cable systems is indispensable. We compared metal cables and superconducting cables, and found that DC superconducting cables are lighter and more compact than metal cables at voltages as high as ±6 kV.

1. Introduction
More electrically driven aircrafts with electric propulsion systems are supposed to require 1-40 MW. In order to realize such heavily electrical aircrafts, in addition to the development of higher voltage electrical systems with motors and generators with much higher power densities than those at present, compatible with low atmospheric pressure at a high altitude, the development of lighter and more compact transmission cable systems is also indispensable. ASCOT is studying superconducting cables for aircrafts, and in this paper we present the results of our feasibility study on superconducting cables for aircrafts.

2. Feasibility Study by following current AC system
As the initial step of this conceptual study, we performed feasibility study to compare metal cables and high temperature superconducting cables. First FS(Feasibility Study) is based on current 230 V AC system and an image of a hybrid aircraft used in our feasibility study is shown in figure 1.

Gas turbine and generator produce 3 phase 230 V AC, which is transmitted by 4 MW transmission cables to 8 motors of 4 MW output. We assumed, for takeoff and landing, the cables carry 100% current of 4 MW power and for cruising 33% current.

We chose cable specifications, as close as current operations. The specifications assumed are shown in table 1. Capacity is 4 MW per circuit. Voltage is 230 V AC line to line as determined by Paschen minimum voltage of 350 V for air [1]. Frequency is 400Hz as is commonly used in present aircrafts. For the application of superconductivity to transmission, DC is more appropriate. However, AC / DC convertor at present is very heavy to be used for on-board transmission. Current is determined by power divided by, voltage multiplied by square root of 3.The length of the transmission line is assumed to be 50 m.

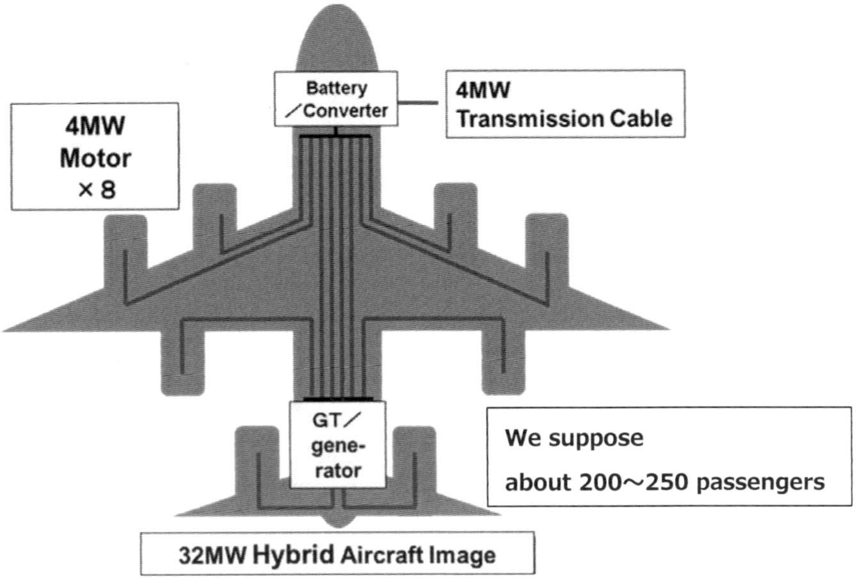

Figure 1. An image of a hybrid aircraft.

Table 1. Cable specifications assumed

capacity	voltage	frequency	current
4MW/cct	230V/Paschen minimum 350 V	400Hz/heavy DC/AC converter	10 kA
length	operation	cooling	
50 m	100% output for 1 hr., 33% for 23 hr.	no-onboard refrigerator	

During 30 min takeoff and 30 min landing, 100% output is assumed. During the cruising of the rest of 24 hours, 33% output is assumed. The refrigerator is heavy, so in this study, no on-board refrigerator is assumed. Sumitomo electric developed 3 phase superconducting cable using 2nd generation wires and tested in the testing field, in a Japanese government project [2]. The verification test and the cable are shown in figure 2 and figure 3. This might be an example for this aircraft cable.

Figure 2. 15m long cable verification system @SEI Feb. 2013[2]

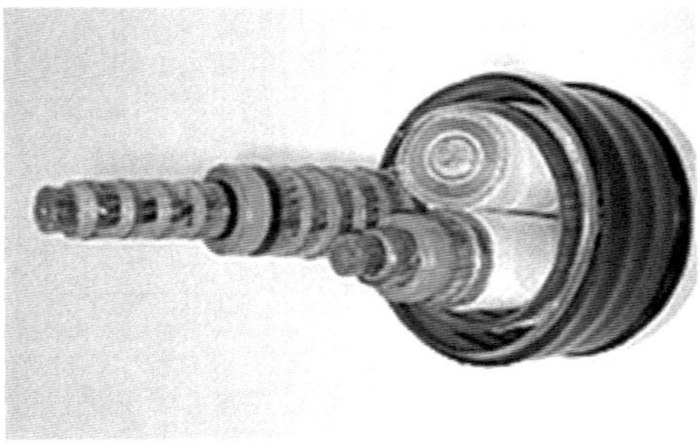

Figure 3. AC Superconducting cable[2]

For a metal cable, we estimated it by following American standard(AS50881) and an example of Cu cable is shown in figure 4. The cable design is as follows.

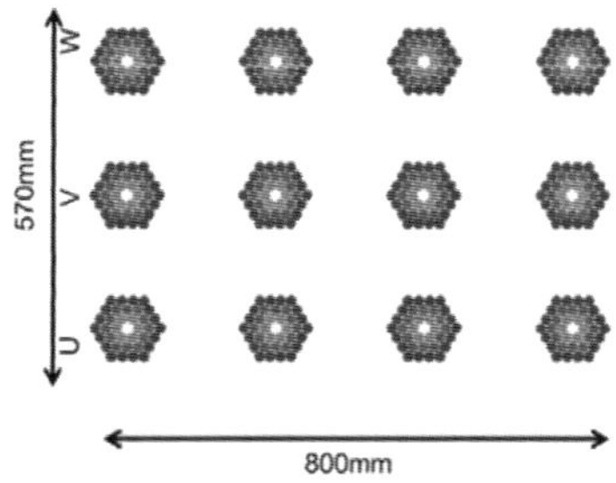

Figure 4. Cu cable example

Wire conductor is composed of 2590 strands. A bundle is composed of 36 wires. They are arranged separately with the separation of one diameter of the bundle. Wire conductor size is AWG 3/0 and wire diameter is 16.5 mm, whose current capacity at 170°C with the surrounding temperature of 100°C is 400 A, following the conditions shown in the reference [3].

Current capacity is reduced due to elevation to 83% at 45000 ft., and due to bundling to 27%, and due to multi bundling to 80%, and finally wire current capacity is 70.6 A. Cross section is 570 mm times 800 mm, and weight is 20.8 thousand kg for 50 m cable.

Next example is Al cable. Wire conductor is composed of 559 strands. A bundle is composed of 36 wires. And an example of Al cable is shown in figure 5. The cable design is as follows.

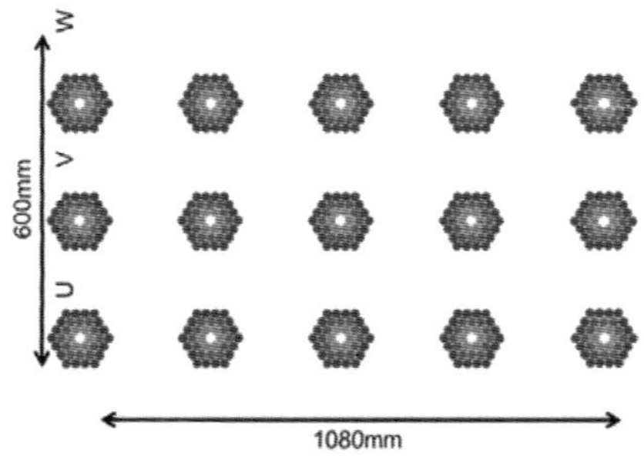

Figure 5. Al cable example

They are arranged separately with the separation of one diameter of the bundle. Wire conductor size is AWG 4/0 and wire diameter is 17.3 mm, whose current capacity at 170°C with the surrounding temperature of 100°C is 376 A.

Current capacity is reduced due to elevation to 83%, and due to bundling to 27%, and due to multi bundling to 68%, and finally wire current capacity is 56.4 A. Cross section is 600 mm times 1080 mm, and weight is 13 thousand kg for 50 m cable.

As for superconducting cable, we estimated 3 types of cable, such as cooled by solid nitrogen, liquid nitrogen with circulation and liquid nitrogen without circulation.

Figure 6 shows solid nitrogen type. On ground liquid nitrogen is poured and cooled by refrigerators or by other measures to, for example, 20K.

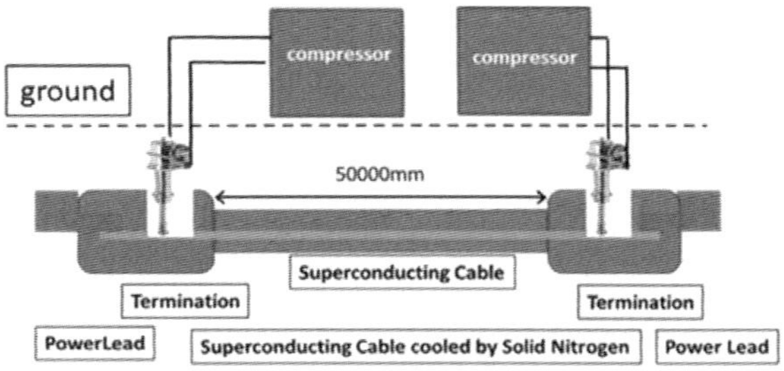

Figure 6. Solid nitrogen type

The cooling system will be removed before takeoff. Cable type is triad, in other words, 3 cores in one cryostat. The power lead is designed as conductively cooled. Maximum operating temperature is set to be 60K as an example. For faster cooling down, thermal conductors are installed in the cryostat. AC loss is assumed to be 1 W/m/p by using lot of wires.

Figure 7 shows a liquid nitrogen cooling type with circulation. The reservoir of liquid nitrogen can store liquid nitrogen consumed during a flight.

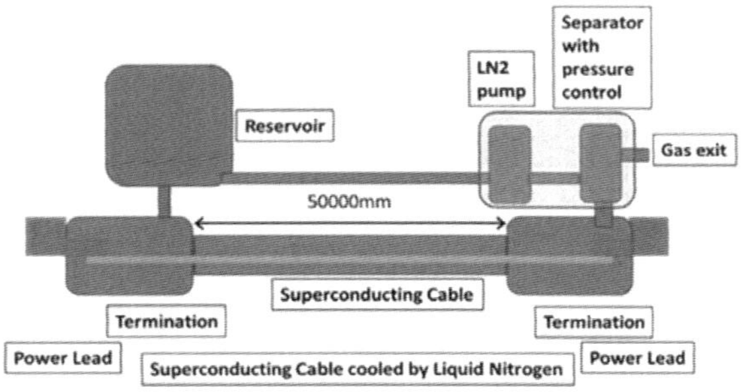

Figure 7. Liquid nitrogen cooling type with circulation

Thicker wall is used because the pressure of liquid nitrogen will be increased due to inclination of the aircraft. Gas nitrogen generated after cooling is moved by the pump to the separator which separates gas nitrogen from the liquid nitrogen and only the gas nitrogen is exhausted thanks to the pressure control.

The cable is assumed to be triad and the power lead is conductive cooling. AC loss is supposed to be 10 W by using a little bit fewer wires to reduce weight.

Figure 8 shows a liquid nitrogen cooling type without circulation. Liquid nitrogen to be consumed in a flight is poured into the cable and termination before takeoff.

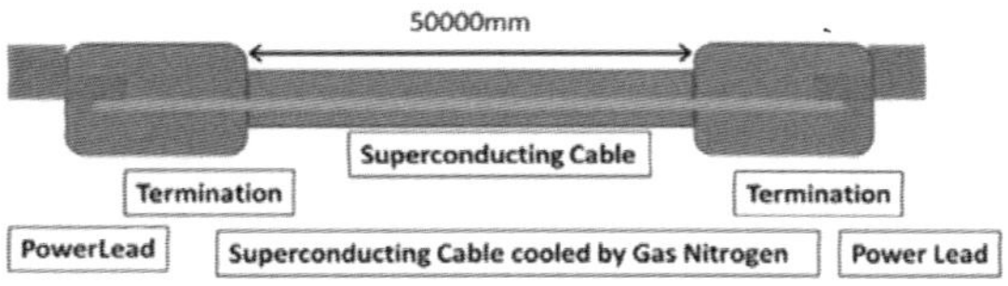

Figure 8. Liquid nitrogen cooling type without circulation

Liquid nitrogen inside is divided by separated rooms to avoid pressure accumulation and large dry area formation of the cable during inclination of the aircraft.

Gas nitrogen generated after cooling is exhausted from each room by a filter. The cable core has layers to increase thermal conductivity. Due to the temperature increase from the liquid nitrogen temperature, the AC loss increases a little bit.

Table 2 shows the comparison between solid, liquid with and without circulation, and Al and Cu cables.

Table 2. Results of FS on cables for 4 MW transmission

case	Solid	Liquid w/ circulation	Liquid w/o circulation	Al	Cu
Cable dia. (mm)	126	108	121	600×1080	570×800
50 m Cable weight (kg)	1020	760	1620	13000	20800
Termination L×dia (m)	1.2×1.8	1.2×0.5	1.2×0.9	-	-
Termination Weight (kg)	5180	840	1580	100	53
Reservoir L×dia (m)	-	2.20.9	-	-	-
Reservoir Weight (kg)	-	1260	-	-	-
Circulator Weight (kg)	-	560	-	-	-
Total weight (kg) (%)	6200 (30%)	3420 (16%)	3200 (15%)	13100 (63%)	20800 (100%)

Cable size and weight are the smallest when it is liquid cooling type with circulation. Termination size and weight are also the smallest in that case. However, this type requires reservoir, whose size and weight are large, and also requires circulator system including a pump, whose weight is not negligible.

Solid cooling type requires a lot of nitrogen, because the sensible heat is smaller compared to the latent heat.

At the bottom, total weights are compared, resulting in that the liquid cooling type with circulation is the lightest and weighs about 15% of the Cu cable.

However, each cable has merits and demerits, so in order to decide which type is the most suitable, more information about the specifications of the cable and layouts of the circuit are required. Some experimental study is also essential.

As the summary of this FS on 230V system, the superconducting cable cooled with liquid nitrogen without circulation is the lightest and the most compact cable for 4 MW capacity with 230 V AC between lines and 50 m length.

3. Feasibility Study with the voltage level of 1.5 kV

Larger transmission capacity as large as 40 MW will be required for the trunk line of wide body aircrafts. For that case, we studied higher voltage applications of ±1.5 kV DC case, which means 1.5 kV line to the earth and 3 kV line to line, and 1.5 kV 3 phase AC case , which means 1.5 kV line to line. This study also turns out to be the comparison between DC and AC, because their current levels are close to each other. The configuration of DC superconducting cable is single concentric and that of AC superconducting cable is triad, which is 3 cores in one cryostat, as are shown in figure 9.

(a) single concentric (DC) (b) triad(3 cores in one cryostat)(AC)

Figure 9. Configuration of superconducting cable

Table 3 shows the comparison between Al cable and superconducting cable for each case. Al cables are estimated by taking cable bending and reduction of current capacity due to multi bundling into account. Superconducting cables are assumed to be cooled by liquid nitrogen without circulation, because it was found to be the lightest.

At the bottom, total weights of cable and termination are compared, resulting in that DC superconducting cable system is the lightest, and weighs about 8% compared to AC Al cable.

Table 3. FS for 40 MW power transmission with ±1.5 kV DC and 1.5 kV 3 phase AC

40 MW system	±1.5 kV DC		1.5 kV(L-L)3 phase AC	
Conductor	Al	SC	Al	SC
Current (A)	**13300**	**133000**	**15400**	**15400**
Cable dia. (mm)	**190 × 1910** 26 cols, 3 rows, (wire dia. **37.5** mm × **39** pairs)	**79** (single concentric)	**600 × 1820** 8 cols, 3 rows,(1 bundle = **17.3 mm** dia. Wire × 36)	**146** (triad(3 cores in one cryostat))
50 m Cable Weight (kg)	**10370**	**360**	**20880**	**2890**
Termination L × dia. (m)	-	**1.2 × 0.9**	-	**1.2 × 1.1**
Termination Weight (kg)	**120**	**1390**	**250**	**2900**
Total weight (kg) (%)	**10490 (50%)**	**1750 (8%)**	**21130 (100%)**	**5800 (27%)**

4. Feasibility Study with higher DC voltages

Because we found that DC is more suitable to reduce transmission system weight, and higher voltage results in lighter weight, we furthermore studied higher DC voltage application, whose voltages are ±3 kV and ±6 kV.

Table 4 is the comparison between Al cable and superconducting cable for each case.

At the bottom , total weights of cable and termination are compared , resulting in that DC superconducting cable system is lighter, even at the high voltage of ±6 kV.

Table 4. FS for 40 MW power transmission with higher DC voltages

40 MW system	±3 kV DC		±6 kV DC	
Conductor	Al	SC	Al	SC
Current (A)	**6700**	**6700**	**3300**	**3300**
Cable dia. (mm)	**110 × 1310** 18 cols, 2 rows, (wire dia. **37.5** mm × **18** pairs)	**78** (single concentric)	**110 × 710** 10 cols, 2 rows, (wire dia. **37.5 mm** × **10** pairs)	**80** (single concentric)
50 m Cable Weight (kg)	**4720**	**330**	**2600**	**270**
Termination L × dia. (m)	-	**1.2 × 0.9**	-	**1.2 × 0.5**
Termination Weight (kg)	**55**	**830**	**30**	**540**
Total weight (kg)	**4775**	**1160**	**2630**	**810**

Figure 10 summarizes the FS of 40 MW DC high voltage cable. The ratio of superconducting cable weight divided by that of Al cable is 17% at ±1.5 kV, 24% at ±3kV, and 31% at ±6 kV.

Therefore, superconducting cable is very effective to reduce the transmission system weight up to very high voltage of ±6 kV.

IOP Conf. Series: Journal of Physics: Conf. Series **1293** (2019) 012073 doi:10.1088/1742-6596/1293/1/012073

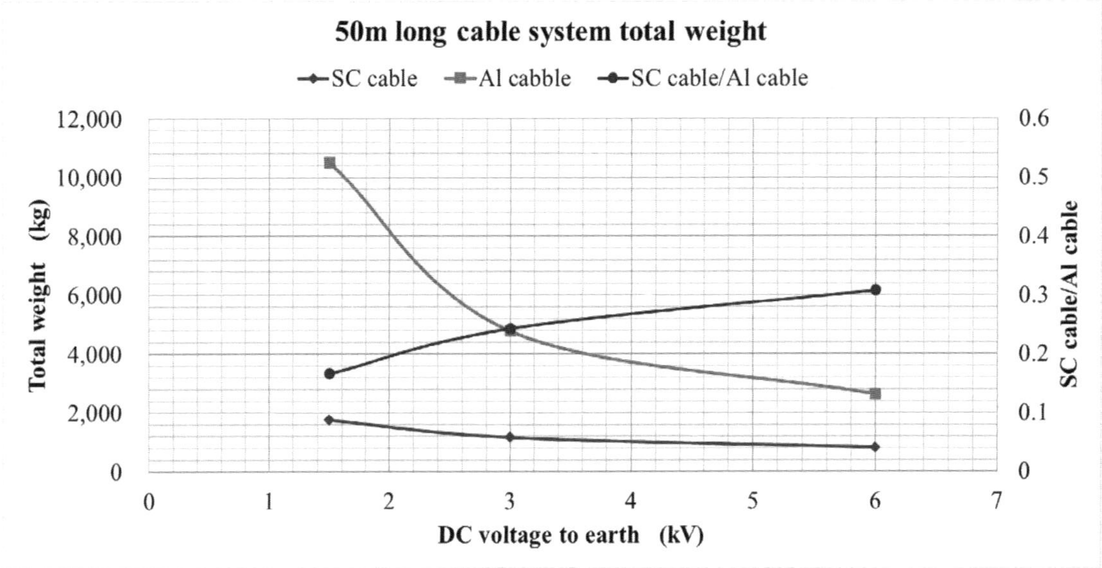

Figure 10. Weight of SC cable/weight of Al cable

5. Future Study

At present, electric power of 1 MW and voltage level of 230V AC are used in B787. However, future more electric aircrafts require 4 MW-40 MW, and application of higher voltages is contemplated.

As for future study, the first challenge is applying high voltage. From the Paschen's law, BD voltage is a function of P times d as is shown in figure 11. P is the pressure of air and d is the gap length between electrodes. When we use air for insulation, the minimum breakdown voltage is 350 V DC, which corresponds to 230 V AC.

For example, at 1bar and 2.5mm, the breakdown voltage is 9kV and at 100mbar and 2.5mm, the breakdown voltage is 1.6kV as is shown in figure 11. In other words, at low pressure of 100 mbar at the altitude of 23500 m, the BD voltage decreases to 1/5.6 of that at sea level. In order to keep the same BD voltage at 100mbar as that of at 1bar, gap must be increased by 10 times of that required at 1bar. Understanding the cable system BD voltage at unpressurized conditions is important.

For lighter cable system, the challenge in case of Al cable is to develop technology to cope with softness at high temperature and to allow light crimp joints. Al will creep at higher temperatures, which means it will be difficult to use it in wire with temperature ratings higher than 150 to 175 C. There are Al conductors used at present, but they are only used on certain portions of the aircraft.

Connectors for Al conductor are heavier than those used for plated copper, and this limits the weight advantage of Al conductor on aircraft to longer runs of wire. Application of Al cable to transmission trunk line in Hybrid Aircrafts requires lots to be developed.

As for superconducting cable, small bending radius and lighter weight are the challenges. Superconducting cable system has to be developed as part of hybrid aircrafts development.

As for future FS, cable system weights should be compared based on the configurations of grids in aircrafts, such as voltage, capacity, length, flight duration, cable branch etc.

For an example, transmission cable system weight is estimated with the cable length as a parameter, resulting in figure 12.

The weight ratio of superconducting cable weight divided by that of Al cable is less than one, when the cable length is longer than 10m. Even for shorter length than 10m, superconducting cable is beneficial, because the voltage of the cable can be lower than that of Al cable for the transmission of the same amount of power, and therefore it will be electrically more reliable.

Consequently superconducting cable system is expected to be beneficial for wide range of application.

Furthermore, the transmission cable system weight study should include the weight of power electronics.

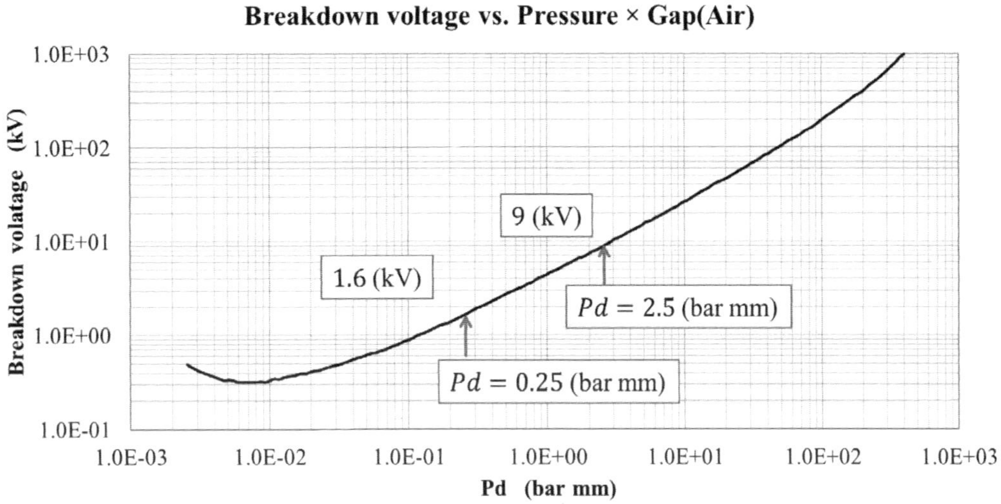

Figure 11. Paschen curve(Air)[1]

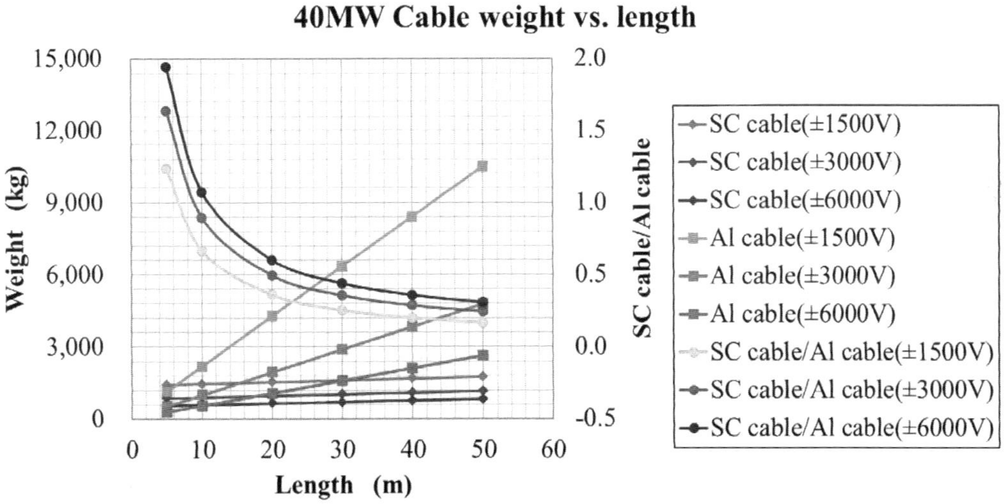

Figure 12. weight vs. length

Acknowledgement

A part of this work is supported by Applied Superconductivity Constellations of Tsukuba (ASCOT).

References

[1] IEEJ 1999 *Handbook of Electrical Discharge* (Ohmsha, Ltd.) p 129

[2] T Yamaguchi *et al* 2014 Large Current and Low AC Loss High Temperature Superconducting Power Cable Using REBCO Wires *SEI Technical Rreview* vol **184** pp 77-83

[3] RIVENC J *et al* 2018 *Proc. Symp. On AIAA/IEEE Electric Aircraft Technologies Symposium*(Cincinnati, Ohio) An Evaluation of Superconducting Power Cables for Airborne Application pp 1-21

IOP Conf. Series: Journal of Physics: Conf. Series **1293** (2019) 012074 doi:10.1088/1742-6596/1293/1/012074

Numerical analysis on the influence of armature winding configuration on AC loss of 10 MW fully superconducting generators of electric aircrafts

M Komiya[1], T Aikawa[1], K Yoshida[1], S Miura[1], M Iwakuma[1], T Yoshida[1], T Sasayama[1], A Tomioka[2], M Konno[2] and T Izumi[3]

[1]Research Institute of Superconductors Science and Systems, Kyushu University, Fukuoka 819-0395, Japan
[2] Fuji Electric Co. Ltd., Ichihara-city 290-8511, Japan.
[3] National Institute of Advanced Industrial Science and Technology, Tsukuba 305-8564, Japan.

E-mail: komiya@sc.kyushu-u.ac.jp

Abstract. 10 MW fully superconducting generators, comprising REBCO superconducting tapes with various armature winding configurations, were designed. The influences of these configurations, such as a distributed-winding and a concentrated-winding configuration, on AC loss, output power density, and output voltage waveform were evaluated. As a result, the short-pitch distributed-winding model exhibited the best properties among all the models. In particular, the AC loss of the armature winding was approximately half that of the concentrated winding. Furthermore, the distortion factor was lower than 10%.

1. Introduction

Currently, the fuel efficiency and emission of aircrafts have attracted attention as an environmental issue. The CO_2 emissions generated by aircrafts will be approximately 4 times greater in 2040 compared to 2006 [1]. However, stringent emission targets have been set by Flightpath 2050 published by ACARE, which requests a 75% reduction in CO_2 emissions per passenger km compared to the year 2000. An electrical propulsion system with electrical motors is an effective solution, as proposed by NASA [2]. The electrical power required to operate the electric fan is provided by generators driven by a gas turbine and/or batteries. However, current conventional rotating machines, composed of iron cores and copper wires, cannot satisfy the very stringent weight and volume constraints imposed by airborne applications. Superconducting technologies have the possibilities of drastically reducing the weight and improving the output power density. According to numerical analyses, superconducting rotating machines can achieve a high output power density of over 20 kW/kg, much higher than conventional rotating machines [3][4]. Recently, fully superconducting rotating machines have particularly attracted attention.

The AC loss generated in the superconducting armature windings of fully superconducting rotating machines is a serious issue because it may cause a severe temperature increase of the armature windings and result in thermal runaway. To realize the fully superconducting rotating machine, it is necessary to reduce the AC loss generated in the armature windings. For this purpose, various winding methods in the armature windings were compared from the viewpoint of the AC loss. In the case of a concentrated winding, the number of turns in one coil will be larger than that of distributed windings. In the result, the influence of the armature leakage reactance on the concentrated winding will become stronger than the distributed winding, and the interlinkage magnetic flux applied to the armature winding will be larger. Therefore, the AC loss will be larger in the concentrated winding than in the distributed winding. In addition, improvement of the sinusoidal waveform of the induced load current can also be expected.

2. Design of superconducting generator

2.1. Properties of REBCO superconducting tape

Table 1 shows the parameters of the REBCO superconducting tapes used for the windings. The superconducting layer is $EuBa_2Cu_3O_y$, a superconducting matrix, and 3.5 mol.% $BaHfO_3$ inclusions. The $BaHfO_3$ inclusions become strong artificial pinning centers to realize a large critical current, I_c, in high magnetic fields. Furthermore, outstanding improvement of the I_c for the thick films was achieved by changing the rare earth material to Eu ratio [5]. The I_c at 77 K under self-field exceeded 380 A.

Table 1. Parameters of the REBCO superconducting tapes.

Width	5 mm
Thickness	112 μm
Substrate	Hastelloy (100 μm)
Buffer Layer	CeO_2 (0.62 μm) + $LaMnO_3$ (0.008 μm) + MgO (0.005 μm) + Y_2O_3 (0.014 μm) + $Gd_2Zr_2O_7$ (0.055 μm)
Superconducting Layer	$EuBa_2Cu_3O_{7-\delta}$ + $BaHfO_3$ [3.5 mol%] (3.6 μm)
Stabilizing Layer	Silver (10 μm)
Number of filaments	10 (Laser-scribed)
I_c at 77 K, self-field	> 380 A

The AC loss of the REBCO superconducting tape was measured using a saddle-shaped pickup coil method [6]. The magnetic field dependence of the I_c, I_c-B, was evaluated using the observed magnetization curves. figure 1 (a) shows the magnetic field amplitude dependence of the AC loss for the non-scribing tapes at 20 K and figure 1 (b) shows the estimated I_c-B characteristics at 20 K. A magnetic field perpendicular to the superconducting tape face was calculated using the incident angle to the tape face. When an AC current is applied to the superconducting tape, magnetic fields perpendicular to the tape face cross each other at both ends of the wire in the process of reversing the

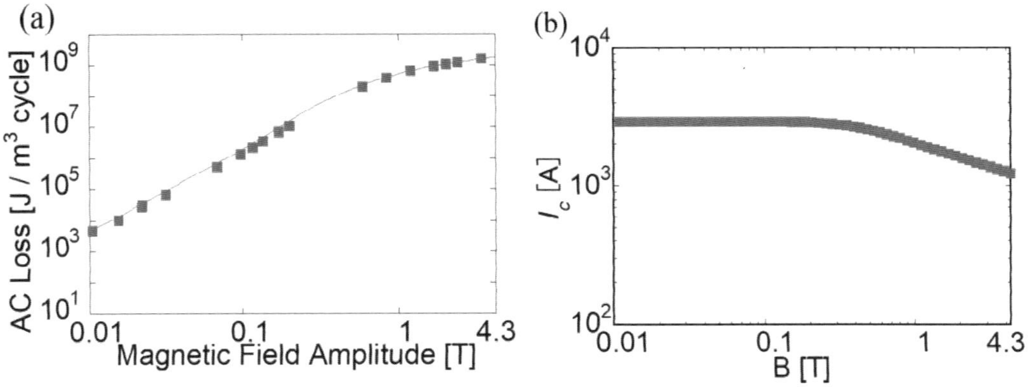

Figure 1. (a) Observed magnetic field amplitude dependence of AC loss, (b) estimated I_c-B characteristic of the REBCO wire.

current. The hysteresis loss is proportional to the square of the width of the superconducting tape. Therefore, thinning the superconducting layer with a laser will reduce the width of inversion and reduce the hysteresis loss [7]. It is assumed that the superconducting tape was divided into a 10-filament structure by laser-scribing. Compared to the non-scribing tape, the shielding current of 1 filament becomes 1/10 in the case of the 10-filament structure. Therefore, it is expected that the hysteresis loss occupying most of the AC loss of the REBCO superconducting tape is reduced to 1/10 by dividing the tape into 10-filaments. We calculated the nonlinear resistance of the stator winding using the *n*-value model. If the tape length is represented as *L*, V_c, the critical voltage, is calculated from the equation: V_c = 10-4 *L* (m). The *n*-value of the REBCO superconducting tape is reported to be approximately 35–40 for a temperature of 20 K and a magnetic flux density of approximately 0–5 T [8]. In addition, I_c is calculated from the observed I_c–*B* curve. As a result, the nonlinear resistance is 0.01–0.03 Ω. This is not the critical value to necessarily affect the calculated data, so we did not consider the nonlinear resistance.

2.2. *Winding method*

We designed three models for the armature winding configurations: short-pitch distributed (SPD) winding, full-pitch distributed (FPD) winding, and concentrated winding. figure 2 shows the armature winding structures of the three models. The U-phase is connected in the order of $U1^+$, $U1^-$, $U2^+$, and so on. This order does not change for each phase.

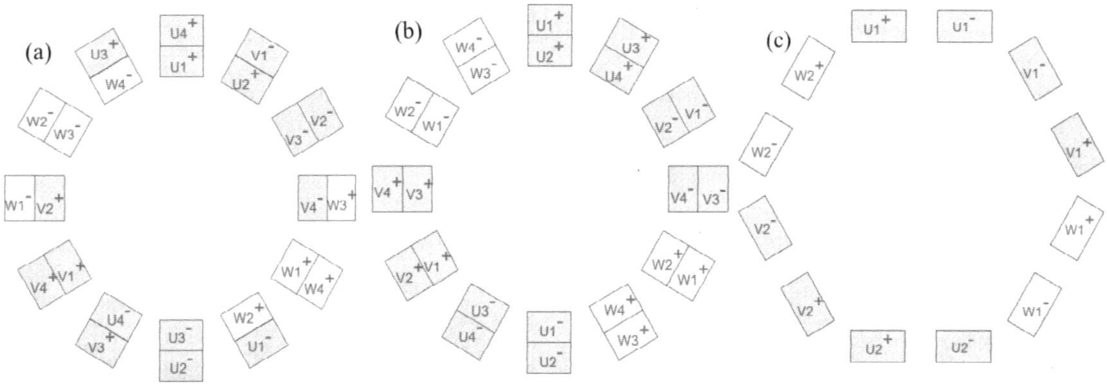

Figure 2. Conceptual diagrams for the three models: (a) Short-pitch distributed (SPD) winding, (b) Full-pitch distributed (FPD) winding, (c) Concentrated winding.

Compared with the distributed windings, concentrated windings do not distribute the magnetic path as much, so the waveform of the induced electromotive force will deteriorate remarkably.

For the SPD winding, the ratio of the winding pitch to the magnetic pole pitch is represented as *β*. The short-pitch factor k_p is as expressed as:

$$k_p = \sin\frac{\beta\pi}{2} \tag{1}$$

When the number of slots in each pole and phase is expressed as *q* and the number of phases is represented by *m*, the distribution factor k_d is expressed as:

$$k_d = \frac{\sin\frac{\pi}{2m}}{q\sin\frac{\pi}{2mq}} \tag{2}$$

Therefore, the winding factor k_w is expressed as:

$$k_w = k_d k_p = \frac{\sin\frac{\pi}{2m}}{q\sin\frac{\pi}{2mq}}\sin\frac{\beta\pi}{2} \tag{3}$$

In this study, for the two distributed winding models, q and m are 12 and 3, respectively. β is 1 for the FPD winding and 5/6 for the SPD winding. Accordingly, k_w of the SPD winding model is less than that of the FPD winding model. The induced electromotive force of the generator is proportional to the winding factor k_w. Therefore, the induced electromotive force in the case of the same number of turns is smaller than that of the FPD winding model for the SPD winding model.
The short-pitch factor k_{ph} for the hth harmonic is expressed as:

$$k_{ph} = \sin\frac{h\beta\pi}{2} \tag{4}$$

In the case of $\beta = 5/6$, k_{ph} decreases significantly when $h = 5$ and 7. Thus, the spectra of the fifth harmonic and the seventh harmonic will reduce in the case of the SPD winding model. Therefore, the waveform of the induced electromotive force of an SPD winding model approaches a sinusoidal wave compared with the FPD winding model. In addition, as the coil-ends are shortened, the SPD winding model has a shorter wire length.

2.3. Design of 10 MW fully superconducting generator

The specifications of the superconducting synchronous generator are shown in Table 2. Four models, SPD winding, FPD winding, concentrated winding, and SPD winding with optimized size, were designed. The former three models were prepared in order to compare the difference in AC loss in terms of the winding method. The models were set to have the same size (diameter and effective length). The last model was prepared to achieve a high power density for the SPD winding by miniaturizing.

Table 2. Parameters of superconducting generator.

Winding Method	SPD	FPD	Concentrated	SPD (opt.)
Output Power	10 MW			
Rated Voltage	6900 V			
Rated Current	780 A			
Number of Poles	2			
Frequency	100 Hz			
Number of Revolutions	6000 rpm			
Operation Temperature	20 K			
Peak Current-to-I_c Ratio	0.7			
Magnetic Flux Density of Gap	1.44-1.55 T			
Effective Length	1200 mm			550 mm
Diameter	560 mm			540 mm
Number of Turns of Armature Winding	12	10	98	30

The load lines of these models of the operation temperature, T_{op} = 20 K, are shown in figure 3 (a). The load lines were determined by a straight line extending from the origin, through the operating point, and to the intersection with the I_c-B curve. The operating point is defined by the value of the current flowing in the field and the armature windings, in a perpendicular magnetic field to the superconducting tape face [9]. All these values were measured during rated operation.

The operating current is represented by I_{op}. The load factor of each winding was defined as the ratio of I_{op} to I_c. The number of turns and the I_{op} were adjusted so that the load factor was approximately 70% for the field windings and armature windings. A schematic of the cross-sectional view of the designed generator is shown in figure 3 (b).

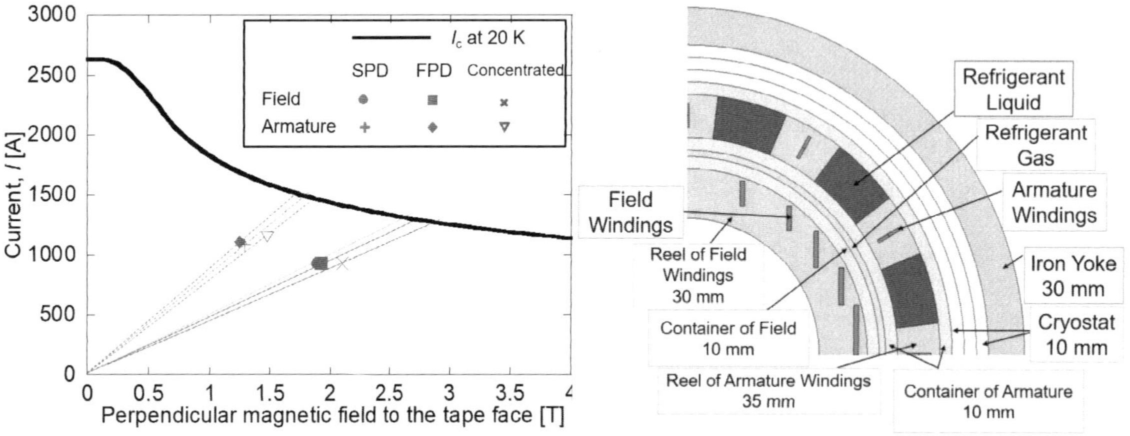

(a) (b)

Figure 3. (a) Load lines in case of T_{op} = 20 K and (b) cross-sectional view in an axial direction of the designed super-conducting generator.

The cryostat, container, and shaft were made of GFRP. The thickness of the parts was changed according to the -component type regarding mechanical strength. For example, the thickness of the container of the field and armature windings was 10 mm. The thickness of the shaft was 30 mm. The distance of the air gap was 5 mm, respectively. Comparing to a partly-superconducting generator, the fully superconducting generator has the advantage of the length of the air gap due to there being no vacuum chamber in the air gap. An iron yoke was placed outside the cryostat to shield magnetic field leakage from the internal magnetic field. For this reason, the iron loss on the iron yoke was not a thermal load of the low temperature cooler. It was assumed that the armature windings were bath-cooled liquid hydrogen at 20 K and the field windings were cooled with helium or hydrogen gas. The iron yoke used was 50 JN 270 made by JFE Steel, and its thickness was set to 30 mm in all models. It was assumed that the cool heat of liquid fuel, such as liquid hydrogen and LNG, can be utilized. The magnetic field at the air gap was set to 1.44 - 1.55 T for all models.

3. Loss and weight evaluation

3.1. AC loss evaluation

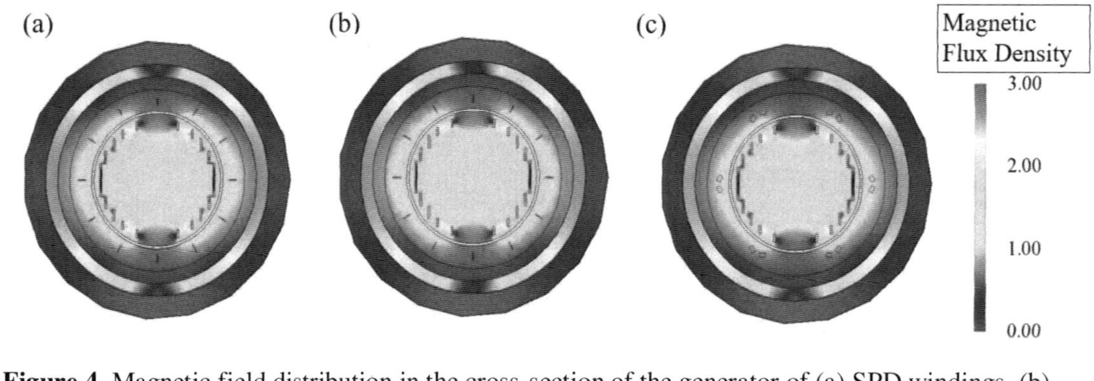

Figure 4. Magnetic field distribution in the cross-section of the generator of (a) SPD windings, (b) FPD windings, (c) concentrated windings.

Figure 4 shows the magnetic field distribution in the cross-section of the generator. Comparing the maximum magnetic flux density of the air gap, SPD model and FPD model were 1.55 T but concentrated winding model was 1.44 T. The current and number of turns of field winding of all models are the same so that the difference of the magnetic flux density of the air gap was due to the armature reaction. The concentrated winding model concentrated the magnetic path more than SPD and FPD winding models. For this reason, the magnetic flux applied to the field windings decreased by the magnetic flux from the armature windings due to the armature reaction. The output power decreased because of decreasing the interlinkage magnetic flux. As a result, the number of turn of armature winding of the concentrated winding increased and generated the output power. In term of the iron yoke, the maximum magnetic flux densities of all models exceeded 2.5 T and the magnetic field of the iron yoke was saturated. In addition, the leakage magnetic flux of all models exceeded 0.5 T. In regards to the relationship between the thickness and weight of the iron yoke and the leakage magnetic flux, we plan to examine it moving forward. We consider replacing the iron yoke to the superconducting shield because of reducing the weight and shielding more leakage magnetic field. In the near future, we will report the experimental results of superconducting shield.

Figure 5 (a) shows the AC loss of each winding and the iron loss of the three models, and figure 6 (b) shows the length of the superconducting tape of each winding.

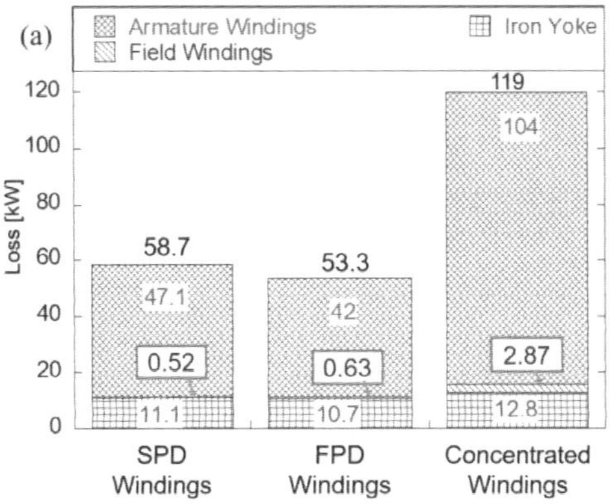

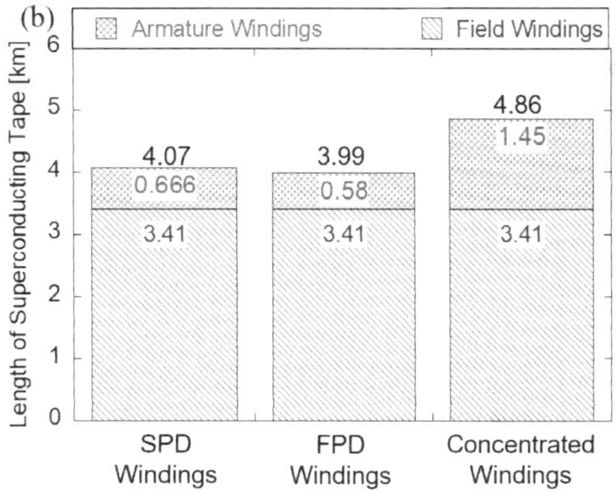

Figure 5. (a) AC loss of each winding and iron loss of the superconducting generator, (b) the length of the superconducting wire of each winding.

As shown in figure 5 (a), the AC loss of the armature winding of the concentrated windings was doubled compared to the AC loss of the FPD windings. This is because the length of the armature winding was approximately twice that compared to the FPD windings, as indicated in figure 5 (b). In the case of the concentrated windings, a large counter electromotive force occurred, as a result, the number of turns of the armature winding increased. Eventually, the AC loss of the concentrated windings became larger. The AC loss of the field winding and the iron loss were small compared to the AC loss of the armature winding.

3.2. The output power density

The iron yoke, cryostat, container of armature, armature windings, reel of armature windings, container of field, reel of field windings, field windings, and shaft were considered when calculating the generator dry-weight. The thickness of each part is shown in figure 3 (b). The previous three models were compared with the SPD (opt.) winding model, which is as small as possible. Figure 6 shows the output power density of all the models.

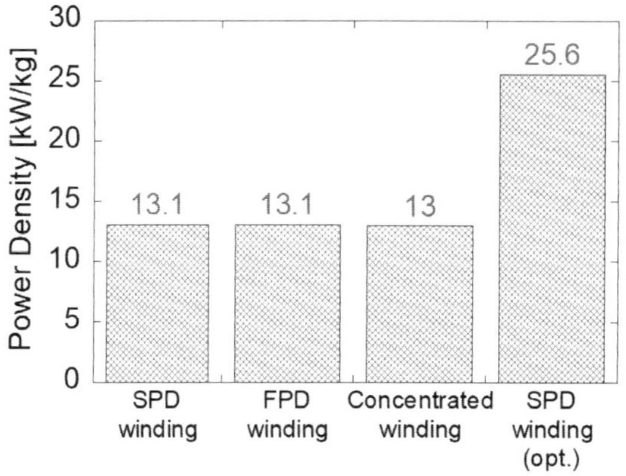

Figure 6. Power density of superconducting generator.

The only difference between the previous three models is the length of the superconducting tape. Hence, the generator weights for the three models were almost the same; thus, the output power densities of the three models were also almost the same, as shown in figure 6. In the case of the SPD(opt.) winding model the effective length and the volume of the iron yoke were both significantly reduced, so the output power density was drastically improved.

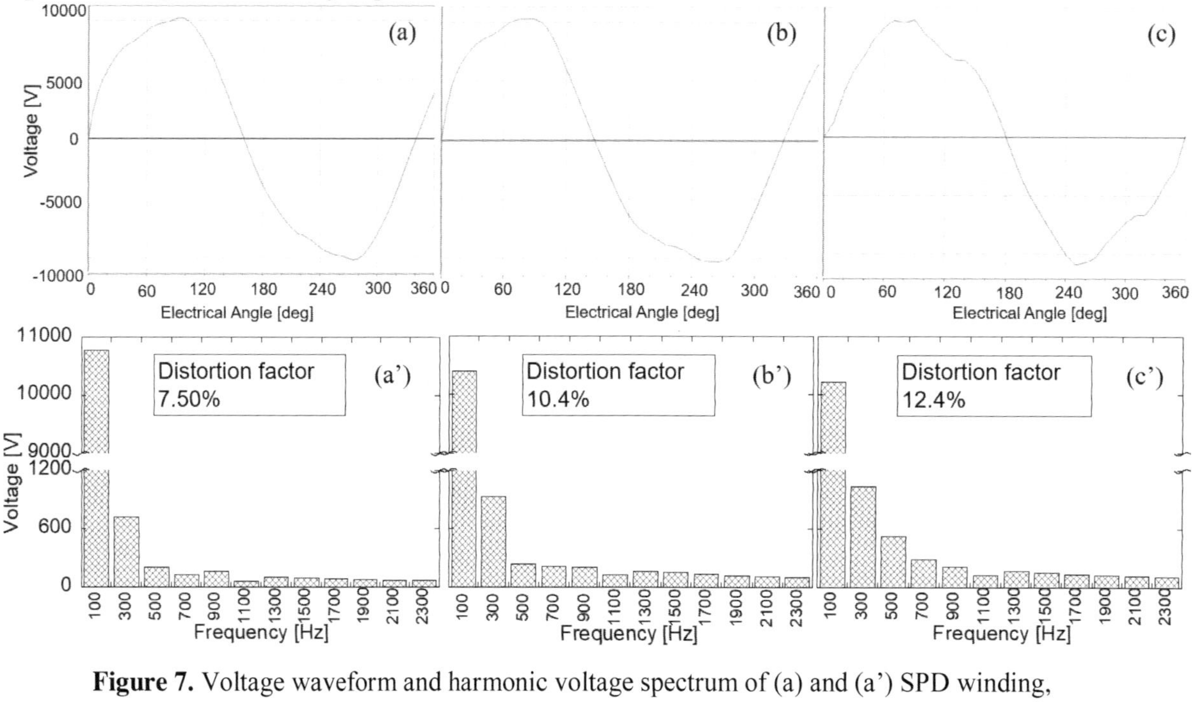

Figure 7. Voltage waveform and harmonic voltage spectrum of (a) and (a') SPD winding, (b) and (b') FPD winding, (c) and (c') concentrated winding, respectively.

3.3. The voltage waveform

Figure 7 shows the voltage waveforms and harmonic voltage spectrums of each model. The voltage waveform distortion is due to the higher harmonic wave. In this research, we calculated the harmonic spectrum of the armature winding. The frequency of the fundamental wave was set to 100 Hz and the odd harmonics until the 12th harmonic wave were considered. In the case of considering only the fundamental wave, distortion of the voltage waveform occurred. The SPD winding model has the lowest distortion factor of 7.50%. Compared to the SPD winding model, the distortion rate of the FPD winding model increased to 10.4%, and the distortion rate of the concentrated winding model increased to 12.4%.

The reason for this difference is the large spectrums of the fifth harmonic and the seventh harmonic. Compared with SPD windings, these were approximately 85% as much as the FPD windings in terms of the fifth harmonic and approximately 60% in terms of the seventh harmonic; approximately 40% as much as the concentrated windings in terms of the fifth harmonic and approximately 45% in terms of the seventh harmonic. By JEC-2130 established by The Japan Electrical Manufacturers' Association (JEMA), the condition of the compliance standard is that the gas turbine synchronous generator has a distortion factor of 10% or less. The SPD winding model satisfies JEC-2130 requirements.

4. Conclusion

With various winding methods for armature windings, the 10 MW-class synchronous generator comprising REBCO superconducting tapes were designed and evaluated by a numerical analysis. The model adopting the distributed winding had a better result in terms of AC loss and output voltage waveform in comparison with the concentrated winding model. In particular, the AC loss of the armature winding was approximately half compared to that of the concentrated windings. Furthermore, the distortion factor of the SPD winding was lower than 10%. Additionally, the output power density of the SPD model with optimized dimensions reached 26.3 kW/kg. Therefore, in terms of AC loss, the output power density, and output voltage waveform, the SPD winding model is the most suitable for aircraft superconducting generators.

Acknowledgments

This research was partially supported by the Japan Science and Technology Agency (JST): Advanced Low Carbon Technology Re-search and Development Program (ALCA) and the Japan Society for the Promotion of Science (JSPS): Grant-in-Aid-for Scientific Research (JP18H03783 and JP17H06931).

References

[1] IPCC 1999 *Aviation and the global atmosphere, intergovernmental panel of climate change* (Cambridge: Cambridge University Press)
[2] Misra A 2012 Technology Challenges for Electric Aircraft *NASA Research Report*
[3] Komiya M, Iwakuma M, Miura S, Aikawa T, Yoshida T, Sasayama T, Tomioka A, Konno M and Izumi T 2018 Design Study of 10 MW REBCO Fully Superconducting Synchronous Generator for Electric Aircraft (*Preprint IEEE Trans. Appl. Supercond.*)
[4] Kong W, Terao Y and Ohsaki H 2018 *J. Phys.: Conf. Ser.* **1054** 012082
[5] Yoshida T, Ibi A, Takahashi T, Yoshizumi M, Izumi T and Shiohara Y 2014 *Phys. C.* **504** 42–46
[6] Iwakuma M, Nanri M, Fukui M, Fukuda Y, Kajikawa K and Funaki K 2003 *Supercond. Sci. Technol.* **5** 545–556
[7] Suzuki K, Matsuba J, Yoshizumi M, Izumi T, Shiohara Y, Iwakuma M, Ibi A, Miyata S and Yamada Y 2007 *Supercond. Sci. Technol.* **20** 822–826
[8] Inoue M, Yamaguchi Y, Sakakibara T, Imamura K, Higashikawa K, Kiss T, Awaji S, Watababe J, T Hiroshi, Yoshizumi M, and Izumi T 2013 IEEE *Trans. Appl. Supercond.* **23** 8002304
[9] Cryomagnetics *Basics of Superconducting Magnets* [Online] Available: "https://www.cryomagnetics.com/wp-content/uploads/reference/Basics_of_Superconducting_Magnets.pdf"

IOP Conf. Series: Journal of Physics: Conf. Series **1293** (2019) 012075 doi:10.1088/1742-6596/1293/1/012075

Experimental and theoretical discussion on step-out characteristics of high temperature superconducting induction/synchronous motor

T Nakamura and D Sekiguchi

Department of Electrical Engineering, Kyoto University, Kyoto 615-8510, Japan

nakamura.taketsune.2a@kyoto-u.ac.jp

Abstract. This paper presents experimental and theoretical results of step-out characteristics of a 20 kW class High Temperature Superconducting Induction/Synchronous Motor (HTS-ISM). A fabricated motor (3-phase, 8-pole, rotor windings: HTS, stator windings: Cu) is investigated, and its maximum quasi-synchronous output is 20 kW at line voltage of 400 V. The step out condition and its behaviour is investigated for different combinations of the input voltage and the drive frequency at 77 K (atmospheric liquid nitrogen). It is shown that tested results of torque versus primary current characteristics agree well with the theoretical expression by compensating with no-load currents. However, the corresponding step-out current differ each other when air-gap magnetic flux density, which is in proportion to (voltage/frequency), increases. Based on electromagnetic field analysis, it could be concluded the reason for the above discrepancy is the magnetic saturation of the iron core. Furthermore, step-out occurs at the phase angle of 45 degree regardless of which is whether or not the step-out current is affected by the above magnetic saturation. These results are really important for the consideration of stable rotation condition of the HTS-ISM.

1. Introduction

One of the advantages of superconducting rotating motor is its high power and/or high torque characteristics. For example, AMSC (USA) has succeeded in testing a 36.5 MW class ship propulsion motor [1]. The group of Kawasaki Heavy Industries et al. (Japan) has also succeeded in testing a 3 MW class motor [2].

The author's industry-university R&D cooperation group has been developing so-called High Temperature Superconducting Induction/Synchronous Motor (HTS-ISM) [3]. Although present mechanical output of our motor is smaller compared to the above reports, one of our main targets is realizing ultimately high power and/or torque density. In order to develop the higher power (torque) density motor, on the other hand, its step-out behavior should be investigated. When the load torque overcomes the generated torque of the motor, which doesn't possess any slip (asynchronous) torque, the motor suddenly stops, and then the motor transits be a dangerous situation. Therefore, so-called overload tolerance for sudden experience of heavy load has to be clarified for the limit design of the motor. Such step-out characteristics are also crucial for the optimum control at the loaded condition.

In this paper, step-out phenomenon and condition are investigated for our fabricated 20 kW class HTS-ISM based on experiment and theoretical expression.

Figure 1. Photographs of a 20 kW class HTS-ISM [3].

Table 1. Specifications of a 20 kW class HTS-ISM [3].

Item	Stator	Rotor
Pole number	8	-
Number of slots	24	34
Inner diameter	160.0 mm	50.0 mm
Outer diameter	265.0 mm	159.4 mm
Length	200.0 mm	206.0 mm
Number of turns	30	-

2. Subject motor

A 20 kW class HTS-ISM is targeted in this study. Detailed explanation of the motor has already been reported in [3], and then basic structure and specifications of the motor is briefly shown in this section. Figure 1 and Table I, respectively, show a photograph and specifications of the developed 20 kW class HTS-ISM. A 3-phase and 8-pole stator structure is adopted and the rated voltage is 400 V. A Bi-2223 squirrel-cage rotor is coupled with conventional (copper) stator, and then the tests are carried out in atmospheric liquid nitrogen (77 K). It should be noted that the motor doesn't possesses any slip (asynchronous) torque [3].

3. Theoretical expression of maximum output and step-out condition

3.1. Expression of torque of HTS-ISM

Rotating principle of the HTS-ISM is similar to that of squirrel-cage induction motor [4]. Therefore, characteristics of the motor can be discussed based on an electrical equivalent circuit that is commonly adopted for describing performance of conventional motor.

Figure 2(a) shows an electrical equivalent circuit of the HTS-ISM, of which r1 is primary winding's resistance and *x* is leakage reactance. V_2' is the HTS rotor winding's voltage drop that is converted to the primary side, and it shows nonlinearity depending upon current ($I_1 = I_2$'). In order to derive intuitively clear expression of the motor's maximum output, exciting circuit is neglected in this study. From this circuit, the torque of the HTS-ISM (t) is expressed as follows [4].

$$\tau = 3\frac{p}{2}\phi_2^{'} I_2^{'} \tag{1}$$

$$\phi_2 = \frac{\sqrt{V_1^2 - x^2 I_2^{'2}} - r_1 I_2^{'}}{\omega} \tag{2}$$

IOP Conf. Series: Journal of Physics: Conf. Series **1293** (2019) 012075 doi:10.1088/1742-6596/1293/1/012075

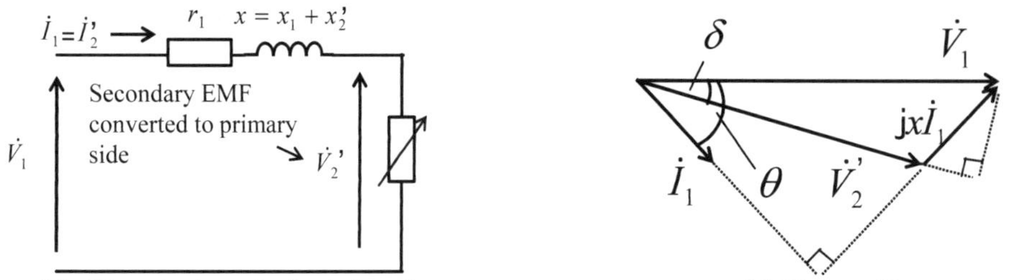

(a) Electrical equivalent circuit. (b) Phasor diagram.

Figure 2. One phase phasor expression of an HTS-ISM.

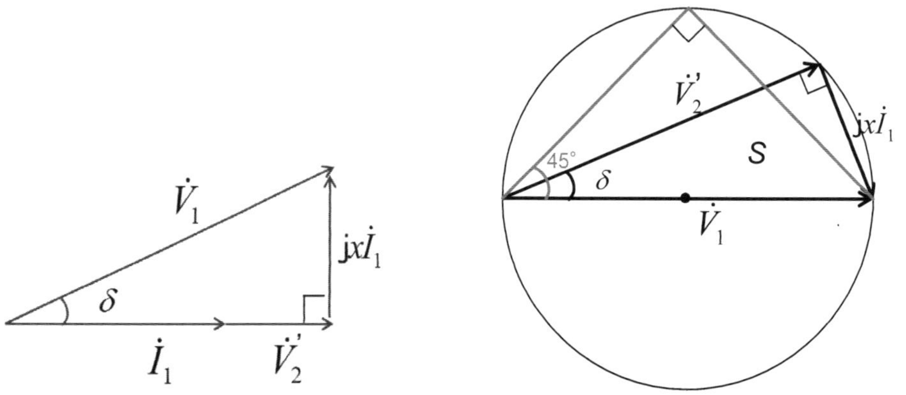

(a) Phasor diagram at maximum output. (b) Variation of maximum output for δ.

Figure 3. Explanation of step-put based on phasor diagram.

where, Φ_2' is secondary magnetic flux that is converted to primary side. As can be seen in equations (1) and (2), t shows its maximum value for the I_2'.

3.2. Maximum output and Step-out condition

Figure 2(b) illustrates the phasor diagram obtained from figure 2(a). Where, θ denotes angle between V_1 and I_1, and δ between V_1 and V_2'. It should be noted that r_1 is neglected for this diagram in order to realize clear discussion. From the diagram, we can express the output (P) as follows.

$$P = 3V'_2 I_1 \cos(\theta - \delta) \tag{3}$$

From equation (3), the following relation must be satisfied for the maximum output (P_M).

$$\theta = \delta \tag{4}$$

Figure 3(a) illustrates the phasor diagram for P_M, and it is expressed as follows.

$$P_M = 3V'_2 I_1 = 3\frac{V'_2 V_1}{x}\sin\delta \tag{5}$$

$$\sin\delta = \frac{xI_1}{V_1} \tag{6}$$

Furthermore, the above P_M will be changed by the mechanical load as shown in figure 3(b), and then the threshold for the stable rotation is determined at δ=45 degree, of which P_M becomes $3V_1^2/2x$ (I_1= $V_1/\sqrt{2}$x. In other word, this angle gives the step-out condition of the motor. In this paper, validity of the aforementioned condition for the 20 kW class HTS-ISM is investigated.

IOP Conf. Series: Journal of Physics: Conf. Series **1293** (2019) 012075 doi:10.1088/1742-6596/1293/1/012075

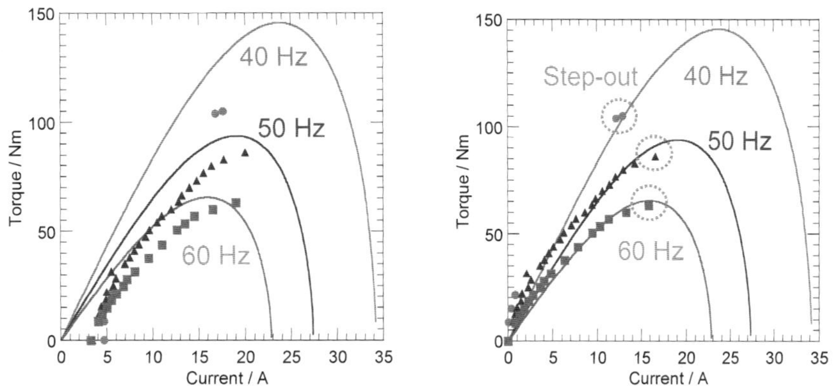

(a) Without no-load current compensation. (b) With no-load current compensation.

Figure 4. Torque versus current curves (77 K).

4. Results and discussion

4.1. Load characteristics

Figure 4 shows torque (τ) versus current (I_1) curves for different frequencies (solid curves: theoretical expression, symbols: tested results). As can be seen in the figure 4(a), although the obtained torque increases with the current before the motor steps out, such values differ from the theoretical expressions. Main reason for this discrepancy would be come from the no-load current. That is, no-load current is included in the current even such current doesn't contribute to the torque. Therefore, the no-load current is simply subtracted from the tested results, and then the results are shown in figure 4(b). As shown in the figure, the tested results agree well with the theory.

4.2. Step-out phenomenon

Step-out currents are marked with open circles in figure 4(b). We can see that the tested step-out current (marked by open circles) agree with the theoretically obtained maximum torque when the frequency is higher. However, discrepancy between the test and the theory becomes distinct by decreasing the frequency. It is known that the air-gap magnetic flux density is in proportion to voltage (V_1)/frequency (f), and then such flux will increases by decreasing the frequency at the fixed value of the voltage. In other words, the magnetic saturation would occur in the iron core when V_1/f increases. Figure 5 shows temporal traces of primary current (I_1) and phase angle (θ) for different voltages at 50 Hz. It should be noted that the initial value of phase angle is about 45 degree for both conditions and the motor is in a step-out mode. Then, the step-out occurs at 176.6 s for V_1= 320 V and 217.2 s for V_1= 400 V. Although both of I_1 and θ shows smooth trace for the case of lower voltage (320 V), such traces vibrate for the higher voltage (400 V). Possible reason for the vibration would also be the magnetic saturation.

In order to validate above assumption, magnetic field analysis (2D finite element method) is carried out for the condition of 400 V/50 Hz as shown in figure 6. Figure 6(a) shows magnetic flux density contour of the core, and its highest value reaches more than 1.5 T. Therefore, the iron core will be in saturation state for this condition. The corresponding current waveforms are also shown in Fig. 6(b), and they are largely distorted due to the above magnetic saturation. That is to say, inclusion of harmonic component due to the magnetic saturation would be the reason for the vibration of current waveform, and also that for the discrepancy of step-out current between the test and the theory.

Furthermore, figure 7 shows step-out angle for various conditions of V_1/f and torque. As can be seen in the figure, the step-out angle always less than 45 degree, as is expected by the conventional theory (figure 3(b)). Namely, the theoretical condition for the step-out angle, fortunately, is always valid to be 45 degree without dependence of the aforementioned magnetic state of the iron core. Based

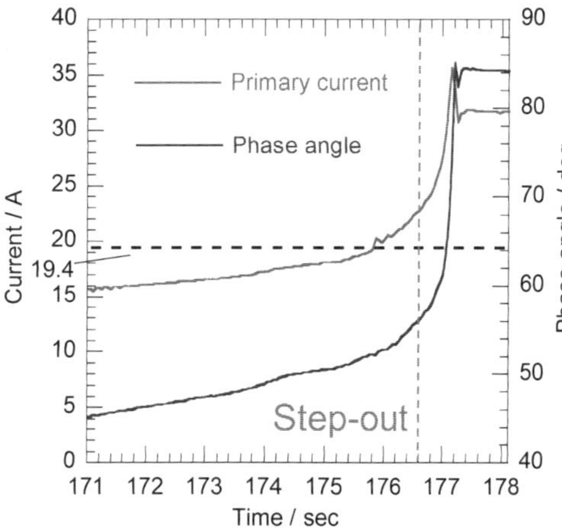

(a) V_1= 320 V.

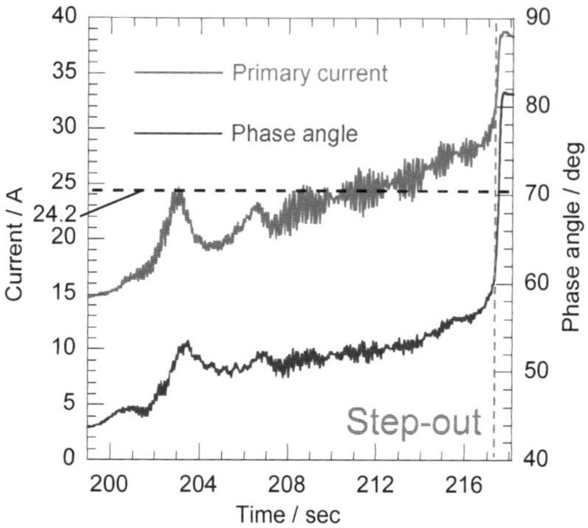

(b) V_1= 400 V.

Figure 5. Temporal traces of primary current and phase angle for different voltages at 50 Hz.

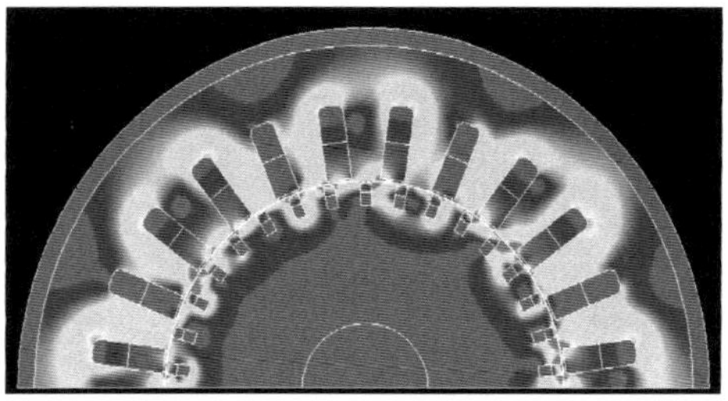

(a) Magnetic flux density contour.

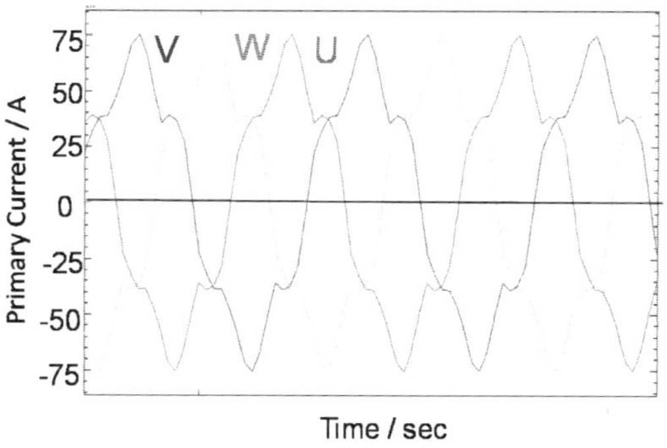

(b) 3-phase current waveform.

Figure 6. 2D electromagnetic field analysis for condition of 400 V/50 Hz.

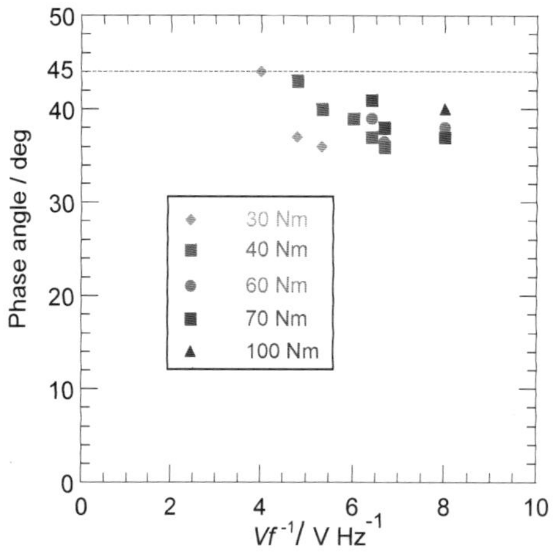

Figure 7. Step-out angle for various conditions of V_1/f and load torque.

on this condition, we have to design or set the operational condition with a safety margin, which depends on applications. This will be one of our future works. It should be noted that the basic mechanism of step-out phenomenon of the HTS-ISM discussed in this paper is same as that of conventional motors.

5. Conclusion

In this paper, step-out characteristics are tested and then discussed based on nonlinear equivalent circuit. It was shown that the tested results of torque versus current curves agree with the theoretical expressions by compensating with the no-load current. Step-out current differs from the theory and the reason for this would be magnetic saturation of the iron core. Even in the above problems, the step-out phenomenon always occurs when the phase angle becomes more than 45 degree as expected by the conventional theory. This results would be effective for the reliable and safe operation of the HTS-ISM.

Acknowledgments
This work was supported by Japan Science and Technology Agency under the program of Advanced Low Carbon Technology Research and Development Program (JST-ALCA) in Japan.

References

[1] Gamble B, Snitchler G and MacDonald T 2011 Full Power Test of a 36.5 MW HTS Propulsion Motor *IEEE Trans. Appl. Supercond.* **21(3)** 1083-1088
[2] Yanamoto T, Izumi M, Umemoto K, Oryu T, Murase Y and Kawamura M 2017 Load test of 3-MW HTS motor for ship propulsion *IEEE Trans. Appl. Supercond.* **27(8)** 5204305
[3] Nakamura T, Itoh Y, Yoshikawa M, Nishimura T, Ogasa T, Amemiya N, Ohashi Y, Fukui S and Furuse M 2015 Tremendous enhancement of torque density in HTS induction/synchronous machine for transportation equipments *IEEE Trans. Appl. Supercond.* **25(3)** 5202304
[4] Morita G, Nakamura T and Muta I 2006 Theoretical analysis of a YBCO squirrel-cage type induction motor based on an equivalent circuit *Supercond. Sci. Technol.* **19(6)** 473-478

IOP Conf. Series: Journal of Physics: Conf. Series **1293** (2019) 012076 doi:10.1088/1742-6596/1293/1/012076

Motor structure and mechanical output density of IPM motor using bulk superconductors as magnetic field

Wataru Akada, Yutaka Terao and Hiroyuki Ohsaki
Graduate School of Frontier Sciences, the University of Tokyo,
5-1-5 Kashiwanoha, Kashiwa, Chiba, 277-8561, Japan

e-mail: y.terao@ohsaki.k.u-tokyo.ac.jp; akada.wataru17@ae.k.u-tokyo.ac.jp; ohsaki@k.u-tokyo.ac.jp

Abstract. Recently, means of transportation with electrical motors such as electric vehicles and ships are rapidly increasing. Various attempts have been made to reduce the size of the motors to increase the freedom of transport equipment system layouts and the energy efficiency of their transportation. Interior permanent magnet synchronous motors (IPMSMs) can realize higher torque than surface permanent magnet synchronous motors (SPMSMs) by using reluctance torque caused by magnetic saliency in addition to magnet torque. In comparison with SPMSMs, IPMSMs can have wide operation ranges by applying control methods such as maximum output control and field weakening control. Therefore, IPMSMs are expected to be used for a variety of application such as electric vehicles, ships and heavy equipment. By replacing permanent magnets with bulk superconductors in the rotors of IPMSMs, the magnet torque increases due to the stronger magnetic field of magnetized bulk superconductors. This leads to an increase in total torque density of the motors and contributes to motor size reduction. In this research, we propose an IPMSM with bulk superconductors with the goal of increasing the mechanical output density. In the proposed IPMSM, cooling space is left inside the rotor to cool the bulk superconductors to their superconducting state. Here, analysis based on finite element method was carried out to verify the utility of the proposed motor. First, the design of the motor with bulk superconductors and cooling space is discussed. It is shown that the total torque can be increased by adjusting the position of the bulk superconductor and modifying the structure of the flux barrier. The maximum total torque shown here was 22500 Nm. Then, the mechanical output density of the proposed IPMSM is compared with a conventional IPMSM. It is shown that the mechanical output density of the proposed IPMSM is 38% higher than the conventional IPMSM. Finally, the magnetic shielding effect, which might be a problem when using bulk superconductors, is discussed. By using electromagnetic analysis, it is shown that the external magnetic flux is weakened by the cooling space to several percent of the magnets' surface and the effect of the shielding current is trivial.

1. Introduction

Recently, for the purpose of higher mechanical output and higher efficiency, interior permanent magnet synchronous motors (IPMSMs) are being used widely [1]. IPMSMs can realize higher torque than surface permanent magnet synchronous motors (SPMSMs) by using reluctance torque caused by magnetic saliency in addition to magnet torque [2]. Higher reluctance torque is realized by proper rotor design, which is essential in utilizing IPMSMs [3]. Also, IPMSMs have an advantage over SPMSMs in that they do not require protection tubes to prevent magnets from scattering. Furthermore, IPMSMs have broad operation ranges by applying control methods such as maximum output control and field weakening control [4]. Therefore, IPMSMs are expected to be used for a variety of application such as electric vehicles, heavy equipment and ships [5] [6].

In this research, we propose IPMSMs using bulk superconductors as field magnets instead of permanent magnets. The stronger magnetic field of magnetized bulk superconductors increases the magnet torque, leading to higher total torque, realizing higher mechanical output (torque × rotational speed) even at low rotational speed. To make use of this, transport equipment which require higher mechanical output, such as heavy equipment and ships, are assumed as the application of proposed IPMSMs.

In this paper, the motor structure of IPMSMs to fully utilize the bulk superconductors is discussed. In the discussion, necessary cooling spaces to cool the bulk superconductors below the critical temperature are considered, making the structure new and original. Then, mechanical output analysis was done to demonstrate the utility of bulk superconductors in IPMSMs by comparing the mechanical output density with conventional IPMSMs using permanent magnets. Finally, electromagnetic field analysis was done to reveal the electromagnetic phenomena occurring in the bulk superconductor in the IPMSM. The shielding current in bulk superconductors might decrease the magnet torque or increase the reluctance torque. The effect of the shielding current is investigated for further uses of bulk superconductors in motors.

2. Design of IPM motor using bulk superconductor

2.1. *Designing method*

Bulk superconductors can be used as strong magnetic flux sources [7]. By magnetizing the bulk superconductor stronger, the magnet torque increases. However, as the motor yoke causes magnetic flux saturation, there is a limit on the magnet torque. Therefore, we consider the design of IPMSMs to maximize the reluctance torque.

2.2. *Motor structure*

A basic motor structure is shown in figure 1, and the motor characteristics are shown in table 1, determined based on past research [8]. Electrical steel of 50A1000 is used for the yoke of the rotor and the stator. To cool the bulk superconductors, a cooling space is set in both sides of the bulk superconductors to flow the cryogen to cool the superconductors [9]. The space at the top/bottom side of the long side of bulk superconductors is 10 mm. The space is occupied by several millimeters of cryostat wall, support plate and evacuated space. Also, the bulk is supported by plates from both side of the space.

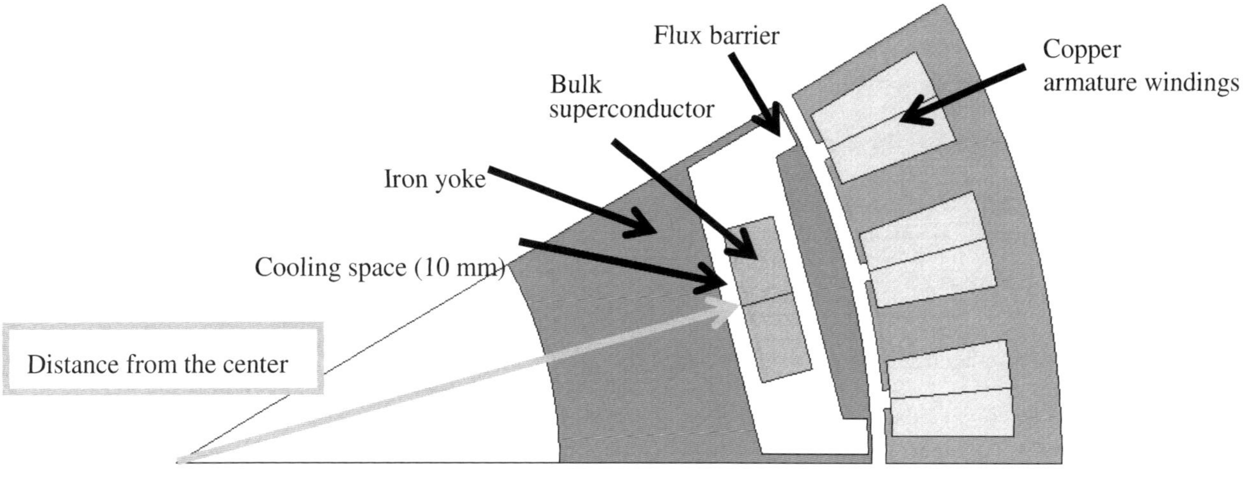

Figure 1. IPMSM with bulk superconductor.

Table 1. Motor specifications.

Parameter	Value
Outer diameter	1000 mm
Gap length	8.3 mm
Effective length	1530 mm
Number of the poles	12
Number of the slots	36
Rotor diameter	784 mm
Size of the bulk superconductor	Thickness 30 mm × Width 85.9 mm
Cooling space	10 mm
Rotational speed	1900 rpm

2.3. Method of analysis

Torque analysis was carried out to find out the motor structures which realized high reluctance torque with bulk superconductors. Here, the distance of the bulk superconductors from the center of the motor and the shape of the flux barriers were changed. An air gap of 10 mm to cool the superconductors was considered in each case.

When deriving the total torque, the permeability of bulk superconductors was defined as μ_0, the permeability of air. A current density of 6.3×10^7 A/m^2 was applied to simulate a magnetization of 2.0 T. When deriving the reluctance torque, the permeability of bulk superconductors was defined as μ_0 and no current density was applied. The permeability of the iron yoke was derived from the analysis of the total torque. The magnet torque was derived by subtracting the reluctance torque from the total torque.

2.4. Position of the bulk superconductor

Firstly, the flux barrier was fixed and the distance of the bulk superconductor from the center of the motor was varied from 311 mm to 326 mm. The total, the reluctance and the magnet torque of the IPMSMs in each case are shown in figure 2, figure 3 and figure 4, respectively.

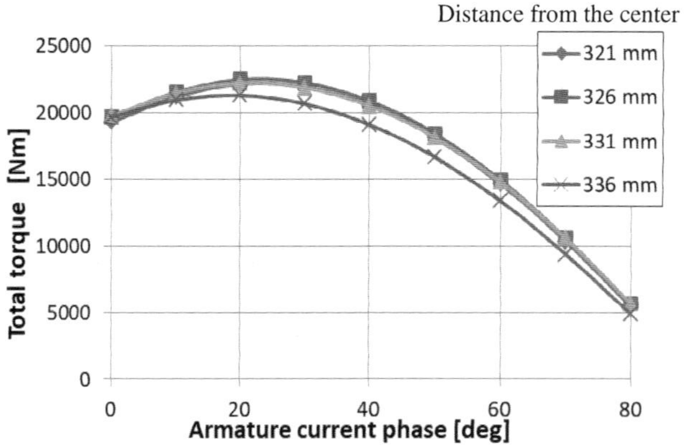

Figure 2. Total torque of the motors as a function of armature current phase.

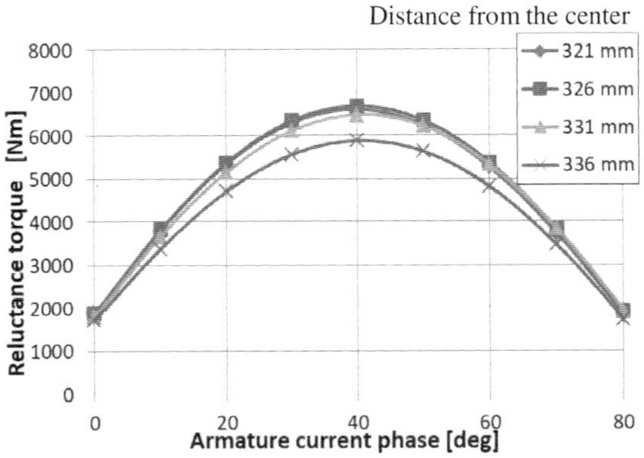

Figure 3. Reluctance torque of the motors as a function of armature current phase.

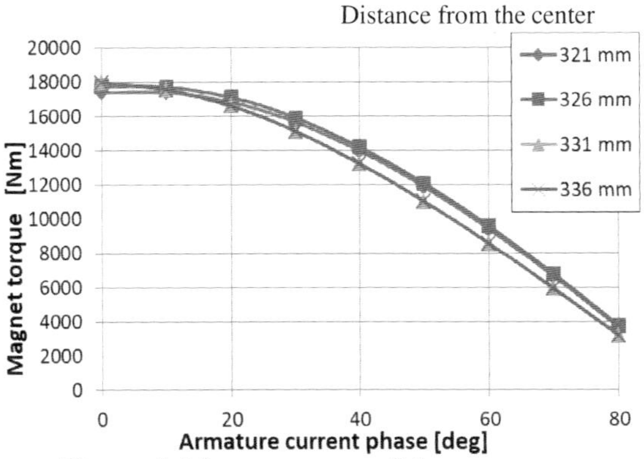

Figure 4. Magnet torque of the motors as a function of armature current phase.

In the analysis, the current phase of the U-phase of the armature current was changed, and the time average of the torque is shown as the result. The standard value of this current phase was set as the current amplitude to obtain maximum torque. From the results, it is shown that when the distance of the bulk superconductors from the center of the motor is 326 mm, the total torque of the motor is maximized due to maximum reluctance torque. Although the maximum magnet torque is higher in 326 mm, the high reluctance torque lead 326 mm to surpass the total torque of the other motors.

2.5. *Structure of the flux barrier*

Secondly, the distance of the bulk superconductor from the center of the motor was fixed as 326 mm, and the shape of the flux barrier was varied as shown in figure 5. The total and reluctance torque of the IPMSM in each model are shown in figure 6.

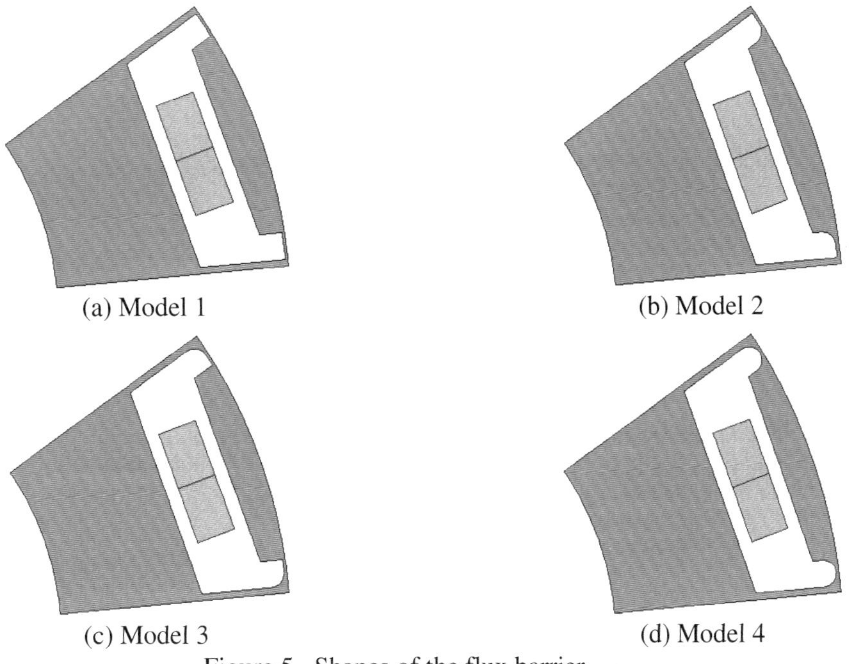

(a) Model 1 (b) Model 2

(c) Model 3 (d) Model 4

Figure 5. Shapes of the flux barrier.

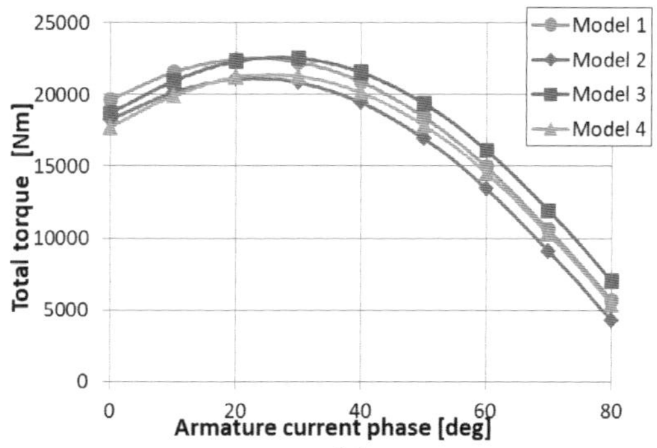

(a)

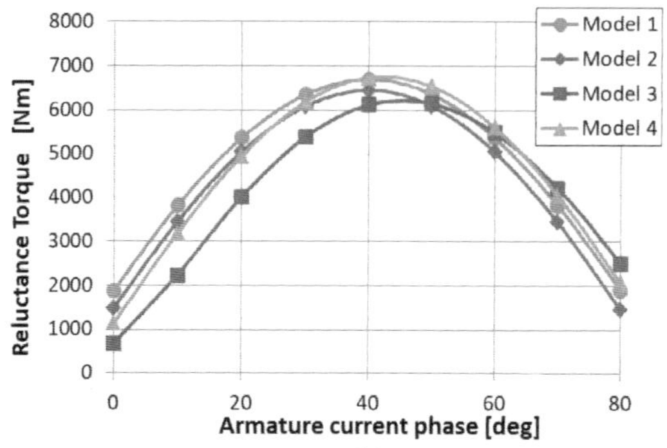

(b)

Figure 6. Total and reluctance torque of the four models of motors.

From the result, it is shown that IPMSM with the rotor of Model 3 had the highest total torque. In this case, Model 3 showed the highest magnet torque. This is due to the curve, concentrating magnetic flux from the bulk superconductor to the edge of the curve, where the reaction with armature current is stronger. This leads to the highest total torque, although the reluctance torque is the lowest. Model 4 showed the highest reluctance torque due to the spread of the magnetic flux, leading to strong magnetic saliency. However, the spread of magnetic flux caused a large decrease of magnet torque and resulted in the lowest total torque. This shows that there is a potential for proposed IPMSMs to show higher total torque by changing the shape of flux barrier, affecting both the magnet torque and the reluctance torque.

3. Mechanical output density of IPM motor using bulk superconductor

3.1. Design of IPM motor using permanent magnet

Similar design of motors was carried out for IPMSMs with permanent magnets. As for permanent magnets, unlike bulk superconductors, cooling space is not needed, the air gap was filled with permanent magnets. So, the permanent magnets used in the analysis were larger than the bulk superconductors. The same rotor shape as the IPMSM with bulk superconductors was adopted. The designed IPMSM is shown in figure 7.

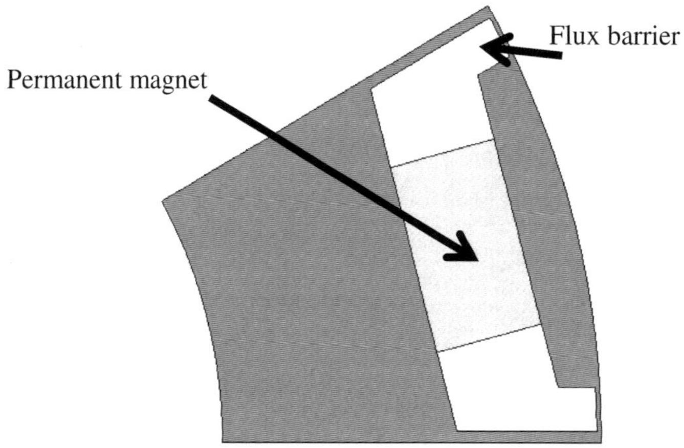

Figure 7. IPMSM with PM used in the analysis.

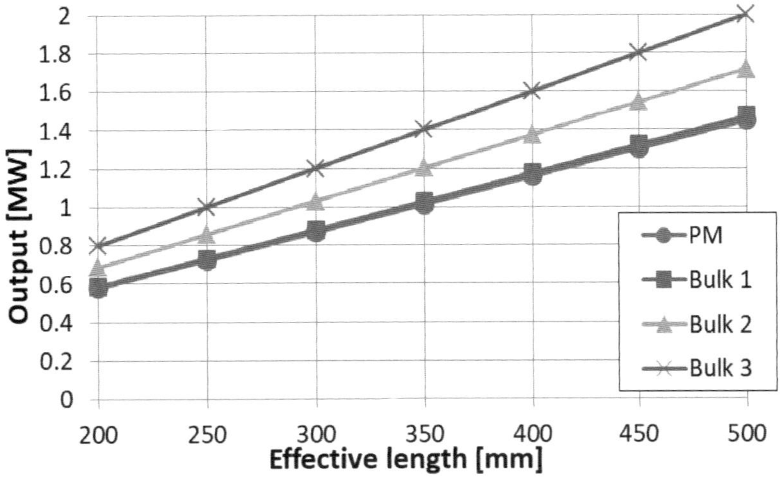

Figure 8. Mechanical output of the IPMSM as a function of the effective length of the motor.

3.2. Comparison of mechanical output density

IPMSMs with bulk superconductors and with permanent magnets were compared by changing the effective length of the motor and aiming for 1000 kW with both motors. The bulk superconductors were magnetized with three current densities, listed in table 2. The permanent magnets used for comparison were magnetized to 1.0 T.

Table 2. The current densities applied to bulk superconductors.

	Bulk 1	Bulk 2	Bulk 3
Current density	6.3×10^7 (A/m^2)	7.7×10^7 (A/m^2)	9.5×10^7 (A/m^2)
Magnetic flux density	2.0 T	2.5 T	3.0 T

The mechanical outputs of the motors with different effective lengths are shown in figure 8. The result shows that the IPMSM with Bulk 1 showed almost the same mechanical output as the permanent magnet machine. When the mechanical output is 1000 kW, the IPMSM with Bulk 3 can reduce the effective length of the motor to 71% against the permanent magnet motor. Finally, the output densities of the motors calculated using roughly-estimated motor masses are listed in Table 3. It is shown that the output density of the IPMSM with Bulk 3 is 38% higher than the conventional IPMSM.

Table 3. The mechanical output densities of the motors.

Magnetic field	Output density
PM	0.47 kW/kg
Bulk 1	0.48 kW/kg
Bulk 2	0.56 kW/kg
Bulk 3	0.65 kW/kg

4. Electromagnetic analysis of bulk superconductors

We analyzed the influence of the magnetic flux density from armature windings on the magnetized bulk superconductors. Superconductors have a nature to expel the external magnetic field [10]. This effect can be modelled with a shielding current flowing inside the superconductor negating the external magnetic field [11]. As bulk superconductors in IPMSMs are exposed to the external magnetic field generated by the armature current, the magnetization might be affected by the shielding effect. In this analysis, we considered the influence of the magnetic flux density of the armature windings on the bulk by comparing the maximal magnetic flux density of the magnetized bulk and the magnetic flux density generated by the armature windings. Finally, the effect on the IPMSM is discussed.

The magnetic flux density of the IPMSM without the bulk superconductors is shown in figure 9. The magnetic flux density at the surface of the bulk superconductor is shown in figure 10, corresponding to the position shown in the red dotted line from A to B in figure 9. We considered the case that the maximal magnetic flux density in the center of the bulk was 2.0 T.

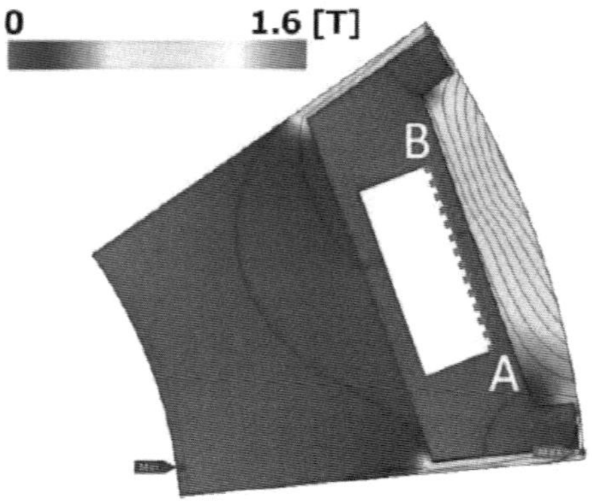

Figure 9. Magnetic flux density of the IPMSM.

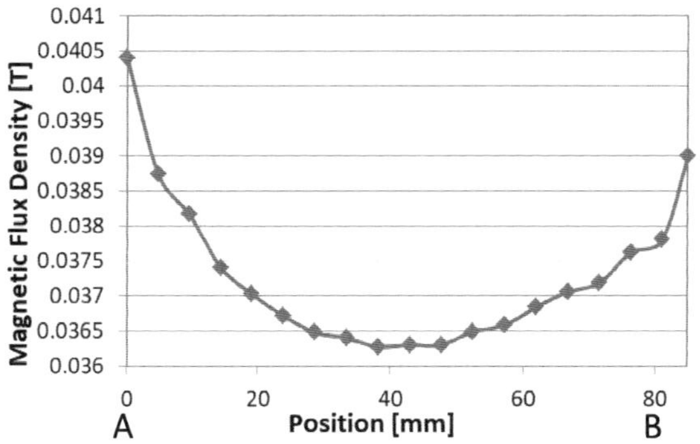

Figure 10. Magnetic flux density of the surface of the bulk superconductor.

As shown in figure 9 and figure 10, the external magnetic field applied to the bulk superconductor was below 0.05 T when the magnetization of the bulk superconductor was 2.0 T. Also, the maximum magnetic flux density of the surface of the cooling space A-B was 1.0 T, which is 50% of the magnetization in the bulk.

Because of the existence of the cooling space for the bulk superconductor, the magnetic flux density from the armature windings, which influences the magnetized bulk superconductor, was severely weakened. This shows that the effect of the magnetic shielding effect of the bulk is small, if at all, and hardly affects the performance of the IPMSM.

5. Conclusion

In this paper, an IPMSM using bulk superconductors as field magnets was proposed. The motor was designed to maximize the reluctance torque, assuming the magnet torque to be compensated for with the magnetization of the bulk superconductors. Then, the mechanical output density of the IPMSM with bulk superconductors was compared with that of a conventional IPMSM with permanent magnets. It was shown that the proposed motor can fulfil the same mechanical output as the conventional motor with an effective length of 71%, showing an increase of mechanical output density depending on the magnetization. Finally, electromagnetic analysis was conducted, investigating the magnetic flux density generated on the surface of the bulk superconductor by the armature current. The result of the analysis was that the magnetic flux reaching to the bulk superconductor is very small due to the air gap made for the cooling. Therefore, it could be considered that the effect of the shielding effect of the bulk superconductor is trivial.

Further optimization is needed for the IPMSM structure by trying other arrangements of design. As with electromagnetic analysis, the magnetic shielding effect can be modelled with shielding current flowing on the surface of the bulk superconductor, leading to a deeper understanding of the electromagnetic phenomena occurring in the bulk superconductors used in the IPMSMs. This may lead to further utilization of bulk superconductors in electric motors.

IOP Conf. Series: Journal of Physics: Conf. Series **1293** (2019) 012076 doi:10.1088/1742-6596/1293/1/012076

References

[1] Okada K, Morimoto S, Sanada M and Inoue Y 2012 *Proc. 15th International Conference on Electrical Machines and Systems (ICEMS)* 13247890

[2] Lusu G and Leila P 2009 *Proc. IEEE International Electric Machines and Drives Conference* 1072917593

[3] Lei X, Mingyao L, Xinghe F, Kai L, and Keman L 2016 *Proc. 19th International Conference on Electrical Machines and Systems (ICEMS)* 16639718

[4] Lindh P, Tehrani M, Lindh T, Montonen J, Pyrhönen J, Sopansen J, Niemelä M, Alexandrova Y, Immonen P, Aarniovuori L and Polikarpova M 2016 *IEEE Trans. Industl. Elec.* **63** 6 3397- 3407

[5] Dezhin D, Kovalev K, Verzhbitskiy L, Kozub S and Firsov V 2017 *IOP Conf. Series: Earth and Environmental Science* **87** 032007

[6] James K, Arijit B and Steven E 2015 *Proc. of the IEEE* **103** 12 2320 – 2332

[7] Ekaterina K, Pavel K, Pavel D and Oleg M 2018 Presented in 20th International Symposium on Electrical Apparatus and Technologies 249

[8] Terao Y, Akada W and Ohsaki H 2019 *IEEE Trans. Appl. Supercond.* **29** 5 5202205

[9] Nakamura T, Itoh Y, Yoshikawa M, Oka T and Uzawa J 2007 *Concepts in Magnetic Resonance Part B* **31B** 2 65–70

[10] Nishi K, Tachi Y, Sawa K, Iwasa Y, Nagashima K, Fujimoto H, Miyamoto T, Tomita M and Murakami M 1999 *Journal of Cyrogenics and Supercond. Society of Japan* **34** 11 640 – 646 (in Japanese)

[11] Ueda H, Morisaki T and Ishiyama A 2004 *IEEJ Trans. PE* **124** 11 1357 – 1364 (in Japanese)

Design of an HTS module coil for a 750 kW-class superconducting wind power generator

O Tuvdensuren[1], B S Go[1], H J Sung[1] and M Park[1]

[1] Changwon National University, Changwon, GN, 51140, Korea

capta.paper@gmail.com

Abstract. Recently, high temperature superconducting (HTS) generators are being studied with the advantages of high efficiency, light weight and small volume compared with the conventional generators. However, the HTS generator requires a large vacuum vessel and has problems such as difficulties in maintenance of the HTS coil. These problems can be overcome through the modularization of the HTS field coil. This paper designs an HTS module coil for a 750 kW-class superconducting wind power generator. The HTS module coil was designed based on the electromagnetic analysis results obtained using a 3D finite element method. For detail structure of the module coil, the heat load analysis and the mechanical stress analysis were performed. As a result, the total heat load was minimized the temperature of the HTS coil could be kept below 30 K. The von Mises stress of the supports was less than the allowable stress of the glass-fiber reinforced plastic material. The design results of the HTS module coil can be effectively utilized to develop large scale the superconducting wind power generators.

1. Introduction

High-temperature superconducting (HTS) generators are more suitable for wind power systems because they can reduce volume and weight compared to conventional generators and thus increase power density [1]-[3]. Many researchers have tried to develop wind power generators with larger capacity, smaller size and lighter weight [4]. However, the HTS generator has its own set problems, such as the need for a huge vacuum vessel and the difficulty of repairing and maintaining the HTS coils. These problems are closely related to the stability of the generators [5]. Furthermore, the HTS generators have high electromagnetic force acting on HTS field coils due to their high current density and magnetic field [6]-[9], and require a current lead (CL) to deliver DC current to the HTS field coil. This CL has a structure exposed to room temperature and causes a large heat transfer load [10]-[12]. Therefore, the modularization of the HTS field coil has been suggested to overcome these problems [13].

The purpose of this paper is to develop a modularized HTS rotor for a 750 kW direct-drive generator, which is smaller in volume and lighter than the UNISON company's U50. The 750 kW-class HTS rotor is composed of 84 module coils. Each module coil consists of an HTS field coil, bobbin, coil support, cryostat and CL, and designed by 3D CAD program. The HTS module coil of the generator was designed based on electromagnetic and heat load analyses, mechanical stress and displacement results obtained using a 3D finite element method (FEM), and the critical current was evaluated by the perpendicular magnetic field density.

As a result, the maximum values of the perpendicular magnetic field and the normal magnetic field of the HTS coils were 2.5 T and 3.9 T, respectively. The total heat load was minimized and a temperature of less than 30 K was achieved in the HTS coil. The von Mises stress of the HTS module coil with the supports was smaller than the allowable stress of the glass-fiber reinforced plastic (GFRP), and displacement was within the acceptable range. The analysis results of the HTS module coil were reflected in the design and discussed in detail. The results will be applied to the design of the HTS module coil to be installed in a 750 kW-class superconducting wind power generator.

2. Design of an HTS module coil for a 750 kW-class HTS generator

2.1 Specifications of the 750 kW-class HTS generator

The 750 kW-class HTS generator consists of rotor and stator parts. The rotor components are HTS field coils, a cryostats and supporting structures, and a rotor body. The stator components are comprised of stator teeth, copper stator coils, magnetic shield. The generator has 84 HTS modules. All the structure of the 750 kW-class HTS generator were drawn by the 3D CAD program as shown in figure 1. Table 1 shows the specifications of the 750 kW-class HTS generator.

Table 1. Specifications of the 750 kW-class HTS generator

Items	Value
Rated output power	750 kW
Rated line to line voltage	780 V
Rated armature current	597.8 A
Rotational speed	25 rpm
Rated torque	0.3 MN·m
Number of poles	84 poles
Diameter of the generator	3.55 m
Weight of the generator	12 ton

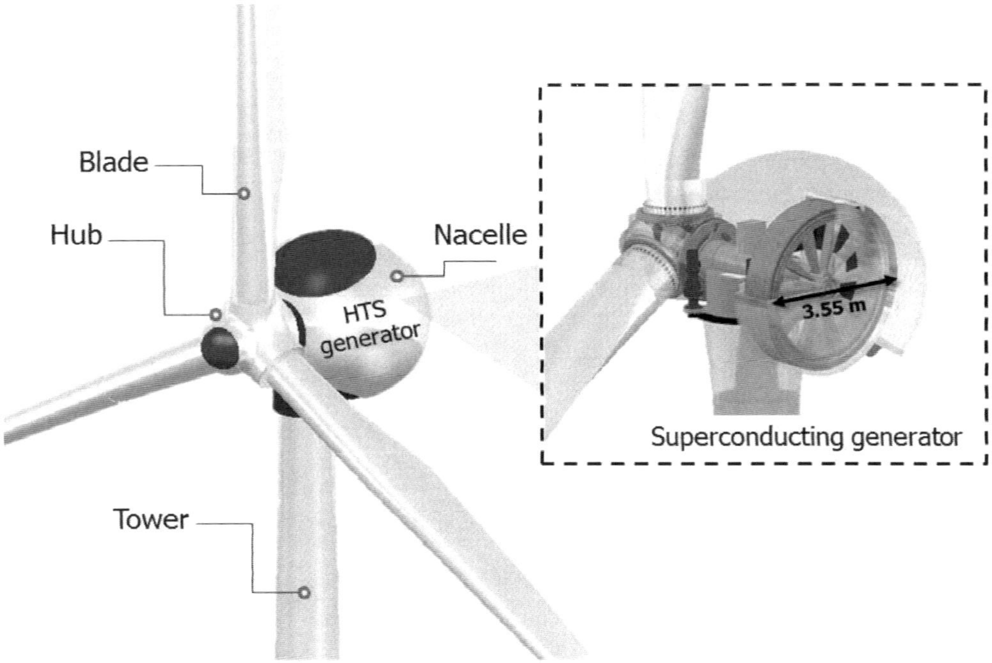

Figure 1. 3D design of the 750 kW-class HTS generator

2.2. Design of the HTS module coil for the 750 kW-class HTS generator

2.2.1 Structural design of the HTS module coil

The HTS module coil consists of HTS field coils, coil bobbins, supports, a copper block, CLs, a cryostat, and a cooling pipe. To prevent iron loss, stainless steel is used as the material of the cryostat. The coil supports are made of GFRP to improve the heat insulation between the cryostat and the HTS coils. The HTS coils are covered with bobbin and bobbin supports to protect the HTS coils from a high mechanical force.

Figure 2 shows the structural design of the HTS module coils. The 84 HTS module coils were divided into 14 parts; each part was mounted on a cryostat with a copper block and cooling pipes. The liquid neon gas flows inside the cooling pipes to cool the HTS module coils. The HTS coils consisted of three double pancake coils (DPC) in race-track form. The HTS wire was coated with a flat conductor tape, and thickness and width were 0.15 mm and 12 mm, respectively.

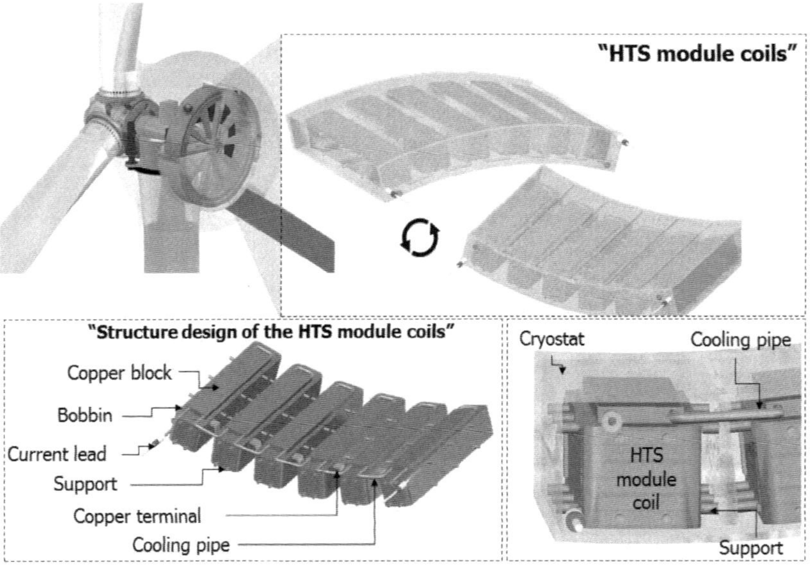

Figure 2. Structural design of the HTS module coils

2.2.2 Structural design of the current lead

The CLs should be designed to minimize the heat load of on the cryogenic system of the HTS wind power generator. Conduction and Joule heat loads of the CLs made of brass were calculated and analyzed. The important factors in designing the CLs are the thermal conductivity of materials and the geometric shapes (length, cross-sectional area and cooling surface area). The minimum heat load and ideal length of the CL can be calculated by equations (1) and (2) [7].

$$Q_{op} = I\sqrt{2\int_{T_L}^{T_H} \rho(T)k(T)dt} \tag{1}$$

$$\frac{L}{A} = \frac{1}{I}\int_{T_L}^{T_H} \frac{k(T)}{\sqrt{2\int_{T_L}^{T_H} \rho(T)k(T)dt}}dT \tag{2}$$

The thermal conductivity k(T) and the electric resistivity $\rho(T)$ are inversely related, according to Wiedemann- Franz law (3). Here, the Lorentz number is $L_0 = 2.45 \times 10^{-8}\ W\Omega K^{-2}$.

$$k(T)\rho(T) = L_0 T \tag{3}$$

where, L is the length of CL, A is the cross-sectional area, and I is the carrying current, T_H and T_L represent the temperature of the cold and warm ends, k(T) is the thermal conductivity, and $\rho(T)$ is the electric resistivity [14]. Brass was used as the material for the metal CL because it has better mechanical characteristics than copper. Figure 3 shows the structural design of the HTS module coil. The CLs, including brass current terminals and current feedthroughs, were designed using a 3D CAD program.

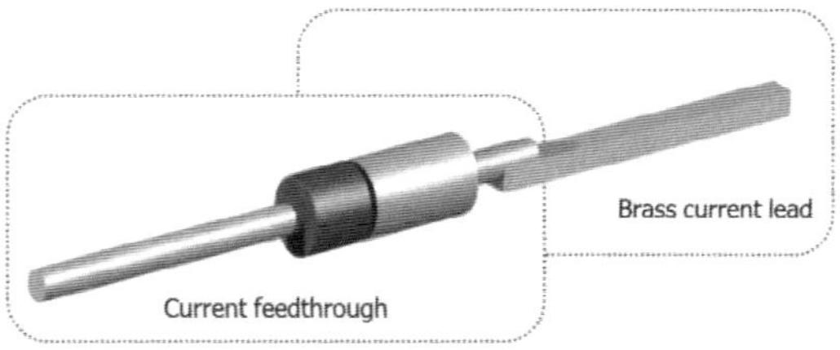

Figure 3. Structure design of the CL

3. Simulation results and discussions

3.1 Electromagnetic analysis

The 750 kW-class HTS generator was analysed using the 3D FEM program. The design specifications of the HTS module coil for the 750 kW-class HTS generator were summarized in table 2. Figure 4 shows the magnetic field distributions of a 1/84 model of the 750 kW-class HTS generator. The maximum value of the perpendicular magnetic field and normal magnetic field of the HTS module coil were 2.5 T and 3.9 T, respectively. The operating current was 397.8 A with 40% margin, and the current was determined by the properties of the HTS wire [15].

Table 2. Detail specifications of the HTS module coil for the 750 kW-class HTS generator

Items	Value
Number of poles	84 poles
Number of HTS coil layers	6 layers
Temperature	30 K
HTS wire width	12 mm
HTS wire thickness	0.15 mm
Turns of HTS coil	136 turns
Effective length of HTS coil	449 mm
Operating current	397.8 A

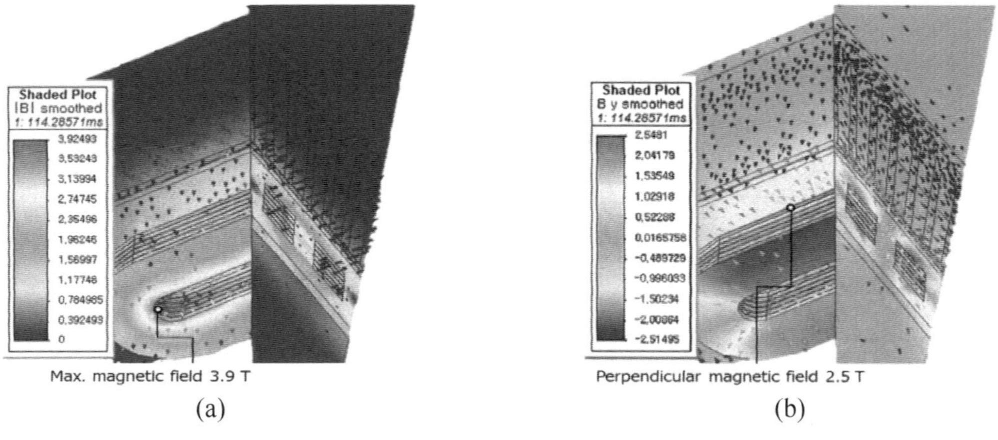

Figure 4. (a) Magnetic field distribution and (b) perpendicular magnetic field of the 750 kW-class HTS generator

As a simulation result, the output voltage and rated torque were 551 V and 0.3 MN·m, respectively. Figure 5 shows the output voltage and torque value of the 750 kW–class HTS generator under the full-load condition.

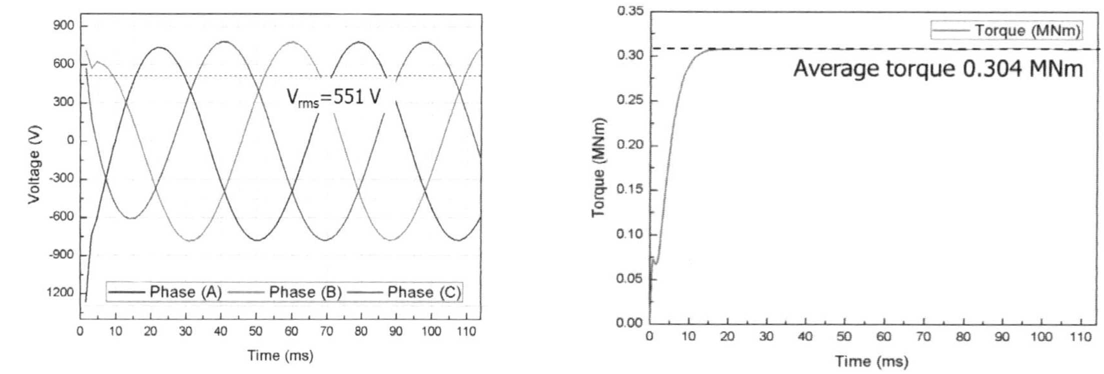

(a) (b)

Figure 5. (a) Output voltage and (b) rated torque of the 750 kW-class HTS generator

3.2 Thermal analysis of the HTS module coil

The total heat load of the HTS module coil, including the Joule heat, the conduction heat load of the CLs and the radiation heat load, the conduction heat load from supports were considered using the 3D FEM program. The heat invasion of the CL was should be close to 0 W which was obtained by changing the thickness of the CL as shown in figure 6. The length and thickness of the CL were 100 mm and 7.7 mm respectively. Figure 7 shows the temperature distribution analysis result of the designed CL. The total heat load of one CL was 20.1 W.

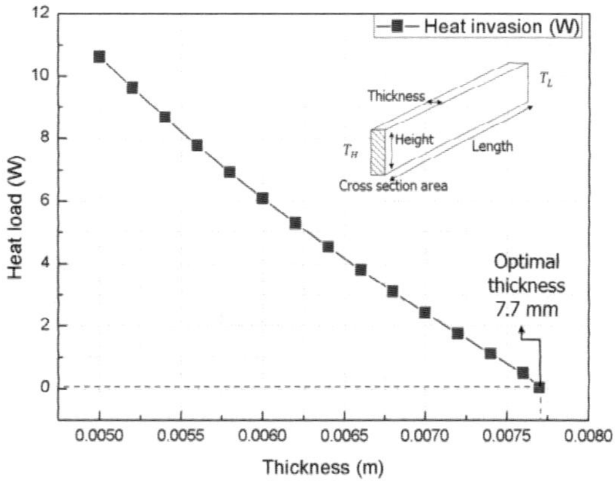

Figure 6. Heat invasion of the CL

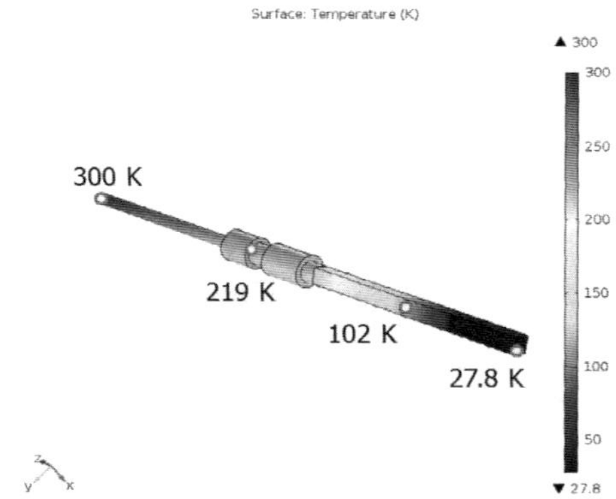

Figure 7. Temperature distribution of the one CL with feedthrough

In order to reduce conduction heat load, GFRP material with low thermal conductivity was used for the support. The length and contact area of the support were 20 mm and 0.078 m^2, respectively. The conduction heat load of the supports in the module coil was analysed by using a 3D FEM program and compared with the calculated result.

The radiation heat load was evaluated by the Stefan-Boltzmann equation using the 3D FEM program. The radiation heat load between the N_i layer of multilayer insulations is estimated as:

$$Q_r = \frac{\varepsilon\sigma}{N_i+1}(T_r^4 - T_c^4) \quad (4)$$

where, ε is the effective total thermal emissivity of the materials. In this paper, the emissivity of 0.01 was applied. σ is Stefan-Boltzmann constant. T_r and T_c are the enclosure temperature and the thermal shield temperature, respectively [16].

For the simulation, the operating current of 397.8 A was applied. The temperature distribution analysis results of the HTS module coils are shown in figure 8. The temperature distribution in the CLs is shown in figure 8 (a). The temperature distributions in the HTS module coils and the cryostat are shown in figure 8 (b) and (d), respectively. The temperature distribution in the supports of the HTS module coil is shown in figure 8 (c).

As a result, the temperature of the HTS coil was 27.1 K. The conduction heat load from the supports was 39.4 W, and the radiation heat load was 7.59 W, the conduction and Joule heat load from the CLs was 40.2 W. The total heat load of the 750 kW-class HTS generator was 87.14 W. Table 3 compares the calculation results with the simulation results. The total heat load obtained by the calculation was 75.05 W, the total heat load obtained by the simulation was 87.14 W, and the temperature was 27.1 K, which was less than 30 K. The AL325 cryo-cooler was used to cool the HTS module coils. To achieve a temperature of 30 K or less based on the capacity map of the cryo-cooler, the total heat load must be less than 140 W, thus the cooling capacity margin of the HTS module coil is 52.9 W.

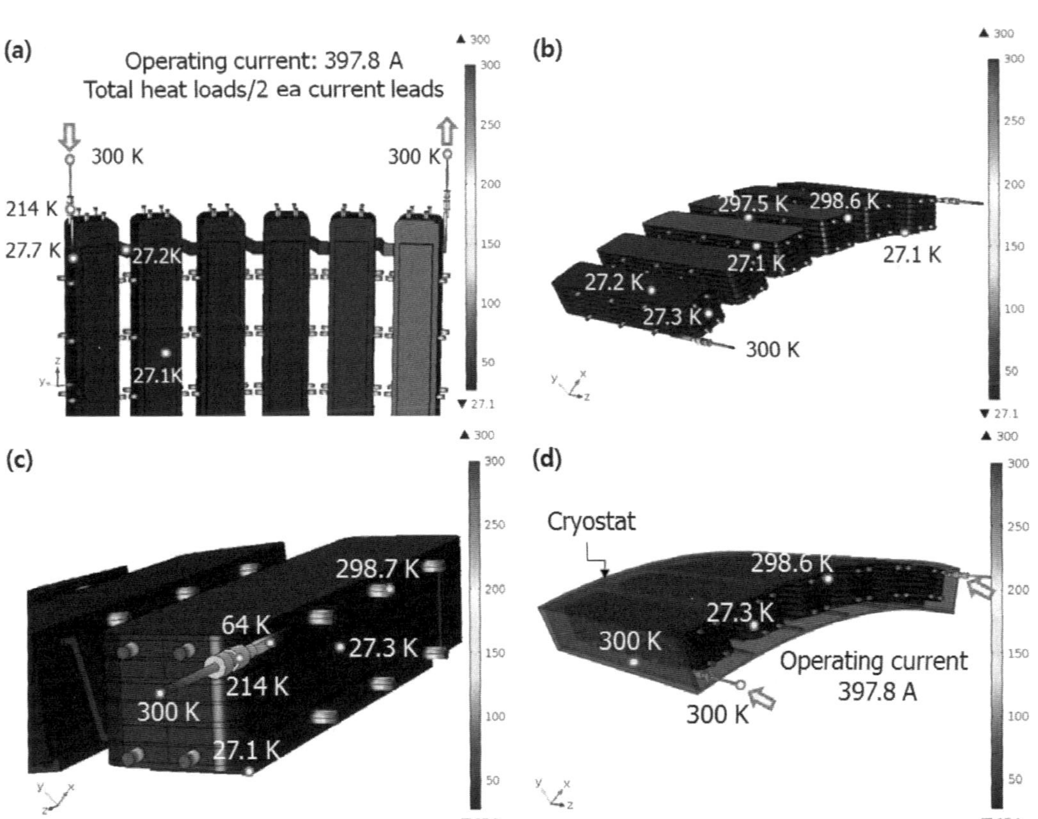

Figure 8. Temperature distributions of the (a) CLs, (b) HTS module coil, (c) supports and (d) cryostat of the 750 kW-class HTS generator

Table 3. The total heat load of the HTS module coil

Items	Calculation result	FEM result
Conduction and Joule heat load from the CLs	37.2 W	40.2 W
Current terminals and joint	0.35 W	0.35 W
Conduction heat load from the supports	32.9 W	39.4 W
Radiation heat load	4.6 W	7.59 W
Total heat load	75.05 W	87.14 W

3.3 Mechanical analysis of the HTS module coil

Torque is generated due to the magnetic field reaction between the rotor and stator parts. When the generator is in a no-load state, the generated magnetic field does not change because the armature current is zero, but the magnetic field generated under the full load condition is influenced by the armature current, thereby changing the electromagnetic force [1]. Therefore, the mechanical stress due to the electromagnetic force of the 750 kW HTS generator was analyzed under full load conditions.

$$\tau = \frac{P}{\omega} \tag{5}$$

where, P is the rated power, τ is the torque, ω is the angular velocity. Tangential force acting on a body moving in tangential direction of the body path is given as follows.

$$F_{tan} = \frac{\tau}{r} \tag{6}$$

where, r is the radius of the rotor. The total torque and the tangential force of the generator were calculated by the equations (5) and (6). The torque and tangential, radial, gravity, centrifugal forces of the HTS module coil were examined. The output torque and the tangential force of one HTS module coil were 3.6 kN and 2.2 kN, respectively. The gravity and the centrifugal forces of the HTS module coils were 1.5 kN and 1.2 kN, respectively.

To analyse the mechanical stress of the HTS module coil, the axial direction and amplitude of the Lorentz force at the HTS field coil are required. The tangential and radial forces of the HTS module coils are shown in figure 9. The maximum amplitudes of the axial direction forces were 15.2 kN (y-axis) and 4.9 kN (x-axis), respectively, as shown in figure 10.

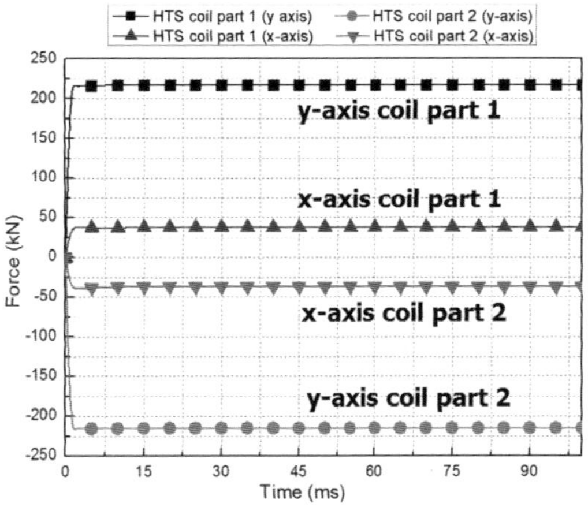

Figure 9. Tangential and radial force of the HTS module coil

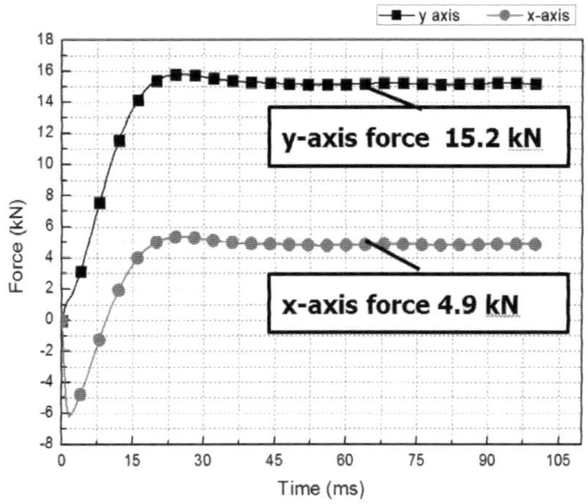

Figure 10. Maximum tangential and radial force of the HTS module coil

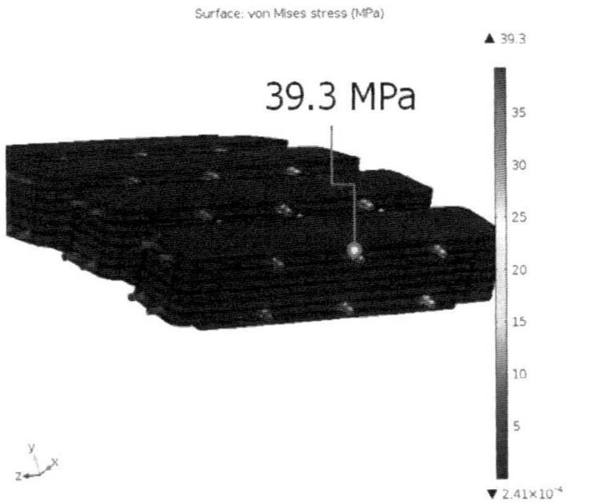

Figure 11. Maximum stress of the HTS module coil

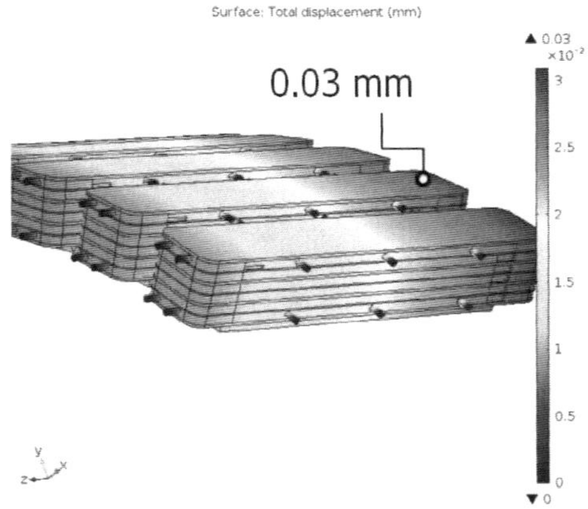

Figure 12. Maximum displacement of the HTS module coil

Based on the force analysis results, the HTS module coil structure was analysed using a 3D FEM program. The simulation results show that the von Mises stress and displacement of the HTS module coil were 39.3 MPa and 0.03 mm, respectively. Figure 11 and 12 show the von Mises stress and displacement of the HTS module coils with support structures. The module coil support was made of GFRP. The tensile strength of the GFRP material was 280 MPa, allowable yield stress was 93.3 MPa. The stress value of the simulation result was 39.3 MPa which is lower than the allowable yield strength.

4. Conclusions

This paper discusses the design of an HTS module coil for a 750 kW-class superconducting wind power generator. Electromagnetic, thermal and mechanical properties were analysed using a 3D FEM program. The required diameter and axial length of the generator were 3.55 m and 0.7 m, respectively. As a result, the maximum values of the perpendicular magnetic field and the normal magnetic field of the HTS module coil were 2.5 T and 3.9 T, respectively. The conduction heat load from the supports was 39.4 W, and the radiation heat load was 7.59 W, the Joule heat load of the current terminal and the joint was 0.35 W, the Joule and conduction heat load from the CLs was 40.2 W. The total heat load and the temperature of the HTS module coil were 87.14 W and 27.1 K, respectively. The AL325 single-stage cryo-cooler was adopted to cool the HTS module coils. Overall heat load of the HTS module coil was less than the cooling capacity of the cryo-cooler, and the cooling capacity margin was 52.9 W. The tangential and radiation forces of the HTS module coils were 15.2 kN and 4.9 kN, respectively. The von Misses stress and displacement of the HTS module coils were 39.3 MPa and 0.03 mm. The von Mises stress of the HTS module coil with the supports was less than the allowable stress of GFRP, and the displacement was within the permissible range. These results can be applied to the development of large scale HTS wind power generator system.

Acknowledgments

This research was supported by Korea Electric Power Corporation (Grant number: R18XA03), and in part by the New & Renewable Energy of the Korea Institute of Energy Technology Evaluation and Planning (KETEP) grant funded by the Korea government Ministry of Knowledge Economy (No. 20183010025270).

References

[1] Go B S, Sung H J, Park M and Yu I K 2017 Structural design of a Module coil for a 12 MW class HTS generator for wind turbine *IEEE Trans. App. Supercond.* **27** 5202405

[2] Sung H J, Park M, Go B S and Yu I K 2016 A Study on the required performance of a 2G HTS wire for HTS wire for HTS wind power generators *Supercond. Sci. Technol.* **29** 054001

[3] Sung H J, Kim G H, Kim K, Jung S J, Park M and Yu I K 2013 Practical design of a 10 MW superconducting wind power generator considering weight issue *IEEE Trans. App. Supercond.* **23** 5201805

[4] Young G P, Woo S L, Jeyull L, Seunghyun S, Young J H, Yoon D C and Tae K K 2016 Operational characteristics of HTS coils with flux diverters in semipersistent mode under alternating magnetic field *IEEE Trans. App. Supercond.* **26** 5204105

[5] Go B S, Sung H J, Park M and Yu I K 2015 Fault characteristic analysis of a modularized HTS field coil-based 12 MW class SCSG *Korean institue of electrical engineers 46th conference (KIEE)* pp 1108-09

[6] Dao V Q, Kim T, Sung H J, Choi J, Park M and Yu I K 2017 Thermal optimization of current leads for HTS DC reactor magnet *International symposium on innovation in information technology and application (ISIITA) international conference 2017* pp 17-21

[7] Tuvdensuren O, Dao V Q, Le T T, Go B S, Park M and Yu I K 2017 Design of a current lead for a large scale superconducting wind power generator *Korean institue of electrical engineers 48th conference (KIEE)* pp 1065-66

[8] Sung H J, Badcock R A, Jiang Z, Choi J, Park M and Yu I K 2016 Design and heat load analysis of a 12 MW HTS wind power generator module employing a brushless HTS exciter *IEEE Trans. App. Supercond.* **26** 5205404

[9] Go B S, Sung H J, Choi J, Park M and Yu I K 2016 Modularization of the HTS field coils for a large-scale superconducting generator *IEEE Trans. App. Supercond.* **26** 5202905

[10] Shafaie R, Kalantar M 2013 Design of a 10 MW class wind turbine HTS synghronous generator with optomized field winding *IEEE Trans. App. Supercond.* **23** 5202307

[11] Go B S, Sung H J, Lim H S, Park M and Yu I K 2016 Design of a large scale HTS generator module coil considering electromagnetic forces *Korean institue of electrical engineers 47th conference (KIEE)* pp 677-678

[12] Tuvdensuren O, Sung H J, Go B S, Park M and Yu I K 2018 Structural design and heat load analysis of a flux pump-based HTS module coil for a large scale wind power generator *J. Phys.: Conf. Ser.* **1054** 012084

[13] Sung H J, Badcock R A, Go B S, Park M, Yu I K and Jiang Z 2016 Desing of a 12 MW HTS wind power generator including a flux pump exciter *IEEE Trans. App. Supercond.* **26** 5206205

[14] Tuvdensuren O, Sung H J, Park M and Yu I K 2018 Thermal analysis of a 1 MW superconducting motor for ship propulsion *Korean institue of electrical engineers 49th conference (KIEE)* pp 841-842

[15] THEVA wire specifications [Online] Available: https://www.theva.com/products/#pro-line

[16] Sung H J, Go B S, Jiang Z, Park M and Yu I K 2016 Heat loss analysis-based design of a 12 MW HTS wind power generator module having an HTS flux pump exciter *Physica C: Supercon. and its appl.* **530** pp 133-137

IOP Conf. Series: Journal of Physics: Conf. Series **1293** (2019) 012078 doi:10.1088/1742-6596/1293/1/012078

Conceptual design of Japan's fusion DEMO reactor (JA DEMO) and superconducting coil issues

K. Tobita [1], H. Utoh [1], R. Hiwatari [1], Y. Miyoshi [1], S. Tokunaga [1], Y. Sakamoto [1], Y. Someya [1], N. Asakura[1], Y. Homma [1], N. Nakajima [2], The Joint Special Design Team for Fusion DEMO

[1] National Institutes for Quantum and Radiological Science and Technology (QST), Rokkasho, Aomori, Japan

[2] National Institute for Fusion Science (NIFS), Rokkasho, Aomori, Japan

E-mail: tobita.kenji@qst.go.jp

Abstract. Goals of Japan's fusion demonstration (DEMO) reactor are to demonstrate (1) steady and stable electric power generation in a power plant scale, (2) self-sufficient production of fuel (tritium), and (3) reasonable availability using a remote maintenance scheme anticipated in a commercial plant. Main design parameters of JA DEMO are a plasma major radius of 8.5 m, fusion output of 1.5-2 GW, magnetic field on the plasma axis of 5.94 T. The superconducting coil system of the reactor consists of a central solenoid (CS), 7 poloidal field (PF) coils and 16 toroidal field (TF) coils. Regarding CS and PF coils, superconducting coil technology on DEMO is basically the same as that on the world largest fusion experimental reactor called ITER. In contrast, TF coils have a technology gap on magnetic energy and the resulting stress between ITER and DEMO due to their size and magnetic field. In particular, the necessity of higher design stress is critical for TF coils, requiring the development of high strength cryogenic steels surpassing the existing ones. The fundamental design strategy to mitigate tolerances in TF coil fabrication is also presented.

1. Introduction

The ongoing core project of fusion development is ITER ("The Way" in Latin) [1], which is an international fusion energy development project and is scheduled to start operation in 2025 and to carry out full burning experiments with actual fusion fuels (deuterium and tritium) as shown in figure 1. In parallel with the construction of ITER, Japan is developing the conceptual design of a fusion **demo**nstration (DEMO) reactor named "JA DEMO" as a post-ITER project realizing electricity production with fusion power. After the completion of the conceptual design of DEMO anticipated in around 2025, the design phase will transit to the engineering design phase in which large or plant scale research and development (R&D) together with detailed DEMO design will be implemented to establish engineering bases toward the decision of DEMO construction [2].

Major goals of JA DEMO are to demonstrate (1) steady and stable electrical power generation, (2) self-sufficient production of fuels, and (3) plant availability extendable to commercialization. Since tritium used in fusion reactors is extremely scarce on the earth, DEMO needs to produce it in a self-sufficient amount in a closed cycle of the plant [3]. For stable electric power generation, fusion output of 1.5 GW or higher, being several times as high as that of ITER, is required for DEMO. For the purpose, a magnetic field higher than that of ITER and toroidal field (TF) coil bore larger than that of ITER are necessary. These requirements lead to technical challenges toward DEMO.

In this paper, design issues on superconducting (SC) magnets in fusion DEMO are overviewed and design strategy to deal with the issues is presented.

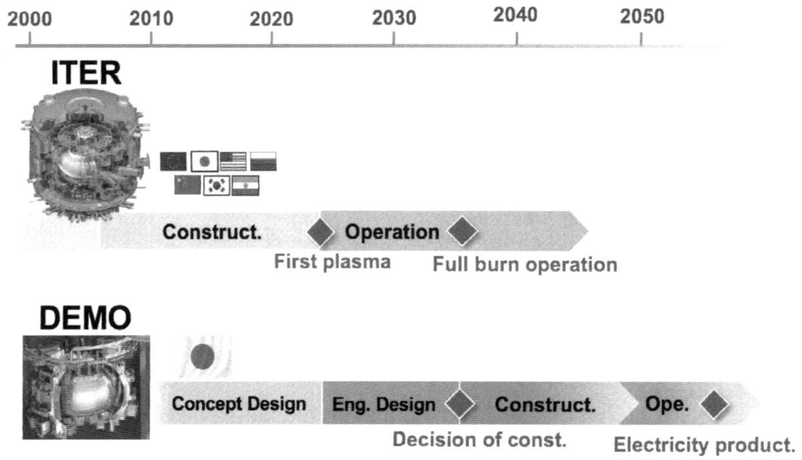

Figure 1. Timeline of ITER project and Japan's fusion DEMO development.

2. Conceptual design of JA DEMO

Design philosophy of Japan's fusion DEMO is to be feasible on the premise of the success of ITER. ITER aims to produce more fusion output from fusion reactions than input power to the plasma. The fusion output (P_{fus}) is anticipated to be about 0.5 GW. Taking account of thermal efficiency of a steam turbine system, a recirculation power required for plasma heating and current drive and auxiliary power in the plant, fusion output needs to be 1.5 GW of higher to demonstrate electric power generation. JA DEMO is designed to generate a fusion output of 1.5-2 GW, eventually producing the net electricity of 0.2-0.3 GW [4,5]. A schematic view of JA DEMO is illustrated in figure 2 and the design parameters listed in table 1. It adopts pressurized water with the water condition of 290-325°C at 15.5 MPa as used in pressurized light water reactors (PWR) to extract thermal power generated in breeding blanket and divertor surrounding the plasma. Use of such water cooling allows us to utilize accumulated knowledge and experience on water handling in light water reactor and thermal power station technology, minimizing technology development for fusion DEMO.

As seen in the figure, the SC magnets consist of a central solenoid (CS), 7 poloidal field (PF) coils and 16 TF coils. The basic specification of these coils is summarized in table 2. The CS is based on Nb_3Sn and vertically segmented into 6 units. The PF coils are made of NbTi conductors except one or two coils being made of Nb_3Sn on the high field side. Regarding CS and PF coils, SC technology on DEMO will be the same as that in ITER in terms of SC conductors, conductor current (~40 kA) and design stress (500 MPa). In the same manner as PF coils of ITER, the PF coils of DEMO will be fabricated with a pancake winding technique on site. As for TF coils, there are technology gaps between ITER and DEMO. The gaps include the yield strength of cryogenic steels, tolerances in TF coil fabrication and fabrication processes. Thus, the following sections focus on design issues of TF coils.

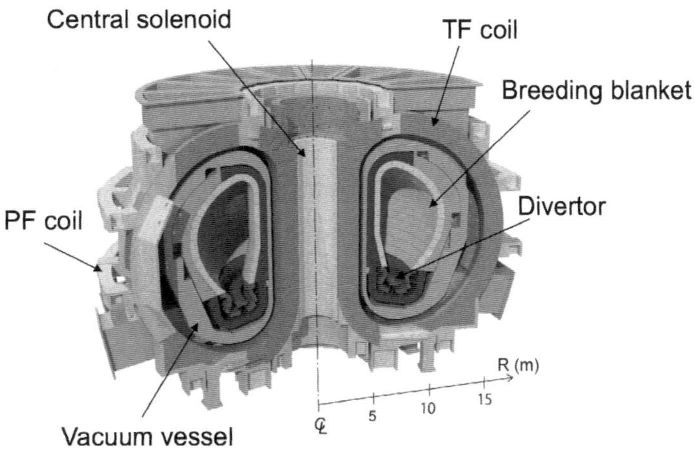

Figure 2. Conceptual view of Japan's fusion DEMO reactor (JA DEMO).

Table 1. Design parameters of JA DEMO.

Parameter	Design value
Major radius, R_p (m)	8.5
Minor radius, a (m)	2.42
Plasma elongation, κ_{95}	1.65
Plasma current, I_p (MA)	12.3
Magnetic field on axis, B_T (T)	5.94
Maximum field, B_{max} (T)	13.7
Fusion output, P_{fus} (GW)	1.46

Table 2. Basic specifications of SC coils of JA DEMO.

	TF coils	CS	PF coils
Conductor	Nb_3Sn	Nb_3Sn	NbTi (partly Nb_3Sn)
N of coils	16	6 units	7
Maximum field	~13 T	~13 T	<13 T
Conductor current	80-100 kA	~40 kA	~40 kA
Maximum stress	800 MPa	500 MPa	500 MPa
Winding scheme	CICC/RP/DP	Rec. CICC /Pancake	Rec. CICC /Pancake

CICC = Cable-in-conduit conductor, RP = Radial plate, DP = Double pancake

3. TF coil issues

Technical specification of TF coils is compared in table 3. Nb_3Sn is considered to be the prime option of SC of the coils, and Nb_3Al and high-temperature SC (HTS) are alternatives. Distinctive difference between DEMO and ITER is magnetic energy and design stress. Including these issues, based on the specifications listed in the table, the following are defined as design issues on TF coils:

1) TF coil size;

2) maximum field (B_{max}) and the total magnetic energy stored in TF coils (W_{mag});
3) design stress;
4) fabrication scheme;
5) tolerances.

Each issue is discussed in the following subsections.

Table 3. Comparison of TF coils between JA DEMO and ITER.

	JA DEMO	ITER
Conductor	Nb_3Sn	Nb_3Sn
N of TF coils	16	18
Maximum field	13.7 T	11.8 T
Total magnetic energy	149 GJ	41 GJ
Ampere-turns (AT) in all TF coils	255 MAT	164 MAT
Conductor current	83 kA	68 kA
Operation temperature	5K	5K
Design stress	800 MPa	670 MPa
Weight of TF coils	11,800 t	5,362 t

3.1. TF coil size

As seen in a comparison of TF coils in figure 3 (a), DEMO requires considerably a large TF coil bore, which is responsible for such a tremendous magnetic energy. Generally, a large plasma volume is advantageous to attain sufficient plasma confinement. In addition, it contributes to increasing P_{fus} as seen in the following scaling.

$$P_{fus} \propto {\beta_N}^2 {B_T}^2 V_p \tag{1}$$

where V_p is the plasma volume, and β_N is the normalized beta defined by $\beta_N = \beta[\%]/(I_p/(aB_T))$ with β= (plasma pressure)/(magnetic pressure = $B^2/2\mu_0$). Here, B_T is the toroidal magnetic field at the plasma axis in Tesla, and I_p is the plasma current in the unit of MA. In addition, B is the total magnetic field which is given by the resultant forces of B_T and the magnetic field produced by the plasma current. In usual tokamak condition, $\beta \approx p/({B_T}^2/2\mu_0)$ with plasma pressure (p).

The plasma size of DEMO is somewhat larger than that of ITER (figure 3 (b)) because voluminous components such as a vacuum vessel, breeding blankets, divertor and support structure are arranged between the plasma and the TF coils as seen in figure 3(c). The problem is that transportation of TF coils with 14 m in width and 20 m in height is unlikely. This means that TF coils for DEMO need to be fabricated on site, which will be an issue in engineering and manufacturing cost.

3.2. B_{max} and W_{mag} of TF coils

As seen in figure 3 (c), the TF coil bore must be much larger than the plasma size, and the resulting magnetic energy in all TF coils become tremendously large for DEMO as expected from the following equation.

$$W_{mag} \approx \int_{V_{TF}} \frac{{B_T}^2}{2\mu_0} dV \tag{2}$$

where V_{TF} stands for the volume of the TF coil bore.

Actually, B_{max} and W_{mag} are determined by a systematic calculation on the coil design with a SCONE (**s**uper**con**ducting coil **e**valuation) code [6] which was developed for superconducting TF coil design for tokamak-type reactors. The main input parameters of the program are a selection of SC wire (NbTi, Nb_3Sn, Nb_3Al or HTS), TF coil size (width, height, thickness, etc.) and conditions of TF coil design (operation temperature, design stress, conductor current, etc.). Using the given input parameters, SCONE first calculate the operation current density of SC wire (J_{op}) and the number of turns of conductor, then solves heat balance and von Mises stress and eventually solves the constitution ratio of SC wire, stabilizer required for quench protection, structural material, coolant and insulator. B_{max} and W_{mag} are quantities determined by given TF coil size and design stress as described in Section 3.3.

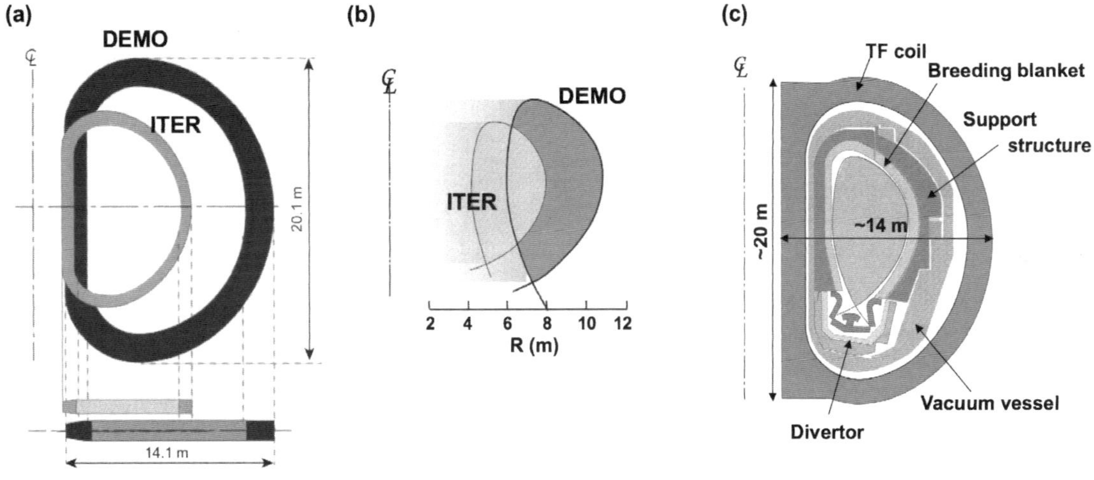

Figure 3. Comparison of (a) TF coil size and (b) plasma size between DEMO and ITER, and (c) schematic cross-sectional view of torus on DEMO.

3.3. Design stress

Figure 4 is an example of the calculation of B_{max} as a function of plasma size (R_p) for TF coil design stress (S_m) of 667 MPa being the design stress adopted in ITER. Generally, B_{max} decreases with R_p in that, as a result of a dramatic increase in W_{mag} with R_p, the fraction of structural material increases and inversely that of SC decreases. For R_p > 8 m, the fraction of structural material exceeds 70% and SC becomes as small as several %.

Previously it was thought that Nb_3Al has advantage over Nb_3Sn to produce a field higher than 13 T. It is true for tokamak reactors with relatively small R_p such as SlimCS [7] and Demo-CREST [8]. SlimCS is a compact DEMO with a **Slim**-sized **C**entral **S**olenoid assuming R_p = 5.5 m, and Demo-CREST is a **DEMO** concept for **C**ompact **R**eversed **S**hear **T**okamak assuming R_p = 7.25 m. Actually, SlimCS and Demo-CREST adopt Nb_3Al as the prime option for the SC of TF coils. An important result shown in figure 4 is that a difference in B_{max} between Nb_3Sn and Nb_3Al becomes small and the difference is smaller than 0.3 T for $R_p \geq 8$ m. Taking account of this finding and the technical maturity of Nb_3Sn, JA DEMO adopts Nb_3Sn as the prime SC option. The figure also indicates an impact of S_m in high field coil design, showing a 1-T increment in B_{max} when S_m is increased from 667 to 800 MPa. From the point of view of reactor system design, the increment in B_{max} mitigates requirements for plasma design parameters, allowing the net electricity production of ~0.3 GW with reasonable design parameters. Therefore, S_m = 800 MPa is favorable for a balanced system design.

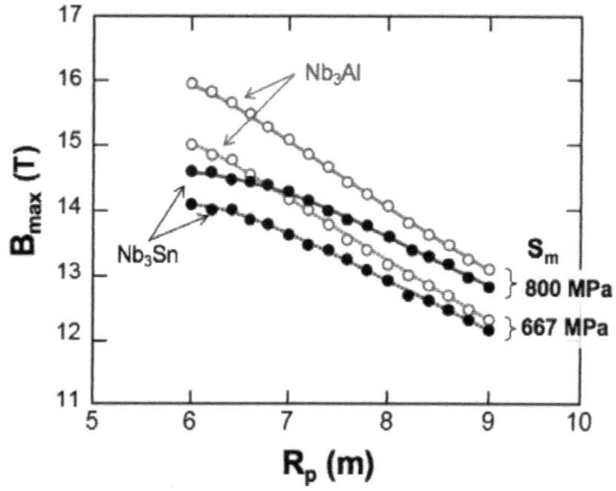

Figure 4. Maximum toroidal field ($B_{\max}$) as a function of plasma major radius for design stress (S_m) of 667 MPa, and increase of $B_{\max}$ when $S_{\rm m}$ is increased from 667 to 800 MPa.

Now that the target of $S_{\rm m}$ is defined, it is reasonable to consider high strength cryogenic steel. Based on the definition of $S_{\rm m}$ with 0.2% yield stress ($S_{\rm Y0.2}$) as follows,

$$S_m = \frac{2}{3} S_{Y0.2} \quad , \tag{3}$$

a cryogenic steel qualified for DEMO TF coil structure with $S_{\rm m}$ = 800 MPa needs to show $S_{\rm Y0.2}$ of well above 1200 MPa. In ITER, main part of TF coil structure is made of strengthen 316LN and partly JJ1 (12Cr-12Ni-10Mn-5Mo-0.2N) is used as high strength cryogenic steel [9]. According to the reference, the 316LN and JJ1 shows $S_{\rm Y0.2}$ of 1078 and 1126 MPa in median and a deviation of 47 and 66 MPa at the temperature of 4 K, respectively. Thus, these steels are not qualified for DEMO TF coil structure. Under the circumstance, Japan is exploring other candidate cryogenic steels for DEMO. Preliminary test indicates that HRX19® (Nippon Steel and Sumitomo Metal) developed for high pressure hydrogen environment may be a potentially qualified steel in that a mechanical test of 30 mmt HRX19® shows $S_{\rm Y0.2}$ = 1570 MPa [10]. Further intensive test on the steel will be continued.

3.4. Fabrication

Considering that situation that TF coils of DEMO will undergo a higher stress than that of ITER and that error field needs to be low enough to stably control plasma, the fabrication method of ITER, namely, the winding with cable-in-conduit conductor (CICC), radial plate (RP) and double pancake (DP), is selected as the prime option. Hence, a series of the fabrication processes require precision machining and welding, leading to problems of a long work period and high fabrication costs. From this point, alternative fabrication methods are under study, including RP grooves with hot isostatic pressing (HIP), CICC and pancake winding, and CICC and layer winding.

3.5. Tolerances

In ITER, an error field requirement of $\delta B/B_{\rm T} \leq 5\times10^{-5}$ for avoiding mode-locking of plasma imposes a tolerance of ±0.023% for the circumference of the winding in the RP groove, a tolerance of ±0.01% for the RP machining and an accuracy of ±0.01% for the conductor length of winding. Here, δB stands for error fields for three modes with toroidal mode number of 1, 2 and 3. A published assessment [11] indicates that a critical error field decreases with R_p, suggesting that TF coils of DEMO need to have smaller tolerances than those of ITER. However, lower tolerances (or higher accuracies) exceeding ITER seem impractical. In reality, the error field requirement should be mitigated utilizing external correction coils [12]. In a plasma physics sense, the target error field that is correctable with external coils is $\delta B/B_T \sim 10^{-4}$. Thus, the question of interest is a consistency between reasonably mitigated tolerances and the target error field.

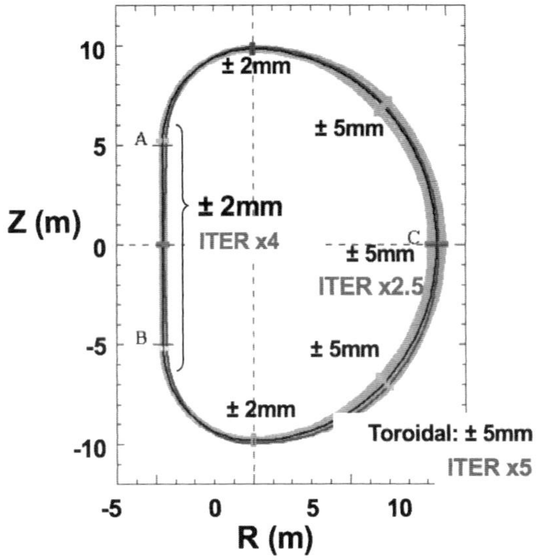

Figure 5. Tolerances assumed in the error field analysis of DEMO TF coils.

In order to resolve the question, a Monte Carlo calculation was carried out giving random dispersions in conductor positions within the tolerances and eventually calculated the resulting error field. Mitigated tolerances are illustrated in figure 5. The tolerances were assumed to be 4 times as large as those of ITER TF coils on the inboard side, and 2.5 times on the outboard side. In addition, the installation tolerance in the toroidal direction was assumed 5 times as large as those of ITER. The calculated result is

$$\frac{\delta B}{B_T} \approx \frac{5.2 \times 10^{-4}}{5.9} \approx 10^{-4} \quad , \tag{4}$$

suggesting that the design strategy of tolerance mitigation in combination with external error field correction can work. In parallel, the design of external correction coils is in progress, indicating that an error field of $\delta B/B_T \sim 10^{-4}$ is reduced to 5×10^{-5} with external correction coils of < 400 kAT.

4. Summary

Japan is working on the conceptual design of fusion DEMO aiming at electricity generation in the next step beyond ITER. This paper mainly focuses on the design issues on SC magnets. Technical challenges on DEMO are mostly due to larger and higher field TF coils. In particular, development of strengthen cryogenic steels is indispensable to generate fusion power enough to produce net electricity. Although the TF coil fabrication technology developed for ITER is matured and will be likely to be applicable to

DEMO, the reliability of TF coil operation under higher field and higher stress environment still remains to be confirmed. Tests to confirm the performance need to be implemented in the engineering design phase anticipated after the completion of the DEMO conceptual design in around 2025. Regarding fabrication aspect, heavy industrial companies greatly concern about tolerances in TF coil fabrication on DEMO. Our recent assessment indicates that, with efficient use of external correction coils, tolerances of TF coils can be mitigated by a factor of 2.5-4.

Acknowledgements

This work was carried out in the framework of the activity of the Joint Special Design Team for Fusion DEMO and partly by the DEMO Design Activity under the Broader Approach. We would like to express cordial gratitude to experts of QST, NIFS, universities and manufacturing companies for their valuable advices and helpful support to the Special Design Team. This work was partly supported by KAKENHI Grant-in-Aid for Scientific Research (C), No.17K07002.

References

[1] https://www.iter.org
[2] Okano K *et al* 2018 *Fusion Eng. Des.* **136** 183
[3] Yamada H *et al* 2016 *Fusion Eng. Des.* **109-111** 1318
[4] Sakamoto Y *et al* 2014 *IAEA 25th Fusion Energy Conf.*, FIR/3-4Rb, St. Petersburg
[5] Tobita K *et al* 2018 *Fusion Eng. Des.* **136** 1024
[6] Utoh H *et al* 2010 *J. Plasma Fusion Res. SERIES* **9** 304
[7] Tobita K *et al* 2007 *Nucl. Fusion* **47** 892
[8] Hiwatari R *et al* 2005 *Nucl Fusion* **45** 96
[9] Hamada K *et al* 2007 *Fusion Eng. Des.* **82** 1481
[10] Ono,Y *et al* 2017 *95th Meeting of Cryogenics and Superconductivity Society of Japan*, 2C-p04 (in Japanese) https://www.csj.or.jp/conference/2017a/2C.pdf
[11] La Haye R J *et al* 1997 *General Atomics Report* GA-A 24000
[12] Scoville T J *et al* 1991 *Nucl. Fusion* **31** 875

IOP Conf. Series: Journal of Physics: Conf. Series **1293** (2019) 012079 doi:10.1088/1742-6596/1293/1/012079

Removal of Scale from Feed-water in Thermal Power Plant by Magnetic Separation -Analysis of Oxygenated Treatment Scale-

M Hiramatsu[1], J Yamamoto[1], Y Akiyama[1], F Mishima[2], S Nishijima[2], H Okada[3], N Hirota[3], T Yamaji[4], H Matsuura[4], S Namba[4], T Sekine[5], Y Kobayashi[6] and M Seto[6]

[1] Graduate School of Engineering, Osaka University, Osaka, Japan
[2] Fukui University of Technology, Fukui, Japan
[3] National Institute for Materials Science, Ibaraki, Japan
[4] Shikoku Research Institute Inc., Kagawa, Japan
[5] Ebara Industrial Cleaning Co., Ltd., Kanagawa, Japan
[6] Institute for Integrated Radiation and Nuclear Science, Kyoto University, Osaka, Japan

E-mail : hiramatsu@qb.see.eng.osaka-u.ac.jp

Abstract. One of the reasons of the deterioration of the power generation efficiency of the thermal power generation is the adhesion of the iron scale generated by the corrosion of the pipe to the inner wall of the feed-water system. Focusing on the magnetic property of scale, we studied on the removal of scales in the feed-water system by high gradient magnetic separation (HGMS) using the superconducting magnet. In this study, we targeted the thermal power plants adopting oxygenated treatment (OT). In order to determine the installation site and the magnetic separation condition of the HGMS system, we analyzed the OT scale and then clarified the aggregation/composite states by magnetic separation experiment. It is concluded that low-pressure feed-water heater drain is suitable for installation site of HGMS system. In addition, it was clarified that homogeneous aggregation particles existed in the OT scale.

1. Introduction

The technology to prevent thermal power plant from deterioration of generation efficiency is needed in order to reduce CO_2 emissions. One of the reasons for the decline is adhesion of the iron scale generated by corrosion on the piping inner wall in the feed-water system. The thermal conductivity of the scale is less than 10% of the piping material, so the scale adhered to the inner wall of the piping causes a drop in heat exchange efficiency. Chemical feed-water treatment such as total volatile treatment (AVT) and oxygenated treatment (OT) is performed in order to suppress the scale generation. However, these methods cannot prevent the scale generation completely. Focusing on the magnetic property of scale, we studied on the removal of scales in the feed-water system by high gradient magnetic separation (HGMS) using the superconducting magnet. In this study, we targeted the thermal power plants adopting OT which is increasing in the power plants with large capacity in recent years.

To determine the magnetic separation condition, analysis of the magnetic properties and particle size of the scale is necessary. The main component of the AVT scale is ferromagnetic magnetite, whereas the main component of OT scale is paramagnetic hematite [1], and also include ferromagnetic magnetite The aggregation state of the mixture of the paramagnetic and the ferromagnetic particles are divided into homogeneous and heterogeneous aggregation. Homogeneous aggregation is the state in which paramagnetic and ferromagnetic particles are separately aggregated, whereas heterogeneous aggregation is the state in which the paramagnetic particles and the ferromagnetic particles are aggregated each other. Since the magnetic separation efficiency depends on the aggregation state of the scale, it greatly influences the determination of the magnetic separation condition. In order to introduce HGMS device to the thermal power plant, it is necessary to minimize the processing cost and the influence of the magnetic field on the surrounding apparatuses. Therefore, it is important to select an appropriate magnetic separation condition. To clarify the aggregation state of the scale, we analyzed the scale sampled from the feed-water in the actual thermal power plant. Then we conducted the small-scale magnetic separation experiment with the superconducting bulk magnet using the actual sampled scale and the mixed sample simulating the actual scale contents. The aggregation state of the actual scale was considered by comparing these results.

2. High gradient magnetic separation (HGMS)

The high gradient magnetic separation (HGMS) is the method of separating an object applying the magnetic force by high magnetic field gradient generated around magnetic filters. The magnetic filter is installed in the magnet bore and magnetized. The magnetic force $\mathbf{F_M}$ (equation 1) and the drag force $\mathbf{F_D}$ (equation 2) act on the separation target particles. The magnetic force $\mathbf{F_M}$ is the force that attracts the separation target particles to the magnetic filter. Drag force $\mathbf{F_D}$ is the force of the resistance to the movement of the particles in the fluid.

$$\boldsymbol{F_M} = \frac{4}{3}\pi r^3 \frac{\chi}{\mu_0}(\boldsymbol{B}\cdot\nabla)\boldsymbol{B} \tag{1}$$

$$\boldsymbol{F_D} = 6\pi\eta r(\boldsymbol{v_f} - \boldsymbol{v_p}) \tag{2}$$

Here, r is the particle radius, χ is the magnetic susceptibility of the particle, $\boldsymbol{B}$ is the magnetic field strength, η is the viscosity of the solvent, $\boldsymbol{v_f}$ is the velocity of the fluid and $\boldsymbol{v_p}$ is the particle velocity. It is possible to estimate whether particles can be trapped by the resultant force of magnetic force and drag force. From the above two equations, it can be seen that the magnetic separation condition greatly differs depending on the particle size r and magnetic susceptibility χ of the scale.

As noted above, homogeneous aggregation and heterogeneous aggregation are considered in the mixture of the paramagnetic and the ferromagnetic particles [2]. In the case of OT scale targeted in this research, homogeneous aggregation means that paramagnetic and ferromagnetic particles are aggregated independently, whereas heterogeneous aggregation means that paramagnetic and ferromagnetic particles are aggregated with each other. In the former case, it is easy to capture the ferromagnetic particles, but to capture the paramagnetic particles, high magnetic field and high magnetic field gradient, corresponding to $(\boldsymbol{B}\cdot\nabla)\boldsymbol{B}$ in equation 1, is required. It is possible to capture paramagnetic particles using a high magnetic field and high gradient. However, as the applied magnetic field becomes higher, the amount of electric power required to start-up the superconducting magnet increases. In addition, since the range of the influence of the magnetic field on the surrounding devices also increases, the large space is required for the introduction site. Whereas in the latter case, the apparent magnetic susceptibility of the particles is larger than that of the paramagnetic particles and the magnetization χ of equation 1 increases. Therefore paramagnetic and ferromagnetic particles can be separated with the relatively low applied magnetic field.

Moreover, in the feed-water system of the thermal power plant, the scale is generated at high temperature and pressure (50 to 290 °C, from 2 to 30 MPa). Hence the particles might be compounded by chemical bonding such as dehydration condensation by heating. In the case that the particles are compounded among the different kinds of particles, dispersion of aggregates due to the flow of the feed-water does not occur, which is the more favorable condition for magnetic separation.

3. Experiment

3.1. Analysis of actual scale

The scale was sampled at 6 sampling points in the feed-water system of the power plant adopting OT. To clarify the magnetic property of the actual scale, the composition was analyzed with Mössbauer spectrometry and magnetization were analyzed by physical property measurement system (quantum design Inc.). Figure 1 shows the sampling points in the feed-water system of the thermal power plant. The sampling points were as follows; (1)outlet of condensate pump, (2)outlet of condensate demineralize, (3)drain of low-pressure feed-water heater, (4)inlet of deaerator, (5)storage tank of deaerator, (6)inlet of the economizer.

Figure 2 and 3 show composition and the magnetization of the actual scales at 6 sampling points respectively. The composition of the scale was different depending on the sampling points. The scale sampled from the low-pressure feed-water heater drain most contained the ferromagnetic magnetite and maghemite compared with the scales of other sampling points. In addition, the sample showed the highest magnetic susceptibility. Therefore, we assumed the introduction of HGMS system into low-pressure feed-water heater drain. In order to clarify the aggregation and composite states of the ferromagnetic and the paramagnetic particles mentioned in the previous chapter, lab-scale magnetic separation experiments by superconducting magnet were conducted. The experiments were conducted both for the actual scale sampled at low-pressure feed-water heater drain and the sample simulating it by mixing iron oxide and hydroxide.

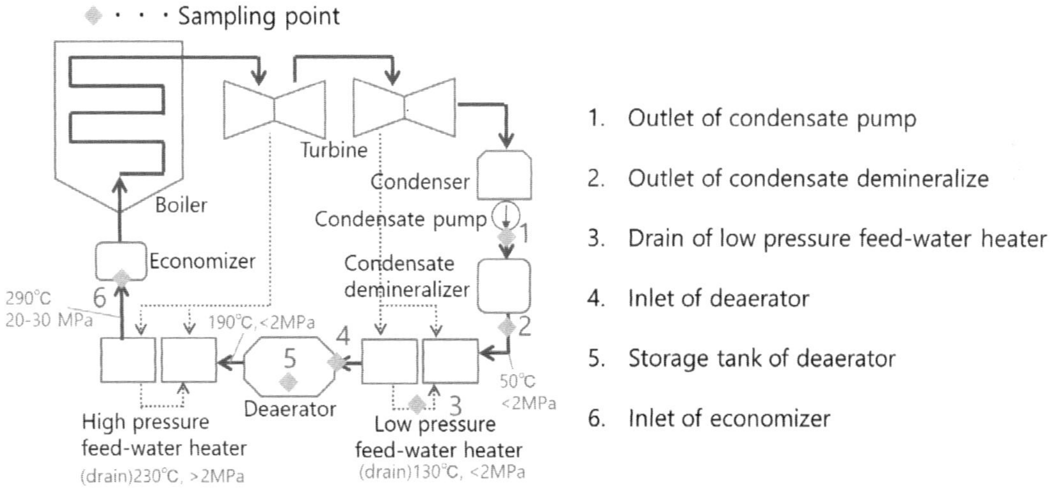

Figure 1. The sampling points in the feed-water system of the thermal power generation.

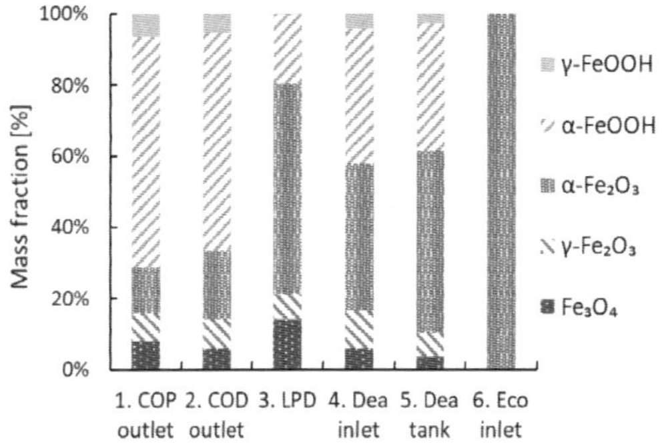

Figure 2. The composition of the actual scales at 6 sampling points estimated by Mössbauer spectrometry.

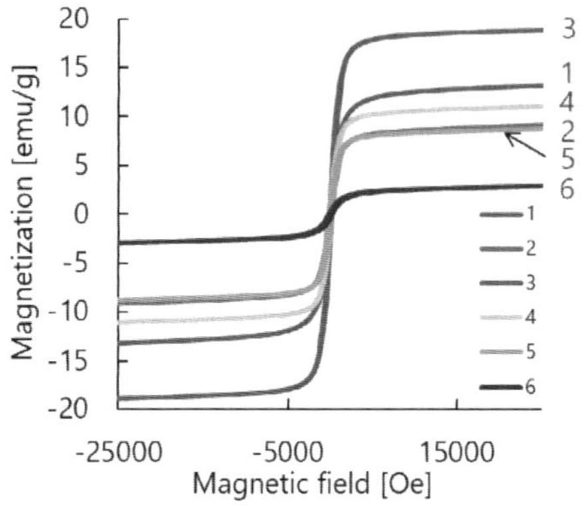

Figure 3. The magnetization curves of the actual scales at 6 sampling points.

3.2. Magnetic separation of actual scale and simulated scale

Figure 4 and 5 show the whole view of magnetic separation experiment and magnetic filter respectively. When we introduce HGMS system in the thermal power plant, we are planning to use the superconducting solenoidal magnet which can apply the strong magnetic field to the large area. However, the purpose of this experiment is the investigation of scale aggregation and small-scale of several hundred mg. Therefore superconducting bulk magnet with GdBaCuO bulk superconductor (Nippon Steel Corp., φ60 mm, Thickness 20 mm) was used to apply the magnetic field. The superconducting bulk magnet was cooled using the GM refrigerator (Aisin Seiki Co., Ltd.) until the temperature of the cryostat reached 30 K. The sample suspension was sent by the constant feed pump (Tokyo Rikakikai Co., Ltd., RP-1000) in the tube flow path. It passed through the magnetic filter on the superconducting bulk magnet in the tube. The particles not captured by the magnetic filter and fluid were recovered in the outlet container.

Table 1 shows the filter conditions and table 2 shows the experimental conditions. 1 g of demister (SUS 430) with a wire linear 0.1 mm was used as a filter (filling rate 23.6%). The flow velocity was adjusted with the constant feed pump and set at 87 mm/s. The simulated scale was prepared by mixing commercial samples simulating the composition of the low-pressure feed-water heater drain (table 3). According to calculation using Equation 1 and 2, the paramagnetic particles (α-Fe_2O_3, α-FeOOH and γ-FeOOH, 0.1 μm- 5.5 μm) in the sample cannot be captured at 87 mm/s. The concentration of suspension is 100 ppm, and the amount of feed-water is 1 L. After magnetic separation, the weight of captured particles and particles not captured was measured to investigate separation efficiency. The composition of the captured scale was also analyzed with Mössbauer spectrometry.

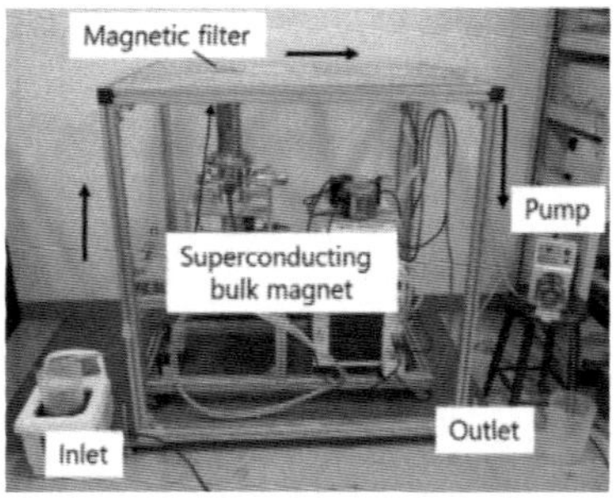

Figure 4. The whole view of magnetic separation experiment.

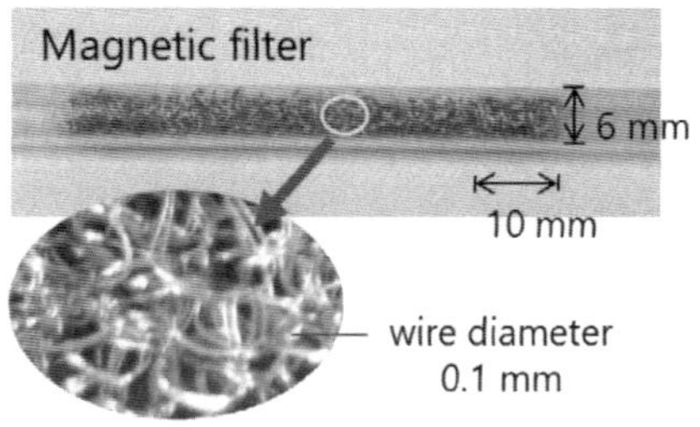

Figure 5. The magnetic filter used in the small-scale separation experiment.

Table 1. Filter condition.

Filter shape	Wavy mesh (Demister)
Filter material	SUS430
Filter wire diameter	0.1 mm
Filter weight	1 g
Filter filling rate	23.6 %

Table 2. Experimental condition.

Flow velocity	87 mm/s
Flow path diameter	6 mm
Applied magnetic flux density	About 3 T
Suspension concentration	100 ppm
Volume of feed-water	1 L

Table 3. The composition of the simulated sample.

	Fe_3O_4	γ-Fe_2O_3	α-Fe_2O_3	α-FeOOH	γ-FeOOH
Company, Brand name	Mitsui Mining & Smelting Co., Ltd., Sample 2	CIK Nano Tek corp.,	Toda Kogyo corp., 180ED	Toda Kogyo corp., Y-313	Kanto Chemical Co., Ltd., 16024-02
Primary particle diameter [μm]	0.1	0.03	0.5	-	-
Composition [%]	14.3	7.00	59.0	19.3	0.400

4. Result and discussion

4.1. The result of magnetic separation

Figure 6 shows the separation efficiency and the composition of captured particles. Figure 7 shows the magnetization curves of the captured particles. The composition of simulated scale before separation was a little different by error of weighing at sample preparation. However both the simulated and actual scale showed the same tendency in the composition and the magnetization.

In this experiment, the flow velocity was set to a condition that paramagnetic particles could not be captured. Nevertheless, paramagnetic particles were captured and total capture rate was about 50%. Two reasons are considered heterogeneous aggregation of particles, or paramagnetic particles involved by the ferromagnetic particles. In order to discuss the former heterogeneous aggregation, zeta potential of each component was measured. Table 4 shows the zeta potential (in the distilled water, pH 5.7) of each component in the sample. The zeta potential at pH 5.7 at which magnetic separation experiment was conducted in all components showed a positive value. Therefore, it is considered that the particles were difficult to aggregate due to electrical repulsion between the particles. However, the zeta potential of hematite shows a slightly smaller value than the other components. It indicates that heterogeneous aggregation may occur partly between hematite and other components. It is considered that the apparent magnetic susceptibility of the particles increased due to heterogeneous aggregation of hematite and ferromagnetic particles and it was captured by the filter. In the latter case, it has been reported that non-target particles were captured in the process of attracting the ferromagnetic particles to the magnet [3]. The ferromagnetic particles aggregate each other in a chain-like form under magnetic field [4]. In addition, the ferromagnetic particles are strongly attracted to the filter and are immediately captured by the filter. Therefore, when ferromagnetic particles with chain-like form were immediately attracted to the filter, part of the paramagnetic particles get caught in the chain and captured together. It is also possible that the magnetic field gradient near the filter increases due to capturing of the ferromagnetic particles to the filter and the paramagnetic particles was captured. However, based on preliminary experiments, its influence is considered to be small.

However, it is considered that aggregation of the particles hardly occurs from the result of zeta potential, showing that all the particles are positively charged. In addition, the paramagnetic particles captured in this experiment was about 30-40 wt.% of total contents. Therefore, it is considered that there were also paramagnetic particles in monodispersed or homogeneous aggregation states.

Also from figure 6, it can be seen that the capture rates showed similar results in actual and simulated scale, indicating the aggregation state of the particles of the simulated and the actual scale were similar, and it is considered that there was almost no composite state in the particles. Figure 7 shows the distribution of magnetic susceptibility and particle size of the actual scale. In this method, the volume magnetic susceptibility of each particle was measured by monitoring the magnetophoresis velocity of the particles, and each particle size was measured by image analysis. In this experiment, the actual scale was sufficiently dispersed in the glycerin solution. A tendency that the distribution of the magnetic susceptibility was divided into high and low ones, confirming the result of magnetic separation experiment.

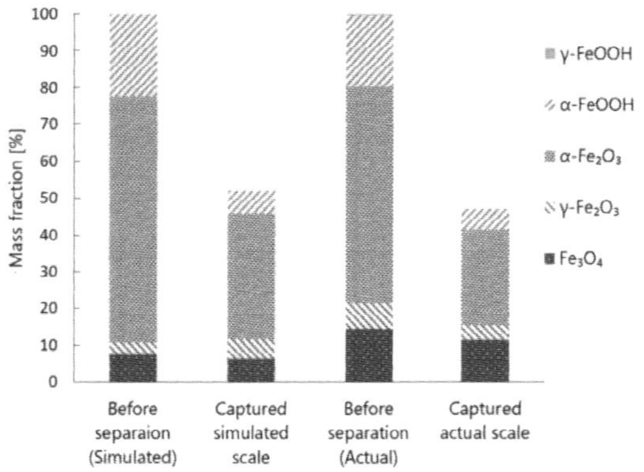

Figure 6. The separation efficiency and the composition of captured particles.

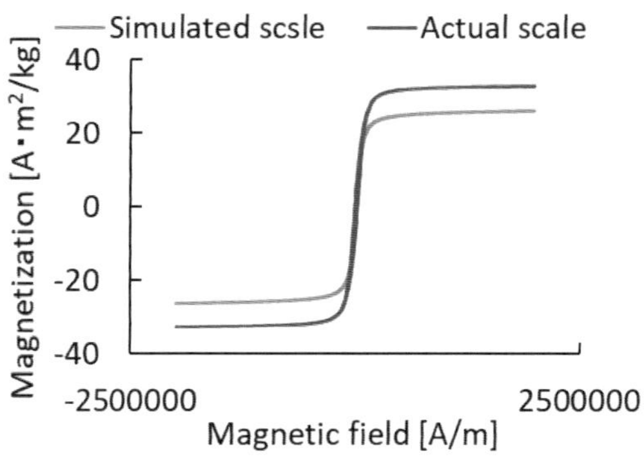

Figure 7. Magnetization curve of captured particles.

Table 4. Zeta potential of each component [mV].

Fe_3O_4	γ-Fe_2O_3	α-Fe_2O_3	α-FeOOH	γ-FeOOH
27.2	26.1	17.0	21.6	29.9

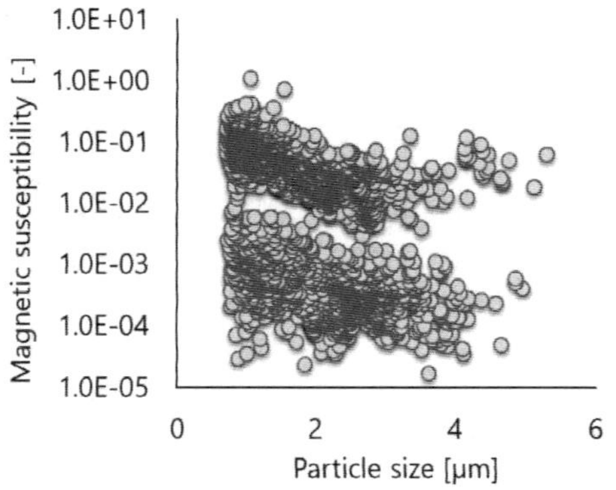

Figure 8. Distribution of magnetic susceptibility and particle size.

4.2. Proposal of HGMS system

The result in section 4.1 showed that there were almost no particles in the composite state in OT scale. Moreover, there were both non-heterogeneous and heterogeneous aggregation particles. In thermal power plants which OT is applied, scale adhering to pipework are removed by chemical cleaning approximately every 10 years. However, one chemical cleaning costs ordinary tens of millions of yen and a three-month shutdown of the power plant. By introducing the HGMS system, it is possible to postpone the interval of chemical cleaning. In order to extend the interval as much as possible, we set it to 80 % as the immediate goal. The reason is as follows. In the previous study, we conducted the magnetic separation experiment simulated the thermal power plant adopting AVT. That experiment was one-seventh of actual power plant and its capture rate was 91.5%. OT scale contain over 80 % of paramagnetic particles and that particle size is smaller than AVT scale, so capturing OT scale isn't easy unlike AVT scale. Nevertheless, 80 % of removal rate is considered to extend the interval of chemical cleaning. Consequently, we need to design a device that can also capture paramagnetic particles in monodispersed or homogeneous aggregation states.

In order to capture paramagnetic particles, large magnetic field its gradient is required. We propose the HGMS system that the ferromagnetic particles and the paramagnetic particles can be captured separately in multiple-stages with different filters. Firstly, ferromagnetic particles are captured with rough mesh filters arranged at the inlet side, and then paramagnetic particles are captured with fine mesh filters arranged at the outlet side. By applying such filter composition, we can capture OT scale with one superconducting magnet system.

5. Conclusion

We analyzed OT scale and clarified the aggregation/composite states by magnetic separation experiment, in order to determine the installation site of HGMS system and the magnetic separation condition in the thermal power plant adopting OT. Composition analysis and magnetization measurement of the scale sampled at six points in the feed-water system were carried out. As a result, the ratio of ferromagnetic particles and the magnetization was the highest in the scale sampled from low-pressure feed-water heater drain. Hence, it is considered that low-pressure feed-water heater drain is the most suitable installation site of HGMS.

In order to clarify the aggregation/composite states of scale, the magnetic separation experiment was conducted using the scale sampled from low-pressure feed-water heater drain and its simulated sample. Both heterogeneous and not heterogeneous aggregation between ferromagnetic and paramagnetic particles are present both in the actual and simulated OT scale. In addition, there are almost no composite state of particles. Therefore, we concluded that it is necessary to design a device capable of capturing single paramagnetic particles with small magnetic susceptibility for removal of the OT scale from feed-water in the thermal power plant.

Acknowledgments

This work is partly supported by Advanced Low Carbon Technology Research and Development Program (ALCA) of JST Strategic Basic Research Programs (JPMJAL1304).

I would like to thank Dr. Tomoyuki Terai in Graduate School of Engineering, Osaka University for measuring magnetization. I would like to thank Mr. Makoto Kawano in Kawano Lab. Inc. for measuring the distribution of magnetic susceptibility and particle size.

References

[1] Tsubakizaki S et al. 2012 *Mitsubishi Heavy Industries technical report* **49** 25-29
[2] Mori S et al. 1994 *shigen-to-Sozai* **110** 325-329
[3] Iwatsuki M and Fukasawa T 1992 *Journal of Japan Society of Air Pollution* **27 (4)** 220-226
[4] Hirota N et al. 2017 *Magnetism and Magnetic Materials* **427** 296-299

IOP Conf. Series: Journal of Physics: Conf. Series **1293** (2019) 012080 doi:10.1088/1742-6596/1293/1/012080

Remediation of Groundwater Contaminated by Heavy Metals Using Magnetic Separation Technique

A J Amosse[1] and Y Akiyama[1]

[1] Graduate School of Engineering, Osaka University, 2-1 Yamadaoka, Suita, Osaka, Japan
E-mail: amosse@qb.see.eng.osaka-u.ac.jp

Abstract. Heavy metals pollution has become one of the most serious environmental problems nowadays. Studies on the environmental pollution indicate that water pollution by heavy metals ions has reached unprecedented levels over the past decade in many countries, with highlights to African countries. Removal of these toxic metals from contaminated water can improve human being life quality and prevent from their adverse effects. This study aims to text the ability of magnetic activated carbon (MAC) and MAC coated with wastes of colloidal silica (MACSi) for adsorption of copper, manganese and zinc ions from water and its separation from water using magnetic separation technique. The adsorption results of MAC and MACSi showed that MACSi was more effective to adsorb the ions than MAC. FT-IR spectrum and XRD pattern of MACSi indicated that silica coating onto MAC surface is in amorphous form.

1. Introduction

Heavy metals are among the wide diversity of contaminants affecting surface and groundwater. Unlike other pollutants, heavy metals are not biodegradable and cannot be decomposed. Therefore, effective methods are needed to remove and detect these toxic metals from water [1]. In recent decades, a large number of researches have been focused on the effective removal methods of heavy metals ions from water. Adsorption method is considered as an efficient and economical technique because of its simple design and facile handling process [2]. Activated carbon has been one of the widely used adsorbent for contaminants adsorption for a long time, but it suffers from a lack of structural and functional stability, low adsorption capacity and limitation to absorb heavy metals contaminants from aqueous solutions [3]. Therefore modification of activated carbon surface with functional groups to enhance its adsorption capacity for heavy metals removal attracts researchers' attention, which is also of great practical value in the future of water treatment.

In this research we focused on colloidal silica as surface modification agent. Colloidal silica is used as a pure polishing agent but from the view point of dispersion stability unused colloidal silica is discharged when they are expired. Colloidal silica possesses an abundant number of silanol groups for surface modification and functionalization [4]. Modification of activated carbon with silane coupling agent has been reported to exhibit high adsorption capacity for heavy metals from wastewater when compared to the materials used traditionally and commercially as adsorbents [5]. In addition, adsorbents with ferromagnetic proprieties have many advantages among others, they can reduce secondary wastes because they can be recovered by the magnetic force and can be used in a high throughput and large scale-treatment.

The objective of present study, is to use non-conventional silica reagent (industrial wastes of colloidal from the expired polishing agent) along with magnetic activated carbon to prepare a low cost ferromagnetic adsorbent and test its ability for heavy metals cations adsorption and separate from solution by using a neodymium magnet.

2. Magnetic separation technique

Magnetic separation technique for a long time has stood out as an interesting and important physical driving force to separate magnetic from non-magnetic components of the mixture [6]. Magnetism is a unique physical property that independently helps in water treatment by influencing the physical properties of contaminants in water. In addition, its combination with other processes enables an improvised efficient treatment technology. The magnetic force, $\mathbf{F_M}$ (equation 1) represents the force that attracts the target particles to the magnet, r is the radius of the particles to remove, μ_0 denotes permeability constant of the vacuum, $\mathbf{M}$ is magnetization of the particles to remove and $\mathbf{H}$ is the strength of a magnetic field. The magnetic field strength to be applied in the water treatment will be determined by the radius and magnetization of the particles to remove.

$$\mathbf{F_M} = \frac{4}{3}\pi r^3 \mu_0 (\mathbf{M} \cdot \nabla)\mathbf{H} \qquad (1)$$

There are many types of magnetic separation for different target pollutants. In 1977 Parker categorized them in four different categories, low intensity dry magnetic separation, low intensity wet magnetic separation, high intensity dry magnetic separation, and high intensity wet magnetic separation, also known as "high gradient magnetic separation" (HGMS) [10]. For large scale treatment in this study, we suggest a low intensity wet magnetic system depending on the magnetism and particle size of the adsorbent with a superconducting magnet or a neodymium magnet shown in Figure 1.

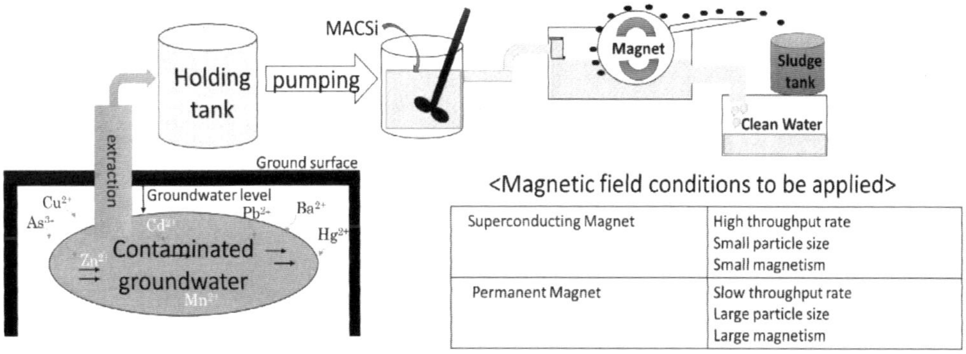

Figure 1. Magnetic separation principle for heavy metals removal from contaminated groundwater.

3. Material and Methods

Copper sulphate pentahydrate ($CuSO_4 \cdot 5H_2O$), manganese sulphate pentahydrate ($MnSO_4 \cdot 5H_2O$) and zinc sulphate heptahydrate ($ZnSO_4 \cdot 7H_2O$) used in this study were of analytical grade (Fuji Film Wako Junyaku, Japan). Industrial wastes of colloidal silica were obtained from a polishing industry (Maeda Material Co. Ltd., Japan). Magnetic activated carbon (MAC) were provided from Osaka Chemical, Japan.

Contaminated water with heavy metals were artificially prepared dissolving the exact amount of salts above mentioned to make solution with 10 ppm of each ion. The concentrations of metal ions were determined by ICP-AES (ICPS-7500 Shimadzu, Japan). FT-IR was obtained as KBr pellet in the range of 4000 cm^{-1} to 400 cm^{-1} using Fourier-Transform Infrared spectroscopy (FT/IR-470 Plus, Jasco, Japan) and X-ray Power Diffraction measurements were performed using PAnalytical Xpert^3 Powder with Cu Kα radiation in the range 2θ from 10 to 80 at room temperature (Malvern Panalytical, United Kingdom).

3.1. Preparation of MACSi adsorbent

The magnetic activated carbon coated with industrial wastes of colloidal silica adsorbent (MACSi) were prepared using two different methods, namely silica coating under room temperature (~24 °C) and under hydrothermal reaction (~120 °C) with the objective to check which method promotes effective interaction between MAC and colloidal silica. Coating of silica onto MAC surface is expected to increase the surface area and promote more active sites for heavy metals cations adsorption as represented in the Figure 2. For room temperature 0.1 g of MAC was mixed with 2 mL of colloidal silica and shook for 24 h under 160 rpm using a shaker (Incubate SHK-101B, Iwaki Glass Co. LTD, Japan) then filtrated and allowed to dry at room temperature before being used.

Under hydrothermal reaction consisted in mixing 0.1 g of MAC with 2mL of colloidal silica, then the mixture was shook for 24 h under 160 rpm using a shaker (Incubate SHK-101B, Iwaki Glass Co. LTD, Japan). Next, the mixture was introduced into a Polytetrafluoroethylene (PTFE) lined stainless steel and introduced to reactor under 120 °C for 24 h. Then the mixture was removed from the reactor, cooled, washed three times with distilled water and dried under vacuum using vacuum oven AVO-200 NS (Misumi-vona, Japan) and it was ready to be used in batch adsorption test.

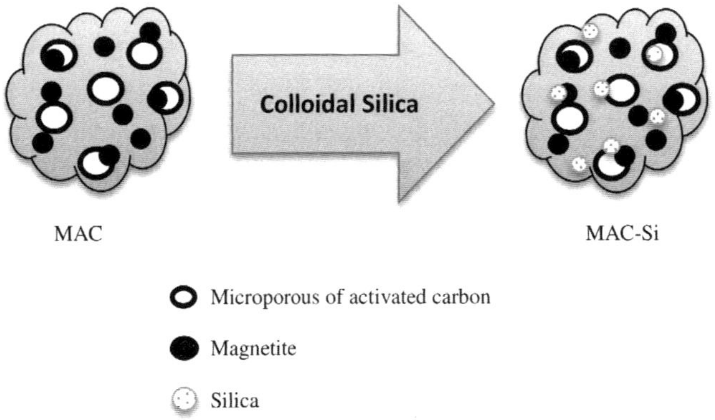

Figure 2. Schematic diagram of MAC surface modification into MACSi.

3.2. Batch Removal Experiments

Adsorption of heavy metals cations, Cu^{2+}, Mn^{2+} and Zn^{2+} was carried out at pH 6 in room temperature (about 23 °C) using both MAC (as standard adsorbent) and MACSi adsorbent prepared under room temperature and hydrothermal reaction. Same quantity of MAC and MACSi, 0.05g was added in two different bottles containing each 50 mL of contaminated water with co-existence of the above mentioned heavy metals cations, each in the concentration of 10 ppm. Then shook for 3h and the adsorbents were separated from the water with a permanent magnet (Neodymium magnet; of 0.5 T).

The adsorption using MACSi adsorbent prepared under room temperature showed excess of silica in the supernatant as shown in Figure 3 (a) suggesting that silica is easily desorbed from the micropores of magnetic activated carbon due to weak interaction between both. So, after the separation the supernatant was not clean enough to be analysed by Inductively Coupled Plasma-Atomic Emission Spectroscopy (ICP-AES). On the other hand, Figure 3 (b) shows the supernatant descendant from adsorption using MACSi prepared under hydrothermal which is clean suggesting effective coating of silica onto magnetic activated carbon surface and the residual heavy metals cations concentration were determined by ICP-AES.

Figure 3. (a) Adsorption test using MACSi prepared under room temperature and (b) MACSi prepared under hydrothermal reaction.

4. Results and Discussions

4.1. Adsorbent characterization

The Powder X-ray diffraction (XRD) pattern of MAC and MACSi prepared under hydrothermal condition represented in Figure 4 shows the characteristics peaks of magnetite, confirming that the cubic phase of magnetite particles are still kept after incorporating silica [13]. The broad peak in 2θ =20 °C in addition to magnetite characteristic peaks in XRD pattern of MACSi prepared under hydrothermal reaction suggest that silica coating is in amorphous form. Same results were observed in the study of Emadi et al. [8].

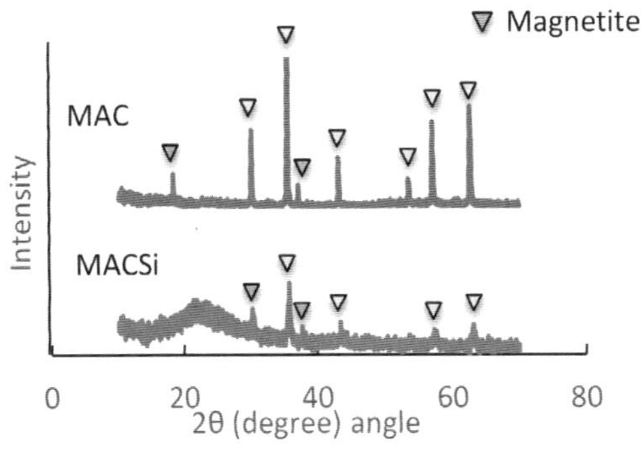

Figure 4. XRD patterns of MAC and MACSi (Prepared under hydrothermal condition).

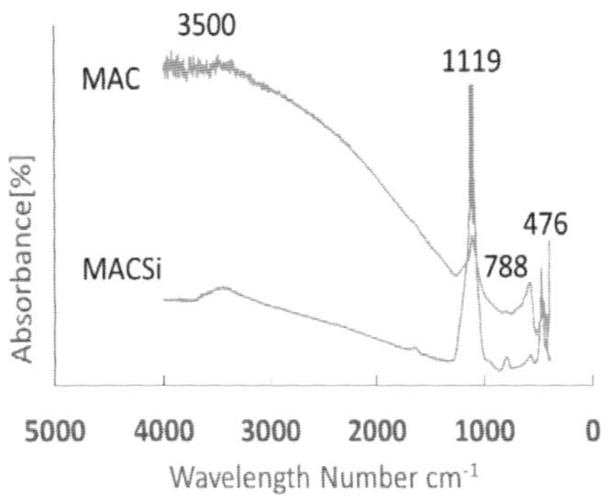

Figure 5. FT-IR results of MAC and MACSi (Prepared under hydrothermal condition).

The FT-IR spectrum represented in Figure 5 correspond to MAC and MACSi. Peaks between 788 and 476 cm^{-1} correspond to both magnetite (Fe_3O_4) and Fe-O-Si vibration. The broad peak in the region of 1119 cm^{-1} in FT-IR spectrum of MACSi indicates the Si-O-Si symmetric stretching of silica phase. The pronounced shoulder in the region of 3500 cm^{-1} in FT-IR spectrum of MACSi indicate OH group present in silica, in form of silanol group (Si-OH). The same stretch vibration were observed in the study of Kong et al., [7]. This suggest that silica was successfully coated onto the surface of MAC.

4.2. Batch adsorption test

The adsorption efficiency (η) was calculated using equation 2, based on ICP-AES measurements, where C_i (ppm) correspond to heavy metal ion initial concentration in the water, C_f (ppm) heavy metal ion residual concentration.

$$\eta = \frac{C_i - C_f}{C_f} x100 \quad (2)$$

The Figure 6 shows the removal ratio of copper, manganese and zinc cations using MAC and MACSi adsorbents, and it suggest that MAC is selective in heavy metals removal as stated in many studies. The removal ratio for copper, manganese and zinc using MAC adsorbent was 83%, 9% and 11% respectively while the removal ratio using MACSi adsorbent was 98%, 87% and 98% for copper, manganese and zinc cations respectively indicating that MACSi adsorbent is effective for heavy metals removal at this pH.

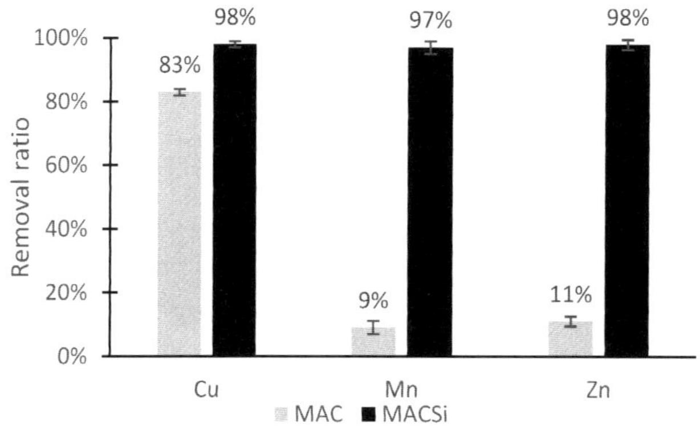

Figure 6. ICP-AES analysis of MAC and MACSi.

This effective heavy metals removal of MACSi can be explained by the enlargement of MAC specific surface area during silica coating which generated more active sites for ions adsorption due to the presence of silica on its surface (Figure 2). The pH of solution is an important controlling parameter for heavy metals adsorption, depending on the pH that occur the adsorption the surface of adsorbent can be negatively or positively charged [9]. As at pH 6 (pH of solution of contaminated water) the surface of silica conjugated in MACSi is negatively charged due to the fact that is greater than its isoelectric point, pH=2 [11]. Thus the negative charges are higher than positive ions as shown in the equation 3, this make effective the electrostatic interactions between the negative charges on the MACSi surface and the tested heavy metals cations.

$$\mathbf{SiOH + OH^- \leftrightarrow Si\text{-}O^- + H_2O} \quad (3)$$

On the other hand, MAC surface is dominated by magnetite at pH 6, which is below magnetite isoelectric point (pH=8) the surface of MAC is positively charged as illustrated in equation 4 [12]. The electrostatic interactions between the MAC surface and heavy metals cations is very weak.

$$\mathbf{Fe\text{-}OH + H^+ \leftrightarrow FeOH_2^+} \quad (4)$$

The significant removal ratio difference between the heavy metals ions using MAC adsorbent it may due to solubility of ions, at pH 6 part of copper ions precipitate as $Cu(OH)_2$ and are physically adsorbed on the microporous of MAC while manganese and zinc precipitate at pH 9.52 and 7.65 respectively.

Table 1: Solubility products and precipitation pH of heavy metal ions

Ions	Solubility product (25°C)	Atomic weight (amu)	Conc. of solution (ppm)	Concentration (mol/l)	Precipitation pH
Cu^{2+}	4.8×10^{-20}	63.55	10	1.57×10^{-4}	6.2
Mn^{2+}	2.0×10^{-13}	54.94	10	1.82×10^{-4}	9.5
Zn^{2+}	3.0×10^{-17}	65.39	10	1.53×10^{-4}	7.7

5. Conclusions

In this study, a new adsorbent for heavy metals cations adsorption (copper, manganese and zinc) were prepared. The method of preparation consisted on coating industrial wastes of colloidal silica onto the surface of magnetic activated carbon. Magnetic character of MAC enables to be recovered from suspension by magnetic separation. Silica coating on MAC surface allows rapid adsorption of heavy metals cations due to the generation of more active sites. The colloidal silica coating was confirmed by XRD technique and it suggests that is in amorphous form. The FT-IR spectrum confirmed the presence of silica by showing Si-O-Si symmetric stretch in the region of 1119 cm^{-1}. Heavy metals adsorption under MAC showed strong affinity to copper with 83% of removal ratio, 9% and 11% to manganese and zinc respectively. Adsorption under MACSi was effective to all heavy metals ions with 98%, 97% and 98% of removal ratio for copper, manganese and zinc respectively. The adsorbent was easily separated from the suspension by a Neodymium magnet of 0.5T. The pH of solution, adsorbent contact time and adsorbent amount dependence are under investigation. Future studies will feature the MACSi adsorption test with a large number of heavy metals and MACSi reusability capacity.

Acknowledgements
Part of this research was conducted as a sponsored research by Japan Society for the Promotion Science (JSPS), No. 17K00598.

References

[1] Momodu M A and Anyakora C A 2010 Heavy metal contamination of groundwater: the surulere case study *Res. J. Environ. and Earth Sci.* vol **2**(1) pp 39-43
[2] Yaoguang W et al 2017 Removal of Pb(II) and methylene blue from aqueous solution by magnetic hydroxyapatite-immobilized oxidized multi-walled carbon nanotubes *J. Colloid Interf. Sci.* vol **494** pp 380-388
[3] Mokhter A M et al 2018 Use of modified colloids and membranes to remove metal ions from contaminated solutions *J. Colloid Interf.* pp 2-19
[4] Hamblin M R, Pinar A and Prow T 2016 Nanoscience in dermatology USA pp 63-64
[5] Kaushal A and Singh S K 2017 Removal of heavy metals by nanoadsorbents: A review *J. Environ. Biotechnol. Res.* vol **6** pp 96-104
[6] Šafařı́k I and Šafaříkováa M 1999 Use of magnetic techniques for the isolation of cells *J. of Chrom. B: Biomed. Sci. Appl.* vol **722** pp 33-53
[7] Kong X et al 2014 Selective removal of heavy metals ions from aqueous solutions with surface functionalized silica nanoparticles by different functional groups *J. Cent. South Univ.* vol **21** pp

3575-3579

[8] Emadi M et al 2013 Removal of zinc from aqueous solutions by magnetic silica sore-shell nanoparticles. *J. Chem.* pp 787682

[9] Berga H E and Roberts W O 2015 Colloidal silica: fundamentals applications *Surfactant Sci. Ser.* vol **131** pp 531

[10] Feng N 2007 et al Heavy metal removal from soils using magnetic separation: 1. laboratory experiments *Clean – Soil, Air, Water* vol **35** pp 362-369

[11] Júnior J A A and Baldo J B 2014 The behavior of zeta potential of silica suspensions *New J. Glass Ceram.* vol **4** pp 29-37

[12] Tombácz E et al 2006 Magnetite in aqueous medium: coating its surface and surface coated with it *Rom. Rep. Phys.* vol **58** pp 281-286

[13] Pasandideh E K et al 2016 Silica-coated magnetite nanoparticles core shell spheres (Fe3O4@SiO2) for natural organic matter removal *J. Environ. Health Sci. Eng.* pp 14-21

IOP Conf. Series: Journal of Physics: Conf. Series **1293** (2019) 012081 doi:10.1088/1742-6596/1293/1/012081

Design and Trial Production of Magnetic Filter for Medical Protein Screening System using High Gradient Magnetic Separation

M. Mori*, M. Kubota, T. Abe, S.B. Kim, and H. Ueda

Graduate School of Natural Science and Technology, Okayama University, 3-1-1, Tsushima Naka, Kita-ku, Okayama 700-8530, Japan.

*Corresponding author. E-mail address: ppbs6nme@s.okayama-u.ac.jp

Abstract. Biomedicine is indispensable for the treatment of various diseases such as cancer and diabetes mellitus. In particular, antibodies utilized by the immune system are believed to represent the future of medical supplies as they offer an effective treatment for diseases with minor side effects. The development of antibody drugs requires a large-scale, sequential, fast separation, and refinement process for the medical proteins. Thus, in the previous study, we proposed a medical protein screening system that is based on high-gradient magnetic separation using a superconducting magnet. In the proposed system, the protein is specifically attached to the magnetic beads and captured by the magnetic force around a magnetic filter generated by a high magnetic field. In this paper, the behaviour of the magnetic beads was investigated using particle trace simulation coupled with electromagnetic and fluid analyses for three types of commercially available magnetic filters. Simulations were performed using the filter stack intervals and number of filters as design parameters. Finally, we designed the suitable filter for capturing the magnetic beads.

1. Introduction

In recent years, the availability of cutting-edge medical technology has led to the rapid development of antibody drugs to meet its ever-increasing demand. It is even anticipated that these drugs will be the most common type of medical care in the future. The antibody drug market is expected to grow by 10% each year and is predicted to expand to $ 8.6 billion in 2025 [1]. An antibody is a type of protein that protects the body by destroying harmful foreign substances. A medicine equipped with this characteristic is called an antibody drug. Maximum benefits and minimal side effects can be realized by utilizing the specificity and immunochemical properties of the antibody and its ability to bind to a specific antigen, thus establishing the future face of medical care. However, this method has disadvantages such as limited and expensive doses, indicating that there will be an increase in the demand with future improvements. The processes of separation and purification are indispensable for the production of antibody drugs. However, it is difficult to manufacture large quantities of inexpensive antibody drugs using the antibody separation and purification techniques that are currently available. Thus, in order to meet the increasing demand, there is an urgent need to develop a device that is capable of separating and purifying the antibodies at a constant, high-speed rate. In the previous studies, we have been working on research and development of an antibody separation and purification device that uses the magnetic force from a superconducting magnet to achieve a high separation efficiency of 1 g/h as compared to an

IOP Conf. Series: Journal of Physics: Conf. Series **1293** (2019) 012081 doi:10.1088/1742-6596/1293/1/012081

efficiency of a few mg/h seen in conventional chromatography [2][3]. We had developed a desktop-type high-gradient magnetic separation system that achieved the trapping and separation of magnetic nano-beads with a high-gradient magnetic force [4-6].

2. High-gradient magnetic separation

The separation of antibodies is a crucial step in the production of antibody drugs. Antibodies attached to a carrier, referred to as affinity beads, are recovered using affinity chromatography and magnetic separation for the non-magnetic and magnetic beads, respectively.

The probability of antibody binding dramatically increases when the bead size is reduced by a tenth as this increases the number of beads by 10^3, while maintaining the same mass and increasing the surface area. In addition, a high-concentration product can be obtained as the magnetic separation method is also capable of separating nano-beads using magnetic force. The purification process using high reaction efficiency is simple and quick. Based on the above, we designed this study investigating magnetic separation using affinity magnetic beads. The magnetic force, $\boldsymbol{F}_{mag}$, on the magnetic beads can be calculated as:

$$\boldsymbol{F}_{mag} = V_p(\boldsymbol{M}_s \cdot \boldsymbol{\nabla})H \tag{1}$$

where V_p and $\boldsymbol{M}_s$ are the volume and magnetization of the beads, respectively, and $\boldsymbol{\nabla}H$ is the magnetic gradient. The magnetic force can be strengthened by increasing the magnetic gradient. Commercially available magnetic separation systems containing permanent magnets can only generate magnetic fields of up to ~1 T. Such low magnetic gradients are unable to effectively capture the magnetic nano-beads. On the other hand, a superconducting magnet can be used to generate a high magnetic field. A superconducting magnet is convenient to operate as it can be easily demagnetized and excited. Figure 1 shows a schematic of the superconducting high-gradient magnetic separation system, while Figure 2 illustrates the magnetic gradient. A tube containing a filter with stacked thin-wire meshes is passed through the bore of the magnet. First, a magnetic field of 3 T is applied to the magnetic filter using a superconducting magnet, which generates a large magnetic gradient around the filter. Second, the suspension containing the magnetic beads that bind to the antibodies is poured into the magnetic filter. This causes the magnetic force to capture the magnetic beads following demagnetization of the magnetic filter. Finally, the magnetic beads are recovered by flowing the flushing solution.

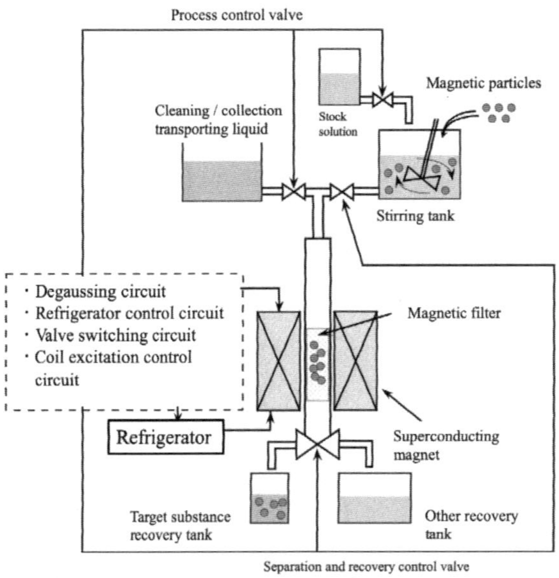

Figure 1. A schematic of the high-gradient magnetic separation system.

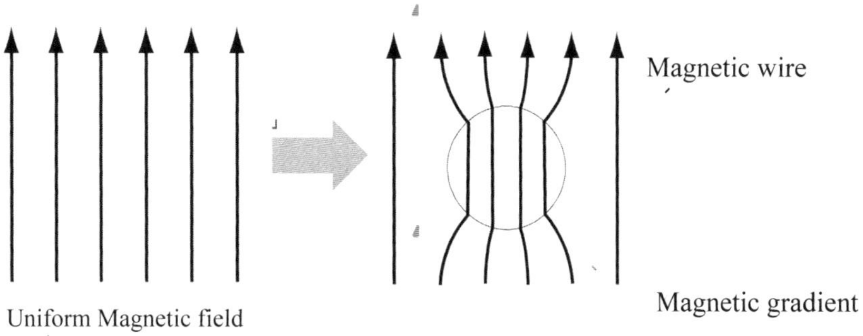

Figure 2. The magnetic gradient generated by the magnetic wire.

The magnetic force of the beads increases with a decrease in the wire diameter as shown in following formula:

$$F_{max} = V_p M_p \left[-4 \frac{\mu_s(\mu_s - 1)}{(\mu_s + 1)^2} \frac{1}{a} \frac{B_0}{\mu_0} \right] \tag{2}$$

where F_{max} is the maximum magnetic force, V_p is the volume of the bead, M_p is the magnetization of beads, μ_s is the relative magnetic permeability of the magnetic wire, a is the radius of the magnetic wire, and B_0 is the external magnetic field. It has also been established in a previous study [7] that the capture ratio of the magnetic beads increases with a decrease in the wire diameter. Based on previous studies, we used the magnetic wire mesh marketed by Kansai Wire Mesh Co. Ltd. to prepare filters that are suitable for the separation of magnetic nano-beads. In our system, the filter consists of a stacked magnetic wire mesh. The parameters of the filter must be optimized to achieve a high capture ratio of the magnetic beads as the capture ratio depends on the arrangement and dimensions of the magnetic wire mesh. In this study, three types of the magnetic wire mesh having a small diameter were selected. The filters in the experiment were designed using the electromagnetic, liquid, and particle tracing analyses methods. Finally, the optimal stack intervals and minimum number of required filter layers were estimated.

3. Analysis of magnetic bead movement

3.1. Analysis model and conditions

The electromagnetic, fluid, and particle tracing analyses methods using COMSOL Multiphysics were applied to a filter with stacked wire mesh, which were selected from the commercially available products of Kansai Wire Mesh Co. Ltd. We have previously confirmed that the capture ratio increases with a decrease in the wire diameter [7]. Thus, three types of wire mesh with a small diameter were selected. Figure 3 shows the analysis model. The S43000 filter was arranged along two dimensions. The periodic boundary condition was applied to the left and right ends of the analytical model. Table 1 shows the specifications of the three types of wire meshes. The wire diameter (*d*) and aperture (*A*) have been defined in Figure 3. The aperture ratio (ε) is determined by the wire diameter and aperture as shown below:

$$\varepsilon = (A/A + d)^2 \times 100 \tag{3}$$

Figure 4 shows the top views of Filter A, B, and C. Table 2 shows the specifications of the particles and suspension. The relative permeability is given by the ratio of the permeability of pure water or magnetic particles to that of vacuum. The particles have specifications corresponding to those of the FG beads (TAS8848N1010) manufactured by Tamagawa Seiki using in the magnetic separation test. In the simulations, the magnetic force distribution was calculated assuming that there is a uniform magnetic field of 3 T along the positive z axis. In addition, for the fluid and particle tracing analyses, it was assumed that the suspension containing the magnetic beads in pure water flows uniformly along the negative z axis. A flow rate of 5 mm/s was assumed, similar to that used in a previous experiment [4].

Table 1. Wire mesh specifications used for the analysis.

	Wire diameter, d (μm)	Aperture, A (μm)	Aperture ratio (%)
Filter A	30	43	34.41
Filter B	35	50	34.41
Filter C	40	62	36.76

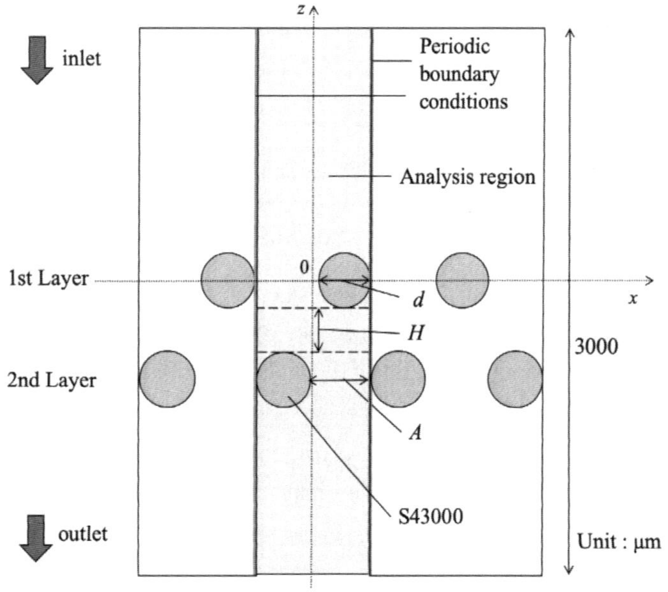

Figure 3. The analysis model, wherein the region of analysis is shown in colour. The periodic boundary condition was applied to the left and right ends of the analysis region.

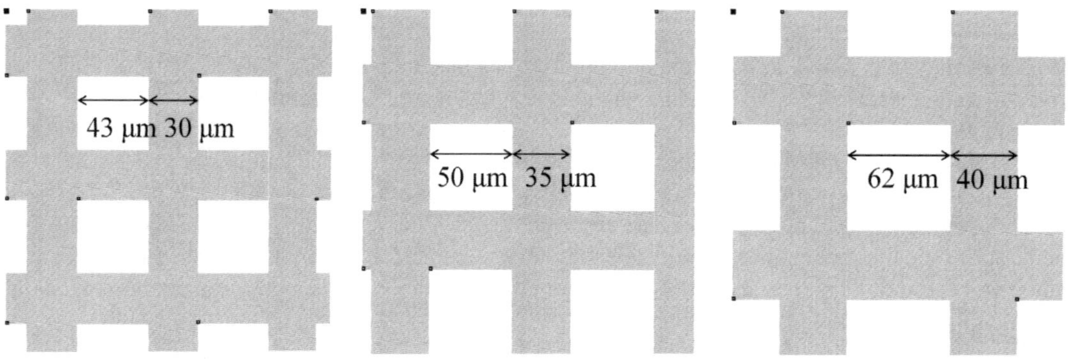

(a) Filter A (b) Filter B (c) Filter C

Figure 4. Top views of the three wire meshes used for the analysis.

Table 2. Specifications of the particles and suspension.

Mass of the particles (pg)	0.01
Diameter of the particles (nm)	200
Relative permeability of the particles	1.01
Relative permeability of the solvent	1
Suspension speed (mm/s)	5
Suspension concentration (mg/ml)	1

3.2. Analysis of the monolayer filter

We first analysed the monolayer filter and verified the capture ratio and range. The capture ratio is defined as:

$$\text{Capture ratio (\%)} = \frac{\text{Captured particles}}{\text{All particles}} \tag{4}$$

The particle traces in the monolayer filter are shown in Figure 5. The capture range (denoted by the red double-headed arrow) is defined as the initial position of the particles finally captured by the magnetic filter. Figure 6 shows the numerical results of the capture ratio and range. The maximum capture ratios of Filter A, B, and C were 20%, 17.6%, 15%, respectively. Filter A had the highest capture ratio and range as the magnetic force increases inversely with respect to the wire diameter (Formula 2). Additionally, the extent of the magnetic force decreases with an increase in the aperture.

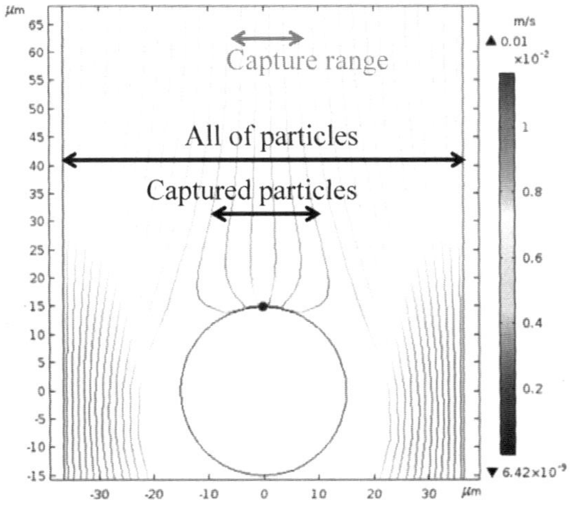

Figure 5. Particle traces for the monolayer filter. The colour scheme of the trace shows the various particle speeds.

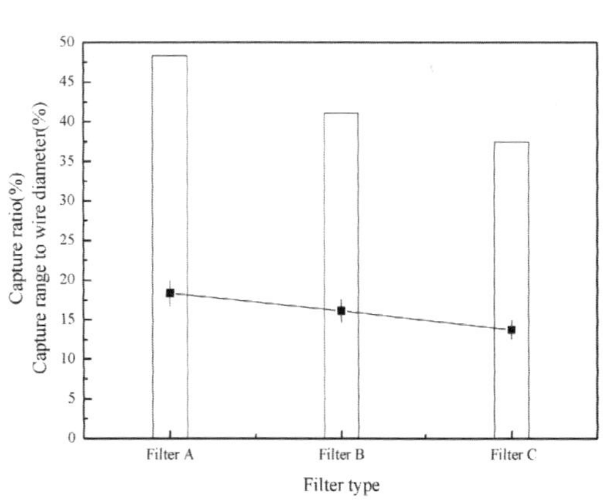

Figure 6. Capture ratios and ranges of filters A, B, and C. The line graph corresponds to the capture ratios while the bar graph shows the capture ranges.

3.3. Analysis of the double-layer filter

The capture ratio was investigated for various stack intervals of the double-layer filter. Figure 7 shows the capture ratio for various stacking intervals. It can be seen from Figure 7 that the capture ratio increases with an increase in the stacking interval, but again decreases beyond a certain interval value. In a small stack interval, the fluid flows through the thin magnetic wires of the first layer with high velocity. This enables the particles to reach the second layer with high speed, thus making the capture of these accelerated particles difficult. The capture ratio subsequently increases with an increase in the filter stack interval. However, there is a decrease in the capture ratio if the filter interval increases by a large margin and exceeds the peak location. Figure 8 shows the traces of the particles released from the same position for stack intervals of 330 μm and 440 μm in Filter A. Figure 9 shows the particle traces that were not affected by the electromagnetic force. It can be seen in Figure 9 that the particle moves gradually along the positive x axis and flows downwards after passing the first-layer filter. This indicates that the particle was not captured when using an interval of 440 μm as it did not pass within the capture range of the next filter.

3.4. Capture ratio in 10-stacked filter and estimated optimum filter

It is necessary to determine the optimal number of layers of the stacked filter to achieve high efficiency in the separation process. The optimal stack intervals with the highest capture ratio were determined in three types of filters in Section 3.3. In this section, the total number of filters required for the capture of all the particles was determined by examining the capture ratio of each layer in a 10-stacked filter with optimal intervals. The stack interval is built by inserting 110-μm-thickness spacers of S43000 between the filters during its production. Thus, the stacking interval was set equal to a multiple of 110 μm. A smaller interval is selected when multiple maximum capture ratios are available for a certain type of filter. This facilitates a faster processing speed leading to high efficiency in the separation process. Figure 10 shows the capture ratio of the particles corresponding to each layer in the 10-layer filter. It can be seen from Figure 10 that the capture ratio of the first and second layers is very high. This indicates that there is already a decrease in the number of particles at the third and subsequent layers, leading to a corresponding decrease in the suspension concentration. The capture ratio of the third and subsequent layers was less than 7%.

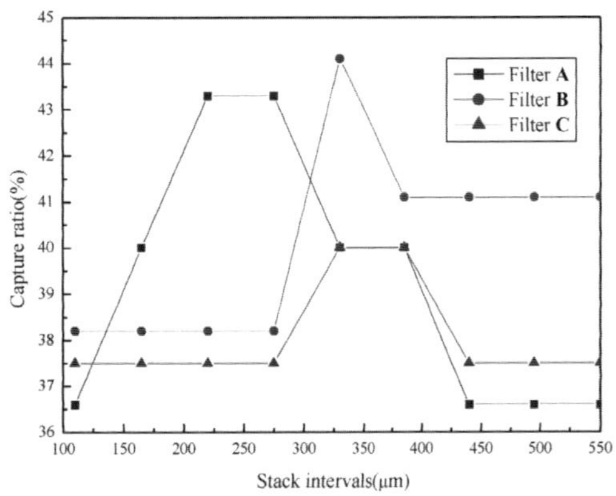

Figure 7. Dependence of the stacking interval on the capture ratio.

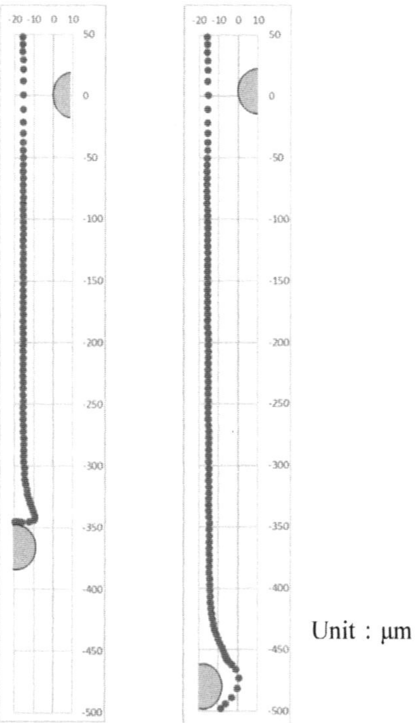

(a) H = 330 μm (b) H = 440 μm

Figure 8. The trajectory of the particles released from the same position for stack intervals of 330 μm and 440 μm in Filter A.

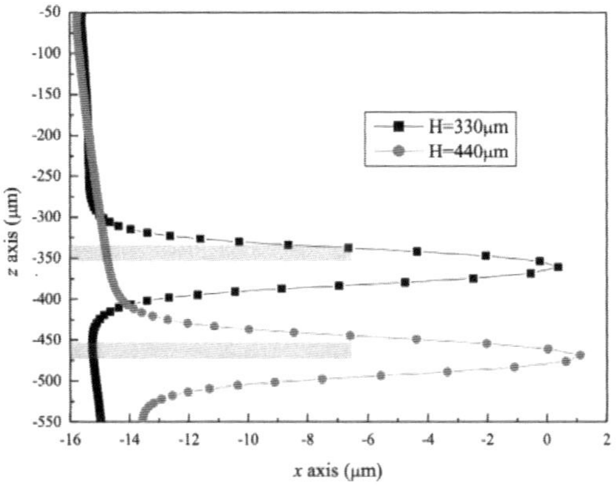

Figure 9. The particle traces unaffected by the electromagnetic force. The bar shows the capture range.

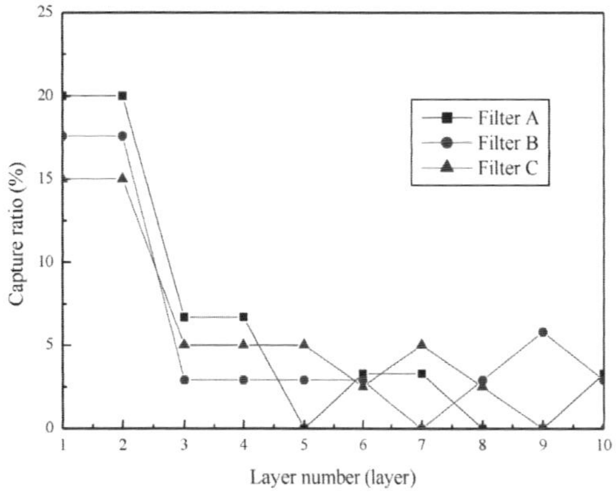

Figure 10. The capture ratios of the particles corresponding to each layer in the 10-layer filter.

Based on the numerical results, the minimum number of filter layers required for capturing all the particles was calculated using the following expression:

$$\text{Required layers (layer)} = \frac{\text{Non-capture ratio of the first and second layers}}{\text{Average capture ratio of the third and subsequent layers}} + 2 \qquad (4)$$

The last term '2' in expression (4) refers to the first and second layers that capture many particles. The first term corresponds to the particles that were not captured in the first and second layers but were captured in the third and subsequent layers. Table 3 shows the optimal stack intervals and the minimum number of filter layers necessary for capturing all the particles. In the cases where two or more optimal stack intervals were estimated, the smaller interval was selected to achieve a higher processing speed. Filter A has the shortest stack interval and the least number of layers among the three type of filters.

Table 3. Estimated number of filter layers and the optimal stacking interval.

	Filter A	Filter B	Filter C
Optimal stack intervals (μm)	220	330	330
Required layers	23	25	25

4. Conclusion

In this study, electromagnetic, fluid, and particle tracing analyses were conducted for the magnetic filters employed in superconducting high gradient magnetic separation experiments. The behaviour of magnetic particles was clarified for a variable number of filters and stack intervals. The capture ratio increased as the filter stack interval was extended and subsequently decreased after crossing a certain peak threshold. For too short stack intervals, the capture ratio is low due to the fluid pressure and increase in velocity around the wire. For large stack intervals that reached or even exceeded the peak value, the fluid drag caused the particles to flow away from the capture range. This shows that it is critical to select an optimal stack interval. Several particles could be captured in the first and second layers of the multi-stacked filters. There is a decrease in the suspension concentration in the third and subsequent layers, with the capture ratio being less than 7% after the third layer. These results were used to calculate the minimum number of filter layers required to capture all the particles. Filter A has the shortest stack interval (220 μm) and the least number of layers (23 layers) among the three type of filters. In the future, the results from this study will be used to prepare the magnetic filter and conduct magnetic separation tests.

References

[1] Market trends in contract drug development and manufacturing market of protein drugs. https://www.seedplanning.co.jp/press/2017/2017090101.html

[2] Ueda H, Agatsuma K, Kajikawa K, Furuse M, Fuchino S, and Ishiyama A 2009 Design and Test of Filter of High Gradient Magnetic Separation System for Trapping Immunoglobulin in Serum *IEEE Trans. Appl. Superoncd.* **19** 2157–2161

[3] Ueda H, Agatsuma K, Fuchino S, Imura T, Furuse M, Kajikawa K, Ishiyama A, Koizumi T Miyake S 2010 Improvement in High Gradient Magnetic Separation System for Trapping Immunoglobulin in Serum *IEEE Trans. Appl. Superoncd.* **20** 949–952

[4] Ueda H, Agatsuma K, Furuse M, Fuchino S, Kajikawa K, Kamioka Y, Iitsuka T, Nakamura S 2014 Performance of Filters in Medical Protein Separation System using Superconducting Magnet *IEEE Trans. Appl. Supercond.* **24** 3700405

[5] Fuchino S, Furuse M, Agatsuma K, Kamioka Y, Iitsuka T, Nakamura S, Ueda H, Kajikawa K, 2014 Development of Superconducting High Gradient Magnetic Separation System for Medical Protein Separation *IEEE Trans. Appl. Supercond.* **24** 4400204

[6] Kajikawa K, Ueda H, Kamioka Y, Agatsuma K, Fuchino S, Furuse M, Iitsuka T, Nakamura S 2014 Development of Degaussing System of Magnetic Filters for High Gradient Magnetic Separation to Improve Recovery Ratio of Trapped Magnetic Nanobeads *IEEE Trans. Appl. Supercond.* **24** 4400305

[7] Kubota M, Mori M, Kim S. B, Ueda H 2018 Numerical Simulation on Behaviour of Magnetic Beads in Magnetic Filter for Medical Protein Screening System using High Gradient Magnetic Separation *IOP Conf. Series: Journal of Physics* **1054** 012085.

Fundamental Study on Cancer Therapy by Blocking Newborn Blood Vessels Using a Rotating Magnetic Field

Makoto Kirimura, Yoko Akiyama

Division of Sustainable Energy and Environmental Engineering, Graduate School of Engineering, Osaka University

E-mail Adress: kirimura@qb.see.eng.osaka-u.ac.jp

Abstract. In this study, we examined a novel cancer therapy by blocking newborn blood vessels with ferromagnetic particles. In this treatment, we try to necrose cancer by accumulating and aggregating the ferromagnetic particles inside the newborn blood vessels. Practically, this method needs to accumulate the particles in the deep part of the body, so the use of superconducting solenoidal magnets is essential. In addition, a design of magnetic field for accumulating the particles selectively in newborn blood vessels is necessary. We designed a rotating magnetic field that can locally accumulate the particles within a spherical range of 10 mm in diameter, 200 mm away from a source of magnetic field. According to the result of experiment and particle trajectory calculation, it was found that the accumulation range of the ferromagnetic particles can be controlled by the rotation frequency of the rotating magnetic field.

1. Introduction

In order to improve the quality of life of patients, some cancer treatments with low side effects and less invasiveness have been studied in recent years. One of them is therapeutic embolization which treats cancer by blocking blood vessels around cancer with microgel[1]. This therapy is expected as a less invasive method with low side effects. In order to treat the diseased part selectively, a skill to guide a catheter to the vicinity of the cancer tissue during angiography is necessary. By the difficulty and the high-risk level of this treatment, the current applications are limited to specific organs such as the liver[2]. To solve this problem, we consider a novel cancer therapy by blocking newborn blood vessels by magnetic force control with multiple superconducting solenoidal magnets. In this therapy, as shown in Figure 1, ferromagnetic particles administered inside the body are accumulated at the target site by the magnetic field control from the outside of the body. After that, blockages of newborn blood vessels are formed around the cancer cells by aggregation of the particles that prevent cancer growth and metastasis.

Figure 2 shows the flow of the therapy we propose. First, the ferromagnetic particles are administered from blood vessels upstream of the cancer tissue with a microcatheter. Next, the administered ferromagnetic particles are accumulated in the newborn blood vessels by the magnetic field control. Thereafter, the particles are aggregated by a uniform magnetic field to block the newborn blood vessel. As a final step, the blockage should be maintained until the cancer tissue is necrosed after removing the magnetic field.

IOP Conf. Series: Journal of Physics: Conf. Series **1293** (2019) 012082 doi:10.1088/1742-6596/1293/1/012082

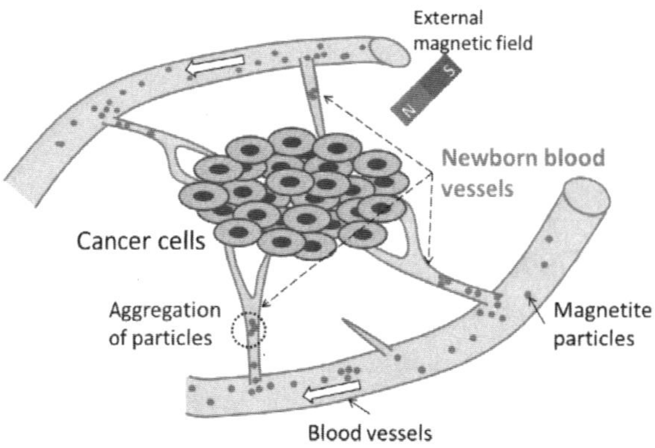

Figure 1. A cancer therapy by blocking newborn blood vessels.

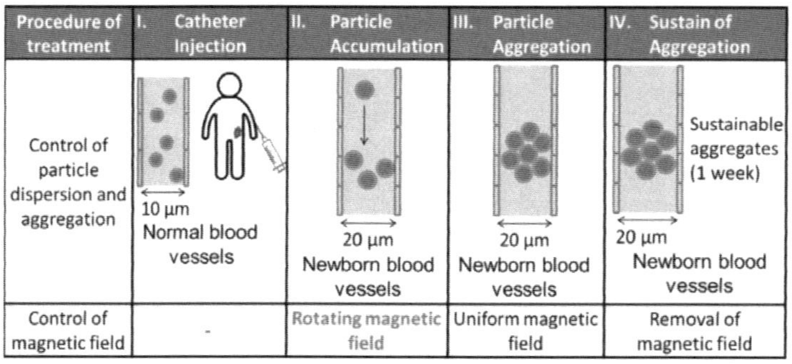

Figure 2. Four-step flow of this therapy.

2. Purpose of this study

In the second stage of this treatment in Figure 2, the particle accumulation in the blood vessels outside the target site causes blood clots. Therefore, in order to apply this therapy to deep cancer in the body, it is necessary to accumulate the ferromagnetic particles only in newborn blood vessels existing deep inside the body.

In our previous research, we simulated ferromagnetic particles trajectories under a rotating magnetic field (RMF) around a human body on the axis containing the target site[3]. In the blood vessel on-axis of the RMF, stable magnetic force acts on the particle in a direction toward the outside of the blood vessel, so the particles show a circular movement along the wall surface of the blood vessel. On the other hand, since a stable magnetic force does not act on the particle in the blood vessel off-axis of the RMF, the particles flow out downstream when the magnetic force temporarily becomes small. As a result, the ferromagnetic particles selectively accumulate on-axis of the RMF[3]. However, the particle accumulation is confirmed only in a simple experiment simulating two blood vessels parallel to the rotation axis of the magnetic field. Thus, the particle accumulation in a more complicated vascular system is necessary for practical use.

We considered an application of this treatment to pancreatic cancer as an example of cancer that develops deep inside a body. The pancreas is about 200 mm from the body surface and the size of the cancer is about 10 mm at the initial stage[4]. Accordingly, we designed the RMF that can locally accumulate the particles within a spherical range of 10 mm in diameter, 200 mm away from a source of magnetic field. Compared with the previous studies[2], we simulated more complicated vasculature closer to practical use. Also, we focused on the frequency of the RMF which is expected to greatly affect the particle accumulation to investigate the relationship between the frequency of the RMF and the accumulation range of the particles.

3. Experiment

3.1. Particle accumulation in simulated organ

In order to investigate the accumulation range of the particles under the RMF as shown in Figure 3, a simulated organ was prepared by filling a syringe (inner diameter: 22 mm) with glass beads (particle diameter: 250 μm). When a fluid flows through this syringe, the flow path (about 57 μm) is formed among the glass beads. Since the size is almost the same as the inner diameter of capillary blood vessels and newborn blood vessels, it can be said that they simulate those blood vessels. In the simulated organ, 10 mL simulated blood in which magnetite particles (particle diameter: 0.10 μm, Fe_3O_4, Mitsui Mining, Japan) were dispersed in 500 ppm flowed in 120 sec. at a constant flow rate of 1.0 mm/s. In the meantime, the RMF was applied to accumulate the particles. After that, the distributions of the particles in the flow direction and in the cross-section perpendicular to the flow were investigated.

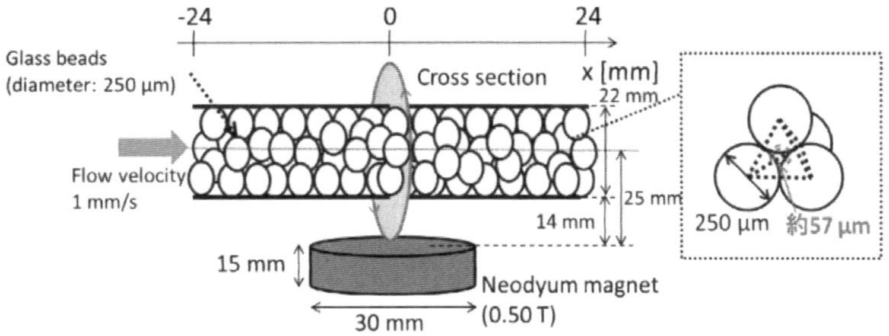

Figure 3. The experimental setup of particle accumulation in the simulated organ by using the RMF.

As simulated blood, a 3.8 wt% gelatin solution was used to simulate the viscosity of blood. The temperature and viscosity of the solution were 293 K and 4.2 mPa·s. A superconducting magnet is necessary for the actual treatment, but as basic research in the small scale, a cylindrical neodymium magnet (φ30 mm × 15 mm, a surface maximum magnetic flux density: 0.50 T) was used. Table 1 summarizes the magnetic field strength applied to the simulated organ. The center of the syringe was set to the central axis of the RMF, and the rotational radius of the RMF was set to 25 mm. The experiments were conducted at the rotation frequencies of 1.2 Hz, 2.4 Hz and 3.6 Hz.

Table 1. The strength of the magnetic field applied to the simulated organ.

The upper side of the organ	2.3×10 mT
On the central axis of the RMF	5.0×10 mT
The lower side of the organ	1.0×10^2 mT

After the experiment, the simulated organ was cooled and solidified at 268 K. The solidified simulated organ was divided into 10 sections (4.8 mm in the width) and a cross-section was observed by an optical microscope to make the images of the distribution of the cross-section perpendicular to the flow direction. Furthermore, in order to quantify the distribution of the particles in the flow direction, after dissolving each section with 8 mol/L hydrochloric acid and diluting them by distilled water, the concentration of iron ion was measured with ICP-AES (ICPS-7500, SHIMADZU, Japan) and converted into the concentration of the particles.

3.2. Particle trajectory in the simulated organ under the RMF

To investigate the accumulation by the RMF, we calculated the trajectories of the ferromagnetic particles flowing into the simulated organ under the RMF.

The forces acting on a particle in the simulated organ are magnetic force by the RMF and drag force by the blood flow. First, a method of calculating the magnetic force when the RMF is applied to the ferromagnetic particle, is described in **Figure 4**. Here, the plane including a rotation axis of the RMF and the initial central axis of a magnet is defined as S1, and the plane including the rotation axis and the central axis of the magnet after t [sec] is defined as S0.

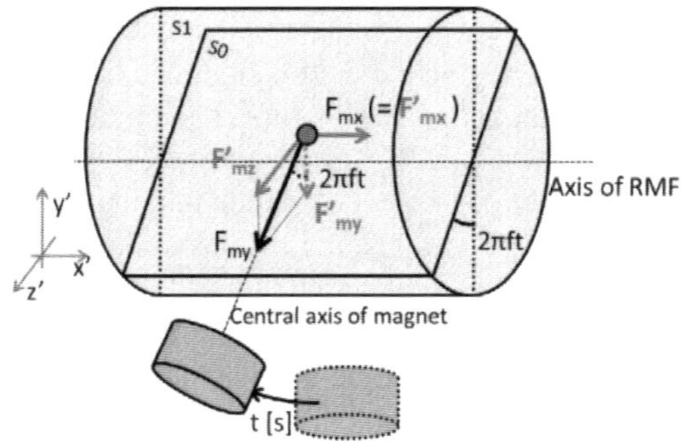

Figure 4. Magnetic force acting on the particle under the RMF.

Firstly, we consider the plane S0 as a reference. Assuming that the rotation axis of the RMF is x-axis and the central axis of the magnet is y-axis, the magnetic forces F_{mx} and F_{my} acting in the respective directions are given by the following equations.

$$F_{mx} = \mu_0 V \left(m_x \frac{\partial H_x}{\partial x} + m_y \frac{\partial H_x}{\partial y} \right) \quad \cdot \cdot \cdot \text{(eq. 1)}$$

$$F_{my} = \mu_0 V \left(m_y \frac{\partial H_y}{\partial y} + m_x \frac{\partial H_y}{\partial x} \right) \quad \cdot \cdot \cdot \text{(eq. 2)}$$

Here, μ_0 is the permeability in vacuum [m·kg/(s^2·A^2)], V is the volume [m^3] of the particle, m_x and m_y are the magnetization [Wb/m^2] of the particle, and H_x and H_y are the magnetic field strength [A/m] in each direction.

Next, assuming that the plane S1 is x'-y' plane and the direction perpendicular to the plane of the drawing is z' axis as shown in Figure 4, the magnetic forces F'_{mx}, F'_{my} and F'_{mz} are expressed by the projection of F_{mx} and F_{my} on the plane S1. So the magnetic forces F'_{mx}, F'_{my} and F'_{mz} are given by the following equations when the rotation frequency is f [Hz].

$$F'_{mx} = F_{mx} \quad \cdot \cdot \cdot \text{(eq. 3)}$$

$$F'_{my} = F_{my} \cos(2\pi f t) \quad \cdot \cdot \cdot \text{(eq. 4)}$$

$$F'_{mz} = F_{my} \sin(2\pi f t) \quad \cdot \cdot \cdot \text{(eq. 5)}$$

In other words, the magnetic force $\boldsymbol{F'_m}$ acting on the particle after t seconds from the start of the application of the RMF can be expressed by (eq. 6) by assuming the unit vectors of x', y' and z'-axis are **i**, **j** and **k**, respectively.

$$\boldsymbol{F'_m} = \mathbf{i}F'_{mx} + \mathbf{j}F'_{my} + \mathbf{k}F'_{mz} \quad \cdot \cdot \cdot \text{(eq. 6)}$$

Next, a drag force $\boldsymbol{F_D}$ acting on the particle by the blood flow was considered. The drag force is calculated by (eq. 7) using the viscosity coefficient η [Pa·s], the fluid velocity $\boldsymbol{v_f}$[m/s], and the particle velocity $\boldsymbol{v_p}$ [m/s].

$$\boldsymbol{F_D} = 6\pi\eta(\boldsymbol{v_f} - \boldsymbol{v_p}) \quad \cdot \cdot \cdot \text{(eq. 7)}$$

By substituting the magnetic force $\boldsymbol{F}_{\mathrm{m}}$ and the drag force $\boldsymbol{F}_{\mathrm{D}}$ calculated by (eq. 1) to (eq. 7) into the equation of motion (eq. 8), the particle trajectory was calculated by solving it by time evolution using Runge-Kutta method.

$$\boldsymbol{a} = (\boldsymbol{F}_{\mathrm{m}} + \boldsymbol{F}_{\mathrm{D}})/m \quad \cdots \text{(eq. 8)}$$

However, we didn't consider the existence of glass beads in order to simplify the calculation, and considered the particle trajectory when simulated blood flowed into the syringe at a uniform flow rate of 1.0 mm/s. The diameter of the particle is supposed to be 5.0 μm which is the secondary diameter after application of the magnetic field, and other necessary values for this calculation are summarized in **Table 2**. Magnetic field distribution and flow velocity distribution were analyzed by finite element method using ANSYS 10.0 (Cybernet system, Japan).

Table 2. Calculation conditions of the particle trajectory.

Particle density	5170 kg/m^3
Particle size	5.0 μm
The saturation magnetization (measured value)	0.55 Wb/m^2
Viscosity coefficient of gelatin solution (3.8 wt%, 298 K)	4.2 mPa·s
Density of fluid	1.0×10^3 kg/m^3
Rotation frequency of the RMF	1.2, 3.6 Hz
Vacuum permeability	1.257×10^{-6} $m\cdot kg/(s^2\cdot A^2)$

4. Results

Figure 5 and Figure 6 show the amount of the magnetite particles accumulated per unit volume in the flow direction under the static magnetic field and the RMF. Comparison between Figure 5 and Figure 6 show that the accumulation range of the particles is more narrowed by using the RMF. This is because the particles tend to accumulate at both ends of a magnet by the large magnetic field gradient under the static magnetic field. Furthermore, it can be seen from Figure 6 that the accumulation range of the particles became narrower with an increase in the rotation frequency. This suggests the possibility of more precise particle accumulation in the flow direction by using the RMF with high frequency.

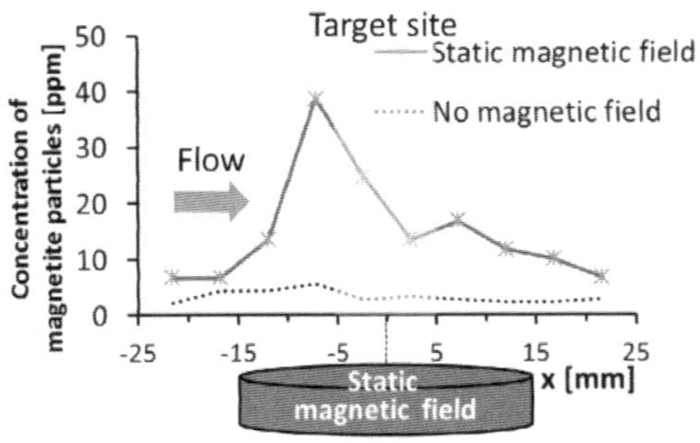

Figure 5. Accumulation in the flow direction (Static magnetic field).

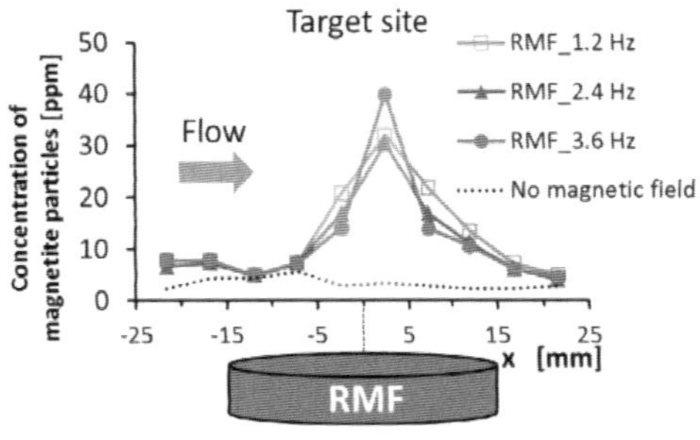

Figure 6. Accumulation in the flow direction (RMF).

Figure 7 shows photographs of the cross-section on the central axis of the magnet after the application of a static magnetic field and the RMF. From the comparison of Figure 7(a) and (b), it can be seen that by utilizing the RMF, the accumulated particles are distributed not only on the installation side of the magnetic field source but also on the entire cross-section. By comparing Figure 7(b)-(d), the particles are accumulated in a part close to the axis of the RMF, and the accumulation range is also extended as the increase in the frequency of the RMF.

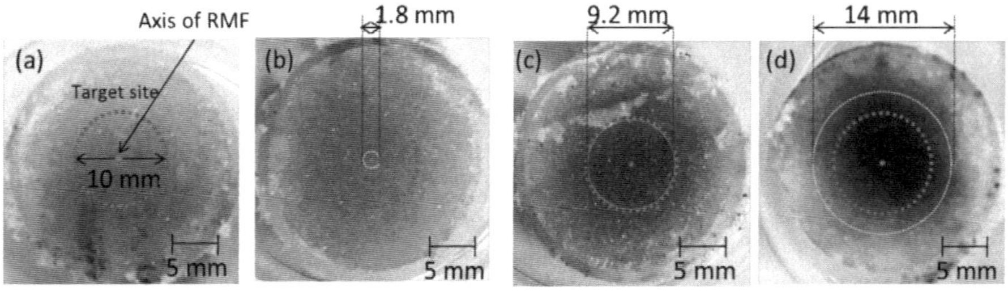

Figure 7. Accumulation of particles in the simulated organ by the RMF
(a)static magnet field (magnetic field source at the bottom) **(b)**1.2 Hz **(c)**2.4 Hz **(d)**3.6 Hz.

5. Discussions

In this part, three results are discussed by using the particle trajectory calculation. We considered a reason why the RMF can make the accumulation range narrower in the flow direction compared with the static magnetic field, the reason why the particles are accumulated on-axis of the RMF on the plane perpendicular to the flow direction, and the reason why the accumulation range is wider by using the RMF with high frequency.

First, the reason why the RMF can reduce the accumulation range of the particles in the flow direction is considered. Figure 8 and Figure 9 show the particle trajectories under a static magnetic field and under the RMF with 1.2 Hz, respectively. From Figure 8, it is shown that the particles flowing in the simulated organ are guided only to the magnet side. On the other hand, in Figure 9, the particles near the side surface of the simulated organ were guided in the outward direction. Furthermore, focusing on the particle introduced from the point closest to the side in Figure 8, the particle is guided to the side surface at 23 mm upstream from the magnetic field source under the static magnetic field. Besides, in Figure 9 the particle is guided on the side surface at 15 mm upstream under the RMF. These results indicate that the accumulation range under the static magnetic field extends to the upstream side. That is, the accumulation range in the flow direction is widened under the static magnetic field because the particles are guided on the further upstream on the side of the simulated organ.

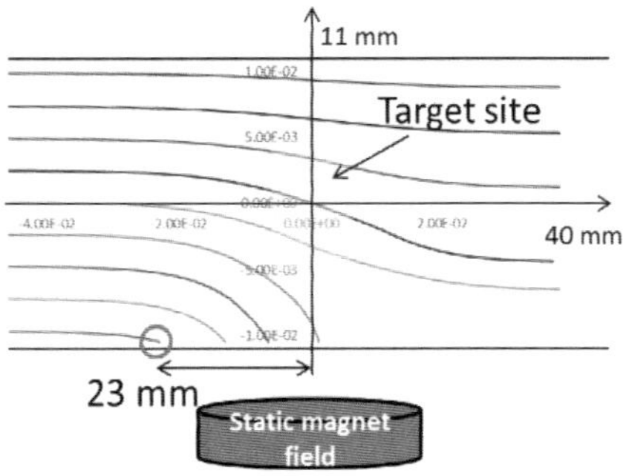

Figure 8. Particle trajectories in the simulated organ under a static magnetic field.

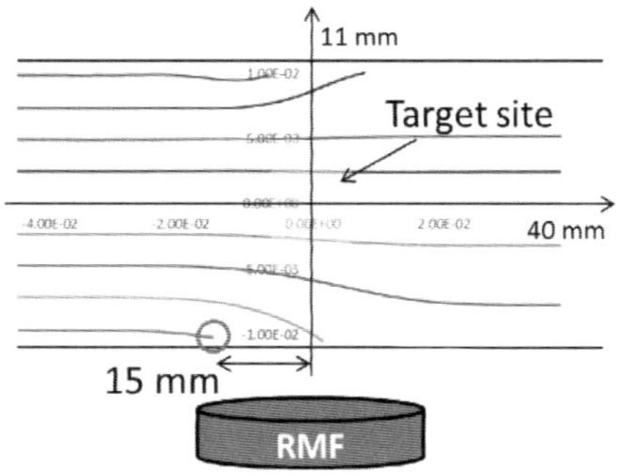

Figure 9. Particle trajectories in the simulated organ under the RMF (3.6 Hz).

Next, the reason why the particles are accumulated on-axis of the RMF on the plane perpendicular to the flow direction is considered. Figure 10 shows the particle trajectories under the RMF with the frequencies of 1.2 Hz and 3.6 Hz. It indicates that the macroscopic particle trajectories hardly changed by the difference of the frequencies, and the particles existing near the side were guided to the outside of the simulated organ. However, the lower diagram in Figure 10 showing the microscopic particle trajectory in the RMF, indicates that the particles on-axis of the RMF periodically move with constant amplitude. Though the particle which is on-axis of the RMF also seems to flow out to the downstream in these Figureures, some particles are considered to show the cyclic motion along the inner wall surface of the glass beads in the actual experiment with the simulated organ[5]. In this case, these particles hardly flow downstream since the drag force becomes small near the inner wall surface of the beads. That is, the reason why the particles are accumulated on-axis of the RMF is the cyclic motion of the particles along the surface of the flow path, that makes the particles hardly flow out to the downstream.

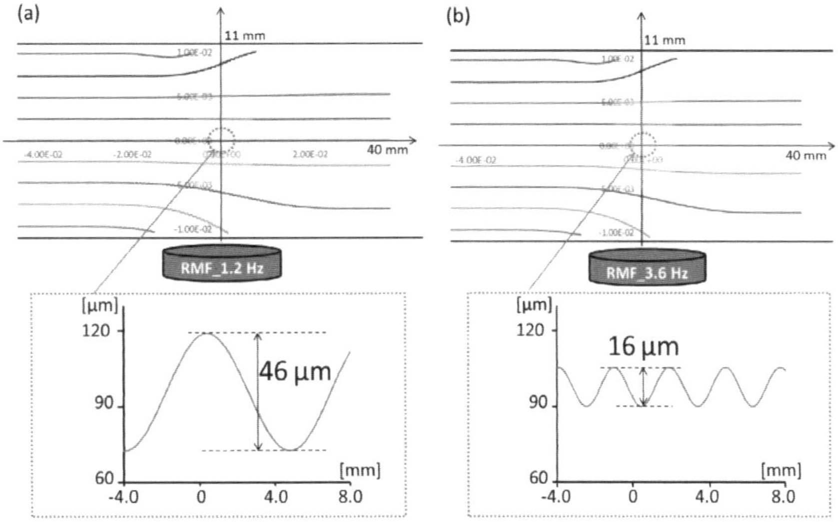

Figure 10. Particle trajectory on-axis of the RMF above the magnetic field source **(a)**1.2Hz, **(b)**3.6 Hz.

Finally, the reason why the accumulation range of the particles becomes wider by using the RMF with higher frequency was considered. Figure 11 shows the microscopic particle trajectories off-axis of the RMF. It can be seen that the amplitude motion occurs at about 180 μm in the cross-section above the magnetic field under the RMF of 1.2 Hz, whereas the amplitude is 45 μm at 3.6 Hz. In the experiments with simulated organ, glass beads form flow path in the inner diameter about 57 μm. These results show that the particles under the RMF of 1.2 Hz have a larger amplitude than the gap of the glass beads. It indicates the possibility that the particles are gradually guided to the outside of the simulated organ at the branch of the flow path. At the same time, the particle flows out gradually to the downstream. On the other hand, under the RMF of 3.6 Hz, the amplitude matches the size of the gap of 57 μm. It indicates that it is easy to continue the periodical motion along the inner wall between the beads staying on the cross-section above the magnetic field source. Combined with this consideration and the discussion on Figure 10, the reason why the accumulation range becomes wider by using the RMF with higher frequency is that the particles near the target boundary are hardly guided to the outside of the simulated organ, and that the particles are likely to show cyclic motion along the wall surface of the flow path between the glass beads.

From above consideration, with respect to particle accumulation by the RMF, when the amplitude of cyclic motion becomes smaller than about 57 μm of flow path width, the particles tend to accumulate. Whereas when the amplitude becomes larger than about 57 μm, the particles tend to be guided to the outside of the simulated organ. Based on this, at the cases of 1.2, 2.4 and 3.6 [Hz], the diameters of the accumulation range in the cross-section are calculated to be 1.2 [mm], 9.2 [mm] and 14.2 [mm], respectively. Also, by comparing the result with Figure 7 showing the results of the experiments with the simulated organ, it can be seen that the sizes of the range almost agree. This suggests that the accumulation range of the particles in the simulated organ is determined by the magnitude relation between the amplitude of the cyclic motion of the particle and the flow path width.

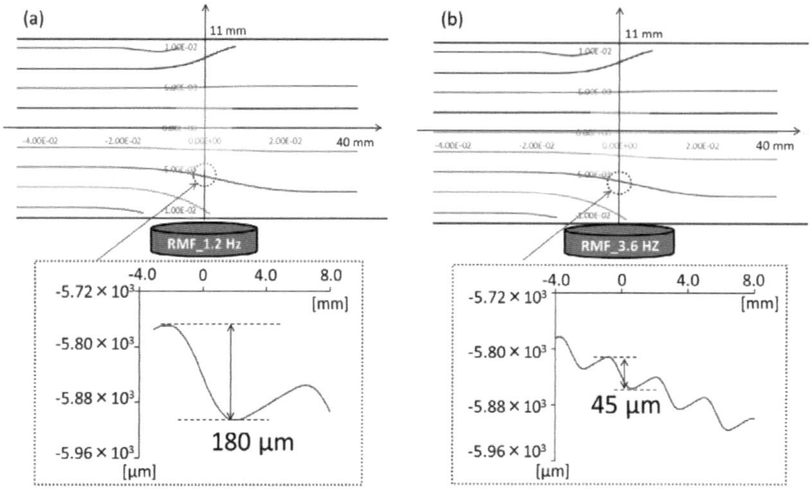

Figure 11. Particle trajectory off-axis of the RMF above the magnetic field source **(a)**1.2 Hz, **(b)**3.6 Hz.

6. Study for practical application by using superconducting magnets

In this chapter, a method for a practical application of the RMF and the specification of superconducting solenoidal magnets for this method are examined. In our previous study[4], we designed the RMF to accumulate ferromagnetic particles (particle diameter: 0.10 μm) in the deep part of the human body. In the above research, it was calculated that a superconducting solenoidal magnet A and B shown in Table 3, can apply a magnetic field product of 8.3 T^2/m respectively to the position 150 mm and 50 mm away from the magnet surface. Here, the magnetic field product means a product of magnetic flux density [T] and magnetic field gradient [T/m].

In the above calculation, the patient was assumed to be a child, whose cross-section of the body was assumed to be an elliptical shape of 200 × 140 mm in long and short axis. In order to apply the RMF in the center of the body, we examined a method by periodically changing the excitation intensity of the four magnets. In this case, a pair of magnets A are arranged in front and behind, and a pair of magnets B are arranged on both sides of the body. However, it is difficult to apply the high-frequency RMF in this setting, because the maximum frequency is 8.9×10^{-3} Hz even in the maximum excitation speed of the superconducting magnet, 0.2 T/s.

Table 3. Specification of superconducting solenoidal magnets for the RMF.

	Superconducting solenoidal magnet A	Superconducting solenoidal magnet B
Superconducting material	Nb_3Sn	Nb_3Sn
Diameter [mm]	190	400
Height [mm]	200	230
Central magnetic field [T]	5.9	4.6

To solve the problem, we propose a new method in this study. As shown in Figure 12, four superconducting solenoidal magnets are arranged around the body, and the RMF is applied by rotating a cylindrical magnetic shielding material with a slit. In this method, the RMF is applied to the body by the leakage magnetic field from the slit. It enables to apply the high-frequency RMF by rotating the magnetic shielding material because it is unnecessary to repeat excitation and demagnetization of the solenoidal magnets.

Based on this, we considered the superconducting solenoidal magnets which are necessary for this method. From the experiment shown in chapter 4, it was found that the magnetic field product required to accumulate the ferromagnetic particles (particle size: 5.0 μm) by the RMF was 0.15 T^2/m on the rotation axis. Because this treatment is also applied to adults, we assumed that the cross-section of the body is an elliptical shape of 440 mm × 320 mm in long and short axis[6]. Accordingly, it is necessary to apply the magnetic field product 0.15 T^2/m to the position 220 mm away from the body surface. Furthermore, considering the thickness of the shielding material, it is necessary to apply the magnetic field product to the part about 300 mm away from the magnet surface. However, since the magnetic field product for this treatment is 0.15 T^2/m, which is much smaller than the magnetic field product 8.3 T^2/m required for our previous study, it can be achieved sufficiently by using the magnets smaller than the magnet A in Table 3.

Furthermore, in order to apply this therapy to cancer of various parts, it is necessary to accumulate the particles selectively at the target site off-axis of the RMF. By controlling the magnetic field strength of each superconducting solenoidal magnet in Figure 12, the particles can also be accumulated at the target off-axis of the RMF. In future work, we plan to design these superconducting magnets and shielding materials.

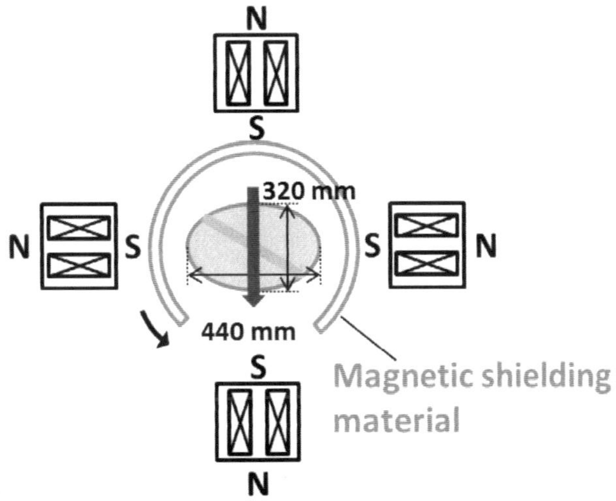

Figure 12. Method of applying the RMF using superconducting solenoidal magnets and a magnetic shielding material

7. Conclusions

In this study, we examined a novel cancer therapy by magnetic force control. The ferromagnetic particles are administered into the body, accumulated in newborn blood vessels existing around cancer, and aggregated to prevent the cancer growth and metastasis. In order to put this therapy into practical use, it is necessary to design a magnetic field to accumulate the ferromagnetic particles only in newborn blood vessels.

In the discussion, the following reasons for the three experimental results were considered by calculating the particle trajectory in the simulated organ under the RMF. Firstly, the reason why the RMF can make the accumulation range narrower in the flow direction compared with the static magnetic field was considered. Secondly, the reason why the particles are accumulated on-axis of the RMF on the

IOP Conf. Series: Journal of Physics: Conf. Series **1293** (2019) 012082 doi:10.1088/1742-6596/1293/1/012082

plane perpendicular to the flow direction was considered. Lastly, the reason why the accumulation range is wider by using the RMF with high frequency were considered.

For practical application, it is necessary to accumulate the particles at a deeper position than this simulation experiment. For that purpose, we use four superconducting solenoidal magnets and the cylindrical magnetic shielding material with a slit arranged around the human body. The magnetic shielding materials are rotated in order to apply the RMF to the human body. In future works, we plan to design these superconducting magnets and shielding materials.

Acknowledgment: A part of this research was supported by Magnetic Health Science Foundation.

References

[1] Horikawa1 M, Miyayama S, Irie T, Kaji T and Arai Y 2015 *American Journal of Roentgenology*, **205-4** 764-73
[2] Yamakado K 2014 *The Journal of the Japanese Society of Internal Medicine* **103-1** 87-92
[3] Chuzawa M, Mishima F, Akiyama Y, Nishijima S 2013 *Physica C* **484** 120-4
[4] Mori T, Mishima F, Akiyama Y, Nishijima S 2017 *Journal of Cryogenics and Superconductivity Society of Japan* **52-2** 98-104
[5] Chuzawa N, Mishima F, Akiyama Y, Nishijima S 2011 *Physica C* **471** 1538-42
[6] Nagai K, Tanoue S, Akahane K, Sato M 2015 *Entertainment Computing Symposium 2015* **2015** 365-71

Development of a contact-less cryogenic rotation mechanism employed for a polarization modulator unit in cosmic microwave background polarization experiments

Yuki Sakurai[1], Tomotake Matsumura[1], Nobuhiko Katayama[1], Teruhito Iida[2], Kunimoto Komatsu[3], Hajime Sugai[1], Hiroyuki Ohsaki[4], Yutaka Terao[4], Yukimasa Hirota[4], Hisashi Enokida[4]

[1]Kavli Institute for the Physics and Mathematics of the Universe (WPI), The University of Tokyo Institutes for Advanced Study, The University of Tokyo, 5-1-5 Kashiwanoha, Kashiwa, Chiba, 277-8583, Japan [2]ispace. inc., Sumitomo Shibakoen Building 10F, 2-7-17, Shiba, Minato-ku, Tokyo, 105-0014, Japan [3]Graduate School of Natural Science and Technology, Okayama University, 3-1-1 Tsushima-naka, Kita-ku, Okayama, 700-8530, Japan [4]Graduate School of Frontier Sciences, The University of Tokyo, 5-1-5 Kashiwanoha, Kashiwa, Chiba, 277-8561, Japan

E-mail: `yuki.sakurai@ipmu.jp`

Abstract. We present the design and the performance of a contact-less cryogenic rotation mechanism used in cosmic microwave background (CMB) experiments. A precise measurement of the CMB polarization is possible to verify the cosmic inflation theory that describes the very beginning (10^{-38} seconds) of the early universe. The polarization modulator, that rotates a half wave plate continuously at the aperture of the telescope, is one of the key instruments in the experiments. In order to reduce noise and systematic uncertainties, the polarization modulator is required a stable rotation with minimal heat dissipation in a cryogenic environment less than 20 K. Thus, we adopted the rotation mechanism that combines completely contact-less bearing and motor, a superconducting magnetic bearing, and a hollow bore synchronous motor. The heat dissipation and the load torque due to the friction can be minimized by avoiding physical contacts. We constructed the prototype of the rotation mechanism and carried out mechanical and thermal performance tests. A continuous rotation test in cryogenic temperature is performed, and it is confirmed that the rotation stability is less than 1% with the rotation frequency between 0.5 Hz and 3.0 Hz. We also conducted a thermal performance test, and obtained the heat dissipation at the rotor of 9.0 mW. We discussed the reduction of heat dissipation using a developed magnetic circuit with improved magnetic field uniformity.

1. Introduction

The Big-Bang theory is established as a standard theory in current cosmology from both theoretical and observational point of view. However, it has several unsolved problems such as the flatness, the horizon and the mono-pole problems. Therefore, cosmic inflation theory [1, 2] has been proposed, which describes rapid expansion of the universe after 10^{-38} seconds from the beginning of the universe. The mysteries of current cosmology can be solved by this theory, and it can be a clue to ultra high energy physics and furthermore ultimate unified theory. This inflation theory is predicted to leave a characteristic polarization pattern to cosmic microwave background(CMB) which was emitted at the Big Bang. The CMB is electromagnetic microwave radiation with a wavelength from μm to mm. It is still observable isotropically from the whole sky today. The CMB polarization observation experiments are actively planned and carried out worldwide toward the experimental verification of the inflation theory. In addition, the CMB polarization signal has sensitivity not only to the inflation but also to the large scale structure of the universe and the mass sum of neutrinos. Thus, the development of experimental instruments related to this experiment has been rapidly advanced.

As the simplest configuration, the instruments of the CMB polarization experiments consist of a telescope and a detector. The incident CMB signals are focused to a focal plane by the telescope, and then they are detected by an array of superconductor detectors. The CMB polarization signal is suffered from the thermal emission from instruments, and thus all instruments are cooled down to cryogenic temperature. In addition, the signal is very weak compared with the detector noise and fake signals from galaxy and atmosphere. Therefore, the critical point of this experiment is how to reduce the noise and systematic uncertainties. One of the key instruments to reduce them is a polarization modulator unit (PMU). It consists of an optical element, a half-wave plate (HWP), and a rotational mechanism. The incident linear polarized signal is rotated after passing through the PMU. As a result of this rotation, the signal is modulated to a high frequency region. This modulation significantly reduces the low frequency noise ($1/f$ noise) and mitigates the differential systematic uncertainties. The PMU has already been installed on ground and balloon experiments and its feasibility has been demonstrated [3, 4, 5]

Installing the PMU has great benefits, while there are systematic uncertainties and thermal noise due to the PMU itself. Thus, the rotation mechanism is required to perform a stable continuous rotation with minimal heat dissipation at cryogenic temperature less than 20 K. It can be realized by a completely contact-less rotation mechanism which can minimize the loss due to the physical friction and the load torque.

We fabricated the prototype system of the rotation mechanism and conducted the performance test at cryogenic temperature. In this paper, we describe the rotational performance and thermal characteristics of the prototype system.

2. Prototype system and experimental setup

We designed and fabricated the prototype system of the rotation mechanism of the PMU. The rotation mechanism has a hollow-bore for mounting the HWP to its inside. The typical diameter of telescope aperture is $\phi \sim 400$ mm in CMB polarization experiment. We designed a slightly smaller inner diameter of $\phi = 380$ mm for the demonstration purpose. The overall design and components follows a small prototype ($\phi = 100$ mm) which has already demonstrated stable rotation at cryogenic temperature [7]. The rotation mechanism consists of a superconducting magnetic bearing (SMB), a synchronous motor, an optical encoder, a cryogenic holding mechanism. Figure 1 shows a photograph and a 3D-CAD design of the prototype system.

The SMB is contact-less bearing, which employs an array of high temperature superconductor tiles as a stator and a permanent magnet as a rotor [6]. The rotor levitates above a stator and spins without physical contact, and thus this technology is suitable for this application. The rotor and the stator are formed into ring shaped using multiple segmented parts due to its large diameter. The stator is formed into a ring shape by combining 20 segmented three seeded superconductor bulks. We employs a YBCO($Y_{1.65}Ba_2Cu_3O_7$) as the material of the superconductor since its transition temperature of ~ 95 K is sufficiently higher than the required temperature of 20 K. We use a ring shaped NdFeB magnet as the rotor. It consists of 16 segmented NdFeB magnets magnetized axially with a magnetic remnance of 1.24 T.

The synchronous motor is used as drive mechanism in the rotation mechanism. It consists

IOP Conf. Series: Journal of Physics: Conf. Series **1293** (2019) 012083 doi:10.1088/1742-6596/1293/1/012083

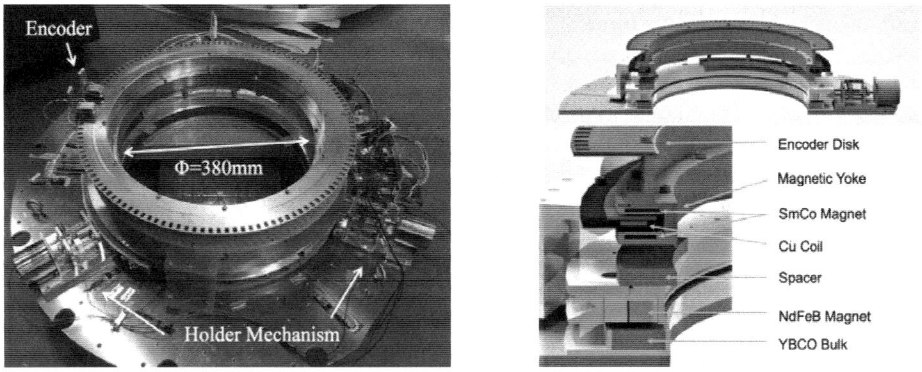

Figure 1. The left and the right panel shows the photograph and the 3D-CAD image of the prototype system of the rotation mechanism.

of 12 poles and 18 phases A copper wire coils and a SmCo permanent magnet are adopted as a stator and a rotor, respectively. The stator coils are sandwiched between upper and lower pairs of rotor magnets. The magnets are surrounded by the C shaped magnetic yoke in order to close the magnetic flux. This design realizes the high efficient motor, and the loss can be minimized. The output of the optical encoder and a cryogenic hall sensor are used as feedback signals in order to control the motor. Three hall sensors are installed inside the stator. The optical encoder is used not only for the feedback signal but also for reconstructing the optical axis angle of the HWP during rotation. An infrared LED and a silicon photo-diode (SiPD) are used as a light source and a detector, respectively. The metal encoder disk with 128 slots is mounted on the top of the rotor. We adopted the incremental encoder to reconstruct both absolute and relative angles, and thus three sets of the LED and SiPD are installed near the stator of the motor.

The cryogenic holder mechanism has two purposes. One is to hold the rotor from room temperature to cryogenic temperature, and the other one is to cool the rotor after the rotation. It consists of a cryogenic stepping motor and a linear actuator. Three mechanisms are installed at 120 degree intervals. A convex arm is equipped in the actuator, and it is designed to match the V groove which is attached to the magnetic holder in the rotor. The stepping motor is possible to drive in vacuum and cryogenic temperature.

The prototype system is installed in a ϕ = 800 mm 4 K cryostat using a Gifford-McMahon cooler with the cooling power of 1.5 W. The whole system is cooled down less than 10 K in about 10 days.

3. Rotational stability

We performed cryogenic rotation test using the prototype system in the 4 K cryostat. The typical rotation frequency of the PMU f is required to be from $0.5 \sim 5.0$ Hz. Also, the rotation stability Δf is required to be less than 1.0% from the viewpoint of systematic uncertainties. The procedure of the experiment is as follows:

(1) Install the prototype system in the cryostat and cool down to a thermal equilibrium.

(2) Operate the cryogenic holder mechanism and release the rotor.

(3) Set the rotation frequency and drive the synchronous motor.

(4) Rotate at set frequency for a fixed time.

(5) Turn off the synchronous motor and stop the rotation.

(6) Close the holder mechanism and cool the rotor to the initial temperature.

We iterate above procedure by changing the setting rotation frequency to 0.5, 1.0, 2.0, 3.0 Hz. The rotation frequency is reconstructed by the encoder signal. The left plot of Fig. 2 shows the reconstructed rotation frequency for each setting frequency. We obtained the rotation stability as $\Delta f \leq 1\%$ for all rotation frequencies. We also tested the continuous rotation for long time of 23 hours, as shown in the right plot of Fig. 2. We did not observe deterioration of rotation even in the 23 hours operation. The 23 hours is not a limitation, and thus longer operation time will be tested according to the requirement of the experiment. From these result, we successfully demonstrated the feasibility of the cryogenic rotation below 10 K.

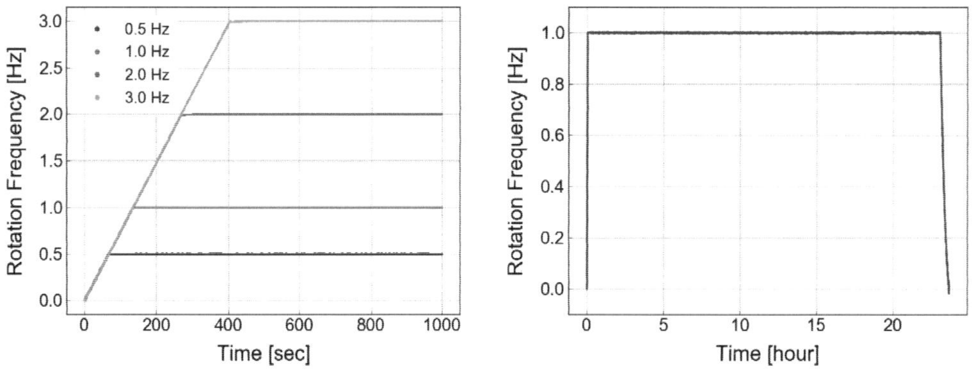

Figure 2. The left plot shows the rotation frequency as a function of operating time for each rotational speed of 0.5, 1.0, 2.0 and 3.0 Hz. The right plot shows the long term operation performance. The continuous rotation is kept for 23 hours.

4. Thermal characteristics

It was confirmed that the requirement of the rotation stability is satisfied using the prototype system. Another important requirement of the PMU is related to the thermal characteristics during the rotation. The temperature of the HWP mounted on the rotor is required to be less than 20 K during the rotation to suppress the thermal loading from the HWP. The thermal simulation model of the prototype system has been built in order to estimate the heat dissipation at rotor with keeping the HWP temperature of less than 20 K. From the simulation, the heat dissipation at the rotor is required to be less than 1.0 mW. We performed the constant rotation test to estimate the heat dissipation at rotor of the prototype system. Since this rotation mechanism is completely contact-less, it is difficult to directly measure the rotor temperature during rotation. Thus, we estimated it using the temperature of the gripper arm of the holder mechanism. The left plot of Fig. 3 shows the thermal profile during the rotation test with setting rotation frequency of 1.0 Hz and the operation time of 5 hours. The procedure of the experiment is the same as that described in the previous section. The temperature of the gripper arm rises following the rotor temperature after the rotation. We measured this temperature rise for each operation time from 10 minutes to 23 hours. We carried out the same procedure in the thermal simulation assuming several constant heat input to the rotor. The temperature rises of the gripper arm from the measurement and the simulation are compared, and then it was confirmed that the heat input to the rotor was 9.0 mW. The experiment and the simulation are consistent within 5%. The right plot of Fig 3 shows the temperature rises of the gripper arm from the experiment and the simulation as a function of the operation time. The estimated rotor temperature is also shown in the plot. The thermal conductivity of the holder mechanism has already been measured, and it is used for the thermal simulation. The validity of the thermal simulation has already been verified, as referred to [10].

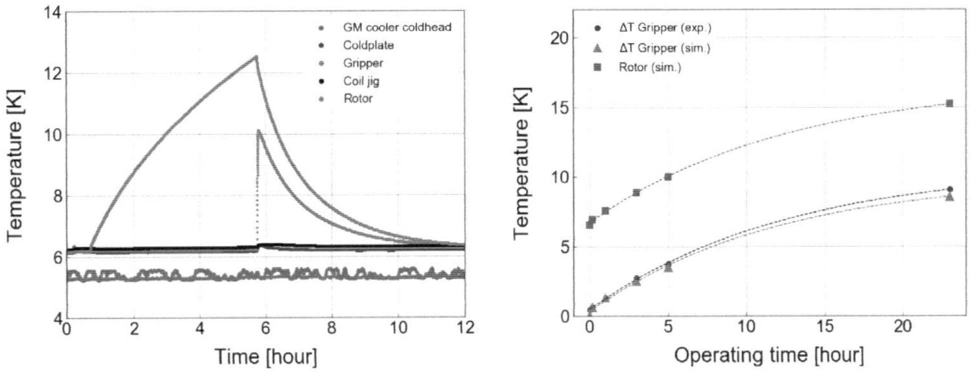

Figure 3. The left plot shows the temperature profile during the rotation and the cooling. The rotation frequency is 1.0 Hz and the operation time is 5 hours. The right plot shows the comparison of the temperature rise of the holder mechanism between the experiment (blue) and the thermal simulation (red). The estimated rotor temperature (green) is also shown.

5. Improvement of the heat dissipation

The obtained heat dissipation at the rotor from the performance test is 9.0 mW, which does not satisfied the requirement of 1.0 mW. Thus, it is necessary to reduce the losses of the rotation mechanism. The major source of the heat dissipation at the rotor is hysteresis and eddy current losses due to the SMB. These losses strongly depend on the magnetic field homogeneity of the rotor magnet. It is defined as

$$\frac{\Delta B}{B} = \frac{B_{\max} - B_{\min}}{\overline{B}}, \tag{1}$$

where $B_{\max/\min}$ and $\overline{B}$ are maximum, minimum and mean values in the magnetic field distribution, respectively. The current rotor magnet is formed by 16 segmented NdFeB magnets with an axial magnetization with respect to the stator. We measured the magnetic field of axial direction at the center position of the magnet in the cross section. The magnetic field distribution of this magnet has spikes due to gaps between the segmented magnets. The magnetic field homogeneity of the NdFeB magnet is 9.4%, and this is the cause of the losses. Therefore, we developed a ring shaped magnetic circuit. It consists of 64 segmented SmCo magnets with a magnetic remnance of 1.17∼1.21T. It is magnetized in the radial direction with respected to the stator. The segmented magnets are sandwiched with ring magnetic yokes in the radial direction. The effect due to the gap between the segmented magnets is alleviated by the magnetic yokes. We measured the magnetic field at the center position and its homogeneity is 2.1%. The NdFeB ring magnet and the developed SmCo magnetic circuit are shown in Fig. 4, and the magnetic field distributions are shown in Fig. 5.

According to the Beans model of the SMB theory, the energy loss due to the hysteresis is modelled as

$$\Delta E \propto \frac{(\Delta B)^3}{J_c}, \tag{2}$$

where ΔE is the energy loss, ΔB is the magnetic field homogeneity and J_c is a critical current density of the YBCO. The hysteresis loss is proportional to the cube of ΔB since J_c is constant

IOP Conf. Series: Journal of Physics: Conf. Series **1293** (2019) 012083 doi:10.1088/1742-6596/1293/1/012083

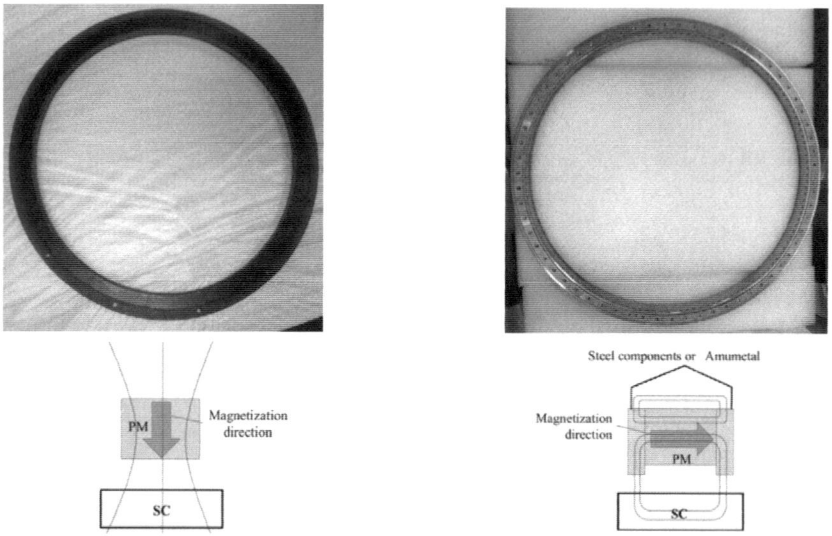

Figure 4. The photograph of the fabricated magnets. The left photo shows the NdFeB ring magnet formed by 16 segmented axial magnetized magnets. The right photo shows the SmCo magnetic circuit formed by 64 segmented radial magnetized magnets and ring shaped magnetic yokes. The conceptual diagrams of the cross section and the magnetic flux lines are also shown below each photo.

value. The energy loss due to the eddy current is proportional to the square of the ΔB. Thus, the hysteresis loss and the eddy current loss are expected to reduce fractions of $(2.1/9.4)^3 = 0.01$ and $(2.1/9.4)^2 = 0.05$, respectively. There are also hysteresis and eddy current losses due to the synchronous motor. However, these losses can be significantly reduced by the optimization of the material selection and the mechanical design. In the prototype system, it is composed of almost metallic material, and the energy loss is not fully taken into account in the mechanical design of each mechanism. In addition, we conducted electromagnetic simulations and confirmed that they have not significant impact compared with the SMB losses. From the above discussion, it is promising to reduce the rotor heat dissipation to less than 1.0 mW.

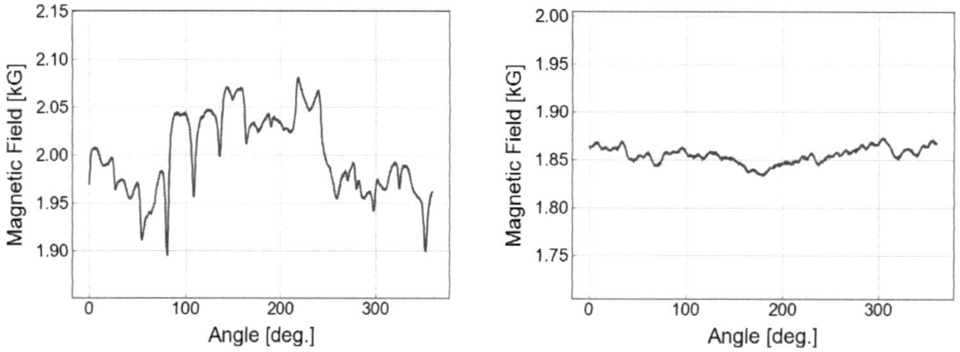

Figure 5. The magnetic field from the NdFeB ring magnet (left) and the SmCo magnetic circuit (rihgt) with respected to the angle of the ring magnet. The magnetic field homogeneities of the NdFeB magnet and the SmCo magnetic circuit are 9.4% and 2.1%, respectively.

6. Conclusion

We are developing a contact-less cryogenic rotation mechanism of the polarization modulator used for the CMB polarization experiment in order to verify the inflation theory. We constructed the prototype system and carried out the performance test.

The stable continuous rotation in cryogenic temperature was successfully demonstrated with rotation frequencies from 0.5 Hz to 3.0 Hz. The rotation frequency was reconstructed with the optical encoder and the rotation stability was less than 1.0%. We also succeed to operate a long term operation of 23 hours.

The requirement of the heat dissipation at the rotor is less than 1.0 mW to keep the HWP temperature less than 20 K. Thus, we carried out the thermal performance test in order to evaluate the rotor heat dissipation. The temperature difference of the holder mechanism before and after the rotation was measured for each operation time. We established the thermal model of the prototype system, and simulated the temperature of the holder mechanism with a constant heat input to the rotor. The measured and the simulated temperatures of the holder mechanism are consistent within 5% with the rotor heat dissipation of 9.0 mW. The estimated heat dissipation does not satisfied the requirement. However, there is much room for improvement. One of the major improvements is to use the developed SmCo magnetic circuit as the SMB rotor, which has better magnetic field homogeneity than the current magnet. Other hysteresis and eddy current losses can be significantly reduced by an optimization of a material selection and of a mechanical design. We will continue to this development to minimize the heat dissipation.

Acknowledgments

This work was supported by MEXT KAKENHI Grant Numbers JP17H01125, JP17K14272, JP15H05441. This work was also supported by JSPS Core-to-Core Program, A. Advanced Research Networks and World Premier International Research Center Initiative (WPI), MEXT, Japan.

References

[1] Sato K 1981 First-order phase transition of a vacuum and the expansion of the Universe *Monthly Notices of Royal Astronomical Society* **195** 46

[2] Guth Alan H 1981 The Inflationary Universe: A Possible Solution to the Horizon and Flatness Problems *Phys. Rev.* D **23** 347

[3] Klein J *et al.* 2011 A cryogenic half-wave plate polarimeter using a superconducting magnetic bearing *Proceedings of the SPIE, Cryogenic Optical Systems and Instruments XIII* **8150** 815004

[4] Kusaka A *et al.* 2014 Modulation of cosmic microwave background polarization with a warm rapidly rotating half-wave plate on the Atacama B-Mode Search instrument *Review of Scientific Instruments* **85** 024501

[5] Charles H *et al.* 2016 Design and development of an ambient-temperature continuously-rotating achromatic half-wave plate for CMB polarization modulation on the POLARBEAR-2 experiment *Proceedings of SPIE - the International Society for Optical Engineering* **9914** 99142U

[6] Hull J 2000 TOPICAL REVIEW: Superconducting bearings *Superconductor Science and Technology* **13** 2 1-15

[7] Matsumura T *et al.* 2016 Design and Performance of a Prototype Polarization Modulator Rotational System for Use in Space Using a Superconducting Magnetic Bearing *IEEE Transactions on Applied Superconductivity* **26** 3

[8] Sakurai Y *et al.* 2017 Estimation of the heat dissipation and the rotor temperature of superconducting magnetic bearing below 10 K *IEEE Transactions on Applied Superconductivity* **27** 4

[9] Sakurai Y *et al.* 2017 Vibrational characteristics of a superconducting magnetic bearing employed for a prototype polarization modulator *Journal of Physics: Conference Series* **871** 012091

[10] Sakurai Y *et al.* 2018 Design and thermal characteristics of a 400mm diameter levitating rotor in a superconducting magnetic bearing operating below at 10K for a CMB polarization experiment *IEEE Transaction on Applied Superconductivity* **28** 4 1-4

IOP Conf. Series: Journal of Physics: Conf. Series **1293** (2019) 012084 doi:10.1088/1742-6596/1293/1/012084

Emulation and experimental analysis of an axial superconductor magnetic bearing

E Rodriguez [1,2], **G Dos-Santos** [1], **Z Huang** [2], **W Wang** [2] **and Z Deng** [2]

[1] Laboratory of Applied Superconductivity – LASUP / Federal University of Rio de Janeiro, P.O.Box 68504, 21941-972 Rio de Janeiro, Brazil.

[2] Applied Superconductivity Laboratory, State Key Laboratory of Traction Power, Southwest Jiaotong University, Chengdu 610031, P. R. China.

elkin@dee.ufrj.br; dosgabrielsantos@poli.ufrj.br; 842650258@qq.com; 2496741148@qq.com; deng@swjtu.cn;

Abstract. Today the high-speed and high-performance systems are each more important for many areas like industry and transport system. The magnetic levitation system can be used to improve the performance. For superconducting levitation, it is important to know the behavior of levitation force. However, its behavior is not completely known and studies about it are necessary. This work presented a description of the test bench and a brief introduction to the new software developed to the magnetic levitation measurement system (SCML-02). The system tested is compound by a superconducting bulk and permanent magnet. This system makes data acquisitions using the load cells and the data processing is built by the new software developed in the LabVIEW. The results are verified and compared by simulations which were done in the software of finite elements and measurement data of the magnetic levitation done at Federal University of Rio de Janeiro - UFRJ. The measurements of the new system show compatible results with the simulations and the measurement data done at UFRJ. Therefore, with this new software, the users can easily to do magnetic levitation measurements through an interface easy to understand.

1. Introduction

High-speed and high-performance industries are developing ways to use the advanced in technology to improve its products, mainly in cost reduction, wear and efficiency. Magnetic levitation has been shown as an important tool, because it allows advantages in reducing losses, maintenance and reliability of the performance [1][2].

Magnetic levitation can be classified in Active Magnetic Levitation (AML), Passive Magnetic Levitation (PML) and Hybrid Magnetic Levitation (HML). These can be found into the industries of transport and the rotary machines. Each levitation technique has advantages and disadvantages and its use depends on the application.

PML is being of interest for not needing a loop control system for levitation. Technological advances and new materials have made these actions possible. Superconducting Magnetic Levitation (SML) and PML using a permanent magnet of $Nd_2Fe_{14}B$ are a good examples. High temperature superconductors (HTSs) may be used in a wide variety of applications such as magnetic bearings, rotating machines and MagLev vehicle due to stability on all axes [3][4][5][6][7]. Permanent Magnet Passive Magnetic Levitation (PMPML) has limitations due to its instability, as shown by Earnshaw's theorem. Nevertheless, this can be used for stability in an axis or the total stability, as in the case of the magnetic top consisting of a couple of the ring-shaped permanent magnets with relative velocity between them [8]. HML uses two or more levitation techniques, in which the technical characteristics each are integrated to reduce the losses and improve the performance [2][9][10].

It is necessary to have suitable tools to identify and study the technical characteristics of the magnetic suspension. These tools can be simulation or test bench. Simulation software is a widely used and reliable way to evaluate real phenomenon and design equipment without an excessive cost. Its results allow getting a good idea of the real product, but in some cases, it is necessary to check these results. Test benches are used to verify or evaluate the results obtained by simulation software [11].

In this context, the test bench at Applied Superconductivity Laboratory (ASCLab) of Southwest Jiaotong University (SWJTU) is under development and improvement aiming to simulate and verify the magnetic suspension prototypes developed. The first generation of magnetic levitation measurement system (SCML-01) was used by measuring the properties between YBCO bulk superconductors and a permanent magnet (PM) and PM guideway (PMG) [12]. An update of this equipment (SCML-02) was accomplished to improve the precision and allow to measure the magnetic properties of single and multi-HTS bulks when moved in three dimensions at one time [13]. These prototypes helped to build an HTS MagLev dynamic test system (SCML-03). SCML-03 is a dynamic test bench that can measure the levitation and guidance forces of a cryostat of the HTS MagLev vehicle when it is moving above of the PMG at 300 km/h [14].

With this in mind, this paper will focus on showing the software update being made on SCML-02. These improvements are looking to develop a test bench with high reliability and precision. More flexibility to programming and movement are also implemented. This new version of SCML-02 can be used to emulate some dynamic magnetic suspension movements. Experimental results showed the behavior of the movement and magnetic force were obtained by SCML-02. These results will be verified through the simulations and experimental results obtained in the Laboratory for Applied Superconductivity (LASUP) of Federal University of Rio de Janeiro (UFRJ).

2. Modeling of the HTS-PM system

There are many formulations that we can apply in the finite elements such as A-Ω, T-Ω, E-Formulation and H-formulation. When the simulations have a superconductor the relationship between electric field "E" and current density "J" is very nonlinear and its structure is also anisotropic. H-formulation works well with that behavior how we can see in [15][16], therefore we choose this formulation to work and this equation were implemented with COMSOL 5.1.

H-formulation can be written using the Faraday's law, Ohm's law and Ampere's Law

$$\nabla\times\mathbf{E} = -\,\mu\frac{\partial \boldsymbol{H}}{\partial t} \tag{1}$$

$$\mathrm{E} = \rho.\nabla\times\mathrm{H} \tag{2}$$

Where ρ is the electrical conductivity and μ is the magnetic permeability of material. Applying the equation (2) in (1) and organizing this equation, we have:

$$\nabla\times(\rho.\nabla\times\mathrm{H}) + \mu\frac{\partial \mathrm{H}}{\partial t} = 0 \tag{3}$$

Comparing this equation with the equation General Form PDE in COMSOL, we can write

$$\mathrm{e_a}\frac{\partial^2 \mathrm{u}}{\partial t^2} + \mathrm{d_a}\frac{\partial \mathrm{u}}{\partial t} + \nabla.\Gamma = \mathrm{f} \tag{4}$$

Where e_a and d_a are represent for tensors and $\boldsymbol{\nabla} = \begin{bmatrix}\frac{\partial}{\partial x} & \frac{\partial}{\partial y} & \frac{\partial}{\partial z}\end{bmatrix}$, H-formulation can be writing how:

$$\boldsymbol{u} = \begin{bmatrix} H_x \\ H_y \\ H_z \end{bmatrix}, e_a = \begin{bmatrix} 0 & 0 & 0 \\ 0 & 0 & 0 \\ 0 & 0 & 0 \end{bmatrix}, d_a = \begin{bmatrix} \mu & 0 & 0 \\ 0 & \mu & 0 \\ 0 & 0 & \mu \end{bmatrix}, \boldsymbol{f} = \begin{bmatrix} 0 \\ 0 \\ 0 \end{bmatrix}, \boldsymbol{\Gamma} = \begin{bmatrix} 0 & -E_z & E_y \\ E_z & 0 & -E_x \\ -E_y & E_x & 0 \end{bmatrix}.$$

$\boldsymbol{E}$ is the electric field vector $\boldsymbol{E} = \begin{bmatrix} E_x \\ E_y \\ E_z \end{bmatrix}$ and it is calculated by Ohm's law. Considering the ρ constant for all axis $\boldsymbol{E} = \begin{bmatrix} \rho.J_x \\ \rho.J_y \\ \rho.J_z \end{bmatrix}$. The relationship of $\boldsymbol{E}$ and $\boldsymbol{J}$ in superconductors is given by $\boldsymbol{E} = E_0.\left(\frac{|J|}{J_{c(B)}}\right)^n$ where E_0 critical electrical field and J_c is critical current, therefore the electrical conductivity is $\rho = \frac{E_0}{J_{c(B)}}.\left(\frac{|J|}{J_{c(B)}}\right)^{n-1}$ [15] and the relationship between J and B was modeled by Anderson-Kim: $\boldsymbol{J_c(B)} = \frac{J_{c0}}{1+\frac{|B|}{B_0}}$ where B_0 and J_{c0} are constant. Figure 1 shows the division of domains used in the simulation.

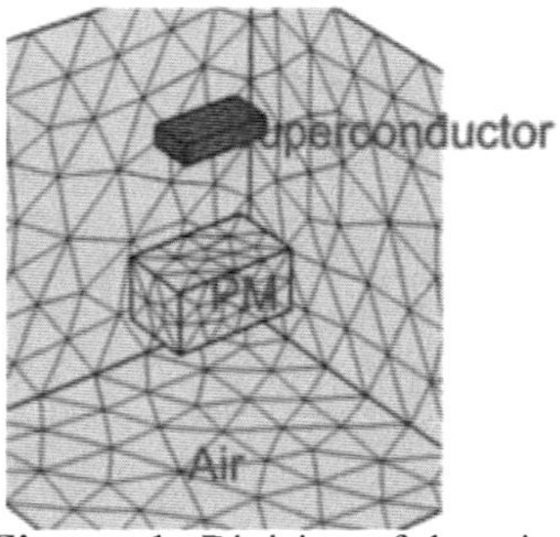

Figure 1. Division of domains.

3. Experimental

3.1. Magnetic levitation measurement system description

This system has three linear actuators, whose position is controlled by servo motors. The hardware of the magnetic levitation measurement system can be simplified, as showed in Fig. 2. This consists by a central processing, servo system used for each axis, the body frame of the SCML-02 and the load cells.

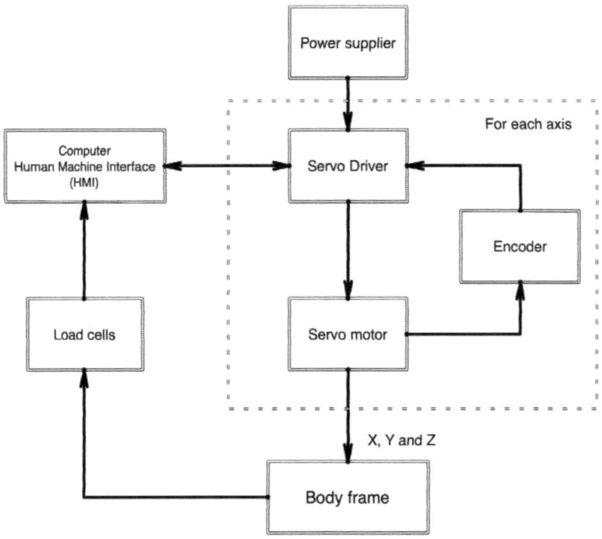

Figure 2. Diagram block summarizing the SCML-02 hardware.

Central processing is developed in LabVIEW software, a graphical language of National Instrument (NI). This has two boards connected to the computer via PCI bus and may be synchronized via the real-time system integration (RTSI) bus. The NI 6036E board is used to acquisition data of the load cells. Motion control 7344 board will be responsible for controlling of movement of the servo motors.

The servo system configuration has the protection of the power supply line, noise filter, servopack of SGDM model and servo motor of SGMAH model. SGDM-04ADA servopack is used on each axis and its power supply is 200 V. Two servo motors are used, SGMAH-04AAA4 and SGMAH-04AAA4C. The difference between them is the use of the brake and it is used on the z-axis.

Body frame of the SCML-02 is shown in Fig. 3. The location of the servo motors of the x, y and z axes can be observed. Underneath the flat table is the counterweight to compensate the inertial force due to the moving parts. The base structure support for load cells shows the position of the load cells to measure the vertical force and the guidance force.

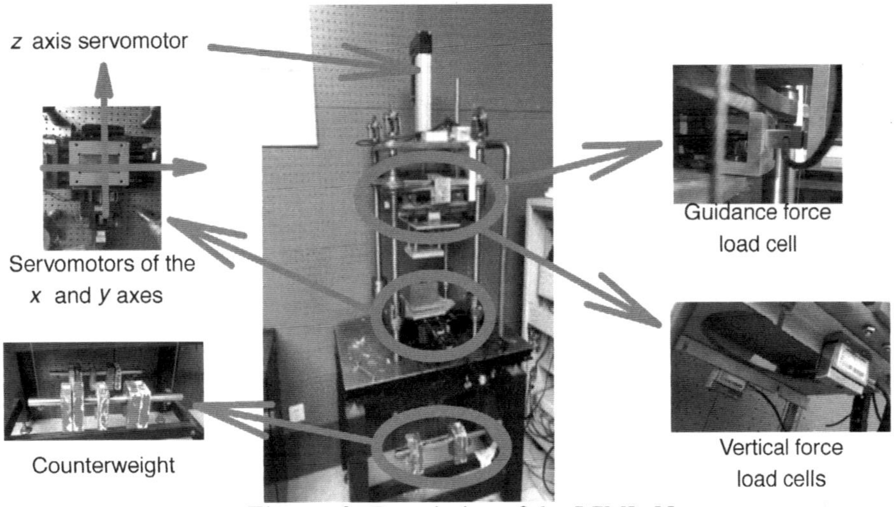

Figure 3. Description of the SCML-02.

3.2. Description of the program and the HMI

The program used to update SCML-02 is developed in NI graphical language called LabVIEW and uses two boards to control the motors and to measure the force. Figure 4 shows the diagram blocks that describe the main program. This can be separated into two main parts. First part is for configuration and checking of the boards. The settings are by reading the file with the basic settings. After setup completed, it begins to check the work of each board. In the end of this stage, it is possible to look for problems with the boards before using the test bench. It can save time and wrong operation. Second part is a wait, where the program is waiting for the user to choose the action. This stage may be separated in four options, some modification of the setup, manual handling of the motors, programming and execution of the test, and exit of the program. HMI of the main program is showed in Fig. 5.

3.2.1. A subsection

The modification of the setup is divided into two options. Changing the parameters of the drivers, such as the speed, acceleration and deceleration of the motor. The other option is to rewrite the file of the basic settings, as well as frequency of the acquisition, number of samples from average of the measure and parameters of linearization of the load cells.

3.2.2. Manual handling of the motor

This subprogram is used to verify the correct movement of the motor and the measurement of the force. HMI of this subprogram is shown in Fig. 6. Two motion options are possible, linear or circular. In the linear motion, the user can control the distance and the axis of the movement. However, in the

IOP Conf. Series: Journal of Physics: Conf. Series **1293** (2019) 012084 doi:10.1088/1742-6596/1293/1/012084

circular motion just *x* and *y* axes can be moved, but the user chooses the radius and the circumference segment to move. This is used to aid for positioning of the target in the operating area for the test.

Load cells data are obtained from each channel and displayed in real time. These can be analyzed before the average or evaluate them with the average. If necessary, they can be saved for a future post-processing.

Information of the motors and position can be showed, as well as the user can remove the TARA before starting the test.

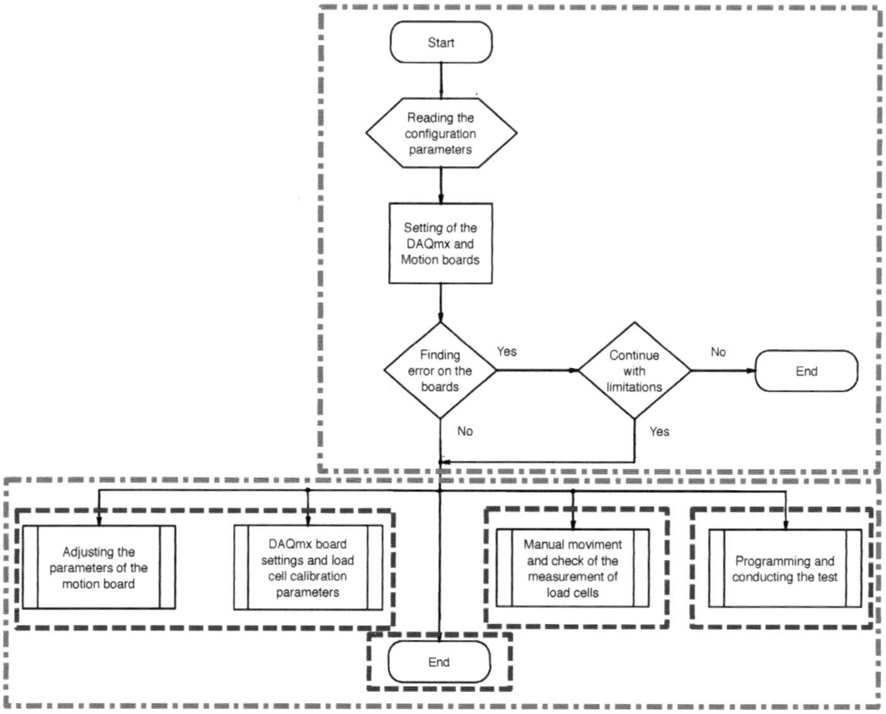

Figure 4. Block diagram summarizing SCML-02 software.

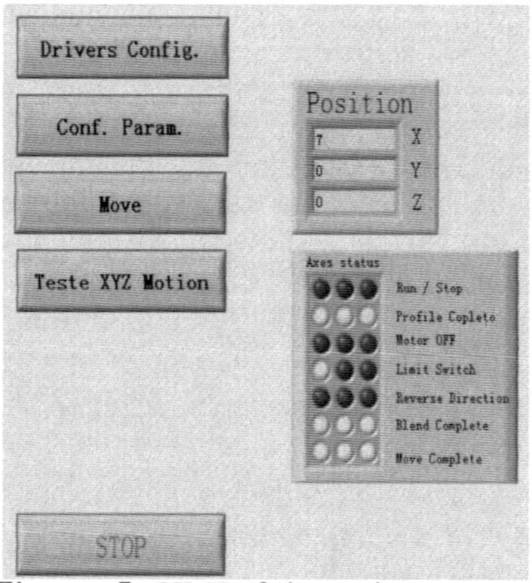

Figure 5. HMI of the main program.

3.2.3. Programming and execution of the test

Figure 7 shows the HMI used for programming and execution of the test. This HMI offers options for motion programming, save the test data, remove the TARA, choose which data to be displayed and position information for each motor. The motion programming button opens a new window with options for building the motion path for each axis and the repetition number of the entire motion, as shown in Fig. 8.

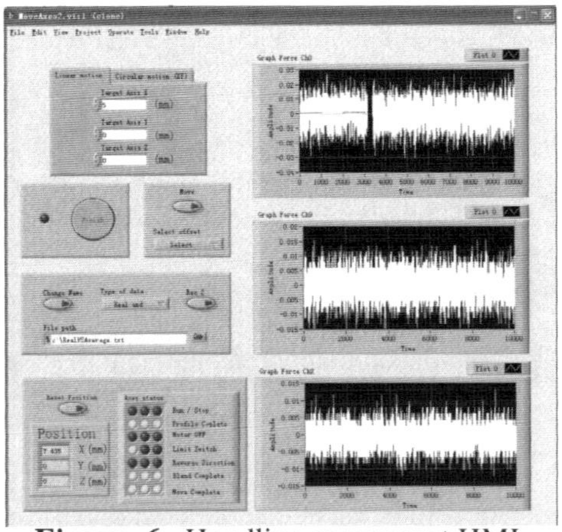

Figure 6. Handling movement HMI.

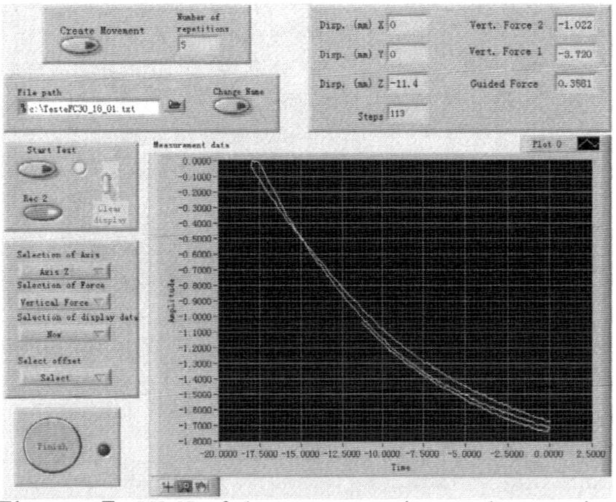

Figure 7. HMI of the programming and execution of the test

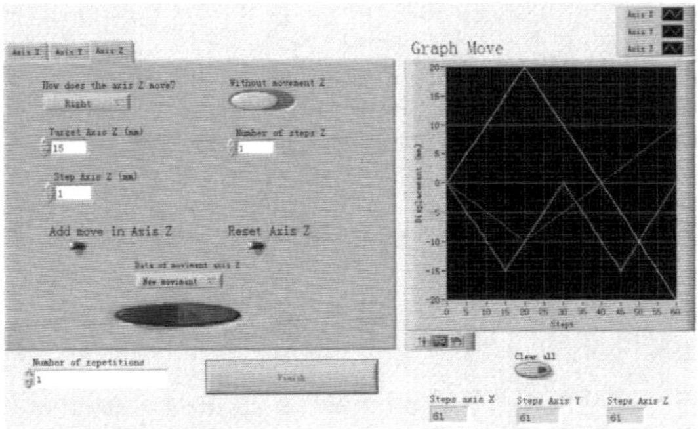

Figure 8. HMI to motion programming.

3.3. Results

In order to validate the results obtained from the tests performed in SCML-02, the test bench of the Laboratory for Applied Superconductivity (LASUP) of Federal University of Rio de Janeiro (UFRJ) was used. For this test, the HTS bulk and the permanent magnet block will be used, where the permanent magnets and superconductors have similar characteristics in each laboratories. Both permanent magnets are the Nd-Fe-B N35 with approximately 0.3 T of magnetic flux density at 10 mm from the face. Dimensions are 47x49x95 mm^3 and 60x50x99 mm^3 for ASCLab-SWJTU and LASUP-UFRJ, respectively. The superconductors used are $Y_1Ba_2Cu_3O_{7-\delta}$ (YBCO) bulks with three seeds and dimension of 64x32x13 mm. Figure 9 presents the test benches assembled in each laboratory.

The test bench was programming to apply discrete displacements of 1 mm along the z axis between desired limits. Average magnetic levitation force is measured after this movement is completed.

In these experiments, the HTS is cooled with liquid nitrogen and obtained the maximum levitation force. For this, the zero field cooling (ZFC) is performed. This procedure requires the HTS is cooled in a position without any external magnetic flux. Therefore, the cooling air gap between HTS and permanent magnet is 90 mm. Figure 10 shows the comparison of the levitation force measured during the ZFC of both test benches and simulations.

(a) (b)

Figure 9. Experimental test bench used to measure force. Details of the permanent magnet and YBCO bulk of the (a) ASCLab / SWJTU and (b) LASUP / UFRJ.

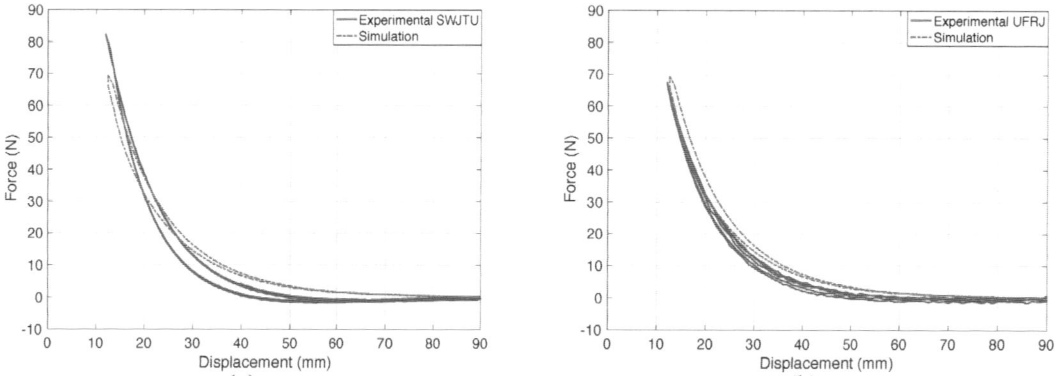

(a) (b)

Figure 10. Comparative of the levitation force between the test benches of the (a) ASCLab / SWJTU and (b) LASUP / UFRJ.

4. Conclusion

This work presented a description of the test bench and a brief introduction of the new software developed to magnetic levitation measurement system (SCML-02). This software is a tool for students with an easy to use and update option for a wide possibility of testing.

Basic structure of the HTS and PM are used to verify the proper operation of SCML-02. The results obtained showed the correct work of the software and hardware. This is confirmed by comparing the results obtained on both test benches. For comparing these results, they were done with basically the same structure. Where the movement of the tests benches were done with the equal distance and the PMs that were used in the tests at UFRJ and SWJTU have the same magnetic density. Besides, those tests were done many times to verify if the results were kept the same. Then with it, the experimental results stay more reliable. After that, the results were compared with the finite element simulations, where it was used H-formulation to represent the behavior of the superconductor.

Moreover this work presented 3D model simulations used to represent the problem. These results showed that simulations represent an approximation the real behavior and good operation of the SCML-02 test bench. The differences between of the experimental results can be variations of the superconducting bulks or problems in the distribution of magnetic flux density. For this work, magnetic flux density was measured in eight points on the PM surface. The magnetic flux density on the surface of the PM will be checked more carefully.

However, more care should be careful consideration of treatment in the simulation to get better results. New experimental tests will be performed to verify the behavior of the SCML-02. For example, the test bench may be tested to obtain the guiding force. For this, the superconductor may be cooled at different height and moved on the *X*-axis. These tests will help identify the scope what SCML-02 can achieve.

Acknowledgements
This work was partially supported by the Science and Technology Partnership Program and China-Latin Young Scientist Exchange Program, Ministry of Science and Technology of China.

References
[1] Kim b S, Park J K, and Daniel K 2015 Integrated dynamic simulation of a magnetic bearing stage for particle free manufacturing process *6th Int Conf on Modelling, Simulation and Applied Optimization* (Istanbul) 15278856. DOI: 10.1109/ICMSAO.2015.7152207
[2] Li Y, Li W, and Lu Y 2008 Computer-Aided Simulation analysis of a Novel Structure hybrid magnetic bearing *IEEE Transactions on magnetics* **44**(10) pp 2283 - 2287. DOI: 10.1109@TMAG.20082000542
[3] Wang B, et al 2015 Dynamic response characteristics of high temperature superconducting maglev systems: Comparison between Halbach-type and normal permanent magnet guideways *Physical C: Superconductivity and its applications* **519** pp 147-152. DOI: 10.1016/j.physc.2015.09.007
[4] Deng Z et al 2017 A High-Temperature Superconducting Maglev–Evacuated Tube Transport (HTS Maglev–ETT) Test System *IEEE Transactions on Applied Superconductivity* **27**(6) 3602008. DOI: 10.1109/TASC.2017.2716842
[5] Deng Z et al 2016 A high temperature superconducting Maglev ring test line developed in Chengdu, China *IEEE Transactions on Applied Superconductivity* **26**(6) 3602408. DOI: 10.1109/TASC.2016.2555921
[6] Mattos L S et al 2016 MagLev-Cobra Operational Tests *IEEE Transactions on Applied Superconductivity* **26**(3) 3600704. DOI: 10.1109/TASC.2016.2524473
[7] Sotelo G G et al 2013 Tests With One Module of the Brazilian Maglev-Cobra Vehicle *IEEE Transactions on applied superconductivity* **23**(3) 3601204. DOI: 10.1109/TASC.2013.2237875.
[8] Azukizawa T, Yamamoto S, and Matsuo N 2008 Feasibility Study of a Passive Magnetic Bearing Using the Ring Shaped Permanent Magnets *IEEE Transactions on magnetics* **44**(11) pp 4277 - 4280. DOI: 10.1109/TMAG.2008.2001490
[9] Mei L, Deng Z, and Liu C 2009 A 5-DOF Magnetic Levitation Motor System with Two Similar Hybrid Magnetic Bearings *4th IEEE Conf on Industrial Electronics and Applications* pp 1215-1219. DOI: 10.1109/ICIEA.2009.5138395
[10] Sotelo G G et al 2016 Tests with a hybrid bearing for a flywheel energy storage system *Superconductor Science and Technology* **29**(9) 095016. DOI:10.1088/0953-2048/29/9/095016
[11] Rodriguez E et al 2016 Designing, simulations and experiments of a passive permanent magnet bearing *International Journal of Applied Electromagnetics and Mechanics* **51**(2) pp 131–149. DOI 10.3233/JAE-150162
[12] Wang J S et al 2000 High Tc superconducting magnetic levitation measurement system (in Chinese) *High Technology Letters* **10**(8) pp 56–58.
[13] Wang S et al 2007 An Update High-Temperature Superconducting Maglev Measurement System *IEEE Trans on Applied Superconductivity* **17**(2) pp 2067-2070. DOI: 10.1109/TASC.2007.899257
[14] Wang J et al 2008 A High-Temperature Superconducting Maglev Dynamic Measurement System *IEEE Trans on applied superconductivity* **18**(2) pp 791-794. DOI: 10.1109/TASC.2008.920568.
[15] Zhang M and Coombs T A 2012 3D modelling of high-Tc superconductors by finite element software *Superconductor science and technology* **25**(1) 015009. DOI: 10.1088/0953-2048/25/1/015009.
[16] Sass F et al 2015 H-formulation for simulating levitation forces acting on HTS bulks and stacks of 2G coated conductors *Superconductor Science and Technology* **28**(12) 125012. DOI: 10.1088/0953-2048/28/12/125012

IOP Conf. Series: Journal of Physics: Conf. Series **1293** (2019) 012085 doi:10.1088/1742-6596/1293/1/012085

Energy losses in magnetic contactless bearings on the base of high-Tc superconducting tapes

I A Rudnev, A I Podlivaev, M A Osipov, D A Abin, I V Anischenko and S V Pokrovskii

National Research Nuclear University MEPhI (Moscow Engineering Physics Institute), Moscow, 115409, Russia

E-mail: iarudnev@mephi.ru

Abstract. High-temperature superconducting tapes are very promising for application in rotary magnetic bearings owing to its excellent magnetic and transport properties as well as possibility to easy create magneto-levitation systems of various geometry. One of problem arising in the rotation superconducting bearings is the presence of the energy loss resulting from the magnetization hysteresis due to inhomogeneity of magnetic field from permanent magnets. In this report we present results theoretical studies of rotation losses in the model superconducting bearing based on commercial high temperature superconducting (HTS) tape. The bearing stator has a cylindrical shape and is made of a multilayer HTSC tape. The bearing rotor is made of permanent magnets located at the same radius relative to the axis of rotation. As a result of numerical calculations carried out in the framework of the critical state model, the rotor geometry was established, at which the value of energy losses was minimal.

1. Introduction

The using of a superconductor/permanent magnet in contactless bearings is promising not only because of the absence of friction forces in such bearings. The absence of direct mechanical contact between moving parts allows efficient use of magnetic bearings in high-speed technology: kinetic drives, gyroscopes, and similar devices (see, for example, [1]).

The manufacture of a superconducting rotor (stator) from a bulk high-temperature superconductor (HTSC) is a rather complicated technological task. First, the HTS material is quite fragile (in this case, it is YBCO perovskite). Secondly, a complex annealing mode is required to ensure uniformity of the superconducting properties of the material. An alternative to a bulk superconductor can be a dial-up superconductor consisting of stacks of HTS tapes [2]. Similarly, a stack of tapes can also be considered short-circuited coils made of HTS tapes, which can be used in various levitation applications. The use of HTS tapes instead of bulk materials has several advantages. The technology for creating a superconducting composite of arbitrary shape made from tapes is simpler. The superconducting characteristics of a stack of tapes are not inferior to bulk material [3]. In addition, HTS tapes have higher mechanical properties than bulk materials.

The manufacture of a magnetic rotor also presents a technological problem. A solid magnet of the desired shape and size should provide an axial symmetry magnetic field. An alternative to a solid magnet (as in the case of a superconductor) can be a dial-up magnet. The typesetting magnet means a mosaic consisting of a large number of small magnets. Each element of the mosaic is a uniformly magnetized ferromagnet having the shape of a rectangular parallelepiped. The disadvantage of this mosaic is the heterogeneity of the magnetic field, especially noticeable at the junctions of neighboring elements. The inhomogeneity of the magnetic field can cause hysteresis magnetization reversal of the superconductor. The magnetization reversal of a superconductor is accompanied by energy dissipation (see, for example, the review [4]) and, therefore, the occurrence of a friction force in a bearing. The subject of this work is a theoretical study of the influence of the inhomogeneity of the magnetic field of a magnet on the friction force in a bearing, in which single-layer and multilayer rings formed from HTS tapes of the 2nd generation will be considered as a superconducting stator.

2. Experimental procedure

The bearing in question consists of a rotor and a stator. The division into "rotor" and "stator" is conditional. It is possible both the rotation of the internal shaft of the bearing with the fixed external part, and the rotation of the external part around the fixed internal axis. For definiteness, in the future, we assume the superconducting part of the bearing is a fixed stator, and the rotor is formed from a set of small permanent magnets. This choice is due, firstly, to the fact that it is technically easier to cool a stationary superconductor. The possibility of cooling allows us to neglect superheat superheating (otherwise thermodynamic instability may occur). Secondly, the rotating massive magnetic rotor around the internal superconducting axis is optimal for creating a gyroscope. Figures 1 and 2 show bearings with quadrilateral and octagonal rotors, respectively. In the following, the number of magnets in the rotor is denoted as N_{mag} (N_{mag} = 4, 8, 16). For calculations, the following sizes of magnetic rotor are selected. The inner border of the rotor (the inner circle indicated by a broken line in Figures 1 and 2) has a radius of 20 mm for both quadrilateral and octagonal and sixteen-square rotors. The outer border of the rotor (the outer circle indicated by a broken line in Figures 1 and 2) has radius of 32.016 mm and 26.34 mm for the quadrilateral and octagonal rotors, respectively.

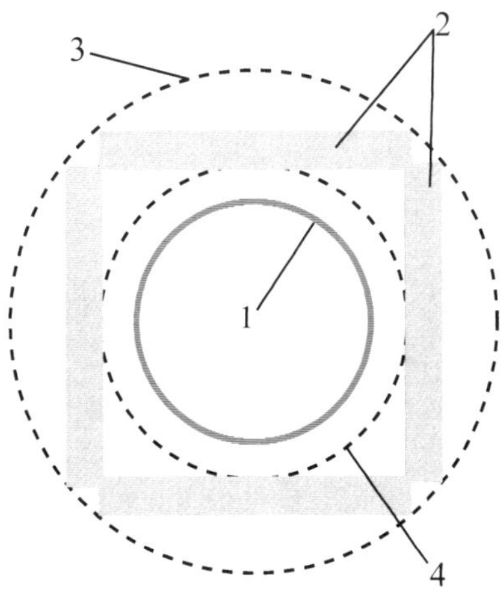

Figure 1. Bearing with a quadrangular magnetic rotor. Top down view.
1 - superconducting stator. 2 - magnetic rotor. 3 and 4 are the outer and inner borders of the magnetic rotor.

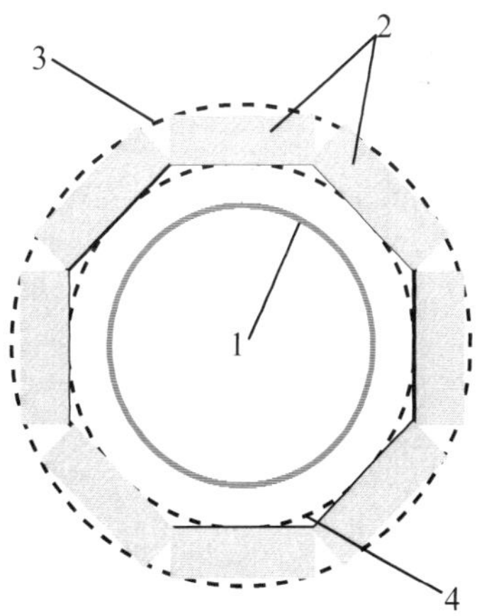

Figure 2. Bearings octagonal magnetic rotor. Top down view.
1 - superconducting stator. 2 - magnetic rotor. 3 and 4 are the outer and inner borders of the magnetic rotor.

The magnet thickness equal to 5 mm and the width of the magnets dZ equal to 12 mm in the direction of the Z axis are also the same for all rotors. When choosing the tape parameters, we focused on quite promising in technical applications 12mm SuperOx tape. The magnetization of each magnet is considered homogeneous. If the density vector of the magnetic moment at a point in space with a radius vector $\vec{r}'$ is denoted $\vec{m}(\vec{r}')$, the magnetic field induction vector at the point $\vec{r}$ created by the magnet (which occupies volume V) has the following form [5]:

$$\vec{B}(\vec{r}) = \mu_0 \int_V \left[\frac{3\vec{n}\cdot(\vec{n},\vec{m}(\vec{r}')) - \vec{m}(\vec{r}')}{|\vec{r}-\vec{r}'|^3} \right] d^3r' \,,\; \vec{n} = (\vec{r}-\vec{r}')/|\vec{r}-\vec{r}'| \tag{1}$$

The vector modulus $\vec{m}(\vec{r}')$ was chosen so that the magnitude of the magnetic field induction of a separate magnet in the center of its surface was 0.3 T.

The usual thickness of the HTS tape is ~ 0.1 mm, so in this paper, just like in [3], we consider the stator winding thickness to be zero. We assume that the critical current density in a single-layer continuous HTS ring is equal to j_c in a two-layer - $2j_c$. The critical current density at the boundaries of adjacent rings perpendicular to the Z axis is zero. The coordinates Z of the boundaries of the stator rings coincide with the boundaries of the bipolar magnets of the rotor. These boundaries have coordinates $Z=0, \pm d_Z/2, \pm 2d_Z/2, \pm 3d_Z/2, \pm 4d_Z/2$, etc.

With a large number of magnets in the rotor ($N_{mag} >>1$) and small width of HTS rings in comparison with their perimeter ($d_p/ d_Z >>1$) the stator curvature can be neglected and replace the problem of determining the density of superconducting current, set in cylindrical coordinates (ρ, ϕ, Z) at $\rho=R_S$ two-dimensional problem in the plane (p,Z), $p = R_S\phi$. In this case the region of definition of magnetic field *and* current density is rectangle $-d_Z/2 \leq Z \leq d_Z/2$, $- d_p/2 \leq p \leq d_p/$. The boundary conditions are periodic both in the angular variable p and in the variable Z (it is assumed that the rotor consists of an infinite number of rings of bipolar magnets adjacent to each other). Replacing a cylindrical problem with a flat one is justified by the fact that, firstly, the ratio $dp / dZ = 8$ is quite large, and secondly, from the practical side, rotors with a large number of magnets ($N_{mag} = 8$ and 16) are of the greatest interest. We consider the quadrangle rotor to qualitatively determine the dependence of the friction force of the bearing on the number of magnets. Moreover, it is for the quadrangular rotor that the effect of the nonuniformity of the external field (in the angular variable) on the distribution of superconducting currents is most pronounced.

The external magnetic field of the magnetic rotor in the stator was calculated by the formula (1) as well as in Comsol Multiphysics. The normal to the stator surface component of the magnetic field on this surface is shown in Fig. 3. The induction was determined for a quadrangular rotor.

The variation of the amplitude of the magnetic field $(B_{max}-B_{min}) / (B_{max} + B_{min})$ with the $Z = 0$ coordinate is 14.8%, 1.3%, and 0.2% for $N_{mag} = 4$, 8, and 16, respectively. The idea of using a non-uniform (periodic) external magnetic field is not new but the feature of this work is the calculation of the HTS tape in a weakly varying field, which, however, affects the superconductor for an unlimited number of periods.

We describe the field dependence of the critical current of a superconducting tape in the framework of a two-exponential model [6, 7], in which the density of surface currents of an HTSC film has the following form:

$$j_c(B) = A_1 \exp(-|B|/\beta_1) + A_2 \exp(-|B|/\beta_2), \tag{2}$$

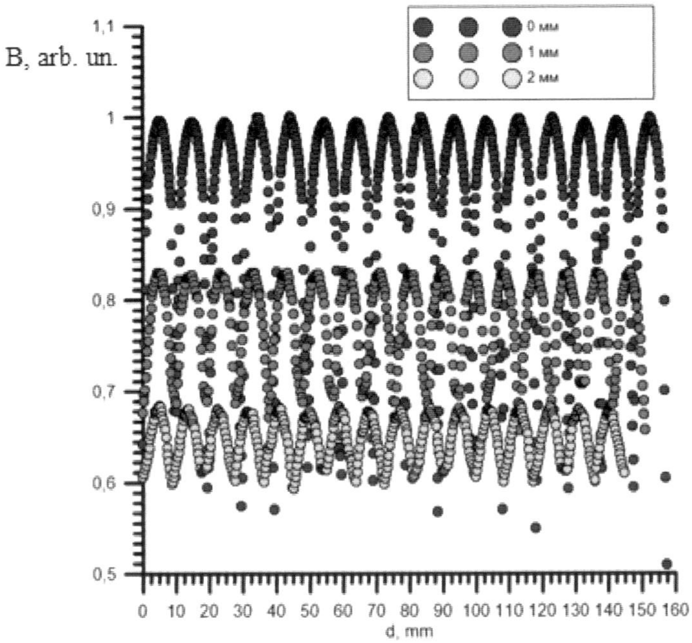

Figure 3. The normal component of magnetic induction along surface on stator (distance from the magnets is 0 (blue), 1 (red) and 2 (green) mm).

with parameters A_1=12.9 kA/m, A_2=13.8 kA/m, β_1= 0.08 T, β_2=1.92 T. The spatial grid step in the presented calculations is $\delta h \sim 0.375$ mm. The calculation was carried out within the framework of the model of the critical state of a superconductor originally proposed in [8], some modifications and applications of which are given in the review [4]. A detailed description of the formulation of the mathematical problem and the algorithm for numerical calculation for the field dependence (2) are given in publications [3, 9], and are not given in this paper due to cumbersomeness.

The mode of changing the external magnetic field was as follows: At the first stage, the magnetic field proportional to the rotor field without rotating the rotor monotonously increased from zero to the value determined by expression (1). Next, the rotor was rotated at 40 rotations (the number of rotations hereinafter referred to as N_{rot}). The corresponding density of superconducting currents was determined. One turn of the cylindrical configuration corresponds to the shift of the magnetic field in the plane problem with respect to the variable p by the magnitude of the period dp. In this shift, the tangential component of the interaction force of the rotor field $B\rho$ with the stator currents (friction force) fp for one stator ring was determined from Ampere's law:

$$f_p = \int_{-d_p/2}^{d_p/2} dp \int_{-d_Z/2}^{d_Z/2} B_\rho(p,Z)\cdot j_Z(p,Z)dZ \tag{3}$$

3. Results and discussion

The dependence of the friction force on number of rotor revolution the quadrangular stator (N_{mag} = 4) is shown in Figure 4. The figure shows that the friction force is established in one turn. For rotors with a large number of magnets (N_{mag} = 8 and 16), the setting also takes place in less than two turns. In this paper, we are not interested in the nature of the transition processes. Therefore, we determine the friction force by averaging this quantity from the third to the tenth turn. In the case of a quadrangular rotor (*Nmag* = 4), the friction force for a single-layer / two-layer stator is 0.013 N and 0.0021 N, respectively. In the case of an octagonal rotor (N_{mag} = 8), the friction force for a single-layer / two-layer stator is 2.01 10^{-6} N and 1.25 10^{-6} N, respectively. At N_{mag} = 16, the friction force for a single-layer / two-layer stator is 1.5 10^{-7} N and 8.1 10^{-8} N, respectively.

The characteristic density distribution of superconducting currents in the stator and the magnetic field of a quadrangular rotor is shown in Fig. 5. The upper part of the figure represents the lines of superconducting currents, the lower part represents the lines of the levels of the magnetic field of the rotor. This figure shows a fragment of a stator with a width of $2dZ$ (the central tape is surrounded on the top and bottom by the halves of adjacent stator tapes). When the rotor is rotated, the pattern is periodically shifted along the p axis.

For a qualitative interpretation of the dependence of the friction force on the number of rotor magnets and turns of the HTS tape on the stator, it is convenient to present the stator as a system of interconnected magnetic pumps. Imagine each cell of the counting grid the size of the stator as an elementary magnetic pump. If we designate the rotor speed as a symbol, the pump frequency of each pump by the external alternating magnetic field will be equal (the phase of the field oscillations for cells with different coordinate p will be different). The moment of the cell induced by the external field is $g \cdot \vec{n}_\rho \cdot h^2$ magnetic. In this expression $\vec{n}_\rho$ is the unit vector normal to the stator surface. The value of g is the density of the magnetic moment related to the density of the surface currents $\vec{j}(p,Z)$ of the stator by the ratio $\vec{j}(p,Z) = \nabla \times (g \cdot \vec{n}_\rho)$ (see review [4] and literature to it). Figure 6 shows the dependence of the density of the magnetic moment of the cell with coordinates $p = 0$, $Z = 0$ on the magnitude of the magnetic induction of the rotor at this point. The upper curve corresponds to a single-layer, and the lower one to a two-layer HTS stator. Dependencies are given for one rotation of the rotor after passing through transients. The magnitude of the magnetic field at a given point varies from the minimum B_{min} = 0.095 T to the maximum B_{max} = 0.128 T.

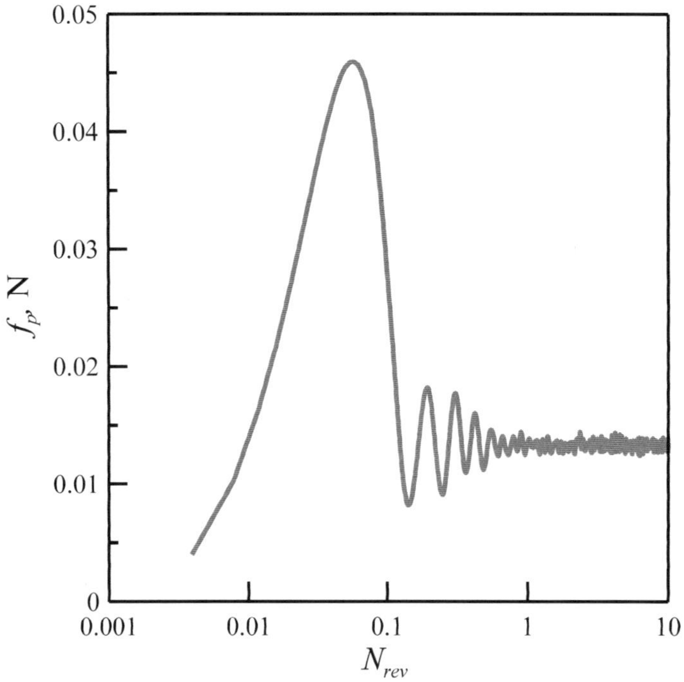

Figure 4. The dependence of the friction force on number of rotor revolution the quadrangular stator.

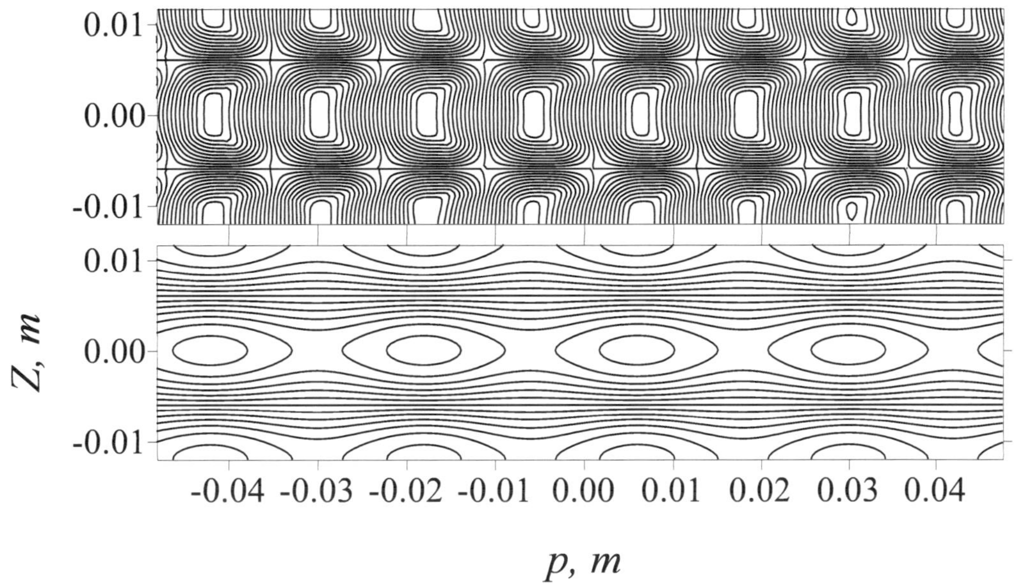

Figure 5. Lines of superconducting currents (upper part of the figure) and level lines of the components of the magnetic field of a quadrangular rotor normal to the stator surface (lower part of the figure). One current line in the upper figure corresponds to the current 5A. One streamline in the lower figure corresponds to a magnetic field induction of 0.02 T.

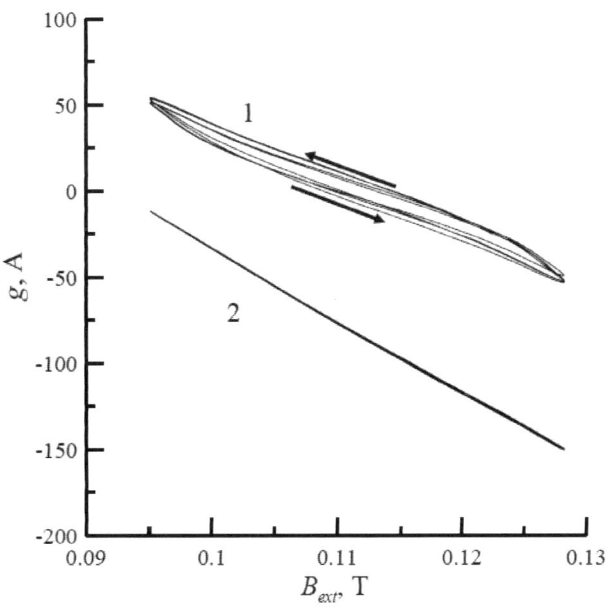

Figure 6. The dependences of magnetic moment density in the point with coordinates p=0, Z=0 on magnetic induction for one turn of rotor: 1 – dependency for one-layer rotor, 2 – dependency for double-layer rotor. The arrays indicate the direction of moving along hysteresis curves.

On the section of the curve corresponding to an increase in the induction of the field from the minimum B_{min} to the maximum B_{max} value, the EMF of a varying field induces currents that reduce the density of the magnetic moment (the induced magnetic field is opposite to the inductive field). As a result, the magnetic moment is minimal at the maximum magnetic field. The shielding process is limited by the finite value of the critical current (otherwise, the shielding would be complete across the entire stator). When the magnetic field decreases from the maximum value at the initial stage, the electromotive force changes sign, and the current density at this point becomes less critical. With a further decrease in the field, the current density passes through zero and again its value reaches a critical value. The amplitude of the change in the magnetic field induction of the quadrilateral rotor on the stator surface reaches 14.8% of the average value. This change is sufficient for the first, in a single-layer stator, the magnetic moment of the unit cell to change sign (see Fig. 6, the upper curve), while the magnetic field is constant. Secondly, the smallness of the critical current with such a change in the field results in a radical reversal of the stator, which leads to a wide hysteresis loop (as a result, to energy dissipation and a large friction force). A two-fold increase in the critical current density in a double-layer stator blocks the process of magnetization reversal (see Fig. 6, lower curve). The area of the hysteresis loop decreases sharply, and as a result, the friction force decreases. Similarly, a significant decrease in the friction force due to an increase in the number of magnets is explained. Variations of the magnetic field amplitude at 1.3%, and 0.2% for N_{mag} = 8 and 16 are not enough to re-magnetize even a single-layer stator.

4. Conclusions

The calculations showed that the force of magnetic friction and the resulting energy losses are not an obstacle in the creation of a combined bearing, the rotor of which consists of more than 8 magnets, and the stator - of several short-circuited layers of HTS tape of 2nd generations. Experimental verification of the output, as well as analysis of the influence of other bearing parameters (eccentric position and misalignment of the rotor axis relative to the stator axis, defects in the HTS stator tapes, etc.) that can critically degrade its characteristics will be carried out separately. It should be noted that the considered configuration of the magnetic rotor and the superconducting stator of HTS tapes has a fundamental advantage compared to the bearings on bulk HTSCs, since it allows for almost unlimited scaling of the device.

Acknowledgments
The study was carried out with a grant from the Russian Science Foundation (project No. 17-19-01527).

References

[1] Miyazaki Y, Mizuno K, Yamashita T, Ogata M, Hasegawa H, Nagashima K, Mukoyama S, Matsuoka T, Nakao K, Horiuch S, Maeda T and Shimizu H 2016 *Cryogenics* **80** 234–37
[2] Kun Liu, Wenjiao Yang, Guangtong Ma , Loïc Quéval, Tianyong Gong, Changqing Ye, Xiang Li and Zhen Luo 2018 *Supercond. Sci. Technol.* **31** 015013–13
[3] Rudnev I A and Podlivaev A I 2016 *IEEE Trans.Appl. Supercond.* **26** 8200104–4
[4] Jooss Ch, Albrecht J, Kuhn H, Leonhardt S and Kronmuller H 2002 *Rep. Prog. Phys.* **65** 651–88
[5] Landau L D, Lifshits E M 1988 *Field theory* M. Nauka 512
[6] Podlivaev A I, Pokrovsky S V, Anischenko I V and Rudnev I A 2017 *Tech. Phys. Lett.* **43** 1136–39
[7] Podlivaev A I, Rudnev I A and Shabanova N P 2014 *Bul. Lebedev Phys. Instit.* **41** 371–374
[8] Bean C P 1962 *Phys.Rev. Lett.* **8** 250-253; Bean C P 1964 *Rev. Mod. Phys.* **36** 31–39
[9] Podlivaev A and Rudnev I 2017 *Supercond. Sci. Technol.* **30** 035021–9

IOP Conf. Series: Journal of Physics: Conf. Series **1293** (2019) 012086 doi:10.1088/1742-6596/1293/1/012086

Evaluation of loss characteristics of superconducting magnetic bearings for LiteBIRD satellite by three-dimensional finite element method analysis

Y Hirota[1], H Ohsaki[1], Y Terao[1], H Enokida[2], Y Sakurai[3], T Matsumura[3], H Sugai[3] and N Katayama[3]

[1] Department of Advanced Energy, Graduate School of Frontier Sciences, the University of Tokyo, 5-1-5 Kashiwanoha, Kashiwa 277-8561, Japan
[2] Department of Electrical Engineering and Information Systems, Graduate School of Engineering, the University of Tokyo, 7-3-1 Hongo. Tokyo 113-8656, Japan
[3] Kavli IPMU, the University of Tokyo, 5-1-5 Kashiwanoha, Kashiwa 277-8583, Japan

E-mail: ohsaki@k.u-tokyo.ac.jp

Abstract. The polarization modulator of LiteBIRD satellite is required to be used by continuously rotating the half-wave retarder under a low temperature environment of about 10 K. Then, it becomes an important issue to reduce heat generation due to rotation loss. Therefore, a rotation mechanism in a cryogenic environment by a superconducting magnetic bearing (SMB) using the pinning effect of the bulk superconductor has been studied. In SMB, because of the pinning effect of superconductor, it is possible to rotate the half-wave retarder with stably floating, and it is possible to avoid frictional heat due to contact. The half-wave retarder is incorporated inside the rotating part of the superconducting magnetic bearing composed of the permanent magnet ring and the iron yoke. At this time, since the permanent magnet ring divides the parallel magnetized permanent magnets into pieces, a gap is formed at the joint of the permanent magnets. For this reason, the magnetic field distribution generated by the permanent magnet ring is not uniform. Although SMB can reduce friction as compared to mechanical bearings, the magnetic properties of the bulk superconductor change due to the magnetic field fluctuation caused by such ununiformity of the magnetic field, resulting in energy loss. In this study, heat generation in bulk superconductor in SMB is evaluated by three-dimensional finite element method. First, using JMAG Designer 17.0, the magnetic field distribution created by the rotor side is obtained by static magnetic field analysis. Next, using the COMSOL Multiphysics 5.3a, by applying the magnetic field distribution created by the rotor side obtained above and rotate the magnetic field, the loss generated in the bulk superconductor is investigated. The resulting loss value satisfied the required value.

1. Introduction

1.1. LiteBIRD and the polarization modulator

LiteBIRD is a medium size artificial satellite planned by JAXA and others with the intention of launching in mid-2020 's. The aim is to observe the polarization of the cosmic microwave background radiation (CMB) and to detect polarization signals [1, 2]. They have traces of "primitive gravity wave" which is evidence of inflation. The appearance of LiteBIRD is shown in Figure 1.

One of the observation instruments installed in this satellite is a polarization modulator, and an optical element called a half wave plate is continuously rotated in a low temperature environment below 10 K. The low temperature environment is realized by a refrigerator for space. But since the refrigerating capacity is limited, it is necessary to suppress the heat dissipation caused by continuous rotation of the half wave plate. In general mechanical bearings, it is difficult to use in low temperature environments because of heat generated by friction. Therefore, it is studied to adopt superconducting magnetic bearing (SMB). Figure 2 and Figure 3 show the appearance of the LiteBIRD polarization modulator targeted in this study.

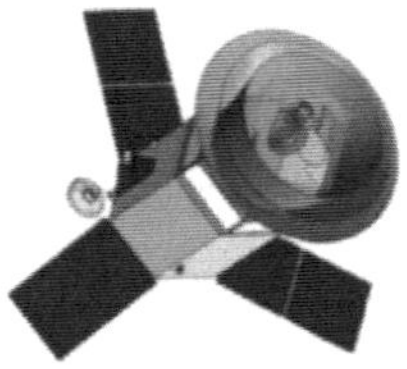

Figure 1. LiteBIRD satellite

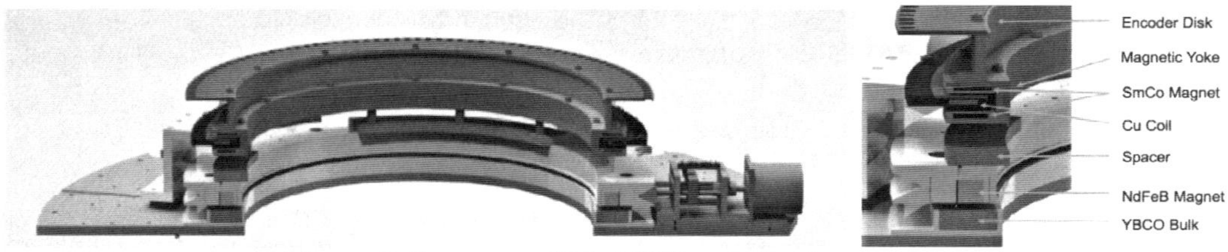

Figure 2. Structure of LiteBIRD polarization modulator

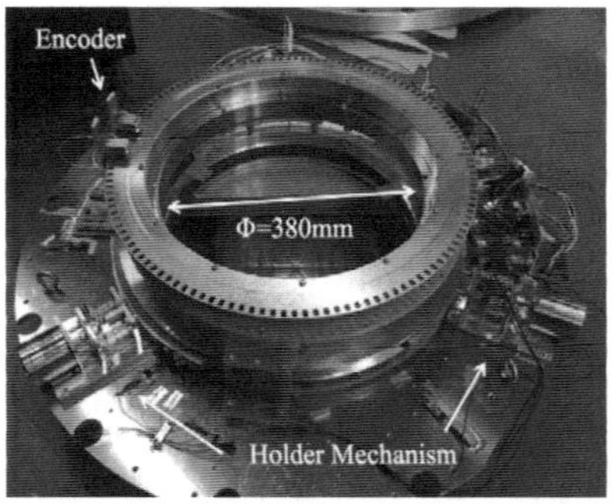

Figure 3. Experimental model of LiteBIRD polarization modulator

Table 1. Requirement of the heat dissipation

[mW]	Rotor	Stator	Total
Drive	< 1	< 1	
SMB		< 3	
Total	< 1	< 4	< 5

1.2. Superconducting Magnetic Bearings (SMB)

Figure 4 show the fundamental structure of the SMB. The rotor side is a ring of Nd-Fe-B magnet magnetized in the radial direction. Also, the PM ring is sandwiched between two iron yoke rings to form a magnetic circuit [3]. Although the iron yoke is not divided, the PM ring is divided into 64 pieces in the circumferential direction. The stator side is YBCO bulk ring. Bulk is divided into 20 pieces in the circumferential direction. The rotor side can stably levitate due to the pinning effect and its height is 3 to 5 mm. The rotor continuously rotates at 60 rpm.

In this structure, the magnetic field created in the circumferential direction by the rotor side is non-uniform. The reasons for this are as follows: 1) there are gaps between the magnets, 2) each magnet is magnetized parallel to the radial direction, 3) an error occurs in the magnetization of the magnet in assembly, as shown in Figure 5. Since such a non-uniform magnetic field rotates over YBCO bulk, an alternating magnetic field is applied to YBCO bulk, eddy current is induced, and the heat loss occurs [4]. The purpose of this paper is to estimate the loss of YBCO bulk in SMB by three-dimensional finite element method analysis.

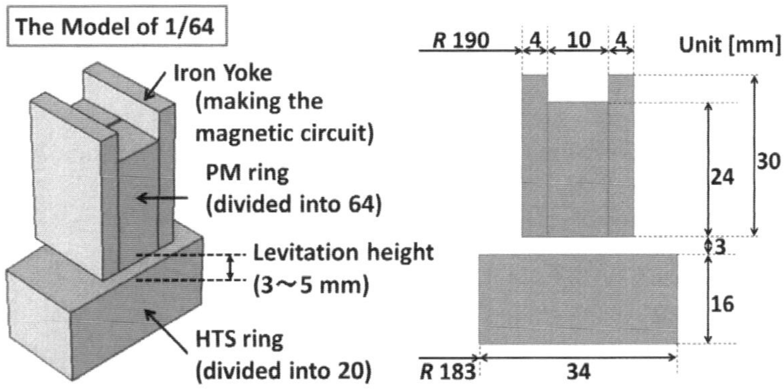

Figure 4. Structure of SMB

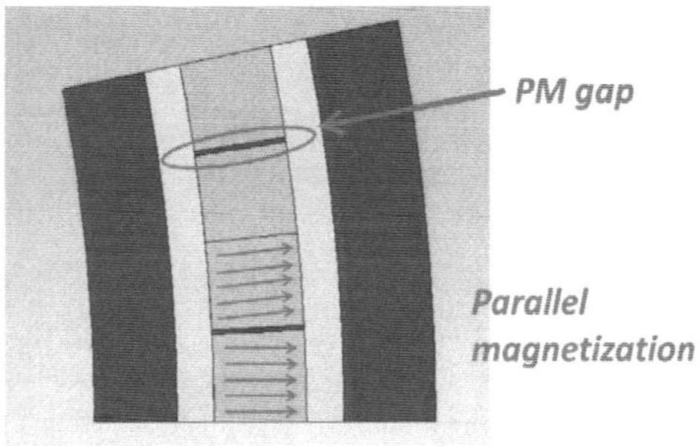

Figure 5. Cause factors of magnetic field inhomogeneity

2. Analysis Procedure

The analysis procedure to estimate the value of the AC loss is shown below.

1) FEM analysis with using conductors instead of superconductors is carried out. The electric conductivity of conductors is varied in the range of 10^4 to 10^{13}. Thus, the electric field distribution in the conductor and the dependence of the loss on the electric conductivity are obtained. At this time, for the equation (1) below, the electric field is determined only by the distribution of the magnetic flux density.

$$\nabla \times \mathrm{E} = -\frac{\partial B}{\partial t} \tag{1}$$

Therefore, the electric field distribution obtained by this analysis is considered to be equivalent to that applied to YBCO bulk.

2) According to the n-value model, the electric conductivity (σ) in the YBCO bulk differs depending on the location. But the minimum value of σ can be estimated based on the electric field distribution obtained in 1) and the equation (2), which is the electric field dependency of the electric conductivity obtained from the n-value model [5, 6].

$$\sigma_{sc} = \frac{J_c}{|E|}\left(\frac{|E|}{E_c}\right)^{\frac{1}{n}} \tag{2}$$

IOP Conf. Series: Journal of Physics: Conf. Series **1293** (2019) 012086 doi:10.1088/1742-6596/1293/1/012086

3) The maximum value of the energy loss in YBCO bulk is determined, based on the minimum electric conductivity in YBCO bulk obtained in 2) and the relationship between loss and electrical conductivity obtained in 1).

By the above flow, the upper limit of the energy loss in YBCO bulk can be estimated.

3. Result and Discussion

3.1. FEM Analysis

We analyzed the 1/64 model using the conductor replaced with the superconductor. The model is shown in Figure 6. Finite element method (FEM) analysis was carried out with *A-V* formulation. The results are shown in Figure 6 to Figure 10. Figure 6 shows the eddy current distribution on the conductor surface. And we saw the eddy current moves with the rotation of the rotor. The electric field distribution applied to the conductor is shown in Figure 7, and from this result, the maximum value of the electric field estimated to be 1×10^{-4} [V/m]. This is necessary for evaluating the heat loss in YBCO bulk. Figure 8 and Figure 9 shows the magnetic flux density distribution. Figure 10 shows the loss as a function of the electrical conductivity: σ. In the region where σ is small, loss increases as σ increases. On the other hand, in the region where σ is large, the influence of the shielding effect by the eddy current becomes dominant and the AC loss is reduced.

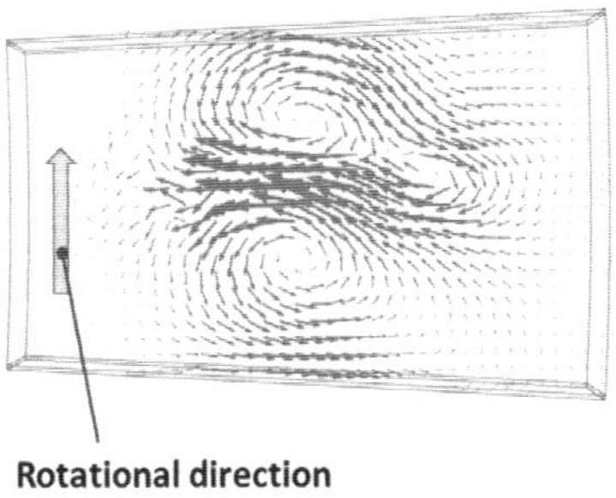

Figure 6. Induced current on the top surface of the conductor

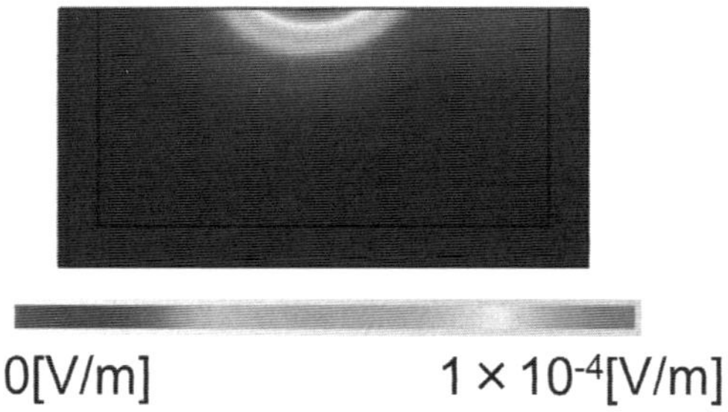

Figure 7. Electric field

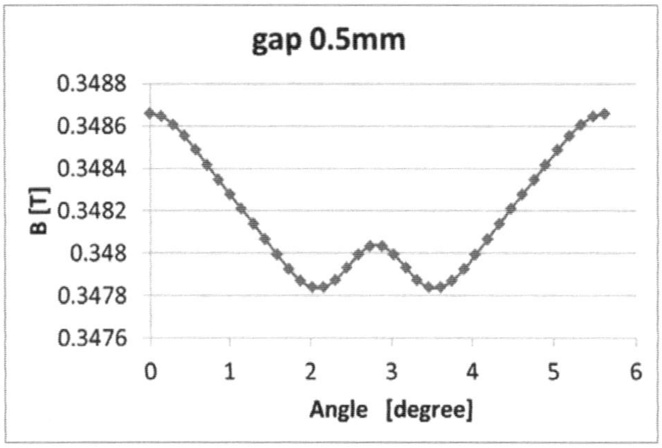

Figure 8. Magnetic flux density distribution in the circumferential direction

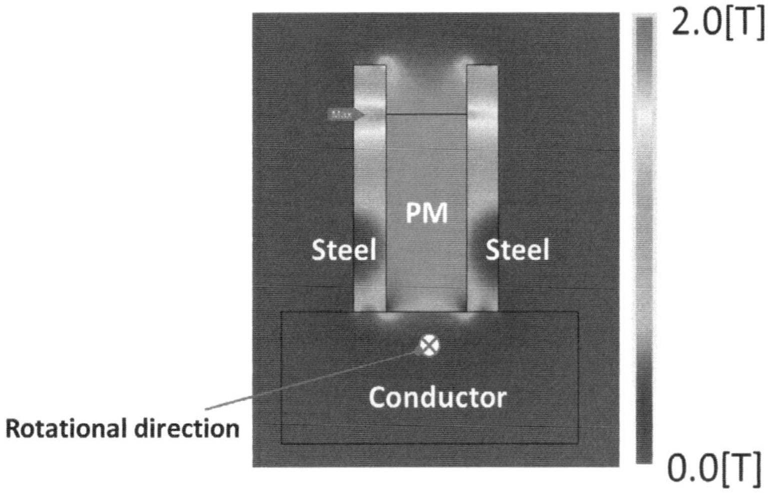

Figure 9. Magnetic flux density distribution

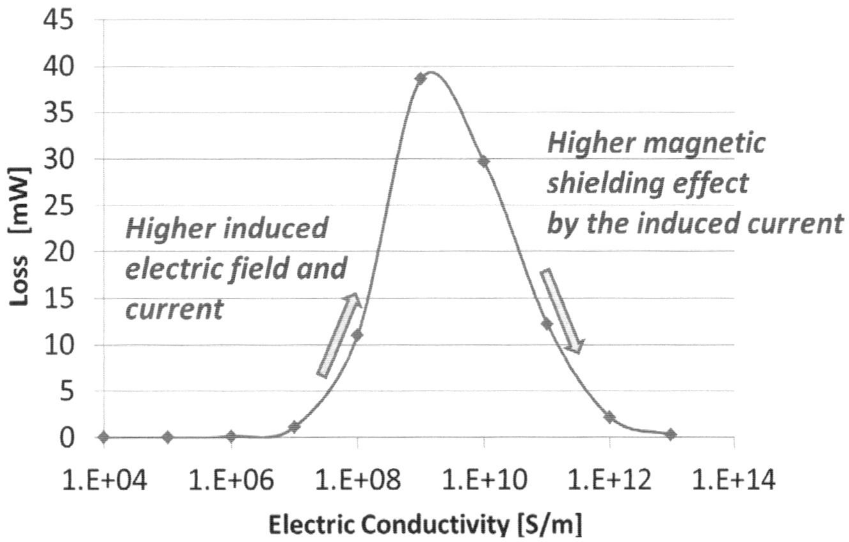

Figure 10. Dependence of the heat loss on the electric conductivity

3.2. Estimation of the energy loss in YBCO bulk

Figure 11 and Figure 12 show how to estimate the upper limit of loss in YBCO bulk using the above results. The parameters used in estimation is shown in Table 2. First, as shown in Figure 9, we estimated the minimum value of the electric conductivity in YBCO bulk: 1×10^{12} [S/m]. This is because equation (2) is a monotonically decreasing function of $|\boldsymbol{E}|$. In Figure 11, $\boldsymbol{E}_{max}$ was set to 1×10^{-4} [V/m] according to the above result and $\boldsymbol{J}_{max}$ was calculated from $\boldsymbol{E}_{max}$. From the results in Figure 10, loss tends to decrease near 1×10^{12} [S/m]. Therefore, the value of 2.2 mW is considered to be the maximum value of loss in YBCO bulk, as shown in Figure 12.

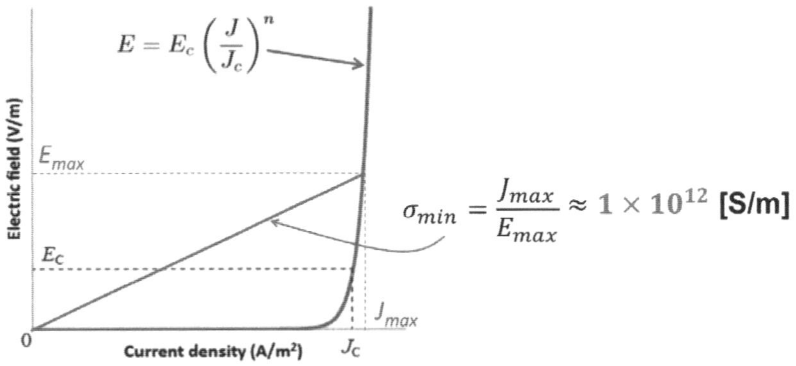

Figure 11. Estimation of the minimum electric conductivity

Table 2. Material parameters of superconductor

J_C [A/m²]	1×10^8
E_C [V/m]	1×10^{-4}
n	8

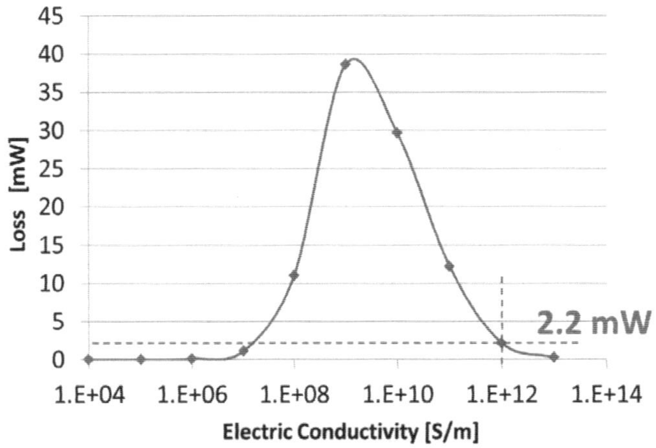

Figure 12. Estimation of the upper limit of energy loss in YBCO bulk

4. Conclusion and Future Works

In this paper, we estimated the AC loss in YBCO bulk of SMB mounted on LiteBIRD satellite. In consideration of the period in the circumferential direction of the SMB, analysis was carried out using the 1/64 model and conductors instead of YBCO bulk. From the results of the analysis, it was possible to obtain the loss as a function of the electrical conductivity. Next, from the maximum value of the electric field obtained by this analysis and the n-value model, the minimum value of electric conductivity in YBCO bulk was estimated. Finally, from minimum value of the electric conductivity and the loss graph, the upper limit of loss occurring in YBCO bulk was estimated. We concluded it as 2.2 mW.

However, we measured the magnetic field distribution in the circumferential direction with a prototype machine to find that it is very different from numerical analysis here. This seems to have a large influence by individual differences of the divided NdFeB magnets. From now on, it is necessary to analyze with considering the actual magnetic field distribution and we need find the method of analysis with the superconducting model.

References

[1] Matsumura T, Kataza H, Utsunomiya S, Yamamoto R, Hazumi M and Katayama N, 2016 *IEEE Trans. Appl. Supercond.* **26** 3602304
[2] Kusaka A *et al* 2014 *Rev. Sci. Instrum.* **85** 024501
[3] Kuehn L, Mueller M, Schubert R, Beyer C, de Haas O and Schultz L 2007 *IEEE Trans. Appl. Supercond.* **17** 2079-2082
[4] Zou J, Ainslie M D, Fujishiro H, Bhagurkar A G, Naito T, Hari Babu N, Fagnard J-F, Vanderbemden P and Yamamoto A 2015 *Supercond. Sci. Technol.* **28** 075009
[5] Quéval L, Liu K, Yang W, Zermeño V M R and Ma G 2018 *Supercond. Sci. Technol.* **31** 084001
[6] Stavrev S, Grilli F, Dutoit B, Nibbio N, Vinot E, Klutsch I, Meunier G, Tixador P, Yang Y and Martinez E 2002 *IEEE Trans. Mag.* **38** 849-852

AUTHOR INDEX